Agrotechnology for Dryland Farming

2nd Revised Edition

Editor
Arvind M. Dhopte
(Retd.) Professor of Botany
College of Agriculture,
Nagpur

Published by
SCIENTIFIC PUBLISHERS (INDIA)

Jodhpur –
5 A, New Pali Road
P.O. Box 91
Jodhpur - 342 001 INDIA

Delhi –
4806/24, Ansari Road
Daryaganj
New Delhi - 110 002 INDIA

ISBN: 978-93-86102-73-7

Visit the Scientific Publishers (India) website at
http://www.scientificpub.com

Printed in India

Dedication

Dedicated to fond memories
and
remembrances of my beloved mother
Late SEETA DHOPTE
who worked in dryland farm and
died in field by thunderbolt hit.
With all odds, she made me
What I am today

KEY - NOTE

Agricultural production has seen a spectacular growth in India during the past 50 years, contributing significantly to the growths and food self-sufficiently. Irrigated areas were largely responsible in bringing about this change from a "begging bowl" situation to self-sufficiency. But even when the full irrigation potential of the country is utilized, nearly 50% of the cultivated area will continue to be rainfed. Rainfed areas support about 40% human population and over 65% cattleheads. Major chunk of nutritive cereals (90%), oilseeds (80%) and cotton (65%) is produced in rainfed conditions.

There are many biophysical and socio-economic conditions in rainfed areas. Soils are highly degraded with poor nutrient status and low water holding capacity. Rainfall, the only source of water for agriculture is scantly in quantity and erratic in distribution. Farmers are poor in terms of resources and depend solely on agriculture and related enterprises for livelihood. The vicious combination of all these factors has made dryland farming a risky proposition leading to lack of investment in rainfed areas and low or non-adoption of technologies.

India is one of the pioneers in dryfarming research. The National Agricultural Research System (NARS), over the past 50 years has addressed the problem of low productivity in drylands through its network of ICAR institute, State Agricultural Universities and All India Coordinated Research Project for Dryland Agriculture. A host of location specific technologies have been evolved for yield stabilisation in dryland farming and conservation of the fragile ecosystem by sustainable use of soil and water resources. The results, though not as dramatic as in irrigated areas have been promising. There was a marked increment in the average productivity of nutritive cereals and oilseeds under rainfed situation. The effect of rainfed agricultural technologies was more visible when they were adopted on a watershed basis. With the Government of India's emphasis on area development on watershed basis, dryfarming technologies hold the key for the sustainable development of rainfed areas.

Despite, spectacular results in some of the watershed projects and their countrywide spread has not been encouraging due to non-adoption of all the components of technologies by farmers. One of the main reasons for this low adoption is their lack of participation in technology generation and dissemination processes. Research related to watershed development needs to be conducted with farmers as focal point in a Participatory Technology Development (PTD) mode and in a farming systems perspective. Therefore, any future efforts in development of dryland agricultural technologies should address the issue at multi-institutional and multi-disciplinary resource- building approach with the farmer at centre stage.

Sound dry farming agricultural technologies along are the answer for the present problem of low productivity (0.8 t/ha) in rainfed agriculture. In order to feed the ever-increasing population, the average productivity of rainfed areas must be increased to at least 1.5 t/ha in the next decade. Besides increasing the productivity, the food and nutritional security of the expanding population is also of great concern. In this context, farming systems perspective assumes a much wider connotation covering livestock, poultry, fisheries, horticulture, and agroforestry besides food grains. A paradigm shift is needed towards attaining the food and nutritional security of teeming millions of human beings by integrating various production components for synergy and with farmers, participation as the core issues. This also calls for adoption of a consortium approach to share knowledge and experience with a view to addressing the problem with a holistic perspective.

Dr. A.M. Dhopte has made an excellent effort in bringing together all the relevant aspects of dryland farming in a realistic frame work. Being a multidisciplinary area of research it is indeed a difficult task to bring together all strategies and applied issues on dryland farming in a single volume. Dr. Dhopte has been able to do this job in a commendable way by integrating the topics in a systematic manner.

I hope the book meets the expectations of all those interested in the subject of dryland agriculture in the developing country and India in particular.

Hyderabad

H.P. Singh
Director
Central Research Institute for Dryland Agriculture
Santoshnagar, Hyderabad 500 059

MESSAGE

Dryland farming necessarily refers to raising of crops with available rain water. Rain water management is primary factor in sustainable agriculture and starts right where rain drops fall on the earth and secondly to harvest every drop effectively with essential agrotechniques. First concept involves *in situ* water conservation with watershed development in a field on a community scale with small holdings or individual watershed on large farms of sizeable area.

Farms are never leveled and have some slope of 0.5 to 10%. Contour farming, formation of live bunds and broad based furrow can be easily adopted for soil conservation and to minimise the run off losses. Rain water is better percolated down the soil and excess run off losses are stored in watershed reservoir during active period of monsoon. With this, we accomplish both water and soil conservation, a must for sustainable and precision farming. Moreover, the stored water can further be used for lift irrigation under occurrence of intermittant drought of short period to save the crop and optimise the productivity.

Next approach for effective rain water use is to adopt microirrigation, organic recycling with loppings and use of various mulches. Clubbing this practices with ideal cropping system as explained in this book shall be more rewarding to the poor farmers to achieve viable and sustainable farming. Continuous contour trench (CCT) for development of plantation under agrohorticulture, agroforestry and other perennial systems leads to uniform *in situ* conservation for rainwater, soil and organic matter.

I must appreciate the hard work and dedication of Dr. A.M. Dhopte, Ex. Professor of Agri. Botany, College of Agriculture, Nagpur and the other authors. I am sure that the publication will definitely be useful to the teachers and researchers serving the two third of our national agricultural land. I wish that the book shall have better patronage in India and abroad.

Akola

G.M. Bharad
Ex. Vice-Chancellor
Dr. Panjabrao Deshmukh Krishi Vidyapeeth
Akola - 444 104 (MS)

FOREWORD

It is needless to emphasize any supportive efforts for making unviable dryland farming a more viable and sustainable enterprise for millions of poor farmers all over the world. This is absolutely essential under conditions when more than 90% land in India is rainfed. Globally, the picture is not different in developing countries. Consortive efforts that were needed in the past, unfortunately are not in hand today. Barrying few examples of few big dams, the efforts towards management of rain water in India were not satisfactory instead of full natural resources. Agriculture sector is not rightly looked into its right perspective as it should have been after 53 years of independence. Quantity and quality of produce need to be monitored right from field production technology and this exactly lacks in previous fifth year plans.

Food quantity is the criterion in rainfed farming though quality can not be ignored. Major share of land asset in India is at the mercy of nature, erratic behaviour, drought and flooding. Any effort to improve living quality of rural folks engaged in rainfed farming with agrotechnology for several crops is worth appreciating. Rainfed crop is cared with every drop of rain water through appropriate measures such as *in situ* conservation of water, micro-watersheds, continuous trenches, restricting run-off, broad based furrow, organic recycling etc.

Second approach is to allow guided flow of water through plants without affecting crop growth to accomplish marginal yields, which a farmer is deprived off, to make agriculture a sustainable enterprise. Pre- and post-planting care is needed for sustainable productivity. Book includes all relevant aspects of rainfed farming and is therefore, a valuable addition for dryland farming. Long outstanding demand has, thus fulfilled the requirements with this book. The approach of the editor is novel and he made an attempt where a farmer lacks to manage the land and he purposefully attempted to provide technologies to monitor the adversities of dryland farming with expertise available in the country.

After going through the contents of the book, I do hope that an editor of this book, Dr. A.M. Dhopte (Former Professor of Agri. Botany) has made full justice to the current problems of dryland in the country and abroad and he deserves appreciation in bringing out such long needed outstanding book for the benefit of scientists, researchers, extension workers, planners and farmers to accomplish marginal productivity from today's unviable non-productive dryland farming of poor cultivators, spread all over the world. To wipe out the tears from eyes of the Indian peasants was the dream of Mahatma Gandhi and such efforts if rightly and timely adopted by the planners and properly executed, I am sure, a dream of

Former Prime Minister of India, Late Lal Bahadur Shastry may come true in days to come. I profoundly, wish, this book is a grand success.

Akola

V.B. Shekhar
Director of Research
Dr. Panjabrao Deshmukh Krishi Vidyapeeth,
Akola (MS)

MONSOON AND RAIN WATER MANAGEMENT — EDITOR'S VIEWS

Economics of any country depends upon agriculture developments and it is commonly approved fact universally. There cannot be progress in India unless growth rates is increased to at least 6.0. As per statistics of 2011, 26 crores of people are totally dependent on agriculture. In past, 70 years after independence, agriculture dependency has raised from 12 crores to 26 crores and out of this 12 crores are peasants and 14 crores are landless labours. This is red signal for economic growth. The people without work is in no way good for any country

Water is a source of life to farm and animals. Satisfactory supply of water to agriculture is essential for progress in agriculture. This has not been attempted in any governance of India during past 70 years. Because of no infrastructure, Dryland farming is 60% in India and 70% in Maharashtra state. The life style of Indian peasant is for sustenance traditionally and it is not planned for income generation at national level. The rainfed farming has become a gambling for crores of people. The crops like sugarcane, wheat and rice have satisfactory source of water supply which is roughly 93%, 85% and 26% respectively. Because of no timely source of water to crops like pulses, oil seeds, sorghum and lesser millets, the production is limited in India as compared to other countries.

This made the Indian economy paralyzed. No ruler has made serious thought over this important issue during past 65 years. The efforts made so far are inadequate to cater the needs of 130 crores of people.

The experts who have made five year plans though were in right direction, were not successfully implemented because of many reasons. The funding system has failed to cope up the urgent needs of the country and now we are at the edge of degradation fast. The execution plan of responsible for poor economy of the farmer. The dryland farmers have become poorer and poorer day by day. He is loaded with loans and suicides have become a common feature among farmers. As per statistics of 2001, 28% of poverty is found among farmers and 43% among landless labours. Large number of farmers take only one crop and they remain workless for rest of the period. It amounts to increase in unskilled and uneducated workless people in the country. The maoist and naxalites are the victims of this systems. The workless hands have opted for such dangerous options. Six crores of our national youth are in the trap of such zone and they become easily vulnerable for sustenance. Under such situation 40% of pleasants are ready to sell their lands without considering nation and nationality. This is because of their uneducation and unskilledness and therefore they always remained under below poverty line. This concept involves two main steps- human resource and economy. The solution is to increase the productivity of dryland farming to make the farmer sustainable economically and save them from being workless. This will definitely lead to dryland farming better both in production and economy. We shall have to wait for next two decades for self sufficiency in present crisis. The Indian governance is failure at both the two ends.

The funding for agriculture researches in India is hardly half percent of total agricultural production as compared to one percent in china. The farmer economy is improved in China due to production and this is a land mark at global level for poverty eradication. The improvement in loan system cannot be a bold solution for viable farming in dryland agriculture but to provide them improved and better technology is the ultimate and absolute system that may help them on sustainable humanitarian ground.

The 3000 farmers in Maharashtra state have made suicides in past five years and this is specially found in Vidarbha region where irrigation facilities are meager and crops are taken on the mercy of rains. Non availability of water at critical stage leads either to total failure of crop or partial production that never allows the farmer to attain self sufficiency. The rainfall is abundant but irrigation sources are meager and therefore they become helpless to supply water during critical stages of the crop. The upliftment of farmer's economy is only possible if irrigation is made available at proper stages of the crop.

The watershed concept is the right choice and necessary steps are taken by the central government before two decades corrupt practices in Indian system however do not allow to have benefit as envisaged by planners. Hopefully, the improvement in governance may allow us to havest the fruits of planners in future.

Various symposium and seminars have been organized at national and international level by reputed societies. Scientist's views are crystal clear and they have emphasised the need to conserve rain water effectively for betterment of human welfare. The concepts that have come in reality are few erected big dams for renewable hydro-electricity and irrigation. This has created some more problems rather solving problems at grass-root level. Huge areas are trapped in high water table leading to flooding effects, salinity, upheld aqualayer and thousands of hectares of land uncultivable. These problems are being tackled at government level now. If one problem is solved, another problem crops- up and this goes on and on keeping real problems unmatched with the requirements. Necessity is the mother of invention. The plans for connecting big rivers in India was the first approach putforth by the scientists in early fifties that could avoid the drought and flooding in some areas of the country. This demanded millions of grants and unfortunately, it could not be implemented by the planners. A march was on in various Vth year plans and slowly, some dams (Big and Small) were erected for two purposes of electricity and irrigation in the country.

First priority was given to the industries primarily to erect industries and barren lands remained neglected till today. The water could not reach to poor farmer's field. Central plains of penninsular India remained neglected and today it is dependant totally on rain water, called dryland. Dryland farming is necessarily a rainfed farming. The yield levels remained marginal in past and it was enough to cater the needs of sizeable population of 60 crores. Now it has increased over a billion and the food basket may be now inadequate to meet the growing demands today. The concept of *in situ* water conservation of rain water is only the solution. How do we implement it effectively ? There are some considerations that should be seriously viewed by the policy makers and I try to illustrate here for the use of readers.

We have set-up innumerable departments and organizations for various purposes like:

1. Water power production
2. Irrigation and soil conservation
3. Management of rivers, harbours and flood control

4. Watershed and drainage management
5. Desalinization and ground water investigations and
6. Experiments with weather control.

The water comes to earth as rains and this water can be used and controlled for the use of humanity. The ability to manage water is one of the man's greatest achievements. Limited progress has been made to manage the water. It remains for the next generation to coordinate these activities to make them more effective and inexpensive. The India can ill - afford the unplanned approach to water management. A more coordinated approach is needed. Average rainfall in India is 550 mm. The world average is around 800 mm. The monsoon rainfall is India's valuable resource which is a unbeatable combination for agricultural development. India is supposed to be the potentially the most productive agricultural country in the world. The question is how to realize this potential. Coordinated nationwide programme can be implemented with primary emphasis on monsoon rainfall. The monsoon water is received during four months period. It therefore should be effectively stored for rest of the year. Certainly, the important reservoir for water is right where the rain drops fall.

The efficiency of soil water storage is determined by water management. The owner of the land where the water falls, has the primary responsibility for water management, both in the quantity stored and the prevention of evaporation losses. This concept must first be established very soon.

Farm storage of water must be in the soil and not on the surface. Surface storage is subject to evaporation and is very wasteful in anything but deep tanks and reservoirs. Studies in the state of Maharashtra have indicated that at least 65% of all water collected back of bunds is lost by evaporation. We wait for rains for nearly eight months in a year and it looks good to have water around. Water held needlessly may represent a sereous loss to the nation in damaged crops, waterlogged soil or wasteful evaporation that could have been used more productively in downstream. This should be remembered by every farmer very critically because life exists where water prevails. The ancient civilization has begun in nearby areas of rivers. The Indus valley civilization was first found near Punjab where five rivers flowed. The ganges valley civilization progressed further. Unfortunately today our farmers do not understand the importance of water and are not after its conservation barrying few cases. We wait for monsoon for eight months but still, when it arrives, conservation part is neglected. Why this happens is the mystery. The run off losses are not controlled and we never care to stop this phenomena. More importantly, it is the right place where it needs to be stored. Excess water goes to the nalla, bunds, ponds, deep crevices and finally it reaches to the river. Precautionary measures are not employed by the farmers. Field ponds were the common approach in areas where paddy cultivation was predominant but in plains, this concept was not practiced.

Each field in India should be considered as primary unit for water conservation. The water management starts right here. The depth, the permeability and the water holding capacity of the soil is considered in relation to the crop and the character of the rainfall. Water moves by gravity, so the slope and the nature of the slope must also be considered. The idea is to plan a system of farming that will enable the cultivator to control the water and put the maximum amount to beneficial use. When a field is prepared to make optimum use of monsoon rainfall, very little additional work is required to make proper use of irrigation, if or when available. How do we plan a field for water management? It is really simple. First determine where excess water should leave the field. Then a drainage system should be planned. Broad contour-bottom sodded drains are used for slopping lands, deep drain ditches are preferred for flat or leveled land. If the land has a long slope, bunds - base terraces may be needed. These should be laid out as a definit part of the drainage system, and must be on an exact contour or graded contour. They may serve as guide lines for contour farming. This practice is mostly followed in hilly areas only but the concept is not followed by farmers of plain lands.

The land is never found leveled. The slope is found some where in the field and contour bunding is the only answer. Live bunds of subabhul, vetiver, grasses, dhancha (*Sesbania rostrata*) can easily be adopted for water conservation and help minimize the run off losses. Better percolation of rain water can be achieved effectively. The terraces or bunds can not be straight unless land has a uniform slope. Property line bunds, road side bunds and nala bunds are usually very wasteful to crop acreage and damaging to productive land and should always be removed when they interfere with a proper field or community water management programme. Deep soils to be used for rice or small grain production may be leveled to a non-erosive grade or to form straight or contour border strips. Border strips are a series of parallel bench terraces constructed across the slope and the land then is leveled, so the water falling on land will flow slowly and uniformly down the strip between the borders.

The borders also serve as guide lines for graded furrow farming of row crops. Border strips are also used for irrigation farming. This is suggested to secure high yields of food grains with monsoon rainfall. Border strips serve the purpose of irrigation and utilised for water conservation.

Graded contour furrow farming is good for steeper slopes. Steeper slopes require deeper furrows. It can hold-up the water until it can be absorbed by the soil or move slowly down the furrows and into the drainage way. The erosion can be controlled and water is stored in the soil for crop use from each rain. The contour furrows control few heavy rains during drought more efficiently. Famine condition will never prevail in India if available water both from irrigation and rainfall are managed effectively on every field.

Consumptive use studies indicate that very few crops mostly perennial one require more water than the monsoon normally supplied during four months' rainy season. India is not a dry desert with insufficient water resources. The only desert in India is along the Pakistan border. The damage to the crop in India is, mostly due to much water rather than not receiving enough. Normally, drought affects the crop on shallow soils. Such piece of land can be efficiently used for fruit crops like Ber, Amla, Sitaphal, Tamarind, Guava, etc. and intercrops of food grains can be cultivated. Strip plantations of these crops can have effective control on soil erosion and can result in conservation of rain water. More rain water results flooding and damages crop growth on millions of hectares of land and also affects more productive land in the world. The probability of such situation is found every year and it affects the soil as well as crops. Large hectares of land is spoiled due to waterlogging, high water table and the accumulation of alkali salts. New irrigation projects are adding to this menace each year. Adequate drainage system is therefore necessary for every field. Soil is precious input in agriculture and million tonnes of soil is lost due to erosion and added to the sea every year.

What a farmer needs to do for water management on his field is explained. Drainage for each field is must. The neighbour farmer complains about this run-off water. All the poor land in the neighbourhood floods and damages crops and soil on all the best lands. The neighbour farmer becomes helpless for this water. This should be the responsibility of organised society. If government programme has to be effective, the programme should start where it begins at local level and then move to the state and national jurisdiction. Very effective implementation of such situation has been achieved in western Maharashtra by forming organised corporate sectors. The "Village pond", Lakes system has been established in Deccan valley tract of western Maharashtra to monitor, manage and distribute available rain water for the needy farmers in critical conditions. This has made the revolution in dryland agriculture in this region. A radical change is seen in past decade and the area which was neglected for years together, is now brought under timely irrigation for better productivity. Protective irrigation can be managed and productivity can be assured. This is a unique example what farmer's community can do on corporate sector. Barren lands have changed to green festivity among farmers due to collective wisdom of the people and good leadership. Such examples can be a land mark for others to follow in central plains for better livelihood, harmony and happy living at village level. Individuals who have common problems of water management should join together to solve their difficulties. The organizations should be endowed with the full authority of Law, as a functioning unit within the block, the district, the state or the federal government. A legislation has come in force in some states of India to deal with such water problems. Water management deserves the undivided attention of enthusiastic and well qualified persons at all levels.

Water management programmes should be coordinated by the government. Around 50 to 100 farmers should join together and form a society for rain water management, arranging watershed in an area of 100 hectares. Expertise can be had from Agri. universities of Govt. Deptt. of Agriculture. A honest leader among the farmer's community is needed to lead the activities in a right and lawfull manner. Unfortunately, farmers are much involved in fighting, confrontation with neighbour farmers for bunds and flood water and this is highly attributed to mostly illiteracy rather greedyness. The training, demonstrations, set examples survey and benefits derived by them need to be brought into the focus of such farmers who are always involved in fighting with neighbour farmers. Unless the sence of brotherhood for mutual benefit is developed within them, the societies though formed may not work and it will lead to poor results. India is sufficient in sunlight, soils, water resources, ideal topography and large forests but overall the crop productivity is poor due to lack of coordination and proper leadership. Honesty, determination and strong nationality lack in most of us and the problems have become much acute and farmers are far behind the world in productivity of food grains in India and developing countries. A radical change can be had provided the people are honest in work and leadership. Many factors can be responsible for such attitude of farmers and the people but some time some day, this approach needs to be adapted if we have to survive in global competition for food and water in days to come.

Population has now increased over one billion and if productivity is not raised from 90% dry land, no one could come to our rescue to feed the people. Begging bowl policy shall not work in future. One can imagine what will happen if food security is remote. Social health can be spoiled leading to anarchy and arson. This situation may be disastrous. A new cell has been launched by federal government on rainfed farming and production of watershed is being promoted at all India level in IXth year plan. Hopefully, the prodigal action may be improved in Future.

India uses about three times a much water than what is really needed by the crop. Due to bright solar radiation during rest of eight months (Excluding rainy days), the ET is higher. Global warming era of millennium shall again add to this and probably, more water could be required to raise the successful crop by 2025 AD. Flood and furrow irrigation system must be stopped to save the water and drip system should be promoted as done in Isrsael and some arid regions of the world. Water management is also required for rice because it is not a swamp grass. Rice likes water but precise water management is needed for maximum yield. Water is only pivotal factor in other managements of nutrients and fertilizers. All other stresses start when water is deficit. India has large water wealth and we have to make the most of it in days to come. The *in situ* conservation is the only key to remedy the problem in dryland farming.

The soil and water are inseparable. The soil resource is old and tried from centuries for farming. Because of heat and monsoon climate, soil forming process in India is most relatively rapid than other temperate countries. The conservation of rain water in America is 90% as against 10% in

India though average rainfall does not vary much (550 and 500 mm/year). During last four decades, the plans for rain water management are not predominantly executed. Paper plans are ideal but reality is unsatisfactory. The water utilization efficiency in India is lowest in the world. Few rainfed crops use 20% of available water and irrigation water efficiency is hardly 10 to 12%. The water resource should be precisely managed. What can be done to increase the water use efficiency? Complete programme of watershed management like reseeding and reforestation on steep or shallow soils, contour farming, strip cropping, cover and green manure crops, graded contour bunds etc. have to be tackeled. Block bunding alone does not contribute to water use efficiency. They hold back water and help control gullies, but the water they hold back does not increase the water available for crops on more than 15% of the total land area. Bunds or broad-based terraces are recommended for farm lands. The campaign to increase moisture use efficiency is much more appealing to the public when it is applied to irrigated land. What is the basic of increasing unit production/unit area? Plant needs water where it is planted and it is almost static. The water need of a plant must be satisfied at its place. Plant roots are searching for water below ground. Competition prevails within plant population. Roots that face horizontally must face more competition than verticle roots. To a plant, soil is just what plant roots can penetrate. At germination, the plant area is low but as it grows to maturity, more and more soil area is covered. The water holding capacity may become more important, not the total quantity but what the plant can extract.

The water requirement of every crop differs and it is dependant on leaf area index. Evapotranspiration (ET) refers to the total amount of water required to raise the crop for entire season. The water use pattern as per percent growth increases to grand period of growth and further decreases in sigmoidal manner. The water deficit brings about the wilting in plants and acute stress terminates into death of tissue. Some crops grow in deficit water and produces marginal. Therefore, we must study every crop and soil and entire irrigation system to supply the water needed for the crop growth. If the root zone water holding capacity of a soil is too low to meet the requirements, the soils need not be irrigated because it is simply a wastage. The ET studies have produced some very useful information and there is now system of irrigation based on crop coefficient. Transpiration and evaporation losses should be separately studied in every crop to know the details of where water goes. Infra-red thermometers have been devised to know the period of water stress in a crop period. The moisture utilization in this country is low. Correcting plant population/ha is the main factor in moisture use. More canopy results in more requirement of water but this thumb rule may not be applicable in field. Several factors are involved in ET of a crop and water requirement.

The factors like radiation, temperature and relative humidity play significant role in crop ET. A complete cover is necessary to utilize maximum solar energy. It is found that narrow spaced crop yields more under dryland conditions. What is harvested by a farmer is the solar energy. India has plentiful of solar radiation and photoefficient species need to be evolved. The algae utilizes 16% of solar energy while terrestrial plants utilize it by only 6%. Unfortunately, very little work is done in this aspect. Normally, dryland crops have less canopy and if more solar harvest is to be attempted, more plants/unit area has to be followed. Theoretical considerations are not experienced in reality under dryland conditions. Crop modelling attempt has been made on certain parameters related to water requirement, but in general biological phenomena can not be put in mathematical equations. Light energy is weak during active monsoon and every inch of space be occupied by leaves in order to make maximum use of limited sunlight. The cloud reduces the light intensity to nearly half (500 W/m^2) and all kharif crops suffer growth during active precipitation. With plenty of available water, a poor crop will use just about as much water as a good crop. This means that with a proper stand and irrigation when needed, yield per unit of water can be greatly increased by careful attention to other things like adapted crops and varieties, planting dates, fertilizers, tillage practices, plant protection, soil amendments, drainage or anything else that will increase production. The water use efficiency can be increased by "good farming" only. The use of some antitranspirants (Radiation reflectant types) are known to have better water use efficiency in crops.

The primary factor limiting the irrigation is the method of irrigation. Drip irrigation is the best system of irrigation under dryland conditions and it saves 40-60% of water. While in furrow irrigation, lot of seed is lost if the bunds are formed before irrigation. Germination is poor and crop stand is inadequate to have expected yield. The water use is very inefficient until plants cover the surface to protect the soil from solar radiation. The studies indicated that sprinkle irrigation is more suitable for vertisol soils if pipes are inexpensive. A computerised schedules of irrigation are prepared for every crop and available water can be monitored as and when necessary. As per expert reports, the agricultural production can be doubled in India provided proper system of water management is adapted by every farmer.

I tried to illustrate the fate of rain water starting from precipitation to river end, importance of *in situ* conservation of water, what approach can be suitable for dryland and how to increase the water use efficiency by plants and assure productivity under unpredicted weather. It is needless to emphasize the importance of rain water conservation. Coordinated and systematic approach on corporate basis can only be the answer to dryland problems. A unique example of village pond concept has been set up in western Maharashtra by devoted leaders among farmers community and such system needs to be adapted by every dryland region to bring another water revolution in the country. It must be remembered that life exists where water prevails.

This book deals with several approaches for dryland questions in India like reclamation of degraded land, enriching the soil with organic matter and construction of watersheds in an area of 100 hectares according to the topography. The use of various mulches, use of field organic waste, selection of suitable drought tolerant and waterlogged tolerant varieties, selection of crops, water requirement, water use efficiency, suitable medicinal plants, best use of waste lands, coordinating the weather forecast and timely precautions to be undertaken, use of antitranspirants and various crop life saving agrotechniques are dealt for the use of readers. How to avoid the yield losses by harvesting the crop according to physiological maturity is suitably explained with photographs. Maintenance of required crop stand is a key factor for assured yield in dryland farming. The subsidies may be accorded for such kind of endeavour by the state governments instead paying farmers after a crop failure.

I suppose, that, this preliminary information may be adequate to have general information a farmer needs to have as primary infrastructure for successful and sustainable agriculture in India and developing countries as well. One should remember that rain water must be stored where it falls. Construction of huge dams, though is a solution to supply the irrigation facility, can not be a final solution. During last fifty years of independence, major break through is not seen in trapping irrigation sources for central plains where dryland is mostly predominant. Every drop of water needs to be conserved if human civilization has to be maintained in orderly and healthy manner. All wisdom is seen only when belly is full. No wisdom can prevail in people with empty stomach. The gravity and magnitude of such serious problem must be understood now without loss of time to maintain civilization on this lonely planet. The chapters, that follow, prepare the reader for rainfed farming with trump-cards opening, one by one from seeding to harvest.

EDITOR

PREFACE

Dryland farming has never been remunerative in India due to several reasons and global scenario is not different in developing countries as well. Geographically, some regions are rainshade and rainfall is highly erratic with limited assured rainfall zones. Regions are classified in various zones according to precipitation pattern and land topography. Plans for water conservation are made in last two decades but rewards are not in focus due to improper implementation. Rain water management has been a top-most priority at Govt. level but still, farmer's socio-economic condition is unsatisfactory even fifty-two years of independance elapse. Drought and flood situations are experienced every year some where and this problem is now being dealt on war-footing by launching a special cell at federal government. Third largest river in the world is Bramhaputra in India and still paucity of water is faced. Is it not a tragedy? What plans we have to utilize this water? Country is poor with all plentiful of resources and potentials. Every inch of land can be brought under irrigation if proper plans are implemented. Having ample resources of water and sunshine hours and favourable topography, the poverty among farmers exists and this is a point of serious concern. Aqua layer is getting down and down day by day due to improper rainwater management at national level. As per recent report of Int. Water Council held at Stockhom on Aug. 14, 2000, Dr. Lester Brown, the president cautioned about paucity of water in future due to dry wells. Countries like India, China, West Asia and America lifted 16000 crore tonnes of water/ year through wells resulting ground water decline heavily. In West Africa, out of every three persons, one is likely to fight for water by 2025 AD. Increasing water and food demand can not be fulfilled for increased population. For every one tonne of food, 1000 tonne of water is required and food is produced only with rain water and lift irrigation.

Recent global statistics indicates ocean salty water to be 94.5% and 1.75% of water is locked as ice on two poles. Only 4% of water is useful for environment and human consumption including animals. In India, 4250 Km^3 water is received through rains but 1880 Km^3 is lost as run-off into the sea and it amounts to 44% of total. The 1258 Km^3 of water gets evaporated and only 423 Km^3 water is percolated in soil which amounts to hardly 10.2% of total. India has 329 m hectares of land with average rainfall of 550 mm. Large quantity of water is not conserved in soil through proper planning. Sufficient grants are allotted for water conservation by Central and State Govt. in every V^{th} year plan but positive results are not in focus. Rural sector is neglected in past and now micro-level planning is essential to have proper impact on agriculture. Run-off and evaporation losses (90%) can be minimised with various systematic effects either in dams, ponds, lakes, watersheds or *in situ* conservation right in fields. One hectare land receives 110 m litres of water through rains and there is a need to conserve 1% of it to fulfill the requirements. In contrast, America and Israel conserves 94% of water in soil inspite of 500 mm average rainfall. In Konkan area of Maharashtra, Average rainfall is 2500 mm but 85% of water is drained back to sea without ground water recharging. Land topography is the limitations for such problem. Who can find the solution? Ground water is not ever lasting if rain water is not recycled. World Bank also envisaged water shortage by 2025 AD leading to social conflicts at global level.

Though soil aqua layer is maintained at some places by construction of heavy dams, some areas face acute crisis of water for drinking and agricultural purposes. Nearly half of ground water is exhausted now and if it is not recycled today, paucity of water is bound to be faced in near future. The Maharashtra State Govt. resolved this situation by making new Act in 2000 to combat likely problems of future. Drinking water is available at 45-60 m depth of soil in some areas and borewells beyond 30 m depth can neither provide pottable water nor useful water for agriculture. It is a danger signal for next progeny to come in millennium if corrective measures are not adopted. Environment

shall come in geopardy because neither the plants nor animals can sustain if water is not provided. ``Life exhists where water prevails". Drought creates barren lands as observed in Rajasthan and some desert areas of the world. Extinction of useful plant species can

not be over ruled. Increased global warming may also lead to aberrent reproductive growth as well. Huge dams lead to upheld aqualayer leaving thousand hectares of cultivable lands non-productive due to water logging.

Promotion of watersheds is the only solution for *in situ* conservation of rain water and this has to be attempted on war footing right now. This can only be a wise attempt to march ahead for successful and sustainable agriculture and protection of environment. Suicidal attempts of farmers in some states of India during last decade are red signal for healthy democracy. Farmers are at the mercy of nature. One English author said "Indian farmers see their God in clouds" and I find this thought to be appropriate even today. Why this happens? Are we inadequate in technology? Is research not enough to resolve this problem? Why are we failure in this front when we rejoice our accomplishment in nuclear technology. Are we failure in transfer of technology? Some may Say "We are not above the nature". Weather can not be monitored. Are huge funds made on this aspects in vain? Scientists are also committed to answer these problems of poor farmers. India's economy is basically dependant on agriculture When nature disposes, we should stand together to fight the bread war. Remote sensing technology is being utilised for agriculture recently.

We are today self-sufficient in food-front only because of green revolution in varieties and hybrids with moderate increase in irrigation resources. A short break of rain during critical period of crop growth brings down the yield levels to marginal or zero level. Bankrupsy of farmer is always at his door steps. Our population is now one billion (1/6 of world) and agriculture shares more responsibility to feed this galloping population. There is therefore a dire need to think upon the problems of dryland farming. The yielding potential in semiarid tropics is uncertain and therefore *in situ* rain water conservation is highly important for assured yields in rainfed farming.

The yields are now plateaued since last two decades and better rain water management can only be the solution to expected crisis. The agrotechniques invented so far have to be transferred to the illiterate farmers. Extension workers should take this challenge through information technology, satellite communication, internet network and taking computers to the village level where no roads and telecommunication exist. The agrotechniques are compiled for successful crop management under biotic and abiotic stresses. True farmer is he who can get the normal yields under abnormal weather. This has been the focus of compiling various agrotechniques to help the farmer. We can certainly have accomplishment in this direction if measures are timely adopted. Stresses can be satisfactorily managed with advance techniques. Losses can be minimised and marginal yields can be assured. Let there be drought, waterlogging or hailstorm damage, total failure can be avoided. This book is an attempt to compile recent and old techniques to monitor crops in semiarid tropics.

After visualising the problems, the relevant aspects and scientists were selected at national levels and agrotechniques were edited.

I have covered my experiences of rain water and dry farming in the beginning for readers. Dryland problems were visualised and the relevant aspects and scientists were selected and exactly framed in systematic manner. The experts compiled various techniques to save the crop in stresses for the welfare of farmers. Experts incorporated their life time experiences and scanned the literature world wide to make the script more comprehensive and highly useful. All the strategies and applied issues on rainfed farming are purposefully dealt to have positive results.

This book deals with seed, soil, watersheds, crop, weed and nutrient management, use of weather forecast, various measures to save the crop under abiotic stresses like drought and flooding, reclamation of degraded land through organic recycling, evapotranspiration, water requirement, water use efficiency, climate and moisture relation, selection of variety and crops, agrometeorological approaches, use of antitranspirants, problematic soils and corrective measures, organic recycling and *in situ* water management, regulation of crop growth and suitable medicinal and aromatic crops for remunerative farming under dryland conditions. Over 90% of land is under rainfed farming. The chapters in this book are written by eminent scientists who devoted their services for rainfed farming in the country. Adoption of these techniques not only results in protection of dryland crops but favour environment as well in larger interest of life and world to prevail in days to come. Every scientist has given full justice to the probable problems of semiarid tropics for use of the readers.

This book could only be edited with the support of all renowned experts in the country. I am aware that the task was tremendous to find the right technology from available literature and it has been brought in limelight for the benefit of farmers. Estimations, predictions and timely operations are only the remedy for assured yields. What to do is given, but when to do is the master key for management of dryland crops. Wrong and delayed operations lead to poor yields. An ideal farmer is he who can take right decision at right time and supply the inputs what crop demands periodically. Risk is always involved in any enterprise and agriculture is not exception. Plentiful techniques are provided to save the crops under odds. One can take the horse to water but not let him drink and this is the task of farmer how best he can use the information.

Suggestions are invited for further improvements in the script. Criticisms are also welcome for better ideas which are not covered in this edition. I gratefully acknowledge the efforts made by all the contributors who have specifically compiled the required information and they deserve compliments for their efforts to ease the job of farmers. I thank them all for commendable task they have made for this book. I specially thank Dr. H.P. Singh, Director, Central Research Institute for Dryland Research, Hyderabad (AP) for writing Key-note, Dr. G.M. Bharad Ex. Vice Chancellor, for writing a message and Dr. V.B. Shekhar, Director of Research, Dr. Panjabrao Deshmukh Krishi Vidyapeeth, Akola (MS) for writing a foreword for this book. I always dreamed Indian farmers without tears and frustration. May I see my dreams come true in near future? Let us hope for the best.

A.M. Dhopte
Nagpur

CONTRIBUTORS

Akhare, B.N.
All India Coordinated Research Project for Dryland Agriculture,
Dr. PDKV, Akola- 444 104 (MS)

Atale, S.B.
Sr. Research Scientist (Sorghum) and Head, Deptt. of Agri. Botany,
Dr. PDKV, Akola- 444 104 (MS)

Babu, M. Raghu
Deptt. of Agronomy, Agricultural College Farm, Bapatla - 522 101 (AP)

Babhulkar, V.P.
Research Associate,
Deptt. of Agri. Chem. and Soil Science,
Dr. PDKV, Akola - 444 104 (MS)

Bharad, G.M.
Ex-Vice Chancellor,
Dr. PDKV, Akola - 444 104 (MS)

Bhardwaj, Ch.
National Research Centre for Soybean,
Indore - 452 001 (MP)

Bhaskar, K.S.
Sr. Scientist (Agronomy),
Central Inst. Cotton Research,
Shankarnagar, P.O. Box 2,
Nagpur - 440 010 (MS)

Billore, S.D.
National Res. Centre for Soybean,
Indore - 452 001 (MP)

Chittapur, B.M.
Professor of Agronomy, Uni. Agri. Sciences,
Dharwad- 580 005 (KS)

Deshmukh, V.N.
Asso. Director, AICRP for Dryland Agri.
Dr. PDKV, Akola- 444 104 (MS)

Deotale, L. Shyam
Assistant Professor, Extension Education Section, College of Agriculture,
Nagpur - 440 001 (Maharashtra)

Deotale, R.D.
Dr. P.D.K.V. Akola, Botany section,
College of agriculture, Nagpur

Deotale, S. Sapna
Associate Scientist, Maharashtra Remote Sensing Applications Centre,
VNIT Campus, South Ambazari Road,
Nagpur (Maharashtra)

Dhankar, S.S.
Deptt. of Agro-Meteorology, CCS, HAU,
Hissar - 125 004 (HS)

Dhopte, A.M.
Retd. Professor of Botany, College of Agriculture, Nagpur
49, Vivekanand Nagar, Khamla Road,
Nagpur - 440 015 (MS)

Dhoble, M.V.
Deptt. of Plant Physiology,
Marathwada Agri. University,
Parbhani - 431 401 (MS)

Ghadekar, S.R.
Asso. Professor, Agrometeorology Section,
Dr. PDKV, College of Agriculture,
Nagpur - 440 012 (MS)

Hegde, D.M.
Director & Project Coordinator (Safflower),
Directorate of Oilseed Research,
Rajendra Nagar, Hyderabad -500 030 (AP)

Hudge, V.S.
Deptt. of Plant Physiology,
Marathwada Agri. University,
Parbhani - 431 401 (MS)

Ismail Syed
Asso. Professor,
Deptt. of Agri. Chem. and Soil Science, MAU,
Parbhani - 431 402 (MS)

Joshi, O.P.
Director, National Res. Centre for Soybean,
Indore - 452 001 (MP)

Kalane, R.L.
Asso. Prof. (Retd.), Deptt. of Agri. Chem. and
Soil Science, Dr. PDKV,
Akola - 444 104 (MS)

Malewar, G.U.
Retd. Professor and Head, Deptt. of Agri.
Chem. and Soil Science, MAU, Parbhani, 4,
Baldhan, Dadarao Layout, Sambhaji Nagar,
Parbhani - 431 402 (MS)

Muthusankarnarayanan, A.
Professor of Agronomy,
Agri. College and Res. Inst. Killikulum,
Vallanad - 628 252 (TN)

Nagdeve, M.B.
All India Coordinated Res. Project for Dryland
Agriculture, Dr. PDKV, Akola - 444 104 (MS)

Pandya, A.V.
MP Pandya Science College, Botany Deptt.
Lunawala, Dist. Panchmahals - 389 230 (Guj.)

Pandian B.J.
Professor of Agronomy,
Agri. College & Res. Inst.
Killikulum (TN)

Patil, J.D.
Professor and Head, Deptt. of Agri. Chem. and
Soil Science, & Ex. Director of Dryland Res.
Solapur. MPKV,
Rahuri - 413 722 (MS)

Rajagopal, V.
Principal Scientists & Head,
Plant Physiology, CPCRI (ICAR)
Kassaragod - 671 124 (Kerala)

Ranganatha, A.R.G.
Directorate of Oilseed Research,
Rajendra Nagar,
Hyderabad - 500 030 (AP)

Ramesh, A.
National Res. Centre for Soybean,
Indore - 452 001 (MP)

Rao, B. Bapuji
Deptt. of Agronomy,
Agricultural College, Bapatla - 522 101 (AP)

Reddy, B.N.
Directorate of Oilseed Research,
Rajendra Nagar, Hyderabad - 500 030 (AP)

Saxena, O.P.
Head, Deptt. of Botany,
Uni. School of Sciences, Gujarat University,
Ahmadabad - 380 009 (Guj.)

Sagare, B.N.
Professor of Chemistry,
Deptt. of Agri. Chem. and Soil Science,
Dr. PDKV, Akola - 444 104 (MS)

Samui, R.P.
Director, Office of the DDGM (Agrimet),
Shivaji Nagar, Pune - 411 005 (MS)

Sheelawantar, M.N.
Professor of Agronomy,
Uni. Agri. Sciences, Dharwad-580 005 (KS)

Sonar, K.R.
Retd. Professor and Head,
Deptt. of Agri. Chem. and Soil Science,
MPKV, Rahuri, 4, Swastic Apt. Savedi,
Ahmadnagar - 414 001 (MS)

Shobhane, M.R.
Deptt. of Plant Physiology,
Marathwada Agri. Uni.
Parbhani 431 401 (MS)

Zode, N.G.
Ex-Seed Research Officer, and Incharge and
Senior Rice Breeder, Paddy Research Station,
Sakoli, Distt. Bhandara. Dr. PDKV,
Akola - 444 104 (MS)

CONTENTS

Chapter - 1

SEED TREATMENT FOR SEED VIGOUR AND VIABILITY IN RELATION TO RAINFED FARMING

A.V. Pandya* and O.P. Saxena

Head, Botany Department, University School of Sciences, Gujarat University, Ahmedabad - 380 009 Gujarat

* Botany Department, M.P. Pandya Science College, Lunawada - 389230 (Panchmahals), Gujarat, India

INTRODUCTION

Deep with in the seed is a dormant miniature plant awaiting the opportunity to continue its growth. It is no exaggeration to say that civilization is based on crop storage. An efficient storage of seeds in the developing parts of the world, will be a major contribution to the solution of a global hunger.

Coles (1973) in his book *'Archaeology by experiment'* goes further by saying.

The status of human group, the organisation of its economy, the size of population, all depend to a certain extent on its success in planned food utilisation and in this, storage is of paramount importance.

Unfortunately, compared to the wealth of information on how crops were grown and cultivated thousands of years ago, that on how they were stored is limited? Each seed faces an uncertain future; the various factors are responsible for life and death of seed in storage. As the green revolution has advanced, there was a need for seed storage to preserve the seed of the improved varieties against the damages caused by poor handling, adverse climatic conditions. Loss of vigour and viability begin while the crop ripens. Following maturation biochemical changes continue and eventually catabolic processes predominate at high rate under high temperature and relative humidity (RH) and deterioration becomes apparent. As the catabolic changes continue with increasing age, the seed vigour and viability is reduced. Man has never been able to stop the ageing process of either seeds or humans, eventually seed and human die. However, the rate of seed ageing can be greatly minimised through proper storage under controlled condition.

Although, such storage are too costly for most agricultural seed lots. They may be extremely valuable for germplasm and certain high value seed stock reservation. But very little progress has been made in field of agriculture crops seed storage (Chin, 1988). There is a sizeable loss of genotype both in quality and quantity due to inadequate storage facilities (Roberts, 1972a). Harrington (1973) derived the interrelationship between seed moisture and storage temperature but these rules of thumb apply only to a condition between 0°and 48°C temperature. Apart from these, many either physical, physiological and biochemical factors not only reduce vigour of the seed but also severely affect the germination potential, productivity and even the life of seeds.

Loss of vigour and viability is a complex one and remains still without any clear picture regarding different segments leading to death. This problem has been tackled differently by different workers (Abdul Baki, 1969; Anderson, 1970; Roberts and Osborne, 1973; Srivastava, 1976; Saxena and Maheshwari, 1980; Saxena *et al.,* 1987) but it is now becoming evident that biochemical changes accompanying loss of viability are well known in many cases but still does not know the loci of seed deterioration. Various factors causing loss of vigour and viability during storage, their impact on the seeds and various modulation methods depending on the type of injury to maintain vigour can be represented diagrammatically as follows (Fig. 1).

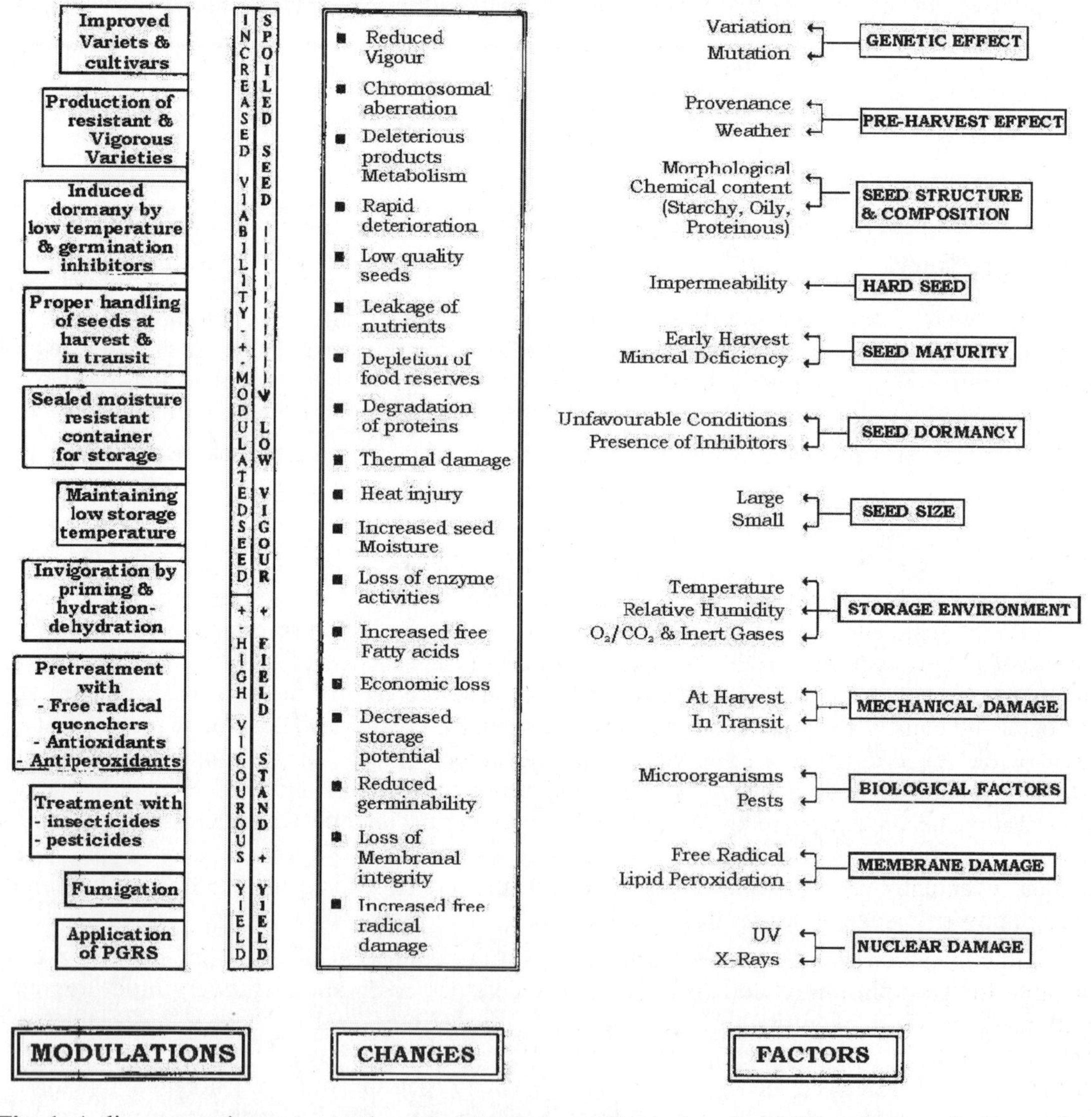

Fig. 1. A diagrammatic representation of major factors affecting seed viability, changes during seed deterioration, and their modulation to enhance seed storage potential

The most visible factors affecting seed vigour and viability is weather, i.e. Relative humidity (RH) and temperature. Seeds of most species may be safely stored for several years by careful control of temperature and RH. In a sealed storage, seed moisture content remains constant. The doubling of the life span of seeds by decreasing the moisture content one percent shows vividly the effects of slight moisture changes, especially at the critical point. Obviously, the thickness, structure and chemical composition of the seed coat affect the rate of water absorption and retention by seeds.

The term ''Seed stimulation'' or ''invigoration'' has been used to describe ways of treating seeds to increase their vigour and germination. The obvious dream of growers for an easy way of seed treatment to stimulate more vigorous germination, greater storability and yield, the willingness of certain persons to promote dubious methods of seed stimulation have created a general distrust of the entire concept. Nevertheless, there are several documented methods of stimulating greater vigour.

Presoaking seed in water has been suggested as a means to speed up germination (Rudrapal and Nakamura, 1988). The basis for the acceleration is uncertain; but it is likely that hydrolytic processes begin during presoaking and the resulting simple sugars can be immediately utilised for synthesis upon germination. Conversely, soaking seeds can be detrimental to germinability and should not be used unless definitely needed (Powell and Matthews, 1978; Mckersie and Tomes, 1980). Prolonged soaking has been found to cause injury to seeds of many species, soaking injury depends on : (i) temperature level, (ii) duration of exposure, (iii) initial moisture level of the seed, (iv) oxygen pressure, (v) Physiological maturity of seed, (vi) Chemical composition of the seed. However, major factors related to soaking injury can be avoided by proper standeredisation before exposure where, there is a minimum chance of damage.

Seed stimulation by substances other than water has been tried with varying success. Plant growth regulators (PGRs), some phenolics and antioxidants are used to modify the crop by changing the rate of the ageing (Basu, 1976; Saxena and Pakeeraiah, 1986; Pandya *et al.,* 1989). Most of these compounds mimic the natural substances present in the plants, such as hormones, metabolites and enzymes. However, the mode of action of both the natural and synthetic chemicals remains largely obscure.

The greatest controversy of all has related to silage additives, i.e. which chemical should or should not be applied on every occasion, irrespective of all other conditions. There are two fundamental points to be observed in relation to any pretreatment which may be applied : (i) Crops are not uniform, either physically or chemically and this variability needs to be reduced to obtain satisfactory results (ii) Pretreatment requires to be correctly applied; in case of chemicals, this implies both correct doses rate and uniform application.

The physiological basis of hardening and chemical treatments are said to involve in a number of physio-chemical changes, e.g. greater hydration of colloids and decrease in limpophytic colloids, higher osmotic pressure, viscosity and elasticity of protoplasm, increase in bound water, triggering of metabolic activity-respiration and photosynthesis, increase protein synthesis with decreasing free amino acids, rise in monosaccharide at the expenses of reserve polysaccharide, stimulation of alcohol and malic dehydrogenase, catalase, phosphatases and ascorbic acid oxidase and increases stress resistance power of seeds (Saxena, 1974; Purohit and Srivastava, 1993). This processes inturn lead to improve germination, vigour, viability and yield of stored seeds (Pandya *et al.,* 1988, 1988a; Saxena 1985; Saxena *et al.,* 1988, 1988a, 1988b; Saxena *et al.,* 1991). However, it is found to be detrimental especially for some thin seed coated crops. This has been overcome by using organic solvents, such as a acetone, dichloromethane, xylene, toluene etc. Which ensure the incorporation of chemical into dry seeds. The solvent (carrier) dry off rapidly after the bulk of the solution has been decanted. Thus, organic solvents assumes significance in avoiding the problem of drying which can be drastically reduced.

However, still the majority of the crop seeds viability and vigour suffer from inadequate storage conditions, container, and from poor chemical treatments techniques, causes seed deterioration faster. For these, selection of the (i) different chemical infusion medium for pre-treatment, (ii) Soaking duration and temperature, (iii) Chemical concentration, (iv) Storage container for packaging and (v) Storage condition are of paramount importance in maintaining the vigour and viability of seed.

2. INFUSION MEDIUM FOR PRETREATMENT

Seed pretreatment claimed to be a method of wide applicability, is to be considered with much caution and a newer perspective. The experimental finding shows the effect of treatment with double the volume (w/v) of the distilled water (DW) and organic solvents on the germination and seedling growth of maize, cotton and peanut seeds following varied soaking duration from 1-12 h, at 25 ± 2°C.

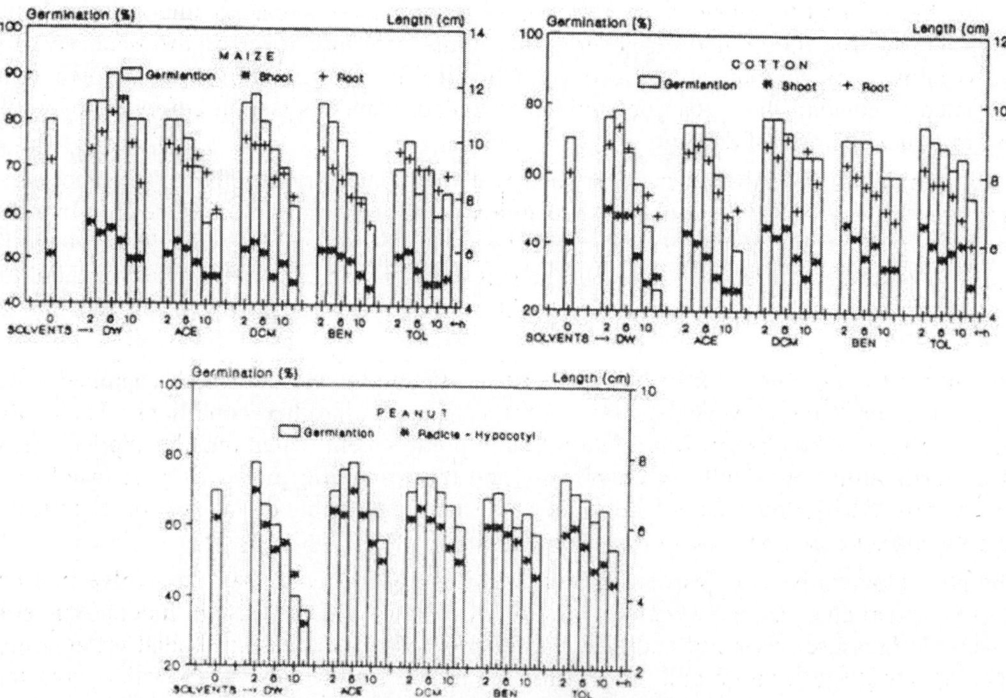

Fig. 2. Germination performance of seeds soaked in different solvents for varying time period (h)

Soaking-drying (DW) observed beneficial than solvents in maize and cotton seeds during early phase of soaking treatment. It may be due to physiological acceleration by carrying out some of the initial steps of germination during soaking (Saxena, 1974). Higher germinability of water soaked seed was found upto 8 h (maize), 4 h (cotton) and 2 h (peanut) followed by sharp reduction in percent germination and seedling vigour with comparison to unsoaked seeds. Webster and Leopold (1977) had shown that the membrane bilayers under a dehydrated state remain in a disrupted phase but on subsequent rehydration resume their normal structure within 20 minutes. Presoaking the seeds for 3 to 6 hr. followed by drying to original weight and moisture equilibration found the least damaging effect on germinability and seedling growth (Chowdhuri and Choudhuri, 1987). Further increase in soaking duration from 12 to 24 hr. has harmful effects. A seed coat rupturing from micropiler end was found in cotton seed after 4 hr. soaking, while in peanut, thin seed coat does not remain intact after 2 hr. soaking. Thus, soaking-drying treatment ultimately reduces the germinability in thin seed coated seeds.

Acetone (ACE) and dichloromethane (DCM) did not show any harmful effect on germination if the soaking duration was restricted to 6 hr. as also shown by Dey and Mukherjee (1988). Furthermore, ACE and DCM are occasionally slightly beneficial by themselves and caused faster germination, while benzene (BEN) and toluene (TOL) found deleterious after 4 hr. soaking. All the organic solvents show more or less reduction in percent germination and growth when exposure is more than 6 hr.

(a) Soaking at varying temperature (Pandya, 1992)

Seed soaking at various temperature viz. 25°, 35° and 45°C and the rate of imbibition shows direct correlation. The period of imbibition by seed is also a period of sensitivity to temperatures (Leopold, 1980). An increase in water uptake was found with subsequent rise in temperature from 25° to 45°C. Seed soaking at higher temperature caused reduction in vigour and growth. This may be due to a rapid imbibition causing soaking injury and motivating thermal damage within seed (Givelberg *et al.,* 1984). Low temperature may decrease the ability of tissue components to withstand expansion during the wetting as noted by Willing and Leopold (1983) and may also make embryonic tissues brittle and hence make them susceptible to the chilling injury. Soaking at moderate temperature (25° ± 2°C) is the protection provided against the chilling injury and thermal inactivation both during pre-and post-storage pretreatment. High vigour, percent germination and seedling length were observed during 6 hr. (Maize), 4 h (Cotton) and 2 hr. (Peanut) soaking at 25°C. Simultaneously, drying at low temperature or in a current of cool air, to safe moisture level provides protection against damgages caused during drying process. Soaking of peanut seeds for more than 2 h at varying temperature reduces the seeding emergence and growth. Therefore, organic solvents can prove better in maintaining vigour within restricted duration of 6 h at 25°C.

Saxena (1979) suggested that embryo can be dehydrated to any stage during phase one (up to 6 h), where only proteins are being made (Probably DNA polymerase), but it is not possible to dehydrate embryo once DNA replication starts (second phase up to 24 h). Osborne (1980) observed that protein synthesis, a prerequisite for correction of biochemical lesions, would start within minutes of imbibition.

(b) Chemical concentration

Studies on application of various plant growth regulators viz. gibberellic acid (GA_3), Kinetin (KIN), etheral (ETH) and abscisic acid (ABA); antioxidants/vitamins, ascorbic acid (AA) and α-tocopherol (TPH); phenolic acids, gallic acid (Ga) and tannic acid (Ta) and manganese chloride ($MnCl_2$); have been applied in a wide range of concentration (10^{-1} to 10^{-7} M) to access their effectiveness at their respective soaking duration, medium and temperature.

Application of higher doses of chemicals were deleterious for the germination and seedling growth of the pretreated seeds of maize, cotton and peanut. Hydration-dehydration (DW) treatment showed increase in per cent germination and seedling growth than unsoaked (UNT) seeds. Moreover, chemicals at very lower concentration (i.e. 10^{-6} and 10^{-7} M) could not show significant increment in germinability and growth, while marginal higher seedling dry weight was observed in KIN, ETH and AA with comparison to other treatments and UNT.

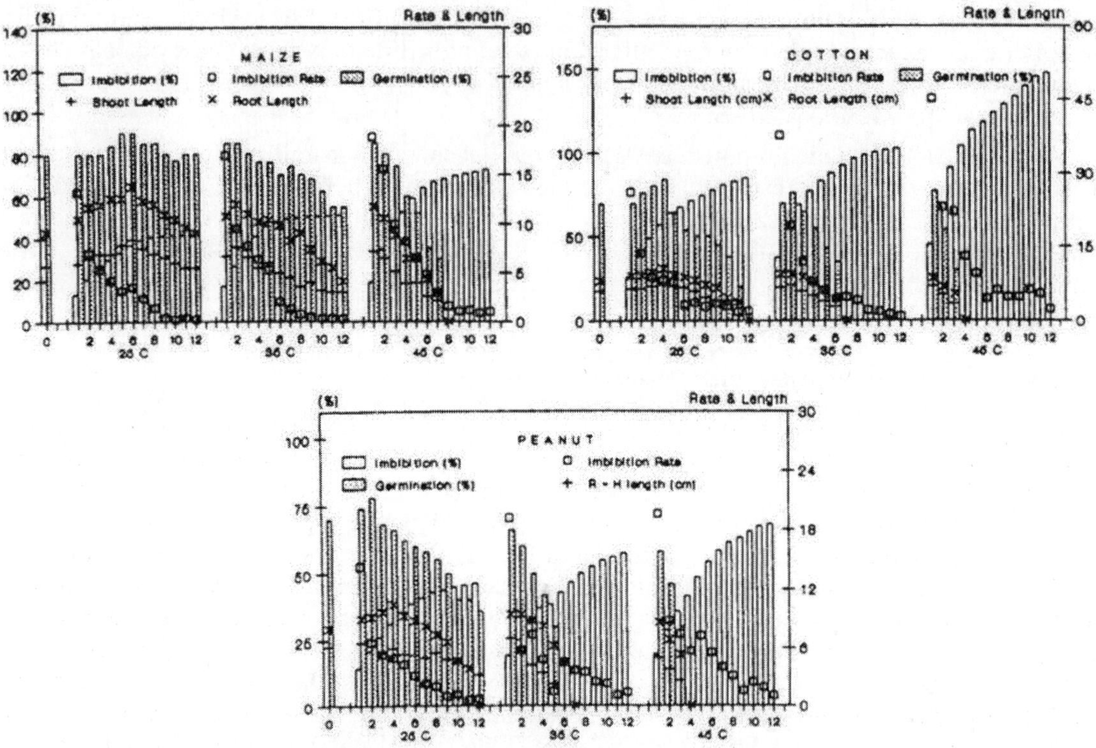

Fig. 3. Germination performance of seeds soaked in water at different temperatures for varying time period (h)

The regulation of plant growth by hormones is dependent on the modulation of endogenous levels of concentration (Hanson and Trewavas 1982). It is evident that pretreatment with GA_3 (10^{-5} M Maize; 10^{-4} M cotton and peanut), KIN (10^{-4} M maize, cotton and peanut) and ETH (10^{-5} M maize, cotton and peanut) showed highest vigour, seedling length and seedling dry weight than DW and UNT. Persson (1988) also observed high vigour and viability of hormones treated seeds. GA_3, cytokinins and etheral are known to enhance the rate of RNA synthesis especially in the transcriptional mechanisms and also increase DNA and protein synthesis (Purohit and Shrivastava, 1993). Etheral can also directly act as a signal for plant to induce genes involved in the biochemical defence against pathogens. Seeds with ABA (10^{-4} M maize and peanut and 10^{-5} M cotton) treatment inhibited the germination at 10^{-1} to 10^{-3} M. However, at lower concentration (10^{-6} and 10^{-7} M) normal growth was found. ABA also promoted the root growth by elongation. The elongation of the slow growing roots is inhibited by both IAA and ABA but the fast growing roots were stimulated when applied at a low concentration (5 x 10^{-9} M) to the maize seeds (Pilet and Saugy, 1987). Seeds with AA (10^{-3} M cotton and peanut; 10^{-4} M maize) showed more dry matter accumulation in seedling parts. Both the phenolic acid found to increase seedling growth parameters than UNT seeds. While $MnCl_2$ did not show much difference than UNT but more growth was found at 10^{-4} M (cotton) and 10^{-5} M (maize) than UNT seeds.

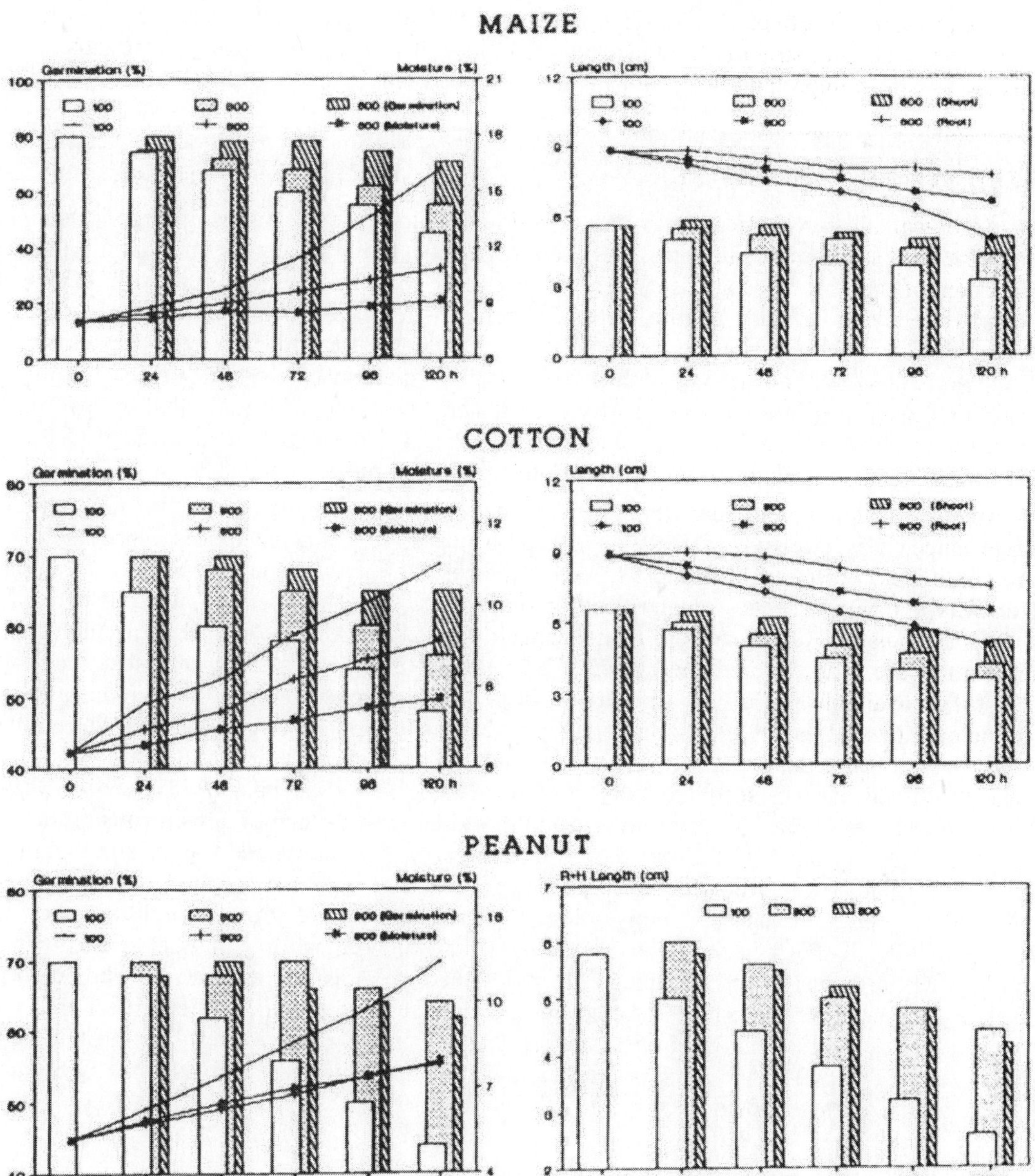

Fig. 4. Germination performance of seeds packed in 100, 300 & 600 gauge polythene bags after accelerated ageing.

Basu *et al.* (1975) found marked increase in per cent germination and seedling length of many crop species with the appropriate concentration of different salts, phenols and vitamins. Seeds treated with α-tocopherol was found beneficial at 10^{-4} M (cotton and peanut) and 10^{-5} M (maize) concentrations. These results corroborate similar finding in other seeds (Basu and Das Gupta, 1974; Basu, 1976; Dey and Mukherjee, 1988).

It is evident that effective doses for pretreatment varied greatly with the cultivars, chemical composition, seed coat structure and sensivity of cultivars to chemicals and endogenous hormones levels are few most important factors for maintaining seed vigour and viability longer.

3. SEED STORAGE CONTAINER

In tropical and sub-tropical areas were humid conditions prevail, moisture-proof container can control the increase in moisture content of the seed and maintain the vigour and viability. However, a selection of suitable package is a compromise between effectiveness and cost. Packages for processed seeds can be made of jute, cotton, paper, plastic, glass, metal, aluminum laminated and polyethylene etc. Each has characteristics that make it suitable for particular type of package or use. (Justice and Bass, 1979). Canning with metal and glass material successfully resist the ingress of moisture although they are the most costly (Freire and Mumford, 1986) and that transporting may cost impaction of seeds against other seeds or hard metal cause mechanical damage to the seeds. Furthermore, jute, paper and cotton bags are poor resistant to moisture uptake.

Amongst different containers, polyethylene bags appear to be the most promising material for storage since, it is effective in resisting the moisture ingress and needs less space for storage. However, it is necessary to select most suitable thickness of polyethylene to resist moisture uptake in different type of crop seed viz. maize, cotton and peanut. There are several methods of measuring the rates of moisture vapour transmission through a material. An easy and reliable method for an individual to use is accelerated ageing (45°± 2°C; 100% RH) of seeds along with test material from the test data available, there was a reduction in per cent germination, seedling vigour and growth with increase in seed moisture content of 100 gauge thickness polyethylene bag (PB) stored seeds. It is a well established fact that within certain limits higher seed deterioration rate was found to increase with increase in seed moisture content (Bass, 1979; Saxena, 1983). Seeds with higher moisture content possessed a heavy invasion by fungi and other microorganism causing the loss of viability (Moreno *et al.,* 1988). Seeds stored in 600 gauge PB maintained vigour and viability with 10% (Maize) and 5% (cotton) loss in per cent germination with the subsequent ingress of 1.1% (maize) and 1.3% (cotton) seed moisture content under hot and humid condition after 120 h. Increase in MC of the seed was considered as a case of permeability of heat sealed containers. Peanut seeds stored in 300 and 600 gauge PB maintained the moisture permeability and deterioration rate. The gaining of moisture also affected by thickness of polyethylene and chemical nature of the seeds. Generally seed containing more starch are more sensitive to moisture than protein and fatty (oily) seeds. PB of 600 gauge are most resistant to moisture with least increase in MC of the seeds (cotton and maize) than 300 and 100 gauge PB (Pandya, 1992).

(a) Storage condition

The apparent influence of storage condition on the moisture content (MC), germinability and seed vigour under ambient (AMB : 32°C ± 5°C) moderate (MOD : 25°C ± 2°C) and low (LOW : 8°± 2°C) temperature stored seeds. Viability and vigour of AMB temperature stored seeds declined rapidly during 36 months of storage of maize (65% germination), cotton [(25% germination (delinted)] and 15% germination (linted) and peanut (no germination)] seeds. However, seeds stored at low temperature maintain the rate of loss of viability and vigour with substantial decrease in growth parameters. Peanut kernels stored in intact pods were more vigorous than stored as naked kernels under any storage conditions (Baskin, 1979; Mehra, 1990) while linted (fuzzy) cotton seeds showed poor storage potential than delinted seeds. Usually lower RH and temperature as a moisture resistant container and lower temperature combination can greatly control the deterioration of stored seeds. (Pandya, 1992).

In recent years an interest in presoaking treatment to seeds as an aid in germination has been revived. Most research on presoaking enhancement of germination using PGRs and other chemicals has been carried out by various researchers from India and abroad, and have reported the beneficial effects of treatment to speed-up germination, growth and yield potential. But at what level of viability of pretreatment showed maximum enhancement in the vigour and viability of seed lots still remains and the whole subject of maintaining quality seeds has been controversial in view of conflicting results. Hence, more critical studies are needed on physiological changes associated with prestorage or presowing, and post-storage or mid-storage treatment which can control the rate of deterioration or modify the vigour during storage. (Pandya, 1992).

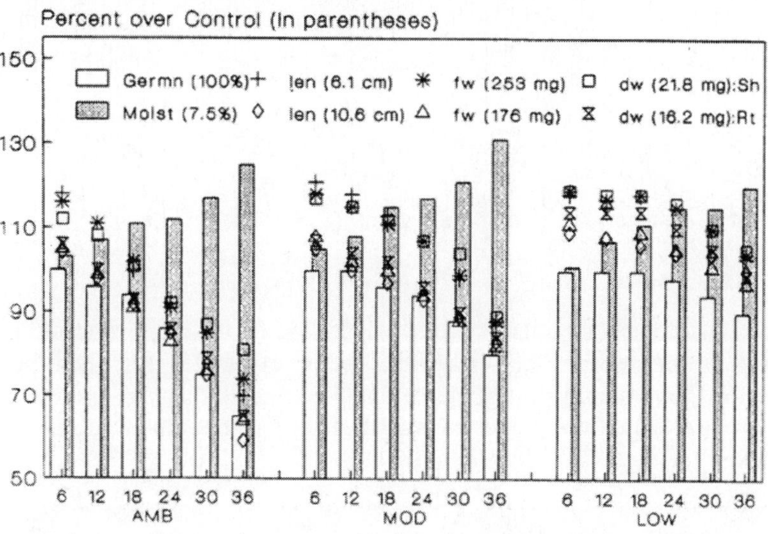

Fig. 5A. Performance of maize seeds stored at different conditions

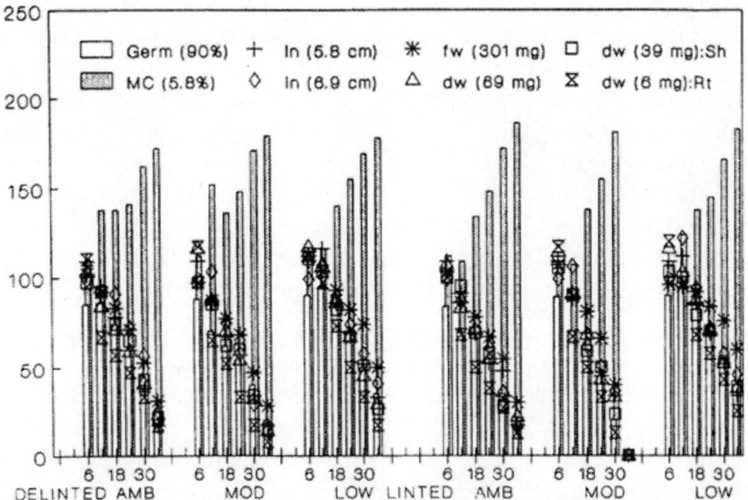

Fig. 5B. Performance of delinated and linted cotton seeds stored at different conditions.

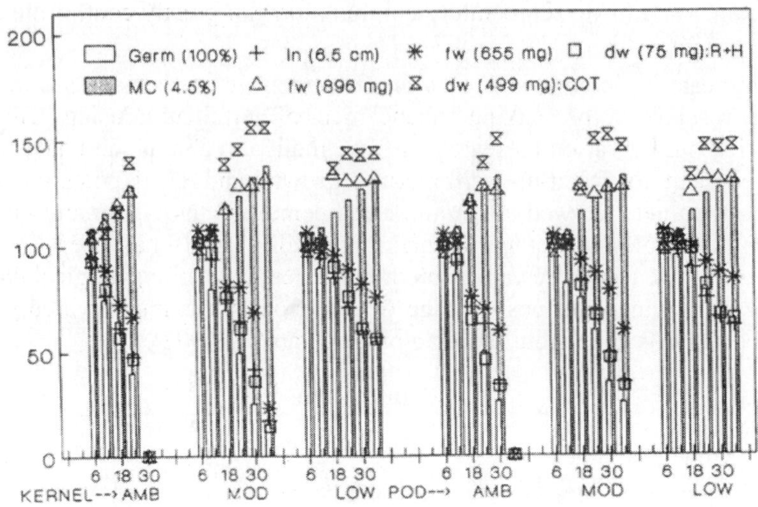

Fig. 5C. Performance of peanut kernels and pods stored at different conditions

(b) Pretreatement in relation to freshly harvested (Pre-storage) seeds

Percent germination, seed moisture content, seedling, growth etc.

Changes in physiological characters are considered to be the major and well established criteria for the seed deterioration studies. Fresh-harvested and pretreated (PGRs, phenolics and antioxidants) seeds of maize (corn), cotton and peanut (groundnut) were stored for 51 (maize) and 36 (cotton and peanut) months in heat sealed polythene bags (PB) at ambient room (AMB) temperature condition along with unsoaked seeds.

4. MOISTURE CONTENT (MC)

The negative correlation between seed viability and moisture has been reported by Ellis *et al.* (1989). Storage temperature and seed MC are the most important factors affecting seed longevity, with seed mc usually more influential than temperature. Therefore, it is essential to preserve the seeds in a suitable container to eliminate moisture to maintain seed viability at safe level (Warham, 1986). There was not much gain in the seed MC, stored in 300 gauge (peanut) and 600 gauge (maize and cotton) PB. Results indicate that PB of 300 and 600 gauge thickness prevent the seeds from gaining extra moisture (Pakeeraiah, 1985, Pandey, 1987). The pretreated seeds of maize, cotton and peanut were stored initially at 7.5, 5.8 and 4.5% MC respectively. There was slow increase in MC of untreated maize (3.0%), cotton (4.2%) and peanut (1.3%) seeds during subsequent storage of 51 (maize) and 36 (cotton and peanut) months at AMB (Pandya, 1992).

Plant growth regulators (GA_3, KIN, ETH, ABA, and AA) phenolics (Ga and Ta) and $MnCl_2$ (in case of maize and cotton) showed more or less similar pattern of moisture uptake as observed in UNT. Seeds stored at AMB showed more deviation in RH and temperature during varying seasonal changes (i.e. 12°C - 45°C; 47 to 94% RH). Here treated as well as untreated seeds of maize and cotton maintained the rate of moisture uptake upto 18 months followed by higher ingress of mc during storage. It was postulated that, higher seed mc caused depletion of reserved storage nutrients (mainly starch, protein and fats) and increased the contamination of pathogens and micro organisms leading to the reduction in seed vigour and viability.

5. GERMINATION PERCENTAGE

There was a definite trend in loss of viability with a gradual increment in storage period of fresh-harvested- pretreated seeds. PGRs and some chemicals along with hydration-dehydration greatly modified the vigour and growth. GA_3, KIN and ETH found to be effective upto 12 (maize) and 6 (cotton and peanut) months followed by rapid decline. Whereas treatment with ABA was most promising with a reduced germination percentage and growth during initial storage period but at the latter stages, it maintained higher vigour and viability longer as compared to DW and UNT. seeds of maize, cotton and peanut. Similarly, in barley, pea and sesame (Pakeeraiah, 1985); Onion, radish, cauliflower and Okra (Pandey, 1987); green gram, black gram, gram, pigeon pea, pea, cow pea and mung bean (Chhabra, 1984; Malhotra, 1990) and peanut soybean, mustard, sunflower, sesame and cotton (Saxena and Maheshwari, 1980; Saxena and Pakeeraiah, 1986; Arya *et al.,* 1990; Mehra, 1990; Pandya *et al.,* 1989; Saxena *et al.,* 1991, 1992) have clearly reported that pre-soaked seeds modified the vigour and showed better seedling performance during early phases but were poor stores as compared to UNT seeds.

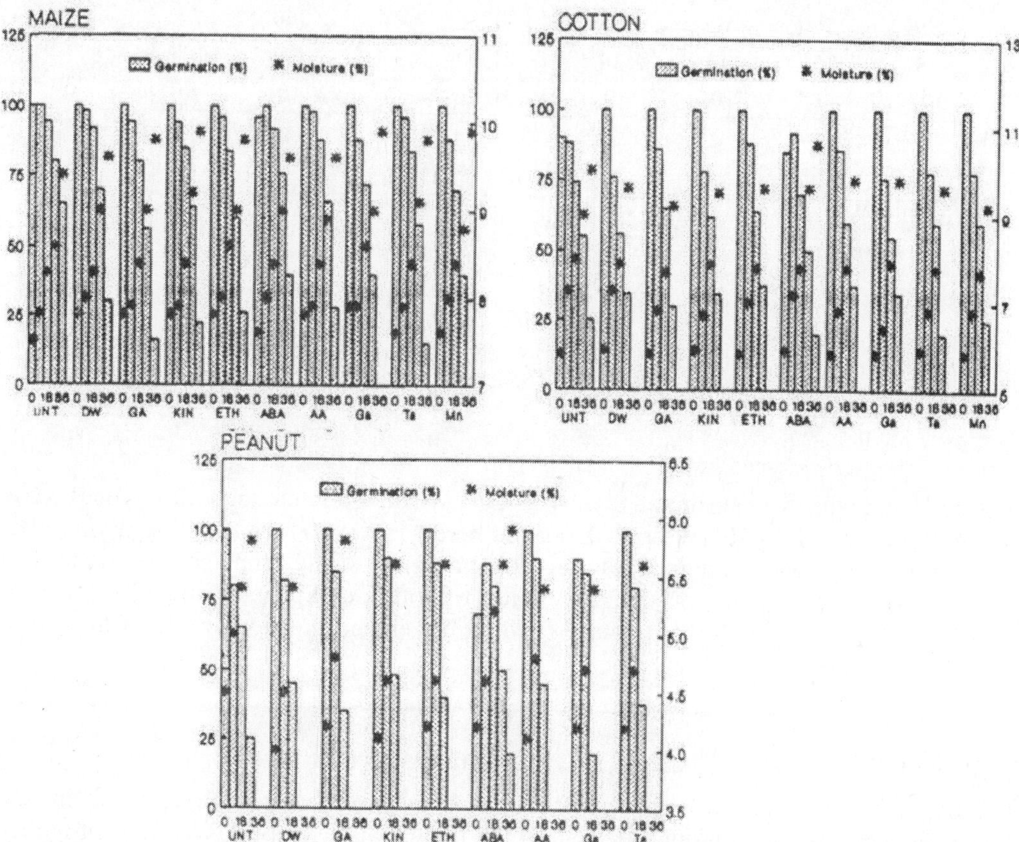

Fig. 6. Germination performance of freshly-harvested treated seeds stored at ambient conditions.

Table 1. Vigour index of freshly-harvested treated seeds of maize, cotton and peanut during storage

MAIZE

Months	UNT	DW	GA	KIN	ETH	ABA	AA	Ga	Ta	Mn
0	168	237	270	223	226	133	268	200	211	191
9	176	224	240	199	208	196	237	165	185	139
18	150	178	169	153	152	166	178	102	133	80
27	112	84	72	71	69	117	91	27	59	21
36	69	15	8	9	10	38	15	0	2	0

COTTON

Months	UNT	DW	GA	KIN	ETH	ABA	AA	Ga	Ta	Mn
0	115	162	259	232	233	105	233	213	219	192
9	115	102	154	100	124	145	131	106	108	105
18	79	56	65	55	61	87	60	43	50	48
27	39	15	13	14	21	37	18	13	7	10
36	7	0	0	0	0	7	0	0	0	0

PEANUT

Months	UNT	DW	GA	KIN	ETH	ABA	AA	Ga	Ta
0	654	712	788	818	808	284	814	585	688
9	422	413	588	655	635	572	623	457	394
18	259	115	69	204	121	450	174	38	72
27	54	0	0	0	0	202	0	0	0
36	0	0	0	0	0	35	0	0	0

The treatment with AA performed similar as GA_3. While no remarkable differences were found between phenolics, $MnCl_2$ and DW or ACE soaked seeds. However, DW soaking-drying was found to be effective during early storage. Prestorage treatment to all the crops showed rapid decline in percent germination and vigour (except ABA) after 6 (cotton and peanut) and 12 (maize) months. Basu and Pal (1979) also found that treatment would not extend the storage life of fresh harvest or even 2-3 month old seeds.

6. SEEDLING PERFORMANCE

Pretreatment with PGRs (GA_3, KIN, ETH and AA), phenolics (Ga & Ta) and DW enhanced the vigour and seedling growth. Vigour index of all the treatments (except ABA) showed higher value than UNT during early stages of storage. In case of maize and cotton, GA_3 and AA treatment scored higher vigour index while in peanut KIN and AA recorded highest vigour which followed by other phenolics, $MnCl_2$, DW and UNT. Khan *et al.* (1988) proposed that GA_3, Cytokinin ethylene and ABA may participate in the control of rice seedling elongation during complete submergence. Action of GA_3 may be closely related to the elongation process showing higher shoot-root/seedling length. It has at least two separate promotive actions on seedling growth. (i) On reserve mobilisation and (ii) On embryo growth. Whereas, cytokinin, ethylene may function by reducing the stress related changes and thus bring about elongation some what indirectly. An enhancement of seed germination by GA_3, Kinetin and Etheral infused *via* ACE was found to be most effective in many crop seeds (Persson, 1988).

The inhibition caused by ABA was counteracted with the advancement of storage period. Pilet and Saugy (1987) found that exogenous application of ABA to maize seedling promoted the root elongation at lower concentration. Finch-Savage and McQuistan (1989) suggested that ABA could make it practically possible to provide the high level of uniform germination required prior to sowing. Further more ABA treatment synchronised the germination leaving no residual effect on growth and could therefore be used as suitable for an osmoticum when priming carrot seeds for subsequent germination.

There was no significant increment found in seedling performance of phenolics and $MnCl_2$ treated seeds. Rapid reduction in germinability and vigour were observed after 18 (maize) and 12 months (cotton and peanut) while untreated seeds maintained their vigour and viability longer. Pretreatment at prestorage stage stimulates the physiological characters but their effect did not persist till the end of storage period and despite early advantages they lost viability and vigour faster. Whereas, ABA considerably slowed down the deterioration process.

Pretreatment in relation to partially-aged (post-storage) seeds

Almost every system within the seed is involved in the series of physiological and biochemical changes leading to loss of viability (Roberts, 1972). The sum effect of process such as loss of membranal integrity and enzymes activities, auto-oxidation of lipids, and degradation of functional and structural macro-molecules like enzymes, membrane proteins (Pandya, 1992, Mehra 1990) and nucleic acids by free radicals, accumulation of mutagenic substances etc. cause a loss of seed vigour and viability with an increase in abnormal seedlings. One of the more popular recent idea to explain the seed deterioration is the apparent failure of cellular repair mechanism, though, some age induced damage can be repaired during an early phase of germination (Berjak and Villers, 1972).

The PGRs and antioxidants are known to minimise such damages and improve the germinability of partially aged seeds. However, it was not possible to repair such injuries once the seed lot(s) reduced to less than 50% viability. In case of partially aged seeds, during hydration phase, the cellular repair system (Villers and Edgcumbe, 1975) could repair the accumulated biochemical lesions in the seeds. Several anti-oxidants are reported to provide some protection against the membranal and free radical damages in the seeds (Basu and Rudrapal, 1980) during ageing.

Pretreated partially aged (65% germination) seeds of maize, cotton and peanut with PGRs (GA, KIN, ETH, ABA and AA) antioxidants (α-tocopherol; TPH) phenolics (Ga & Ta) and $MnCl_2$ were stored in 300 (peanut) and 600 (cotton and maize) gauge PB for 36-69 (maize); 20-25 (cotton) and 18-51 (peanut) months as the viability of seed lots fails to 65% germination at AMB temperature to study their ultimate effect on physiological characters.

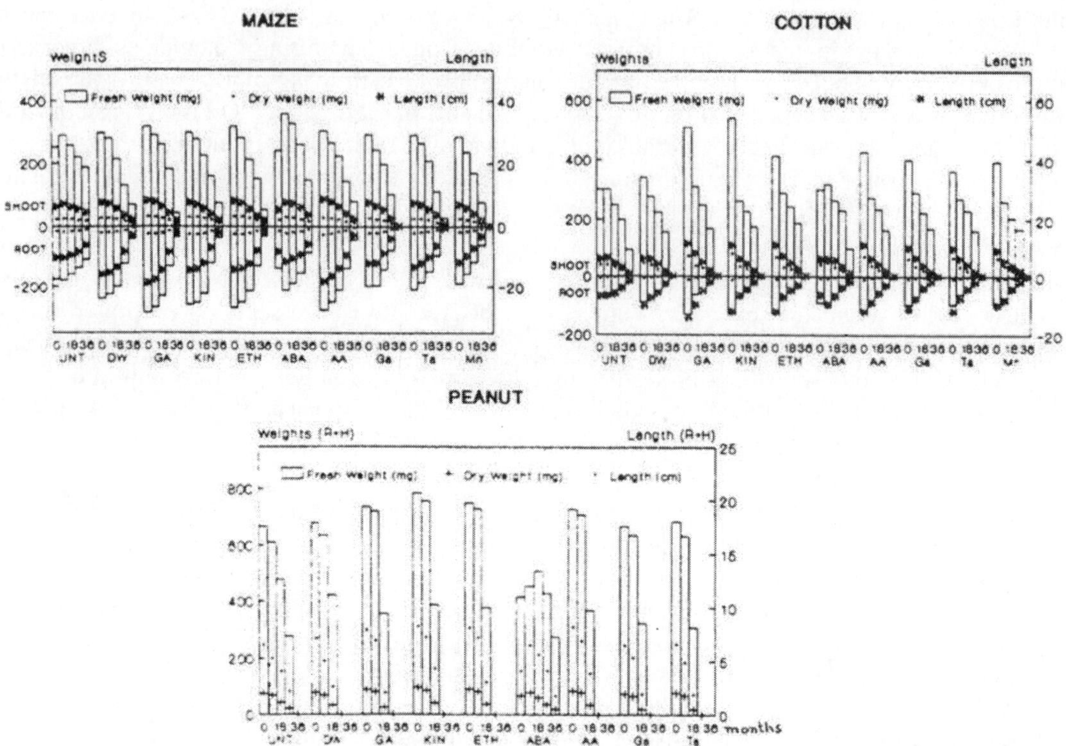

Fig. 7. Germination performance of freshly-harvested treated seeds stored at ambient conditions

Table 2. Vigour index of partially-aged treated seeds of maize, cotton and peanut during storage

MAIZE

Months	UNT	DW	GA	KIN	ETH	ABA	AA	Ga	Ta	TPH	MnCl2
36	70	110	164	157	151	75	179	142	148	185	110
45	16	77	128	126	125	129	141	103	112	138	75
54	0	35	74	79	73	103	87	61	66	63	41
63	0	0	21	26	29	47	35	23	27	30	6
69	0	0	3	3	4	20	6	2	4	9	0

COTTON

Months	UNT	DW	GA	KIN	ETH	ABA	AA	Ga	Ta	TPH	MnCl2
36	73	116	190	176	165	67	168	128	146	142	119
45	34	67	128	119	116	137	113	77	90	96	74
54	2	20	72	67	59	85	67	32	39	48	28
63	0	2	39	34	29	53	35	10	16	25	11
69	0	0	24	21	15	32	12	3	5	11	2

PEANUT

Months	UNT	DW	GA	KIN	ETH	ABA	AA	Ga	Ta	TPH
36	26	33	46	49	44	19	44	37	35	45
45	7	12	21	23	19	27	14	14	18	25
54	0	0	5	11	5	13	12	0	4	12
63	0	0	0	0	0	4	0	0	0	3
69	0	0	0	0	0	0	0	0	0	0

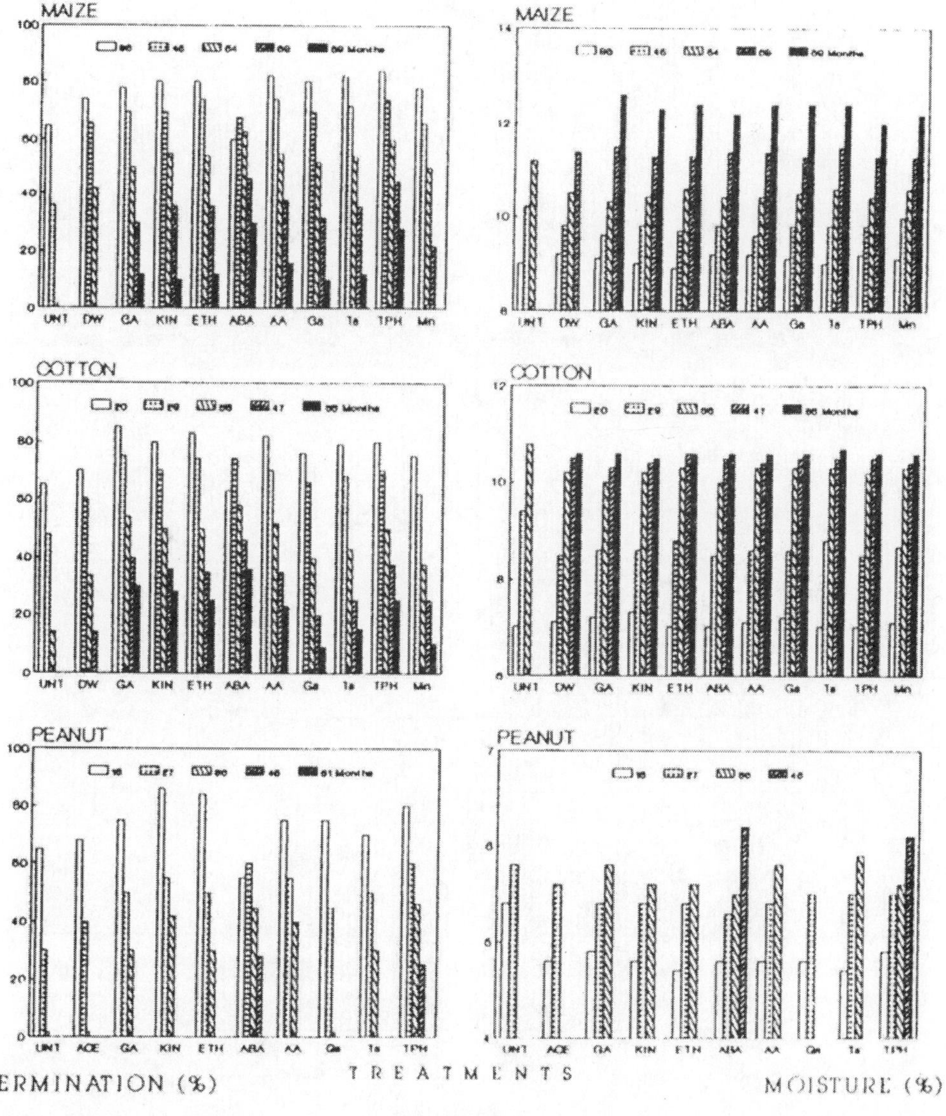

Fig. 8. Germination and seed moisture content of partially-aged treated stored seeds

7. SEED MOISTURE CONTENT (MC)

Seed MC increases at a controlled rate in case of treated as well as untreated seeds from 9.0 to 12.4% (maize), 7.0 to 10.8% (cotton) and 4.7 to 6.2% (peanut). Storage of seeds in a suitable seed containers of suitable size greatly minimised the rate of moisture uptake during prolonged storage at uncontrolled conditions.

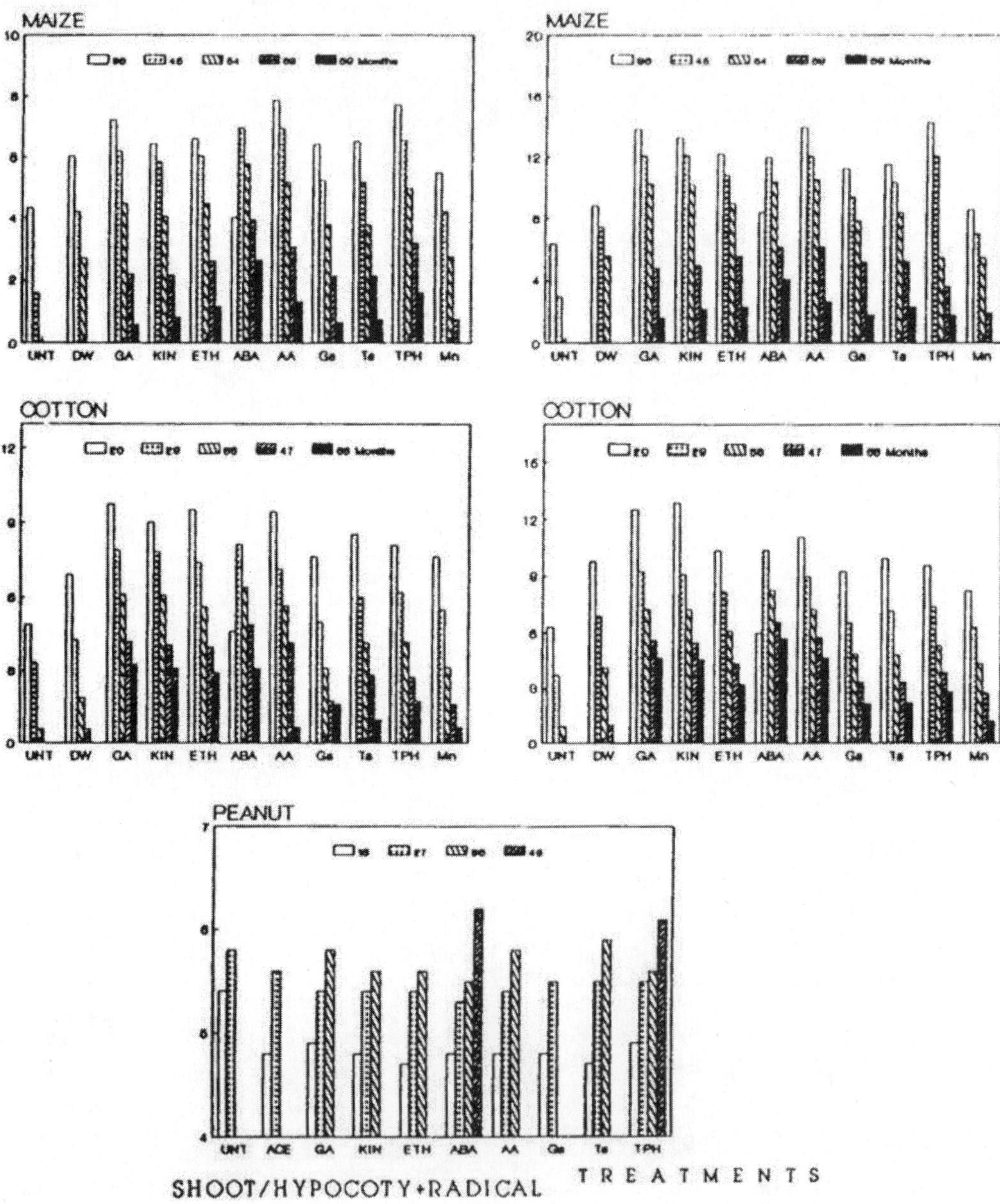

Fig. 9. Germination performance (seedling length cm) of partially aged treated stored seeds

The minimum moisture uptake was observed in case of DW (2.6%), TPH (2.8%) and ABA (3.0%) treatments after 36 months of storage. Pretreated maize seeds found more resistant to higher moisture level, they are viable at MC of 12.4-12.6% however, UNT lost germinability rapidly at 11.0% MC.

Pretreated seeds maintain their viability longer, while UNT lost faster after 12 (maize), 18 (cotton) and 9 (peanut) months of storage with about 2.2% (maize), 3.8% (cotton) and 0.2% (peanut) ingress of moisture. Peanut seeds found most sensitive ingress of moisture than cotton and maize seeds.

PGRs, antioxidants, phenolic acids and DW increases the storage potential of aged seeds at their respective moisture levels. TPH and ABA were found to be most effective treatments in maintaining the vigour and viability.

Germination percentage

Pretreatment with PGRs (except ABA), phenolics and TPH showed 10-20% increment in per cent germinability of aged maize and cotton seeds. The longevity of the seed is mainly determined by the pre-storage deterioration level. Pre-soaking with hormones enable the seeds to withstand adverse conditions reduce and/or reverse the hostile effect which would have otherwise been manifested in unsoaked aged seeds. It has been reported that endogenous level of hormones decreases during ageing and the loss of ability to produce, essential hormones which govern germination was a fundamental ageing process reducing the seed vigour and ultimately the viability (Harrington, 1973a; Pakeeraiah, 1985).

Petruzzelli and Toranto (1985) showed that GA_3 and ethaphon could withstand viability loss under adverse conditions, possibly by correcting metabolic lesions resulting from the deterioration of hormonal and enzymatic mechanism (Saxena and Pandey, 1986; Arya *et al.*, 1989; Chandak *et al.*, 1990; Pandya *et al.*, 1993). Pretreatment with Ga & Ta showed higher germination and storability over UNT of maize and cotton, while same showed poor response in peanut seeds. Research finding shows that GA_3, KIN, ETH, AA and α-tocopherol treatments maintained viability upto 36 (maize and cotton) and 18 (peanut) months over DW/ACE and UNT, where it declined to 0% within 12, 18 and 9 month respectively.

ABA showed slight reduction (5-10%) in germination initially followed by subsequent improvement in germinability (upto 9 months). This treatment greatly minimised the loss of vigour and viability. Treatment with free radical controlling agents and anti-oxidants greatly reduced the cellular damage and improved vitality of the seeds (Ramamoorthy and Basu, 1984, Tittel, 1984; Dey and Mukherjee, 1986). Ascorbic acid and α-tocopherol were found to be more suitable free radical scavenger protecting the seeds from free radical and/or radiation induced damage (Tewary and Mookherjee, 1984).

Seedling performance [Fig. 9; Table 2]

The vigour of seed at the time of storage is an important factor that affects their storage life. Vigour and viability can not be differentiated during storage experiments, especially in seed lots that are rapidly deteriorating or at midway of the viability level. Pre-treatment to the aged-stored seeds of maize, cotton and peanut remarkably enhanced the vigour and seedling growth with the stimulation of the germination rate, shoot and root/radical + hypocotyle in maize cotton and peanut with the corresponding increments in their fresh and dry weights. GA_3, KIN, ETH, AA and TPH treatments significantly recovered the losses caused during ageing. ABA treated seeds performed better than any other treatment but at the later stages of storage. Presoaking also retarded seed deterioration in *Phaseolus aureus* (Agrawal, 1983). Joshua and Heydeckar (1973) also reported than KIN in lettuce seeds resulted in rapid and uniform germination. Similarly GA_3, KIN and etheral also enhanced the vigour (Nakamura 1980; Pandya *et al.,* 1993a; Powell and Matthews, 1988) also reported increases in seedling growth by soaking-drying treatment with GA_3 in cotton, Chickpea, wheat, lettuce and peas. Etheral (Ethephon), KIN and GA_3 also markedly increased germinability of sunflower seeds (Mohammad *et al.,* 1982; Mehra, 1990).

The sum total of those properties of the seeds which determined the level of activity and performance of the seed lots during germination and seedling emergence, is known as seed vigour. Vigour index and aged pretreated seeds of all the three crops was found to be higher than UNT during subsequent increase in storage period. In maize 2.0-2.5 fold increase was observed in seeds treated with GA_3, KIN, ETH, AA and highest with TPH. Soaking-drying along and with $MnCl_2$ showed 2.0-2.6 fold higher value over UNT. GA_3 caused most marked effect in increasing the vigour of partially aged cotton seeds where as KIN recorded highest vigour index in peanuts seeds. The treatment with ABA reduced the vigour index initially but had highest value among all after 12 (maize and cotton) and 9 (Peanut) months of storage. Thus, the modultion of vigour and viability during post-storage pretreatment with PGRs and antioxidant found to be most effective than pre-storage treatment.

8. CONCLUSIONS

From the foregoing, it is evident that infusion of growth regulating substances via water is beneficial to invigorate the seeds. However, for the seeds of thin seed coat it would be more appropriate, if the PGRs and other substances are infused via organic solvent to avoid drying injury. Acetone and dichloromethane were better media within restricted exposure duration. Six hours soaking duration and moderate temperature (25°C) was found to be most suitable both for aqueous and dry permeation of chemicals. Both the factors prevent the seeds from soaking and chilling injuries. Furthermore, effective doses of PGRs and antioxidants appear to be more effective within the range of 10^{-4} M to 10^{-6} M.

Polyethelyne bags of 300 and 600 gauge considerably controlled the ingress of moisture. Seeds storage at low temperature provides higher storability than moderate and ambient temperature.

Seed treatment to the partially aged seeds is found to be beneficial for long term storage and invigoration of physiological parameters. However, treatment to the freshly harvested seed lot resulted in better seedling performance and vigour but seeds were poor storer as the viability was reduced rapidly in comparison to untreated seeds.

Problem of maintaining seed viability and vigour is a complex one and remain still without any clear picture regarding different segments leading to the seed deterioration. As deterioration proceeds, storability, potential vigour and performance capability are affected. Severe deterioration results in poor germination, vigour and plant establishment even under favourable conditions. Seed pre-treatment is claimed to be a method of wide applicability and used to increase vigour and viability of seeds stored under adverse conditions. However, the willingness of certain persons to promote dubious methods of invigoration have created a general distrust of the entire concept. Nevertheless, there are several precautions for minimising soaking injury to obtain satisfactory results in maintaining vigour and viability, although, some chemical treatments have not been found worthwhile or suitable.

Present article deals with the results of experiments on selection of proper infusion media, their soaking duration and temperature; chemical concentration for pretreatment; suitable container, storage condition and the effect of pre-storage and post-storage chemical treatment for maintaining vigour and viability longer.

It is evident from the study that infusion of chemical *via* water (maize, cotton) is beneficial to invigorate the seeds but for most of the thin coated seeds (eg. Peanut) organic solvents (especially acetone, dichloromethane) are found to be better infusion medium. Soaking duration of 6 h (I phase of germination) at moderate temperature (25 ± 2ºC) was most effective for aqueous and dry permeation of chemicals. Both the factors prevent the seed from soaking and/or chilling injuries. Furthermore, chemical concentration ranging from 10^{-4} to 10^{-6} M appears to be more effective for maize, cotton and peanut seeds. Polyethylene bags of 300 (peanut) and 600 (cotton and maize) gauge considerably controlled the ingress of moisture. Seed stored at low temperature provides higher storability than moderate and ambient temperature.

GA_3, kinetin, etheral, ascorbic acid and α - tocopherol maintained high vigour and seedling growth parameters. Pretreatment to the freshly harvested (pre-storage) seeds resulted in better vigour and seedling growth, but were poor storers as the viability was reduced at faster rate in comparison to partially aged (post-storage) treated seeds. Not only that, pre-treated aged seeds also regained the vigour and viability which were lost during ageing. Abscisic acid (ABA) also proved better in both pre-and post-storage pre-treatment for maintaining the viability longer.

REFERENCES

Abdul-Baki A.A. 1969. Relationship of glucose metabolism to germinability and vigour in barley and wheat seeds; Crop Sci. 9 : 732-737.

Agrawal B. 1983. Effect of pre-harvest, environmental condition and chemical pre-treatments on seed viability; Indian J Plant Physiol. 25 (2) : 197- 202.

Anderson, J.D. 1970. Physiological and biochemical differences in deteriorating barley seed : Crop Sci 10 : 36-39.

Arya, V, Pandya A.V., Saxena O.P. 1989. Performances of presoaked stored partially aged seeds of mustard under field conditions : Proc. Nat Symp: Current trends in Biological Sci, Vallabh Vidyanagar (Dec. 1-3, 1989), India PSC - 14 : 22.

Arya V, Pakeeraiah, T. Chandak, N. Malhotra, A. and Saxena O.P. 1990. Performance of aged seeds of gibberallic acid; Expanding Horizons in Botanical Research; Proc. All. India Bot. Seminar, Bikaner 25 : 218 - 222.

Baskin, C.C. 1979. Relation of certain physiological properties of peanut seed to field performance and storability; Doctoral Thesis, Mississippi state Univ. Mississippi State, MS.

Bass, L.N. 1979. The physiological and other aspects of seeds preservation; in the plant Seed, Development, Preservation and Germination PP 125-170 eds I Ruberstein, R.L. Phillips, C.E. Green, B.G. Geogenbach (New York : Academic Press).

Basu, R.N. and Dasgupta M. 1974 Control of seed deterioration in wheat (*Triticum aestivum* L.); Indian Agriculturist 18 : 285-288.

Basu, R.N., Bose T.K., Chattopadhyay K, Dasgupta M, Dhar M., Kunda O, Mitra R, Pal P., and Pathak G 1975. Seed treatment for the maintenance of vigour and viability; Indian Agric, 19 : 91-96.

Basu, R.N. 1976. Physico-chemical control of seed deterioration : Seed Res. 4 (1) : 15-23.

Basu R.N., and Pal. P. 1979. Physico-chemical control of seed deterioration in rice; Indian J. Agric. Sci. 49 : 1-6.

Basu R.N., and Rudrapal A.B. 1980. Iodination of mustard seed for the maintenance of vigour and viability : Indian J. Exp. Biol. 18 : 492 - 493.

Berjak, P. and Villiers T.A. 1972. Ageing in Plant embryos, II : Age induced damage and its repair in early germination; New Phytol 71 : 135-144.

Chandak, N., Thanki A. and Saxena O.P. 1990. Quantification of germination of maize using accelerated ageing technique : Intern. Symp. Environmental Influence on seed and Germination Mechanism, Recent Advances in Research and Technology. 102 : 140-141. (Jodhpur : Jan 27 - 29, 1990)

Chhabra D. 1984. Seed deterioration studies in wheat and mung; M Phil Dissertation, Gujarat Univ. Ahmedabad India.

Chin, H.F. 1988. Storage and vigour ; Seed Sci & Technol 16 (1) 1-4

Chowdhuri S.R. and Choudhuri M.A. 1987. Effects of presoaking and dehydration on germination and early seedling growth performance of two jute species under water stress condition; Seed Sci & Technol 15 : 23-33.

Coles, J. 1973. Archaeology by experiment ; Hutchison & Co Ltd. (fide)

Dey G. and Mukherjee R.K. 1986. Deterioration changes in seeds during storage and its control by hydration - dehydration pretreatments; Seed Res 14 (1) : 49 - 50.

Dey -Pathak G. and Mukherjee R.K. 1988. Invigoration of dry seeds with physiologically active chemicals in organic solvents; Seed Sci & Technol 16 : 145-153.

Ellis R.H., Hong T.D., and Roberts, E.H. 1989. A comparison of the low-moisture content limit to the logarithmic relation between seed moisture and longevity in twelve species ; Ann. Bot. 63 : 601-611.

Finch-Savage W.E., and Mc. Quistan C.I. 1989. The use of abscisic acid to synchronize carrot seed germination prior to fluid drilling; Ann Bot 63 : 195-199.

Freire, M.S. and Mumford P.M. 1986. The efficiency of a range of containers in maintaining seed viability during storage; Seed Sci & Technol 14 : 371-381.

Givelberg A, Horowitz M. and Poljakoff-Mayber A. 1984. Germination behaviour of *Solanum nigrum* seeds : J. Exp. Bot. 35 : 588 - 598.

Hanson J.B. and Trewavas A.J. 1982. Regulation of plant cell growth : The changing performance : New phytol 90 : 1-18.

Harrington, J.F. 1973. Packaging seed for storage and shipment ; Seed Sci. & Technol. 1 : 701-709.

Harrington J.F. 1973a. Problems of seed storage; in Seed Ecology pp 251-263 ed. W. Heydecker (London : Butterworths)

Joshua A. and Heydecker W. 1973. Incorporation of Kinetin in dry lettuce (*Lectuca sativa* L) seeds to permit at higher temperature; in Seed Ecology pp 55 ed W Heydecker (London : Butterworths).

Justice O.L. and Bass L.N. (eds) 1979. Packages and packaging materials; In Principles and practices of seed storage pp 142 - 168 (London : Castle House Pub. Ltd. 1979).

Khan, A.A., Thakur, R., Akbar M., Hille Ris. Lambers D., and Sensu D.V. 1988. Hormonal regulation of elongation in floating rice during submergence; Crop Sci. 28 : 121-127.

Leopold A.C. 1980. Temperature effects on soybean imbibition and leakage; Plant Physiol 65 : 1096-1098.

Malhotra A. 1990. Storage studies on some pulse crops; Doctoral Thesis, Gujarat Univ. Ahmedabad India.

Mckersie B.D. and Tomes D.J. 1980. Effect of dehydration treatments on germination, seedling vigour and cytoplasmic leakage in wild oats and birds foot trefoil; Can J. Bot. 58 : 471-476.

Mehra (nee Arya) V. 1990. Physiological changes accompanying loss of viability of some oil seeds; Doctoral Thesis Gujarat Univ. Ahmedabad India.

Mohammad S., Sunder R.S., Dutt. K.V.L.N. 1982. Effect of growth regulators on germinability of sunflower; Field Crops Abst 3 (12) : 9222.

Moreno E., Benavides C. and Ramirez J. 1988. The influence of hermatic storage on the behaviour of maize seed germination; Seed Sci & Technol. 16 : 427-434.

Nakamura S. 1980. Effect of fusiccoccin in the germination of agricultural seeds, 19th ISTA Cong. Vienna (June 6-13) Reprint No. 32 SV : 1 -6.

Osborne D.J. 1980. Senescence in seeds, in Senescence in plants pp. 13-37 ed. K.V. Thimann (Florida : CRC Press Inc.)

Pakeeraiah, T.1985. Presowing and storage studies in barley, pea and sesame; Doctoral Thesis Gujarat Univ. Ahmedabad India.

Pandey N. 1987. Seed deterioration studies in some vegetable crops; Doctoral Thesis, Gujarat Univ. Ahmedabad India.

Pandya A.V., Chandak N., Singh G. and Saxena O.P. 1988. Storage potential of hormone - pretreated wheat seeds; Ist Reg. Conf. APPSC, Kathmandu, Nepal (Oct. 3-7) G-59, p. 43.

Pandya, A.V., Thanki A., Chandak N. and Saxena O.P. 1988a. Germination and growth behaviour of pretreated aged cotton seeds under controlled and natural conditions, 57th Ann. Body Meet. SBC (Abst. 443). Oct. 9-12 New Delhi.

Pandya, A.V., Arya V., and Saxena O.P., 1989. Effect of plant growth regulators and chemicals on dry matter accumulation and crop productivity of partially aged seeds of cotton, In proc. Nat. Sem. Agric. Biotechnol. Navasari pp. 80-85 (March 7-8, 1989).

Pandya A.V., 1992. Viability studies on pretreated stored seeds of peanut, cotton and maize; Doctoral Thesis, Gujarat Univ. Ahmedabad, India.

Pandya, A., Arya, V., Malhotra A. and Saxena O.P., 1993. Seed storage potential of presoaked seeds of cotton (*Gossypium hirsutum* L.), mustard (*Brassica juncea* L.) and cow pea (*Vigna unguiculata* L); Seed Res (special Vol. 1) pp 300- 305 eds S.P. Sharma, K. Kant, A. Varier, Published by Indian Soc. Seed Technol, Div. Seed Sci. & Technol, IARI, New Delhi, India.

Pandya, A.V., Malhotra, A.S., Saxena O.P., 1993 a. Performance of partially aged seeds of cotton and maize : Presoaking seed treatment under field condition; The 3rd Conf. of Assoc of P1 Physiologist of SAARC Coun. (APPSC) and Annual Conv. of PGRSI on the Pl. physiol. & Biotechnol in relation to improving plant productivity (April 9-11, 1993) Bot. Dept. Guj. Univ. Ahmedabad, India. Abst 5.9

Persson B. 1988. Enhancement of Seed germination by Plant growth regulators infused via acetone; Seed Sci. & Technol 16 (2) : 391- 404.

Petruzzelli L. and Taranto G. 1985. Effects of permeation with plant growth regulators via acetone on seed viability during accelerated ageing; Sci. & Technol. 13 : 183 191.

Pilet P.E., and Saugy M. 1987. Effect on root growth of endogenous and applied IAA and ABA : A critical re-examination; Plant Physiol. 83 : 33-38.

Powell, A.A. and Matthews, S. 1978. The damaging effect of water on dry pea embryos during imbibition; J. Exp. Bot. 29 : 1215-1229.

Powell A.A. and Matthews, S. 1988. Seed treatments : Developments and Prospects : Out look on Agriculture 17 (3) : 97-103.

Purohit S.S., and Shrivastava, P. 1993. Mechanism of hormone action : The 3rd Conf. Asso. Plant Physiologists of SAARC Countries (APPSC) and PGRSI on Plant Physiol & Biotechnol in relation to improving plant productivity (April 9-11, 1993, Ahmedabad, India) Abst 4.36.

Ramamoorthy, K. and Basu, R.N. 1984. Control of seed deterioration in groundnut (*Arachis hypogaea* L.) by hydration- dehydration treatments; Plant Physiol & Biochem 11 (2) : 148- 152.

Roberts, E.H. (ed.) 1972. Cytological, genetical and metabolical changes associated with loss of viability; In viability of seeds pp 253-306.

Roberts E.H. (ed) 1972a. Storage environment and the control of viability; In Viability of Seeds pp 14-58 (London : Chapman & Hall).

Roberts B.E. and Osborne D.J. 1973. Protein synthesis and viability in rye grains : In Seed Ecology pp 99-114 ed W Heydecker (London : Butterworths)

Rudrapal D. and Nakamura S. 1988. The effect of hydration-dehydration pretreatments on egg plant and radish seed viability and vigour; Seed Sci. & Technol 16 : 123-130.

Saxena O.P., 1974. Presowing hardening treatments and crop production : A Review : Planta 4 : 1 - 11.

Saxena, O.P., 1979. Physiological aspects of seed deterioration; In current Advances in Plant Reproductive Biology pp 328-348 ed C.P. Malik (New Delhi : Kalyani Publishers)

Saxena O.P. and Maheshwari D.C. 1980. Biochemical aspects of viablity in soybean : Acta Botanica Indica. 8 : 229-234.

Saxena O.P. 1983. Physiological parameters associated with accelerated ageing studies in some crops : In. Ann. Conf. Adv. Bot. (May. 28-28, 1983) pp 27 HAU, Hissar.

Saxena O.P. 1985. Presowing seed treatments : A review : In. widening Horizons of Plants Sciences pp 199-217. ed C.P. malik (New Delhi : Cosmo Publication)

Saxena O.P. and Pakeeraiah T. 1986. Seed deterioration studies : Indian Rev Life Sci. 6 : 189-214.

Saxena, O.P. and Pandey N. 1986. Role of gibberellic acid in maintaining viability of onion seeds; 22nd Hort Congr. California, Davis (Aug. 10-18, 1986) Hort. Sci. 21 (3) : 832.

Saxena O.P., Singh G., Pakeeraiah T. and Pandey N. 1987. Seed deterioration studies on some vegetable seeds; Acta Hort. 215 : 39-44.

Saxena, O.P., Arya V., Pandey N. and Maheshwari D.C. 1988. Performance of the aged seeds under the influence of bioregulants : In. Proc. Pl. Growth Reg. Soc. Amer Meeting, San Antonio pp 194-204.

Saxena, O.P., Arya, V., Maheshwari D.C. and Broker H.C. 1988a. Performance of aged seeds under the influence of bioregulants : 416. In. 13th Intern. Conf. Pl. Growth Substances (July 17-22, 1988) Calgary

Saxena, O.P., Maheshwari D.C., Pandey N. and Arya V. 1988b. Low temperature storage and use of bioregulants to maintain viability of soybean, peanut and onion; In. Int. Conf. Research in Plant Sciences and its Relevance to Future pp 39 Delhi Univ. Delhi India.

Saxena, O.P., Thanki A. and Pandya A. 1991. Seed deterioration - A Review : In Current concepts in Seed Biology, Prof. S.N. Dixit Commemoration Volume Chap. 9 : 133-145 (Gorakhpur, India)

Saxena O.P., Pandya A. and Arya V. 1992. Role of PGRs on improvement of seed germination vigour and yield potential of some soil seeds at varying levels of viability; In. Vistas in Seed Biology Vol. II pp. 161-171 eds T. Singh and P.C. Trivedi Univ. of Rajasthan (Jaipur : Printwell, India)

Srivastava A.K. 1976. Physiological Studies on soybean seed viability during storage and its practical application; Seed Res. 4 : 56-61.

Tewari B.K. and Mookerjee A. 1984. Effect of pre-and post- treatment of ascorbic acid on wheat (*Triticum aestivum* cv *sonalika*) and maize (*Zea mays* cv ganga) seeds exposed to gamma radiation; Indian J. Plant Physiol 25 : 258- 265.

Tittle C. 1984. Improving the storability of seeds by chemical treatment; Field Crop Abst. 38 (5) : 2718 pp 300 (Abst).

Villiers T.A. and Edgecumbe D.J. 1975. On the cause of deterioration in dry storage; Seed Sci. & Technol. 3 : 761-774.

Warham E.J. 1986. A comparison of packaging materials for seed with particular references to humid tropical environments; Seed Sci & Technol. 14 : 191-201.

Webster B.D. and Leopold A.C. 1977. The Ultrastructure of dry and imbided cotyledons of soybean; Amer J. Bot 64 : 1286-1293.

Willing R.P. and Leopold A.C. 1983. Cellular expansion at low temperature as a cause of membrane lesions : Plant Physiol. 71 : 118-121.

Chapter - 2

WATERSHED CONCEPT AND YIELD OPTIMIZATION WITH LAND USE PLANNING UNDER DRYLAND FARMING

K.S. Bhaskar

Retd. Principal Scientist (Agronomy), Central Institute for Cotton Research, Post Bag. No 2, Shankar Nagar Post Office, Nagpur- 440010 MS

INTRODUCTION

The rapidly increasing population of the country, shrinking good quality land resources for crop production and increasing concern for environmental degradation, creates an urgency for continuously enhancing the productivity of the land in the country. Soil, water, climate, vegetation, nutrients and energy are the basic natural resources needed for agricultural production. Because of population pressure, these natural resources are shrinking very fast. Since, agricultural development is not possible on deteriorating natural resources base, conservation of natural resources needs urgent attention; not only to fulfill the present need but for posterity also. Future challenges of food security are intimately linked with the environmental security. Increasing population would demand a continuing increase in agricultural productivity and production ensuring the quality of the resource base is atleast maintained, if not improved. The scenario requires development of strategies that take into account the potential and limitations of the natural resources for their sustainable use coupled with improving productivity and production. The strategies should be need based and obviously, have to be specific for a given agro-climatic region or sub-region (Anonymous, 2000b).

With the current food grain production of about 206 Mt, India needs to produce an additional 5-6 Mt of food grain annually in the next decade to meet the requirement of an estimated population of nearly 1120 Million in 2010 A.D (Table 1). This high population also necessitates significant increase in the production of other agricultural products like oil seeds, cotton, animal products, fruits, vegetable, fodder and timber etc. In the past, more attention was rightly given on pulses, oilseeds and cereal production, but for making balanced diet a reality, live stock, horticulture, poultry, fishery etc. are also have to be given needed encouragement. The added problem is that in the coming years, more food and other agricultural commodities have to be produced under conditions of diminishing per capita arable and abiotic stresses. In other words, the need for more food and other agricultural commodities has to be met through higher yields per unit of land, water, energy and time. This obviously is uphill task, but there is no escape from accepting the challenge and achieving the goal. Meeting the challenging task necessitates judicious utilisation of our natural resources through adoption of sustainable management practices at watershed basis. The important questions to be addressed are national food scarcity, nutritional security, maintenance of soil health, enhancement of soil productivity and leaving a good heritage for the posterity.

Table 1. Food requirement in 21st Century (Million tonnes)

Crop	Year		
	1996-97	2001-2002	2006-2007
Rice	81.2	89.8	98.8
Wheat	65.2	72.5	80.7
Coarse grain	30.7	32.6	34.4
Pulses	15.9	18.4	21.5
Oil	6.5	7.9	9.5
Sugar	14.5	16.8	19.6
Vegetable	77.3	91.7	108.5
Fruits	40.4	52.6	69.1

Beside this, highly intensive agriculture would be needed to meet the growing demand of ever increasing population. It would call for more efficient management of inputs like seeds, fertilizer, water, plant protection measures, weed management and labour. Resource management has direct social and economic implications therefore future activities have to originate from and should address the social and economic needs of the farmers. Farmers should be the nucleus of all research and developmental activities. Bottom up approach, involving the farmers in the project planning, execution, monitoring and feed back would facilitate better understanding of farmers problem and proper use of resources available at the field in watershed. So far, adequate informations are lacking on characterisation and evaluation of soil and water resources and climatic parameters at micro level, which are very essential for efficient land use planning and resource management, inventorization, characterisation and monitoring of natural resources.

Development of proper land use plans for each agro econological zone, considering their resource base, potential productivity, risk factors and social acceptability at micro level has to be attempted. It would help in creating essential infrastructure to support the system for yield maximisation without causing ecological problems, the goal of the sustainable agriculture production can be achieved only with rational and scientific management of plant nutrients as also other resources. Water is one of the most important crucial input for improving agricultural productivity. Expansion of irrigation network therefore, has been the key strategy in the agricultural development. Even, if India's estimated irrigation potential of about 114 Million hectares is achieved about 20% of total cultivated area would still remain as rainfed. However, considering the kind of investment needed for the targeted expansion of irrigated area, any significant progress in this direction is not in sight. Consequently, the vast rainfed area has to be managed through watershed approach to improve the productivity of cotton, pulses, cereals, horticultural crops, livestock and fishery etc. (Anonymous, 2000b).

Irrigated area has a potential to produce 10 t of food per hectare per annum. Better management of irrigated area (about 43 m ha) alone can thus produce about 400 mt of food grains. Irrigation system management and enhancement of water productivity by about 10% from the current level has to be achieved. Cost effective drainage technology for saline and water logged conditions needs to be developed to reduce the cost of reclamation considerably. Development of technology for efficient and so far utilisation of poor quality waters for crop production should get top priority as the share of water availability to agriculture is likely to reduce considerably. Efficient irrigation system like drips, sprinkler and fertigation have to be given emphasis in the years to come. (Anonymous, 2000).

Climate plays a key role in agricultural production through rainfall, sunshine, temperature and humidity. Rainfall is the main source of soil moisture. It must be conserved and protected at any cost, first at field, farm and finally at watershed level. Excess runoff should be collected, stored and recycled to crop as life saving irrigation as and when there is a need of water to crop. Water storage in the water harvesting pond will not only satisfy the water need of plants but at the same time it will increase the recharge capacity of irrigation wells also. The importance of temperature and humidity in enhancing plant nutrient availability and absorption and also the role they play in disease and pest infestation is well documented. Development of proper methodologies for improvement of agrometeorological advisory services and their efficient use in mitigation of adverse effect of aberrant weather conditions on agricultural production system has to be attempted. Weather forecasting can also help in planning the nutrient addition and also timely and effective control of pests and disease through adoption of integrated plant protection management system (IPPMS) consisted of chemicals, biological and mechanical control measures of plant enemies (Anonymous, 2000b).

The land use planning is integral part of rainfed farming. Different types of soils and its appropriate use for sustainable dryland farming is the need of the day. Various multivariate regression analysis expands the scope for proper land use and can help rainfed farming viable with confidence. Centrally sponsored micro watrershed projects explored at Kutki & Saongi on cotton and sorghum in Maharashtra are described for the benefit of readers.

Food security is the country's prime need and increase in food production is pre requisite to it. If, natural resources are properly managed for the human need, but not the greed, nutritional security and balance diet would pose no problem. Environmental problem is a global one. Therefore, global thinking and local actions are of great importance. If nature is not properly cared for, it would take appropriate action. Therefore, the slogan should be "Love those whom, he loves" (Anonymous, 2000b).

2. WHAT IS WATERSHED?

Watershed is defined as "Drainage area on earth's surface from which runoff resulting from precipitation flows fast through a single point into a stream, a river, a lake or ocean. Magrath and Doolette (1990) defined the watershed as "The area of land contained within a drainage divided above a certain specified point of stream. Watershed size vary considerably and each watershed contains number of sub or micro-watershed or in other way any watershed is a part or a micro-watershed of larger one. Depending upon the field situation and topographys each field of a farmer is micro-watershed for implementation of planned programme.

The prime object of the watershed management is human welfare through sustainable production and environmental upgradation. In rainfed watersheds, the objective would in - siti moisture conservation development and utilization of natural resources, for productive, stable, sustainable and equitable agro-eco-system. In short, water management in rainfed areas is an adoption of appropriate rain water management programmes over the entire area of the water.

The watershed being the geohydrological unit, the sustainable production of food, feed, fuel, fibre, fodder and timber would be a product of scientific utilization and management of natural resources of lands and water without any further deterioration of environment. The land resources comprising of soil, water and associated plants, animals involving the total eco-system are being mismanaged with the result as much as 57 per cent land in India is degraded. The increasing human and livestock population, ecological imbalances and environmental degradation, regional imbalances in production systems and non judicious utilization of resources have forced to manage problems within the boundaries of watershed leading to sustainable production of food, feed, fruit, fiber, fuel, fodder, timber and shelter.

The watershed approach aimed at achieving rational utilization of all the natural resources for sustainable production of biomass with the least damage to the environment, is the only technically sound, economically viable and socially accepted strategy for the conservation, development and management of land and water resources in the country.

3. CONCEPT AND NEED OF WATERSHED MANAGEMENT

Watershed based agricultural research and resource management approach for efficient crop production is being prioritized to replace the traditional system of sporadic experiments and for the same time obtaining greater effectiveness and higher resource use efficiency. During recent past, watershed based holistic and integrated development approach has emerged as a potential approach for the development of rainfed areas, which can lead to higher productivity and sustainability in agricultural production. The productivity of rainfed crops largely depends on soil-site and elimatic characteristics of the cropped area. Each crop requires a specific set of climatic, soil-site characteristics and management inputs for optimum production. But in majority of cases, the soils are cultivated to different crops influenced by other factors including social ones. This ignoring the soil-site requirement of particular crop, leads to sub optimum yield or complete failure of the crop(s). This necessitated to properly characterised and quantify such soil-site parameters at watershed basis for suitability assessment of commonly grown rainfed crops or a cropping system as a whole.

Production ability of the land is deliminated by its soils, land form conditions, and the climate which is turn in conditioned by the nature of land use and the management of land. It is therefore, prerequisite to have indepth knowledge of land resources endowment and its potential for planning economic development. In land use planning, questions often asked are : is there sufficient land to meet future food and agriculture production, where are those potential lands, what land use are suitable, what is the level of technology required, what is the risk of degradation and measures required for minimizing the risk and what are the research, extension education needs for production optimisation. The present study will give the answers of these questions only at watershed based research and development. All such measures are possible at watershed basis and this will be a unit of every kind of natural resource management. This is the main reason that Government of India, various state departments and NGO's are investing crores of rupees for the development of watersheds in the country. It is true that there is no other alternative, except this for the overall development of rainfed areas of the country. Watershed is only the unit, where zero runoff and zero soil loss concepts could be developed efficiently.

Pre requisite for watershed management

This watershed management implices judicious use of soil and water resources in the areas of catchment, therefore, for giving land treatment land use recommendations, following data would be needed (Varma, 2000).

Watershed:	Size, shape, slope and relief
Soil:	Geology, physiography, physical and chemical characteristics

Hydrology:	Drainage pattern, stream density, moisture pattern
Cover conditions:	Grass type, stocking density, associated vegetation
Climate:	Seasonality, rainfall, temperature, humidity, wind velocity, sunshine, frost, fog and snow
Geographical situation:	Altitude, latitude and longitude
Social and legal:	Government, private, clan, tribe states

Keeping this in view, Government of India has started watershed development programme in the country as under:

National watershed development program for rainfed areas

The centrally sponsored scheme for National Watershed Development Project for Rainfed Areas (NWDPRA) was launched in 1990-91 and covers 25 states and 3 union territories. The project aims to achieve twin objectives of sustainable production of biomass and restoration of ecological balance in the vast rainfed areas of the country. It specially focuses on:

(i) Conservation, upgradation and utilisation of natural endowments like land, water, plant, animal and human resources in a harmonious and integrated manner with low cost, simple, effective and replicable technology

(ii) Generation of massive employment and

(iii) Reduction of inequalities between irrigated and rainfed areas.

It is targeted to cover an area of about 28 lakh hectare in all the community development blocks of the country where less than 30 per cent arable area is under assured means of irrigation with an outlay of Rs. 1,080 crore. The requisite integrated watershed development plans of the identified areas have been developed and are being executed with active participation of watershed community. The project is being implemented by the participating state governments through cent per cent funding from the centre comprising 75 per cent grant-in-aid and 25 per cent loan to the state.

Salient achievements in the field of integrated watershed development are as follows:

(i) Unified command approach of multi disciplinary team has proved to be an effective organisational set up for implementing the watershed management development programme

(ii) Contour farming in both black and red soils of different crops

(iii) Different farming systems based on contour, vegetative barriers in watershed development programme has shown substantial in situ moisture conservation and yield increase and

(iv) Cropping intensity has increased significantly following the integrated approach of farming

In order to strengthen the activities of watershed development in the country, a full fledged division of rainfed farming system has been created in the Ministry of Agriculture, Government of India.

4. LAND RESOURCE AND SOIL DEGRADATION IN INDIA

A large variety of soils spreading all over Indian sub continent are derived from a wide range of geographical rocks and parent materials interacting with soil forming factors like climate, vegetation and topography for varying time period. Out of 329 M ha of geographical area of the country, it covers 105.5 M ha red soils, 73.5 M ha of black soils, 58.4 M ha alluvial soils, 11.7 M ha of laterite soils and 30.0 M ha hill and tarai soils (Singh, 2000b). The capacity of soil to produce is limited by intrinsic characteristics, agro-ecological setting and land and water management. The nature and distribution of different kinds of soils have been surveyed by state and central soil survey organizations under the over all coordination of National Bureau of Soil Survey and Land Use Planning of the ICAR and soil maps at 1:250000 scales have been prepared. Detailed map at 1:50000 scale for district level planning and 1:10 000 scale for watershed level planning have also been developed on demand for selected areas.

Both due to intrinsic features and human interventions, the soil resources of the country are being constantly exposed to different forces of degradation. The reported statistics of land degradation in India varies widely. The potential human induced land degradation is reported to be as high as 187.7 M ha. (Paroda, 1998, Anonymous, 2000). Atleast 57.1% of geographical area of country suffers from different forms of land degradation and water erosion accounts for 65% land degradation (Table 2).

Plate 2.1. There is land, but no soil due to faulty crop cultivation exposed to severe soil erosion.

So far, the research and resource organizations in India has characterised the soil and prepared the soil maps of the country (1:1 Million scale), state (1:250000) and districts (1:50000) depicting extent of degradation (1:4.4 Million). Thematic maps depicting soil depth, texture, drainage, calcariousness, salinity, pH, slope and erosion have been published. Based on soil correlation work, about 180 soil series have been given the status of 'established series' in addition to 78 soil series given tentative status. Much efforts have gone to delineate 21 agro ecological zones and 60 sub zones for the country based on physiography, soil climate, length of growing period and available water content and mapped on 1:4.4 Million scale. Digitisation of physiography and soil resource base on 1:50000 scale for 14 state have been completed. Fertility status of Indian soils has also been characterised. Soil fertility maps of N, P, K, S, Zn, Cu and Mn have also been prepared. Guidelines to fertilizer application based on soil test for targeted yield have been developed.

Plate 2.2. Cotton cultivation on shallow soil not profitable due to shallowness and low moisture holding capacity of soil.

Table 2. Extent of soil degradation in India

Degradation type	Area affected (M ha)	
	Total	% of Geographical area
1. Water erosion	**148.9**	**045.3**
a. Loss of top soil	132.5	040.3
b. Terrain deformation	016.4	005.0
2. **Wind erosion 013.5 004.1**		
a. Loss of top soil	006.2	001.9
b. Loss of top soil/ terrain deformation	004.6	001.4
c. Terrain deformation/over blowing	002.7	000.8
3. **Chemical deterioration**	**013.8**	004.2
a. Salinisation	010.1	001.1
b. Loss of nutrients	003.7	003.1
4. **Physical deterioration**	**011.6**	**003.5**
Water logging	011.6	003.5
Total affected area	187.7	057.1

Source : Paroda (1998)

Climate characteristics

In our country, considering as a whole, 74% of the mean annual rainfall is being received during June to September. It ranges from less than 300 mm in Western Rajasthan to over 3000 mm in parts of NE India. The undependable and erratic rainfall introduced an element of risk, uncertainty and instability in crop production. The results of research have shown that adoption of suitable measures can remedy the situation to a maximum extent. Climate is of basic importance in determining the suitability of land area for agriculture including suitability for different crops and the relative merits of various types of farming systems. There are 3 most important components that are:

- Temperature
- Precipitation and
- Light.

Water resource of India

Water is life and one of the most important natural resource for the over all development of a nation. The source of all water is precipitation. On an average, the annual rainfall over the Indian subcontinent has been estimated at 1200 Million ha m. On this basis, the annual precipitation including snow fall is estimated at 400 Million ha m (4000 km^3). Rain water is a crucial resource and key input in rainfed agriculture. The success of agricultural enterprise depends largely on how efficiently and effectively we conserve and store the rain water and how wisely we use this precious resource at farm level. Out of 400 Million ha m annual rain water, hardly around 150 Million ha m enters in the soils and sizeable amount of 180 Million ha m goes waste as run off (Anonymous, 2000). The distribution of rainfall with respect to different rainfall zones is given in Table 3. It is estimated that about 300 M ha m rainfall is generated during June to September, while another rainfall of about 100 M is found during rest of the year (Anonymous, 2000).

Net irrigated area : The net irrigated area in the country is 53.5 M ha, through tanks 3.1, canal 17.1, tubewells 17.9, other wells 11.9 and other sources including drip and sprinklers 3.5 M ha. Now, there is an ample scope for rain water harvesting and recycling. Through rain water harvesting significant area could be increased under irrigation where total rainfall exceed 700 mm per annum.

Per capita water availability : The water resource is limited and need its proper management. The per capita water availability is decreasing, it was 5.3 thousand m^3 during 1955, reduced to 2.5 during 1990 and it is estimated that during 2025, it will be 1.5 thousand m^3 only. So water is very precious and use it judiciously. This alarming situation is not only in India, the similar trend is also in China, Pakistan, Bangladesh, Nepal and USA. The position of UK is comparatively better.

Table 3. Distribution of geographical area in different rainfall zones in India

Rainfall zone (mm)	Geographical area (M ha)	Rainwater available (M ha m)
100-500	52.07	15.62
500-750	40.26	25.16
750-1000	65.86	57.63
1000-2500	137.24	205.86
> 2500	32.57	95.73
Total	328.00	400.00

Source : Anonymous (2000)

Why rain water management is essential

Water is essential to support about 17% world human population and 15% livestock (about 453 million) from 2-3% area, 4% fresh water on 1% forest. We need water and land use more judiciously to ensure : Food security, Lively hood security Health security and Ecological security.

However, the conservation and management of natural resources are possible and more effective of watershed basis. In this view, in words of Dr. Kofi Annan, U.N. Secretary General that "by 2025, two third of the world's population is likely to live in countries with moderate or severe waters shortage". In this connection rainfed crop production is mainly affected by soil degradation and decreasing availability of water for irrigation. The efforts are being made in this chapter to utilize soil and water resources at field level within the boundary of the watershed for natural resources' conservation and crop production at sustainable basis.

Irrigation potential : The per capita per annum water availability for irrigation is presently at the level of 2001 m^3 will be reduce to the stress level of 1700 m^3 in the next 2-3 decades. The ultimate irrigation potential of the country has been estimated at 139.5 M ha comprising of 58.5 M ha from major and medium schemes, 15 M ha from minor irrigation schemes and 66 M ha from ground water exploitation. The present irrigated area in the country is about 53 M ha. However, even after achieving full irrigation potential, nearly 50% of the total cultivated area will remain rainfed. The present allocation of water for agriculture is about 85% of the developed water resources, and it is likely to be reduced by 10-15% to meet the ever increasing demand for drinking water, industrial use and the irrigation in agriculture. On an average, 34.8 per cent irrigated area is only under cotton in the country. In order to realise the full potential of cotton production on sustainable basis, efficient management of rain water is very very essential. Once, rain water is properly managed at field/farm level, the problem of soil erosion is automatically minimised to a maximum extent.

There is a big question of poor economic returns from major irrigation projects and the negative effects on environments resulting in soil degradation, water logging and salinisation in many canal command areas. In India, the percentage of area damaged by poor water management is highest in the world (36%). Grave concerns are being expressed by many international and national organisations about the ill effects of poor management of irrigation on ecology of the command areas of many irrigation projects. The life of many irrigation reservoirs has been reduced by accelerated erosion in the catchment areas & increased rate of sedimentation than estimated by the planners.

5. PROBLEM OF LAND DEGRADATION IN INDIA

Soil is viewed as a medium of main plant growth for providing suitable environment for germination of seed, supplying adequate quantity of water to growing plants, providing oxygen to respirating roots and giving sufficient strength to plant needs which are primarily the physical factors in addition to supply of nutrients to plants. All soils are not equal in influencing crop growth because of one or the other physical limitations. The soil having any kind of such limitation which adversely affect plant growth is termed as Problem Soil. A soil may be having some physical, mechanical problem or chemical problems or both depending upon types of limitations imposed on it. Most of the soil have some kind of growth limiting problem depending upon the nature of the soil. The soils, areas, where plant production often encountered are -acid soils, halomorphic soils (saline, saline alkali soils), red yellow and laterite soils, medium and deep black soils, calcareous soils, desert soil, hilly soils and marshy wet soils. In addition to these soils, there are highly degraded areas which are affected by water erosion over the years. The present scenario of land use and degraded land areas in India are given in Table 2.

The fast disappearance of prime lands for urbanisation and increasing rate of land degradations through various processes is making the chances of extension of agriculture to new areas rather remote. As against 0.48 ha of agricultural land availability per capita in 1951 and 0.20 ha in 1981, it has been reduced to 0.15 ha by the end of this year (2000). Besides the decrease in area per capita, which is of major concern, it is the decline in quality of land environment which is of even greater concern for the quality of life in future. The land use statistics of India shows that the forests and grass lands which constitute the along of a nation are woefully inadequate. There is hardly 0.80 ha forest area per capita as against 1.0 ha in the world. Moreover, these lands are highly degraded and low in productivity. The wastelands are so much eroded, impoverished and denuded that they have become threat to adjoining agricultural lands, because of accelerated erosion that they are subjected to. The abnormally high rate of increase of animal population (2% per annum from 1951) is threatening these fragile environments further because of over grazing by the animals. India houses 15% of world's animal population, while it share only 2% of the world's geographical area. It is a matter of great concern that out of 30.4 M ha reporting area, 187.7 M ha is affected by problems of soil erosion by water, wind and various other types of soil degradation such as salinity, alkalinity, acidity, ravine and gullies etc. (Table 2). The degraded lands form 57.1% of the total geographical area of the country (Paroda, 1998).

Status of land degradation : The total area affected due to various land degradation types in the country is 187.7 Million ha which is 57.1 per cent of the total geographical area of the country (Paroda, 1998).

Soil and nutrient loss : It is a matter of great concern of every citizen of India that every year almost we are loosing 5.3 Mt very good fertile soils and due to that 80 Lakh tonnes major soil nutrients such as nitrogen 29 Lakh tonnes, phosphorus 26 Lakh tonnes and potash 25 Lakh tonnes only due to soil erosion and with this result, about 30 Lakh tonnes gross loss of food grains every year we are loosing in the country.

Major drought incidents : Drought in simple terms is the dryness due to deficient rainfall or shortage of water for an extended period, a season, a year or several years over a particular region. Droughts are devastat crops and arises in difficult times leading to famine, malnutrition, epidemics, forced migration, economic devastation, loss of lives and social conflicts.

To overcome on these problems, as discussed earlier, efficient rain water management is must to save the country from any further disaster. On an average, about 20 Million peoples are affected by nominal or severe drought condition every year. The current drought of the year 2000 is expected to become one of the biggest droughts of this century. As of now, 11 states have reported drought conditions of varying magnitude. Almost 5 crore people have been affected seriously in the states of Gujarat and Rajasthan alone. In Gujarat, 17 out of 25 districts, and in Rajasthan, 26 out of 32 districts are reeling under severe drought condition. The entire cattle population of about 4 crore is faced with acute shortage of fodder and water. So, this year drought surpassed the earlier drought of 1979, 1982 and 1987 (Bose and Saxena, 2000).

6. CAUSES OF SOIL DEGRADATION

Both natural and man made factors contribute to soil deterioration but the human interventions have assumed far greater significance in recent years. The following important factors associated with a host of agrarian, silvi-pastoral, socio economic developmental, administrative, political and organisational activities are responsible for soil degradation (Singh and Hazra, 1996) as given below:

- Erosion leading to heavy loss of top fertile soil
- Increased leaching of nutrients, particularly in the high rainfall areas
- Negative balance between addition and withdrawal of plant nutrients
- Agriculture on sloping, marginal and fragile land involving in appropriate cultivation methods
- Overgrazing by animals
- Removal of vegetative cover through indiscriminate deforestation and shifting cultivation with shorter fallow cycle, thereby exposing the soil to the action of erosive forces
- Water logging, salinisation and alkalisation in the irrigated areas due to faulty use of irrigation water
- Excessive withdrawal of underground water, resulting in lowering of water table and consequently the moisture deficiency
- Pollution due to agro-chemical, untreated industrial wastes, town sewage and other toxic substances
- Loss of land due to urbanisation, settlements, mining, air fields, road construction and other developmental needs and
- Inappropriate cropping and fertiliser application systems causing nutrient imbalances, nutrient deficiencies and deterioration of physical, chemical and biological properties of soil.

Problem Scenario and Need of Integrated Watershed Development

The degraded land, water and vegetation resources of the watershed are critical parameters which provide the life support to human beings and livestock in the watershed. The natural phenomenon and human activities are solely responsible for resources and environmental degradation causing non sustainability of productive systems. The need for integrated watershed development is crucial in many respect as detailed below :

- It has been observed that the total soil loss every year is of in order of 5334 Mt, out of which nearly 3254 Mt (61%) is transferred from one location to another, about 533 Mt (10%) is deposited in river beds, reservoirs and storage structures and about 1547 Mt (29%) is washed away into the sea. Due to poor management of the land resources, annual soil loss is as high as 16.55 $t.ha^{-1}$, while it should normally range from 4.5 to 11.2 $t.ha^{-1}$.
- Substantial areas in India periodically experience droughts leading to considerable loss of agricultural production, livestock and vegetation besides causing misery to people inhabiting these areas. Similarly, in almost all parts of the country floods commonly occur due to heavy storms or excessive flush, runoff causing enormous loss of life, property and vegetal cover. The attempts of mitigating the suffering of the people at the time of both events is generally aimed at provision of adhoc relief through income generating activities and supply of essentials for a short time from the government and non governmental agencies without integrating various activities for land and water conservation in a watershed.
- Development of water resources, surface and ground water in the rainfed watersheds are very poor. Most of the precipitation goes to river as runoff and no integrated efforts are made to harvest it for irrigation purposes.

- In the canal command areas, the culaba system has been taken as the developmental unit. In such areas, where the pumping sets are also the means of irrigation, the dark and grey blocks are increasing due to over exploitation of the ground water. Integrated recharge activities would be planned on watershed basis for water balancing in these areas to cope with the requirements of ground water available for irrigaion purposes.
- The basic requirements of food, feed, fibre, fuel, fodder and timber are to be met on priority for the growing human and livestock population. It is only possible by integrated development of watersheds and co-ordination of various developmental activities into it for effective and sustained productivity.
- Most of the biological raw materials are produced in watersheds and are being used in cottage, small scale and heavy industries. The production of such raw materials and byproducts should be encouraged by suitable land use planning in industrial watersheds.
- The employment of landless and agriculture labourers in rural areas is a primary task in order to check their migration towards cities. The watershed development activities can provide enough and round the year opportunities of employment to such class of people in their locatlities.
- The environmental degradation and ecological imbalances are increasing in rainfed areas causing loss of production system. In order to improve the ecosystem, integrated watershed development would be taken up on priority basis.

Farmers participation for watershed management

No watershed development activity could be completed without active participation of the farmers. So far, the participation of farmers was less even at lower level planning and the land use at village level. Therefore, participation of farmers is must for effective and productive planning and execution of watersheds. Development planners of yester year concentrated more on the large river valley projects, requiring huge capital investments but with poor long terms returns. They also imbalance the geological, hydrological cycles, natural vegetation and the wild life. The focus has now shifted towards minor irrigation with micro level planning. Watershed is an geophysical, hydrological unit, where it starts from ridge ling and ends through one natural rain water disposal system. Natural resources viz., soil, water, crop, livestock, horticulture, agroforestry, fishery and human beings are managed scientifically so as to bring prosperity to the entire village. The primary activity is to conserve the soil and water in to a percolation well or water harvesting farm/ village pond at a non erodible speed by various soil conservation measures. This collected and stored water can then be recycled at critical stage of crop growth during dry spells, or used for supplemental irrigation of rainfed crops. In many parts of the country, the harvested and stored water is used for drinking water purpose also, The experience of model watersheds in black and red soils found that it raised the productivity of rainfed crops, ground water level and could provide supplemental irrigation. In deep black soil it was feasible to go even for double cropping due to high AWC.

Present management of soil and water resources

Soil and water resources in India, specially as in Bundelkhand Region are grossly mismanaged through undesirable activities such as : indiscriminate falling of forest trees, deforestation of slopy land, shifting cultivation, over grazing of pastures, improper cropping pattern and practices for undulating lands, bringing marginal land under the plough, plugging of natural drains while levelling the land for irrigation or for the construction of roads and railway lines, use of land for brick making and neglecting the mined land. These practices not only enhance the quantity of run off, but also add to its velocity and thus increase the land deterioration manifolds.

As regards water resources, these are equally mismanaged. Of the total run off approximating to 180 M ha m, only about 15 M ha m is stored in various reservoirs and tanks, of which about 5 M ha m is lost by evaporation. The utilisation through diversion works and direct pumping aggregates to 15 M ha m which is more than that from storage work. Thus, 150 M ha m of water still flows into rivulets and rivers and is ultimately lost to the ocean.

This 150 M ha m of surface run off should be a boon to our countrymen, but it is proving a curse, as soil erosion by run off is the most common and severe form of land degradation. Out of 175 M ha of land under cultivation, 120 M ha suffers from water erosion. According to Swaminathan (1973), the annual loss of soil is about 6000 Mt and that of nutrients is 2.5 Mt of nitrogen, 3.3 Mt of phosphorus and 2.6 Mt of potassium

7. SOIL, WATER & CROP MANAGEMENT

The major issue relating to lands already under cultivation are sustainability of agriculture, and not at the subsistence level of low production; rather continuous growth at enhanced level of production so that their negative effect on environment is minimised. For enhancing productivity and sustainability, it is essential to identify the chemical, physical and biological limitations which restrict crop growth. Appropriate technologies are needed to remove these limitations. It hardly needs to be stressed that the main limiting factor like in most Indian soils is the nutrient deficiency, low out put of fertiliser and manures, imbalanced use of fertilisers and lack of consideration for the nutrient removal and loss in various forms besides the low efficiency of rain and irrigation water. Specially in Bundelkhand region, where 78% of the cultivated area is under drylands and major irrigated area is also not so assured under irrigation, it is the most important aspect to look for appropriate dryland technologies developed for these areas. For dryland areas increased efficiency of rain water use is essential and it can be achieved in the following ways:

- To retain precipitation in situ and minimise the run off
- To reduce evaporation in relation to transpiration
- To use drought tolerant crops and their varieties that fit in the rainfall pattern
- To use varieties of crops with higher water use efficiency
- To harvest and recycle the run off water
- To use watershed concept for maximisation of rain water harvest and its intensive use
- To adopt appropriate drainage methods where canal irrigation is practised.

Infact, it is a combination of all these that ensures best results with minimum degradation of environment. It hardly needs to be emphasised that the most critical factors of rainfed areas are the moisture stress and nutrient deficiency stress and an integrated technology is needed for removing these stresses to ensure sustainable agriculture.

Land use planning as land capability class.

In most of the fields, the crops are grown as per market demand of as per the family requirements and not as per land capability, which exerts pressure on marginal lands, resulting in further degradation. However, as per land capability classification class I to III lands can be put under arable cropping, alley cropping, agro horticulture system for sustainable agriculture. The dryland horticulture crops like guava, custard apple, phalsa and pomegranate can survive with minimum irrigation under extremely dry conditions. The crop diversification is the need of the hour for sustainable crop production in SAT against the vagaries of weather, pests and market fluctuations.

The class IV and beyond lands must be used for pasture and range purposes. These lands can be used to grow forages, perennial trees for fuel wood. Marginal lands should be established with Cenchrus sp., and Stylo grass seed which can sustain extreme droughts. Tree framing can be taken up on class V lands with closer planting and proper pit filling. The class VII and VIII lands with extreme slope, marshy conditions must be exclusively reserved for wild life and rural man as buffer zone, otherwise they may damage field crops and domestic animals

Alternate land use

The highly degraded lands specially from soil and water conservation view point and also lands with greater slope per cent (8% and above) should invariably put to alternate land use. This refers to the use of land with such practices which is primarily aimed at conserving the sites with production benefits other than the normal arable cropping. Here conservation of natural resources is of paramount importance. This alternate land use is in fact the most appropriate land use for such situation. The studies at Gaharawa watershed have clearly indicated the usefulness of such measures in containing soil and water loss and improving soil productivity.

Efforts made so far for soil and water conservation in India

According to Bhumbla (1975), the total estimated agricultural land treated under soil conservation programme upto 1968-69 was 9.4 M ha. The fourth five year plan target was to cover an additional area of 5.39 M ha of agricultural land and 0.45 M ha of non agricultural land, with an estimated outlay of 158 crores. Also, soil conservation programmes in the selected river valley projects covered an area of 0.84 M ha, with a total expenditure of Rs. 22 crores. Bali (1980) estimated that 23 M ha area had been treated with soil and water conservation practices at a total cost of about 639 crores during the period 195152 to 1978-80. In fifth five year plan, some important developments in dryland research took place and these affected the policy of soil and water conservation. In each state, one ORP (Operational Research Project) was started by the Indian Council of Agricultural Research to transfer and test the technology evolved by research. The first ORP with British collaboration, known as Indo-UK Dry Faming Project was started at Indore in 1947. The project successfully demonstrated the management of land and water resources on watershed basis. Subsequently, a number of other ORPs were also started by the ICAR and located at Soil Conservation Research Centres and at Agricultural Universities. In some states, e.g. Sukhomajari ORP (near Chandigarh), ORP at Guna (M.P.) and at Sidhi (M.P.) and at Chitradurga (Karnataka) all had successful stories. The work of these ORPs proved beyond doubt that soil and water conservation on farmers fields could be achieved only on watershed basis. Therefore, during the last two years of fifth five year plan, an emphasis was laid on the integrated watershed management of 276 watersheds as part of the soil conservation programme (Shankarnarayan and Shankar, 1984).

In 1983-84, an attempt was made by the Govt. of India and the ICAR to identify critical watersheds and bring the development workers and the scientists together to prepare a Master Plan for the development of model watersheds and execute it. Forty seven watersheds were identified in 16 states, and the plans were executed in 46. Excellent work was done in some of these, and the same was recognised at national level by agencies such as National Productivity Council, Indian Farmers Fertilizer Co-operative (IFFCO) and Krishak Bharati Co-operative (KRIBHCO). These agencies gave National Awards to Mettemari Watershed (Karnataka), Chavela Watershed (A.P.) Nartora Watershed (Raipur, M.P.). Later on, the Department of Agriculture and Cooperation launched the National Watershed Development Project for Rainfed Areas (NWDPRA) covering almost the same states. These activities were in micro and mini watersheds covering 500-2000 ha. Simultaneously, another programme on watershed development was initiated with the help of the World Bank and the Union Dept. of Agriculture. To start with, four large watersheds covering 25,000 to 30,000 ha were selected at Maheshwaram (A.P.), Kobbalnala (Karnataka), Manoli Maharashtra and Peruamala (Sehore, M.P.) During 7th five year plan, 4400 watersheds were covered under the National Watershed Programme. In 8th five year plan, massive programme of soil and water conservation, land reclamation and integrated land development for minor irrigation etc. were undertaken. Ravine reclamation programmes in M.P., U.P. and Rajasthan states were also undertaken from time to time, particularly under the antidacoity campaign.

Cropping pattern choice

Weather/climate as a natural resource defines the choice of crops and cropping patterns in a given region (Singh, 2000a). Country has been delineated in to 4 rainfall zones viz. very low, low, medium and high rainfall zones considering 500, 500-750 mm, 750-1150 mm and 1150 mm and above rainfall respectively. Agro-climatic characterization of the country based on moisture, thermal and sunshine regimes was completed. Agricultural production strategy was developed for Agro-Met advisories based on weather forecasts. Short duration drought resistant pulses such as mungbean, mothbean, cowpea and cereals of 10-12 weeks duration such as pearlmillet and minormillet are suitable in arid zone. In semi-arid regions, where length of the humid period is more, rainy season crops are grown in soils that have capacity of holding less than 150 mm of water. Additional post rainy season crops can be grown on conserved soil moisture, in soils that can hold more than 200 mm. In soils with 150-200 mm capacity, inter cropping is possible. In sub humid areas, where humid period is more than 12 weeks duration and the rainfall is twice that of PET, rice-based cropping system is suitable as other crops can't tolerate water stagnation (Table 4).

Table 4. Choice of cropping pattern for different growing period

Length of growing period	Cropping system that can be adopted
75 days	Perennial vegetation, monocropping of short duration pulses
75 – 140 days	Monocropping
140-180 days	Intercropping
>180 days	Double cropping

Source : Singh, G.B. (2000a)

Sustainability of watershed

Watershed projects have now been implemented for more than a decade. Evaluation reports have clearly shown that these can not succeeded without full participation of project beneficiaries and careful attention to issues of social organisations. This is because, the costs and benefits of watershed interventions are location specific and unevenly distributed among the peoples affected. Unfortunately, most projects have failed to generate sustainability because of the failure of the government agencies to involve the people. On the whole, villages with projects operated by non government agencies (NGOS) or in collaboration between NGOs and Government agencies performed significantly better than those with purely government projects, which in many cases, did not perform better than control villages with no project. Most government watershed development investments have yielded disappointing results given the vast resources allocated to date.

A survey of 70 villages of Maharashtra and Andhra Pradesh covering several watershed development programmes showed that a most programmes, increase in agricultural production did not last for more than two years. Structure was abandoned because of lack of maintenance and there were no mechanisms for looking after common lands. Farmers were not convinced about the need to contribute which would happen only when they make the decisions about that kind of measures are introduced on their plots. Often in government projects, farmers do not have this choice and technological norms are too inflexible. The very fact that farmers are unwilling to contribute towards the cost of works shows that they have little faith in the effectiveness of the programme.

8. CONCEPT OF SUSTAINABILITY

Sustainable agriculture, depending upon the objectives means differently to different agencies and researchers. According to FAO (1989), sustainable agriculture involves successful management of resources for increased agricultural production to satisfy challenging human needs, while maintaining or enhancing the environment and natural resources. In a predominantly agriculturally driven Indian economy, sustainable agriculture could more aptly be defined as the one that over the long term enhances the environmental quality and resource-base on which agriculture depends, provides for human needs of food, fodder and fibre is economically viable and enhance the quality of life for farmers and society as a whole (Anonymous, 1989). To realise sustainable agriculture, it is important :

1. To maintain soil health and quality
2. Practice crop rotations mutually beneficial in thriving on the finite soil resources
3. Maximise benefits from nutrient cycles and flows
4. Minimise soil loss
5. Harness biological nitrogen fixation (BNF) potential fully
6. Protect ground water from contamination through agro chemicals including fertilizer nutrients
7. Maintain rich heritage of genetic bio diversity and most important of all
8. Apply the chemical fertilizers in conjuction with bio solids (consisting of FYM, crop residues, industrial and urban wastes) and biofertilizers in such a way that crop yeids and nutrient use efficiency get maximised without any deterioration of soil, water and air quality (Goswami and Rattan, 2000)

Approaches of integrated watershed development

Integrated watershed development is an area development programme for sustained production of food, fuel, fodder and timber according to requirements of the watershed, scientific utilization of resources, improvement of environment and restoration of balanced eco system. In planning of developmental activities of a watershed, resources inventory, local requirements, proper land use and national perspectives are to be considered in detail (Ambedkar, 1996).

Plate 2.3.1. A corner of field in low land. It can well be constructed in watershed. Find such spot in your field and arrange for watershed to have supplementary irrigation. **2.** Watershed on an area of half hectare land at down hill in a field of 50 hectares.

- Watersheds provide living opportunities to inhabitants including land holders, landless labourers and village artisans. Hence, the basic approach for development should be employment oriented
- The production system should not only be limited to edible commodities, but it should also include production of raw materials for local as well as for national consumption.
- Participatory management system and effective involvement of local communities in watershed development should be ensured at each stage so that proper utilization and maintenance of land resources and community assets would bring prosperity in the area.

- The watersheds located in various parts have different socio economic as well as hydrological and topographical characteristics. In planning of each watershed, basic need of the people has to be looked into according to production capability of watersheds.
- Watershed development is based on conservation cum production technology. Hence engineering and vegetative measures combined together with in situ moisture conservation and rainfall harvesting play an important role. Unless foundation work of mechanical/ vegetative measures are done in a watershed, no other activities would sustain the productivity and harnessing of utilizable capabilities of resources.

Thus, it is evident from the above account that integrated watershed development is the only solution for proper conservation of land and water resources and availability of basic requirements to the Millions of people living in rainfed and drought prone areas of the country.

9. WATERSHED PLANNING AND DEVELOPMENT

Biodiversities and imbalanced ecosystems are gradually increasing in rainfed watersheds. Poor vegetal cover, scrub forests, overgrazed pastures and denuded soils with retarded productivity are the result of over exploitation and mismanagement of resources by human beings and livestock. The restoration is possible by creating resource literacy and integration of various conservation cum production technologies in accordance with land use capabilities and requirements of people inhabiting the watershed.

Holistic approach for conservation and management of land resources for making available food and firewood for 843.93 Million population and fodder for 419.58 Million livestock is certainly to be based on watershed development with integration of all activities related to farming system, livestock development and maintaining the proper ecosystem for healthy growth of flora and fauna (Ambekar, 1996). In planning of a watershed for sustainable conservation-cum-production system, the following points should be considered.

1. Village based planning of watershed development would be the main thrust in the new Panchayati Raj system, but drainage line control for water harvesting can not be taken up unless the whole watershed lying in different villages is selected at one time. Smaller watersheds can be planned within the village boundaries but bigger ones should be delineated on the basis of maps of different villages taking ridges as natural boundaries of the watershed. Modalities for such planning should be worked out on the Nyay panchayat level.
2. In rainfed watersheds, planning should be based on land use capability classification of the area in which integrated approach of production system, cropping, agroforestry, agrohorticulture, tree farming on arable lands and silvipastorals, silvihorticulture, dryland horticulture and afforestation on non arable lands after proper moisture conservation. The ICRISAT, Hyderabad has analysed the benefit cost ratio of various cropping systems which are given below

Farming System	Benefit : Cost ratio
Agro forestry	1.73 : 1
Agro horticulture	5.78 : 1
Tree farming	2.08 : 1
Silvi pastoral	2.56 : 1
Farm forestry	18.60 : 1

With the above results, it is advisable to plan various farming systems in watershed development projects.

3. Moisture conservation plays an important and crucial role in rainfed farming. Moisture conservation *in situ* consisting of farming measures i.e. summer ploughing, formation of dead furrows, strip and contour cropping and mechanical vegetative measures i.e. contour bunding, submergence bundhis and small gully checks have proved very effective in increasing productivity and cropping intensity in rainfed areas. Productivity of different crops increased to the tune of 60 to 289% in Tejpura (Jhansi) 219 to 472% in Randhar and 81 to 141% in Sheetalpur projects of watershed development. Cropping intensity also increased from 83% to 156%.

4. Success stories of watershed projects in the country show that runoff harvesting through various soil conservation measures and recharge of ground water play an important role in providing life saving irrigation and in increasing productivity. Giving one or two life saving irrigations to a crop at certain critical stages of moisture stress during its growth cycle is most effective in rainfed farming. Water harvesting bundhies are the cheapest source of irrigation in rainfed watersheds. Impounding of water along the bundhies is submerged area has a greater effect in improving moisture regime of soils in upstream and recharging of ground water through storage in the vicinity along with irrigation facilities either by bundhi or by dug wells in downstream fields. Cost of irrigation development by bundhi is Rs. 4000 ha^{-1}, while it is Rs. 15000 and Rs. 40000 ha^{-1} by state tubewells and canals respectively.

Integrated watershed development works executed in Kharaiya Nala watershed of Jhansi, reveal an appreciable increase in recharge status of ground water. It is observed that ground water level is now between 7.5 to 9.5 m in summer, while it was between 12 to 14 m before implementation of the project during the same season.

The farmers are effectively utilizing ground water by construction of wells. At present there are 1060 seepage wells in the project while their number was only 20 in the year when the project was started. The life saving irrigation percentage is 64.3 in comparison to pre project percentage of 10.9 only.

Runoff harvesting being the most important activity in rainfed farming, construction of check dams, bundhis, sunken ponds and dugout ponds should be given top priority. Unless a network of water storage structures in drainage lines are established in rainfed watersheds, successful cultivation and mitigation of drought hazards would never be achieved.

5. In hills, undulating topography, shallow soils, steep terrain and torrential rains are prone to severe erosion and land slides unless they are properly managed. Watershed based farming system coupled with mechanical and vegetative soil conservation measures i.e. contour bunds, bench terrace, grassed waterways, check dams and diggis at appropriate locations can retain maximum rainfall within the slopes and safety disposing off excess runoff from slopes to foot hills with non erosive velocities.

Diggis play an important role in providing irrigation water to terraces for agriculture as well as for horticulture. The check dams constructed in perennial streams can also be utilised for irrigation purposes by erecting pipeline conveyance system upto terraces and then constructing open channels. To avoid choking of pipeline with silt and debris, higher diameter should be preferred.

Effect of reclamation

The practices which have been recommended for gulley reclamation are (1) Closure to grazing and other biotic interference, (2) Construction of a peripheral bund to divert run off from gully heads, (3) Levelling and contour bunding on shallow ravine and (Soil from a shallow V-shaped channel on the upstream of the bund is utilized for its formation), (4) Stabilization of bushes and bunds with sidding, safe disposal of excess run off through grassed ramps in the bunds, (5) Bench terracing in medium gullies and (6) Gully plugs/check dams and afforestation of deep gullis. The strategy is to reclaim shallow and medium gullies for cultivation, while deep gullies put to permanent vegetative cover of grasses and trees. The gulley plugs should be composite structures of earth and brick masonry for spillway. The detailed specifications about the design and construction of the checkdams and bench terraces are described by Tejwani and Dhruvanarayan (1960).

The benefit cost ratio for ravine reclamation for agricultural use has been reported by Shankarnarayan and Shankar (1984), for (i) humps having a slope upto 6 per cent as 1.86 with manual labour and 1.89 with bull dozer and manual labour, and for (ii) shallow and medium gullies as 0.86. However, the indirect effects/ benefits of gully reclamation can not be assigned a monetary value and therefore benefit, cost ratio alone is not relevant in evaluation of gully reclamation works. The resource gain analysis (value of treated land, value of untreated land, cost of land treatment) of ravine reclamation in various categories of land based on 1977 prices were found to be (1) Rs. 2325 ha^{-1} for class II (ii) Rs. 1350 for class III, (iii) Rs. 2150 loss ha^{-1} for class IV to VI land, (iv) nil for class VII land, and (v) average for all classes Rs. 406 ha^{-1} (Seckler, 1978). Rai (1970) evaluated the ravine reclamation work in districts of U.P. bordering Rajasthan and reported a net profit of Rs. 56 ha^{-1}.

Bunding on Alluvial Soils

Results of field/properly bunding in alluvial soil tracts have been some what satisfactory. Many evaluation studies have been conducted and reported regarding return on investment of about Rs. 250 ha^{-1} (in sixtees/seventees) on contour bunding programme could not lead to very high direct benefits, particularly when levelling of land was also involved, and bunds on all the four sides of the fields had to be constructed, instead of a single bund on a contour line/grade. Still beneficial effects of soil moisture conservation by construction of bunds and levelling of land have been extensively demonstrated in the semi arid alluvial plains of U.P., where 35,63 and 98 per cent increase in the yield of both Kharif and rabi crops has been obtained by bunding alone, levelling alone and bunding cum levelling respectively (Khan, 1962). In Madras state, bunding increased yields of setaria, cotton and sorghum by 18, 11 and 17 per cent respectively. Rehmankheda farm in U.P. badly eroded soils showed continuous increases in yields for the last 10 years to the extent of over 900 per cent (Tamhane and Bali, 1963). Tamhane et al. (1967) reported that contour bunding resulted in yield increases of 24.5 and 25.2 per cent in case of rabi sorghum and pearlmillet respectively; in Maharashtra state, yield increases of 35.6 and 25.4 per cent in case of sorghum and pearlmillet respectively, in Tamil Nadu state, yield increases of 21.4, 15.0, 19.7 and 13.9 per cent in case of wheat, gram, maize and perlmillet respectively, in Punjab, Tejwani (1968) found that bunding levelling and bunding cum levelling increased crop yields by 35, 63 and 98 per cent, respectively, in semi arid plains of U.P. Patel and Sohoni (1969) observed that contour bunding not only increased crop yields in dry farming areas of Maharashtra state, but also enhanced the water level in wells. Soil conservation measures on 1600 ha of marginal land of the farmers, led to 16 per cent increase in yield ha^{-1} (1.36 q ha^{-1}) of Kharif crop (Rai, 1970). Agrawal et al. (1970) reported that the yield of bejhar (mixed crop of gram and barley) was increased 3 and 13 percent by bunding alone at Rehmankhera and Lucknow centres. Kulkarni and Raheja (1970) studied the impact of contour bunding on the productivity of important crops during 1950-51 to 1962-63, in Maharashtra state and reported a substantial increase in the productivity of sorghum, pearlmillet, cotton and groundnut, but observed that benefits of contour bunding comes only after a couple of years. This proves the adverse effect of levelling due to shifting of productive top soil.

Plate 2.4. 1. Eroded soil with grassy growth due to low land. Convert such patch of field in watershed with two objectives. a. to stop erosion and b. to arrest runoff losses, **2.** Nalla bunding example to utilize rain water for protective irrigation and allow increase water table under ground.

10. EFFECT OF LEVELLING AND BUNDING ON CONSERVATION OF SOIL AND WATER

Many studies have been conducted to measure the effect of field levelling and bunding on the silt load and rate of run off. At Chandigarh, the rate of sediment was reduced from 80 tonnes ha^{-1}

annum^{-1} to 6 tonnes ha^{-1} annum^{-1} within 5-6 years of treatment (Mishra *et al.,* 1975). As reported by Nema and Kamanver (1976), on treatment of a 89 ha catchment of Vasad, the loss of 1422 tonnes of sediment per year was prevented. At Dehradun, 62 per cent reduction in run off and 40 per cent reduction in peak discharge was observed by Ram Bau et al. (1980). While at Vasad, the amount of rain water retained in the catchment treated with bunding and check dams was 53 per cent more than in the untreated catchment. Patnaik (1977) observed that sediment load of 0.18 ha m/ km^2 (before treatment) to 0.02 ha m/Km^2 after treatment

These results indicate that field bunding and levelling of land with alluvial soils reduced the run off, floods, and sediment load considerately. However, the picture was not so rosy in black soils regions. A recent survey by the M.P. Deptt. of Agriculture on the effectiveness of contour bunds in increasing the yield of wheat crop has shown that the effect of bunding was strongly negative. Similarly, a study on the effect of bunding on the yield of kharif crops in Bellary area indicated that contour bunding decreased the yields of sorghum, cotton and safflower both as a result of water stagnation and delay in cultural operations (Shankarnarayan and Shankar, 1984). In western M.P., black soils with an annual total rainfall of 800 to 950 mm were growing good crops of sorghum, pigeonpea, maize and groundnut in kharif under rainfed conditions and were subjected to a slight sheet erosion, particularly where the land slop was greater than 1 per cent. Levelling and field bunding (under the name of contour bunding) made the fields quite unsuitable for raising kharif crops due to waterlogging and water stagnation. Not only this, when excessive run off overtopped the field bund, the concentrated water started running through a breach in the field bund and the result was formation of a gully, which became deeper and wider with time leading to manifold losses of soil and water as compared to those before soil conservation works were undertaken. Further, one breach in a bund initiated a chain of breaches down the stream, thus creating havoc in the watershed. As regards the effect on crop production, the entire area got converted into kharif fallow rabi cropped one. The yield of rabi crops was limited by the amount of profile stored moisture and impoverishment of bare in soil during the rainy season, particularly on the washes in the field, from where almost whole top soil was removed, and kankar layer in the sub soil exposed. Situation was worse on black soil with higher annual rainfall (greater than 1000 mm) in Southern Central M.P., field bunding in such areas led to "Haveli System" of farming, a gross wastage of both land and water resources. Anybody can guess how unhappy those farmers would have been after getting the soil conservation work done on their fields. Thanks to our system in which soil conservation service employees were not responsible to realise the charges/cost of works executed by them on farmers' field on loan. The only excuse the soil conservation service had for its failure was the lack of adequate provision for the maintenance of their works.

This is an example of transfer of technology evolved by research, in a vitiated form. There is need to plan and implement soil conservation measures in a much more scientific manner, taking into consideration not only the need to check erosion but also to promote crop growth.

Urgency to control the disaster

No nation can afford such disastrous losses. It is not surprising that the neglect of soil led to the disappearance of once flourishing civilisations. It is generally agreed that a man needs 0.05 ha of land for housing, roads, waste disposal, power supply, etc. and 0.4 ha for production of food. In contrast, the availability of agricultural land in 1991 was only 0.21 ha per person and would get reduced to 0.15 ha per person in 2000 A.D. Therefore, the urgency to conserve our land resources can not be over emphasised, simply because land resource is shrinking (as soil erosion is a continuous process) and population is continuously increasing.

Rain water management in watershed

Sources of all water is precipitation. Rain water a crucial resource, is key input in Indian Agriculture. The success in agricultural enterprise depends largely on how effectively we conserve rain water and how wisely we use water resources at farm level.

Khasi Kaintia hills in the extreme east is one of the rainiest area in the world with an annual average of the order of 1000-10500 mm. In the extreme west in Rajasthan, there are areas receiving less than 100 mm. Almost the entire country east of the longitude 79ºE and the west coast has normal annual rainfall of more than 100 mm. Most of the west coast gets more than 2500 mm. Rainfall in the peninsular India excluding coastal belts is generally in the range of 500-700 mm. In the north west India it decreases to less than 300 mm west of Jodhpur. In this zone more than 80% annual rainfall is received during June to September. On an average, over space and time average annual rainfall over the Indian sub continental has been estimated to 1200 m ha m. On this basis, the annual precipitation including snowfall is estimated 400 m ha m in (4000 km^3). Distribution of rainfall with respect to different rainfall zones is given in Table 5. About 300 M ha m rainfall is generated during June to September, while another 100 M ha m is during rest of the year. About 20 M ha m of runoff is brought from outside India. Thus making of 420 M ha m of water resource for exploitation and use.

Table 5. Area distribution in India according to rainfall (1982-83)

Rainfall Zone (mm annum^{-1})	Net sown area (Mha)	Net irrigated area (Mha)	Net rainfed area (Mha)
Very low (<mm)	14.3	3.9	10.4
Low (500-700 mm)	33.9	10.5	23.4
Medium (750-1150 mm)	48.2	15.7	32.5
High (>1150 mm)	42.4	8.9	33.5
Himalayan and Shivalik Mountains	3.00	0.9	2.1

Water conservation : Water conservation is a vital to agriculture specially in rainfed situation. In water conservation, storage of water in the soil as such through mechanical methods or man made structures and effective utilization of given quantity of water to the plants are important aspects of the water conservation. The detailed studies on rainwater harvesting, *in situ* moisture conservation and its management for sustainable crop production on shallow, medium deep and very deep soils were conducted on watershed basis at Central Institute for Cotton Research, Nagpur during *Kharif*, 19952000 Water conservation in the real sense deals first : in-situ moisture conservation and second Water harvesting and recycling.

Indigenous rain water management practices.

The study was carried out during Kharif 2000-01 to 2002-03 at Thugaon micro watershed, district Amravti, Maharashtra covering four (Thugon, Mardi, Pathergao and Karla) villages. The main object was to identify the indigenous water management practices and its documentation to assess the awareness of indigenous as well as improved water conservation practices by watershed farmers and their impact on yield improvement in various crops on different topo-sequences under cotton based cropping system at watershed basis. The information was gathered from records, magazines and journals and also with the group discussions and in intreaction among watershed farmers. The yield data were collected after conducting on farm trials on different topo sequences with varying levels of moisture conservation treatment. Study revealed that cultural practices such as intercultural operations, cropping pattern, crop rotation and sequence cropping, use of cotton sticks and sorghum stubbles, stone bunding, mud bunding, use of agave on field boundaries for water saving, use of *matka* system for micro-irrigation etc. were found helpful in soil and water conservation (Wasnik et al. 2003). Further results indicated that 46, 43 and 32 percent awareness was recorded among watershed framers who adopted efficient moisture conservation practices like tillage and seedbed preparation, proper crop rotations and efficient inter-cropping systems. Through adoption of soil and water conservation technologies at watershed basis, maximum yield of 7.5, 36.75 and 8.75 q/ha of seed cotton, sorghum and pigeon pea was recorded, respectively at valley bottom followed by lower plain and the upper plain. This showed that soil and water conservation measures done at watershed basis by the farmer have immense value and use and provide the suitable path for rapid development of viable cotton – based production system for different topo – sequence.

In - situ moisture conservation

When and where to grow rainfed cotton a meteorological approach.

Soil moisture is the main yield limiting factor in rainfed crop production and due to that the crop growth and subsequent development of the plant is affected considerably. Rainfall availability and its distribution play an important role in crop planning and sustainable crop production under rainfed situation. Cotton being deep rooted and long duration crop required enough soil depth with high available water capacity of soil. Our experience show that the crop growth and seed cotton yield of rainfed cotton varied from to genotype soil-to-soil, genotype and management-to-management. Cotton being most important cash, commercial and fiber Kharif Italic, crop, the average yield of cotton in Vidarbha region is 170 kg lint ha^{-1} only, which is considerably low and below the national average yield of 321 kg ling ha^{-1}.. However, the yield of rainfed cotton could be increased significantly by assessing and managing the soil moisture movement in different soils. In this view, field studies were conducted on shallow (<45cm deep) black cotton soil of Central Institute for Cotton Research, Nagpur during Italic *kharif*, 1997-2004. The climate of the region is hot, semi-arid and sub-tropical. The mean annual rainfall at the station is 1156 mm. The mean maximum temperature during summer is 43 Degree C and the mean minimum 10.9 Degree C in winter. The weekly and monthly rainfall availability, annual rainfall distribution pattern and soil moisture status under cotton and fallow soil at different crop development stages was estimated. Crop calendar was developed and 75 % rainfall probability was evaluated.

Based-on, occurrence and regular monsoon and rainfall availability, 23rd Met. Week was evaluated as the best time for sowing of rainfed cotton in the region. Rainfall sufficiency and 75% probable rainfall was found as the best for sowing for rainfed cotton, getting proper seed germination and sufficient plant stand. Severe soil moisture stress was observed during 40th Met Week when total and 75% probable rainfall almost goes down to its minimum level. Under such circumstances, life saving irrigation to cotton crop becomes essential to meet out the evaporative demand for the survival of the crop. If life saving irrigation is not applied to the cotton crop at this you stage, the crop will not mature and you will not get economic yield. Further, results on soil moisture movement and its availability under bare and cultivated land clearly indicated the difference in total moisture availability, its path of movement and also varied with change in soil depth. Soil moisture availability was highest at 60cm soil depth and the lowest at 15 cm soil depth of the soil surface. The trend of soil moisture showed that as the rate of evaporation increases, the percent soil moisture was decreases. However, the soil moisture movement was found associated with rate of evaporation and the amount of rainfall. It was affected due to change of soil depth, both under fallow and cropland. At initial crop growth stage, soil moisture under cotton crop was comparatively less and it was increased simultaneously towards mid-season stage and after that it declined towards crop harvesting stage, but situation was just reverse in case of barren field.

It may be inferred from the study that soil depth plays an important role in moisture retention, its period of availability and movement in Vertisol. Amount of soil moisture coupled with soil depth was found helpful in proper plant growth, total biomass production and finally seed cotton yield of rainfed cotton.

11. IN-SITU MOISTURE CONSERVATION

In the rain water management, the first approach is to utilise the water where it falls. Appropriate land treatment and *in-situ* moisture conservation, agronomic practices have been developed to conserve the rain water in the soil as such. Some of these include off season tillage, soil and organic mulching, dead furrow, compartment bunding, graded boarder strip and, inter plot water harvesting. Broad bed and furrow (BBF) is another system adopted in cotton cultivation found suitable in deep vertisols. The effect of ridge and furrow in conservation of soil moisture mainly in shallow to medium deep soil is found beneficial in improving soil moisture by about 2-3 per cent which was found effective in increasing the seed cotton yield of about 3 q (Bhaskar *et al.,* 1998).

Effect of moisture conservation measures on seed cotton yield of popular genotypes

The response of *in situ* moisture conservation measures on well adopted genotype, LRA-5166, AKA-8401 and NHH44 was evaluated for 3 years. Results indicated that the response of ridge and furrow method of moisture conservation was highest in case of LRA-5166 and NHH-44, which was able to increase seed cotton yield by about 3 q, but in case of AKA-8401, the response of broad bed furrow was highest and able to influence seed cotton yield by about 4 q (Table 6).

Table 6. Effect of moisture conservation measures on seed cotton yield (q ha^{-1}) of popular genotypes. 3 (1995-1997) years mean

Moisture conservation measures	Genotypes		
	LRA-5166	AKH-8401	NHH-44
Control	10.58	10.61	10.60
Ridge ad furrow (RF)	13.60	13.97	13.80
Broad bed furrow (BBF)	12.00	14.43	12.29

Source : Bhaskar et al. (1998)

Effect of tillage operations on seed cotton yield

The beneficial effect of opening of furrow at the end of August + application of 5 t ha^{-1} of FYM increased soil moisture by about 2-3 per cent with similar additional yield benefit. Application of FYM @ 5 t ha^{-1} at the time of sowing found beneficial in increasing soil moisture as well as seed cotton yield.

Water harvesting : Water harvesting is a centuries old practice adopted by the farmers and relegated in many part of the country due to availability of irrigation. Now, need is felt to adopt this practice. However, recent research efforts to stabilize dryland production have generated renewed interest in the use of water harvesting terminology. For developing the early technology to the small and marginal farmers, water harvesting ponds at filed level of size 18 x 18 x 3 m^3 were digged in shallow, medium deep and very deep black cotton soils representing the land area at Agriculture Research Farm, Panjari of the Central Institute for Cotton Research Nagpur (Table 7).

Roof top water harvesting is considered and found essential by the city planner, Architect, land users colonisers state and Central Government Department and made provision for rain water collection from roof top of the Govt. building, Hotels, Hostels, Guest Houses, Schools etc. and stored in the ponds and recycled the same for washing of floors, car, irrigation in lawn, kitchen garden etc. This will help in reducing the huge amount of city water used for drinking purpose.

Water storage : The harvested rain water was stored in the water harvesting ponds on soils of varying depths for a period of about 3-3.5 months starting from first week of July uptill time of irrigation, that is first week of October.

Table 7. Rainwater storage and area to be irrigated in soils of varying depths

Soil	Rainwater collected from 1 ha cotton field (litres) Mean of 5(1995-96 to 1999-2000) years	Area to be irrigated (ha)*
Shallow (<45 cm deep)	3.5-4.00	0.3-0.4
Medium deep (45-90 cm deep)	4.5-5.0	0.4-0.5
Very deep (>90 cm deep)	5.5-6.0	0.5-0.6

Source : Bhaskar & Kairon (2000)
* Area to be irrigated is mainly depend on method of irrigation (by flood or bed or drip irrigation), accordingly area can be increase or decrease.

12. SOIL MOISTURE STRESS

It has been experienced that cotton crop suffers from moisture stress some time from very early stage to peak boll development stage. It is highly susceptible to drought at 2 to 4 leaf stage. This happen mainly due to less soil moisture available in the soils, mainly in shallow profile depth, high temperature coupled with high wind velocity compell the soils to come in moisture deficiency, and due to that cracks are formed on the upper surface which makes the plant life difficult. The available moisture in the soil get evaporated and increased soil temperature make short of moisture to plants, if rains are not occurs, plants mostly dried at this stage. But, after 4-5 leaves, there is no danger to cotton plants and it can tolerate dry spell of about 15-20 days or so. To overcome this situation, following measures are essential.

i. Keep the fields free from weeds
ii. Light interculture operations like Bakhar (Dora) is essential to conserve the moisture, and soil moisture at lower zone through capillary will protect the plants.
iii. 30-40 days advance sowing of cotton under drip have been found suitable and remunerative.
iv. Sowing of cotton on side of ridge in shallow to medium deep soil have been found suitable for conservation of soil moisture.
v. Farm yard manure @ 5 t ha^{-1} at the time of sowing conserved the soil moisture and helped in increasing soil moisture as high as 3 per cent.

The second moisture stress in cotton mostly occurs during last week of September to first week of October, at this time, cotton crop reaches to peak boll development stage, at this time temperature is increases, and the evaporative demand of the plants increases and deep cracks are developed. Under such situation, cotton crop needs irrigation, if stored rain water is available, one irrigation @ 4 ha cm could be given as life saving found beneficial and increased seed cotton yield significantly.

Economization of irrigation water and its use through microirrgation system

Microirrigation (Drip and sprinkler system) is a modern and efficient irrigation system for saving water, fertilizer, labour and time. Microirrigation system is applicable for all type of crops depending on type and systems of planting. It is found suitable for the crops where irrigation is needed daily, alternate day or system is not needed to shift frequently. This system is also applicable where farmers need to irrigate more and more area within short time during day or night on or undulating topography also. This system is also suitable where irrigation water is limited mainly for the condition where rain water is harvested and stored in water harvesting pond.

In the central zone, utilization of micro irrigation systems particularly drip systems, to establish a premonsoon hybrid cotton will substantially help in improving the yield of cotton, wherever limited water is available during summer. However, a complete package of practice to reap the benefits of drip irrigation, particularly plant geometry, fertigation, herbigation, water delivery rates, pest management etc. need to be standardized to offset the high initial investment in instabilizing drip irrigation. Field experiments at CICR, Nagpur, and elsewhere have established, the superiority of drip on sprinkler system in economizing water and improving water use efficiency.

Table 8. Water use efficiency under irrigation systems

Irrigation System	Seed cotton Yield (q ha^{-1})	Water applied (ha/cm)	Water use efficiency (kg ha^{-1} cm)
Drip	12.07	13.63	88.5
Sprinkler*	11.45	11.81	96.9
Furrow	11.99	19.72	77.9

*At 0.6 IW/CPE (6 cm of water at 100CPE)

Source : CICR Annual Report, 1996

Very little work has been done on the evaluation of drip for cotton in India. Padmakumari and Sivanappan (1979) observed a clear advantage of drip irrigation on seed cotton yield over surface method, under Coimbatore conditions. Subsequently, while developing the drip technology for cotton they observed that irrigation of cotton at a rate of replenishment of more than 75 per cent of Eo tended to reduce the yield resulting in curvilinear response. Similarly, prolonged interval between two irrigation tended to reduce the yield. Contrary to the behavior of seed cotton, the total dry matter production showed a linear response to replenishment levels as well as a prolonged irrigations interval. The yield per unit water application was higher under low to medium level of replenishment induced early maturity.

Studies conducted at MPKV, Rahuri, indicated that with drip irrigation the cotton yield can be increased by about 34 per cent. In another study conducted at the same place, the yield increase was 27 per cent with water saving of 53 per cent (Benke, 1996).

Similarly, an on going trial at Akola, indicated that 50 per cent of fertilizer can be saved when applied through drip as liquid fertilizer without any reduction in the yield. But when the recommended dose was applied through drip, the yield increased by 28 per cent (Benke, 1996).

In Gujarat, systematic research for finding out the economic feasibility of drip in cotton in different agroclimatic conditions was initiated in 1995. An ongoing experiment at the Cotton Research Station, Surat indicates that with drip a water saving to the tune of about 40 % can be achieved without affecting the yield. In Hybrid-8 through drip the seed cotton yield increased by about 25% with a water saving of about 50%. A curvilinear response was observed upto 70%, replenishment of evaporated water from the open pan. The ultimate economic return showed that with the use of drip, the farmer can get about Rs. 2000.00 per hectare additional income and about Rs. 50,000 more for the same quantity of water used as in the surface method. In Navasari area of South Gujarat, a heavy rainfall zone where kharif cotton is on the decline due to waterlogging problems, rabi cotton was found to have good prospects. In the rabi cotton, drip irrigation increased the seed cotton significantly.

Based on the review of available literature on drip irrigation, the following broad conclusions can be drawn.

1. Response of cotton to drip varies under different soil and agroclimatic conditions.
2. Under conditions of non constraints of irrigation water and in non problematic conditions the yield response may be poor, though considerable saving in irrigation water can be achieved.
3. Under constraints of good quality irrigation water, but with enough land availability, drip may be a boon to bring more area under irrigation and improve the overall economy of the farmers with the same quantity of water used as in the surface method.
4. In salt affected soils, but with availability of either good or moderately saline water drip can most profitably be used both for increasing cotton productivity and maintaining soil health.
5. Fertigation through drip can save on fertilizers
6. Under high water table conditions with poor quality under ground water, irrigation through drip to maintain shallow and restricted root zone is a better proportion than the surface irrigation.

Under rainfed and erratic rainfall conditions, use of drip and sprinkler technology can boost the yield many fold. Appropriate ET replenishment has to be worked out for different agroecological situations. The cotton seed/lint yield follows a curvilinear relationships while the total dry matter production follows a linear response, with increasing moisture requirement through drip.

Irrigation efficiency

Is the percentage of irrigation water that is available for consumptive use by the crops. When the water is delivered and measured at the farm headgate it is called farm irrigation efficiency. When it is measured at the field or plot it may be designated as field irrigation efficiency. However, irrigation efficiency of 70 to 80 % is considered good.

Scope of Drip Technology

Maharashtra and Gujarat put together contribute to more than 40% of the total cotton area in the country. But in both these states about 67% and 97% of cotton crop is grown under rainfed conditions. In these areas, generally the rainfall is erratic. In Gujarat, the Government has announced 85% subsidy for farm ponds. If the rainfed cotton growers of this state take advantage of this scheme and adopt drip irrigation, the cotton productivity can be increased many folds. Secondly, Gujarat state has maximum cotton area under salt affected soils. Many of the cotton growing areas fall in the saline soil categories as this crop is relatively more tolerant to salinity. Both in saline and normal soils, in considerable area, the underground water is either very less or saline or both. In such conditions, use of drip technology will greatly improve the existing lint productivity, and also the overall production as more area can be irrigated with the available water. Farmers of North Gujarat, have already put into use this technology especially for their seed production plots. About 400 ha of seed plots of cotton in Sabarkantha district are under drip. This is also adopted by many growers in other parts of North Gujarat, Kutch and Saurashtra area. Thus, the drip technology can boost the productivity and overall production of cotton, in addition to betterment of soil and plant health especially in problematic conditions.

Apart from yield increase, water and fertilizer saving, drip has been especially useful in problem, soil and water conditions. In the sodic soils of Thiruchallapalli in Tamil Nadu, used sodic water through drip in cotton and harvested 10% more yield than surface method. Moreover, under drip irrigation, about 1/3 water is used than surface method. Sprinkler irrigation is equally important effective and useful for increasing area under irrigation and to get uniform germination of crops. Sprinkler can be used in any crops of medium height of the sprinkler sets. Sprinkler irrigation in cotton crop, should not be done after boll bursting mainly due to spoil of the fibre quality of cotton.

Response of rainfed cotton to life saving irrigation of harvested rain water in different toposequences

Detailed field study for rain water management practices on different toposequences by Bhaskar et al. (2003) at Thugaon micro water-shed was carried out. Three years (2001—2003) results showed that the response of cotton to drip in different soils occurring on different toposequences was different in relation to yield per plant, total biomass per plant and finally seed cotton yield qha^{-1}.

Under conditions of non-constraints of irrigation water and in non-problematic, conditions, the yield response of cotton was found poor through considerable saving in irrigation water, but in case of conditions where water is limited, the response of life saving irrigation water, water but in-situ moisture conservation was found very helpful and effective in increasing the seed cotton yield by 5-8 q/ha. Under constraints of good quality irrigation water, but with enough land availability, drip and sprinkler may be a boon to bring 50-100% more area under irrigation and improve the overall economy of the farmers with the same quantity of water used as surface irrigation. In salt affected soils, but with the availability of either good o moderately saline water, drip can be most profitably used both for increasing cotton productivity and maintaining soil health. Fertigation through drip can save 40-50 per cent of fertilizer requirement. Under high water table conditions with poor quality underground water irrigation through drip to maintain shallow and restricted root zone was a better proposition than the surface irrigation. In shallow, sloppy degraded soils, in-situ moisture conservation practices such as ridge and furrow system, use of the organic or plastic mulch was also found very useful over the other management practices by reducing runoff water and soil loss with 15-25% increase in productivity of rainfed cotton crop.

Critical stages in cotton for irrigation

The various critical plant development stages has been identified in cotton, among these, peak flowering and peak boll development stages are very critical, and if, life saving irrigation is not provided at this time, the biomass and seed cotton yield decreases drastically.

Recycling of harvested rain water : Results from the field experiments conducted at the Institute during kharif, 1995-96 to 1999-2000 on shallow, medium deep and very deep soils at watershed basis, indicated that one irrigation of harvested rain water @ 4 ha cm is applied at peak boll development stage, not only increased biomass, but also increased seed cotton yield significantly in shallow to medium deep soils. The effect of irrigation on seed cotton yield was more in medium deep over other soils.

Plate 2.5.A. Cultivation of cotton on totally eroded soil having surface layer purely of sand. Note growth in month of Sept. (Warora takuka of Chandrapur dist. in Maharashtra)

Plate 2.5.B. A view of water harvesting pond on shallow soil 5-8 q ha^{-1} seed cotton yield could be increased by recycling of harvested rain water

Effect of rain water on seed cotton yield in soils of varying characters: The results from 5 years research study indicated that the maximum seed cotton yield of 18.15, 18.03 and 15.03 q ha^{-1} was recorded in shallow, medium deep and very deep soils respectively, when irrigation @ 4 ha cm of rain water at peak boll development stage was given to cotton crop. Cotton yield before and after this stage decreased the seed cotton yield due to irrigation (Table 8). Irrigation effect on seed cotton yield was almost equal in shallow to medium deep soil, while, it was less in very deep soil. This showed that the yield of the cotton could be increased by about 4 q in shallow and 5 q in medium deep soils by giving one irrigation to cotton at peak boll development stage, while about 2 q only in very deep soil over no irrigation. This showed that shallow to medium deep soils are very responsive to irrigation compared to very deep. Very deep vertisols can be cultivated without irrigation in the region. Further, it may be concluded that the yield of seed cotton could be increased by about 25% in shallow to medium deep soils due to irrigation through rain water over very deep soil (Table 9).

Table 9. Effect of irrigation through harvested rain water on the yield of rainfed cotton on soils of varying depth. 6 (1995-2000) years mean yield

Treatment	Seed cotton yield (q ha^{-1})			% yield increase over very deep soil	
	Shallow * (<45 cm deep)	Medium deep (45-90 cm deep)	Very deep (>90cm deep)	Shallow	Med. deep
I 0 Control	14.17	13.60	12.72	11.40	06.9
I 1. Irrigation @ 4 ha cm water at flowering stage	16.03	15.20	12.63	26.92	20.45
I 2. Irrigation @ 4 ha cm water at early boll development stage	16.74	16.98	14.04	19.23	20.94
I 3. Irrigation @ 4 ha cm water at peak boll development stage	18.15	18.03	15.03	20.86	20.0
I 4. Irrigation @ 4 ha cm water at early boll development stage and peak boll development stage	16.65	16.68	13.44	23.92	4.1

Source : Bhaskar & Kairon (2000) Note :

- No irrigation was given in shallow soil during 1995, as no rain water was available.
- During 1997, no runoff was recorded due to soaking rains.
- During 1999 due to continuous rains no moisture stress was occurred.

Aquaculture in water harvesting ponds

Now, it become not only possible, but economical also, to cultivate fishes in the stored rain water for a period of about 3-4 months time. Fingerlings of Rohu, Katala, Mrigal and Catfish were raised in the water harvesting ponds during last 3 years. Within 3-4 months time, catfish gained the weight of 300-700 g per fish from finger lings of 3-4 g weight, while, other breeds were found comparatively very less in weight gain. With the adoption of this practice, during 3-4 months time, stored water could be utilized for fish production and after that when there was a need of the water to cotton crop, same quantity of water was recycled to the crop.

Economic significance : With the adoption of watershed based research, approach, the additional income of Rs. 15000-20000 could be obtained from sale of fishes and sum of Rs. 10000-

15000 as additional income from cotton could be realised from same piece of 1 ha land. However, by harvesting and recycling of rain water, sum of Rs. 25000-35000 as additional income was realised from same piece of land through rain water management.

Impact assessment of watershed-based Agro-Production technology.

Cotton-based cropping systems in most of the area suffered form resource constraints and also confronted with a number of problems is rainfed tracts of Maharashtra. The productivity of the system in rainfed agro-eco-system watershed areas is considerably low as compared to irrigated agro-ecosystem. In rainfed system, watershed management has been accepted as the most rational approach in preventing deterioration of agro-eco systems, restoration of degraded lands; use of rainwater and runoff harvesting. Efficient management has helped in improving overall productivity in rainfed areas. The soil and water conservation measures such as growing of grasses on boundaries, mulching, forestation, ridges and furrow methods, deep tillage, strip cropping, sowing of crops across slope, land leveling, contour farming, collection of excess runoff water in harvesting ponds, etc are some of the important measures considered for improving the land use efficiency and crop yields at watershed basis. Considering all these facts, a comprehensive programme of technology transfer at watershed basis on farmers field at Thugaon micro-watershed in Amravati district was launched by Central Institute for Cotton Research, Nagpur under NATP Rainfed-agro-eco-system during 2000-2004. For assessment of the technology transfer programme, an instrument was developed to record knowledge and adoption of watershed-based technologies. This instrument was used before start and after implementation of the programme in selected villages and farmers under Thugaon micro-watershed. The study revealed that the watershed farmers had higher enhancement of knowledge, adoption and improvement in yield levels in cotton-based crops after implementation of the programme. The score for adoption of technologies ranged from 35-51 per cent before programme initiation and 50-102 per cent after the implementation of programme. However, the composite mean score was increased to the tune of 27.39 and 71.83 per cent, respectively, in case of knowledge and adoption of practices. It was remarkable that due to on-farm trails, an increase of 129,56,67,172,73, 44.71 and 114.29 per cent yields, in cotton, jowar, pigeonpea, wheat and gram was recorded, respectively. Further, it indicated that the yields obtained in on-farm trials were comparatively higher than the state and the districts average yields. This showed that the watershed farmers who were not covered under the programme were also taking interests in adoption of cotton based watershed technologies. However, concerted extension efforts are required to popularize the watershed-based cotton production technology among the villagers.

While comparing the adoption of watershed-based cotton production technologies, which are 30-35 per cent of the actual adoption, and due to that it has created a yield gaps in cotton crops. Further, at watershed basis, the marginal productivity of cotton is because of several associated problems like undulating topography, shallow and sloppy soils associated with low water holding capacity and less growing period. Erratic and unpredictable rainfall and its poor management needs sound cotton production technologies at watershed basis. At the same time, socio-economic conditions of the farmers is of paramount importance in such a high-risk prone rainfed farming situations as it regulates the decision-making and adoption of improved technology. To overcome of this situation watershed based toposequence management is being considered as holistic approach to bridge the yield gaps due to efficient moisture management practices, increased yields with higher gross returns, improved socioeconomic conditions etc. of the watershed farmers. Integrated toposequence based cotton production technology at watershed basis has been developed for the farmers by Central Institute for Cotton Research, Nagpur under NATP Rainfed Agro-eco-system and conducted on farm trials in 40 farmers field at Thugaon micro-watershed in Amravati district,

Maharashtra. The study has, therefore, been planned to probe into the socio-economic issues in terms of socio-personal, socio-economic, psychological and communication characteristics of the farmers on watershed basis and its overall impact on socio-economic condition of the farmers. It was observed that due to implementation of watershed based cotton production technology, it not only increased the yields of the cotton and cotton-based crops, but also helped in increasing monitory returns ranges from Rs. 8511 in upper plain in case of cotton sole to Rs. 31941 in lower plain when soybean was intercropped with cotton and chickpea was taken as sequence crop under protective irrigation after the harvest of cotton and soybean. Due to implementation of this programme, considerable increase has been noticed in number of wells and water harvesting ponds and percolation tanks with improvement in ground water status. The increased water availability in wells, ponds and bundies has enabled farmers to provide life saving irrigation to about 10 per cent additional areas. It has also improved the retention of soil and water and stopped its further degradation. It has also helped sustaining water percolation, reducing run-off and soil loss and improved water and use efficiency, developed better environmental conditions and empowered the farmers to take decisions to follow proper crop and water management practices at watershed basis for farmers prosperity and sustainability in rainfed cotton production.

Additional benefits due to watershed based research technology

- The recharge capacity of irrigation wells has been reported to increased by about 1 meter
- Soil loss reduced to lowest limit of about 2-3 t ha^{-1}.
- Not a single drop of water and a soil particle left unattended.

13. LAND EVALUATION

Land evaluation work different locations on watershed basis was conducted 1st at Kutki Research Station of Dr. Panjabrao Krishi Vidyapeeth, Akola and 2nd at Sangi near Nagpur under UNDP, watershed Management programme by the National Bureau of Soil Survey and Land Use Planning, Nagpur; and 3rd at Thugaon, District Amaravati under NATP and its Research Farm Panjari of The Central Institute for Cotton Research, Nagpur.

Detailed field experiments on rain water management, in-situ moisture conservation runoff and soil loss, cropping system water harvesting and storage, recycling, soil-site suitability evolution, micro irrigation etc. Detailed location wise results are given as under:

Research work done at Kutki Research Station of Dr. PDKV Akola on Land Evaluation

Land Evaluation For Rainfed Cotton

Land evaluation is the assessment of land performance when used for specific purpose. Soil depth, texture and drainage determine the performance of any crop on identified soil unit (Davidson, 1984). While evaluating the cotton yield performance, Bhaskar (1987) reported that, deep to very deep & dark clayey soils are highly suitable for rainfed cotton cultivation in Vidarbha Region of Maharashtra. As such, land evaluation provide a rational land use relationship between land use and land giving estimate of required input and projected output (Anonymous, 1983). The present study was undertaken with a view to evaluate the soil suitability for cotton cultivation at Agricultural Research Station (Punjabrao Krishi Vidyapeeth), Kutki District Wardha Maharashtra. The detailed soil survey of the farm was carried out and seven soil mapping units were identified and mapped. Brief description of soil mapping units is given is Table 1. In general, soils of the farm are developed from weathered basalt and basaltic alluvium. They are very dark greyish brown to dark greyish brown in colour shallow to deep, clay loam to clay in texture and slightly alkaline in reaction with high base saturation.

Table 10. Soil and their brief description

No.	Mapping unit	Brief description
1.	Kt4 md2 B2	Kutki 4 clay. shallow on 1-3% slope with moderate erosion-Lithic Ustorthent.
2.	Kt5 md3 B2	Kutki 5 clay. Moderately 1-3% slope with moderate erosion – Vertic Ustochrept
3.	Kt2 md4 A1	kutki 2 clay. Deep on 0-1% slope with slight erosion – Typic Chromustert.
4.	Ktl md5 B2	Kutki 1 clay. Very deep on 1-3% slope with moderate erosion – Typic Chromustert.
5.	Ktl md5 B2	Kutki 1 clay. Very deep on 1-3% slope with moderate erosion – Typic Chromustert.
6.	Ktl md5 B2	Kutki 1 clay. Very deep on 1-3% slope with moderate erosion – Typic Chromustert.
7.	Ktl md5 B2	Kutki 1 clay. Very deep on 1-3% slope with moderate erosion – Typic Chromustert.

Source : Bhaskar (1988).

The ultimate of the area is semi-arid tropical. The mean annual rainfall is 86 mm with 60 rainy days. Crops yield data have been collected from different soils for the year 1981 to 1984 under rainfed condition. The test crop was cotton B -1007.

The identified soil units of the farm have been evaluated using revised storie soil Index rating (SIR) Method (Storie 1978) which is a product of four major factors, viz. character of physical profile (A), surface texture (B), slope (C) and condition other than those in factor A,B, and C like drainage, nutrient (fertility) level and erosion (X), etc. Rating in present has been worked out as under :

$$\text{Soil Index Rating} = A \times B \times C \times X.$$

Data indicated that the highest (39%) Soil Index Rating (SIR) was recorded in Typic chromusterts and lowest (15%) in Lithic Ustorthents (Table 11).

Table 11. Effect of soil rating and yield contributing characters on the yield of cotton.

S. No.	Soil Mapping	Soil rating	Yield (q ha^{-1})	Plant height (cm)	Number per (average)		
					Branches	primary	Root Length (cm)
1.	Kt4 md2 B2	15	4.4	36.8	4.2	8.0	22.2
2.	Kt5 md2 B2	25	4.7	51.2	10.8	10.6	35.6
3.	Kt2 md2 A1	31	6.2	70.4	10.2	11.4	42.6
4.	Ktl md2 B2	33	6.6	73.0	11.6	11.8	46.8
5.	Ktl md2 B2	35	9.3	76.0	12.2	19.2	50.6
6.	Ktl md2 B2	37	8.8	74.4	11.8	15.8	47.8
7.	Ktl md2 B2	39	9.7	79.8	12.6	19.6	51.8

Source : Bhaskar et al. (1988).

Highest (9.7qh^{-1}) cotton yield was recorded in Typic chromusterts and lowest (4.4 q ha^{-1}) in Lithic ustorthents (Table 8). The yield in Typic chromusterts ranged from 6.2 q ha^{-1} to 9.7 q ha^{-1}. The higher (100%) cotton yield in Typic chromusterts may be attributed to soil depth, high water and nutrient holding capacity of soils less exosion and less slope.

The cotton yield data (y axis) from different soil types were correlated with the SIR (X axis) and the relationship was linear. The correlation coefficient (r=0.89) is significant at 1% level. The regression equation worked out is $Y = 0.234\ x - 0.088$. Where Y is the yield of cotton in qha and X is the soil Index Rating.

Cotton being deep rooted crop, requires good soil conditions for proper root and shoot development which can be significantly related to soil Index Rating. The yield contributing characters like plant height (cm), number of branches primary roots and root length per plant (cm), relatively supported the cotton yield in all the soils.

Cotton Response to Soils of Varying Depths

Results obtained from another field study, conducted at the same Research Farm on four soils, namely shallow (7.5 to 22.5 cm deep), medium (22.5 to 45 cm deep), deep (45 to 90 cm deep) and very deep (more then 90 cm deep). Soils were identified, characterized and grouped into four category. All soils were developed on basalt and basaltic alluvivour. Soil based crop yield data on cotton for eight years (1980-1987) have been collected and interpreted with soil depth and depth wise soil group. Results obtained from the study are very encouraging and found suitable for soil suitability evaluation and sustainable land use planning for the devclopment of rainfed agriculture in the country.

Table 12. Yield (q/ha) evaluation of cotton to different soils.

Year	Soil				Rainfall (mm)	No. of r ainy days
	Shallow (7.5 to 22.5- cm deep)	Medium (22.5 to 45- cm deep)	Deep (45 to 90- cm deep)	Very deep (90 cm and above		
1980	3.69	4.18	–	7.52	–	–
1981	2.26	5.71	8.72	10.02	694.8	37
1982	3.86	–	5.74	9.17	606.6	36
1983	4.31	5.24	6.79	8.58	865.1	44
1984	2.50	–	6.93	8.85	304.3	25
1985	2.26	7.83	7.89	13.57	1065.6	76
1986	0.87	2.77	8.85	10.67	1310.0	35
1987	3.34	6.41	6.74	8.26	437.6	26
Average	2.88	5.36	7.38	9.58	660.5	29
S. E.	0.37	0.65	0.40	0.62	---------	--------

Source : Bhaskar et al. (1989)

Table 13. Economic (Rs/ha) evaluation of cotton in different soils.

Year	Soil			
	Shallow	Medium	Deep	Very deep
1980	2140	2424	NG	4361
1981	1310	3311	5057	5811
1982	2238	NG	3329	5318
1983	2499	3039	3938	4976
1984	1450	NG	4019	5133
1985	1310	4541	4576	7870
1986	505	1606	5133	6188
1987	1937	3717	3909	4790
Average	1674	3106	4280	5556

NG = Not Grown, Price = RS 580/quintal.

Source : Bhaskar et. al. (1989)

Under uniform set of management, that is, one deep ploughing followed by 2 to 3 harrowings and the recommended dose of fertilizer at 75 kg nitrogen and 50 kg P_2O_5 per hectare, sowing improved variety 'B-1007', with 3 weeding, 3 hoeings and 3-4 sprayings of pesticide. Very deep soils produced the highest (9.58 q/ha) yield of cotton seed, while deep, medium and shallow soils produced 7.38, 5.36 and 2.88 quintals per hectare respectively.

It is evident that very deep, deep and medium-deep soils are capable of producing additional cotton yield of 6.70,4.50 and 2.48 quintals per hectare respectively, which amounts to an increase of 232.6, 156.3 and 86.1 percent respectively over shallow soils (Table 8) under similar management practices. Higher yields in deep to very deep soils are attributed to higher clay content, water and nutrient holding capacity of these soils as compared to shallow medium-deep soils.

Soil Depth And Economic Returns.

Table 8 indicates that the average economic returns realized from shallow, medium, deep and very deep soils are Rs. 1,6784, Rs. 3,106, Rs. 4,280 and Rs. 5,556 respectively. Economically, very deep, deep and medium deep soils are more remunerative by 232, 156, and 85 percent respectively over shallow soils. Shallow soils having depth less than 22.5cm are not suitable for cotton cultivation and thus they may be put under shallow-rooted cover crops. Medium deep soils (22.5-45.0 cm) though they are better than shallow soils, should be profitably put under other climatically adopted crops like sorghum, pigeonpea and other leguminous and cover crops. Deep soils (40-90 cm) and very deep (more than 90 cm) soils produce the highest yield of cotton and are most suitable for Vidarbha region of Maharashtra.

Effect of Soil Depth on Cotton Yield.

Land evaluation study on the yield of rainfed was conducted at the same Research Station Kutki of Dr. P.D.K.V. Akola Maharashtra, by Bhaskar (1997) in order find out the effect of soil depth on the yield of rainfed cotton.

Soil-based crop yield data for four years (from 1981 to 1984) were collected analysed and correlated with the soil depth (Table 9).

Table 14. Soil depth, cotton yield and productivity rating.

Soil depth (cm)	Cotton yield (q/ha)				Average	Productivity rating	Crop productivity Rating
	1981	1982	1983	1984			
20	3.70	-	5.03	-	4.30	45	Medium
42	3.90	-	5.42	-	4.66	48	Medium
63	-	5.49	-	7.00	6.24	64	High
100	-	3.85	-	9.35	6.59	68	High
123	10.02	-	8.59	-	9.31	96	Very high
136	-	8.60	-	8.91	8.75	90	Very high
139	-	9.74	-	-	9.74	100	Very high

Source: Bhaskar et. al. (1987).

The maximum cotton yield (9.74 q/ha) of the farm was taken as a standard yield as suggested by Ableiter (1937). The variation in the yield was expressed as percentage of the yield taking the value of standard yield as 100. The same was graded into five crop productivity potential (CPP) class. The marginal productivity rating takes into account minimum possible yield target that is needed to be harvested for covering the expenditure on input for the management of a given crop. Data (Table 7) indicated that the yield data were above marginal i.e. CPP was medium at 20 to 42cm soil depth, high at 63 to 100cm depth and very high at 123 to 139 cm depth.

Correlation studies revealed that there was a significant correlation of cotton yield with soil depth ($r = 0.96$ and $t = 7.850$) with a regression equation of $Y = 3.17+0.0044\ X$. It can, therefore, be concluded that 123 to 139 cm soil depth for cotton cultivation at the farm and surrounding area was highly suitable and cotton can also be grown successfully in the soil depth range of 63 to 100 cm. Soils below 63 cm depth will give medium yield and such soils should not be brought under cotton cultivation. Such soils may be cultivated for shallow rooted crops.

Response of Cotton and Sorghum to Soil Parameters

For realization of rainfed agricultural potential, the major constraint is lack of available moisture. The moisture capacity of Vertisols is high and that of associated soils is relatively low. The soils are usually put to monocrop cultivation more particularly during post-rainy season. With the introduction of improved agricultural practice, it has become feasible to have double cropping system in the dependable rainfall area of Vertisols and associated soils. A case study was undertaken to identify the soil units, their characteristics and productivity behavior of different crops under uniform set of management practices.

Detailed Soil Survey of Agricultural Research Farm, Kutki, of Panjabrao Krishi Vidyapeety, Akola, was carried out (Gaikawad and Bhaskar 1988) and 7 mapping units were identified and mapped. Farm soils were characterized and classified as Lithic Ustorthent, Vertic Ustochrept, Typic Chromustert and Typic Ustochrept. They are developed on weathered basalt and in alluvium. They are shallow to very deep, clay loam to clay in texture and field trials on sorghum (CSH-9) and cotton (B-1007) were conducted during the rainy season of 1986 and 1987, on different soil units with uniform levels of management under scientific and farmer's management practices.

Crop Response to Soil

Gaikwad and Bhaskar, (1988) reported that the average highest yields of sorghum grain (35.0 q/ha) and cotton (9.3 q/ha) were recorded in Typic Ustrochrept and Typic Chromustert, respectively under uniform set of management practices. On comparing the yield under both scientific and farmer's management practices, soil belonging to Typic Ustochrent and Typic Chromustert were found highly productive as seen from the additional yield of sorghum grain 17.3 q/ha and seed cotton grain 6.6 q/ ha constituting an increasing of 98 and 357 per cent, respectively Similar trend of result on sorghum (Tamgadge et al 1986) and cotton (Bhaskar et al. 1987) were also reported.

Since climatic factors were considered constant, yield differences are attributed to various soil parameters. Soil depth seems to be the most important yield influencing factor. Soil having the depth > 45 cm were found highly suitable for cotton (9.3 q /ha) and sorghum (28.2 q/ha) crops as compared to medium deep soils (22.5 -45.0 cm) and shallow soils (< 22.5 cm) where the yields are low. The site and soil characteristics were correlated with crop yields and their effects were tested statistically using yield Prediction Model of the type given below :

$$Y = a + b1\ X1 + b2\ X2 + b3X3 + \ldots\ldots\ldots\ldots\ldots\ldots + e,$$

Where (Y = Yield, X1 + X2 .... Are explanatory variables, a intercept, b's are coefficients are e random error). The factors like slope, concretion and exchangeable Ca are not significant ; CEC creates a problem of multicollinearity. Hence these are excluded in the model. The final multivariate equation comes to $Y = 1.8035 + 0.0667\ X1 + 0.05087\ X2 + 1.6139$ for cotton and $Y = -4.4917 + 1.0666\ X1 + 0.03927\ X2 + 2.8903\ X3$ for sorghum, where X1 = % clay ; X2 = % of soil moisture and X3 = soil depth (m). It is observed that yield of cotton and sorghum was positively and significantly influenced by the factors like soil depth, clay and soil moisture.

II. Research Work Done At Saongi Water Shed Near Nagpur By National Bureau of Soil Survey and Land Use Planning, Nagpur on Land Evaluation.

Soil- Site Suitability Evaluation for Cotton and Sorghum

The study was undertaken to evaluate production potential of different soils as per their capabilities with cotton and sorghum as test crops, in order to provide a suitable base for soil based agro-technology transfer.

The experimental site - a micro watershed measuring 74 ha near village Saongi in Nagpur district of Maharashtra State (India) is situated between 20 Degree 1'15'' N Latitude and 78 Degree 56'41''E Longitude. The area represents foothill slopes and alluvial plan land forms. The area receives 1125 mm of rainfall and represents sub-tropical, dry and subhumid climate. The soil moisture regime is Ustic and temperature regime is Hyperthermic.

A detailed soil survey of area was carried out as per procedure outlined in the Soil Survey Manual (IARI, 1970 and USDA Hand Book No. 18. 1951) on a base map of 1:4000 scale. Agronomic trials on different soils were conducted using sorghum and cotton a test crops during 1987 to 1989 (July - December) under scientific and farmer's management practices to find out the production potentials of the soils.

Six soil series were identified tentatively with 22 mapping unit. The soil and other thematic maps (soil depth, erosion, stoniness surface – texture, clay content) were prepared. A brief description of each soil is given below :

Saongi 1. - belongs to : fine loamy, skeletal, missed, hyperthermic, calcareous, very shallow family of Lithic Ustorthents.

Saongi 2. - belongs to : fine loamy, mixed, hyperthermic, calcareous shallow family of Typic Ustorthents.

Saongi 3. - belongs to : fine, monotmorillonitic, hyperthermic, moderately deep family to Lithic Ustochrepts.

Saongi 4. - belongs to : fine, monotmorillonitic, hyperthermic, deep family of Typic Ustochrepts.

Saongi 5. - belongs to : fine, montmorillonitic, deep family of Vertic Ustochrepts.

Saongi 6. - belongs to : very fine, montmorillonitic, hyperthermic, deep family of Typic Chromusterts.

The yield of different crops, as influenced by site and soil characteristics and management levels, were evaluated; their trend as influenced by different parameters (highlighted below) indicates similar response for cotton and sorghum although their magnitude varies :

Soil depth : Yield of the crops increased with increasing soil depth.

Slope : As slope increased, yield of the crops decreased.

Erosion : With increasing erosion hazards, yield of the crops decreased.

Stoniness : With increasing stoniness, yield of the crops decreased.

Surface : As the texture become coarser, yield of the crops decreased.

Texture

Clay content : The yield of the crops increased as the amount of clay increased.

Soils: The data indicated that the mean maximum grain yield of sorghum was recorded in Udic Ustochrepts (54.5 q ha^{-1}) followed by Typic Chromusterts (39.6 q ha^{-1}) and the mean minimumin Littic Ustorthents (13.5 q ha^{-1}). Similarly, the mean maximum seed cotton yield was recorded in Udic

Ustochepts : (20.1 q ha^{-1}) followed by Typic Chromusterts (13.5 q ha^{-1}) and the mean minimum from Lithic Ustorthents (8.1 q ha^{-1}). The yield responses in Lithic Ustochrepts and Typic Ustochrepts for both the crops are of medium range.

Land capability : The data showed that the mean yield of cotton and sorghum in class II, III,& IV lands were recorded in the order of 13.5, 11.4 & 9.6 and 39.2, 32.7 & 26.4 q ha^{-1}, respectively.

Since climatic factors were considered constant, yield difference are attributed to various site and soil parameters. These were correlated with crop yield and their effects were tested statiscally by using : $Y = a + bx$

Where ; Y = Yield, x = Soil parameter, a = Intercept and b = Regression Coefficient

It became evident that yields of cotton and sorghum were positively and significantly influenced by the factors, such as soil depth, clay and available water content. This enables us to predict crop performance on similar soils occurring elsewhere.

From the study, it can be inferred that :

1. Deep to very deep, clayey soils are found to be economical for cultivation.
2. The highest crop response to management was recorded in Udic Ustochrepts followed by Typic Chromusterts.

3. Based on yield prediction models, soil depth, clay content and available water capacity were found to be important soil parameters influencing the crop yields.
4. By adopting scientific management practices, the yield of cotton, and sorghum can be increased by 7.0 (120%) and 23.0 (150%) q ha respectively over farmers' practices.
5. This technology will provide a suitable base for soil-based agro-technology transfer on similar soils occurring elsewhere.

The challenges for the food security are linked to the land resources its quality and management. Watershed approach is well considered as an approach of area planning of natural resources (land, water and plants) to achieve high productivity, while sustaining or enhancing our natural resource base. Land evaluation is the process of determining the potential of land for alternative use and is pre requisite for land resource management to manage land parcels for better land uses like cropping etc. based on potentials. Land evaluation is the recent concept of increasing production potential of soils based on soil-site characteristics management levels and climatic parameters. Result of the case studies conducted at Saongi watershed, near Nagpur showed that in the upper sector, the major soil constraints were shallow depth, low length of growing period (LGP) and low plant available capacity (PAWC) and hence are not suitable for cotton, soybean and pigeonpea with low yield level and B.C. ratio.

Vertic ustochrepts were found suitable for sorghum (32.3 q ha^{-1}) and soybean (13.6 q ha^{-1}). Valley floor soils with high PAWC (126.2 mm) and high LGP (120-180 days) highly supported cotton, sorghum and soybean with high yield and high B.C. ratio.

Approach Towards Sustainable Land Use Planning

Land use planning appears to be rational response to the challenge. However, it is clear that attempts to meet the challenge are not keeping pace with the escalating severity of the problem. Land use depends on land users, individual companies and communities, the act first to meet primary need as to avoid imminent calamity. Beyond this, they act to reduce risk as in the expectations of gain in the longer or shorter term. Hence land use planning is expected to make a major contribution to the realization of sustainable development as it can facilitate the allocations of land to the use that provide the greatest sustainable benefit (FAO, 1976). The soil subjected to expletive agriculture for many years can not regenerate by itself so as to maintain its productivity. Proper plant growth can be attained only by creating optimum conditions in soils. In view of this a soil resource inventory for land use planning of Yerla watershed around Nagpur, Maharashtra has been under taken.

The present trend of population and potential production, suggested land use planning with due consideration of expected yield are computed for carrying capacity. This denotes that the present land use and rate of production of present market price, it may Rs 4122/- while the consumption at present population and live stock as per standard diet Rs 6427/- which showed the production is surplus and with due consideration of growth rate both for human and animal at average, present rate of production may be sufficient for 20 years (2017-18) where as land use, present market price and expected yields per unit area may be sufficient for another 34 years (2051-52) because of production Rs. 14586/-.

It is concluded that the suggested land use plan for the watershed may prove tools for the development of watershed village in particular and/or nation as a whole (Nimje *et al.*, 2000).

Soil Suitability Evaluation For Cotton In Relation To Sustainability.

In another fields study conducted at Saongi water shed, the soil-site characteristics of soils was evaluated as per FAO (1976). Soils-site parameters were correlated with seed cotton yield and a sustainable soil based land use path was developed, as detailed below.

Soil-site parameters in Relation to Cotton Yield

A significant correlation between soil parameters and cotton yield (R^2 =0.89) was recorded and yield prediction models was developed as Y =1.6032 + 0.0012 soil depth + 0.0096 AWC = 1.1430 CEC = 1.0202 clay.

Soil Series in Relation to Cotton Yield

Saongi 5 and Saongi 6 are considered as S_1 (highly suitable) as they produced maximum (18.34 q ha^{-1}) seed cotton yield. While Saongi 4 as S_2 (moderately suitable), Saongi 3 as S_3 (marginally suitable) and Saongi 2 as N1 (currently not suitable) produced mean yield of 14.67, 10.67 and 7.54 q ha^{-1} respectively.

Soil Suitability vis-s-vis Sustainability

Both sustainable Yield Index (SYI) and the soil suitability increased with the increase in effective soil depth. The degree of soil sustainability also increased accordingly and was found closely related with soil-site characteristics which is evident from the soil series sequence path from highest to lowest in order of :

- Saongi 5 Vertic Ustochrepts (89) Saongi 6 Typic Chromusterts (71) Saongi 4 Typic Ustorthents (26).

Soil Suitability Sequence Path

- Saongi 5 (100) Saongi 6 (89.2) Saongi 4 (75.8) Saongi 3(55.2) Saongi2 (38.7)

Soil Suitability Class

- S1 (Very deep, silty clay S_2 (medium deep, clay loam to silty clay-loam) S_3 to N1 (shallow to-very shallow and silty clay-loam to sandy caly-loam).

Soil Survey for Cotton Yield Maximization

- It is seen that seed cotton yield increased with increase in effective soil mass which followed the sustainable soil class sequence path from highest to lowest in order of :
- Very deep (18.34 q $ha^{-1)}$ Medium deep (12.60 q ha^{-1}) shallow (7.54 q $ha^{-1)}$

It may be concluded from the study that the yield of cotton could be increased from very deep - 60 to140% by giving cotton in medium (45 to90 cm) soils. Sustainable yield in trend was found associated with higher seed cotton yield in Saongi 4, 5 and 6 soils which as are evaluated S_2 and S_1.

Soil depth, AWC, CEC and clay content were found yield influencing parameters. Soil-based agro-technology developed on watershed basis was found economically viable. The research findings can be extrapolated to similar soils occurring elsewhere for yield prediction and sustainable land use planning.

Evaluation Of Suitable Cropping Systems. (Saongi watershed)

The identified soils of the Saongi micro-watershed were tested with different cropping systems during 1989—90 and 1990—91 under uniform set of management practice to evaluate the suitable cropping system for each soil under irrigated conditions of Vidarbha Maharashtra.

Efficient Cropping System

The yields of various crops expressed as wheat grain equivalent for the sake of comparison since the cropping system involves number of crops including vegetables are given in Table 7.

Table 15. Suitable cropping systems for different soils of Saongi watershed near Nagpur,

Soil	Suitable cropping system	Mean Yield (qha^{-1}) Grain	Straw	Wheat grain equivalent	Net Returns (Rs ha^{-1})	Cost Benefit ratio
Saongi 2	Soybean-gram	9.8 - 9.5	16.7 – 3.8	36.8	10483	1.7
	Soybean – wheat	9.8 - 14.9	18.6- 13.4	39.9	9904	1.2
Saongi 3	Soybean – wheat	10.5-22.8	18.9 -31.9	52.5	15466	1.6
	Soybean – wheat	26.7 -20.7	45.4 – 18.8	54.9	15400	1.7
Saongi 4	Cotton – Onion	12.3-112.5	13.5 –Nil	84.3	26761	2.4
	Sorghum – wheat	35.9 -21.6	61.0 -28.0	68.1	21234	2.3
	Soybean -wheat	12.4 -25.6	21.1 – 35.8	60.1	18792	2.3
Saongi 5	Cotton – Tomato	16.2 – 149.4	21.1 – Nil	111.9	34166	2.1
	Sorghum –Wheat	48.2 - 27.2	89.2 -38.1	91.3	31504	3.3
	Soybean - Wheat	16.0 – 35.3	27.2 -53.0	81.0	27989	3.3
Saongi 6	Sorghum – Wheat	50.9 – 29.7	91.6 -41.6	97.0	33992	3.5
	Sorghum - Gram	50.9 – 9.0	91.6 – 4.5	73.9	35889	3.5

Source : Bhaskar et. al. (1995).

Sorghum @ 350 q^{-1}, straw @100 q^{-1} ; soyabean @ 850 q^{-1}, straw @100 q^{-1}

Wheat @ 450 q^{-1}, strae @80 q^{-1} ; cotton @ 1200 q^{-1}, sticks @ 50 q^{-1}

It is seen from the data (Table 15) that the cotton-tomato produced the highest (111.9 q ha^{-1}) Wheat grain equivalent and net profit (Rs. 34166) in Saongi -5 (Vertic Ustochrepts) followed by Sorghum- Wheat with 97.0 q ha^{-1} wheat grain equivalent and Rs. 33992 net profit in Saongi -6 (Typic Chromusterts) and the lowest wheat-grain equivalent (19.9 q ha^{-1}) and net profit (Rs.4,851 ha^{-1}) in Saongi-2 (Typic Ustorthents).

Out of all the cropping systems, only five were tested on all soils. Of these sorghum-wheat and sorghum-gram for Saongi-6 (Typic Chromusterts) and Saongi-4 (Typic Ustochrepts) and soyabean-gram and soybean-wheat for Saongi-2 (Typic Ustorthents) were found highly suitable and remunerative cropping systems. Further, it was observed that soybean being a legume increased the available soil N to the extent of 60 kg ha^{-1} while sorghum and cotton resulted in a net loss of N (40 and 70 kg $ha^{-1)}$.

Deep to very deep, clayey soils are found to be most economical. This helps in extending cropping patterns to similar soils occurring in the same agro-ecological region.

Management Level In Relations To Soils.

It is very clear from the data (Table 15) that the yields under scientific managements levels influence yields patterns of both (Sorghum and Cotton) the crops significantly in all the soils by creating a big yield gap ranging from 9.6 to 22.0 q ha $^{-1}$ in sorghum and from 2.5 to 7.9 q ha^{-1} in a case of cotton over the farmers level in different soils, mainly due to cultivation of these crops under scientific management practices : When we compare the yield pattern of sorghum in different soils, it ranges from 20.8 q ha^{-1} in Lithic Ustorthent to 46.3 q ha^{-1} in a Vertic Ustochrepts, under scientific, and from 11.1 to 24.3 q ha^{-1} in same soils respectively under farmers management practices. However, almost similar yield pattern in case of cotton under both the management levels has been recorded. Study shows, that under scientific management practices, by adopting soil-based crop production technology at watershed basis has created a ample scope in increasing the yields of principal crops (Table 16).

Table 16. Comparative performance of crops in different soil under scientific and farmers management levels.

Soil Classification	Scientific Management		Farmer Management		Yield gap	
	Sorghum	Cotton	Sorghum	Cotton	Sorghum	Cotton
Lithic Ustorthents	20.8	7.6	11.1	3.1	9.6	4.5
Typic Ustorthents	36.3	11.2	17.3	4.6	19.0	6.6
Lithic Ustochrepts	33.9	9.1	18.3	4.9	15.5	4.2
Typic Ustochrepts	38.9	11.0	21.2	8.5	17.7	2.5
Vertic Ustochrepth	46.3	13.0	24.3	8.0	22.0	5.0
Typic Chromusterts	43.3	12.1	32.9	4.8	19.4	7.9

Source : Bhaskar et. al-(1995)

Yield and Economic Significance of Soil-Site Character on Soughum and Cotton Economic significance of soil series

Amongst 5 soil series, Sawangi-5 produced the highest (Rs. 10,540 and12,070 ha $^{-1}$) economic return followed by Sawangi-3 (Rs. 8,640 and 10,795 ha^{-1}) from sorghum and cotton crops, respectively.

The yield as affected by soil depth, texture, erosion, land capability and management levels is given as follows :

Soil depth Vs crop yield

The yield of both the crops, sorghum as well as cotton, increases with increasing the soil depth. The maximum yield of Sorghum (48.0 q ha^{-1}) and that of cotton (13.0 q ha^{-1}) was recorded from very deep soil. The minimum yield of these crops was 22.4 q ha^{-1} and 8.6 q ha-1 respectively from shallow soil.

Fineness of the texture is found to cause the higher production in these soils. Sorghum and cotton yield was higher by 37.5 and 21.6 percent respectively in silty clay, over gravelly silty clay soils. Sorghum : The maximum yield (48.0 q ha^{-1}) was recorded in slightly eroded, followed by moderately eroded (31.6 q ha-1) and the minimum in severely eroded (22.4 q ha^{-1}) soils.

Cotton : The maximum yield (18.0 q ha^{-1}) was recorded in slightly eroded, followed (9.1 q ha^{-1}) by in moderately eroded and the minimum in severely eroded (8.6 q ha^{-1}) soils.

Land capability classes Vs crop yield

Sorghum : The highest grain yield (47.3 q ha^{-1}) was recorded in class II, followed by class III (33.7 q ha^{-1}) and 22.4 q ha^{-1} in class IV lands. The yield of sorghum in class II and III lands was increased by 24.9 and 11.3 q respectively over class IV soils.

Cotton : The maximum (13.1 q ha^{-1}) yield was recorded in class II, followed (10.0 q ha^{-1}) by class III and (6.9 q ha^{-1}) in class IV soils. The yield of cotton in class II and III soils was increased by 6.2 and 3.1 respectively over class IV soils.

Table 17. Sustainable Soil Suitability Path.

Soil	Sustainable Yield index	Sustainability index	Sustainable Class
Saongi 5-Fine, montmorillonitic, hyperthermic, deep family of Vertic Ustochrept.	89	100	51
Saongi-6 : Very fine montmorillo-nitic, hyperthermic, deep, family of Typic Chromustert.	71	82.0	51
Saongi-4 : Fine montmorillonitic hyperthermic moderately deep,family of Typic Ustochrepts.	60	75.8	52
Saongi-3 : Fine montmorillonitic,hyperthermic, moderately deep, family of Lithic Ustochrepts.	40	55.2	53
Saongi-2 : Loamy, skeletal mixed hyperthermic, calearious very shallow family of Typic Ustorherts.	26	38.7	N1

Source : Bhaskar et. al (1995)

Sustainability Concept

It is very clear from the data (Table 9) that the production potentials of vertisols and associated soils depend on various soil-site characteristics and management in puts. Soils depth, AWC, CEC and Clay content of the soil play an important role and has close relationship with crops yield. Detailed soil survey of the area is must for the management of natural resource at watershed basis. Priority should be given first to soil and rain water management, because it takes care of all the parameters. Based on, yield prediction model was developed.

Yield prediction model.

Y = 1.6032 +0.0012 Soil depth +0.0096 AWC +1.1430 CEC -1.0202 Clay.

Y = a+b1 x 1 + b2x 2 +b3 x3.....................+ e R^2 =O.89

Where,

Y = yield, x1,x2 and x3 explanatory variables

a = Intercept

b = Regression coefficient

e = Random error

Saongi Watershed

Runoff and soil loss in soils of varying characters as affected by cropping system. Runoff and soil loss in different crops on different soil under simple and graded bunding was determined. The soil loss under pigeonpea was observed to be 8.4 t ha^{-1} year $^{-1}$ which needs to be controlled in view of the shallow depth soil. The soil needs to be put under strip cropping along with suitable combination of crops. Sorghum under graded bunding resulted in soil loss of 7.0 t ha^{-1} year^{-1}. Soil loss under graded bunding with sorghum crop is more than permissible value of 5 t ha^{-1} year$^{-1,}$ which calls for measures to bring down the soil loss below permissible limit by undertaking vegetative control measures in association with graded bunding. Under graded bunding pigeonpea and cotton resulted in lowest soil loss on Typic chromusterts. Similarly on Typic Ustochrepts cotton (H-4) 3-4 t ha^{-1} year^{-1} soil loss (Table 18).

Table 18. Runoff and soil loss potential in different crops under graded bunding

Crop	Soil	Rainfall (mm)	Conservation practices	Runoff (mm)	Soil loss (t ha^{-1} hear $^{-1}$)
Pigeonpea	Lithic	1037	Bunding	138	8.4
Sorghum	Typic Chromusterts	1037	Graded bunding	124	7.0
Pigeonpea sole	Typic Ustortherts	1037	Graded bunding	110	4.1
Cotton sloe	Typic Ustochrepts	1037	Graded bunding	180	3.4
Pigeonpea + cotton intercropped	Typic Chromusterts	1037	Graded bunding	165	2.8

Source : Draft report Saongi watershed(1993)

Soil – Site Suitability Evaluation For Wheat

To Evaluate the soil-site suitability for irrigated wheat, different soils of Saongi micro watershed, detailed soil survey of the Sawangi micro watershed was done and soils were identified, charactersied and soil series were developed. Based on these parameters the soil – site suitability evaluation study for irrigated wheat in soils of sawangi micro watershed was carried out by Bhaskar et at (1996).

Field trials with wheat on Saongi 2 to 6 were conducted during Rabi 1987- 88 to 1989-90. The climatological data such as rainfall and temperature during crop development stages have been collected and used for assessing climate suitability. Soil- site parameters such as depth, available water capacity, clay content, cation exchange capacity, organic carbon, erosion and slope were expressed in terms of degree of limitations (0-4). The suitability of each soil was evaluated by number and kinds of limitation (s) as per FAO Framework (1976), and Dent and Young (1981) by considering S1 as highly suitable, and N1 as unsuitable (Table). The influence of individual soil parameter on yield of wheat was evaluated for sustainable potential.

A single parametric mode was tested.

$$Y = a + bx$$

Where, Y = yield ; a = regression coefficient, and x = soil parameter.

The parameters were correlated and multiple regression equation and correlation matrix were computed by using multivariate analytical model.

Y = a+bX1+b2X2+b3X3...............+e

Where, Y = yield ; X1X2X3..........Xn are variables ; 'a'

Intercept, 'b' s are regression coefficient, and e' = random error.

Evaluation of Climate and Soils

The optimum range of temperature for proper growth and development of wheat was found to be 15° to 20°C during germination, 8° to 10°C during vegetative stage, and 20° to 25°C during ripening stage. The prevailing temperature in the area is classified as moderate (2) during germination, and severe (3) during vegetative and ripening stage (Table 19).

Table 19. Criteria used for assessing the suitability of climate and soils for wheat.

Characteristics	Degree of limitation				
	No (0)	Slight (1)	Moderate (2)	Severe (3)	V. Severe (4)
Sufficiency of Energy					
Temperature (°C)					
Germination	15-20	12-18	10-15	8-12	5-10
	20-25	20-25	25-28	28-30	30-35
Vegetation growth	8-10	10-12	12-15	15-20	20-30
Ripening stage	20-25	22-25	22-30	25-35	35-40
Sufficient of water					
Soil depth (cm)	V. deep	Deep	Med deep	Shallow	
Sufficiency of Air					
Drainage	Well drained	Mod well drain	Imperfectly drained	Poorly drained	Ponded
Condition Affection Germination					
Texture	Sil, l	Sl, C, Sicl	Sl, fecl	Sl	Ls,s
Sufficiency of Nutrient					
OC(%)	V. high	High	Med	Low	V. low
CEC (com (p+) kg^{-1})	V. high	High	Med	Low	V. low
Erosion hazards	None	Slight	Mod	Severe	V. Severe

Source : Bhaskar et al. (1996).

Further, it has been observed that this temperature sets from tillering stage to milk formation and continues upto dough stage. When sufficient winter rains occur and bless the crop temporarily by increasing the humidity thereby lowering down the maximum and minimum day temperatures. High temperature and low humidity coupled with high wind velocity at grain filling lead to physiological drought (forced maturity) resulting low yields. Results are in agreement with Das and Kaira (1990).

However, due to this unstable and aberrant climatic (temperature) characteristics, highly suitable (S1) soils such as Saongi 4.5 and 6 were evaluated as moderately suitable (S2) although they produce comparatively high, (23.5 to 29.4 q ha^{-1}) yield. Where as, Saongi 3 produced 18.7 q ha^{-1} grain yield was evaluated as marginially suitable (S3), and Saongi 2 (14.5q ha^{-1}) as N1 (Non suitable) for wheat cultivation (Table 2)

Factors Affecting Wheat Yield.

Depth; The highest (29.4q ha^{-1}) grain yield of wheat was recorded from the soils of 101 cm depth followed by (25.9q ha^{-1}) 65-90 cm depth, and the lowest (14.5q ha^{-1}) from the soils of 16-20 cm depth Bhaskar et at. (1989) reported that deep (40-90) cm to very deep (> 90 cm) soils produced the maximum yield of cotton in vertisols.

Soil depth influences the wheat yield either through storage of available water or through root proliletion. It is evident from a significant correlation between soil depth and yield (r=0.749). The regression equation is computed as Ye =14,6980 +0.107 depth (cm).

Table 20. Soil-site characteristics.

Slope (%)	Drai-nage	Ero-sion	Tex-ture	Depth (cm)	AWC at rooting Depth (mm)	Clay (%)	CEC Cmol(+) Kg^{-1}	OC (%)	Yield (q ha^{-1})	Production Efficiency (%)
Saongi 2: Loamy, mixed, Typic Ustorthents										
3-10	Wd	e2-e3	Gscl	16-20	13-36	30.7-35.5	27.3-31.8	0.50-0.51	14.5	13
Saongi 3: Fine, Lithilc Udtorthents										
3-5	Wd	e2	Gscl	35-48	70-149	40.4-44.6	36.1-41.2	0.66-0.71	18.7	17
Soangi 4: Fine, Typic Ustochrepts										
1.5	Wd	e1	Sci,cl	65-90	168-179	48.4-45.5	43.1-46.8	0.73-0.74	25.9	23
Saongil 5: Fine, Vertic Ustochrepts										
3	Wd	e1	Scicl	101	204	55.6	49.3	0.63	29.4	26
Saongi 6 : Very fine, Typic Haplusterts										
1-3	Md	e1	C	150	311	60.1	53.7	0.68	23.5	21

Source : Bhaskar et al. (1996).

It indicated that the yield of wheat increase in soil depth. The most optimum range of soil depth was found to be in between 65 to 104 cm.

Available water capacity (AWC) : Data revealed that the highest (29.4q ha^{-1}) grain yield of wheat was recorded with 204 mm AWC, followed (25.9q ha^{-1}) by168-179 mm AWC. The lowest yield (14.5q ha^{-1}) was with 13-36mm AWC. Further, it may be stated that the yield of wheat was increased with increasing AWC. Similar results have been reported by Gaikawad et at. (1980) and

Bhaskar (1995). The correlation between wheat yield and AWC (r=0.7459) was significant. The regression equation described as : Y =14.6732 + 0.0474 x AWC. The optimum range of AWC was from 170 to 200 mm.

CLAY : The maximum (29. 4 q ha^{-1}) grain yield of wheat was recorded from the soils with 55.6 percent clay closely followed by (25.9q ha^{-1}) 48.4 – 51.4 per cent clay. The lowest yield (14.5q ha^{-1}) was with 30.7 -35.5 per cent clay (Table 2). The significant effect of day content on the yield of cotton and sorghum have also been reported by Gaikawad and Bhaskar (1988), Naidu et. at. (1988), and Gaikawad et at. (1990).

Further, there was a significant correlation between clay content and wheat yield (r=0.8319) with regression equation Ye =0.5174 x clay (%) - 2.3486. The suitable limit of clay content was found to be in between 48 and 56 per cent (Table 2).

CATION EXCHANGE CAPACITY (CEC) : Data indicated that the maximum (29.4 q ha^{-1}) yield was recorded from the soils having 49.3 CEC followed closely (25.9 q ha^{-1}) by 43.1 -46.8 CEC. Further it was observed that the yield was increased with increasing CEC. The correlation (r =0.8708) and regression equation Ye = 0.5032 x CEC -3.2198 was computed to find out the suitable limit of CEC for optimum wheat production. The most favourable CEC limit ranges from 43 to 53 Cmol (p+) kg^{-1}

ORGANIC CARBON (OC) : The significant (29.4 q ha$^{-1)}$ grain yield of wheat was recorded from the soils having o.63 per cent OC followed (25.9 q ha^{-1}) by o.73 to 0.74 OC, and the lowest (14.5 q ha^{-1}) from 0.50 – 0.51 per cent OC (Table). The correlation (r=0.5890) and regression equation Ye = 38.697 - 3.5846 x OC (%) shows that OC has significant effect on wheat yield. The yield is in agreement to Sharma and Sharma (1991).

EROSION : Yield trend showed that with increasing erosion, the wheat yield was decreased. Similar results were obtained by Massee and Waggoner (1985). It was highest (29.4 q ha^{-1}) under slightly (e1) and lowest (14.5 q ha^{-1}) under moderately (e2) to severely (e3)eroded soils (Table 2). Further, correlation (r =0.5586) and regression equation Ye =30.0 - 4.9286 x q ha^{-1} x Erosion showed eroded soils are found suitable (Table).

SLOPE : Yield trends indicated that the crop yield decreased with increasing degree of slope. The maximum (29.4 q ha^{-1}) wheat yield was recorded in soils having 1-3 per cent slope (Table). This explains the existence of negative correlation (r = - 0.5604) between the yield and slope parameter. The regression equation is worked out to be Ye=28.4161 -1.9095 x slope.

TEXTURE : Yield trend indicated that the wheat yield was increased with increasing fineness of soil texture (Table). The maximum (29.5 q ha^{-1}) yield of wheat was recorded in fine (Sicl) texture; closely followed (25.9 q ha^{-1}) by very fine (Sic) texture and the minimum (14.5 q ha^{-1}) from gravelly, skeletal, sandy clay loam soils. Medium (S1,1) to line (Sicl, CI) textured soils were found suitable. The yield was decreased by 50 per cent when the soil texture was changed from clay – loam to gravelly sandy clay loam (Table 2). Yield changes with soil texture have been reported by Naidu et al. (1988). Yield reduction in very heavy soils may be contributed to high (>60%) clay content; increased AWC, which directly reduce the oxygen essential for root development.

PRODUCTION EFFICIENCY (PE) : Production efficiency is a function of soil fertility, productivity and capability (Bhaskar et al. 1987). The maximum (26%) PE was recorded from Saongi 5, followed (23%) by Saongi 4 and the minimum (13%) in Saongi 2. High PE in Saongi 5, may be attributed to 1-3 per cent slope, sufficient rooting depth, fine texture, high AWC, clay CEC, and OC. Similar results have been reported by Bhaskar (1995) and Naidu et al. (1986).

Soil –Site Parameters And Wheat Yield: Interrelationship of various soil-site characters, influencing wheat yield was studied by using parametric approach for which a correlation matrix was worked out. It showed that soil depth, AWC,CEC, clay a OC have significant positive correlation while erosion and slope have significant negative correlation with yield. However, soil-site characters are most important yield influencing as it is evident from high degree correlation coefficient (R2 =0.9628) (Table). The findings are similar to De La Rosa et al. (1981) and Gaikawad et al. (1990). The soil-site characters, correlations and degree of extend have been ranked and arranged from highest to lowest order as below :

Soil : AWC > clay> CEC>OC. > Erosion

depth >Slope

AWC : Clay >OC > Erosion > Slope

CEC : Clay > OC > Erosion > Slope

Clay : OC > Erosion > Slope

OC : Slope > Erosion

Erosion : Slope

Based on these parameters, a multivariate analytical model was developed by substituting experimental values.

Yield = 82.2392 – 0.2700 x Depth -0.0211 x AWC - 2.0850 x CEC + 4.4644 x Clay - 18.3214 x OC +9.8673 x Erosion.

It is seen that soil-site characters are very important as they significantly influence the wheat yield. Texture is one of the most important permanent physical characteristic directly related to AWC and nutrient holding capacity. Soil depth per se supports the plant biomass and is responsible for moisture storage capacity.

Table 21. Soil-site suitability classification for wheat.

Soil	Soil – Site Parameters									
	Clim-ate	Topo-graphy	Drain-nage	Depth	Tex-ture	Stoni-ness	Base sat	Sali-nity	Nutrient availability	Overall suitability
Saongi 2: Typic Ustorthents	3	2-3	0	3	2	1-3	0	0	4	N-N1
Saongi 3: Lithic Ustochrepts	3	2	0	2-3	2	2	0	0	2	S3
Saongi 4: Typic Ustochrepts	3	1	0-1	2	0-2	0	0	0	1-2	S2
Saongi 5: Vertic Ustochrepts	3	0	0	0	0	0	0	0	0	S2
Saongi 6: Typic Haplusterts	3	0	1	0	2	0	0	0	1	S2

Source : Bhaskar et al. (1996).

Table 22. Correlation matrix for wheat.

Soil Parameters	Correction						
	Yield	Depth	AWC	CEC	Clay	OC	Erosion
Depth	0.7488**						
AWC	0.7457**	0.9709**					
CEC	0.8416**	0.9533**	0.9762**				
Clay	0.8319**	0.9603**	0.9747**	0.9979**			
OC	0.5819**	0.5243**	0.6084	0.6731*	0.6579*		
Erosion	-0.5586	-0.7425**	-0.7987**	-0.8207**	-0.8398**	-0.5393	
Slope	-0.5604	-0.6999*	-0.7704**	-0.8145**	-0.8254**	-0.5977	0.9792**

*Significant at 5% level; **significant at 1% level.

LAND EVALUATION : Thugaon Micro - Watershed Land Evaluation For Cotton – Based Cropping Systems In Different Toposequences.

Comparative yield performance of cotton- based cropping system was evaluated at Thugaon micro- watershed District Amravati, Maharashtra, under uniform set of management level in different toposequences. They were developed from weathered basalt and alluvium . The soils of the toposequences were very shallow to very deep as per toposequence with high base saturation. The climate of the area was semiarid tropical. The mean annual rain fall was 866 mm. The test crops were cotton, sorghum and pigeonpea. The size of the experimental plot was 100 x 40 m. in each toposequence. The gross yield of different crops in different toposequences found varied, it was lowest (5.07 q/ha) in upper plain and the highest (52.82 q/ha) in valley floor, which was higher by 412% and 800 % in the lower plain and the valley of the floor, respectively over upper plain (Table). The higher yield of the crops was found associated to deeper soils, having fine texture, high organic matter with higher soil moisture retention capacity and better fertility status (Bhaskar et al 2001). The evaluation of toposequence- based crop yield performance could be a better choice for soil selection and its management for higher yield potential and better development of rainfed agriculture at watershed basis for sustainable land use planning in the country (Table 23).

Table 23. Soil types and gross yield of crops in different toposequences at Thugaon micro – watershed

Topo-sequence	Soil Characteristics	Yield (q ha^{-1})			Gross yield	% yield increased in lower to valley floor over upper plain
		Cotton	Sorghum	pigeonpea		
Foot hill	Extremely Shallow to very shallow (<25 cm deep). 10-15% slop, very severe erosion and calcareous.	-	-	-	Not suitable for cultivation	-
Upper plain	Shallow to slightly deep (25-50cm), gravelly sandy clay loam, gently slopping (3 to 5%) to strongly	1.36	2.50	2.01	5.87	-

	slopping (5 to 10%), severe to very severe erosion, moderately to strongly stony and calcareous.					
Lower plain	Moderately deep to deep (50-100 cm), gravelly sandy clay to clay lome, gently slopping (3-5%), moderate erosion, calcareous and slightly stony to stony.	4.66	19.17	6.25	30.08	412.44
Valley floor	Deep (>100cm), silty clay loam, gently slopping, moderate erosion and slightly stoney.	7.50	36.57	8.75	52.82	800

Source : Bhaskar et al. 2001.

Technological Impact

1 Ridge and furrow system is considered the best among all the moisture conservation practices which is being adopted by some farmers in larger area in their fields.

2 Inter cropping of greengram, soyabean and sorghum as a strip crop in cotton is being adopted by the farmers to reduced runoff, improved soil moisture and soil protection.

Technology Accepted by the Watershed Farmers

- Timely sowing with 2-3 seeds hill^{-1} with proper placement.
- Application of fertilizer through ring method in cotton.
- 3-4 spray of insecticides/pesticides in place of 10-12.
- Sowing of cotton greengram and soyabean as intercrop across the slope.
- Cultivation of greengram on upper and soyabean on lower toposequence.
- Two plants per hill of cotton on upper to lower toposequences and one plant on bottom toposequence.
- Ridge and furrow at the end of August and first fortnight of September.

Conclusion

The toposequence based in-situ moisture conservation and rain water management practices showed the eye opening results for increasing quantity, quality and gross return of cotton and cotton based crops in different toposequences under rainfed situation. Efficient use of rain water as protective irrigation at watershed basis is the need of the hour. Technologies developed at watershed basis were found environmentally sound, economically viable and socially acceptable and considered only the viable alternative for the development of rainfed agriculture in the country. It may be inferred from the study that based on 4 years results technologies developed at the Thugaon micro-watershed could be transferred to another area occurring elsewhere under similar soil and climatic condition in the country.

14. CONCLUSIONS

An increasing food, feed, fodder, fuel, fibre and fruit demand by growing population can be met by adopting scientific land use of our finit and precious natural resources. Soil, water, climate, vegetation, nutrients and energy are the basic natural resources needed for agricultural production. Developing the system towards sustaining the productivity of the natural resources is, therefore, an essential issue not only for the present circumstances but for the next century too and this will lead to continued human posterity. Rainfed farming is the backbone of Indian agriculture, as it contributed by 75 per cent (143 m ha) of our cultivated area. The low productivity of these drylands is due to poor fertility, aberrant weather and short growing season. The fortunes of the majority of the farmers are linked with the progress and development of this appropriate system of farming at watershed basis. It contributes about 42 per cent to country's food store. The possibility of obtaining maximum production per unit area and per unit time through adoption of watershed-based approach is seems to be the right approach, by way of evaluating, correlating and utilising natural resources at watershed basis as per their land capability. Watershed will be a unit of every kind of natural resource management as 57% land in India is degraded. It is true that there is no other alternative except watershed based research and development of rainfed agriculture in the country.

The concept of watershed-based agricultural research and resource management approach for efficient crop production is being prioritiesed to replace the traditional system of sporadic experiments and for the same time obtaining greater effectiveness and higher resource use efficiency. During recent past, watershed-based holistic and integrated development approach has emerged as a potential approach for the development of rainfed areas, which can lead to higher productivity and sustainability in agriculture production. Watershed-based agricultural research and development is the only technically sound, economically viable and socially acceptable strategy for the conservation, development and management of land and water in the country. The production ability of land is deliminated by its soil, land form conditions and the climate which is turn in conditioned by the nature of land use and its management.

For sustainable land use planning, questions often asked are : is there sufficient land to meet the future food and agricultural production, where are those potential lands, what land use are suitable, what is the level of technology required, what is the risk of degradation and measures required for minimising the risk and what are the research, extension education needs for production optimisation. The present study will give the answers of these questions through water-shed-based research and development to convert gray area into green.

Rain water is a crucial resource and key input in development of rainfed agriculture. The success of agricultural enterprises depends largely on how efficiently and effectively we conserve and store the rain water, and how wisely we use this precious resource at farm level. Out of 400 Million ha m annual rain water, hardly around 150 Million ha m enters in the soils and about 180 Million ha m goes waste as runoff. This showed a ample scope for rain water conservation, harvesting and recycling in the country (Plate 2.5). It is a matter of great concern of everybody that every year we are loosing 5.3 M t good fertile soil and due to that 80 lakh tones major soil nutrients such as nitrogen 29 lakh tonnes, phosphorus 26 lakh tonnes and potash 25 lakh tonnes and due to that about 30 lakh tonnes gross loss of food grains is there in the country.

In the same time, on an average 20 million people and 4 million cattle are affected by nominal or severe drought every year. This showed that natural resource management at watershed basis will not only protect and conserve the soil and water, it will also help in evaluating the soils, cropping and farming system, which will increase overall production of the watershed farmers at sustainable basis.

It is hoped that the present investigation may provide a useful information having applicabilities in other area with similar climatic conditions which has immense value to the farmers, land users decision makers and researchers about the food security in the region. This would help in meeting the daunting challenge of raising fiber and food grains production to a substantial scale and to conserve the natural resources for sustainable land use planning.

Therefore, the slogan should be "**Love those whom he loves**".

REFERENCES

Agrawal, O.P., Goel, K.N. and Singh, J.P. (1970). Role of leveling and bunding in moisture conservation and increasing crop yields. Paper presented at All India Seminar on Dryland Farming held at IARI, New Delhi, Jan. 22-24, 1970

Ambekar, V.W. (1996). Integrated watershed development of Bundelkhand (Tomar, R.S. and Srivastava, A.K. editors) Indian Grassland and Fodder Research Instt. Jhansi. P. 50-57.

Anonymous (1983) Guidelines : Land Evaluation for Rainfed Agriculture FAO, Rome.

Anonymous (1989). American Agronomy Society News, 1989.

Anonymous (2000b). Natural Resource Management Strategies. Frank Note. Fertilizer News. 45(2) : 11-12.

Anonymous, (2000). Frank note Natural Resource Management strategies. Fertilizer News 45(2) : 11-12.

Anonymous, (2000a). Vision, 2020. Natural Resource Management Research, Div. of Natural Resource Management, ICAR, Krishi Bhawan, New Delhi. February, 2000. P. 8.

Bali, J.S. (1980). Soil and water conservation on agricultural land. Key paper presented at the National Symposium on soil conservation and water management in 1980s, held during March 12-13, 1980 at the Central Soil and Water Conservation Research and Training Institute, Dehradun.

Benke, S.D. (1996). Drip Irrigation system on cotton. Paper presented at Expert Meet on Water Management Practices with special reference to Drip Irrigation System in Cotton, Directorate of Cotton Development, Mumbai.(Maharashtra.)

Bhaskar, K.S, Lal, S, Challa, O and Madavi S.H. (1987). Effect of soil depth on cotton yield. J. Maharashtra Agric. Univ. 12:139—140.

Bhaskar, K.S. (1995) Response of pearlmillet, greengram and sesamum to the levels of fertilizers in different soils for land use planning. Ph. D. Thesis, Agri. Univ. Agra.

Bhaskar, K.S. , Sawaji, B.V. and Kairon, M.S. (1998). Rainwater management for rainfed cotton production in Vertisols, First Intl. Agronomy Congress on Agronomy, Environment and Food Security for 21st Century, New Delhi, 23-27 November, 1998, Abstract, pp 30-31.

Bhaskar, K.S. ; Mendhe, P.N. and Wasnik S.M. (2003) Soil and water conservation management practices in yield maximization of rainfall cotton, National Seminar on extension for efficient irrigation water management and water conservation, December 13-14 (2003). Mahatma Phule Krishi Vidyapeeth, Rahuri, Dist. Ahmadnagar (M.S.).

Bhaskar, K.S. Gaikawad S.T. and Bhoyar, S.M. (1989). Grow cotton on deep soils in Vidarbha Region of Maharashtra : Indian Fmg 39 : 4-5.

Bhaskar, K.S. Gaikawad S.T. and Madavi, S.H. (1988). Evaluation of soil based crop yield efficiency of Vertisols for land use planning J. Maharashtra Agric. Univ. 13 : 325—360.

Bhaskar, K.S., Gaikawad, S.T. and Anantha Rao D. (1996). Soil-site availability evaluation for wheat. A case study, Agropedology. 6 (1), 1996 : 89—94.

Bhaskar, K.S., Gaikawad, S.T. and Anantha Rao, D. (1995). Systematic Soil Survey for sustainable cotton production. National workshop on Soil inventory for perspective land use planning OCT. 15—18 (1995) N B S S & L U P Nagpur, page 139—141.

Bhaskar, K.S., Giri, J. D., Naidu, L.G.K. and Dent, D (1990). Application of FAO Framework for land evaluation : Case studies in India, Indo-UK Work-shop : New Approaches to Land Use Planning, NBSS, Nagpur, Sept. 24-27 & at Bangalore Sept28-29.

Bhaskar, K.S., Kairon, M.S. (2000). Watershed based rainwater management approaches for development of cotton-fish farming system : A case study. Intl. Conf. on Managing Natural Resourse. for Sustainable Agrl. Prodn. in the 21st Century, Feb 14-18, 2000, IARI New Delhi. Extended summary, Resource Management 3: 1191-1192.

Bhaskar, K.S., Lal, S., Challa, O., Warade, L.K. and Madavi, S. H. (1987) Land evaluation for cotton J. Indian Soc. Soil Sci. 36 : 195—197.

Bhaskar, K.S., Lal, Sohan., Challa O., Warade, L.K. and Madavi S.TL. (1988). Land evaluation for cotton. J, Indian Soc. Soil Sci. (36) : 195—197.

Bhaskar, K.S., Mendhe, P.N. and Rap, MRK (2006). Rain water management for sustainable cotton production,presented. International workshop on R and D frontiers in water and waste water management. January 20—21 2006. National Environmental Engineering Research Institute, Nagpur.

Bhaskar, K.S., Mendhe, P.N., Rao, M.R.K. and Suryavanshi, M.R. (2005). Soil moisture availability and its manangement for rainfed cotton production in Vidarbha. (Abstract), National Seminar on Land Resource Appraisal and Management for Food Security. 5-6 April 2005. N. B.S.S. and LUP, Nagpur.

Bhaskar, K.S., Wasnik, S.M., Barbde N.P. and Mendhe, P.N. (2001). Comparative performance of cotton-based cropping systems in different toposequences on watershed-basis for sustainable land use planning. 66th Annual conservation : Oct. 30-Nov. 2, 2001, National Seminar on Development in soil science, 2001 (Abstract). National Bureau of Soil Survey and Land Use Planning, Nagpur.

Bhumbla, D.R. (1975). Address to the officer trainees at the closing function of the XXXII regular course. Central Soil and Water Conservation Research Institute, Dehradun. Soil Conserve. Digest 3(1) : 91-93.

Bose, S. and Saxena, N. (2000). Drought, the Indian scenario. Employment News Weekly 25 (9) : 1

Das, D.K. and Kalra, N. (1995). Adjustment to weather variation through cropping systems and fertilizer use. Fert. News 40 : 11—21.

De La Rosa, D.F. Cardona, F. & Almorja, J. (1981). Crop yield predictions based on properties of soils in Sevilla, Spain, Geoderma 25 : 267—274.

Dent, D and Young, A (1981) Soil Survey and Land evaluation. George Alien Unwin, London.

FAO (1976) A Framework of Land Evaluatin. Soils Bull. 32. FAO, Rome.

Food and Agriculture Organization, Rome, Italy, (1989).

Gaikawad, S.T. and Bhaskar, K.S. (1988). Crop production potential of Vertisols and associated soils : A case study. Trans. Intl. Workshop Swell—Shrink Soil, Oct. 23-28 NBSS, Nagpur; pp 226-228.

Gaikawad, S.T. and Bhaskar, K.S. (1988).Crop production potential of Vertisols and associated soils : A case study, classification management and use potential of Vertisols and associated soil. A case study classification management and use potential of Swell Shrink Soil, Oct. 24-28, 1988. Transaction, International workshop S. well-shrink soils, National Bureau of Soil Survey and Land Use Planning, Nagpur.

Gaikawad, S.T., Bhaskar, K.S. and Sehgal, J.L. (1990). Evaluation of site and soil characteristics and management levels on the yield of crop in different soils in a micro-watershed : A case study. Trans. 14th Intl. Cong. Soil Sci. Kyoto, Japan, Aug. 12-18, 1990 V : 525-526.

Goswami, N.N. and Rattan, R.K. (2000). Ecofriendly and efficient integrated nutrient management in sustainable agriculture. Intl. Conf. Managing Natural Resources for Sustainable Agricultural Production in 21st Century, IARI, New Delhi. Feb. 14-18, 2000 pp. 41-47.

Khan, A.D. (1962). Measurement of increase in productivity by adopting soil and water conservation practices. J. Soil Wat. Conserv. India, 10(3,4) : 25-33.

Kulkarni, G.A. and Raheja, S.K. (1970). Effect of contour bunding on production and productivity of important cereals and cash crops in Maharashtra State. Paper presented at "All india Seminar on Dryland Farming" held at IARI, New Delhi, Jan. 22-24. 1970.

Massee, T.W. & Waggoner, S.T. (1985). Productivity losses from soil erosion on dryland crops. J. Soil Water Cons. 40 (5) : 497.

Mishra, P.R., Kaushal, R.C., Dayal, S.K.N. and Prabha Shankar (1975). Annual Progress Report. Central Soil and Water Conservation Research and Training Institute, Dehradun.

Naidu, L.G.K., Verma, K.S. & Jain, S.P. (1988). Soil family : A key in Soil Taxonomy for agricultural interpretations J. Indian Soc. Soil Sci. 36 : 192—194.

Naidu, L.G.K.. Verma, K.S.. Jain, S.P.. Rana, K.P.C. & Sindhu, G.S. (1986). An appraisal of the production potential of dominant soils of Delhi, Teerritory J. Indian Soc. Sci. 34 (3) 558— 567.

Nema, J.P. and Kamanver, H.K. (1976). Ann. Progve Rep. CSWCRTI, Dehradun.

Nimje, M.B., Tamgadge D.B. and Wankar, W.V. (2000). Soil resource inventory for land use planning : A case study. Intl. Conf. On managing natural resources for sustainable agricultural production in 21st Century Extended Summaries Vol. 3 Resource Management pp. 1112-113.

Padamakumari, O and Sivanappan, R.K. (1979). An advance irrigation method for better yield in cotton. Madras Agric. J. 66(3) : 173—176.

Paroda, R.S. (1998). Plenary lecture, Natural Resource Management for Sustainable Agriculture; A New Paradigm. First Intl. Agronomy Congress, 24th November, 1998. P. 1-27.

Patel, R.G. and Sohoni, D.K. (1969). Long term economic benefits of soil conservation programme. J. Soil and water conservation, India 17 (1,2) : 22-26.

Patnaik, N. (1977). Soil Conservation works in river valley projects (RVP). Soil Conservation Digest 5 (2) : 53-61.

Rai, P (1970). An evaluation of ravine reclamation programme in U.P. (1952-68) planning Research and Action Institute, Lucknow. Pub. No. 380 (44-46).

Rai, P. (1970). An evaluation of reclamation programme in UP (1952-68). Planning, Research and Action Institute Lucknow. Pub. No. 380 (44-46).

Ram Babu, Agrawal, M.C. Vishwanathan, M.K. and Joshi, P. (1980). Paper presented in National Symposium on Soil conservation and Water Management in 1980s, held at CSWCRTI, Dehradun.

Saxena, N.C. (2000). Rehabilitation of degraded lands in India through watershed development Intl. Conf. Managing Natural Resources for Sustainable Agricultural Production in 21st Century IARI, New Delhi, February 14-18, 2000 pp. 25-30.

Seckler, David (1978). A Guide to the Economic Evaluation of Soil and Water Conservation projects. Central Soil and Water Conservation Research and Training Institute, Dehradun.

Shankarnarayan, K.A. and Shankar, Vinod (1984) *Grasses and Legumes for Forage and Soil Conservation* . Indian Council of Agril. Res. New Delhi.

Sharma, K.R. & Sharma, P.K. (1991). Soils-site suitability for wheat in different Agro-climatic Regions of Punjab, Agropedology, 1 : 65—73.

Singh, D.P. and Hazra, C.R. (1996). Soil water resources and land degradation problem management of water and soil resources with special reference to Bundelkhand region. Resource conservation and Integrated Development of -Bundelkhand. Tomer, P.S. and Srivastava, A.K. (ed). IGFRI, Jhansi pp. 17-25.

Singh, G.B. (2000a). Natural resource management, a must for sustaining agriculture. Indian Farming 49 (11) : 4-8.

Singh, G.B. (2000b). Overview of Natural resources in India. Intl. Conf. Managing Natural Resources for Sustainable Agricultural Production in the 21st Century Souvenir, New Delhi, India February, 14-18, 2000. P. 1-8.

Storie, R.E. (1978), Storie Index Soil Rating Spl. Pub. 3201, Div. of Agric. Sci. Univ. of California, U.S.A.

Swaminthan, M.S. (1973). Our Agricultural Future. Sardar Patel memorial lecture. All India Radio, New Delhi, pp. 1-54.

Tamgadge, D.B. Gajbhiye, K.S. and Gaikwad S.T.(1986). Evaluation of crop productivity potentials of soils. J. Indian Soci. Soil Sci. 34:551—557.

Tamhane, R.V. and Bali, Y.P. (1963). Soil Conservation. Indian Farming (Dec. 1963) : 33-40.

Tamhane, R.V., Khemchandani, H.T. and Kulkarni, G.A. (1967) Critical Review of the Assessment of Soil Erosion Measures in India. Ministry of Food, Agriculture, Community Development and Cooperation, Govt. of India, New Delhi.

Tejwani, K.G. (1968). Conservation of soil moisture- the way to success under dry farming conditions. Indian Farming, XVIII (5) : 19.

Tejwani, K.G. and Dhruvanarayan, V.V. (1960). Reclamation of small and medium size gullies and ravines in Gujarat. J. Soil Water Conserv. India, 3(2.3) : 26.

Varma, A. (2000). Harnessing the potential of Animal husbandary. Fertilizer News, February 2000, 45(2) : 47-48.

Verma, G,.P., Panwar, O.P.S. and Tomar, P.S. (1996). Soil and Water Conservation. Resource Conservation and Integrated Development of Bundelkhand. Tomer, P.S. and Srivastava, A.K. (ed). IGFRI, Jhansi pp. 1-16.

Wasnik, S.M. and Bhaskar, K.S. (2004). Assessment of watershed—based technology transfer among watershed farmers under Thugaon micro-watershed. Presented, National Symposium on changing world order—cotton Research, Development and policy in context, August 10-12 2004, Acharya N.B. Ranga Agricultural University, Rajendranagar, Hyderabad. A.P..

Wasnik,S.M., Bhaskar, K.S., Mendhe, P.N. and Barbde, N.P. (2003). Indigenous technologies knowledge : Documentation awareness and extension strategy for transformation of moisture conservation practices of water shed-basis. Abstract, National Seminar on extension strategy for. Efficient Irrigation water management and water conservation, Dec. 13-14 (2003). Mahatma Phule Krishi Vidhyapeeth, Rahuri, Dist. Ahmadnagar. (M.S.)

Chapter - 3

DIFFERENTIAL PROPERTIES OF SOILS, NUTRIENT AVAILABILITY AND THEIR MANAGEMENT UNDER SEMI-ARID REGIONS

G.U. Malewar* and Syed Ismail**

* Head, Department of Agril. Chem. & Soil Science (Retd.),
Marathwada Agricultural University, Parbhani - 431 402 (India)
** Soil Scientists, Sorghum Research Station,
4, Baldhan, Dadaro Layout, Sambhaji Nagar, Parbhani - 431 402 (India)

Soil-plant-climate system integrates the extent of response to the growth and development of organisms including crops over and above the usual agricultural inputs. Thus, agro-climatic zone or agro-ecological concept is gaining significance in deciding the suitability of crop for specific agro-eco unit for assured and maximum production potential. In dry land agro-ecosystem, climatic difference is the main cause of inter annual variability of crop yields. Water deficit environments are unfavorable for growth and development of rainfed crops and often produce low crop yields. Because of climatic variations, there is likelihood of differences in physico-chemical properties of soils within the semi-arid regions of the country which intern decides the availability of nutrients to the crops. The of nutrients in soils developed under semi-arid climate is meager and scattered. In this chapter, an attempt is made to delineate the area of semi-arid region, characterize the soils formed in varied semi-arid type of climate variations, factors affecting nutrient availability, nutrient deficiencies and nutrient utilization by crops. Nutrient management aspects are also covered.

1. GEOGRAPHICAL DISTRIBUTION AND CLIMATIC CHARACTERIZATION OF SEMI-ARID REGIONS OF INDIA

A vast area of dryland occurs in India having hot as well as cold deserts, delineation of which has been attempted by several workers. The semi-arid tropics as defined by Troll (1965), has a mean annual information on properties and availability temperature >18°C with rainfall exceeding evapotranspiration, 2-4.5 months of the year. Total rainfall varies considerably from 400 to 1900 mm. Katyal *et al* (1997) referred 'semi-arid' to climatic classification of typically dry areas with rainfall ranging from 500 to 1000 mm and evaporative demand in excess of rainfall for a greater part of the year. According to Sehgal *et al.* (1992), semi-arid conditions prevail over a third (116.4 m ha) of the total geographical area (329 m ha) and are concentrated in the western peninsula (Fig. 1). Subsequently, Sehgal and Sharma (1994) divided the Indian territory into four categories : arid land covering the north western portion of Thar desert bordering Pakistan (5% of the overall dry areas of the country), Semi-arid land adjoining these areas (25%), Sub-humid (dry) land (30%) and Transitional land with adequate moisture supply (40%). Categorically, the rainfall ranges from 500-1000 mm in semi-arid region, however, potential evapotranspiration (PET) varies between 1400 and 2000 mm resulting in water deficit of the magnitude of 700 to 1200 mm. By and large, the climate of semi-arid region is characterized by hot and dry summers and cool winters. The moisture regime is ustic, while the soil temperature regime is hyperthermic. Physiographically, semi-arid conditions occur in northern plain and central high lands including Aravallis, Central (Malwa) highlands, Gujarat plains and Kathiawar peninsula, Deccan plateau in Maharashtra, Telangana and Karnataka and Eastern ghats of Tamilnadu uplands (Table 1). These semi-arid eco-regions are distinctly formed from alluvium, basaltic and lateritic parent material resulting in deep loamy, medium/deep/shallow black and shallow to deep loamy red soils. Within the semi-arid ecoregions, minimum range of precipitation (500-800 mm) was observed in northern plain and central high lands including Aravallis followed by central (Malwa) high lands, Gujarat plains and Kathiawar peninsula. Maximum potential evapotranspiration (1600-2000 mm) was in central (Malwa) high lands, Gujarat plain and Kathiawar peninsula followed by northern plains and central highlands including Aravallis (1400-1900 mm). The constraints of various semi-arid ecoregions also differed significantly particularly in sandy to clay texture, low AWC, salinity and sodicity, drainage, prolong dry-spells and high runoff during storms and low nutrients content i.e. N, P and zinc. Further, Sehgal et al (1992) also estimated on an over all basis about 158 districts in 11 states including Delhi and Union Territory of Diu in semi-arid climate (Table 2). Out of 30 districts of Maharashtra, 20 were categorized as semi-arid conditions. The growing period for rainfed crops in these area lasts for 90-150 days, which is associated with water storage capacity of soils. High available water capacity of deep black soils spread out to a large semi-arid area is attributed to high clay content with montmorillonitic clay.

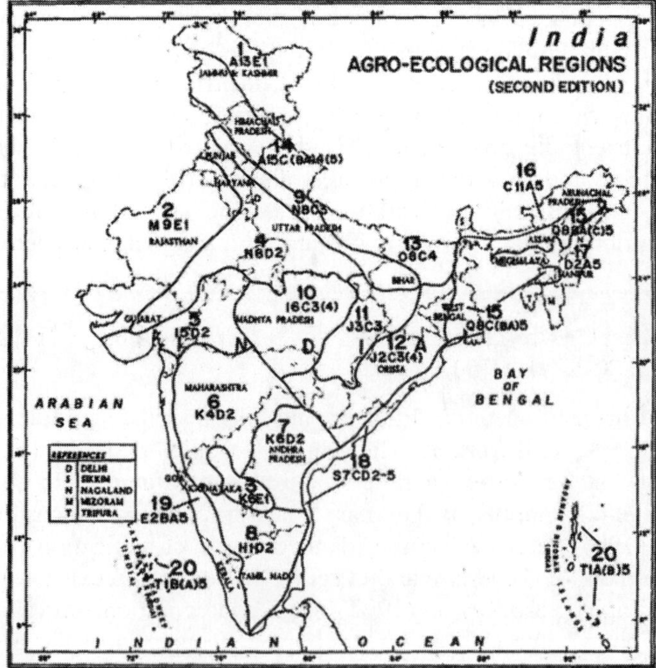

Fig. 1. Arid and semi-arid ecoregions of India (Sehgal *et al.,* 1992)

Table 1. Edaphic features of representative semi-arid regions of India.

Semi-arid ecoregion	Soil type	Soil group taxanomy	Climate	Precip-ition (mm)	Potcutial evapotrans-piration	Water deficit (mm)	High crops	Constraints
Northern plains and central high lands including Aravallis	Deep loamy alluvium derived soils with medium AWC and shallow to medium sandy to loamy grey brown soils with low to medium AWC.	Ustochrept Ustopsamments Haplustalf Ustorthents Ustifluvents	Hot semi arid	500-800	1400-1900	700-1000	Millets, Pulses, Wheat, Maize Sugarcane & cotton under irrigation	Coarse texture low AWC impread drainage loading to salinity and sodicity
Central (Malwa) high lands, Gujarat plains and Kathiawar peninsula	Medium and deep fine loamy and clay black soils with medium and high AWC	Ustochrept Ustopsamments Chromusterts Ustorthents	Hot Semi arid	500-1000	1600-1200	800-1200	Millet, Wheat Pulses	Drought, poor drainage in low laying areas, salinity in irrigated areas.
Deccan plateau	Shallow and medium loamy to clay black soils with low to medium AWC inclusion of deep clay black soils with high AWC	Ustochrept Ustorthents Chromusterts	Hot Semi arid	600-1000	1600-1800	800-1000	Millet, pulses sugarcane and cotton under irrigation	Prolong dry spells, high runoff during stormns and erosion, N,P and Zn deficiencies
Deccan (Telangana) plateau and Eastern ghat	Medium red loamy soils with low and medium AWC Deep clay black soils with medium to high AWC	Ustochrept Rhodustalf Plinthustalf Ustorthents Chromusterts	Hot Semi arid	600-1000	1600-1700	700-800	Millets, oil seeds, rice, cotton and sugarcane under irrigation	High runoff and erosion, subsoil salinity & sodicity in black cotton soils, N,P, Zn deficiencies
Eastern ghats (Tamilnadu) and Deccan (Karnataka) plateau	Shallow to deep loamy red soils with low to medium AWC, inclusion of deep, loamy black soils with high AWC.	Ustropepts Haplustalf Rhodustalf Pellusterts	Hot Semi arid	600-1000	1300-1500	400-700	Millets, pulses, oil seeds, sugarcane and rice under irrigation	Coarse texture low AWC, draught, high runoff, and erosion, N,P and Zn deficiencies

Table 2 : List of districts covered under semi-arid regions of India

State	Representing districts
Andhra Pradesh	Chittoor, Cuddapah, Rangareddy, Guntur, Karimnagar, Khammam, Krishna, Kurnool, Mahaboobnagar, Nalgonda, Nellore, Prakasam, Medak, Warangal, East Godawari, West Godawari, Adilabad, Nizamabad (18).
Gujrat	Amreli, Bhavnagar, Junagarh, Panchmahals, Kheda, Vadodara, Bharuch, Rajkot, Surat, Ahmedabad, Sabarkantha, Mehsana, Surendranagar (13).
Haryana	Faridabad, Gorgaon, Jind, Karnool, Kurukshetra, Mahendragarh, Rohtak, Sonepat (8)
Punjab	Amritsar, Kapurthala, Patiala, Songrur, Ludhiana, Firozpur, Faridkot (7).
Karnataka	Belgaum, Bider, Dharwad, Gulburga, Raichur, Bijapur, Bangalore, Chitradurga, Chikmanglur, Shimoga, Hassan, Kolar, Mandya, Mysore, Tumkur (15).

Madhya Pradesh	Dewas, Dhar, West Nimar, East Nimar, Indore, Jhabua, Reewa, Ujjain, Shajapur, Bhind, Datia, Gwalior, Morena, Shivpuri (14).
Maharashtra	Akola, Ahmednagar, Amaravati, Aurangabad, Bhir, Buldhana, Dulia, Jalna, Jalgaon, Kolhapur, Latur, Nanded, Nasik, Osmanabad, Parbhani, Pune, Sangali, Satara, Solapur, Yeotmal (20).
Rajasthan	Ajmer, Alwar, Bharatpur, Bhilwara, Dholpur, Durgapur, Jaipur, Sawal Madhopur, Tonka, Udaipur, Bundi, Banswara (12).
Tamilnadu	Dindigual Anna, Chengalpattu, Coimbatore, Dharmapuri, Kamarajar, Kannyakumari, Madurai, North Arcot, Periyar, Pudukkottai, Romanathpur, Salem, South Arcot, PN Tevar, Thanjavur, Tiruchirapalli, Tiruneiveli, Chidombaranar, T. Sambuvarayar (19).
Uttar Pradesh	Agra, Allahabad, Farrukabad, Badaun, Banda, Bulandshahar, Etah, Elawah, Fattepur, Ghaziabad, Hamirpur, Hardoi, Jalaun, Jaunpur, Jhansi, Kanpur, Lucknow, Moradabad, Mathura, Meerut, Varanasi, Mujaffurnagar, Pratapgarh, Raebareilly, Lalitpur, Shajahanpur, Unnao (28).
Delhi	----
Diu and Daman	----

2. VARIABILITY IN SOIL PROPERTIES AND AVAILABILITY OF NUTRIENTS

Soil, water, plant and atmosphere are the most important resources, which are critical, not only for agriculture but also for the existence of life. Among these, soil is the most precious one, since this finite resource provides foothold for most plants. The crop culture based on soil and climate requirement has been an age-old practice now coinciding with the recent concepts like soil-site suitability and agro-climatic zone/agro-ecological unit. Soil being the main component of overall production system, reacts with crop characteristics, climate and management in determining the final productivity.

The soils are the outcome of a combined action of the climate and vegetation on parent rock as conditioned by topography over a period of time, within any ecosystem. Depending on the variation in one or more soil forming factors or processes, a great variety of soils with variegated properties are formed. Considering the spatial variability of soils, the spatial and temporal variability of climate and diversity of crops, different soils would behave and diversity of crops, different soils would behave differently with respect to water availability for crops due to differences in their water retention and release characteristics. Soil management practices for a given crop under specific climatic condition would differ with soil. In dry land ecosystem, soils are fragile and land use capability can be lost rapidly by mismanagement. Thus, it is necessary that they are to be used rationally as per their capability after proper characterization and classification. Since, the contribution of soil to crop productivity is only a part of an integrated effect of many variables, i.e. climate, moisture availability, plant type etc, the role of soil factors in this regard needs to be understood for each agro-climatic situation considering the interactive effect of all factors. The specific phytoclimatic conditions exist in dry land agro-ecosystem that leads to impairment of normal physiological and biochemical processes of plant. Such condition arises when precipitation (P) remains less than potential evapotranspiration (PET) around 3/4 period of the year. Soil moisture control section (SMCS) remains dry for more than 150 days in a year resulting in typic and aridic ustic sub-group of the ustic and torric/cold aridic moisture regimes. Soil temperature regime ranges within hyperthermic and megahyperthermic while at places it is cryic, mesic and thermic, both iso and non-iso. Sehgal and Sharma (1994) discussed the physico-chemical properties of dominant soil series developed under semi-arid climate in India (Table 3).

Table 3. Physico-chemical characteristics of some typical soils in Semi-arid regions of India.

Eco-region	Dominant soil	Soil taxonomy	Soil characteristics						B.S. (%)	E.S.P (%)
			Clay (%)	pH	EC (dSm^{-1})	OC (%)	$CaCO_3$ (%)	CEC m.eq/ 100 g		
Northern plain and central High lands including Aravallis (Hot semi-arid eco-region)	Kanali/ Kapurthala/ Punjab	Typic Ustorchrept	17.7	7.9	0.6	0.19	1.1	9.2	95	—
	Zarit viran/ Kanal/Haryana	Typic Natrustalf	30.5	9.7	3.0	0.17	5.7	13.5	100	84
	Ghabdan/ Sangrur Punjab	Salic Natrustalf	26.0	9.6	2.5	–	7.6	11.5	99	72
	Chomu/ Jaipur Rajasthan	Typic Ustip-samment	6.1	7.8	0.2	0.05	–	6.0	98	11
Central Malwa High Land, Gujarat Plain & Kathiawar peninsula (Hot Semi-arid)	Sarol/Indore Madhya Pradesh	Typic Chromustert	55.8	8.1	0.1	0.30	4.7	49.7	98	9
	Kamaliakheri/ Indore M.P.	Vertic Ustochrept	63.2	7.2	0.2	0.54	Tr	61.9	98	–
Deccan plateau (Maharashtra) (Hot Semi-arid)	Paragon/ Ahmednagar, Maharashtra	Lithic Ustorthents	16.7	7.9	<0.2	0.15	30.0	29.8	100	–
	Sawargaon/ A'nagar Maharashtra	Vertic Ustorthents	39.0	7.8	0.4	0.35	34.0	40.0	100	–
	Barsi/Solapur Maharashtra	Typic Chromustert	70.2	8.4	0.5	0.49	1.6	61.7	100	9
Deccan (Telangana) plateau and Eastern Ghats (Hot Semi arid)	Kasireddipalli/ Medak A.P.	Typic Pellustert	61.9	9.2	0.3	0.39	6.9	59.8	100	14
	Patancheru/ Medak Andhra Pradesh	Udic Rhodustalf	32.7	7.0	–	0.60	–	17.2	99	–
Eastern Ghat (Tamil nadu) Uplands and Deccan (Karnataka) Plateau (Hot Semi arid)	Tyamagondalu/ Bangalore/ Karnataka	Oxic Palustalf	45.3	6.3	–	0.45	–	8.8	81	–
	Palathurai/ Coimbatore Tamil nadu	Typic Haplustalf	21.8	8.5	–	0.76	5.4	23.4	84	–

The edaphic factors and soil properties influence the plant growth. Considerable variations in soils of varied semi-arid regions were observed in clay content, soil reaction, soluble salts, organic carbon, calcium carbonate, cation exchange capacity, base saturation and salinity and sodicity (Table 3). The clay content varied in the range of 6.1 to 70.2 percent in Typic ustipsamment and Typic chromustert, respectively. These variations in clay are attributed to the parent material rather than semi-arid climate. The pH and EC of soils varied greatly from moderately acidic (6.3) to highly alkaline (9.7) and low (0.1 dSm^{-1}) to high (3.0 dSm^{-1}), respectively. Northern plain and central high lands semi-arid region is appeared to be more prone to the development of salinity and sodicity. It is mainly attributed to the alkaline parent material, canalization, high water table and semi-arid hot climate. Lowest organic carbon content (0.05 to 0.19) was registered in northern plain and central high lands, however, highest organic carbon was observed in eastern ghat (Tamilnadu), central (Malwa) and Deccan plateau (Maharashtra) except Paragon series of Ahmednagar district.

Table 4. Edaphic and physico-chemical characteristics of soils of different agroclimatic zone of Marathwada.

S.No.	General/Physical characteristics	Scarcity zone (3)	Assured rainfall (44)	Moderately high Rainfall zone (7)
1	Location of zones	Bhir, Osmanabad	Osmanabad, Parbhani, Aurangabad	Nanded and Parbhani
2	Average annual rainfall (mm)	600-700	700-900	900-1300
3	Soil group	Medium to deep black	Shallow and medium to deep black	Deep black
4	Parent material	Basalt	Basalt	Trap rock
5	Crops grown	Kharif-Bajra Udid, Paddy Rabi-Wheat, gram Jowar, Safflower	7.2 - 8.6 (8.0)	
6	pH	7.9 - 8.3 (8.1)	0.18 - 3.23 (0.70)	8.00-8.3 (8.1)
7	Organic carbon (%)	0.42 - 0.60 (0.51)	0.10 - 22.0 (3.25)	0.48-0.72 (0.58)
8	Calcium carbonate (%)	0.25 - 11.50 (4.16)	9.4 - 61.2 (41.7)	0.32-6.45 (1.87)
9	Clay (%)	42.0 - 56.8 (49.3)	0.28 - 4.40 (0.86)	42.0 - 55.8 (51.0)
10	Available Zn (.mg kg^{-1})	0.44 - 0.70 (0.59)	6.6 - 28.6 (13.7)	0.48 - 0.92 (0.72)
11	Available Fe (.mg kg^{-1})	6.6 - 15.0 (11.0)	13.3 - 65.2 (23.1)	8.4 - 14.0 (10.8)
12	Available Mn (.mg kg^{-1})	14.9 - 24.0 (19.6)	1.2 - 7.4 (3.3)	19.2 - 24.0 (21.1)
13	Available Cu (.mg kg^{-1})	1.6 - 1.7 (1.63)		1.6 - 3.2 (2.1)

Figures in parentheses indicate average values.

This pattern of organic carbon is attributed mainly to the extent of precipitation received at the soil-site. It is also observed that the decreasing trend in organic carbon was noticed with reducing precipitation of the agro-ecological semiarid regions. Highest calcium carbonate was present in Deccan plateau Pargaon (Lithic ustorthents) and Sawargaon (Vertic ustorpepts) which may probably be associated with alkaline earth metal containing basaltic parent material. CEC more or less followed the trend of clay accumulation in the respective soil series. Malewar (1979) assessed the variation in soil properties in various agroclimatic zones of Marathwada region of Maharashtra who indicated increasing accumulation of organic carbon with increasing precipitation from scarcity to moderately high rainfall zone (Table 4). Katyal and Sharma (1984) confirmed that with climate changing from humid to arid, a sharp fall in organic matter occurs. Subsequently, Katyal and Sharma (1991) elucidated the association between organic matter levels of soils and their pedoclimate (soil moisture regimes). Tropical rainfed soils of arid and semi-arid region manifesting, respectively, aridic and ustic regimes were typically low in organic matter and high lime. Compared to 2 to 4% organic carbon content of semi-arid soils of temperate environment, semi-arid tropical (SAT) soils contain less than 1 per cent organic carbon (Virmani et al. 1982). However, $CaCO_3$ content was relatively more in soils under scarcity zone and lower content of $CaCO_3$ in soils under moderately high rainfall zone. These differences in organic carbon and $CaCO_3$ are mainly due to climatic complex (hot semi-arid to sub-humid). Similarly, available zinc, iron, manganese and copper also found to be lowest in scarcity zone against assured rainfall or moderately high rainfall zone. matter were associated positively with Zn According to Malewar *et al* (1998), soils of and Mn. Availability of Zn, Mn, B and Mo semi-arid area of northern Marathwada increased with their total contents in soils, were rated as moderately alkaline, calcare-Available Zn in semi-arid zone was of lower ous and low in organic carbon. They further magnitude than sub-humid and moderashowed wide variation in soil physical tely high rainfall zone. Malewar (1999) and properties and nutrient availability within Katyal and Sharma (1991) indicated wide-the various soil groups of semi-arid area spread deficiency of Zn, Fe and B in soils of (Table 5). The mean values of bulk density, semi-arid regions in Maharashtra and porosity and water holding capacity (WHC) India. Sporadic patches of Mn and Cu varied with the type of soil groups deficiencies were exhibited. Further, indicating low bulk density and high Malewar (1997) summarized that sulphur porosity and WHC in lithic ustorthents and would be a limiting nutrient in soils of Typic Haplusterts. The nutrient availability western India particularly in area of N, P and K was however highest under oilseeds and pulses.

Table 5. Important physico-chemical characteristics of the soils of semi-arid area of Northern Marathwada region of Maharashtra

Soil groups and No. of samples	pH	EC (dSm^{-1})	$CaCO_3$ (gkg^{-1})	Clay (%)	Org.C ($g\ kg^{-1}$)	Bulk density ($Mg\ m^{-3}$)	Porosity (%)	WHC (%)	Avail. N ($kgha^{-1}$)	Avail. P_2O_5 ($kgha^{-1}$)	Avail. K_2O ($kgha^{-1}$)
Typic Haplusterts (30)	7.50- 7.98 (7.79)	0.18- 0.62 (0.33)	37.50- 110.0 (63.70)	52.64- 68.42 (59.42)	2.3- 8.3 (5.7)	1.17- 1.33	46.04- 56.23	42.36- 60.06	144.25 266.96 192.64	17.02- 44.80	246.08 544.10
Vertic Ustochrepts (38)	7.45- 7.96 (7.70)	0.13- 0.78 (0.37)	36.50- 110.0 (66.60)	31.92- 52.63 (39.82)	3.2- 7.4 (5.5)	1.16- 1.52 (1.30)	42.65- 57.74 (50.68)	32.35- 56.73 (48.75)	147.39- 272.83 198.04	8.96 - 30.46 (19.80)	455.84 609.98 546.63
Typic Ustorthents (19)	7.40 - 7.96 (7.71)	0.12- 0.69 (0.34)	46.00- 99.0 (73.70)	25.30- 43.75 (29.44)	5.2- 8.3 (5.7)	1.22- 1.43 (1.31)	45.29- 55.48 (50.51)	31.05- 52.53 (43.31)	111.98 238.33 174.07	11-20- 26.88 (17.99)	318.00 511.84 426.55
Typic aplusterts (30)	7.42 - 8.00 (7.77)	0.18 - 0.42 (0.32)	38.50 - 114.0 (91.80)	17.58 - 31.64 (22.54)	0.9 - 5.8 (3.1)	4.14 - 1.29 (1.22)	51.70 - 57.74 (54.29)	37.71- 53.93 (44.28)	78.40- 191.29 138.26	8.94 - 17.94 (13.65)	346.08 530.08 (423.31)

Figures in parentheses indicate average value

Vertic ustochrepts followed by Typic Various workers analysed the soils of Haplusterts. Patil and Shingate (1982) semi-arid regions including dryland assessed the soil properties of drought forming areas of Andhra Pradesh represenprone area of Pune region and agro-climaticting black clay Vertisols, red-laterite and zones of Maharashtra. These soils were sandy soils of Gaunter and Prakasham neutral to moderately alkaline, normal in district (Subbiah and Manickam, 1987; soluble salt and high in $CaCO_3$. The organic Venkateshwarlu *et al.* 1995, Jagannadham carbon varied between 0.27 to 0.92 per cent *et al.* 1995; Prasadrao *et al.* 1995 and in these soils. Available Fe, Mn, Zn, Cu, B Madhavi *et al.* 1996). Most the black soils and Mo were deficient in the soils with were neutral to slightly alkaline, normal to varied magnitude indicating widespread critical in soluble salts, low in N, P, organic deficiencies of Zn and iron. Chavan *et al* carbon but high in available K. Vertisols of (1980) further reported rainfall and organic these regions were more prone to zinc and iron and at some sites in Mn. Red/laterite soils, however, low in N and P and K was moderate to high. Among micronutrients, Zn was deficient in all red soils but fairly rich in Cu, Mn and Fe. Sandy soils of south Karlapalem (Guntur) were neutral to slightly alkaline, normal in salt content, low in organic carbon and available N, P and K. Micronutrient like Cu, Zn and Fe were adequate but indicated Mn deficiency.

In Gujarat also, more or less similar situation was observed in nutrient availability in various agro-climatic zones. Gundalla *et al.* (1996) reported that alluvial sand or loamy sand were low in available N whereas lower pediment and plains of Saurashtra region and Northern and northeastern part of state exhibit medium available N but sub-humid tropical soils were high in available N. Dominated soil group in semi-arid area of Gujarat was deficient in phosphorus with low P fixation capacity (Patel, 1996). Kalyansundaram *et al.* (1996) indicated high K status of exchangeable K in middle Gujrat semi-arid agroclimate zone containing alluvial sandy loam to sandy clay loam (Goradu) soil dominating group along with black soils, hill soils etc. The exchangeable K in semiarid belt varied from 0.33 to 1.44 C mol (±) kg soil and reserve K (acid soluble) ranged between 5.26 and 10.91 c mol (+) kg^{-1}. The K saturation varied from 1.95 to 2.80 per cent and observed to be more than the critical limit (1 per cent). The semi-arid belt of middle Gujrat also showed sulphur deficiency to the extent 42.0 per cent and available S varied in the range of 0.9 to 171.7 with an average of 17.7 mg kg^{-1} (Meisheri and Patel, 1996). The overall scenario of micronutrient research carried out in Gujarat provides useful information of their deficiencies and management either cropwise or regionwise based on soil characteristics (Dangarwala and Patel, 1996). They reported 24 and 8 per cent soils were rated as deficient in zinc and iron, respectively. However, deficiency of Mn and Cu was quite meager. Accordingly to Katyal *et al.* (1997), nutrient deficiencies have multiplied in Indian soils and crops. N and P deficiencies affect 65 to 70 % of the Indian soils. Further, micro and secondary nutrient deficiencies became the creeping sickness of soils and crops across diverse region. Malewar and Ismail (1997) reviewed the sulphur scenario of western India and reported incidence of S scantiness amounting to alarming rate and sulphur deficiency was more prone to rainfed oilseed and pulse growing belts.

There are several soil properties, which regulate the nutrient availability by way of direct or indirect influence. Organic matter accumulation in soil decides the extent of beneficial effect on improvement of physical, physico-chemical and biological properties of soils. Thus, organic matter is a key variable and provides several plant nutrients through its decomposition. Nearly 90 to 95 per cent N, 45 to 50 per cent P and 98 per cent S reside in organic matter. Similarly, about 70% of available Zn and Cu are in the organic form. Contribution of organic P to availability is variable ranging from 15 to 25 per cent. A decline in organic matter will multiply nutrient deficiencies due to serious suppression in their availability. Malewar and Ghonsikar (1984) explained the greater deficiency of micronutrients cations in soils of semi-arid regions in terms of their poor organic matter level. In addition to being generator of essential plant nutrients, organic matter aids availability by complexing nutrient elements. Complex elements remain more available because of their protection from immobilisation. A striking improvement in nutrient availability in soils of semi-arid region is possible by organic manure's addition regularly. Palaniappan *et al* (1978) studied the influence of environmental factors on organic C, total N, available N, exchangeable N and nitrate N. Close relationship of organic C and total N was established with rainfall and altitude. It was estimated that increase in 100 mm mean annual precipitation gave rise in organic matter content to the extent of 0.17 per cent whereas every increase in 100 m elevation, there was an increase of 0.31 per cent of organic matter. On the contrary, mean increase in 10°C temperature, there is 2-4 fold increase in rate of decomposition of organic matter. Bharambe (1976) pointed out differences in C:N ratio in soils under forest and cultivation. The C:N ratio of 17.05 and 11.8 was observed in forest and cultivated soils, respectively. However, little or no differences were observed in C:N ratio under semi-arid region of Maharashtra. The average C:N and C:P ratio varied from 9.2 to 14.1 and 5.3 to 7.9 within different agroclimatic zones of Maharashtra i.e. semi-arid, to transitional zone with high rain fall (Patil and Sonar, 1994). The wide C:N ratio of forest soil of sub-humid region is attributed to continuous fall of leaves and other plant material with rather slower rate of its decomposition. Whereas narrow C:N and C:P ratio explained the faster rate of organic residue decomposition and formation of desiccated immobilized N and P in soil.

Soil reaction (pH) also controls solubility and consequently availability of nutrients in soils. With rising pH, deficiencies of Zn, Fe, Mn, Cu and B will increase and that of Mo and P decrease. The availability of N, total N and organic carbon decreased with increasing pH of soil (Palaniappan *et al* 1978). Occurrence of Zn deficiency is a very common phenomena in saline-alkali soils due to alkaline soil reaction. Most of thc micronutrients in alkaline pH range get precipitated into insoluble form. Presence of native lime in the profiles of soils of semi-arid regions aggravates incidence of Fe deficiency. In contrast, it is the strongly acid condition which inhibits P and Mo availability (Goswami and Sahrawat, (1982).

Soil texture influences the occurrence of nutrient deficiencies. Dominance of clay size particles (<0.002 mm diameter) was positively related with the abundance of several plant nutrients in soil. A negative relationship between availability of Zn and Cu and content of coarse (Sand) fraction in soil was established (Malewar and Randhawa 1978 and Katyal and Sharma, 1991). Further, Tandon (1995) pointed out the expansion in S deficient area with soil texture becoming coarser. Besides inherent property, sand fraction possess limited surface area for nutrient exchange on account of this disability, nutrient ions in sandy soils move freely into soil solution which may be liable to leaching. Apart from micronutrients, high sensitivity of N (Mainly NO_3^-) to leaching is universal in soils.

In addition to climatic, environmental and pedogenic factors which are native to semi-arid tract, there are several man made interventions which have contributed significantly to mounting nutrient deficiencies. Accumulation of nutrients arising from their regular additions adversely influences the absorption of others through negative interactions. High build up of P due to regular dressings is known to depress Zn and S availability. Excessive liming induces Fe deficiency.

Land degradation is a leading cause of overall pathetic nutrition of Indian crops. The harmful effects of land degradation (Water and land erosion) spring up from fertility depletion and poor physical environment for plant growth. Soils of semi-arid regions are typically erosion prone due to sparse vegetation cover promoted by extended periods of aridity and over grazing by cattle. Soils of rainfed semi-arid regions are nutrient hungry and erosion holds the major responsibility for this condition. Apart from erosion, land degradation is set in motion by excessive wetness, hard setting of soils, perched water table and continual use of agricultural chemicals, due to net enrichment of soil with some toxic factors or change in soil physical or biological environment.

Crops and cropping systems in semiarid regions affect physical, physicochemical and biological properties (Malewar, 1995; Malewar *et al.,* 1999 and Reddy, 1997). Grasses influence the buildup of soil organic carbon and thereby improve the physical properties like bulk density, porosity, infiltration rate, and soil aggregation. Among the various oilseed based cropping systems, cotton-groundnut proved to be superior over others in improving bulk density, porosity, water stable aggregates and organic carbon in soil. The productivity of inter-cropping systems (Cotton + soybean, sorghum + pigeon pea) can be significantly increased through improvement of physical, physico chemical and microbial activity by incorporating organic residues like PMC, glyricidia, FYM and wheat straw @ 5 t ha^{-1}, 20 days prior to sowing. The productivity of inter cropping (Cotton & black gram) can be further enhanced through recycling of crop residue of black gram through post harvest tillage coupled with incorporation of organic source with wide C:N ratio, wheat straw (5 t ha^{-1}) 20 days prior to sowing.

3. SOIL PROPERTIES AND ABSORPTION OF NUTRIENTS AND THEIR EFFECT ON CROP GROWTH

Soil physical environment particularly soil physical properties decide the magnitude of successful management of soil resource in arable farming system. Soil physical requirement is crop specific and thus, each crop behaves differently under varied set of soil conditions. Basically, soils also differ significantly in imparting different physical properties due to large differences in soil forming elements and related processes. Further, solid, liquid and gaseous matter present in the different proportions in soil, control its physical and dynamic properties. These properties, in turn affect the availability of water and nutrients to the plant. Soil studies in terms of their different dynamic properties were tried to relate with crop behaviour and production in a given environmental condition.

Water retention and transmission characteristics of a soil determine the amount of water available to plant roots. The differences in soil moisture energy relations and pore size distribution observed in the soil were reflected in water transmission characteristics viz., saturated and unsaturated conductivity and soil water diffusivity. Soils identical in moisture content but differing in rate of water movement would significantly influence the amount of water taken up by a given plant species. Moisture retention and release characteristics of clay soils of Parbhani revealed that though soils were clayey, yet there was variation in the moisture content held at tension of 0 to 1500 kPa. The moisture held at these tensions as a function of depth was constant, erratic or showed an increasing trend. This emphasizes the importance of relative distribution of the different size particles class on the moisture retention at different tensions (Bharambe, 1981). In his compilation, it was observed that water transmission behaviour of soil clay, transmitted water more slowly than those with lesser clay content. The diffusivity and captivity of water were less in soils with higher clay than those with lower clay content. Similarly soils with higher clay had lower hydraulic conductivity values and suffered a lesser reduction in these values on reduction in moisture content compared to soils with lesser clay content. The presence of free lime in soil has been found to adversely affect the water diffusivity, weighted mean diffusivity, penetrability and captivity of medium black soils of Gujarat. This has been presumed to be due to unfavorable changes in the distribution of micro and macro water filled pores, total water filled pores and total water filled porosity (Pathak and Patel, 1982). While studying the influence of lime on physical properties of silt loam soils of Vertisols of Saurashtra region of Gujarat, Patel *et al* (1992) observed increasing swelling with increased $CaCO_3$ and water content. Soil moisture largely controls soil fertility and productivity either directly or indirectly by regulating soil air and temperature regimes. In two extremes of soil moisture conditions, several chemical reactions are either much retarded or hindered regulating the supply of nutrients to crops. Ammonification and nitrification processes were retarded in dry soils.

Consequently, nitrate reduction and supply to crop was limited. A moisture content of 50-75% of water holding capacity is optimum for ammonification and nitrification. At a given P status of soil, the availability of P was less in dry soils but was much increased under waterlogged conditions as in case of rice. This is mainly because of the reduction of the insoluble iron and aluminium phosphate compounds to soluble iron and aluminium phosphate compounds at high moisture levels under highly reducing environmental conditions. The oxidation-reduction processes as a consequence of different moisture regimes, influenced the availability and hence the uptake of micronutrients by rice plant (Table 6). At a given exchangeable K, its availability depends on soil moisture conditions. A decrease in soil moisture content results in reduction of diffusive flux and also availability of K. Consequently, K may become critical in soils of marginal K status when K supply is just enough under optimum conditions. At a given water content, diffusion depends mainly on K concentration in soil solution. By implication, the effect of reduced water content can be compensated by an increase in K concentration in soil solution. The K availability was restricted in dry soils and the decrease in yields in dry years remained least in soils receiving high P and K. This is because of the fact that for a given diffusive flux soil water content is inversely proportional to the K concentration. In a compilation, Subbarao (1987) explained that at each of K levels tested, K uptake increased with increase in water content and reduction in K availability caused by low water content was compensated by increasing the K concentration in solution. In a field, experiment conducted on clay soils at Navsari, Mehta (1982) observed that bulk density values and resistance to penetration was considerably lesser nearer the roots, while depletion of moisture was greater. Hydraulic conductivity was marginally affected. At lower depth, bulk density and resistance to penetration were greater while hydraulic conductivity and depletion of moisture were lower. The axial and lateral pressures exerted by the roots are attributed as cause for such modification in physical properties of soils. Patel *et al* (1983) and Polera *et al* (1984), in pot and field experiment in groundnut observed that leaf area, number of flowers and gynophores, haulm and pod yield and root dry weight, relative leaf water content and leaf water potential decreased as the soil moisture stress increased from 0-1/3 to 0/14 bar (green house) and 0-0.5 to 0-20 bar (field). Further, they observed moisture stress at any of the growth stage to significantly decrease shoot and pod yield including decline in accumulation of N, P and K.

Table 6. Uptake (ha^{-1}) of micronutrients by rice under different moisture regions

Water treatment	Black soil				Red soil			
	Zn (g)	Cu(g)	Fe(kg)	Mn(kg)	Zn (g)	Cu(g)	Fe(kg)	Mn(kg)
Saturation	687	201	5.1	5.8	513	208	4.8	17.1
Standing water (5 cm) throughout	494	162	12.1	11.1	306	126.	7.7	17.3
Alternate flooding and drying	659	194	4.2	7.2	369	152	4.2	13.3

Source : I.V. Subbarao, 1987.

A significant difference in N, P and K content and uptake was observed in the groundnut varieties. In case of semi-arid ecosystems, the quality of arid, moist and humid periods, the quantum of moisture storage over the period were found to be different between two areas (Aurangabad and Solapur). These variations in the quality of growing periods within a ecosystem could be effectively used for crop planning specific to an area (Challa *et al* 1999). Basically, available soil moisture and storage capacity varies with soils and it depends mainly on soil volume, texture, kind and amount of clay minerals of soils. On the basis of P/PET ratio, the duration of dry and moderately dry period in Aurangabad area was more (35 weeks) than Solapur area (31 weeks). In spite of higher annual rainfall in Aurangabad area, the moist and humid periods which could sustain the crop growth were only 17 weeks while they were 21 weeks in Solapur area. On the contrary, the humid period in Aurangabad area was 14 weeks and it was 8 weeks only in Solapur area. This clearly indicates that crops in Aurangabad area during this period had practically no moisture stress and this may not be true in Solapur area. The longer the crop growing period in Solapur area as compared to Aurangabad area could be ascribed to the distribution of rainfall under soil-site conditions.

Soil moisture controls the soil temperature

Soil and ambient temperatures influence the nutrient availability and uptake by plants. The rates of chemical reactions, in both the cells as well as the soils, increased with increasing temperature upto a point, and decreased thereafter following the Vant Hoff's rule. Ammonification and nitrification take place at soil temperatures between 5° and 40°C, being more rapid around 35 to 40°C. A classical example of the effect of soil temperature was that of iron chlorosis in roses in cool winter months which however, disappears as temperatures shoot up. The availability of Zn and P was adversely affected by low temperatures. The P uptake in corn was 2 to 4 fold greater at 25°C than at 18°C soil temperature (Table 7). If the early growth of rabi crop is caught in cool winter months, P and Zn deficiencies become aggravated. Iron deficiency occurs at low and very high temperatures. The differential uptake at varying temperatures was because of differential absorption of mineral nutrients by living membrane. The solute absorption is much reduced at low temperature and the salt accumulation increased with increasing temperature but the magnitude of such absorption for all the ions is not same. Uptake of N, P and K in rice was maximum at 32/20°C and decreased on either side of this range. Similarly, increase in temperature increased P, Ca, Mg and to some extent Mn in potato.

Table 7. Effect of Temperature on P uptake by corn

Air temperature °C		18		25			25
Soil temperature °C		18		25			18
	Po	P22	P44	Po	P22	P44	P22
Yield g/pot	0.89	1.02	1.59	3.63	639	10.23	4.45
P uptake mg/pot	2.73	3.81	5.39	5.38	11.97	20.91	7.39
Root length cm/pot	2.27	3.09	4.50	13.20	15.70	11.82	8.59

Sources : Subbarao I,V.1987

Soil structure is the key to soil fertility. Bulk density, soil aeration, moisture release pattern, uptake of nutrients, growth and yield of crops depends upon structure. Subbarao (1987) compiled data on the influence of bulk density on rice at Kharagapur and Rajendranagar (Hyderabad). The studies revealed that optimum bulk density for puddled rice was found to be1.5 g/ml at which highest grain yield was obtained along with highest root density. Similarly, the uptake of phosphorus by daincha increased upto 1.5 g/ml bulk density and then decreased. Karle *et al* (1995) indicated that the compaction of black clay by 8, 12 and 16 passes of 400 kg roller and 0.3 g/cm^3 for uncompacted soil (Table 8). The final infiltration rate was decreased from 6.0 to 4.0 cm/ha by 8 passes of the roller and further decreased to 3.0 cm/ha by increasing number of passes to 16. The grain yield of paddy (PBN-1) was increased significantly by compaction of black clay soil.

Table 8. Effect of compaction of puddled black clay soil on physical properties and paddy grain yield.

Passes of Roller	Bulk density (mg/m^3)	Final infiltrate rate (cm/hr)	Paddy (PBN-1) grain yield ($q\ ha^{-1}$)
0	1.00	6.0	11.5
4	1.12	4.6	15.8
8	1.22	4.0	21.7
12	1.32	3.4	19.3
16	1.43	3.0	24.2

Source : Karle *et al* (1995)

Soil aeration

The composition of soil air differs from that of atmospheric air in 2 or 3 respect. Soil air contains 0.25 per cent CO_2 against that of 0.03 per cent CO_2 in atmospheric air. However, 20.6 and 20.96 per cent O_2 was present in soil and atmospheric air, respectively. Thus, it is clear that CO_2 in soil air may be 10 to 1000 times more than atmospheric air but 0_2 is slightly lower in soil air than that of atmospheric air. As a result, several soil processes and properties of soils are affected by presence of varied magnitude of aeration (Table 9). The impact of well presence of oxygen in soil (oxidised), results in more availability of N and S to the plants. Whereas Fe and Mn are available adequately in waterlogged soils as in case of rice rarely approaching toxic limits in certain cases. Phosphorus availability is usually low in rainfed crops and highly available to low land rice in the same soil. Lack of sufficient O_2 or high amount of CO_2restricts the nutrient uptake by roots of a large number of crops. The order of reduction in nutrient absorption by maize from a silty loam due to restricted aeration is K > Ca > Mg > N P. At one hand, increasing oxygen percentage increased the nitrate nitrogen but on the other hand the CO_2 in soil increases P availability as a result of CO_2-P complex in soil due to the action of carbonic acid on less soluble phosphates. The partial pressure of CO_2 similarly influences Ca relationships. The distribution of Ca^{2+} between OH^{-1}, CO_3^{-2} and HCO_3^{-1} can be expressed as a function of the partial pressure of CO_2.

$$CaCO_3 \xrightarrow{CO_2.H_2O} CaHCO_3$$

Directly, this reaction is significant in liming and indirectly on soil fertility. The significance of this reaction lies in the availability of iron to groundnut in black calcareous soils. Severe iron chlorosis has been observed with increase in $CaCO_3$ content of these soils on groundnut (Table 10). Though, DTPA extractable Fe and metabolic Fe were far above the critical limits in soil and plant, severe chlorosis, papery thin small leaves and lastly death of plant was the symtomology in groundnut (Subbarao *et al* 1987).

Table 9. Changes in the state of oxidation of soil constituents.

Element	Normal form in oxidised soils	Reduced form in water logged soil	Remarks
C	CO2	CH4	
C	–	Complex aldehydes	
N	NO_3	NH2, N2	Nitrate greatly available in oxidised form, sulphide injury in waterlogged soils.
S	SO_4	H_2S	
Fe	Fe^{3+}	Fe^{2+}	Fe and Mn highly available in reduced soils.
Mn	Mn^{3+}	Mn^{2+}	May become toxic in highly reduced environment in the presence of organic matter

4. SOIL -NUTRIENT MANAGEMENT

As pointed out, major soil groups in semi-arid region are : Vertisol, Alfisol, Inceptisol and Entisols. In one or the other way, soils of these groups poses problem of water and wind erosion. Further, Vertisols also showed variation in moisture availability, poor internal drainage, narrow workable moisture range, lack of opportunity for interculture, moisture loss through cracks and swelling causing impeded drainage, low organic matter, multinutrient deficiencies etc. Alfisols, Inceptisol and Entisols are highly permeable, prone to erosion, low soil depth, undulating topography, low productivity and nutrient deficiencies.

Table 10. Availability of iron in relation to $CaCO_3$ content of soils in groundnut crop.

$CaCO_3$ in soil (%)	DTPA extractable iron (.mg kg^{-1})	Metabolic (active) iron in plant (.mg kg^{-1})	Extent of chlorosis
3.0	17.7	23.4	Normal
5.0	13.0	19.8	Newly emerged leaves slightly chlorotic
9.0	11.9	19.0	Intervinal chlorosis extended to lower leaves also.
16.0	8.8	14.4	Severe chlorosis and stunted growth
20.0	7.8	9.9	Severe chlorosis, papery thin, small leaves

Source : Subbrao I.V., Narasimha Rao, P., Balaguravaiah, D, Narasimha Rao,Y and Adinarayana, V. PARS, Lam (A.P.), 1987.

Nutrient Management

Deficiency of nitrogen in almost all the rainfed semiarid region is more widespread. Maximum effort on nutrient management are naturally focused on fertilizer N management. Since fertilizer N exhibits a strong tendency to exit readily the soil-plant continuum, fertilizer N use research has emphasised more on partitioning of fertilizer N between crop use (fertilizer N recovery) and misuse (fertilizer N loss). Applied N undergoes various pathways of transformation in soil and its rate is governed by several factors. The losses of applied nitrogen through leaching, runoff, denitrification, volatilization and immobilisation are of great practical significance. The nitrogen recovery ranged between 30 to 50 per cent. The N use efficiency be improved through split doses, appropriate selection of N fertilizers in relation to crop and soil, placement of fertilizers, urea coating with neem cake/neem oil/sulphur, premixing of urea with moist soil, use of urea super granules/ briguetts, balanced fertilization are some of the field level techniques which showed great promise for enhancing its efficiency under various soil-crop situations. Deep black Vertisols may be conducive for ammonia volatilization following surface application of ammonium and ammonium forming fertilizers. Because of high pH, Vertisols are suspect to loss of urea and ammonium fertilizers through ammonium volatilisation. Similarly, high clay content on receiving water get water logged and nitrate present could be lost by denitrification. Saturation of soil profiles provokes denitrification and loss of N. Runoff in case of high intensity of rainfall displaces fertilizers N from the site of application. High leaching conducts N to depth beyond the root reach. Thus, with shallow soils, N use efficiency of crop is expected to decrease with increasing rainfall and will be lower than that with deep soils (Malewar, 1996). The KNO_3 lost twice as much N than urea when rainfall coincides with fertilization. Agronomic superiority of divided application of N in 2 or 3 splits to a single application is evidenced. Placement of fertilizer N adds to the efficiency of divide application because it curbs losses of N through surface runoff and ammonia volatilisation (Katyal, 1989 and Katyal and Gadalla, 1990). Placement behind plough proved superior over broadcast in N utilization by rice crop (Talekar and Dongle, 1993). Absorption, primarily is a function of accessibility of NH_4N or NO_3^- -N to plant roots. This explains why top dress application to grown up crop with abundant root system was utilized more efficiently than basal application to just beginning to grow seedling the scant roots (Table 11).

Table 11. Effect of Method of fertilizer application on recovery yield and nutrient uptake by rice and sorghum.

Method/Time application	Rice		Recovery		Sorghum	
	Grain yield (q ha^{-1})	Total uptake (kg ha^{-1})	N	P	15 N recovery (%)	15 N loss (%)
Broadcast	29.9	52.1	28.0	8.4	-	-
Placement behind plough	34.2	61.6	42.1	12.2	-	-
CD @ 5%	3.6	5.0			-	-
Basal					51	49
Top dressing					78	22

Source : Rice (Talekar and Dongale, 1993) Sorghum (Hong *et al.* 1992)

A survey of Indian agriculture shows that the crop recovery of added P varied from 5 to 35% on single crop basis with mean of 15-20%. Tandon and Kanwar (1984), in their analysis showed that sorghum tended to respond to P application more frequently when raised on Alfisols than on Vertisols in rainfed conditions. Shallow depth encouraged higher crop response to P. Due to high clay, $CaCO_3$, and high pH of Vertisols. Phosphate management assumes more significance for improving efficiency of added phosphates. The outcome of phosphate research in the form of tips for fertilizer P use are summarized by Chavan and Dongle (1996) as under (1) The water soluble phosphatic fertilizers be applied either by spot application, in bands or placement by seed-cum fertilizer drill so as to reduce the P fixation and to maintain higher P availability in soil for crop growth, (2) Avoid broadcast application of water soluble phosphates, (3) The water insoluble P fertilizers like rock phosphate should be applied by broadcast method on soil surface followed by its incorporation for better utilization in strongly acidic soils. It's particle size should be fine and it should be applied 3-4 weeks prior to sowing for higher benefits, (4) Rock phosphate along with SSP or partially acidulated RP could be used on neutral to alkaline soils, (5) P applied to first crop also increase the yield of succeeding crops due to residual P fertility, (6) so far as phasing of P application is concerned higher emphasis be given to fertilize rabi-summer crops while the kharif crop like rice be grown under residual P or with partial P application, (7) use of P solubilising organisms may be made to make soluble the native added P and increase the efficiency of P and (8) The combined use of N and P fertilizers helps to improve their efficiency.

On large scale, potassium, secondary and micronutrients have not become so critical in increasing the productivity of rainfed crops (Katyal, 1994), though sporadic responses of potassium, sulphur and micronutrient fertilizers are evidenced under dry land crops and that to high value crops like cotton, chillies, groundnut and other pulses and oilseed crops.

Plate 3.1. Soil profiles and root growth of Agroforestry trees. Profile Depth 300 cm. Photographed at Bhopal (MP)

More loss of organic matter is evidenced elsewhere whether by erosion or high temperature in semi-arid tropical regions. This further adds up to impoverishment of soil resource of several elements essential for plant growth. A decline in organic matter multiplies nutrient deficiencies and its fall by 2/3 symbolizes a serious suppression in nutrient availability. Moreover, the presence of organic matter in soil is a single property which controls the physical, chemical and biological properties of soils. Thus, role of organic matter is major in nutrient management and soil sustainability. Presently, the use of organic manures in rainfed agriculture is totally neglected which needs to regularised by its either annual application (@ 5-10 t ha^{-1}) or once in 2-3 years so as to upkeep the soil properties above the critical level particularly in soils of semi-arid regions of rainfed agriculture. Results of many experiments (Table 12) revealed that in corporation of subabul lopping and sunhemp @ 5 t ha^{-1} and sorghum stubble and safflower stalk significantly increased the infiltration rate, water content at field capacity, organic carbon, available N and P as compared to inorganic fertilizers (Badanur *et al* 1990, Bellaki and Badanur, 1994, Reddy 1997, Karle *et al* 1995).

Table 12. Changes in soil physical and chemical properties with the addition of organic residues in a vertisol

	Treatments	Infiltration rate (cm ha^{-1})	Bulk density (mg m^{-3})	Field capacity (%)	Organic carbon (%)	Available nutrients (kg ha^{-1})		
						N	P_2O_5	K_2O
T1	Application of fertilizer @ 50 kg N and 25 kg P2O5 ha^{-1}	0.5	1.3	33.4	0.55	135	27.50	300
T2	Fertilizer application @ 25 kg N and 22.5 kg P2O5 ha^{-1}	0.5	1.3	33.3	0.53	128	27.20	305
T3	Sunhemp at 5 t ha-1	0.8	1.3	37.0	0.61	150	28.25	310
T4	Subabul at 5 t ha^{-1}	0.8	1.3	37.0	0.66	148	28.75	315
T5	Sorghum stubbles at 5 t ha^{-1}	1.6	1.2	39.5	0.63	120	28.80	335
T6	Safflower stalks at 5 t ha^{-1}	1.5	1.2	33.7	0.61	126	26.25	330
T7	Farmyard manure at 5 t ha^{-1}	0.6	1.3	36.4	0.68	145	28.00	320
	C.D. at 5%	0.09	0.03	0.72	0.03	4.25	1.52	8.08

Source : Badanur *et al.* (1990).

5. FURTHER RESEARCH PROJECTIONS

1. Considering poor resource capability of rainfed farmer of semi-arid tropical region, low cost nutrient input technology with maximum efficiency is required to be developed. Systematic research approach is necessary to develop quantitative relationship between judicious fertilizer use, rain water availability and harvest and soil type.
2. Practically, meager data under Indian condition are available on nature, distribution and characterisation of soil organic matter developed under semiarid regions. Thus, these aspects of soil humus under different semi-arid agroecological regions, sub-regions soil groups, cropping systems and management levels need to critically studied. This would facilitate to further evolve suitable soil organic matter management systems for various semi-arid regions in the country.
3. Integrated nutrient management and supply system models are to be developed for various semi-arid regions, considering availability and replacement of inorganic component to organic nutrient base, residue management including their recycling, agro-industrial waste product locally available and biofertilizers as component of the nutrient supply system.
4. Watershed approach through soil and water conservation practices need to be developed. Due attention on rain water harvesting and its recycling coupled with nutrient management would be a priority area of research.

REFERENCES

Badanur V.P., Poleshi C.M. and Balachandra K. Naik. 1990. Effect of organic matter on crop yield and physical and chemical properties of Vertisol. Journal Indian society of Soil Science. 38 : 426-429

Bellakki M.A. and Badanur V.P. 1994. Effect of crop residue incorporation on physical and chemical properties of Vertisol and yield of sorghum. Journal Indian society of Soil Science. 42(4) : 533-535.

Bharambe, P.R. 1976. A comparative study of quartzitic soil of Nagpur district under continuous cultivation and afforestation. J. Maharashtra agric. Univ 1 (Addi.) 285-287.

Bharambe, P.R. 1981. Soils of Marathwada (Ed) Department of Agril. Chem. & Soil Sci, Marathwada Agricultural University, Parbhani.

Challa, O. Reddy, R.S. and Shivprasad. 1999. Assessment of quality of growing period in arid and semi-arid ecosystems for crop production. Agropedology 9 : 63-70.

Chavan, A.S. and Dongle, J.H. 1996. Transformation of major fertilizer nutrients in soil and their efficient management.Workshop Org. ADMS/IFFCO/FAI(WR), Pune.

Chavan, D.P., Bangar, A.R. and Shingate, A.K. 1996. Zn, Mn, B and Mo distribution and soil profiles of different agroclimatic zones of Maharashtra. J. Maharashtra agric. Univ. 5(3) :183-189.

Dangarwala, R.T. and Patel, K.P. (1996). Micronutrient Research in Gujarat In : Soil Research in Gujrat Agriculture a compendium GAU, Anand Campus, pp 65-76.

Gundalla, J.D., Golakiya, B.A., Hadvani, G.J., Hadiyal, T.M. and Qureshi, A.U. 1996. Nitrogen research in Gujrat. In soil Research in Gujarat Agriculture. A compendium. Gujrat Agril. Univ. Anand campus. pp. 23-32.

Jagannadham, J., Bhanuprasad, V., Seshagiri Rao, M. and Pillai, R.N. 1995. Nutrient status of red soils of Kandukur division of Andhra Pradesh. Andhra Agric. J. 42(1-4) : 74-76.

Kalyansundaram, N.K., Patel, V.R. and Patel, B.A. 1996. Potassium; An Emerging fertilizer nutrient in intensive agriculture of Gujrat. In soil research in Gujrat Agriculture - A compendium. GAU, Anand Campus, pp 43-52.

Karle, B.G., Awasarmal, B.C. Khating, A.E. and Raut, V.V. 1995. Highlights of Research (1982-1992) of AICRP on Improvement of Soil Physical condition, Marathwada Agricultural University, Parbhani, India, pp 1-60.

Katyal, J.C. 1989. In soil fertility and fertilizer management in semi-arid Tropical Indian (Christianson, B.Ed) ICRISAT, Hyderabad, A.P. IFDC, Musde shoals, ALA, USA.

Katyal, J.C. 1994. Nutrient Management in soils of arid and semi-arid regions. Bull. Indian Soc. Soil Sci. 16 : 78-91.

Katyal, J.C. and Gadalla, A.M. 1990. Pl. Soil 121 : 21.

Katyal, J.C. and Sharma, B.D. 1984. Commun. Soil Sci. pl. Anal. 15, 1065.

Katyal, J.C., Sharma, K.L. and Reddy M. Narayana. 1997. Balanced fertilizer use in Semi-arid soils. Fert. News 42(4) : 59-69.

Katyal, J.C., Sharma, K.L. and Sriniwas, K. 1997. In TSI/FAI/IFA symposium on sulphur in balanced fertilization, New Delhi pp 1-12.

Madhavi, A., Seshaiah, B.V., Subbiah, G.V. and Pillai, R.N. 1996. Nutrient status of turmeric growing soils of Guntur district in Andhra Pradesh. Andhra Agric. J. 43(1) : 35-39.

Malewar, G.U. 1979. Distribution of Zn, Fe, Mn and Cu in different agroclimatic conditions of Marathwada region. Res. Bull. Marathwada agric. Univ. 9(1) : 18-19.

Malewar, G.U. 1995. Micronutrient availability as influenced by cropping patterns in Marathwada region of Maharashtra. Journal Maharashtra agricultural University 20(3) : 330-333.

Malewar, G.U. 1996. Soil nutrient management for sustainable productivity. Invited paper presented at workshop on Management of soil for sustainable productivity. Org. ADMS/IFFCO/FAI(WR) 17-18, Jan, 196. Pune

Malewar, G.U. 1999. Increasing importance of micronutrients in sugarcane production. Special lecture delivered at 15th Annual cane development conference, VSI, Manjari 17-18 September, 1999.

Malewar, G.U., Dhamak, A.L. and Syed Ismail 1998. Inter relationship of forms of Fe with properties of soils of semi-arid area of Northern Marathwada. J. Maharashtra agric. Univ. 23 (2) : 199-201.

Malewar, G.U., Ganure, C.K., Rudraksha, G.B. and Syed Ismail 1999. Impact of varied oilseed based cropping systems on soil fertility and soil nutrient balance. J. Maharashtra Agric. Univ. 24(2) : in press.

Malewar, G.U. and Ghonsikar, C.P. 1984. Soils of Maharashtra and their zinc chemistry. J. Maharashtra agric. Univ. 9(1) : 1-7.

Malewar, G.U. and Syed Ismail. 1997. Proceeding of sulphur in balanced fertilization-western India. TSI/FAI/IFA seminar. New Delhi.

Meisheri, M.B. and Patel, V.R. 1996. Sulphur research in Gujarat. In soil research in Gujarat Agriculture a compendium GAU, Anand campus pp. 53-64.

Palaniappan, R., Krishnamoorthy, K.K. and Ramanathan, G. 1978. Carbon-nitrogen relationship under varied environmental conditions in Hill regions of Tamilnadu. Journal Indian society of Soil Science. 26(1): 79-80.

Patel, C.L., Padalia, M.R. and Babaria, N.B. 1983. Journal Indian society of Soil Science. 53(5) : 340-345.

Patel, C.L., Raval, Y.M. and Patel, B.K. 1992. Journal Indian society of Soil Science. 40(4) : 361-363.

Patel, M.L. 1996. Phosphorus related research in soils of Gujarat. In soil Research in Gujarat Agriculture. GAU, Anand Campus, pp 33-42.

Pathak, S.R. and Patel, C.L. 1982. Journal Indian society of Soil Science. 30(3) : 366-368.

Patil, J.D. and Shingate, 1982. Micronutrient status of soils from drought prone area of Pune region. J. Maharashtra agric. Univ. 7(3) : 216-218.

Patil, J.D. and Sonar, K.R. 1994. J. Maharashtra agric. Univ. 19(2) : 169.

Polera, K.B., Patel, C.L. and Pathak, S.R. 1984. Int. Legume Research. 7(2) : 95-100.

Prasada Rao, G. Reddy, A.V. Pillai, A.V. Reghavulu, P and Appa Rao, D. 1995. The Andhra Agric. J. 42(1-4) : 68-70.

Reddy, G.R. 1997. Management of swell-shrink soil through tillage, intercropping and crop residues. Ph.D. Thesis, Marathwada Agricultural University, Parbhani, India.

Sehgal, J.L. Mandal, D.K. Mandal. C. and Vadivelu, S. 1992. Agro-ecological Regions of India 2nd Ed. Oxford & IBH Publishing Co (P) Ltd, New Delhi.

Sehgal, J. and Sharma, J.P. 1994. Soil and climatic resource characterization of dry land ecosystems in India for sustainable agriculture. In soil Management for sustainable Agriculture in Dryland Areas. Bull. Indian Society of Soil Science, 16 : 12-25.

Subbiah, G.V. and Manickam, T.S. 1987. Nutrient status of Vertisols in dry farming areas of Andhra Pradesh. Andhra agric. J. 34(3) : 290-99.

Subba Rao, I.V. 1987.Factors affecting soil fertility. Presented paper at Training course on soil testing and fertilizer use. APAU, Hyderabad pp 47-80.

Subba Rao, I.V. Narsimha Rao, P. Balaguravaiah, D. Narsimha Rao, Y. and Adinarayana, V. 1987. Compiled in factors affecting soil fertility and presented paper at Training course on soil testing and fertilizers use. APAU, Hyderabad pp. 47-80.

Talekar, J.K. and Dongle, J.H. 1993. Comparative study on fertilizer placement behind plough and surface broadcasting for transplanted rice. J. Indian Soc. Soil Sci. 41(2) : 372-373.

Tandon, H.L.S. 1995. Sulphur fertilizers for Indian Agriculture - A Guide Book (Ed) FDCO, New Delhi.

Tandon, H.L.S. and Kanwar, J.S. 1984. Res. Bull of International crop Research Institute for the semi-arid Tropics Patancheru A.P.

Troll, C. 1965. Seasonal climates on the earth In : World Maps of Climatology, Berlin, Springer Verlag, p 28.

Venkateshwarlu, M., Seshagiri Rao M., Bhanuprasad, V. and Pilai, R.N. 1995. Nutrient status of black soils of Kandukur division in Prakashan district of Andhra Pradesh. Andhra Agric. J. 42 (1-4) : 88-90.

Virmani, S.M. Saharawat, K.L., Burford, J.R. 1982. In Vertisols and Rice soils of Tropic Tran. 12th int. Congr. 2,80, New Delhi.

Chapter- 4

MICRONUTRIENT DEFICIENCY PROBLEMS IN MAHARASHTRA AND ITS CORRECTIVE MEASURES UNDER DRYLAND FARMING

V.P. Babulkar* and B.N. Sagare**

* Asstt. Prof. of Soil Sci. and Agril. Chemistry, College of Agril, Nagpur, Dr. Panjabrao Deshmukh Krishi Vidyapeeth. Akola - 444 104 (MS)

** Professor of Soil Science and Agril. Chemistry (Retd), Deptt. of Soil Science and Agricultural Chemistry

INTRODUCTION

The mineral nutrient elements are classified as macro secondary and micronutrients based on relative amounts that are normally present in plants. Nitrogen, phosphorus, potassium are primary macronutrients: calcium magnesium and sulphur are secondary and zinc, iron, manganese, copper, iron and molybdenum are micronutrients. To this list of micronutrients chlorine, cobalt, sodium, vanadium silicon, selenium which are required for a particular species of plants are also recently added. Nickel (Ni) has been newly incorporated to the list of the important essential micronutrients (Sahu & Samant, 2006).

The importance of micronutrient has been realized during the past four decades when widespread micronutrient deficiencies, particularly of zinc and boron were observed in most of the soils in our country, where intensive agriculture has been practicised. Micronutrients are not only important for better crop productivity but also essential for sustaining human and animal health. It is important to note that minerals are trace elements, which are essential in human nutrition, are not necessarily same that are essential in plant nutrition and vice versa (Ratan *et al.,* 2009). Inadequate consumption of any micronutrient results in adverse metabolic disturbance leading to sickness, poor health, impaired development in animal and human health. Dietary micronutrient deficiencies affect a large part of human population. The world health organization estimates that globally some two billion people are affected by iron deficiency and that some 750 million people suffer from iodine deficiency. Also zinc deficiencies are increasingly recognized as an important public health problem (Ratan *et al.,* 2009).

With the introduction of green revolution, use of chemical fertilizer although contributed 40% of crop production. Continues use of NPK fertilizers without adequate organic manures seriously jeopardized the soil health and environment (Sahu & Samant, 2006). The chief source of micronutrients used by farmers are organic matters available in farm house. These include FYM,

Compost, dungs of various animals, poultry mature, green manure and crop residues in farm fields (Table 19).

1.PHYSIOGRAPHIC REGIONS

The state can be divided into three broad physiographic regions i.e. Western Konkan coast, Western ghats and North Deccan plateau.

Konkan coast is narrow coastal strips on the western part of Sahyadris with longitudinal distance of 500 km and having with of about 15-20 km. Western ghats commonly known as Sahyadris, formed the western age of the deccan plateau with several basaltic lava flows at the height of 1500 m; separating the plateau and coastal low lands. Towards east of Sahyadris, there is a vast stretch of Maharashtra plateau occupying a distance of 700 km and it is subdivided in to three sub-regions:

(a) Upper Maharashtra (Deccan) plateau.

(b) Lower Maharashtra (Deccan) plateau and

(c) Lower Maharashtra plateau (metamorphic)

2. SOILS OF MAHARASHTRA STATE

Konkan coast consists of Sindhdurg, Ratnagiri, Thane and Raigarh districts, having rock formations of Deccan trap, granite, genisses and laterite. Soils on gently sloping lands are aquic, and/or swell shrink characteristics, saline in creeks and nonsaline at other places. Soils of valley are moderately deep, well drained, loamy to clayey, neutral and slightly eroded. Major constraints of this region are shallow depth of soil, no moisture retention, excessive runoff, severe erosion and high amount of Fe and Al that decreases the availability of plant nutrients. However, these soils have great potential for growing export oriented crops like mango, cashew, pepper etc.

Western ghats include hilly terrian of basaltic origin in the districts of Dhule. Nasik, Ahmadnagar, Pune, Satara, Kolhapur, Sangli and Sindhdurg. Soils are very shallow, loamy, acidic and some what excessively drained. Major constraints are erosion, limited root zone, low Moisture and nutrient holding capacity, and with moderate to strong acidity. These soils are under forest vegetation, grasses and minor millets and have congenial environment to grow a horticultural and plantational crops.

Upper Maharashtra plateau includes Nasik, Aurangabad, Jalana, Ahmadnagar, Bid, Latur, Pune, Satara, Sangli and Part of Kolhapur districts having basalt as a parent material. Soils observed on steep slope are very shallow to shallow loamy skeletal and slightly acidic. Soils on gently sloping lands are clayey, well drained, slightly to moderately saline, slightly alkaline and calcareous. Major limitations are erosion, soil depth, acidity and low water holding capacity. These soils are cultivated to minor millets, pearlmillet, cotton, sorghum, sugarcane, onion and horticultural crops like grapes. oranges and banana where irrigation facilities are available.

Lower Maharashtra plateau is as a part of Dhule, Jalgaon, Jalana, Aurangabad, Ahmadnagar, Solapur, Osmanabad, Nanded and majority tehsils of Buldana, Akola, Washim, Amravati, Wardha, Yeotmal and part of Nagpur districts. Soils on moderate slope are very shallow to shallow clayey to clayey skeletal neutral to alkaline, with moderate erosion. Soils on plains and vallies are moderately deep to deep, neutral having high water holding capacity. These soils are cultivated to sorghum, cotton, soybean, groundnut and horticultural crops like citrus. Shallow rooting depth, salinity and sodicity problem in very deep soils and high runoff due to low permeability are the major limitations of these soils.

Lower Maharashtra plateau includes Bhandara, Gadchiroli and part of Chandrapur and Nagpur districts, of Vidarbha region. Soils on slopy land are shallow, loamy with severe erosion and low water holding capacity, whereas soils of alluvial plains are very deep clayey moderately well drained high water holding capacity and calcareous. Major constraints are low water and nutrient holding capacity and steep slopes causing erosion. Paddy is among the predominant crops grown in Chandrapur, Bhandara and Gadchiroli districts, whereas citrus, guava, cotton, soybean, sorghum, greengram and wheat are the major crops in Nagpur districts.

3. CAUSES OF DEFICIENCY OF MICRONUTRIENTS

Geochemically the black soils (Vertisols and its associated soils) which are extensive in Maharashtra is expected to be adequate in micronutrients required for normal growth of plants, as these soils are developed from basalt rich in divalent cations. Red Paddy soils developed from mixed parent material (Granite gneisses sandstone limestone) are comparatively low in micronutrients. Deficiency of micronutrients may appear due to following reasons under dryland conditions.

i. Depletion due to erosion and leaching as major areas under dryland farming are highly susceptible to erosion. In India, annual soils loss is estimated to be about 6000 M. tonnes and obviously micronutrients loss through runoff water and soils.

ii. Continuous use of micronutrient free NPK fertilizers in dryland agriculture and diminishing the use of organic matter, FYM, compost and green manures etc.

iii. Use of high yielding varieties, adoption of intensive systems of farming and cropping and use of high doses of fertilizers increased proportionately the micronutrient harvest from soil. The hidden hunger of micronutrients is manifested only when the major nutrients are present and supplied in adequate quantities.

iv. The increased crop production means heavy demand for plant nutrients in the soil and it results in rapid depletion of macro and micronutrients unless regularly replenished. Replenishment of micronutrient in the soil has unfortunately not received much attention in dryland crops. Consequently the deficiencies of micronutrients have become evident and that of Zn, Fe and B in particular are being fairly wide spread in Maharashtra (Anonymous, 1995).

4. MICRONUTRIENT STATUS OF SOILS

Micronutrient status of bench mark and established soil series of Maharashtra are presented in Table 1.

Table 1: Micronutrient status of soils of Maharashtra (after Patil and Sonar, 1994)

Soil Series	Agro-climatic zone and sitrict	pH	CEC (meq 100 g^{-1})	$CaCo_3$ (%)	ZTPA extractable nutrients (ppm)			
					Zn	Fe	Mn	Cu
Tadpangari	IV A. Parbhani	8.3	54.9	6.4	0.66	10.9	23.2	2.4
Sayala	IV. A. Parbhani	8.2	61.0	4.7	1.44	21.3	35.8	3.4
Ranjani	IV. A Parbhani	7.9	59.1	8.8	0.72	12.2	35.3	2.5

Aroli	V.B. Nagpur	8.0	61.1	1.3	0.72	11.0	34.9	2.7
Karla	V.B. Nagpur	6.8	54.1	0.2	1.0	11.7	63.0	3.8
Linga	V.B. Nagpur	8.2	51.2	5.8	0.68	26.0	27.8	1.7
Kirnapur	V.B. Nagpur	8.1	45.2	2.2	0.80	12.3	31.3	3.0
Shendwade	III. Dhule	8.0	63.1	6.6	0.70	10.4	20.4	2.7
Hated	IV. B Jalgaon	8.0	59.9	1.2	0.70	16.5	19.6	2.4
Hingona	IV. B. Jalgaon	8.7	50.1	8.0	0.80	6.1	35.1	2.3
Rastapur	III. Ahmednagar	8.3	50.1	14.4	0.70	12.0	14.6	2.4
Vadgaonamli	III. Ahmednagar	8.7	60.1	13.7	0.58	9.7	14.9	1.3
Otur	III. Ahmednagar	8.4	60.0	7.7	0.70	9.5	24.8	3.2
Nimone	III. Ahmednagar	8.5	60.6	7.5	0.70	11.4	27.3	3.1
Dholwad	III. Ahmednagar	9.1	61.1	7.5	1.06	11.3	32.5	3.6
Umbraj	III. Ahmednagar	8.2	63.1	6.6	1.08	12.3	25.7	3.8
Sawargam	III. Ahmednagar	8.7	62.0	5.2	0.78	14.9	17.8	3.6
Barshi	III. Solapur	8.2	68.9	1.8	0.88	14.9	42.5	6.1
Wadegaon	II. Pune	7.9	62.0	0.4	0.92	15.7	17.3	3.2
Kandegaon	II. Sangli	8.8	65.0	6.2	0.70	12.1	34.8	3.1
	Mean	8.2	58.4	6.3	0.87	13.1	28.9	3.0

4.1 Zinc

DTPA extractable Zn constant of swell-shrink soils of Maharashtra varied between 0.58 to 1.70 ppm with average values of 0.87 ppm (Table 1). The soils of Vadgaonamli series (Ahmadnagar district) arc poorest in DTPA zinc, whereas Kandegaon series (Sangali district) contained maximum amount of DTPA Zn. On an average, 33 per cent of soil samples of Akola and 12 percent from Nagpur districts were low in DTPA Zn (Table 2). Widespread deficiency of Zn was observed in Patur tahsil of Akola district (Badhe *et al.,* 1983). Most of the soils of Maharashtra are poor to marginal in available Zn on the basis of critical limits (O.6 to 0.8 ppm) suggested by Lindsay and Norvell (1978).

Table 2. DTPA Zn in Nagpur and Akola districts

District	No of Samples analysed	Percentage		
		Low (< 0.6)	Med. (0.6 -1.2)	High (> 1.2)
Akola	413	33	51	16
Nagpur	333	12	50	38
Patur Taluka (Washim, dist.)	25	72	28	

Deep black soils have higher values of available Zn (0.40 -1.40 ppm) than medium (0.40-1.20 ppm) and shallow soils (0.40-1.10 ppm). Zn released from the weathering of primary minerals is washed down from the shallow soils (situated at higher position) and deposited in lower position in deep soils (Dakhore and Bang, 1982). Maximum percentage of soil samples (29%) from shallow category were found to contain available Zn in deficiency range followed by medium (16%) and deep black soils (8%) soils. Recently, Naphaded *et al.,* (1998) reported that 40-60 percent soils of Vidarbha region are deficient in DTPA Zn and 9-14 percent in DTPA Fe.

4.2 Iron

DTPA Fe in swell-shrink soils of Maharashtra varied from 6.1 to 26 ppm (mean values of 13.1 ppm). Soils of Linga series (Nagpur district) contained highest amount available Fe, whereas Hingona series (Jalgaon district) is least in available Fe, content (Table 1). Available iron content appears to be sufficient in the light of critical values (4.5 ppm) proposed by Lindsay and Norvell (1978).

On an average, 45 and 22 percent soils (Table 3) of Akola and Nagpur districts, respectively and that of 64 % soils of Patur Tahsil (Washim district) were found low in available Fe (Badhe *et al.,* 1983).

Table 3. DTPA Fe in Akola an Nagpur districts

District	No of Samples analysed	Percentage	
		Low (< 4.5 ppm)	High (> 1.2 ppm)
Akola	315	45	55
Nagpur	333	22	78
Patur Taluka (Washim dist)	25	64	36

4.3 Manganese

Available manganese content in swell shrink soils of Maharashtra varied from 14.6 (Rastapur series of Ahmednagar district) to 63 ppm (Karla series of Nagpur district) with mean values of 28.9 ppm. In the light of critical levels of 1.2 ppm of DTPA Mn (Lindsay and Norvell, 1978). Soils are well supplied with available Mn (Table 1).

Mn status of Nagpur soils (Table 4) is high, however, 21% soil samples of Akola district were found deficient. However, 57% sample in akot tahsil of akola district where low in available Mn status (Badhe *et al.,* 1983).

Table 4. DTPA Mn in Nagpur and Akola

District	No. of samples analysis	Percentage		
		Low < 1 ppm	Med 1-2 ppm	High > 2 ppm
Akola	305	21	28	51
Nagpur	333	-	-	100
Akot Tahsil	91	57	35	8

4.4 Copper

DTPA copper content in swell-shrink soil of Maharashtra ranged from 1.3 to 6.1 ppm with mean value of 2.91 ppm and well supplied with available Cu. Barshi series of Solapur district is highest, wheres Vadegoanamli (Ahmadnagar district) is lowest in available Cu. soils of Akola and Nagpur districts are found to be high in DTPA Cu. (Table 5).

Table 5. DTPA extractable Cu in Nagpur and Akola districts

District	No. of samples analysis	Percentage	
		Low <0.2 ppm	High > 0.2 ppm
Akola	389	-	100
Nagpur	333	2	98

4.5 Boron

Vertisols and associated black soil (swell-shrink) of Vidarbha region (Maharashtra) contained B in the range of 0.28-3.14 ppm and 89 per cent soil samples are above critical limits (0.5 ppm of Boron). However, 11% of soil sample are found to contain available 'boron' in deficiency range (Atre *et al.,* 1989). Deficiencies and toxicities might be encountered at the same time, hence the use of 'Boron' requires great caution. Soil (Parbhani and Basmat talukas of Parbhani district) are deficient in available boron content.

4.6 Molybdenum

Molybdenum status of vertisonal and its associated soils (Swell shrink soil) of Maharashtra is about 0.45-0.50 ppm (Sagare *et al.,* 1980) which considered as moderate to sufficient for most of the dryland crops.

5. DETECTING MICRONUTRIENT DEFICIENCIES

Deficiency symptoms appear when supply of a micronutrient is too low and the plant can no longer function properly. However, plants require micronutrient before showing the micronutrient deficiency symptoms (hidden hunger). Hidden hunger refers to a situation, in which a crop need as a particular micronutrient before showing deficiency symptoms.

5.1 Plant Analysis

Plant and soil analysis is widely used as means of detecting micronutrient deficiency. Under conditions of hidden micronutrient deficiency, plant analysis is the most effective method for diagnosis of micronutrient disorders because neither plants exhibit deficiency of symptoms nor soil text predicts situation.

Table 6. Critical limit of dryland crops

Micronutrients	Crops	Plant part	Age (days)	Soil type 2	Critical limits (ppm)
Zinc	Wheat	Top half	30	Medium black	20.0
	Rice	Top half	30	Medium black	15.5
	Linseed	Top half	35	Deep black	15.9
	Maize	Top half	35	Mixed red and black	14.0
	Soybean	Top half	35	Mixed red and black	40.0
	Blackgram	Top half	35	Mixed red and black	38.6
Copper	Wheat	Top half	30	Medium to black	7.50
Iron	Sorghum	4th and 5th leaf	20	Medium to black	44.5
Molybdenum	Rice	3rd leaf	35	Medium to black	1.80

(Source: Rathore *et al.,* 1995)

The concept of critical micronutrient concentration in plant parts is the basis for diagnosing the critical concentrations of nutrient in plant or soil. It refers to a level at or below which develops deficiency symptoms or causes significant and / or 5 to 10 per cent reduction in its yield as compared to optimum. Critical limits established for micronutrients in plant are given in Table 6.

Limitation of plant analysis for diagnosing micronutrient deficiencies is that by the time a plant shows deficiencies. A reduction in the yield occur and is usually too late to correct the deficiencies. Moreover, Plant analysis indicates the micronutrient status in plant parts and is unable to assess the status for subsequent crop growth.

5.2 Soil Analysis

Soil analysis is very useful and a convenient diagnostic tool for quick and timely assessing micronutrient availability. This helps in estimating whether or not a soil can supply adequate amount of micronutrient for optimum plant production. On the basis of soil test values, the average critical limits of drylands crops may be as follows.

Table 7. Critical limits of micronutrients in soil for dryland crops

Micronutrient	Critical limits (ppm)
Zinc	0.6
Copper	0.2
Boron	0.5
Iron	4.5
Manganese	1.2
Molybdenum	0.02

The soil can be tested in soil testing laboratory for micronutrient analysis and on the basis of soil tasting values, the amount of micronutrient fertilizers to be added can estimated from the following three-tier system.

Table 8. Limits of micronutrients status in soil (ppm) for suggesting micronutrients fertilizer

Available nutrient in soil (ppm)	Low	Medium	Sufficient
Iron	<2.3	2.3-4.5	>4.5
Manganese	<1.0	1.0-2.0	>2.0
Copper	<0.3	0.3-0.5	>0.5
Zinc	<0.3	0.3-0.6	>0.6
Molybdenum	<0.3	0.3-0.5	>0.5
Boron	<0.03	0.03-0.06.	>0.06
Recommen dations	20% more than Recommended	Recommended dose	No need to apply

5.3 Requirement of micronutrients for dryland crops

Requirement of hybrid sorghum for micronutrients was maximum as compared to other dryland crops (Table 9).

Table 9. Uptake of micronutrients (g ha^{-1}) by different crops

Crop	Variety	Zn	Cu	Fe	Mn
Sorghum	CSH5	346	200	437	789
Cotton	AHH 468	242	140	197	328
Groundnut	JL24	252	179	305	559
Green gram	Vaishakhi	130	119	137	159
Black gram	T9	133	120	144	171
Cowper	Pusa	196	155	121	207
Safflower	-	172	182	195	333
Wheat	Kalyan sona	121	160	167	224
Pearlmillet	BJ 104	111	142	152	284

(Source: Phirke. 1983)

Micronutrients content and uptake (Table 10) by groundnut (cv SB XI) grown in vertisols are given in Table 10.

Table 10. Micronutrient requirement of groundnut

Micronutrient	Content (ppm)	Uptake (g ha^{-1})
Zinc	40-60	270-321
Iron	60-70	375-520
Manganese	8-10	53-84
Copper	4-5	13-29
Molybdenum	2-3	12-20

(Source: Sagare *et al.*, 1988)

6. DIFICIENCIES AND CORRECTIVE MEASURES

For correcting the micronutrient deficiencies various micronutrients can be applied to soil or through foliar sprays.

6.1 Zinc

Zinc is related in regulation of oxidation reduction potential within cells, growth promoting substance i.e. auxin and sulphydryl compounds such as cysteine. Zn deficient leaves are narrow and small, Lamina chlorotic, veins green, dead spots developed all over the leaves (old leaves) including vein, tips and margins. In case of cereals, the internodes become very short and results into stunted growth. A pale yellow chlorosis develops between the margins and midvein near the base of leaflet. In cotton, there is bronzing on older leaves starting from leaf tips and extending towards the base of leaf. In pulse crop, older leaves show interveinal marginal chlorosis, which lateron turns into brownish spots.

In case of deficiencies, zinc may be applied through zinc sulphate @ 10 to 40 Kg ha^{-1} depending on soil status at the time of sowing of dryland crops. When the zinc deficiencies appear in the standing crop, foliar spray @ 0. 5 per cent zinc sulphate plus 0.25 percent lime is found more effective and for recovery of deficiencies two to three sprays at an interval of 15 days may be used. Foliar application of Zn is found more effective and is mainly used for tree crops. Chelates and natural organic complexes are suitable for quick recovery of seedlings.

Table 11. Zinc containing fertilizers

Name of fertilizer	Formulae	% Zn
Zinc sulphate	$ZnSO_4.\ H_2O$	35
Zinc sulphate	$ZnSO_4.\ 7H_2O$	23
Zinc oxide	ZnO	78
Zinc corbonate	$ZnCO_3$	52
Zinc phosphate	$Zn(PO_4)_2$	51
Zinc chelates	Na_2Zn EDTA	14
	NaZn H. EDTA	9

Studies indicated that amongst the various carriers, zinc sulphate is the most efficient and cheaper source of correction of zinc deficiencies. Zinc sulphate is most effective when applied in soil below the seed prior to sowing or transplanting. Application of Zn by dipping rice seedlings in zinc oxide (4%) was observed effective in correcting zinc deficiencies of rice. Field experiments conducted on black soils having DTPA Zn in 0.38 ppm showed that 37.5 Kg zinc sulphate gave significantly higher yield over 12.5 and 25 kg zinc sulphate per hectare. Further, application of zinc to the soil was found better rather than spray (Anonymous, 1995).

Table 12. Effect of zinc and boron on yield of sorghum and sunflower

Treatments	Yield (q ha^{-1})	
	Sorghum	Sunflower
Control	18.20	3.21
Reco. NPK	30.9	9.90
Reco. NPK + zincated super phosphate (0.60% Zn)	35.5	12.5
Reco. NPK+zinc sulphate 10kg ha^{-1}	37.30	13.96

(Source: Dakhore and Gawande. 1990)

Increase in the grain yield of sorghum by 14.90 and 20.70 and that of sunflower seed by 26.30 and 41.40 per cent over NPK was recorded (Table 12) due to zincated super phosphate and zinc sulphate, respectively.

Besides soil and foliar application. Zn deficiencies can be cured by quoting seeds with micronutrient powders, soaking and dipping seeds in micronutrient solution and suspension and this has largely been attempted to economise Zn use. The integrated effect of organic manures enriched with zinc proved superior to inorganic Zn fertilizers applied to rice-wheat sequence (Table 13). Maximum rice yield was recorded with FYM @ 5 t ha^{-1} enriched with 5 or 10 kg Zn ha^{-1} through zinc sulphate alone. Under dryland farming, the application of Zn in combination with FYM will be more effective in relation to better soil physical environment. Order of yield response to Zn application in deficient black soil was observed as sunflower > sorghum > gram> soybean (Anonymous. 1995).

Table 13. Influence of Zinc enriched organic manures on the grain yield of rice and wheat crops in black clay soil

Treatments	Rice			Wheat		
	Grain yield (q ha^{-1})	% increase	Uptake of Zn (g. ha^{-1})	Grain yield (q ha^{-1})	% increase	Uptake of Zn (g. ha^{-1})
Control	22.5	-	107	28.6	-	137
10 Kg Zn ha^{-1}	34.7	12.2	250	34.7	6.0	444
10 t FYM ha^{-1}	30.6	8.1	185	34.6	5.7	603
5 t FYM ha^{-1}	26.9	4.4	95	30.0	2.0	82
5 t FYM + 5 kg Zn ha^{-1}	34.6	12.1	139	33.9	5.3	384
5 t FYM + 2.5 kg Zn ha^{-1}	29.1	6.6	104	31.3	2.7	187
2.5 t FYM + 5 kg Zn ha^{-1}	32.3	9.8	122	33.3	4.7	174
2.5 t FYM + 2.5 kg Zn ha^{-1}	31.5	9.0	109	31.8	3.2	161
CD (0.5%)	6.8	-	-	NS	-	-

Source : Rathour *et al.,* (1995)

6.2 Iron

Best known role of iron is as catalytic agent in enzymic activity. The characteristic symptoms of iron deficiency is pale yellow interveinal chlorosis, which develops on young leaves. Eventually, white necrotic areas develop within the chlorotic tissues and whole leaf turns pale yellow to white in colour.

Table 14. Amount of iron present in Iron carries

Name of fertilizer	Formulae	%Fe
Ferrous sulphate	$FeSO_4 \ 7H_2O$	19
Ferric sulphate	$Fe\ (SO_4)_3.\ 4H20$	23
Ferrous oxide	FeO	77
Ferric oxide	Fe_2O_3	69
Ferrous ammonium Phosphate	$Fe\ (NH_4)\ PO_4.\ H_2O$	29
Ferrous ammonium sulphate	$(NH_4)_2SO_4.\ FeSO_4\ 6H_2O$	14
Iron chelates	Na Fe EDTA	5-14
	Na Fe H EDTA	5-9

Iron deficiencies are mostly observed at high pH calcareous soil in a arid and semi-arid region. Iron is applied through iron sulphate @ 10 to 40 kg ha^{-1} or as a foliar spray @ 0.5 per cent ferrous sulphate with 0.25 percent lime. Soil application of iron on alkaline soils is less effective because of water soluble iron that is converted in to insoluble Fe $(OH)_3$. Therefore, foliar spray of iron is most efficient way of correcting iron deficiencies in alkaline soils. Order of yield response to Fe deficient black soil is observed as gram > sunflower > soybean > sorghum > wheat at (Anonymous, 1995). The various iron carriers with per cent of iron are given in Table 14.

Application of Fe either through $FeSO_4$ or Fe-EDTA and pyrite in presence or absence of FYM enhanced the yield arid Fe uptake of groundnut grown on calcareous soils of Pune. The chelated form of Fe was found more effective than non-chelated form and response was better in presence of FYM (Anonymous, 1995) @ 30.

Addition of FYM (67 g per pot) and dilute sulphuric acid separately recorded the higher DTPA-Fe in soil and dry matter, grain yield and iron uptake by rice grown on incept sols was increased. Use of spent wash slurry and spent wash compost influenced the yield, concentration and uptake of micronutrient in respect of maize. Iron chlorosis is a major problem in all calcareous soils in Maharashtra and demonstrated the utility of soil and foliar application of $FeSO_4$, for controlling iron chlorosis at Padegaon (Anonymous, 1995).

6.3 Manganese

It acts as a catalyst in important enzymatic and physiological reactions in plant. In manganese deficiency, leaves are chlorotic main and smaller veins are dark green forming a chequered appearance. In sorghum, the symptoms are developed on young leaves. Initially, small discrete chlorotic lesions develop between the veins, usually about midway along or towards the tip of the blade. In wheat, leaves become chlorotic with white streaks between the veins.

Availability of Mn in soil is influenced by high pH which favours the formation of less organic complexes of Mn and activity of soil micro organisms which oxidise soluble Mn to unavailable form that reaches to maximum extent. Commonly the manganese is applied through soil @ 10 to 40 kg ha^{-1} in the form of manganese sulphate or as a foliar spray @ 0.5% manganese sulphate plus 0.2 per cent lime. The various manganese carriers with percentage of manganese is given in Table 15.

Table 15. Amount of Manganese present in Manganese fertilizers

Name of fertilizer	Formulae	%Mn
Manganese sulphate	$MnSO_4 . 4H_2O$	26-28
Manganese oxide	MnO	41-68
Manganese	$MnCl_2$	63
Synthetic chelates	Mn EDTA	5-12
Manganese feats	---	10-35

Inadequate Supply of Mn in soil reduces the grain yield even at higher level of fertilizer application. In a field experiment conducted on black clay soil (pH 7.8, DTPA Mn 7.4 ppm), the addition of 10 kg Mn ha^{-1} through $MnSO_4$ resulted into increase in wheat yield over control. Further, soil application, of 10 kg Mn ha^{-1} was statistically at par with 3 foliar sprays of 1% $MnSO_4$ applied at 15 days interval (Rathore *et al.*, 1995).

6.4 Copper

It is a constituent of several enzymes which take part in oxidation reduction reactions and in the formation of vitamin-A in plants. The deficiency symptoms appeared on young leaves are interveinal chlorosis, rosting, permanent wilting, and leaf detaches easily. In sorghum, symptoms appear on leaves which are still within the whorl and young leaves. These leaves turn pale green and interveinal chlorosis developed at the base of the blades. In wheat, the new leaves are pale green in colour and show loss of turgor. Copper deficiency may cause "Wither tip" in fruit trees.

Table 16. Amount of copper present in copper fertilizers

Name of fertilizer	Formulae	%Cu
Cuprite	Cu_2O	89
Copper sulphate	$CuSO_4.\ H_2O$	35
Basic copper sulphate	$CuSO_4.\ 3Cu(OH)_2$	13-53
Chalkanthite	$CuSO_4.\ H_2O$	25
Copper ammonium phosphate	$Cu\ (NH_4)\ PO_4.\ H_2O$	32
Capper chelate	Na_2Cu ECTA	13

The mobility of copper in soil solution, often decreases with increase in pH and its supply to plants is correspondingly decreased. A wide range of compounds can be used to supply Cu through soil and spray. However, usual Cu source is copper sulphate ($CuSO_4.\ 5H_2O$) @ 10-40 kg ha^{-1} or spray of 0.1% $CuSO_4$ plus 0.05 per cent lime. Application of copper through foliar sprays is generally confined to emergency treatments of deficiencies identified after planting. Soil and foliar application are effective but soil applications are more common. The various sources of copper fertilizers and there contents are given in Table 16.

6.5 Boron

It increases permeability in cells membrane and facilitate, carbohydrate transport. In boron deficiency young leaves of terminal buds lose colour at the base, break down starting from the base and death of terminal buds is seen. The characteristic white or transparent lesions develop in interveinal tissues on young leaves.

Table 17. Amount of in boron fertilizer

Name of fertilizer	Formulae	%B
Borax Power	$Na_2B_4O_7 : 10H_2O$	11
Bond Acid	H_3BO_3	17
Sodium Pentaborate	$Na_2\ B_{10}O_{16}.10H_2O$	18
Sodium Petraborate	$Na_2B_4O_{7.}\ 5H_2O$	14-15

Soil pH and liming affect the availability of boron and is generally less available to plants with increase in soil pH arid its availability is markedly dropped when pH is greater than 6.5. Boron deficiency can be corrected with soil application of 5 kg Borax ha^{-1} which remains effective for longer period. Foliar application of B @ 0.2 percent borax is often beneficial, if the deficiencies are observed in standing crop. Foliar application of boron is generally practiced for fruit crops in combination with pesticides. Response of groundnut to boron application in black clay soils was reported by Dakhore and Gawande (1990). When boron was applied through borax @ 5 kg ha^{-1} and boronated super phosphate @' 0.2 per cent boron, it enhanced the pad yield of groundnut by 20 percent over NPK alone. List of boron containing fertilizer is given in Table 17.

Effect of boron Was studied (Anonymous, 1995) on vertisols deficient in boron and it was found that application of 50 kg P_2O_5 + 5 kg borax ha $^{-1}$ through boronated superphosphate gave significantly higher yield of cotton, groundnut and sunflower. The order of yield response to-boron was gram > sunflower > soybean > sorghum> wheat.

6.6 Molybdenum

It is related to nitrogen utilization and nitrogen fixation and is a constituent of nitrate reductase and Nitrogenese enzymes. The deficiency is markedly evident in leguminous plants. The lower leaves are light green, golden yellow to orange and dead spots developed all over the areas except veins. Affected areas extrude resinous gum through lower surface.

Plant availability of MO, unlike other micronutrients increases with depressing soil acidity. Crops that are very sensitive to insufficient molybdenum are legumes, cruciferrous crops and citrus. A perceptible increase in the yield of cauliflower grown in Vertisols with molybdenum @ 0.75 kg Mo ha^{-1} applied through spray and soil was reported by Sagare and Badhe (1980). Excessive amount of Molybdenum is toxic especially to grazing cattle and sheeps. The molybdenum deficiency is generally controlled by soil application of sodium molybdate @ 0.5 to 2 kg ha^{-1}. Various sources of molybdenum are given in Table 18.

Table 18. Percentage of molybdenum in molybdenum fertilizers

Name of Fertilizer	Formulae	%Mo
Ammonium molybdate	$NH_4 16\ MO_7O_{24}.\ 2H_2O$	54
Sodium molybdate	$Na_2M_oO_4.\ 2H_2O$	39
Molybdenum Trioxide	MoO_3	66

Table 19. Micronutrient content of the Organics (Source: Sahu & Samant, 2006)

Sr.No.	Organics	Micronutrients (mg/kg)					
		Fe	Zn	Mn	Cu	B	Mo
1	FYM	1788	34	137	2.5	4.6	2.1
2	Pig Manure	1200	50	70	8.9	3.5	1.4
3	Poultry manure	1400	90	210	7.1	5.0	2.0
4	Rural compost	3600	85	200	14	10.0	2.0
5	City compost	7000	400	560	150	15.0	9.0
6	Goat/Sheep manure	6500	2570	150	61	46.00	3.5
7	Sewage/sludge	8000	2459	262	643	9.0	6.0
8	Rice straw	225	30	700	4.5	8.0	0.1
9	Wheat straw	190	25	90	3.4	5.0	0.1
10	Green manure	140	17	80	3.2	20.0	0.2

7. ECONOMICS OF MICRONUTRIENTS

The micronutrient products are available in most of the firms supplying chemicals and fertilizers or these products may be supplied after putting the demand with the agencies. The cost of micronutrient application may differ with the type of micronutrients to be applied and way of its application. However, approximate cost of two sprays is around Rs. 500 ha^{-1}.

Correction of micronutrient deficiencies is not a simple job because of very small amounts required for application and an addition of excess doses may also prove toxic to crops. Therefore, a blanket application of micronutrients is not advisable. Foliar fertilization is effective, when given during seed filling period in grain crops. The nutrients in the leaf accumulated during early growth stages migrate to grains/fruit during later stages of development. This foliar nutrient depletion affects photosynthesis in "flag leaf" in particular, which supplies the major portion of photosynthesis to the grain in crops (Kannan, 1990). Adoption of suitable crop and soil management practices, aiming of sustainable agriculture in problematic soils of the state is a basic need of today. In this context, a blanket recommendation of 0.5% for maintaining the good level of organic carbon with regular addition of organic manure in any available form and or green manuring is almost important. It not only improves the physical properties of all types of soils but also contributes towards balanced plant nutrition improving the quality of crop yield.

Zende (1995) concluded that micronutrients can be applied as a single nutrient, in multiple combination, with organic matter in different forms, along with amendments of N and P fertilizers or with biofertilizers etc. Use of multiple micronutrients in dryland agriculture will be always better for yield as well as for economical feasibility.

REFERENCES

Atre, A.H., Guhe, Y. S. and Bhake, R.M. (1988). Surface and depth wise distribution of water soluble boron in relation to soil properties in soils of Vidarbha region of Maharashtra, Res. Highlights, ISSS Chapter, PKV, Akola. 3(1): 1-7.

Badhe, N.N., Patil, D.B. and Wankhade, S.G. (1983). Annual Project Report of "Micronutrient status of soils in Vidarbha and to prepare contour map and data Bank". Dr. P.D.K.V. Akola, MS, India.

Challa, O., Vade Velu, S. and Sehgal, J. (1995). The soils of Maharashtra for Optimizing land use. NBSS. Nagpur, Publ.. 54. pp. 25-49.

Dakhore, R.C. and Bang, S.K. (1982). Zinc status of soils of Central Res. Station, PKV, Res. J. 6 (2): 12-16.

Dakhore, R.C. and Gawande, D.R. (1990). Agronomy and soil science Research Review Committee Report, 1990. Dr. P.D.K.V., Akola, MS, India.

Kanan, S. (1990). Role of foliar fertilization on plant nutrition. Crops Enhancers of Nutrient Use. Acad Press Inc. pp. 313-348.

Lindsay, W. and Norvell, W. A. (1978). development of a DTPA soil test for Zn, Fe, Mn and Cu. Soil Sci. Soc. Amer. J. 42 : 421.

Naphade, P.S., Sarode, P.V. and Patil, D.G. (1998). Report, Research Review Committee. Agronomy and Soil Science, 1997-98. pp. 50-53. Dr. P.D.K.V. Akola. MS, India.

Patil, Y.M. and Sonar, K.R. (1994). Status of major and micronutrients in swell-shrink soils of Maharashtra. J. Maharashtra agric. Univ. 19 (2): 169-172.

Phirke, V.N. (1983). Effect of different crop rotations on chemical properties of Vertisols., M.Sc. (Agri.) Thesis, PDKV, Akola.

Rathore, G.S., Khamparia, R.S., Gupta, G.P., Dubey, S.B. Sharma, B.L., Tomar, V.S. (1995). Twenty Five Years of Micronutrients Res. In Soils and Crops of M.P., JNKKV, Jabalpur, Res. Bull. pp. 1-98.

Rattan, R.K.,Patel, K.P., Maujiah, K.M. and Datta, S.P.(2009). Micronutrient in soil, Plant and Human Health. J. Indian soc. of soil sci. 57: 546-558

Sagare, B.N. and Badhe, N.N. (1980). Effect of molybdenum alone and in combination with N and P on yield, uptake and ascorbic acid content of cauliflower, PKV Res. J. 4 : 28-30.

Sagare, B.N., Rewatkar, S.S. and Ghawghave, P.B. (1988). Micronutrient harvest by peanut as influenced by S and P application in vertisols. Ann. Plant. Physiol. 2 (2): 187-192.

Sahu, S.K. and Samant, P.K. (2006). Micronutrient management through organic farming. Orissa Review, April. 2006: 57-58.

Singh, M.V., Patel, K.P. and Ramani, V.P.(2003) Crop response to secondary and micronutrients in swell – shrink soils. Fertiliser News 48: 55-56.

Zende, G.K. (1995). Lead paper presented at the seminar on. "Management of problematic soils", organised by Akola. Chapter, ISSS, Dr. PDKV, Akola, Octo. 11-12. 1996.

Chapter - 5

MANAGEMENT OF SLOWLY PERMEABLE SOILS UNDER DRYLAND CONDITIONS

R.L. Kalane and V.K. Kharche

Deptt. of Agri. Chemistry and Soil Science, Dr. Panjabrao Deshmukh Krishi Vidyapeeth, Akola, 444 104 (MS)

* Prof. of Chemistry & Head, Soil Science, Dr. P.D.K.V. Agri. University, Akola - 440 104 (M.S.)

INTRODUCTION

India has about 108 million hectares of rainfed area which constitutes nearly 75% of the total 143 million hectares of arable land. India has about 47 million hectares of dry lands out of 108 million hectares of total rain fed area. Dry lands contribute 42% of the total food grain production of the country. Thus, dry lands and rainfed farming will continue to play a dominant role in agricultural production. Dryland generally receives less than 750 mm of rainfall in arid and semi-arid regions. However, in the present Indian context, dryland means unirrigated rainfed area. The emphasis in rainfed agriculture is often on disposal of excess water, maximum crop yield, high level of fertilizer inputs and water erosion constraints.

Vertisols and associated soils, *viz.* Inceptisols and Entisols in peninsular India are commonly known as black soils with clay content is (predominance of montmorillonite exhibiting swell shrink characters) ranging between 40-60% or more. These black soils are predominantly clay in texture (61.1%) in Maharashtra (Challa *et al.,* 1995). These clay soils are often characterized by low air capacity or drainable porosity (10 or less) following poor aeration and exhibiting ODR (Oxygen Diffusion Rate) value less than its critical limit (33.3×10^{-6} g O_2 ms^{-1}). These soils are also slow to very slow in respect of infiltration rates (<0.5 mm hr^{-1}) and are susceptible to water logging events during the rainy season under dryland conditions, as a result, crops suffer from poor aeration, and insufficient supply of oxygen.

Low drainable porosity and slow infiltration rates are the soil physical constraints which limit the crop production both under irrigated and dryland conditions. To overcome the soil physical constraints of crop production, it is of great importance to have the knowledge of management of these clay soils in the light of improving drainable porosity (macroporosity) and thereby enhancing the infiltration rate through incorporation of FYM, farm residues and amendments like gypsum. Secondly, by way of providing drainage through the land treatments viz. ridges and furrows, broad bed furrow, (BBF), raised and sunken beds.

For proper understanding of management aspects of these clay soils susceptible to waterlogging and facing a problem of poor aeration and insufficient oxygen supply to the crops, it is highly essential to have the back ground knowledge on the aspects viz.

1. Vertisols and associated soils
2. Climatic conditions
3. Characterization of black soils
4. Soil aeration
5. Soil respiration and aeration requirement
6. Soil air composition
7. Soil air and plant response.
8. Approaches for management.

1. VERTISOLS AND ASSOCIATED SOILS

The major area of Vertisols and associated soils are located in Australia (70.5 m ha), India (70 m ha), Sudan (40 m ha), Chad (16.5 m ha), and Ethiopia (10 m ha). These five countries contain over 80% of the total area (250 m ha) of Vertisols in the World (Dudal, 1965). In India, substantial Vertisol areas occur in Maharashtra, Madhya Pradesh, Gujarat, Andhra Pradesh, Karnataka and Tamil Nadu (Murthy, 1981). Most of these receive 500 to 1300 mm of annual rainfall, concentrated in a short period of 3 to 3.5 rainy months interspersed with droughts. Crop yields in this area are low and vary from year to year. Vertisols and associated soils, *viz.* Inceptisols and Entisols occur in peninsular India covering an area of about 72.9 m ha. They are generally known as 'black soils'. They constitute about 22.2% of the total geographical area of the country. The distribution of these soils in nine States in India is given in Table 1.

Table 1. Distribution of black soils in India

State	Area (m ha)
Maharashtra	29.9
Madhya Pradesh	23.0
Gujarat	8.2
Andhra Pradesh	7.2
Karnataka	6.9
Tamil Nadu	3.2
Rajasthan	2.5
Orissa	1.3
Bihar	0.7
	82.9

Source : R.S. Murthy (1981)

Maharashtra, having an area of 30.76 m ha comprising 10 per cent of the total geographical area of India is endowed with wide variations in physiography, climate, geology and natural vegetation which are responsible for development of a large variety of soils. The spatial distribution of major soil scapes (land form units) of vertisols and associated soils reveals that 84.1 per cent of state geographical area (25.8 m ha) is covered by Vertisols and it's associated black soils (Murthy *et al.,* 1982). Entisols (36.77%), Inceptisols (30.9%) and Vertisols, (26.2) are the predominent soils followed by Alfisols (5.6%) (Challa *et al.,* 1995). Typic Ustorthents, Vertic Ustropepts. Vertic Ustochrepts and Typic Haplusterts are the major sub-groups of Vertisols and its associates.

2. CLIMATIC CONDITIONS

Evaporation exceeds precipitation in the low rainfall region in all months while in the medium rainfall region, precipitation is in excess during August and September. In the high rainfall region, it is in excess from July to September. When a ratio of actual evapotranspiration potential evapotranspiration was calculated, the growing season (AE/PE>50%) was below 20 weeks for low rainfall regions, 20-30 weeks for the medium rainfall zones, and above 30 weeks for the high rainfall regions. In the low rainfall regions, the shallow black soils are cropped during the monsoon whereas the deep black soils are sown with post-monsoon crops. In the medium rainfall regions intercropping is practised, and in the high rainfall regions sequence cropping is possible. Some 15 m ha of vertisols exist in the last group, but almost 12 m ha are fallowed during the monsoon.

The vertisols region of Maharashtra has the monsoonic climate of semi-arid and sub-humid type with annual rainfall ranging between 400-1250 mm, mostly occurring during June to September in 45-60 rainy days. This region may be classified into two climatic sub-regions (Virmani *et al.,* 1981 and Challa *et al.,* 1995).

3. CHARACTERIZATION OF BLACK SOILS

Shankaranarayan and Sarma (1981) pointed out that the black soils characteristically have a clay content of 40-60% or more. The clay is dominated by montmorllonite, which is responsible for the problems and potentials of these soils. The soils swell, shrink and crack and are also easily dispersed. They have a high moisture retention capacity and potential fertility.

Infiltration rates of Vertisols vary from almost infinity to extremely low, due in part to the high coefficient of expansion and contraction. Large soil-volume changes accompany changes in soil-water content. Shrinkage curves of the Houston Black clay surface soil indicate volume changes of up to 40% by drying from field capacity to -15 bar water potential (Johnston and Hill, 1945). As drying proceeds, shrinkage cracks develop. These cracks may be 5-7 cm wide and 60-90 cm deep at the end of a dry growing season with row crops. While the cracks are open, the water intake rate is very high. Expansion of the swelling clay soil begin immediately upon wetting, with an attendant decline in infiltration rate. As the soil approaches saturation, the infiltration rate drops to 2-3.5 cm day^{-1} (Ritchine *et al.,* 1972). Therefore, precipitation rates in excess of about 2.5 cm day^{-1} will result in runoff after the swelling clay soils are near saturation. If the soil remains saturated with the infiltration rate of about 2.5 cm day^{-1}, poor internal drainage contributes to soil-aeration problems, with a consequent detrimental effect on plant growth. This situation occurs most frequently in the early spring growing season and particularly in the humid Vertisol region.

Infiltration of rain or irrigation water is initially high in dry soil, reaching values of 50-80 cm h^{-1}, if the surface soil has a well-developed mulch or a fine tilth. Infiltration is very uneven : the surface soil is thoroughly wetted, whereas the sub-soil receives water only through the cracks. Terminal infiltration rates, after the cracks have been sealed by swelling of the soil, can be extremely low, below 1.0 mm h^{-1} (Jewitt *et al.,* 1979). Vertisols have a low terminal infiltration rate (about 0.2

mm hr^{-1}). During the rainy season, water may create pond if drainage is inadequate and then crop suffers from water logging.

Vertisols regions of Maharashtra State

Drought prone, dry semi-arid area with undepedable and low rainfall	Rainfed, Semi-arid to sub-humid area with dependable rainfall
Mean annual rainfall is 750 mm (450 - 750 mm)	Mean annual rainfall ranges between 750-1250 mm or more
Irregular raining pattern	Reliability of rainfall occurrence at shorter intervals is higher
Frequent occurrence of long and repeated drought periods	Short periods of drought
Rainfall covers only 40-50 per cent of annual PET	Rainfall covers 50-60 per cent of Annual PET
Length of growing period (LGP) ranges from "90-150 days.	Length of growing period ranges from 150-180 days
Western Maharashtra dry zone, and western part of CMP zone.	Central Maharashtra plateau Zone, Central and Eastern Vidarbha Zone.
Total area 10.8 m ha.	Total area 15.07 m ha.

4. SOIL AERATION

The process of soil aeration is one of the most important determinants of soil productivity. Plant roots adsorb oxygen and release carbon dioxide in the process of respiration. In most terrestrial plants (excepting such specialized plants as rice), the internal transfer of oxygen from the parts above the ground (leaves and stems) to those below the ground surface (roots) cannot take place at a rate sufficient to supply the oxygen requirements of the roots. Adequate root respiration requires that the soil itself be aerated, that is to say, that gaseous exchange take place between soil air and the atmosphere at such a rate as to prevent a deficiency of oxygen and an excess of carbon dioxide from developing in the root zone. Soil micro-organisms also respire, and, under conditions of restricted aeration, might compete with the roots of higher plants (Stotzky, 1965). Gases can move either in the air phase (that is, in the pores which are drained of water, provided they are interconnected and open to the atmosphere) or in dissolved form through the water phase. The rate of diffusion of gases in the air phase is generally greater than in the water phase, hence soil aeration is dependent largely upon the volume fraction of air-filled pores.

Impeded aeration resulting from poor drainage and waterlogging, or from mechanical compaction of the soil, can strongly inhibit crop growth. In particular, the problem of soil compaction seems to have worsened in recent decades, along with the growing trend to use larger and heavier machinery and the tendency to trend over the field repeatedly for such purposes as fertilization and pest control. Moreover, with greater use of fertilizers and irrigation, shortage of nutrients and water have been obviated in many places so that more it is aeration which has become a major limiting factor to the attainment of maximal productivity. It seems likely that root systems are commonly restricted in extent by the progressive decrease of aeration in the deeper regions of the soil profile. Poor aeration can decrease the uptake of water and induce early wilting. According to Kramer (1956), restricted aeration also causes a decrease in the permeability of roots to water.

Anaerobic conditions in the soil induce a series of reduction reaction, both chemical and biochemical. Included among these reactions are denitrification (the processes by which nitrate is reduced to nitrite, then to nitrous oxide and eventually to elemental nitrogen: $NO_3 \rightarrow NO_2 \rightarrow N_2O \rightarrow N_2$ manganese reduction from the manganeous form); iron reduction from the ferric to the ferrous; and sulfate reduction to form processes are toxic to plants (e.g. ferrous sulfide, ethylene, as well as acetic, butylic, and phenolic acids).

5. SOIL RESPIRATION AND AERATION REQUIREMENTS

The overall rate of respiration due to all biological activity in the soil i.e. the amount of oxygen consumed and the amount of carbon dioxide produced by the entire profile – determines the *aeration requirement* of the soil. It is important to acquire quantitative knowledge of the aeration requirements of different crops and soils in varying circumstances if we are to devise means to ensure that these requirements are indeed met. However, the information is difficult to obtain, as the rate and spatial distribution of soil respiration, as well as its temporal variation, depend on numerous factors, included among which are temperature, soil wetness, organic matter, and the time-variable respiratory activity of both macro- and micro-organisms (Alexander, 1961) *Anaerobiosis*, or oxygen stress, will occur in the soil whenever the rate of supply falls below the demand. This condition can develop quite quickly, since the storage of oxygen in the soil is generally rather low in relation to the quantity required for soil respiration. To illustrate, let us consider a soil with an effective root zone depth of 60 cm and 15% air filled porosity, containing 90 litre of air under each square meter of soil surface. With an initial oxygen concentration of 20% in the gaseous phase, the storage of oxygen can be calculated to be 18 litres, equivalent to about 25 g. If the oxygen requirement for soil respirations of the order of 10 g/day, per square meter of ground, the initial oxygen reserve in the soil would last only 2½ day. Oxygen stress symptoms would probably begin even earlier. However, these figures are given only to provide an order of magnitude, since in actual conditions the aeration rate probably varies between wide limits. Plant growth probably depends more upon the occurrence and duration of periods of oxygen deficiency than upon average conditions (Erickson and Van Doren, 1960).

Soil respiration values reported by various investigators have varied widely. Papendick and Runkles (1965), working under laboratory conditions, found respiration rates of 1.7×10^{-11} mole/cm^3 sec, equivalent to 0.75 mole of oxygen per cubic meter per 12-hr day, or 24 g/m^3 day. At the other extreme, Grable and Siemer (1968) reported values 10,000 times higher. Under filed conditions, respiration values are likely to be smaller than those found in growth chambers.

6. COMPOSITION OF SOIL AIR

This is relatively stable, due to the establishment of local equilibrium between two processes : Oxygen consumption and Carbon dioxide production on the one hand, and gaseous exchange with surroundings on the other. In a well aerated soil, the gas phase contains 18 to 20.5% Oxygen, but this can drop to 10% after rain, or even as low as 2% around roots or in poorly structured zones. As carbon dioxide is produced within the soil, its concentration there is higher than in the open air (0.03%), and increases at the expense of Oxygen concentration. On average it is in the range of 0.3 to 3%, but may reach 5 to 10% or higher (e.g. in the rhizosphere, or following recent incorporation of plant residues), and increases with depth. Nitrogen concentration is very close to its free atmosphere level (about 79%). Various other gases can be found in the Soil : N_2O, argon, H_2, H_2S, methane, etc. (the later three under strongly anaerobic conditions).

In a well aerated soil, the composition of soil air is close to that of the external ("open") atmosphere, as the oxygen consumed in the soil is readily replaced from the atmosphere. Not so in a poorly aerated soil. Analyses of the actual composition of soil air in the field reveal it to be much more variable than the external atmosphere. Depending on such factors as time of year, temperature, soil moisture, depth below the soil surface, root growth, microbial activity, pH, and above all the rate of exchange of gases through the soil surface, soil air can differ to a greater or lesser degree from the composition of the external atmosphere. The greatest difference is in concentration of carbon dioxide (CO_2), which is the principal product of aerobic respiration by the roots of higher plants and by numerous macro-and micro-organisms in the soil. The CO_2 concentration reaches levels which are ten or even one-hundred times greater.

As CO_2 is produced in the soils by oxidation of carbonaecous (organic) matter, an increase of CO_2 concentration is generally associated with a decrease in elemental oxygen (O_2) concentration (though not necessarily to an exactly commensurate degree, since additional sources of oxygen may exist in dissolved form and in easily reducible compounds). Since the O_2 concentration of air is normally about 20% it would seem that even a hundred fold increase of CO_2 concentration, from 0.03 to 3% can only diminish the O_2 concentration to about 17%. However, even before they begin to suffer from lack of oxygen *per se*, some plants may suffer from excessive concentration of CO_2 and other gases in both the gaseous and aqueous phases. In more extreme cases of aeration restriction, the O_2 concentration can fall to near zero and prolonged anaerobic conditions can result in the development of a chemical environment characterized by reduction reactions such as denitrification, the evolution of such gases as hydrogen sulfide (H_2S), methane (CH_4), and ethylene, and the reduction of mineral oxides (such as those of iron and manganese). Another interesting difference between the atmosphere and soil air is that the later is characterized by a high relative humidity, which nearly always approaches 100% except at the soil surface during prolonged dry spells.

7. SOIL AIR AND PLANT RESPONSE

Plant response to deficient soil aeration depends on the plant and on certain environmental factors, particularly temperature. A wide spectrum of plants, from hydrophytes to mesophyts to xerophytes, have adapted to different levels of soil aeration. Hydrophyts do well with permanent flooding while mesophytes and xerophytes require an increasing amount of air for well adoption. Most of the negative phenomena associated with deficient aeration are caused solely by lack of oxygen for metabolic processes.

Table 2. ODR values for emergence of various crops (x 10^{-10}g/cm^2/min)

Plant	ODR_L	$ODR_{0.5}$	ODR_C
Barley	25	17	8
Oats	30	20	12
Beans	33	25	12
Wheat	40	–	8
Maize	40	27	16
Tomato	40	30	25
Sugar-beet	50	30	13
Rye	50	33	12

ODRL = limiting value; ODR0.5 = half limiting value and ODRc = critical value.

These values vary, however, with the crop varieties.

ODR requirement of various crops are given in Table 3.

Many plants, e.g. wheat, carrot and onion, germinate well even at low oxygen concentration while plants such as rice and cucumber preserve germination ability even in completely anaerobic conditions. There are also plants, such as tomato, grasses etc. which show an increase in germination ability with an increase in oxygen concentration throughout practically the whole range of concentration found in soil air. Threshold Oxygen Diffusion Rate (ODR) values for emergence of various crops are reported in Table 2.

In plants which can survive anaerobic conditions, the root growth has been observed to decrease with a decrease in soil aeration, e.g. tomato, maize, barley, wheat, pea, sugar-beet and mustard. Under anaerobic conditions the root mass of flood-sensitive plants decreased earlier and to a higher degree than the mass of the shoot. A completely anaerobic condition stops root growth entirely within 2-3 minutes. This has been observed in cotton and soybean. When the stress did not last longer than 30 minutes the growth rate subsequently returned to normal. But this condition lasting 3 hours for cotton and 5 hours for soybean killed the tap roots. Roots of plants which are very sensitive to anaerobic conditions, such as tobacco and legumes, die after a few days of flooding. But the adverse effect appears after 1-2 hours and permanent damage may appear after 24 hours, especially on warm sunny days. Roots growing in an environment with restricted oxygen supply are thicker, shorter and less branched. In the case of wheat and desert shrubs, ODR values of 40-50 and 30-35 were respectively found to be limiting and critical. In general, root initiation reduces or ceases when the ODR in soil is in the range 18-23 while an ODR above 23 seems to permit relatively good root growth. Shoot growth is reduced with reduced soil oxygen. The general tendency for plants under adverse conditions to lengthen but not to fill in is commonly observed under field conditions.

Table 3. Oxygen Diffusion Rate (ODR) requirement of various crops.

Crop	ODR (x 10^{-10} g/cm^2/mn)
Paddy	2.0
Maize	20.0
Mustard	22.9
Wheat	22.9
Soybean	28.4
Peas	39.5

8. APPROACHES FOR MANAGEMENT OF SLOWLY PERMEABLE SOILS

Management aspects of clay soils susceptible to water logging, low in drainable porosity and slow infiltration rates are followed through two approaches as :

a. Increasing drainage porosity and/or infiltration rates through incorporation of FYM, farm residues and gypsum.

b. Providing drainage through land treatments like ridges and furrows, broad bed furrow (BBF) and raised and sunken beds.

(c) Enhancing Soil Organic Matter (SOM)

Soil organic matter is of central importance in maintaining soil quality and also has potential for carbon sequestration in soils. Soil organic carbon (SOC) levels are influenced by management practices. Many management practices that are effective in increasing SOC are also effective in improving crop yields. The actual amount of soil carbon that can be stored is dependent on the cropping system (management practices), soil type and climatic conditions, as well as the initial soil carbon level of the site. SOC promotes soil structure by holding the soil particles together as stable aggregates, improves soil physical properties such as water holding capacity, water infiltration, gaseous exchange, root growth and ease of cultivation.

There are a wide range of management options and cropping systems that can increase SOC levels. It can be raised by direct additions of organic materials, composts, manure and other recycled organic materials. Green manuring, crop residue recycling have considerable potential for improving SOC status. Increased carbon input in agro-ecosystems can be achieved in a number of ways such as selection of high biomass producing crops, residue recycling or residue retention by lessened tillage intensity, application of organic materials (e.g. animal manure, compost, green manure etc.), adoption of agro-forestry systems, intensification of agriculture through improved nutrient and water management practices, reducing fallow and changing from monoculture to rotation cropping. Soil carbon loss could be decreased by adopting conservation agriculture and minimizing soil disturbance, checking erosion through reduced tillage intensity, and using low quality organic inputs. Since no single land management strategy in isolation may be adequate to enhance carbon sequestration, it is important to evaluate the combination of various land management strategies.

Technological options that have been found to be efficient for soil C sequestration in Indian agro-ecosystems include integrated nutrient management and manuring, soil crop residue incorporation, mulch farming and/conservation agriculture, agroforestry systems, grazing management, choice of cropping systems and intensification of agriculture. Integrated nutrient management (INM) involving addition of organic manures/ compost along with inorganic fertilizers results in improved soil aggregation and greater carbon sequestration specially in macro-aggregates (Benbi and Senapati, 2009). Incorporation of organic manures include decomposition of organic matter where roots, hyphae and polysaccharides bind mineral particles in to micro-aggregates and then these micro-aggregates bind to form C rich macro-aggregates. This type of C is physically protected within macro-aggregates. The free primary particles are cemented together into micro-aggregates by persistent binding agents characterized by humification of organic matter and stimulate accumulation of C in aggregates. Cover crops improve carbon sequestration by enhancing soil structure, and adding organic matter to the soil.

Kharche et al (2010) observed gradual changes in physical properties of sodic swell shrink soils of Maharashtra. Bulk density after two years field experimentation decreased while hydraulic conductivity increased due to gypsum as well as its combined use with spent wash press mud compost. Hydraulic conductivity improved from 0.45 to 0.81 cm h^{-1} during reclamation. Replacement of sodium due to addition of gypsum gradually helped in flocculation process resulting in better aggregation, which causes decrease in bulk density and improvement in hydraulic conductivity of soil. The reduction in bulk density was also due to regeneration of structure caused by improved aggregation with the addition of organic matter present in compost. This suggests that the integration of chemical amendment and organic conditioner along with bio-inoculants was found more beneficial in improving air-water relationship in sodic soils.

The magnitude of increase in hydraulic conductivity due to gypsum is very high in sodic soils of alluvial plains in northern states of India (Verma *et al.,* 1985) while the improvement in black sodic swell shrink soils is very slow. Sagare *et al.* (2001) also reported the lower extent of increase in hydraulic conductivity of sodic Vertisols from 0.41 to 1.05 cm h^{-1} due to gypsum. Kharche et al (2010) further revealed that the added advantage of increase in hydraulic conductivity observed in their study under the combined use of compost with gypsum can be ascribed to the dissolution of calcite which releases more calcium in addition to the calcium from gypsum causing flocculating effect. Use of spent wash press mud compost improves the efficiency of reclamation of calcareous cracking clay sodic soils and also can act as an alternative to gypsum.

(d) Managing degraded black saline-sodic soils

The saline-sodic cracking clay soils of Deccan plateau in semi arid regions of India are difficult to manage owing to their dual problem comprising of sodicity coupled with very slow permeability. Kharche et al (2012) observed that despite a low level of sodicity (ESP 5.0-11.5), soils of Purna valley in Vidarbha region of Maharashtra in central India pose severe drainage problems due to low hydraulic conductivity. Very high amount of clay in these soils (50-70%) owing to its predominant smectitic nature leads to increase in bulk density within polygonal cracks resulting into hard and compact soil structure. As an effect of climate change the rainfall has been observed to gradually decrease over a period of time in this area and the rainy days are drastically reduced with prolonged dry spells coupled with increase in temperature. This is causing further increase in soil degradation and there has been increase in pH in subsoil associated with increase in ESP towards lower horizons with concomitant decrease in exchangeable Ca/Mg ratio.

Pal et al (2012) reported that these vertisols of dry climates have poor drainage, but they show no salt-efflorescence on the soil surface as evidence of the threat of soil sodicity. These soils also do not qualify as salt-affected soils as per the United States Salinity Laboratory criteria; however, the sHC of their sub-soils is adversely affected by clay dispersion caused by exchangeable magnesium (Vaidya and Pal, 2002). This point suggests that the saturation of vertisols, not only with Na^+ ions but also with Mg^{2+} ions, blocks small pores in the soil. They further reported that, the current lower limit of a 15 ESP by the United States Salinity Laboratory for all soils is arbitrary and necessitates the evaluation of a lower ESP limit. Pal et al. (2006) advocated a value of sHC less than 10 mm h^{-1} instead of ESP or SAR. As regards their management therefore, improving drainage porosity is most crucial.

Although the use of chemical amendments like gypsum, iron pyrite has been advocated for their reclamation, the on farm available crop residues and green manuring have shown promising results. The crops having potential of biomass addition to the soils like green gram, chickpea, pigeon pea act as cover crops and enhance the process of soil reclamation and also improve the efficiency of chemical amendments. The hydraulic conductivity and mean weight diameter of sodic black soils has been found to improve due to growing of green manuring crops in between two rows of cotton and subsequently burying them in situ. This has also increased the soil organic carbon stock which is often low under only gypsum application. The widespread deficiency of nitrogen, sulphur, zinc in these soils can also be overcome by green manuring besides reclamation. These soils thus need simultaneous attention in respect of drainage, pH, Ca/Mg ratio, ESP, emerging deficiency of nutrients and soil organic carbon levels so as to enhance and sustain their quality.

9. INCREASING DRAINAGE POROSITY AND/OR INFILTRATION RATE

Incorporation of FYM, farm residues helps in enhancing the macroporosity of soils with low drainage porosity and thereby infiltration increases resulting in avoiding the problem of crops suffering from poor aeration and insufficient supply of oxygen.

Kalane *et al.* (1995) reported incorporation of FYM wheat straw and green manuring supplying 50% N combined with 50% NPK through fertilizers showed increase in infiltration rates respectively to 2.39, 1.33 and 1.90 cm hr^{-1} from 0.70 cm hr^{-1} under control.

Badanur *et al.* (1990) reported that incorporation of sorghum stubbles @ 5 t ha^{-1} or safflower stalk @ 5 t ha^{-1} enhanced infiltration rates respectively to 1.5 cm hr^{-1} and 1.6 cm hr^{-1} from 0.5 cm hr^{-1} under the plots receiving only nutrients through fertilizers on Vertisols. Kalane *et al.* (1994) found significantly highest yield (14.75 q-ha^{-1}) of soybean under incorporation of gypsum @ 5t ha^{-1} in Vertisols charactersized by poor aeration porosity (8.45%) and slow hydraulic conductivity class (8.05 x 10^{-5} cm^{-1}) s followed by the yield (14.06 q ha^{-1}) at par under the treatment FYM @ 5t ha^{-1} and lowest under control (9.98 ha^{-1}). This may be due to better drainage conditions avoiding waterlogging under gypsum and FTM incorporation treatment in the plots. They further, reported ODR value (29.56 x 10^{-6} g O_2 $m^{-2}s^{-1}$) which is lower than its critical value 33.3 x 10-6g O^{-2} s^{-1} sunder control.

10. MEASURES FOR PROVIDING DRAINAGE THROUGH LAND TREATMENTS TO AVOID WATER LOGGING

Kharif crops generally suffer or even fail due to excess of moisture in the root zone. The interchannel area on flat lands should be suitably treated to get rid of the active root zone of excess water. The practices which have been found suitable are : (a) narrow ridge and furrow system, (b) broad ridges and furrow system, (c) graded furrows and (d) raised and sunken bed systems.

Planting of upland *kharif* crops on ridges, laid out along a general slope less than one per cent on vertisols of low permeability in central parts of Madhya Pradesh have been observed to increase soybean, sorghum and maize yields (Tomar *et al.,* 1978, Gupta *et al.,* 1979) by 30 to 100 per cent (Table 4).

Table 4. Effect of ridge planting on grain yield of soybean, maize and sorghum grown on vertisol at Jabalpur (MP)

Crop/year	Slope (%)	Grain yield (kg/ha)		Increase (%)
		Flat	Ridge	
Soybean (1973)	0.3	1890	2900	53.4
	0.6	1740	2650	52.3
	1.2	2390	3030	26.7
Maize (1977)	0.3	1601	3376	106.0
	0.6	1704	3247	88.0
	1.2	2642	3644	38.0
Sorghum (1977)	0.3	2638	3608	38.0
	0.6	2445	3754	54.0
	1.2	2717	3553	30.0

Work at ICRISAT and Indore centre of AICRPDA has demonstrated that 90 to 240 cm broad ridges (depending on crop and row spacing) made with furrow openers attached to tool bar are easier to maintain and work with and are as satisfactory as narrow ridge and furrow systems.

Observations in the Indo-U.K. Operational Research Project at Indore (Verma, 1982) have shown that furrows opened on a grade of 0.2 to 0.3 per cent serve as surface drains and let out excess water safely. The furrows should have a cross section of 0.10 to 0.15 m^2 and be spaced at 8 to 10 m intervals depending upon crops and slope of land. Such furrows can be easily constructed by a farmer with a piece of wooden plank has been tied.

The system developed at J.N. Agricultural University, Jabalpur (Gupta *et al.,* 1978) consists of an array of raised and sunken beds laid out side by side. Runoff from raised beds planted to an upland crop is collected in the adjacent sunken beds planted to an upland crop is collected in the adjacent sunken beds supporting a rice or any waterlogging tolerant crop. Generally bed width is kept equal to calculated drain spacing, whereas the elevation difference between raised and sunken beds is maintained at 20 to 30 cm. The rainfall and soil characteristics are the deciding factors for the specification of beds. About 6 to 9 meter wide raised beds and 6 m wide sunken beds with 25 to 35 cm elevation difference have been observed to significantly improve grain yield of upland crops at Jabalpur (Table 5) while 8 m wide raised beds and 4 m wide sunken beds with elevation difference worked successfully presents the data performance of configurations with of 15-20 cm have at Indore. Table 6 on the relative different land respect to crop performance and field water balance (Anonymous, 1989-90) under relatively low seasonal rainfall situation.

- a system of resource conservation, management and use based on small watersheds,
- a well designed semi permanent or permanent broadband and furrow system or well graded flat system.
- land smoothing and installation of drains to ensure effective surface drainage and runoff disposal from the field.
- performing the primary tillage operations for both post-rainy as well as rainy season crops immediately after harvesting the previous seasons crop, when the soil has adequate moisture and is friable;

Table 5. Seed yield of Soybean (kg/ha) grown on raised beds and that or rice grown in sunken beds at Jabalpur

Width of raised beds (m)	Years				
	1979	1980	1981	1982	1983
6	1373	2962	1830	2230	2340
9	1315	2751	2931	2003	2410
12	1233	2621	1484	2161	2334
15	1197	2387	1587	2200	2305
Conventional method (Control)	Crop failed	2229	1050	1054	817
Rice in sunken beds	Crop failed	5100	Crop failed	2030	1400
Annual rainfall (mm)	478	1531	674	1246	1444

Source : Annual progress reports of All India Coordinate Research Project for Improvement of Soil Physical Conditions. Deptt. of Soil Science. J.N. Krishi Vishwa Vidyalaya, Jabalpur (M.P.)

The components of improved land management strategy for vertisols include.

- shallow cultivation of the land (only the beds) whenever an effective rainfall (20 mm) is received.
- use of minimum tillage and stubble mulch residue management;
- application of moderate amounts of N and P fertilizer, followed by dry sowing about one week before the expected date of onset of the rainy season.
- use of high yielding varieties and other recommended agronomic practices;

Table 6. Influence of different land configurations on soybean seed yield and field water balance at Indore during 1989.

Land configurations	Rainfall (mm)	Runoff (mm)	Deep percolation (mm)	Flux to root zone (mm)	Seed yield (kg/ha)	Increase in seed yield (%)
Flat	534	11	80	45	1778	–
Raised beds	534	Nil	78	101	1999	12.4
Sunken beds	534	Nil	112	108	1834	3.2
NBF (1.05 m)	534	5	63	39	1831	3.0
BBF (1.5 m)	534	3	65	18	1881	5.8

(Kanwar, 1981 and Kampen, 1982)

Using the above techniques, Vertisols have been managed successfully at ICRISAT to produce high and stable yields even under very fluctuating rainfall conditions. Various aspects of this approach for the management of Vertisols have been discussed in detail by El-Swaify *et al.,* (1985), Kanwar (1981), Kampen (1979) and Kampen (1982).

Table 7. Grain yields under improved and traditional technologies on a Vertisol at ICRISAT in 12 successive years

Year	Rainfall during cropping period	Grain yield (t ha^{-1}) Improved system double cropping		Traditional system, single crop	
		Sorghum/ Sequential Maize	Chickpea intercropped pigeonpea	Sorghum or Chickpea	
1976-77	708	3.20	0.72	0.44	0.54
1977-78	606	3.08	1.22	0.38	0.87
1978-79	1089	2.15	1.26	0.56	0.53
1979-80	715	2.30	1.20	0.50	0.45
1980-81	751	3.59	0.92	0.60	0.56
1981-82	1073	3.9	1.05	0.64	1.05
1982-83	667	3.27	1.10	0.63	1.24
1983-84	1045	3.05	1.77	0.84	0.48
1984-85	546	3.36	1.1	0.69	1.23
1985-86	477	2.70	0.73	*	0.84
1986-87	585	4.45	0.38	0.37	1.27
1987-88	841	4.26	1.35	0.80	0.92
Mean	776	3.20	1.10	0.60	0.80
SD	200	0.70	0.40	0.20	0.30
CV (%) (0.05)	26	21	34	27	38

* No crop sown.

In addition to its direct benefit to crop yield, improved soil management has been more effective than the traditional rainy season fallow system in reducing runoff and soil erosion (Pathak *et al.,* 1985). On an average, the improved system reduced annual runoff to one-third, soil loss to one-eleventh and peak runoff rate to one-half when compared with traditional rainy season fallow system. As shown in Table 8, the performance of the broadbed and furrow system was consistently superior to the traditional system in reducing annual runoff, soil loss and peak runoff rate. In 1977 when rainfall was very low and moisture conservation was crucial, the broadbed and furrow system conserved most of the annual rainfall, producing an annual runoff of only 1 mm. In contrast, when removal of excess water by drainage was crucial, during the high rainfall years of 1975 and 1978, this system produced substantial runoff of 162 and 273 mm, respectively. The better control on runoff in the improved system was due to good crop cover during the rainy season, periodic land smoothing, slope adjustment of controlled runoff velocity as it flows through many furrows and disposed through grassed waterways.

Plate 5.1. 1. Ill managed canal leading to seepage spoiling adjacent land. It creates slowly permeable soils unfit for cultivation. **2.** Mis-managed canal made another problem of swampness in down hill lands (4 m down of canal). Only grasses may grow around the area.

Table 8. Annual rainfall, runoff, soil loss, and peak runoff rate for a cropped vertisol with broadbed and furrow system and a traditional monsoon fallow system (1975-80)

	Broadbed and furrow at 0.6% slope,				Traditional flat, monsoon fallow		
Year	Rainfall (mm)	Runoff (mm)	cropped Peak runoff Rate ($m^3.S^{-1}.ha^{-1}$)	Soil loss ($t.ha.^{-1}$)	Runoff (mm)	Peak runoff Rate ($m.^3S.^{-1}ha^{-1}$)	Soil loss ($t.ha.^{-1}$)
1975	1041	162	0.06	1.39	253	0.15	5.21
1976	687	73	0.09	0.98	238	0.16	9.20
1977	585	1	0.01	0.07	53	0.06	1.68
1978	1125	273	0.11	2.93	410	0.15	9.69
1979	690	73	0.08	0.70	202	0.15	9.47
1980	730	116	0.06	0.97	166	0.11	4.58

In the both, the traditional and improved management systems, the sediment load per unit runoff decreased as the crop cover increased. The broadbed and furrow system performed better in the disposal of excess water than the flat system because of lower values of sediment load in the runoff. This is highly desirable on vertisols which often have waterlogging conditions when disposal of runoff is essential with least soil loss.

Plate 5.2. Soil profile at Bhopal (M.P.) 3 m. Depth from surface.
Note various layers of soil in alfisol under very shallow conditions.

The data presented clearly show that the improved management system for vertisol is superior to the traditional monsoon fallow system both in increasing productivity and in enhancing the effective use and conservation of soil and water resources.

For slowly permeable soil and deep Vertisols, the recommendation of broad bed and furrow has been cited by Varade (1992) focusing that wide spread adoption of these practices by the farmer is lacking. Kalane *et al.* (1990) reported significantly highest grain yield (2530 kg ha^{-1}) of Hy. Sorghum under vertical mulch on vertic Ustochrept and it was at par with that (2354 kg ha^{-1}) broad bed furrow (BBF) because of better drainage condition under these land treatments. Kalane *et al.* (1993) obtained highest grain yield of 18.4 q ha^{-1} from plot with 6 m drain spacing on Vertisols characterized by poor aeration porosity (9.1%) and slow permeability (4.694 x 10^{-5} cm S^{-1}) and it was at par with 9 m drain spacing. ODR values were observed significantly lowest at 10, 20 and 30 cm depths under control with no drain.

Kool *et al.* (1992) reported improvement in infiltration characteristics of Vertisol by Deep cultivation and application of chemical amendments up to depth of 25 to 40 cm. Generally chemical like gypsum and pyrite reclaim the soil to a depth of 20 to 40 cm and therefore, influence on properties of the soil in lower layers is negligible. Even application of pyrite did not cause any appreciable changes in the properties of soil in lower layers. Kool *et al.* (1994) observed reclaim of sodic Vertisol by providing filter drain to a depth of 1 m in the Nimar region of Madhya Pradesh at 9, 12 & 15 m spacing by growing cotton on raised bed and rice in sunken beds (1 : 1 ratio).

Waterlogging or water stagnation is a more serious problem than erosion on flat lands with less than 0.5 per cent slope particularly on deep black soils which have low basic infiltration rate. The effective solution to the problem is possible only if provision for both total land drainage and crop drainage is made. The land treatment described have been found useful. Safe disposal of runoff water from areas between waterways and from fields between two crests to a grassed waterway is essential. For this, open drainage channels on a grade are to be constructed at suitable intervals from ridge to grassed waterway. The cross section of drainage channels will depend upon the rate of runoff and the area of its catchment. Depending upon the difference in the bed level of the grassed waterway and those of fields to be drained, rams or suitable fall structures have to be provided. Oswal (1994) cited for management of slowly permeable soils by placing tile drains at 0.9 - 1.2 m depth and 9-21 m spaced.

CONCLUSIONS

Vertisols and associated soils particularly clay in texture are low in macroporosity and slowly permeable, as a result these clay soils are susceptible to water logging during kharif season. Crop production on such soil is limited due to poor aeration and insufficient supply of oxygen. These slowly permeable soils can well be managed through approaches viz. (i) incorporation of FYM, farm residues, gypsum etc. (ii) land treatments like ridges and furrow, broad bed furrow and raised and sunken bed (iii) by providing subsurface drainage.

REFERENCES

Alexander, M. 1961, 1977. "Introduction to Soil Microbiology", Wiley, New York.

Anonymous, 1989-90. Annual progress report of AICRP dryland agric. JNKV College of Agril. Indore.

Badanur, V.P., C.M. Poleshi and Balachandra, K.N. 1990. Effect of organic matter on crop and physical and chemical properties of vertisols 38 (3), pp. 426-429.

Benbi, DK and Senapati, N. (2009) Soil aggregation and carbon and nitrogen stabilization in relation to residue and manure application in rice-wheat systems in north-west India. *Nutrient Cycling in Agro-ecosystems* 87: 233-247.

Challa, O, Sehgal, J.L. S.S. Vadivelu 1995. In Abstract National Workshop Soil Resource Inventory for Perspective Land Use Planning Oct 15-18, 1995, Nagpur.

Challa, O., Vadiveu, S. and Sehgal, J. 1995. Soil of Maharashtra for optimising land use, NBSS, Publ. No. 54.

Deshmukh, A.S., R.L. Kalane, P.G. Babrekar, P.T. Kharkar and A.K. Tiwane 1993. Soil moisture regime, oxygen diffusion and yield of soybean on Vertisols with surface drainage of varying drain spacings. Ann. Plant Physiol. 7 (2) pp 225-229.

Dudal, R. 1965. Dark clay soils of tropical and subtropical regions. Agricultural Develop paper No. 83, Rome Italy : FAO.

El-Swaify, S.A., Pathak, P., Rago, T.J. and Singh, S. 1985. Soil management for optimized productivity under rainfed conditions in the semiarid tropics.

Erickson, A.E., and Van Doren, D.M. 1960. The relation of plant growth and yield of soil oxygen availability. Trans. Int. congr. Soil Sci., 7th, Madison, Wisconsin 3, 428-434.

Grable, A.R. 1968. Advances in Soil Science 1 : 1-64 E.G. Effects of bulk density, aggregate size, and soil water suction on oxygen diffusion, redox potentials and elongation of corn roots. Soil Sci. Am. Proc. 32. 180-186.

Gupta, R.K., Sharma S.G., Tembe, G.P. and Tomar S.S. 1978. A new approach in farming system for problem soils of Madhya Pradesh JNKV Res. J. 12 pp. 73 - 79.

Gupta, R.K., Tomar S.S. and Tomar A.S. 1979. Improved soil management practices for maize and sorghum grown on vertisols of central India J. Agron and Crop Sci. 148 pp 479 - 483.

Jewitt, T.N., Law, R.D. and Virgo, K.J. 1979. Vertisol soils of the tropics and subtropics : their management and use. Outlook on Agriculture. 10 (1) : 33 - 40.

Johnston, J.R. and Hill, H.O. 1995. A study of the shrinking and swelling properties of Rendzing. Soils Sci. Soc. Am. Proc. 9 : 24-29.

Kalane, R.L., A.S. Deshmukh and P.T. Kharkar 1995. Effect of gypsum, sand, FYM and farm residues on soil moisture regime, oxygen diffusion and yield of soybean on vertisols. In extended summaries - National Seminar on Recent Development in Soil Science P.A.U. Ludhiana Nov. 2-5, 1995 pp 47-48.

Kalane, R.L., B.N. Sagare, Y.S. Guhe and R.A. Marathe 1990. Moisture use pattern and yield of Hy. Sorghum under different moisture conservation practices on vertic Ustochrepts. Ann. Plant physiol 4 (1) pp 1-8.

Kampen, J. 1979. Watershed management and technology transfer in the Semi-Arid Tropics. Proc. Int. Symp on Development and Transfer of Technology, ICRISAT, Patancheru, P.O. A.P. India pp 111-120.

Kampen, J. 1982. An approach to improved productivity on deep vertisols. ICRISAT Information Bulletin II, Patancheru, P.O. A.P. India.

Kanwar, J.S. 1981. Problems and potentials of the black soil of India : Some suggestions for an action plan. In proceedings of the seminar on Management of Deep Black Soils for Increased Production of cereals, pulses and oil seeds, New Delhi 21 May 1981, ICRISAT P.O. Patancheru, (A.P.), India.

Karmer, P.J. 1956. Physical and Physiological aspects of water absorption In "Handbuch der pflanzen physiologie, Vol. III, Pflanze and Wasser, pp. 124-159. Springer - Verlag, Berlin and New York.

Kharche, V. K, Patil, S. R., Belur, S. V. and Ghogare, N. S. (2010) Integrated use of gypsum, spent-wash press-mud compost and bio-inoculants for reclamation of sodic swell shrink soils of Maharashtra, *Journal of Soil Salinity and Water Quality* 2(2), 110-115.

Kharche, V. K., Kadam, Y. B., Konde, N. M., Katkar, R. N., Taley, S. M. (2012) Soil degradation assessment by modified diagnostic criteria for site specific resource management in Purna valley of Vidarbha region in Maharashtra, In: Proc. State Level Seminar, 'Soil Resource Management for Resilient Agriculture under Changing Climate', March 16-17, 2012, MAU, Parbhani, pp76.

Kool, Y.M., N.C. Shrivastava, R. Gupta and D.H. Ranade 1994. Improvement in Crop yield with provision of filter drainage system. J. Ind. Soc. Soil Sci. 42 (3) pp. 351-356.

Kool, Y.M., Shrivastava, N.C. and Gupta R.K. 1992. 5th Int. Workshop Drainage, WAPD Lahore Pakistan pp 1-14.

Murthy, R.S., Bhattacharjee, J.C., Lande, R.J. and Pofali, R.M. 1982. Distribution, characteristics and classification of Vertisols. Transactions, 12th International congr. Soil Sci. Symposia paper No. II pp 3-22.

Oswal, M.C. 1994. Soil Physics, Publ. Oxford & IBH Publ. Co. Pvt. Ltd. New Delhi P. 193.

Pal D.K., Wani, S.P., Sahrawat, K.L. (2012) Vertisols of tropical Indian environments: Pedology and edaphology, *Geoderma* 189-190 : 28–49

Pal, D.K., Bhattacharyya, T., Chandran, P., Ray, S.K., Satyavathi, P.L.A., Raja, P., Maurya, U.K., Paranjape, M.V. (2006). Pedogenetic processes in a shrink–swell soil of central India. *Agropedology* 16, 12–20.

Papendick, R.I. and Runkles, J.R. 1965. Transient state oxygen diffusion in Soil. I The case when rate of oxygen consumption is constant soil sci. 100, 251-261.

Pathak, P., Miranda, S.M. and El-Swaify, S.A. 1985. Improved rainfed farming for the semiarid tropics. Implications for soil and water conservation. In soil Erosion and Conservation (eds.) S.A. El-Swaify W.C. Modenhauer and L. Andrew. Soil conservation Soceity of America, Ankeny, Lowa, pp 338-354.

Ranade, D.H., Shrivastava, N.C., Gupta, R. and Verma S.K. 1992. J. Ind. Soc. Soil, Sci. 40 356.

Ritchie, J.T., Kissel D.E. and Burnell E, 1972. Water movement undisturbed swelling clay. Soil science soc. Am. Proc. 36 : 874- 879.

Sagare, B. N., Babhulkar, V. P. and Vaidya, P. H. 2001. Effect of different amendments on the properties of sodic Vertisols and yield of sorghum in the Purna valley, Maharashtra. *Agropedology*, 11: 61-65.

Shankaranarayana, H.S. and Sarma, V.A.K. 1981. Soils in India pp 41-70 in bench mark soils of India, Nagpur, NBSS & LUP (ICAR).

Stotzky, G. 1965. Microbial respiration. In ''Methods of soil Analysis'', pp. 1550-1569. Monograph 9, Am. Soc. Agron. Madison, Wisconsin.

Tomar, A.S., Tomar, S.S., Gupta, R.K., Sinha, S.B. and Motiranani D.P. 1978. Effect of surface drainage and system of planting on soybean yield Ind. J. Agric. Sci. 48 pp 38 -40.

Vaidya, P.H., and Pal, D.K., 2002. Microtopography as a factor in the degradation of Vertisols in central India. *Land Degradation and Development* 13, 429–445.

Varade, S.B. 1992. Soil management for sustainable productivity. J. Indian Soc. Soil Sci. 40 (1) pp 25-38.

Verma, G.P. 1982. Land and water management in Semi arid rainfed areas : Review of Soil Research in India part II 12th Int. Congr. Soil Sci. New Delhi Feb 8-16, 1982 pp 666-675.

Verma, S.K., Gupta, R.K. and Dubey, S.K. 1985. Comparative effectiveness of some chemical amendments in reclamation of sodic Vertisol. *Journal of the Indian Society of Soil Science* 33 : 948-95.

Viramani, S.M., Sivakumar, M.V.K. and Reddy, S.J. 1981. Rainfall probability estimates for selected locations of semi arid India. ICRISAT, Res. Rep. I. Hyderabad, India, p. 128.

Chapter - 6

ORGANICS IN REHABILITATION OF DEGRADED LANDS

V.N. Deshmukh, B.N. Akhare and M.B. Nagdeve

All India Coordinated Research Project for Dryland Agriculture, Dr. Panjabrao Deshmukh Krishi Vidyapeeth, Akola-444 104, India

INTRODUCTION

When land becomes degraded, its productivity declines unless steps are taken to restore the productivity and check further losses. Degraded land consumes labour and capital inputs into production. However, assessment of degradation in qualitative term is not an easy task. There are no historical records and/or other means to create the past to compare it with present.

Dudal (1977) has empahsised the urgent need for a better understanding of the risks of soil degradation and their geographical distribution as well as a better knowledge where soil degradation occurs. Soil degradation is defined by FAO/UNEP Expert Consultation Groups : "The result of one or more processes which lessen the current and/or potential capability of soil to produce (quantitatively and/or qualitatively) goods and/or services" (FAO, 1977).

Soil quality is defined as the capability of soil to produce safe and nutrition crops in a sustained manner over the long term without impairing natural resource base or harming the environment. Large extent of land is affected by salinity, alkalinity, acidity and waterlogging. Unscientific land use practices have also led to increase in the area affected by soil erosion, toxicities and deficiencies of a number of nutrient elements. The amelioration or reclamation of problem soils involves the application of soil amendments like gypsum/lime to improve fertility status. Another alternative that could be thought to improve their fertility status is incorporation of green manures, compost, other biological products or to grow tolerant crops or trees, pave way for reclaiming these soils indirectly. Though it is a long down process, the advantages derived are highly sustainable.

Table 1. Status of degraded lands in India by different agencies

Sr. No.	Source	Estimated area (M. ha)
1	National commission on Agriculture	175.00
2	All India Soil and land use survey	173.65
3	National Bureau of Soil Survey and land use planning	166.00
4	Society for promotion of Wastelands Development	129.60
5	National Remote Sensing Agency, Hyderabad	053.30

Source : Dagar *et al*. 1994

Out of 328.73 M.ha of India's geographical area, about 57.1 per cent are considered degraded or problematic land (Sehgal and Abrol, 1994), where crop/forage production is very poor or uneconomical. Though the estimates of degraded and problematic lands significantly vary in different survey reports (Degar *et al.* 1994), (Table 1) it has been estimated that about 141.25 M.ha suffer from erosion, 9.08 M.ha are salt affected, 8.53 M.ha are waterlogged. (Table 2).

Table 2. Extent of degraded and problematic lands in India

Category of degradation	Area M.ha	Occurrence
Erosion (water and wind)	141.25	Water erosion in humid and subhumid regions. Wind erosion in hot and cold deserts.
Acid soils	4.50	North eastern Himalayas, higher attitudes of western and middle Himalayas, high rainfall areas of peninsular and coastal regions.
Reverine torrents and stream banks	2.73	Costal regions, interior mountainous region and steep slopes in Himalayan regions.
Alkali soils	3.58	Mainly arid and semiarid regions of Indo Gangetic plains and peninsular regions.
Saline Soils (including coastal sandy soils)	5.50	Arid and semiarid and coastal regions.
Waterlogged area	8.53	Lowlying areas
Shifting cultivation	4.91	Mostly in north-eastern regions.
Ravines and gullies	3.97	Mostly in Gujarat, Madhya Pradesh and Uttar Pradesh.
Barren hill ridges or rocks	2.70	Distributed in several states
Total	177.67	

In India, it has been estimated that about 60 per cent of the land area is undergoing degradation due to one or another cause. Among the different categories, lands under rainfed cultivation have the highest problem, followed by deserts, irrigated crop lands and salt affected lands in the decreasing order. Soils occurring in arid environments undulation terrain is a familiar feature of rainfed region. The biological potential of many soils has diminished to the point that no more/further deterioration is likely to take place under existing subsistence cultivation methods. Excessive pressure of man and his livestock joinly with improper land use are the key elements supporting the present state of wide-scale degradation of the Indian land mass. Erosion by water (and wind) is the leading (47%) cause of rampant soil degradation. Loss of vegetation cover promoted the long period of aridity and openness of lands to degradation by man and over grazing by livestock makes drylands highly susceptible to erosive forces of water (and wind).

Alkali soils formed under the influence of Na_2CO_3 and $NaHCO_3$ are characterised by high pH (>8.5) and excess exchangeable sodium percentage (ESP>15) throughout the soil profile and adverse soil physical properties. Barren alkali soils are highly deficient in organic matter, N, Zn, low in Fe and Mn and adequate in P and K. The major alkali area is situated in the Indo-Gangetic plains.

In arid and semiarid regions in particular soils, development is characterised by high salt levels in the soil profile. Depending on specific conditions one or more of a number of ions (Na^+, Cl^-, SO_4^{2-}) may be present within the root zone in high concentration thus adversely affecting crop growth.

Acid soils are formed due to one or more reasons viz., leaching due to heavy rainfall at higher altitude, origin of soil from acid parent material like granite and use of acid forming fertilizer as ammonium chloride and ammoium sulphate. The major processes involved in the formation of acid soils are laterisation of varying degree, podzolization in areas with subtemperate climate and marshy conditions with sufficient amount of partly decomposed organic matter. All along coast wherever the mangrove ecosystems have been denuded for agriculture, particularly in Bay islands and Kerala, patches of acid sulphate (black cat) soils are predominant (FAI, 1985). These soils have low soil pH, sometimes as low as 3.5.

Further about 8.5 M.ha area is under waterlogged condition. The survey conducted by the Central Ground Water Board has revealed that an area of 36.36 M.ha is temporarily waterlogged during the monsoon period. Problem of short or long duration flooding have been increasing largely due to deforestation in catchment area, destruction of surface vegetation, change in land use, reduction in measures for water control and soil conservation (terracing, contour ditches, etc.) and increased urbanization. Waterlogged and saline soils also occur in areas where water is stored in tanks, rivers and lakes. These soils are marginal owing to their physicochemical properties which are unsuitable for plant growth (Singh *et al.* 1994).

Shifting cultivation (jhumming) is still prevalent in the hill areas of Andhra Pradesh and north-eastern states of India, where patches of dense forest are cleared and crops are raised without any management. After three or four years of cultivation when the yield decreases, new forest area is cleared for the purpose, leaving the old area exposed to the vagaries of nature resulting in soil erosion. Forest cutting, burning and dibbling of seeds caused nearly 4.0 t ha^{-1} yr^{-1} of soil material to slide/roll down to foot hill (Sehgal, 1995).

The soil quality and its productive capacity must be enhanced beyond preventation (Status quo) through soil building measures, e.g. presenting soil erosion and enhancing development of the rooting depth, replenishing nutrients harvested in crops through judicious use of mineral fertilizer and organic amendments and effective nutrient recycling practices, encouraging biological activity of soil fauna and improving soil organic matter content. The land use or farming system to be adopted must be "soil restorative" rather than "soil depletive" or "soil degrading". In addition, soil should not be misused as a dumping ground for toxic wastes.

2. PROCESS, FACTORS AND CAUSES OF DEGRADATION

In India, it has been estimated that about 57 per cent of land area is undergoing degradation due to one or other cause. Processes of soil degradation are mechanisms responsible for decline in soil quality. Process leading to soil degradation may be physical, chemical or biological (Fig. 1). Purely environmental processes such as leaching and erosion occur with or without human interference. Other processes such as acidification, salinisation and alkalization are generally triggered by excessive population pressure on land to meet the competing demands of rapidly growing population. Human induced degradation occurs when land is poorly managed or where natural forces are so powerful that there is no means of management that can check its progress. The direct/indirect human interference has caused the soil to degrade due to natural or anthropenic factor (Fig. 2). Natural factors include climate, hydrology, terrain, parent material and vegetation. Among important anthropogenic factors are population density, land use, development of roads, waterways and disposal of industrial and urban waste.

The causes of soil degradation are anthropogenic perturbations related to socio-economic pressure and natural like soil depth, clay mineral, etc., (Fig. 3). The causes and factors are especially rempant in many developing countries of Asia, Africa and tropical America.

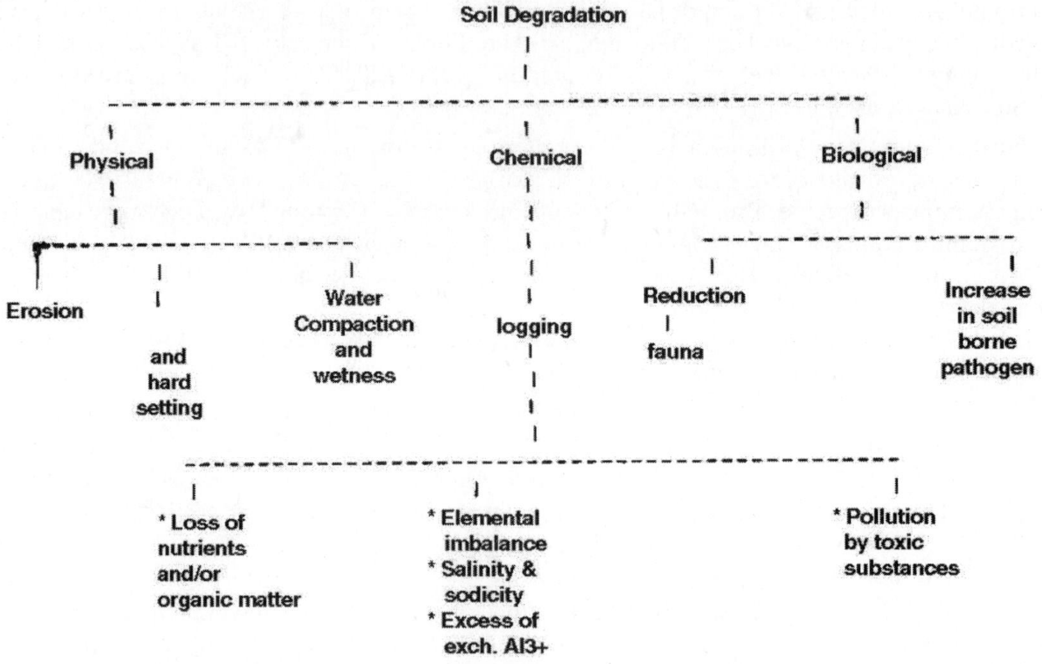

Figure 1: Types/processes of soil degradation

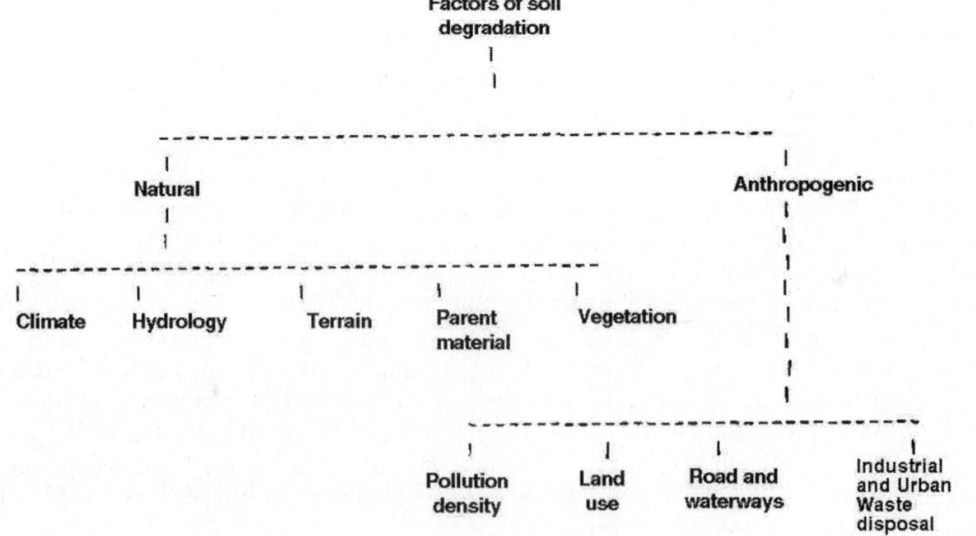

Figure 2: Factors responsible for soil degradation

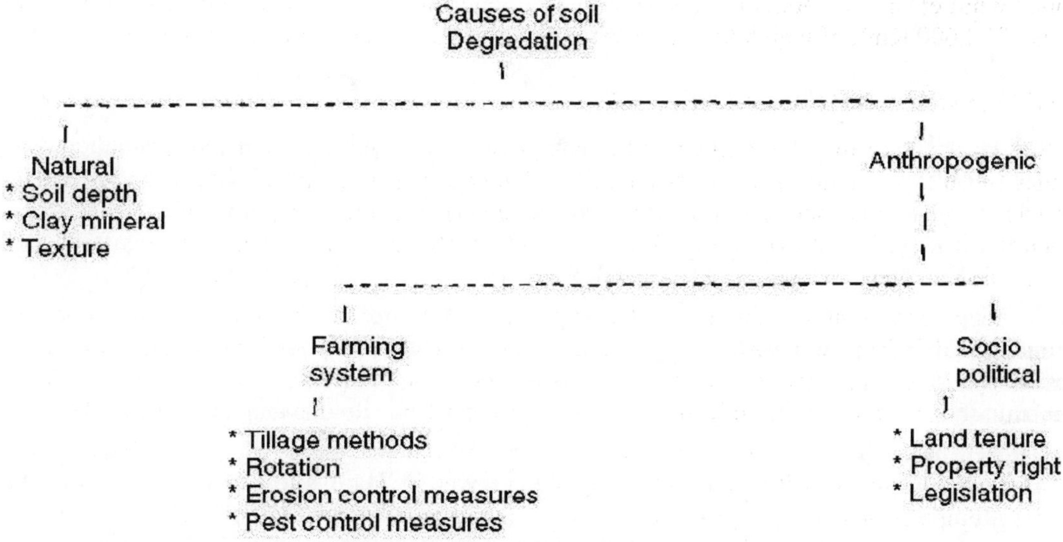

Figure 3: Causes of soil degradation

Types of Soil Degradation

Physical process or types

(a) Soil Erosion

Soil erosion is the wearing away, detachment and removal of top soil from one place and its deposition at another approximately 6000 million tonnes of soil through the forces of striking and moving carrying about 8.4 million tonnes of plant water, blowing winds, strong waves, snow nutrients are lost from 80 million ha of and gravity. It is estimated that cultivated land every year through erosion.

A survey and monitoring of the soils of Solapur district (Maharashtra State), after 75 years of interval showed that nearly 17 per cent of the land with a soil depth of more than 50 cm was converted into shallow soils and very shallow soils (with less than 50 and 25 cm depth, respectively).

The black cotton soils in Peninsular India (covering 73 M.ha) are highly erosive because of high runoff (42-50% of total rainfall), indiscriminate deforestation removing protective canopy, erosive slope and low permeability. It has been observed that the soils under normal cultivation (cotton and soybean) on slopes of 2 to 3 per cent have been eroding at the rate 8.4 t ha^{-1} $year^{-1}$ (Anonymous, 1996).

The red soils covering about 71 M.ha are another major soil group subject to severe water erosion. Most red soils (Alfisols, Ultisols and Oxisols) having low infiltration rate due to crusting problem, suffer from surface runoff and erosion by water. The lateritic soils suffer from rill erosion losing almost 40 tonnes ha^{-1} year in the absence of adequate soil conservation measures (Sehgal and Abrol, 1994). About 4 million ha^{-1} of land is affected by shallow, medium and deep gullies on bank of rivers in northern India. (Yadav, 1984).

Wind erosion is caused by strong winds mainly in arid and semi-arid regions, through processes like saltation, suspension and surface creep which are directly related to the wind velocity.

In India, wind erosion is moderate to severe in arid and semiarid regions of the northwest, covering an area of 28,600 Km^2, of which 68% is covered by sand dunes and sandy plains (Gupta, 1990).

(b) Soil Compaction and Hardsetting

Soil compaction may be defined as the compression of a mass of soil into a smaller volume. Changes in bulk properties are accompanied by changed in structural properties, like thermal and hydraulic conductivity and gaseous transfer characteristics thus in turn affect chemical and biological balances. In a world, the soil environment changed in such a way to affect all soil processes to a greater or lesser extent, depending on the degree of compaction.

The degree of compaction is usually expressed in terms of dry bulk density, porosity or penetration and resistance. Excessively compacted soils suffer from poor aeration and low hydraulic conductivity. Less than excessive compaction is not necessarily detrimental to crop growth as has been found in long term studies of crop response to compaction (Raghavan and McKyes, 1983).

Hardsetting soils are that set to a hard, structureless mass during drying and are thereafter difficult or impossible to cultivate until the profile is rewetted. Hardsetting of cultivated soil usually involves slumping, which is a process of compaction (i.e. increase of bulk density) that occurs without the application of an external load. Soils that undergo permanent hardening upon drying and do not soften upon rewetting (e.g., as can occur with laterization of cleared forest soils) are not classified as hardsetting. Soils such as sodic soils with a high proportion of dispersible clay are particularly noted for hardsetting behaviour. However, the presence of dispersible clay is not a necessary condition for hardsetting, since Young (1987) has observed hardsetting in a soil with no dispersible clay. Hardsetting soils have a clay mineralogy dominated by hydrous mica (illite) and/or Kaolinite. Hardsetting is dominantly observed in soils with low concentrations of organic matter and low activity clays. Soils prone to compaction and hardsetting are susceptible to accelerated runoff and erosion.

(c) Waterlogging

The estimate regarding the extent of waterlogged area vary a great deal, however, it is thought to be substantial under estimated to 8.53 M.ha (Anonymous, 1992). Waterlogging, in the form of seasonal or perennial surface ponding or rise in water-table upto the active root zone reduce the crop yields due to poor root aeration and flooding over extended period may even wipe out the crop. Due to the pattern of heavy concentrated unpredicted monsoon rains, a large area is affected by drainage problems during monsoons which is further exagerated by unchecked irrigation during the rainy weather. Most of the salt affected lands are also prone to waterlogging. Some localized waterlogging problems are also due to excessive seepage of water from canals as is evident in Indira Gandhi and Jayakwadi Commands. In some selected commands, 3.7 to 47.6 per cent of the potential irrigated areas are found waterlogged and 0.8 to 38.1 per cent is salinized (Singh, 1992).

Waterlogging also occurs in areas where water is stored in tanks, rivers and lakes. These waterbodies bear large slat loads rendering them marginal to unsuitable for irrigation. Waterlogged and saline soils usually occur in low rainfall areas (except some coastal soils) where fresh water supplied are scarce and more often the ground waters are brackish and at times close to the soil surface.

The major physical change that can be defined as soil degradation associated with soil wetness is the loss of strength in the soil, which encourages soil erosion and, facilitates soil compaction and dispersion of soil aggregates. When raindrops, wheels, animal feets or other energy- imparting instruments, are applied to the soil, the detachment of particles and/or puddling may occur. This represents true degradation of the soil physical condition. Sharma and DeDatta (1985) documented that puddling decreased bulk density, percolation and saturated hydraulic conductivity of the surface layer of dry compacted soil. Slaking is another manifestation of degradation by soil wetness. Excessive wetness created by flooding has thus been described as a soil degrading agent. Excessive wetness degrades soil physically and chemically. The oxidation-reduction (redox) potential decreases in the most dramatic and quantifiable short-term change that occurs in a soil as a result of waterlogging. Aerated soil characterized to have redox potentials in the range of +400 to +700 mv. Waterlogged soils may have redox potentials as low as -300 mv. Below +400 mv., soil is characterised as moderately reduced, and at -100 mv. as highly reduced. When oxygen is displaced and excluded from the soil by waterlogging, other oxidized components of the soil become electron acceptor and get reduced. The most easily reduced compound is nitrate. Ponnamperuma (1972) provides a very complete review of the work on this subject upto the 1970.

2.1.2 Chemical Degradation/ Deterioration

Chemical deterioration of soils occurs through a number of processes; the main agents under Indian conditions, appear to centre around the accumulation of slats and/or pollutants and the loss of nutrients and/or organic matter.

(a) Soil Salinisation and Alkalization

Salt accumulation is one of the most serious forms of land degradation affecting approximately 10% of the total land surface of the globe. The problem occurs in varying intensities in more than 120 countries, extending to all the continents and is more prominently witnessed in the arid and semiarid areas, though many coastal areas in the humid climate are also affected due to the ingress of sea water. Agricultural production is drastically reduced in most salt affected soils which are either lying barren or produce very poor crop yields owing to their inhospitable properties.

According to Szaboles (1977), an area of about 951.1 M.ha is estimated to suffer from this problem at the globle level. In India, out of an estimated area of 173.64 M.ha of total degraded land, approximately 10.6 M ha are affected by soil salinity/ sodicity (Suman, 1994), though some workers have put this figure as high as 26.1 M.ha (Yadav, 1986). Some salt affected soils are formed due to natural processes including release of electrolytes by weathering, their transportation mostly by water and subsequent accumulation *in situ* or elsewhere primarily caused by impeded drainage. Other soils developed due to prolonged use of poor quality groundwaters for irrigation in the arid and semi arid areas (Yadav, 1982), faulty management of irrigation water has resulted in wide spread occurrence of the twin problem of water logging and soil salinity in several irrigation projects, particularly in the arid and semi-arid areas, (Joshi and Agnihotri, 1984).

(b) Loss of Nutrients and/or Organic Matter

Use of adequate amounts of plant nutrients for optimum crop yields and profitability is a key to environmental protection. Although excessive fertilizer use does create environment problem, the main concern is the degradation of soil due to mining of soil fertility caused by the inadequate replenishment of nutrients removed by more harvests. In India, the soils are already very poor in nutrient content and malnourished.

In 2000 AD, removal of N, P and K will be about 32 million tonnes and the consumption is expected to be 18 million tonnes (as against 13.5 m tonnes in 199495) creating a gap of 14 million tonnes. Though organic manures, biological nitrogen fixation, etc. will contribute to some extent, a major portion of this nutrient will be mined from soil.

The major soil productivity constraints in soils are low organic matter status, low available N, P, Ca, Mg and S and toxicity of Fe, Al and Mn coupled with deficiency of Zn, Cu, B and Mo etc. Such unproductive soils can be made productive by more application of liberal quantities of organic matter (Pedro and Salinas, 1981). But in Garhwal region of Himalaya, the scope of organic amendment is also restricted due to paucity of sufficient quantities of farm and animal wastes and geographical constraints.

Addition of organic amendments like FYM, compost or even pressmud increased not only the availability of native nutrient but also increased availability of fertilizer (Narwal *et al.* 1990; Singh and Sarkar 1992).

(c) Pollution by Toxic Substances

There is an increasing concern about occurrence of trace elements in the environment in concentrations which can be harmful for animal health. Many fertilizers contain varying amounts of trace elements such as F, As, Cd, Co, Cr, Hg, Mo, N, Pb, etc. Incidental addition of these elements may lead to their accumulation in soils and ground water aquifers. But this will take place if contaminated fertilizers are marketed.

In most of the areas, the water supply for irrigation is problematic due to the poor resources of irrigation water, particularly in the arid region of northern India. Content of heavy metal increased with increasing of irrigation with sewage water as well as refinery effluents. The increase in Cd, Pb, and Ni concentration in berseem crop was more with sewage water than with refinery effluents. DTPA extractable Cd, Cr, Ca, Fo, Mn, Ni, Pb and Zn contents in the soil increased significantly with increasing number of irrigations with sewage water and refinery effluents (Singh *et al.* 1991).

Continuous application of sewage effluents makes the soil alkaline and increases the salt contents. Bicarbonates and chloride ions are at toxic level. Although sewage effluents and sludges are rich in nutrients, the toxicity due to micro nutrients and other heavy metals which they generally possess problem, has to be seriously taken into consideration before their application in soils. (Maiti *et al.* 1992).

Use of irrigation water containing higher concentration of toxic elements results in the enrichment of the top soil with the toxicants (Anderson and Nelson, 1976) and their entry into food chain may cause health hazards of the animals and human beings. Use of smelter's diluted effluent for irrigation has resulted in several hundred fold increase in DTPA-extractable Zn in the surface (0-15 cm) layer of the soil (Totawat, 1991).

2.1.3 Biological Degradation

Biological degradation of soil as described herein refers to the impairment of or elimination of one or more "significant" population of microorganisms in soil, often with a resulting change in biological processing within the associated ecosystem. Decrease in activity and diversity of soil fauna are ramification of biological degradation. Because of prevailing high soil and air temperatures biological degradation of soil is more severe in the tropics than in the temperate zone. Biological degradation can also be caused by overgrazing, excessive mining of nutrients, indiscriminate and excessive use of chemicals and soil pollutants.

A discussion on degradation of land would be incomplete without a mention of biological degradation, although a detailed discussion on this issue; reader is referred to detailed treatises on the subject (Sims, 1990; Fansey and Lal, 1990; Curry and Good, 1992).

3. ORGANIC AMENDMENTS IN REHAB- ILITATION OF DEGRADED/PROBLEM

Soils

Organic manures play a vital role in maintenance of physical, chemical and biological characteristics of the soil and replenish both macro and micro nutrients for crop production. In absence of organic matter, the soil is a mixture of sand, silt and clay. Organic matter is widely termed specifying the material that arises from the debris of green plant, animal residues and extra matter that are deposited on the surface and admixed to a variable extent with soil mineral component. The soil itself is defined as a dynamic natural body of mineral and organic constituents. The dynamic nature is solely due to the activity of macro and micro organisms supported by the organic matter contained in the soil.

In India, plenty of organic resources are available for recycling. The major sources are livestock and human wastes including Farm Yard Manure (FYM) and poultry manure; organic residues, including crop residues, tree wastes (litter, etc.) and aquatic weeds, green manure, urban and rural wastes (either as such or composted) including sewage sludge, microbial preparation, vermicompost, by-products of agro-industries such as press mud; fly-ash, oil cakes, slaughter house wastes, marine 1993; Sree Ramulu, 1994, Bellakki and wastes tank silt, etc. (Table 3.1).

There are many studies on the use of organic manures/agricultural wastes, in agriculture and their influence on the physico-chemical and biological characteristics of the soil as well as on crop yields (Bhat, 1991, Patel and Singh, 1991, Maiti *et al.* 1992; Palaniappan and Natarajan, Badanur, 1994; Pal, 1995; Das *et al.* 1995, Raman et al. 1996) but the information related to problematic soils is limited. The beneficial effects of organic amendments in the re-habilitation of problem soils and for crop production has special significance in a country like India, where a large extent of the arable land is already degraded.

Table 3.1. Average Chemical Composition of Organic Wastes and Manures (Per cent)

Organic Waste/Manures	N	P_2O_5	K_2O
	1. *Human habitation wastes*		
Night Soil	1.2 - 1.5	0.8	0.5
Human urine	1.0 - 1.2	0.1 - 0.2	0.2 - 0.3
Town compost	1.5 - 2.0	1.0	1.5
Sewage	15 - 30ppm	3.04 - 6ppm	10-20 ppm
Activated sludge	3 -6	3.0 -3.5	0.5 - 0.7
Settled Sludge	1.5 - 4.5	0.7 -4.0	0.3 - 0.6
	2. *Animal Wastes*		
i. *Cattle-shed wastes*			
Cattle dung	0.3 -0.4	0.1 - 0.15	0.15-0.20
Cattle urine	0.8	0.01-50.02	0.5-0.7
FYM/Farm compost	0.5 - 0.6	0.15 - 0.2	0.5- 0.6
ii. *Wastes/byproducts from slaughter house and fallen animals*			
Dried blood	10 - 12	1 - 2.6	0.8
Meatmeal	10.5	2.5	0.3
Bonemeal (raw)	3.0	20.0	-
Bonemeal (steamed)	-	22.0	-

Hoof, horn meal	10 - 15	1.0	-
iii. *Other animal wastes*			
Sheep/goat droppings	0.70	0.51	0.29
Paultry litter	3.03	2.6	1.4
Bird guano	7 -16	8 - 12	2 - 3
	3. *Crop wastes*		
Wheat straw	0.43	0.17	0.91
Rice straw	0.40	0.26	1.16
Oat straw	0.55	0.24	1.39
Barley straw	0.44	0.19	1.07
Sorghum straw	0.40	0.23	2.17
Bajra straw	0.65	0.75	2.50
Bean straw	1.57	0.74	1.62
	4. *Agro-industrial wastes/by-products*		
i. *Oil cakes*			
Castor-cake	5.5-5.8	1.8	1.0
Coconut-cake	3.0-3.2	1.8	1.7
Cotton-seed cake;			
-undecorticated	3.9 -4.0	1.8	1.6
-decorticated	6.5	2.8	2.1
Groundnut cake			
decorticated	7.8	1.5-1.9	1.4
undecorticated	4.5	1.7	1.5
Karanj-cake	3.9-4.6	0.9-1.0	1.3
Linseed-cake	5.5	1.4	1.2
Mahua-cake	2.3-2.6	0.8	1.8
Mustard-cake	4.5	1.5	-
Neem-cake	3.2	1.0	1.4
Niger-cake	4.8	1.8	1.3
Rape seed-cake	5.1	1.8	1
Sunflower-cake			
undecorticated	4.8	1.4	1.2
decortocated	7.8	2.2	2.0
ii. *Husks and bulls*			
Rice husk	0.45	0.25	0.45
Groundnut hulls	1.75	0.45	1.50
iii. *Sugarcane wastes*			
Bagassee	0.25	0.12	-
Molasses :			
Sulphitation	0.24	0.078	2.84
Carbonation	0.17	0.03	2.8
Press-mud	1-1.5	4 - 5	2.7
iv. *Tea wastes*			
Tea leaves (used)	2.6-3.6	0.3-0.4	1-1.9

v. *Tobacco wastes/byproducts*			
Tobacco wastes	0.05-1.0	0.8	0.8
Tobacco seed cakes	4-4.5	7 -15	5-5.3
vi. *Cotton, wool and silk wastes*			
Cotton dust	1.5-2.0	0.5-0.6	0.6-0.8
Wool waste	4 - 7	-	1 - 5
Silk worm waste	10	–	–
vii. *Tannery refuse*			
Leather waste	7	0.1	0.2
viii. *Sawmill waste*			
Sawdust	0.24	0.20	0.40
	5. *Fisheries and Marine wastes*		
Fishmeal	4 -10	3 - 9	0.3-1.5
Seaweed	1-2	1 - 2	2 - 7
	6. *Other organic wastes*		
Forest litter (Mixed)	1.51	0.18	0.57
Water hyacinth	2	1	2
Tank silt	0.3	0.3	0.3

* Data are taken from "Organic Manures" by A.C. Garg, M.A. Idnani and T.P. Abrahar, *Tech. Bull* (Agric) Indain Council. agric. Res. 32 and classified by Dr. O.P. Vimal of the Division of Soil Science and Agricultural Chemistry, Indian Agricultural Research Institute, New Delhi.

3.1 Livestock Wastes and Crop Residues

The cattle and buffalo population over the years showed increasing trend. On an average 7% increase occurred every five years, the population of the above animals during 1990 would be around 290 millions. The annual manurial potential of bovine excreta alone is 3.442, 1.307 and 2.214 million tonnes of N, P_2O_5 and K_2O, respectively. It cannot be said with any certainty as how much is used for fuel and how much for manure.

In India, a major part of crop residues is used as animal feed. While some find use as fuel, some may make housing materials in rural areas. Thus, only a small part of the crop residues would be available for direct incorporation into the soil or composting or by way of mulch.

In addition to cereals and millets, pulses, oilseeds and sugarcane are also important producers of crop residues (Table 3.2). It is well to recognise that the availability of crop residues in future will increase as a result of higher crop production.

Reclamation of sodic soil is the partial or complete removal of exchangeable sodium on the clay complex by the more favourable Ca ions, supplied through the soluble, calcium salt. The replaced sodium is often removed beyond the root zone by ponding of water on the soil surface. This can be accomplished by chemical or organic amendments (which directly or indirectly furnish, divalent cations for replacement of exch.-Na). The amendments are materials that directly or indirectly, through chemical or microbial action, furnish divalent cations (usually Ca^{++}) for replacement of exch-Na. Acids and acid forming materials react with $CaCO_3$ invariably present in sodic soils to form soluble Ca slats which in turn replace Na from exchange complex. The effectiveness of an organic material in reclaiming sodic soils depend on the amount of CO_2 produced and reduced conditions through drop in redox potential under submerged conditions. In a field study it was observed that 50% of total gypsum required (50% GR), pyrite equivalent to Gypsum on sulphur basis and FYM @ 30 tonnes ha^{-1} were equally effective increasing rice yield (Table 3.3) when a period of 30 days submergence prior to planting of rice was allowed. The reduction in pH and ESP was, however, the largest in gypsum treated soils followed by pyrites (Swarup, 1994). Enhanced availability of nutrients in submergence, FYM treated plots was chiefly responsible for better rice growth.

Table 3.2 : Annual production and nutrient potential of bovine excreta and crop residues

Source-Crop pro	Annual excretion/residue duction (million tonnes)	Nutrient Potential (million tonnes)		
		N	P_2O_5	K_2O
Wet dung	1227.80	1.842	1.227	0.614
Urine	800.00	1.600	0.080	1.600
Rice	106.01	0.61	0.26	1.70
Wheat	80.99	0.40	0.21	1.04
Sorghum	21.04	0.08	0.06	0.30
Pearl millet	15.58	0.11	0.06	0.30
Maize	12.50	0.07	0.09	0.19
Total pulses	13.70	0.22	0.07	0.27
Sugarcane	40.92	0.14	0.05	0.24
Total	2318.54	5.07	2.11	6.19

Source : Gaur and Sadasivam (1993)

Table 3.3 Effect of soil amendments and duration of submergence prior to rice planting on soil properties and yield of rice and wheat grown on sodic soil.

Treatment	Duration of submergence (days)	Yield (t ha^{-1})		Soil properties after			
		Rice	Wheat	Rice pH	harvest ESP	Wheat pH	harvest ESP
Control	0	0.08	0.005	10.3	82.8	10.3	82.5
	30	0.18	0.007	10.1	81.0	10.2	81.4
Gypsum	0	5.19	2.01	9.5	37.5	9.5	37.0
	30	5.58	2.07	9.3	35.2	9.3	33.1
Pyrites	0	4.81	1.81	9.6	39.6	9.6	39.2
	30	5.43	1.97	9.4	37.5	9.5	37.8
FYM	0	3.29	0.98	9.9	72.5	9.9	71.8
	30	5.52	1.28	9.8	62.0	9.9	62.1
Rice husk	0	2.70	0.66	10.0	74.5	10.0	73.5
	30	4.52	0.99	9.9	70.1	9.0	70.0
LSD Amendments		0.23	0.23				
(P = 0.05) submergence		0.13	0.13				
Interaction		0.29	0.26				

Source : Swarup (1994)

Earlier experiments conducted as CSSRI, Karnal showed that gypsum applied at rate of 12-15 t ha^{-1} (50% GR of 0-15 cm soil) was sufficient enough to initiate reclamation process in rice based cropping system. Field studies (1989- 90 to 1991-92) further showed that gypsum dose could be reduced from 50 to 25% GR when FYM @ 20 t ha^{-1} was applied (Table 3.4). Reduction in pH and ESP and improvement in crop yields also supported the above conclusion. The result thus suggested that gypsum @ 25% GR alongwith FYM @ 20 t ha^{-1} is better choice than hitherto used GR of soil @ 50% for successful productivity of rice and wheat at maximum level (Swarup, 1995).

Table 3.4 Effect of soil amendments on the grain yield of rice and wheat in sodic soil

Treatment (Amendments)	Grain yield (t ha^{-1})							
	Rice				Wheat			
	1989	1990	1991	Pooled	1989-90	1990-91	1991-92	Pooled
Control	2.98	4.55	5.23	4.25	0.20	1.00	1.18	0.79
Gypsum 25% GR	5.10	5.23	5.32	5.22	1.67	2.22	2.33	2.07
Gypsum 50% GR	5.44	5.46	5.34	5.41	1.99	2.31	2.42	2.24
FYM@ 20t ha-1	4.05	5.20	5.30	4.85	1.42	2.00	2.16	1.86
Gypsum 25% GR + FYM@ 20t ha-1	5.78	5.76	5.83	5.79	2.14	2.78	2.82	2.58
Gypsum 50% GR + FYM @ 20 t ha-1	6.13	6.01	5.90	6.01	2.36	2.82	2.91	2.70
LSD (P = 0.05)	0.33	0.41	0.40	0.36	0.35	0.36	0.34	0.35

Source : Swarup (1995)

In a field experiment conducted on Vertisol in Madhya Pradesh, India, different organic amendments at five tonnes ha^{-1} have decreased the bulk density and penetration force and increased the hydraulic conductivity and infiltration rate resulting in higher yields of paddy and wheat (Tomar *et al.* 1992). The results recorded in the study are shown in table 3.5.

Table 3.5 Effect of organic amendments on structural improvement of soil at flowering stage of paddy (average of 5 years).

Treatment	Bulk density (g cm3)			Penetration force kg/cm2			Hydraulic conductivity (cm hr -1)	Infiltration rate (mm hr-1)
	Depth (cm)			Depth (cm)				
	0- 15	15-30	30-45	5	10	15		
T1 FYM	1.15	1.25	1.40	1.87	3.44	5.36	0.65	3.50
T2 Rice husk	1.16	1.27	1.40	2.34	3.77	5.80	0.59	3.30
T3 Saw dust	1.18	1.29	1.44	2.81	3.85	5.78	0.42	3.00
T4 Control	1.19	1.30	1.44	3.87	4.41	6.06	0.27	2.53

Source : Tomar *et al.* 1992

The incorporation of fresh crop residues as mulch has moderating effect on the hydrothermal conditions of the soil. The crop residues not only regulate the soil temperature and surface evaporation but also improve the structure of the soil leading to better aeration and enhanced water holding capacity. Dev and Bhardwaj (1993) reported the beneficial influence of sugarcane trash and mustard straw on the physical properties of an acid silty clay loam soil when the residues were applied for two years prior to sowing wheat (Table 3.6).

Table 3.6 Effect of crop residue incorporation on the physical properties of an acid silty clay loam soil.

Residue	Bulk density		Pore space		Water holding capacity (%)
	0-7.5 cm	7.5-15 cm	0-7.5 cm	7.5-15 cm	0-15 cm
No residue	1.25	1.36	52.4	48.5	46.4
Sugarcane trash	1.16	1.27	55.3	51.8	52.0
Mustard straw	1.19	1.30	54.1	50.1	49.9
C.D.at 5%	0.04	0.04	1.7	2.3	1.7

Source : Dev and Bhardwaj (1993)

In a greenhouse experiment conducted on Aquic Natrustalf at Karnal, the soil is representative of large areas of sodic soils occurring in the Indo-Gangetic alluvial plain of northern India. Gypsum was applied at the rate of 50% GR of soil and *Sesbania aculeata* green manure and farmyard manure were applied at the rate of 0.5 per cent soil on oven-dry basis. The initial pH of the experimental soil was 10.2 and decreased to 9.8, 9.3, 9.0 and 8.9 in control, gypsum, gypsum + GM and gypsum + FYM treatment, respectively (Table 3.7). The magnitude of decrease was more during 30 days after transplanting that at 60 and 90 days of crop growth. This was expected as a result of reclamation process operating under submerged conditions. Submergence of sodic soil is known to build up pCO_2 (Swarup, 1981) which cause dissolution of $CaCO_3$ invariably present in these soils. The Ca thus get replaced adsorbed Na from the exchange complex. Because of these process, exchangeable Ca+Mg increased even in control treatment (Table 3.7). The pH and exchangeable Na of soil further significantly decreased with gypsum application. When manures were applied with gypsum, marked decrease in pH and ESP and increase in exchangeable Ca+Mg took place.

Loganathan (1990) reported that application of organic amendments viz., saw dust, groundnut shell powder, coir dust and FYM, each at 2.5 and 5.0 t ha^{-1} improved the soil characteristics like infiltration rate, total porosity and hydraulic conductivity of red soil with hard pan. The details are given in Table 3.8.

Table 3.7 Effect on soil properties (Average of three replications).

Treatment	Soil pH			ESP	Exch. Ca + Mg [cm mol (p+)kg^{-1}]
	30 days	60 days	90 days	90 days	90 days
Control	10.1	10.0	9.9	78	2.6
Gypsum	9.6	9.4	9.3	36	5.6
Gypsum + GM	9.3	9.1	9.0	28	6.6
Gypsum + FYM	9.2	9.1	9.0	28	6.7

Source : Swarup (1991)

Plate 6.1. 1. Soil cracking after irrigation during rabi season leading to poor crop stand. Peculiar property of vertisol, **2.** Addition of farm yard manure and crop residues (Note dried leaves after harvest of soybean) can improve the soil property. Application of FYM at plant site is better rather than broadcasting.

Dadhwal and Katiyar (1989) reported that FYM application @ 40 t ha^{-1} increased organic carbon content and water holding capacity (from 0.23 to 0.50% and 12.1 to 20.1%) whereas the bulk density and $CaCO_3$ content decreased. Further they reported that an application of FYM at 20 t ha^{-1} is adequate for the initial establishment and better growth of *Eulaliopsis binana* in degraded mine spoil lands. Application of paddy straw, rice hulls, proved beneficial in the reclamation of alkali and saline-alkali soils and reduced the salinity status (Agarwal and Gupta, 1968).

Table 3.8. Physical properties of soil as influenced by organic amendments

Treatments	Infiltrate rate mm ha^{-1}	Total porosity	Hydraulic conductivity [cm ha^{-1}]
Control	9.7	42.0	1.08
Gypsum 200 kg ha^{-1}	8.4	42.4	1.28
Gypsum 500 kg ha^{-1}	9.7	43.2	1.34
Saw dust 2.5 t ha^{-1}	6.2	43.2	1.37
Saw dust 5.0 t ha^{-1}	9.8	43.9	1.26
Groundnut shell powder 2.5 t ha^{-1}	12.7	43.8	1.57
Groundnut shell powder 5 t ha^{-1}	11.1	43.8	1.65
Coir dust 2.5 t ha^{-1}	13.3	44.3	1.73
Coir dust 5 t ha^{-1}	17.6	44.1	2.03
FYM 2.5 t ha^{-1}	21.6	43.6	2.34
FYM 5 t ha^{-1}	10.8	44.0	1.68

Source : Loganathan (1990)

Singh *et al.* (1987) reported that plantation of five to six years age increased the nutrient status of soil but could not balance the loss as a result of clear felling of trees. Mixed plantation with number of species were beneficial for enriching the soils. Handa and Sur (1987) reported the benefit of mechanical bunding, minor land leveling and provision of structure for safe disposal of surplus water on agricultural fields, thereby conserved more nutrients, stored more water in profile from rains and decreased physical degradation of agricultural fields due to rilling and gullying and these resulted in increase of pearl millet fodder yield.

Incorporation of residues of sorghum, mungbean, *Leucaena leucacephala*, perennial pigeonpea and groundnut shell and FYM equivalent to 40 kg N ha^{-1}, phosphorus and sulphur extractability increased with decomposition reaction period and substantial portion of added P and S had been mineralized by the 60 and 30 days of incubation (Das *et al.* 1995). They further concluded that residue incorporation in soil enhanced soil quality which has hitherto remained much more unaccounted. Aggarwal *et al.* (1995) reported that long term green manuring increased water content in soil profile and that organic manured plots had greater root mass and deeper penetration of maize roots. Pandey *et al.* (1985) from Kanpur reported that incorporation of maize, rice, sorghum and wheat straw in soil prior to growing rice showed improvement in bulk density, infiltration rate, stability index and water retention at 0.33 bar when soils were sampled after harvest of succeeding wheat.

A field experiment was conducted on an acid Inceptisol of Sikkim at ICAR Research Complex Farm Tedong at an altitude of 1350 m above msl continuously for three years (1988-90). Four levels of goat manure (0, 20, 40 and 60 t ha^{-1} containing 50% dry matter, C, N, P, and K at 23, 1.2, 0.7 and 1.1 per cent respectively) were applied with three levels of N (0, 70, and 140 kg ha^{-1}) through urea in

12 treatment combination without P and K fertilizer in RBD with three replications. Nandi [*Setaria sphacelate* (Schum)] the perennial grass was the test crop. After three years of application of manure and nitrogen, manure increased the pH, organic carbon, cation exchange capacity and decreased the amount of exchangeable Al^{3+}exponentially (Table 3.9).

Nitrogen application tended to decrease pH, and increase exchangeable Al^{3+} (Pati Ram and Singh, 1993). Thus, in long term forage production system, fertility level of acid soils can be improved through use of manure.

Mulching involves the laying of plant material on the surface of the soil to reduce evaporation and erosion, to smother weeds and for temperature control. Both sand and clay soils need conditioning to make them productive. Working plant material into soil improves its texture and also by acting as manure improves the nutrient content. In high rainfall areas, ground cover of mulch is effective in controlling erosion. Several species of aquatic weeds that can be used as manure are *Pistia stratotes*, *Hydrialla Verticillata, Salvinio* spp; *Eichhorina Crassipes* and Azolla. (Gopal and Sharma, 1981, Gupta and Lamba, 1976). Minhas and Gill (1985) reported that maximum reduction in waterloss (23.7%) resulted from subsurface placement of oat straw @ 7 t ha^{-1}, followed by the treatment involving its application on the surface, incorporation into surface soil and tillage alone, where in the evaporation was reduced by 160; 12.8 and 5.5% respectively, over control. A mulch cover of crop residues has been found to be an effective way to control soil erosion (Lal, 1984). In India, residues or straw of many crops like wheat, rice, mustard and cotton (which could be used as organic mulch to check both water and soil erosion and also to conserve moisture) are wasted or not used properly. Proper use of these will help in ameliorating the problem soils.

Table 3.9. Effect of manure and nitrogen on soil pH, exch Al^{3+}, organic carbon and CEC

Manure (t ha^{-1})	Level of N (Kg ha^{-1})					
	0	70	140	0	70	140
	pH			Exch	Al^{3+}	(mg kg^{-1})
0	4.8	4.8	4.6	29.0	39.2	37.5
20	5.0	4.9	4.9	3.1	4.8	6.1
40	5.1	5.1	5.1	1.0	0.8	1.0
60	5.2	5.2	5.1	0.8	0.8	0.8
Initial	4.8	–	–	34.4	-	-
	Manure	N	Manure XN	Manure	N	Manure XN
CD (P=0.05)	0.07	0.03	NS	5.93	NS	NS
	Organic C (%)			CEC [cmol (p^+) Kg^{-1}]		
0	1.2	1.1	1.2	6.6	5.7	5.9
20	1.6	1.5	1.5	7.9	7.6	7.9
40	1.8	1.8	1.8	8.5	8.6	8.5
60	2.1	2.1	2.0	9.8	9.5	10.3
Initial	0.9	-	-	4.7	-	-
	Manure	N	Manure XN	Manure	N	Manure XN
CD (P=0.05)	0.20	NS	NS	1.25	NS	NS

Source : Pati Ram and Singh (1993)

Table 3.10. Characteristics of some important green manure crops

Green manure crop	Green matter (t ha -1)	Moisture (%)	Nitrogen content	Phosphorus content	C:N ratio
			(% of dry matter)		
Cultivated annual legumes					
Astragalus sinicus (Milk-vetch)	–	–	3.20	–	13.1
Cassia mimosoides	4.7	74	2.99	–	–
Crotalaria juncea (Sunnhemp)	16.5-25.0	73	2.0-3.12 (2.86)	0.18-0.40 (0.34)	13.5-24.0 (16.1)
Cyamopsis tetragonoloba (Cluster bean, guar)	7.0-8.0 (20.0)	60	2.30-3.35 (2.80)	–	12.7-24.0 (17.3)
Indigofera anil (Indigo)	6.8	74	3.66	–	–
Melilotus alba (Senji)	28.6	–	0.57	–	–
Phaseolus aureus (Mung/green gram)	7.7-8.0	82	2.21-2.96	–	16.1
P. trilobus (Pillipesara)	5.3-18.3	79	1.10-2.47	–	–
Vigna unguiculata (Cowpea)	10.0-15.0	85	1.60-3.89 (2.69)	0.22-0.43 (0.28)	10.3-23.2 (17.1)
Sesbania aculeata/ (Dhaincha) bispinosa/cannabina	14.8-25.0 (20.0)	78	1.79-4.16 (2.62)	0.21-0.40 (0.32)	8.6-28.8 (16.4)
S. speciosa, sesban (Sesbania)	7.8-60.0	78	2.43	–	–
Trifolium alexandrium (Berseem)	15.5	80	0.43	–	–
T. incarnatum (Crimson Clover)	–	–	2.20-2.95 (2.50)	–	13.5-15.9 (15.0)
Perennial legumes					
Cassia hirsuta	5.0	81	2.52	–	–
Desmodium gyroides	1.4	72	3.35	–	–
Gliricidia sepium	30-110	75	2.50-4.30 (3.49)	0.13-0.29 (.022)	10.42
Indigofera tinctoria (Indigo)	10-12	–	–	–	–
Leucaena leucocephala	–	–	2.42-3.50	0.13-0.27	12.7
Sesbania punctata	3.7	73	2.42	–	–
S. rostrata	3.5	74	–	–	–
Wild annual legumes					
Aschynomene americana	8.9	78	3.14	–	–
Calopogonium mucunoides (Calopa)	4.5-5.0	74	–	–	–
Cassia tora	5.2	71	2.13	–	–
C. occidentalis	4.3	78	2.80	–	–
Lathyrus sativus	–	82	4.67	–	–
Tephrosea purpurea (Wild indigo)	3.5-6.0	70	3.46	–	–
Azolla pinnata (Aquatic)	–	–	0.95-6.57 (4.46)	0.06-1.74 (0.38)	9.0-12.9

Values in parenthesis are average

Plate 6.2. 1. Heap of stover after soybean threshing. Mix it in soil to improve water holding capacity, **2.** Field tillage in rabi season after soybean harvest. Note crop residues in field. Addition of organic residues increases nutrient availability to next crop.

3.2 Green manuring and green Leaf manuring

Green manuring is a cheap alternative to the use of fertilizer nitrogen, help to correct soil conditions. The process also makes a positive contribution to the maintenance of soil organic matter content at a satisfactory level. Green manuring is the practice of incorporating *in situ* plants grown specially for use as green manure. Green leaf manuring is the practice of collecting green leafy material from perennial shurbs and trees growing in different locations and using them as manure in field where crops are to be raised. Biomass production of green manure crops and their N and P contents vary widely according to the species, their growth stage, environmental conditions, soil fertility and crop management practices. Panse *et al.* (1965), Abrol and Palaniappan (1987) and Singh *et al* (1991) listed large number of green manure crops (Table 3.10). Green manure crops are usually raised with objective of supplying nitrogen and correct the soil conditions. Green manure crops are also grown as cover crop to avoid erosion of the soil by the overflow of water and protects the soil from wind erosion. *Sesbania aculeata* (dhaincha) and palm leaves correct the alkali soil while *Argemone mexicana* and *Tamarindus indica* has buffer effect when applied to sodic soil.

Besides legumes, several other plants may be used as green manure. For example, the water hyacinth (*Eichhorina* crassipes) which grows wildly in stagnant water (see chapter on weed) and sea weeds may be used for manuring crops in coastal areas. *Argemone mexicana* a common winter weed of waste lands have also found useful as green manure (Singh, 1984). Application of five tonnes ha^{-1} *Argemone mexicana* was as good that of *Sesbania aculeata* in reclaiming calcareous sodic soils (Krishnamurthy, 1984). Some of the trees and shrub species which are extensively used for green manuring, their nitrogen contents and time taken for decomposition are given in Table 3.11.

Table 3.11 : Nitrogen content and decomposability period of some green leaf manure

Species used for green Stem to leaf ratio manure		N content (% dry matter)		Decomposability period (weeks)
		Leaf	Stem	
Glyricidia sepium	1:1.9	3.98	1.84	6
Calotropis gigantea	1:0.8	3.52	1.40	4
Croton sparsifiorus	1:1.2	3.85	2.10	4
Cassia siamea	1:3.8	2.42	1.54	4
Thespesia populnea	1:2.3	3.28	1.56	6
Delonix regia	1:3.9	2.27	1.47	6
Pongamia pinnata	1:1.3	3.71	1.89	6

Source : Krishnamurthy (1984)

Organic materials act as binding agent in holding the primary soil particles together to form soil aggregates, Improvement in soil aggregation by green manuring has been reported by several workers (Sharma and Singh, 1970; Chaudhary and Bajwa, 1979; Jiao *et al.* 1986; Liu, 1988). Boparai (1982) observed that green manuring in wetland rice increased water stable aggregates, between 0.1 to 0.5 mm; by 68% over non-green manuring. Chaudhary and Bajwa (1979), Jiao *et al.* (1986) & Liu (1988) reported that green manuring increased the water stable aggregates, decrease modulus of rupture in paddy soils (Table 3.12). Joshi *et al.* (1990) reported that the settling percentage of wet soil aggregates after rice harvest was 10.8 in the *Sesbania* treated plots compared with 14.2 in the organic fertilized plots. Effect of land effectiveness of green manuring did not last long unless continuous additions were made every year.

In addition to adding organic matter, green manure crops, due to their extensive and sturdy root system, exert a favoural effect on soil permeability and lowering of bulk density. Sharma and Singh (1970) Boparai (1982) and Liu (1988) reported a marked decreased in bulk density of soil in green manured plots under wet land rice (Table 3.12). This effects of green manuring, however, decreased with time.

Table 3.12 : Effect of green manuring (GM) on physical properties of soil

Reference and treatment	Soil aggregate (%)		Modulus of rupture (bar)	Hydraulic conductivity (cm ha^{-1})	
	>0.25 mm	>0.1 mm			
Chaudhary and Bajwa (1979)[a]					
-GM + gypsum	4.5	15.3	4.3		0.49
-GM + gypsum	13.7	25.7	2.4		0.65
Jaio *et al.* (1986)[b]	5-2 mm	2-1 mm	1-5.0 mm	0.5mm	Total
- GM	9	5.9	4.0	3.4	22.3
+ GM	34.5	13.6	2.3	5.1	55.5
	Soil aggregates 0-10 cm layer (%)		Maximum water holding capacity (mm)	Aeration (%)	Bulk density (g/cm^3)
Liu (1988)[c]					
– GM	3.6		43.6	36.5	1.48
+ GM	9.0		46.5	45.9	1.33

a : Pot study on sodic soil

b : 3 year field study

c : 3 year field study on saline coastal soil.

An attempt to study the possibility of organic recycling of *Leucaena leucacephala* (Lamk) de Wrt) planned in association with winter sorghum grown on a Typic Chromustest at Solapur, India, was made by Narkhede and Ghugare (1987).

Leuceana produced a dry matter of 1.97 t ha^{-1} which contained 3.86% N. Thus, 76 kg N ha^{-1} was added by leucaena loppings; the loppings were applied on the soil surface. The addition of loppings increased available nitrogen in the surface 0-15 cm soil by 44.5% over control (Table 3.13). The addition of leucaena loppings increased grain and straw yield and moisture use efficiency and left more moisture in soil at sorghum harvest.

Table 3.13. Effect of organic recycling on sorghum, organic carbon and moisture content in soil

Characteristics	Control	Leucaena loppings added
Sorghum		
Grain yield t ha^{-1}	1.2	1.4
Stover yield t ha^{-1}	3.3	3.6
Moisture use efficiency (kg grain mm -ha^{-1})	5.4	6.7
Soil		
Organic carbon (%)	0.42	0.47
Moisture content in 50 cm soil depth (mm)		
At sowing	232	234
At flowering	175	172
At harvest	132	136

Source : Narkhede and Ghugare (1987)

Plate 6.3. 1. Tillage after soybean harvest. Note crop organic litter on surface soil during tillage, **2.** Broad based furrow (BBF) at 6 m distance for rain water conservation and better drainage.

On Vertic Inceptisol, continuous application of leucaena loppings for seven years increased organic carbon content, field capacity, hydraulic conductivity and water use efficiency in cotton + green gram intercropping (Anon, 1994). The increase in organic content in soil was 31.3 per cent (Table 3.14). Similar was effect of farmyard manure on physico-chemical properties. Application of *Leucaena leucacephala* loppings @ 4.5 t ha^{-1} for eight years alone or in combination with chemical fertilizer had a favourable effect on organic carbon content and field capacity (Table 3.15) under monocropping system of cotton.

Table 3.14. Effect of leucaena loppings, FYM on physico chemical properties and water use efficiency in cotton-mungbean intercropping

Treatment	Organic carbon (%)	Field capacity (%)	Hydraulic conductivity (cm ha^{-1})	Water use efficiency (Kg ha^{-1} mm^{-1})
Control, no fert. no manure	0.48	34.86	0.79	1.39
Recom. full dose to cotton (50:25 kg NP ha^{-1})	0.55	34.69	0.87	1.53
25 kg N ha^{-1} through leucaena loppings	0.63	36.33	0.57	1.49
25 kg N ha^{-1} through FYM	0.66	37.86	1.04	1.53
Initial value (1987)	0.46	35.90	–	–

Source : Anon. (1994)

Table 3.15. Effect of addition of leucaena loppings (LL) on soil properties

Treatment	Organic Carbon (%)	Field capacity (%)	Available N (Kg. ha^{-1})	Cumulative addition to loppings (t ha^{-1}) from 1985-896
Control	0.56	38.0	121	0
Leucaena loppings @ 4.5 ha -1	0.73	39.0	149	36
L.L. + 1/2 recom. fert. dose (25: 12.5 N.P.kg ha^{-1})	0.73	39.0	150	36
L.L. + 3/4 recom. fert, dose	0.73	40.0	158	36
Full recom. fert. dose	0.51	37.0	130	0
Initial Value (1988)	0.32	35.5	148	–

Source : Anon, (1994)

Table 3.16. Effect of green manuring on soil properties and respiratory capacity of soil at rice harvest (three year mean)

Treatment	Organic C %	WHC * kg kg^{-1}	Bd mg m^{-3}	CO_2 evolved (mg 100 kg^{-1} g^{-1} soil)
Sunnhemp	0.90	0.52	1.32	112
Sesbania	0.89	0.52	1.34	110
Cowpea	0.90	0.53	1.33	110
Sunnhemp + N 40	0.82	0.50	1.35	-
Sesbania + N 40	0.84	0.50	1.36	-
Cowpea + N 40	0.82	0.50	1.35	–
N 60	0.78	0.48	1.36	99
N 90	0.80	0.48	1.37	99
N 120	0.78	0.48	1.37	100
CD (P=0.05)	0.06	0.023	0.02	6

* Water holding capacity, ** Bulk density Source : Datta and Bhardwaj (1995)

Datta and Bhardwaj (1995) reported that substantial usefulness of summer grown green manure legumes for rice-wheat cropping on a productively poor and structurally degraded silty clay loam soil. Compared with the original status green manuring brought about increases in organic carbon, water holding capacity and higher respiratory activity as evaluated by CO_2 evolution from soil (Table 3.16). Further they reported that sunnhemp was found to be better green manure than Sesbania.

Bellakki and Badnur (1994) reported that water infiltration rate and water retention characteristics of a Vertisol were increased and bulk density was reduced significantly with sorghum stubbles alone or in combination with *Leucaena leucoephala* loppings over that with fertilizer application. Leucaena loppings alone or in combination with sorghum stubbles increased water stable aggregates significantly (Table 3.17).

The beneficial effects of legumes in crop rotation, intercropping in enhancing soil fertility and its usefulness to succeeding cereal are well known (Sharma and Bajpai 1989). In addition, legume based cropping systems maintain high level of organic matter and high microbial load in soil as compared to cereal based cropping systems. Bhandari *et al.* (1989) reported that green manuring of cowpea or *Sesbania* sp. prior to sowing of maize added substanied amount of total nutrients (Table 3.18). The green manured plants during their decomposition released organic acids which helped in making available the unavailable soil nutrients.

Bhardwaj and Dev (1989) reported that incorporation of easily decomposable plant materials in soil causes increased and rapid production of CO_2. For this reason, green manuring has been suggested as an important practice for the reclamation of alkali soils. Swarup (1986) studied the effect of green manuring with *Sesbania aculeata* grown form 50 days and of decomposition time on the yield of rice grown on sodic soil. Green manuring significantly increased the rice yield, contributing N at 111.6 kg ha^{-1}. Continuous submergence improved the sodic soil, and green manuring further decreased pH and exchangeable sodium.

Table 3.17. Influence of crop residues, green leaf manure on the properties of Vertisol

Treatment	Organic C %	Bd g/cm^3	Infilltration rate ms^{-1} x105	WSA % >25mm
Incorporation of L.L. @ 5 t ha^{-1}	0.68	1.24	0.21	48.8
Incorporation of S.S. @ 5 t ha^{-1}	0.55	1.17	0.40	50.6
Incorporation of S.S.+ L.L. (50: 50) @ 5 t ha^{-1}	0.58	1.27	0.36	51.1
Incorp. L.L. @ 5 t ha^{-1} alternate year starting from first year	0.56	1.35	0.21	45.0
Incorp. S.S. @ 5 t ha^{-1} alternate year starting from first year	0.57	1.21	0.36	46.1
Incorp. S.S.+L.L. (50:50) @ 5 t ha^{-1} alternate year	0.58	1.19	0.37	45.8
Incorp. of S.S.+L.L. (25:75) @ 5 t ha^{-1} alternate year	0.54	1.25	0.33	46.4
Recommended dose of fert.	0.50	1.37	0.14	42.4
No fertilizer	0.43	1.35	0.15	43.2
	0.012	0.057	0.080	0.85

L.L. = leucaena loppings; S.S.= Sorghum stubbles Bd = Bulk density, WSA = Water stable aggregates
Source : Bellakki and Badnur, 1994.

Table 3.18. Effect of green manuring and fertilizer nitrogen on organic carbon content and available nutrients after maize harvest

Treatment	Organic carbon (%)	Available nutrients (kg ha^{-1})		
		N	P	K
N0	0.55	154	17.9	178
N60	0.55	155	17.6	178
N120	0.56	158	16.9	179
Cowpea, 4 t ha^{-1}	0.57	169	19.9	190
Cowpea + N60	0.58	167	19.7	189
Sesbania, 6 t ha^{-1}	0.57	168	19.7	189
Sesbania + N60	0.57	167	19.7	189
C.D. (P=0.05)	0.007	4.7	1.22	4.0

Source : Bhandari *et al.* 1989

Krishnamoorthy *et al.* (1985) reported that green manuring could partially reduces the fixation of applied inorganic P substitute the gypsum required to reclaim fertilizer. In a calcareous sodic soil, alkali soils and this practice was more incorporation of Sesbania plant material economical than using gypsum alone. reduced an initial ESP by 36% over a 10 Application of 5 t ha^{-1} of green manure was week incubation period, whereas barely better than that of gypsum at 7.5 t ha^{-1} and straw resulted in only a 10% reduction other chemical ameliorants in the (Singh, 1974). In a field experiment, reclamation of sodic soils. Yash Pal *et al.* application of gypsum [20 t ha^{-1}] alongwith (1995) reported that sesbania green green manures to a sodic soil resulted in a manuring helps in solublizing native P and marked reduction in soil pH and ESP compared to that upon application of gypsum alone (Swarup, 1987).

Yadav and Agarwal (1961) could observe complementary effects of gypsum and Sesbania green manuring; the former facilitating base exchange reactions and the latter improving physical characteristics. Increase in both permeability and removal of soluble salts due to green manuring with *Sesbania* were attributed to an extensive and sturdy root system, which renders the leaching process much easier (Rao, 1985). It is interesting to note that root growth of *Sesbania* also caused a marked decrease in ESP and an increased porosity (Swarup, 1987). Organic acids and CO_2 liberated by the roots of *Sesbania* helped in solubilizing native calcium carbonate and gypsum applied to the alkali soils.

Green manuring can improve both the physical and biological properties of soil. The data from a few studies indicated that the physical parameters more likely to be affected by green manuring are aggregate stability and bulk density. Several studies have shown that green manuring caused a marked reduction in redox potential and a significant reduction in soil pH. Rarely, soil organic matter levels will be increased by green manuring, however, it has a favourable effect on the availability of P, K, Fe and Mn in the soils. Green manuring hastens the reclamation of saline and sodic soils and thus proved an important management practice for their reclamation.

3.3 Agro-industrial wastes and other wastes

Oil cakes of non-edible and edible like neem (*Azadirachta indica*), Karanj (*Pongamia glabra*), mahua (*Madhuca butyracea*), castor (*Ricinus communis*) rape seed (*Brassica campestris*), linseed

(*Linum usitatissimum*), and cotton seed are widely used as organic manure as these contain high amounts of plant nutrients. Neem cakes have been found to retard nitrification (Patil, 1972) as cake contains the alkaloids, nimbin and nimbidin. Similarly Karanj seed cake and the bark of tree possess nitrification inhibitor properties. Mahua is a potent nitrification inhibitor equal in efficiency to nitropyrin (N-serve) in retarding the nitrification of ammoniacal nitrogen and increasing rice yield (Saharawat, 1981). The utility of neem cake blended urea for increasing the yield of upland rice was known (Surve and Daftardar, 1985).

Waste product of sugarmills, pressmud from sulphitation process can be used for reclaiming sodic soils. Gypsum and pressmud treatments reduced pH, ESP and increased exchangeable Ca of soil at all levels of application (Table 3.19). However, it was significantly higher with gypsum treatments than with pressmud. The beneficial effects of pressmud is to supply calcium on decomposition of organic matter (Patel and Singh, 1991).

Table 3.19 : Effects of amendments on soil properties (after rice)

Treatments	pH	Ec dSm^{-1}	ESP	Exch. Ca [$cmol(p^+)\ kg^{-1}$]
Initial Value	10.2	1.2	55.3	4.3
Control	9.8	1.5	46.6	5.1
Gypsum -25% GR	8.8	1.0	28.9	6.8
Gypsum -50% GR	8.0	1.4	14.6	8.1
Pressmud - 25% GR	9.1	1.1	31.5	6.2
Pressmud - 50%	8.7	1.2	25.3	7.1
GR Pressmud - 75% GR	8.7	1.3	28.0	7.5
C.D. (P=0.05)	–	0.1	1.67	0.35

Source : Patel and Singh, (1991)

The pressmud being organic in nature brings marked changes in the soil properties when used for a number of years. Higher level of pressmud increased the initial low N to medium N and medium P to high P level. The movement of N and P after 3 years of experimentation was observed upto 30 - 45 cm soil layer. (Kapur and Kanwar, 1989). Jagannathan *et al.* (1987) observed an increased hydraulic conductivity and the decreased bulk density by the application of coirpith and pressmud treated plots. Addition of sulphitation pressmud to plant cane had a significant residual effect on the increased availability of N, P and micronutrients in the soil (Yaduvanshi and Yadav, 1996). Further they reported that an economy of 75 kg N ha^{-1} may be achieved due to residual effect of pressmud. Combined application of FYM and fly-ash proved to be beneficial in augmenting proliferation and activity of microorgani-sms in acid soil (Lal *et al.* 1996). Patil (1993) has reported 74.3 per cent increase in the pod yield of groundnut with the application of 30 t ha^{-1} of flyash as compared to control yield. The beneficial influence of flyash was in reducing the strength of soil and in maintaining favourable soil physical properties. The result of studies carried out by the College of Agriculture, Raichure, Karnataka on the influence of amendments on soil crust strength are shown in Table 3.20.

Application of flyash as mulch on the soil surface with and without FYM has been found beneficial for improving physical properties of heavy deep black cotton soils from Nagpur region of Maharashtra State, (Anonymous, 1994). Zalawadia and Raman (1994), reported that with 75 per cent of the recommended fertilizer application with the distillery waste water, the sorghum yield was at par with 100 per cent recommended dose of fertilizer application with normal water irrigation. Application of sewage effluents were reported to be beneficial in increasing crop yield and reduced fertilizer requirement (Jayaraman, 1988).

Flyash is chemically a ferroalumino silicate and major solid waste of industrial area. Huge amount of flyash is formed due to low grade of coal and inferior model of electrostatic precipitator used. Flyash application increased pH, organic carbon as well as available P and K of the acid soil (Table 3.21). Microbial activity was also higher in flyash amended soil (Lal *et al.,* 1996).

Table 3.20 : Effect of different amendments on soil crust strength

Amendment (20 t ha^{-1})	Unconfined crust strength (kg cm^{-1}) at 15 DAS		
	1988	1988-89	1989
Control	2.23	3.36	2.60
Flyash	0.98	1.94	1.95
FYM	1.03	2.48	2.90
Sand	1.68	2.88	2.10
Paddy hust	0.53	0.95	-
C.D. (P=0.05)	0.28	1.14	NS

Source : Patil (1993)

Raman *et al*. (1996) reported that application of gypsum reduced the soil pH and increased the EC. Flyash and pressmud did not have any effect on pH and EC. The organic carbon content significantly increased due to pressmud application. Application of all the three materials enhanced the infiltration rate and water stable aggregates. (Table 3.22).

The current annual rate of loss of arable land due to soil degradation is estimated at 5 to 7 M ha and it may increase to 10 M ha by the year 2000 (Dudal, 1982). There is constant and continuous decrease in availability of per capita land. This situation leads to a consideration that per unit production has to be increased. The decline in the quality affects about 35% of the land surface. This discussion lead us to the fact that serious attention need to be paid to soil productivity.

Table 3.21 : Effect of Flyash on pH, organic carbon and microbial activity in post harvest soil samples

Flyash (% w/w)	pH	Organic carbon (%)	CO_2 evolution dg kg^{-1}
0	4.72	0.39	13.72
4	4.85	0.38	13.44
8	5.01	0.36	16.32
16	5.37	0.41	15.92
32	5.64	0.44	13.42
CD (P=0.05)	0.07	0.02	–

Source : Lal *et al*. (1996)

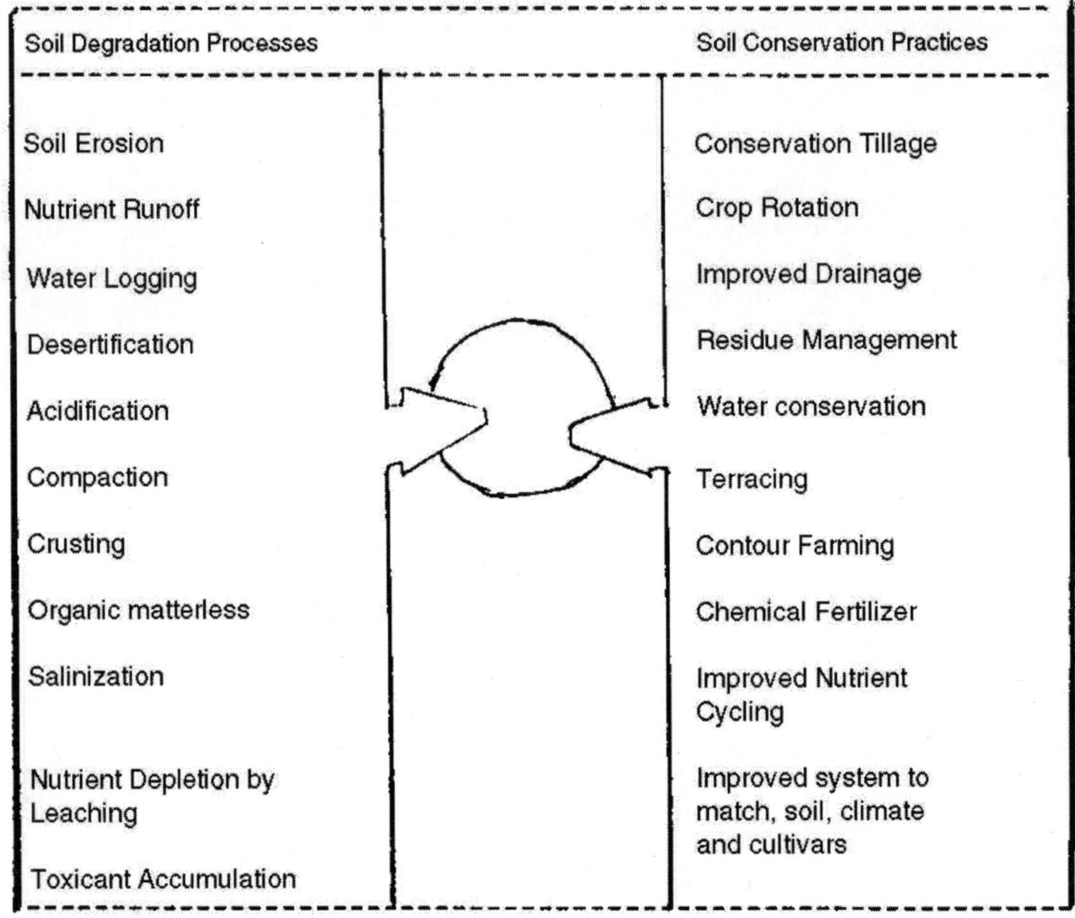

Fig. 4. Relationship of soil degradative processes and soil conservation practices

Usually the relationship between soil erosion; nutrient losses, organic matter depletion and beneficial effects of conservation and management practices occur simultaneously. This is often neglected. It is important to realise that potential productivity of a particular soil at any point of time is the result of on-going degradative processes and applied management practices. On the best agricultural soils, a high level of productivity can be maintained with few management practice. On marginal soil with limited capability such as highly sloping, course textured, shallow, poorly structured and poorly drained soils, the management practices need a package. A combination of more practices is necessary to maintain productivity. Organic waste or residue offer the best possible measure for restoring the productivity of severely ended soils.

It is clear from Figure 4 that the productivity of soil is a dynamic process, particularly organic matter content is one of the most important factors.

Table 3.22: Effect of industrial waste on soil pH, EC, organic carbon infiltration rate and water stage aggregates (three years pooled data)

Treatments	pH * (1:2.5)	EC dSm-1	Organic C (g kg^{-1})	Infilt. rate (in 10-6 ms^{-1})	Water-stable aggregate (70.25mm)
Control	8.1	0.45	7.3	0.89	58
Gypsum, 6 t ha-1	7.9	0.89	8.7	1.42	58
Gypsum + RS	8.1	0.90	8.8	1.72	64
Flyash, 6 t ha-1	8.2	0.48	7.2	1.28	65
Flyash + RS	8.3	0.49	8.7	1.42	72
Pressmud, 6 t	8.3	0.47	10.8	1.75	75
Pressmud + RS	8.3	0.47	10.9	2.08	80
C.D. (P=0.05)	–	0.2	1.0	0.11	5.5

RS = Rice Straw @ 5 t ha-1, * Average of three years
Source : Raman *et al.* 1996

CONCLUSIONS

The state of land degradation being as it is, agricultural scientists will have crucial role in generating and promoting a successful technologies for rehabilitation of degraded lands and for preventing further degradation of already degraded lands.

The soil organic matter is the most dynamic component of the soil. It has profound influence on the fertility and productivity of the soil. Agricultural wastes including cereal straw/stalk/stover, animal wastes, etc. are available in large quantity and contain valuable plant nutrients. The continuous use of organics maintain the organic matter content and fertility status of the soil. Organics not only support sustained crop production but also ameliorate problematic soils.

Green manuring hastens the reclamation of saline and sodic soils, thus proved an important management practice for their reclamation. Similarly, by- products of agro-industries viz., pressmud, gypsum, flyash have played a significant role in the problemed soils. Sustainability in agriculture cannot be fulfilled without people's education, information, training and effective participation.

REFERENCES

Abrol, I.P. and Palaniappan S.P. (1987). Green manure crops in irrigated and rainfed lowland rice based cropping system in South Asia. In: *Green Manure in Rice Farming. Intnal. Rice Res.* Inst., Los Banos, Philippines. pp.71.

Aggrawal, G.C; Sekhon N.K; Sidhu, A.S. and Mahant (1995). Plant water status of maize and succeeding wheat grown on organically amended soils. J. Indian Soc. Soc. Soil Sci. **43**(2): 152-155.

Anderson, A. and Nelson K.D. (1976). *Swedish J. agric. Res.* 6:151.

Anonymous (1994). Annual Report. AICRP for Dryland Agriculture, Dr. P.D.K.V., Akola.

Bellakki, M.A. and Badanur, V.P. (1994). Effect of crop residue incorporation on physical and chemical properties of a Vertisol and yield of sorghum. *J. Indian Soc. Soil Sci.* **42**(4): 533-35.

Bhandari, A.L., Sharma, K.N., Kapur, M.L. and Rana, D.S. (1989). Supplementation of N through green manuring for maize growing. *J. Indian Soc. Soil Sci.* **37**(3): 483-86.

Bhat, A.K; Beri, V. and Sidhu B.S. (1991). Effect of long-term recycling of crop residues on soil productivity. *J. Indian Soc. Soil Sci.* **39**(2): 380-82.

Boparai, B.S. (1982). Ph.D. Thesis, Indian Agricultural Research Institute (IARI), New Delhi.

Chaudhary, M.R. and Bajwa, M.S. (1979). *J. Res.* (Punjab Agric. Univ.) **16**: 7-13

Curry, J.P. and Good, J.A. (1992). Soil Fungal Degradation and Restoration, *Advances in soil Sci.* **17**:171-215.

Dadhawal, K.S. and Katiyar V.S. (1989). Effect of farmyard manure on growth, yield and chemical composition of *Bhabar* in a mine spoil. *J. Indian Soc. Soil Sci.* **37**(2): 412-15.

Dagar, J.C., Singh, N.T. and Singh, G. (1994). India agroforestry options for degraded and problematic soils in India. In: Singh, P., Pathak P.S. and Roy, M.M. (eds) *Agroforestry Systems for Sustainable Land Use.* Oxford and IBH Publishing Co. Pvt. Ltd., New Delhi, pp. 96-120.

Das, S.K., Sharma, K.L., Srinivas, K., Narayana Reddy, M. and Singh OM. (1995). Phosphorus and sulphur availability in soil following incorporation of various organic residues. *J. Indian Soc. soil Sci.* **43**(2): 223-28.

Datta, N. and Bhardwaj, K.K.R. (1995). Nitrogen contribution and soil improvement by legume green manuring in rice-wheat cropping on an acid clay loam soil. *J. Indian Soc. Soil Sci.* **43**(4): 603- 07.

Dev. S.P., Bhardwaj, K.K.R. (1993). Effect of crop wastes and nitrogen levels on physical properties of soil in wheat-maize sequence. *Ann. Agric. Res.* **14**: 116-18.

Dudal, R. (1977). World Food Conference, Resolution, F.A.O., Rome.

FAI (1985). Soil of India and their Management, New Delhi. pp. 445.

Fansey, N.R. and Lal, R. (1990). Soil Wetness and Anaero-biosis, *Advances in Soil Sci.* **11**:173-186.

FAO (1977). Assessing soil Degradation, *Soil Bull.* 34, Rome.

Gaur, A.C. and Sadasivam, K.V. (1993). Theory and practical considerations of composting organic wastes. In: Thampan, P.K.(ed). *Organics in Soil Health and Crop Production.* Peekay Tree Crops of Development Foundation, Cochin 682-020. pp.1-22.

Gopal, B. and Sharma, P.S. (1981). Water Hyacinth. The Most Trouble some Weed of the World. Hindasia Publishers, Delhi.

Gupta, O.P. and Lamba, P.S. (1976). Some aspects of utilization of aquatic weeds. In: *Aquatic Weeds in South East Asia* (eds.) C.K. Varshney and J. Rzoska, Dr. W.Junk, B.V. The Hague, pp. 361- 67.

Gupta, J.P. (1991). Sand dunes and stabilization. In: Technologies for wasteland development. (I.P. Abrol and V.V. Dhruvanarayana, Eds), ICAR Publ., New Delhi, pp. 59-68.

Jagannathan, S., Mayalagu, K. and Peer Mohamed, M. (1987). Effect of amendments on soil physical properties *J. Indian Soc. Soil Sci.* **35**(4): 726-27.

Jaio, B., Gu, R. and Zhang. X. (1986). "*Chinese Green Manuring*". Chinese Agric. Publ. House. Beijng, China.

Jayaraman, S. (1988). *Livestock Advisor*, **13**(12): 3

Joshi, P.K. and Agnihotri, A.K. (1986). An assessment of the adverse effects of canal irrigation in India. *Indian J. agric. Econ.* **39**: 528-536.

Joshi, P.K. and Agnihotri, A.K. (1986). An assessment of the adverse effects of canal irrigation in India. *Indian J. agric. Econ.* **39**: 520.

Joshi, R.C., Hookip, D.D. and Singh, K.N. (1990). In: "International Symposium on Natural Resources Management for Sustainable Agriculture". pp. 61-62, New Delhi.

Handa, M.S. and Sur, H.S. (1987). Effect of land modifying measures on erosion, nutrient, water storage and yield of pearl millet fodder. *J. Indian Soc. Soil Sci.* **35**(3): 480-86.

Indian Agriculture in Brief (1992). Directorate of Economics and Statistics, Deptt. of Agric. and Co-operation, Ministry of Agric., New Delhi.

Kapur, M.L. and Kanwar, R.S. (1989). Effect of sulphitation and carbonation PMC and FYM on the accumulation and movement of N, P and K. J. *Indian Soc. Soil Sci.* **37**(1): 56-60.

Krishnamurthy, K. (1984). *A Manual on Compost and Other Organic Manures*. Today and Tomorrow Printers and Publishers, New Delhi.

Krishnamoorthy, V.S., Ramanathan, G. and Vadival, K. (1985). Reclamation of alkali soils. Paper presented in *National Seminar on Salt Affected Soils*, TNAU, Soil Salinity Research Centre. Trichy. pp. 107-11.

Lal, R. (1984). Soil erosion from tropical arable lands and its control. *Advances in Agron.* **37**: 183-248.

Lal, J.K., Mishra, B. and Sarkar, A.K. (1996). Effect of flyash on soil microbial and enzymatic activity. *J. Indian Soc. Soil Sci.* **44**(1): 77-80.

Liu, C. (1988). In: *Green Manure in Rice Farming*. pp. 319-31. Int. Rice Res. Inst. Los Banos, Laguna. Philippines.

Loganathan, S. (1990). Effect of certain tillage practices and amendments on physico-chemical properties of problem soils. *Madras agric. J.* **77** : 204-08.

Maiti, P.S., Sah, K.D., Gupta, S.K. and Banerjee, S.K.(1992). Evaluation of sewage sludge as a source of irrigation and manure. *J. Indian Soc. Soil Sci.* **40**(1): 168-72.

Minhas, P.S. and Gill, A.S. (1985). Evaporation from soil as affected by incorporation and surface and sub-surface placement of oat straw. *J. Indian Soc. Soil Sci.* **33**(4): 774-78.

Narkhede, P.L. and Ghugare, R.V. (1987). Organic recycling in Drylands-II: Effect of Organic recycling of *Subabhul* on yield, nutrient uptake and moisture utilization by winter Sorghum. *J. Indian Soc. Soil Sci.* **35**(3): 417-20.

Narwal R.P., Gupta, A.P. and Antil, R.S. (1990). Efficacy of TSP and Mussoorie rock phosphate mixture incubated with sulfitation process pressmud. *J. Indian Soc. Soil Sci.* **38**(1): 51-55.

Pal, S.S. (1994). Use of pine litter as an organic amendment for an Inceptisol of Garhwal region of Himalaya. *J. Indian Soc. Soil Sci.* **43**(3): 448-52.

Palaniappan, S.P. and Natarajan, K. (1993). Practical aspects of organic matter maintenance in soil. In Thampan, P.K.(ed.). *Organics in Soil Health and Crop Production*. Peeday Tree Crop Development Foundation, Cochin, pp. 23-42.

Panse, V.G., Abraham, J.P. and Lellavathi, C.R. (1965). *Green Manuring of Crops*, Bull. No.2 pp.84 ICAR, New Delhi.

Pande, S.P., Hari Shankar and Sharma, V.K. (1985).Efficacy of some organic and inorganic residues in relation to crop yield and soil characteristics. *J. Indian Soc. Soil Sci.* **35**(1): 179-81.

Patel, K.P. and Singh, B. (1991). A comparative study on the effect of gypsum, pressmud and pyrites on leachate composition, soil properties and yield of rice and wheat on sodic soil. *J. Indian Soc. Soil Sci.* **39**(1): 154-59.

Patil, N.D. (1972). Neem oil as nitrification inhibitor. *Fertil. News*, **17**(9): 37.

Patil, C.V. (1993). Project on agricultural utility of flyash. Agril. College, Richur, Karnataka, India.

Pati Ram and Singh, K.A. (1993). Effect of continuous application of Manure and nitrogen fertilizer on some properties of acid Inceptisol. *J. Indian Soc. Soil Sci.* **41**(3): 430-33.

Pedro, A.S. and Salinas, J.G. (1981). Advances in Agron. 33: 279.

Poonnamperuma, F.N. (1972). The chemistry of submerged soils. *Advances in Agron.* **24**: 29-96.

Raghavan, G.V.S. and McKyes, E. (1983). Physical and hydraulic characteristics in compacted soils. *J. Terramechanics* **19**(4): 235- 42.

Raman, S., Patel, A.M., Shah, G.B. and Kaswala, R.R. (1996). Feasibility of some industrial wastes for soil improvement and crop production. *J. Indian Soc. Soil Sci.* **44**(1): 147-50.

Rao, D.L.N. (1985). "*Sesbania* for Green Manuring". Brochure No.6 p.7, Central Soil Salinity Research Institute, Karnal, India.

Sehgal, J. and Abrol, I.P. (1994). Soil Degradation in India - Status and Impact, Oxford and IBH Publishing Co. Pvt. Ltd. New Delhi.

Sehgal, J. (1995). Land resources appraisal for land use planning to meet challenges of 21st Century. *J. Indian Soc. Soil Sci.* **43** (4): 504-28.

Sehrawat K.L. (1981). Karanja (*Pongamia glabra*) as a Source of nitrification inhibitors. *Fertil. News* **26**(3):29-33.

Sharma, C.P. and Bajpai, P.D. (1989). Effect of legumes on succeeding wheat crop and soil properties at different levels of nitrogen. *J. Indian Soc. Soil Sci.* **37**(3): 475-82.

Sharma, P.K. and De Datta, S.K. (1985). Puddling influence on soil, rice development and yield. *Soil Sci. Soc. Amer. J.* **49**: 1451-1457.

Sharma R.P.R. and Singh, A. (1970). Effect of growing winter legumes on certain properties of an alluvial soils of northern India. *Indian J. Agric. Sci.* **40**: 45-53.

Sim, G.K. (1990). Biological degradation of soil. *Advances in Soil Sci.* **11** : 289-320.

Singh, N.T. (1974). Physico-chemical changes in sodic soils incubated at saturation. *Plant & Soil* 40: 303-34.

Singh, N.T. (1984). Green manures as source of nutrients in rice production In: *Organic Matter and Rice.* IRRI Los Banos, Philippines, pp.217-28.

Singh, N.T. (1992). Dryland salinity in Indo-Pakistan subcontinent. In: Dregne, H.E.(ed). Degradation and Restoration of Arid Land Cent. Arid and Semi-arid Land Stud. Texas Tech. Univ. USA. pp.179-248.

Singh, S.B., Das, P.K., Banerjee, S.K. and Singh, B. (1987). Effect of vegetation covers on the nutrient status of soils. *J. Indian Soc. Soil Sci.* **35**(2): 232-37.

Singh, Y., Khind, C.S. and Singh, B. (1991). Efficient management of leguninous green manures in wetland rice. *Advances in Agron.* 45:135-89.

Singh, N.T., Dagar, J.C. and Singh, G. (1994). Conservation of soils against waterlogging and salinization. *Indian J. Soil Conservation* **22**(1-2): 112-24.

Singh, R.R., Singh, V., Shukla, A.K. (1991). Yield and heavy metal contents of berseem as influenced by sewage water and refinery effluent. *J. Indian Soc. Soil Sci.* 39(2): 402-04.

Singh, K.P. and Sarkar, M.C. (1992). P availability in soils as affected by fertilizer, P, sodium silicate and farmyard manure. *J. Indian Soc. Soil Sci.* **40**(3): 762.

Sreeramulu, U.S. (1994). Utilization of sewage and sludge for increasing crop production. *J. Indian Soc. Soil Sci.* **44**(4): 525- 32.

Surve, S.P. and Daftardar, S.Y. (1985). Effect of some *Neem* (*Azadirachta indica*) and Karanj (*Pongamia glabora*) products on utilization of nitrogen by upland rice. *J. Indian Soc. Soil Sci.* **33**(1): 182-86.

Swarup, A. (1985). Yield and nutrition of rice as influenced by pre-submergence and amendments in a highly sodic soil. *J. Indian Soc. Soil Sci.* **33**(2): 352-57.

Swarup, A. (1986). Effect of preplanting submergence and green manure on yield and soil improvement. *Int. Rice Res. News* **11**(5): 39.

Swarup, A. (1987) *Biol. Fertil. Soils* 5: 203-08.

Swarup, A. (1991). Effect of gypsum, green manure, farmyard manure and zinc fertilization on the Zn, Fe and Mn nutrition of wetland rice on sodic soil. *J. Indian Soc. Soil Sci.* **39**(3): 530- 36.

Swarup, A. (1994). Chemistry of salt affected soils and fertility management. In: Rao, DLN, Singh N.T., Gupta, R.K. and Tyagi, N.K. (eds). *Salinity Management for Sustainable Agriculture*, C.S.S.R.I., Karnal, India, pp. 18-40.

Swarup, A. (1995). Management and balanced use of inputs in achieving maximum yield in salt affected soils. *Fertil. News* **40**(11): 39-47.

Szabolcs, I. (1977). Proc. Indo-Hungarian Seminar on "*Management of Salt affected Soils*", Karnal, pp.20-30.

Tomar, S.S., Tembe G.P. and Sharma, S.K. (1992). Effects of amendments on soil physical properties and yield of crops under rainfed upland conditions of paddy-wheat cropping. In: Rai, M.M. and Verma, L.N. (eds) *National Seminar on Organic Farming*, College of Agriculture, Indore, M.P., India, pp. 116-19.

Totawal, K.L. (1991). Effect of effluent form zinc smelter on metallic cations in soil and crop in cropping sequences. *J. Indian Soc. Soil Sci.* **39**(3); 542-48.

Yadav, J.S.P. (1982). In: "Review of Soil Research in India" *Indian Soc. Soil Sci.* pp. 623-34.

Yadav, J.S.P. (1984). Presidental address -48th Annual Convention. *J. Indian Soc. of Soil Sci.* **32**(1): 1-8.

Yadav, J.S.P. and Agarwal, R.R. (1961). *J. Indian Soc. Soil Sci.* 9: 151-56.

Yaduvanshi, N.P.S. and Yadav, D.V. (1996). Residual effect of pressmud cake with nitrogen on cane yield, juice quality and soil nitrogen in sugarcane ratoon. *J. Indian Soc. Soil Sci.* **44**(1): 158-160.

Yashpal, Vig, A.C. and Milap-Chand (1995). Effect of P fertilization and *Sesbania* green manure incorporation on phosphate availability. *J. Indian Soc. Soil Sci.* **43**(4): 673-81.

Young, I.M. (1987). Soil strength and hardsetting behaviour of some structurally unstable British Soils, Ph.D. thesis, Aberdeen Univ. Aberdeen, U.K.

Zalawadia. N.M. and Raman, S. (1994). Effect of distillery waste water with graded fertilizer levels on sorghum yield and soil properties, J. Indian Soc. Soil Sci. **42**(4): 575-79.

Chapter - 7

IMPROVEMENT OF CROP STAND UNDER DRYLAND CONDITIONS

*N.G. Zode**

Seed Research Officer, Seed Technology Research Unit, Dr. Panjabrao Deshmukh Krishi Vidyapeeth, Akola - 444 104 (MS)
* Present address: Incharge & Breeder, Paddy Research Centre, Substation Sakoli. Distt. Bhandara (MS)

Seed is a basic input in agriculture. Without good seed, the investment on fertilizer and water will not pay the dividend (Dr. M.S. Swaminathan). Crop performance is entirely determined by a complex genotype and environmental interaction. The main reason for low yields in dryland farming is the poor plant stand, the outcome of which can be radically altered by the quality of seed. Numerous factors interact to determine the number of healthy autotropic plants obtained from planting a given sample of seed i.e. seedling establishment. Under field conditions, physical, chemical, and biotic properties of soil; the seed and seedling vigour and seed treatments to increase vigour, reduce barriers affecting seedling establishment as well as method of sowing, seed rate and depth of seeding shall influence the plant stand. In addition, each of these factors interact with the environmental factors like water, temperature, oxygen and light that regulate the rate and percent germination.

In normal situation i.e. under favourable conditions of factors like water, air, temperature and light, the plant stand in the field shall be in proportion to seed germination percentage which is reported by seed testing laboratory. However, under dry-land situation, it becomes very complex phenomenon. The interaction of soil type, soil moisture availability and seed quality ultimately result in poor stand establishment. To overcome this problem, the enormous research has been done in the country and abroad. Efforts are made to review important findings in brief, which may enable the farmer to harvest dividends of his efforts.

1. FIELD FACTORS INFLUENCING SEEDLING ESTABLISHMENT

The important field factors affecting seedling establishment are soil type and seed bed moisture availability.

1.1 Soil type

The different types of soils differ in their properties like bulk density and water holding capacity. The shallow soils (black or red) are coarse to medium structured, well drained but poor in water holding capacity. Whereas, medium and deep black soils are rich in nutrients with very good water holding capacity. In soils with good water holding capacity, the moisture supply rarely limits germination, while in poor soils, drying occurs fast and seeds always face water scarcity resulting scanty seed germination. Another important soil property which differs with soil type is the soil strength. It influences the seedling emergence greatly. As the soil becomes dry, soil strength increases, restricting the seedling emergence, though there is sufficient soil moisture to continue cell division and elongation in seedling. Seedlings growing horizontally under the crust reflect this phenomenon.

Heavy soils, especially clay and clay loams, are likely to develop a crust on the surface following heavy rains. It is formed due to increase in soil strength by soil compaction and loss of soil structure. Crusting increases the resistance to shoot and root penetration and is almost frequently a problem in stand establishment on fine structured soil with poor soil aggregation and low organic matter. The emergence of seedling through soil crust is governed by seedling size, number of seedlings at a hill and spacing between seeds. Soil cracking patterns the strength of crust around the emerging seedling. Multiple seedlings in a group are able to rupture the soil crust. The small seeds like mustered or sesame are more prone to affect by soil crust. If the vigourous cracking in soil occurs, it may help in seedling emergence. However, in the event that a surface does form the crust and meagre possibility of cracking has it should be broken up with spike tooth harrow or a rotary hoe before the seed starts to germinate.

The soluble salts and salinity of soil also influence the seed germination and field emergence due to osmotic and in some cases ion toxicity effects. Field crop establishment in arid and semiarid soils, where salts were more prevalent, heavy rates of fertilizers in the seed zone may pose the similar problem.

Crops exhibit a wide salt tolerance range among and within the species. Saline soils contain more than 0.2% soluble salts and have conductivity greater than 4 dSm^{-1}. The soil salinity at low value of 5 dSm^{-1} had no effect on seed germination but at 20 dSm^{-1} it reduced germination in tomatoes and capsicum. In flax and cauliflower there was variable response depending on a variety. In lolium, differences for salinity tolerance prevailed between species. Card Well (1984) stated that conductivity greater than 6 dSm^{-1} reduces emergence rate and final emergence percentage of sorghum, sugarbeat, field bean, pigeonpea and cowpea. Salt tolerance is usually less during germination than for established plants. The emergence is related to the combined stresses of moisture and osmotic potential. Emergence declines rapidly when combined forces exceed 0.8 MPa.

1.2. Seed bed Preparation

The well prepared seed bed achieves better soil-seed contact, and more precise planting depth altimately results in higher field emergence. Traditional seed bed preparation techniques of ploughing, clout crushing, harrowing twice or thrice before planting generally result in firm seed bed giving good soil-seed contact and uniform, rapid emergence under favourable temperature and moisture conditions. Now a days, because of fast mechanization, by use of tractor and rotavator, it is easy to prepare desired seed bed. Minimum tillage seed bed preparation frequently results in lower plant population when older equipments which are not designed for rough field condition are used.

1.3. Soil moisture

About 100 to 125% moisture must be imbibed by seeds of crops for proper germination. The maximum seedling emergence was observed at -0.15 and -0.3 bars in sandy loam and clayey soils respectively. More than 80% of viable seeds of lentil, niger and wheat germinated at -6.0. -4.0 and-1.0 bar respectively in clayey soil (Rajput and Gupta, 1978). This indicates that soil moisture conditions exert a dominant influence on stand establishment because of modifying effects of soil moisture on soil properties (see soil type). Soil structure, soil water potential and seed-soil surface contact determine the rate of moisture uptake by seed in soil. Increasing seed surface contact with liquid water decreases germination time and increases per cent germination. Any factor reducing soil hydraulic conductivity (determined by pore diameter) or seed soil water contact (determined by aggregate size and matrix potential) reduces rate of water uptake and delays seed germination. Card Well (1984) suggested general guidelines for preparing seed beds that obtain in crow zone, a mean soil aggregate size of 1/5 the seed diameter to ensure rapid water uptake and germination. Many crop seeds are able to imbibe sufficient water at moisture near or slightly below the permanent wilting point to initiate germination but do not elongate. Aksenova *et al.* (1993) reported enhanced germination in conditions of low water availability, by treating seeds with various ionic and non-ionic surfactants. Insufficient aeration presumably limits germination at very high soil moisture conditions, since diffusion of oxygen in air is 10,000 times greater than in water at 20ºC.

2. SEED FACTORS INFLUENCING SEEDLING ESTABLISHMENT

Seed is the fertilized matured ovule consisting of embryo, a living organism embeded in supporting or the food storage tissue, which are surrounded by protective coat. Since ancient times, seed has been important for agriculturist. The great Hindu Scriptures like "Manu Smruti" states *Subijam Sukshetre Jayate Sampadyate"* i.e. good seed in good soil yields abundant. When we talk of a 'quality seed', we quite often presume high genetic purity, high in vigour and germination, free from seed borne diseases and higher yielding ability.

Good quality seed has a great role in crop production primarily through improvement of plant stand. Feistritzer (1978) rightly pointed out that the use of good quality seed combined with complementary inputs viz. fertilizers, irrigation and plant protection chemicals etc. has potential for rapidly and substantially increasing crop production. The use of good quality seed of improved varieties has been mainly responsible for success of green revolution in India.

However, it is observed that except in case of hybrids, most of the farmers use saved seed of varieties from previous year/ season's produce. The seed replacement rates (approximately) currently followed are 100% in case of hybrids, 10% in case of rice, maize, sunflower, and safflower, 13% in case of pulses and cotton, and 8% in case of sorghum, pearl millet, wheat and groundnut. It is very less in case of black gram and green gram.

Table 1: Comparison between certified and uncertified seed of wheat and Moong

Seed category	Wheat yield (q/ha)	Off type (%)	Moong yield (q/ha)	Off type (%)
Uncertified	37.3	12.0	3.0	1.9
Certified	40.6	0.6	4.3	0.8

(Source : Yadav *et al.,* 1988)

In the experiments conducted by National Seed Project (ICAR) it was observed that productivity of farmer's saved seed was less than that of certified seed, even in self pollinated crops like wheat and often pollinated crops like sorghum. The comparative performance between certified and uncertified seed of wheat and green-gram (Tables 1 & 2) revealed superiority of certified seed over uncertified seed.

Table 2. Comparative performance of different categories of wheat

Seed category	Yield (q/ha)	Germination %	Physical purity
Uncertified	59.8	93.00	97.2
Certified	64.1	98.00	99.3

(Source : Agrawal, 1988)

At present about 70.0 lakh quintals of certified quality seed of different crops is being distributed in India through both public and private sectors. Also the large quantity of notified and not notified improved varieties are marketed under truthful label. Still there exists a very wide gap between total seed requirement and quality seed produced.

In general, independantly and interacting with each other, genetic quality, physical quality, physiological quality and seed health quality constitute a over all seed quality.

2.1. Genetic Quality

It represents the inherited differences for specific plant characteristics, such as yield, lodging, plant height, disease and pest resistance etc. It can be assured only by rouging out individual plants which deviate from the variety grown for seed production and following desired isolation distance, through field inspections, grow out tests and laboratory tests. As per the seed certification standards, confirmation of genetic purity is compulsory for certification of hybrid seeds of notified hybrid varieties. Further to ensure the genetic purity of improved varieties of cotton and soybean the Governing Board of Maharashtra State Seed Certification Agency has decided to conduct grow out test before issueing the certification tag. The National Seed Project is emphasizing on grow out test for nucleus and breeder seed of all the notified varieties.

2.2. Physical Quality

When you buy seed, you expect to receive good seed and not a mixture of chaff, sand, clay, weed seed etc. The pure seed is referred to the species in question or found to predominate in test and shall include all varieties and cultivars of that species. The structures of intact seeds like achenes, and similar fruits, schizocarps and mericarps with or without perianth and regardless of whether they contain a true seed; as well as pieces of their seeds or fruits that are more than one half their original size shall be considered as pure seed. In India, with a promulgation of seed Act 1966, the seed certification system is accepted on voluntary basis for notified kinds or varieties of crops. The Indian minimum seed certification standards and certification and labeling rules emphasize on reducing all adulteration with foreign matter and weed seeds. If a lot is labelled as specific variety, it should have 99.00% purity for foundation seed, 98% for composits, synthetics, multilines etc. and 95% for hybrids (Indian minimum seed certification standard, 1988). The other species seeds and inert matter percentage should be as low as possible (below 2%). The percentage of seed of other species should be almost negligible or nil (below 0.1 per cent).

3. SEED PHYSIOLOGICAL QUALITY

3.1. Seed Viability

Because of its role in stand establishment, seed germination is a key to modern agriculture. The seed germination determined in seed testing laboratory is to indicate the probable capability of seed lot for field emergence as well as to compare the germination value of different seed lots. Seed germination in laboratory has a very strong positive correlation with field emergence i.e. higher the germination in laboratory more will be the field emergence. The 'word viability' is also used synonymously with germination capacity and is probably highest at the time of physiological maturily. The deterioration in viability occurs with the ageing of seed depending upon the environmental conditions it receives. The good quality seed of crops like wheat, sorghum, pearlmillet etc. has a viability span of more than 12 months but less than 24 months. However, in case of cotton, pigeonpea, gram, black gram, and green gram the value more than 24 months from harvest is also considered.

The research at Dr. Panjabrao Deshmukh Krishi Vidyapeeth, Akola, showed that black gram and green gram had the germination above 90% even after 64 months of storage under ambient but insect free storage conditions of Akola.

3.2. Seed and seedling vigour

Seed vigour is just one of the several estimates of quality and is not a absolute term but is a relative measure. Differences in vigour are only revealed in practice when germination test tells to indicate emergence differences in the field. Estimates of vigour give assurance as to the reliability of emergence. The laboratory germination of seeds describes the percentage of seeds that are capable of producing normal seedlings under optimum conditions in laboratory. However, the field conditions are always more unfavourable than laboratory conditions and lack of vigour results in delayed and/or reduced seedling emergence. The longer the time required from the start of germination to the emergence of seedling; the most vulnerable stage in any plant life cycle, the greater is the chance of mortality from either physical stresses or biotic stresses. The table 4 indicates that the vigour differences also are seen in the ageing of seeds in the storage even among the lots with similar high levels of germination. (Powell and Matthews, 1984).

Table 3. Storage period in months for which germination has been maintained above minimum seed certification standards (MSCS)

Sr. No.	Crop (MSCS %)	Variety/Hybrids	Storage life (M)
1.	Sorghum (75%)	(i) CSH-1	22
		(ii) CSH - 5	17 to 22
		(iii) CSH - 9	18
2.	Wheat (85%)	(i) HD 1553	20 to 29
		(ii) Lok -1	24
		(iii) HD 1593	22
		(iv) HD 2189	20
		(v) Arjun	20
3.	Pigeonpea (75%)	(i) BON-L	46
4.	Gram (85%)	(i) Chaffa	53
		(ii) Phule - 1	54
		(iii) G-5	45
5.	Green gram (75%)	S-8	64

6.	Black gram (75%)	(i) T-9	64
		(ii) TAU - 1	52
7.	Cotton (65%)	(i) AHH- 468	36
		(ii) H4	50
8	Pearlmillet (75%)	(i) ICMS7703	35
		(ii) MH-179	22

(Highlights of Research. Seed, Technology Research Unit, Dr. PDKV Akola.)

Ageing, both before harvest and in the storage after harvest, is the major determinant of the germination level and the vigour of a seed lot. This is illustrated in Fig. 1 for two hypothetical seed lots, 'A' (high vigour), 'B' (low vigour) and two storage conditions, 'A 1' (less adverse) and 'A2' (more adverse).

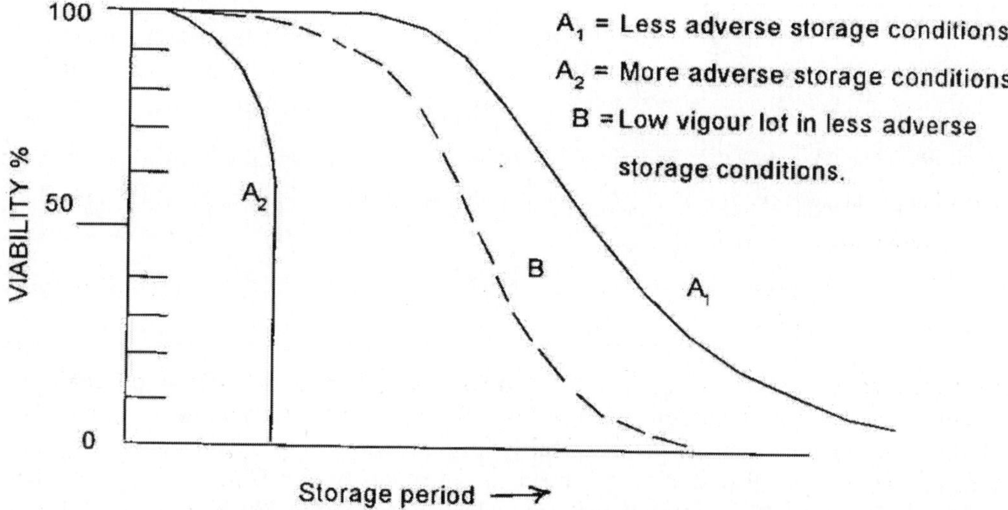

Fig. 1. Effect of storage conditions on seed viability.

Abundant evidences suggest that differences in emergence of seed lots exhibiting high levels of germination in the laboratory (above 80%, see Table 4) arise because there can be considerable differences in seed age that are not revealed in germination tests. Matthews (1980), has shown this by accelerated ageing test and controlled deterioration test.

An ageing of the seed in general has been attributed in part to the accumulation of free radicals having deleterious side effects, produced by metabolic processes and induced by ionized radiation. The short lived radicals produced during auto-oxidation can be damaging to proteins, enzymes and other biological compounds in their proximity. As seed deterioration progresses, cellular membranes loose their selective permeability, permitting the cytoplasmic metabolites to leach into the intercellular spaces. Membrane degradation occurs from both hydrolysis of phospholipids by phospholipase and phospholipid auto-oxidation, therefore in the strictest sense, it might be considered a result of ageing rather than cause.

The phenomenon of imbibition damage can occur in field, specially in conditions of high soil moisture and low temperature. Rapid inhibition can damage cell on the surface of the normally living cotyledons of grain legumes such as peas, soybeans, phaseolus, beans, chickpeas, and cowpeas (Powel *et al.,* 1984). Seed conditions and genotypes within the species also influence the occurrence and severity of imbibition damage. Cracks in seed coats arising from processing damage such as the mechanical handling of dry seed can lead to rapid imbibition. Two other aspects of seed condition that can influence imbibition damage are ageing and excessive dryness. Seeds that have been aged either by conditions during production and storage or through prolonged storage show much greater sensitivity to rapid imbibition (Powel, 1984). Seeds dried to less than 5% moisture content can be severely damaged by imbibition unless this is done slowly in a humid atmosphere (Ellis *et al.* 1982).

Table 4. Range of per cent germination and emergence in field sowing of commercially available seed lots of eight crops.

Crops	Number of Lots compared	Range among lots	
		Lab. germi%	Field emergence %
1 *Pisum sativum* (Vining peas)	80	>80	8-85
2 *Pisum sativum* (combing peas)	17	76-80	7-85
3 *Glycine max* (soybean)	18	86-96	22-90
4 *Vicia faba* (field beans)	14	83-99	63-97
5 *Phaseolus Vulgaris* (green bean)	30	75-100	34-93
6 *Allium cepae* (onion)	12	77-92	44-72
7 *Brassica oleracea* (Brussels sprouts)	15	71-78	26-72
8 *Brassica campestris* Turnip	30	88-99	36-76

Vigour differences in the seed available to growers are clearly of great significance to crop establishment. In case of low vigoured seeds, if low seed rates or even equal to recommended rates are adopted, there will always be the possibility of reduction in yield because of poor stand establishment. It is also possible that stands made up of plants grown from low vigour seeds may yield less than those grown at the same density from seeds of high vigour (Matthews, 1993). Robert (1986) found that yields were lower in crops established late in the season from low vigour seed of spring wheat compared to high vigour seeds sown at similar densities, similarly revalidated seeds of sorghum hybrids were found low in vigour as compared to fresh seed lots. And though revalidated seeds had germination above certification standard, it proved poor yielder than fresh seed, mainly because of failure in crop stand establishment. (Anonymous, 1998).

Table 5 Pooled data for three years on seed germination (Laboratory germination) field emergence (initial plant stand)

Seed Category	Germination %	Field emergence per hectare	Final stand per hectare	Yield/ hectare qtl.
Fresh	94.68	1.38 lakh	1.08 lakh	29.99
Revalidated	84.4	0.99 lakh	0.81 lakh	24.11

Thus, it will only be the vigour in the seed already at acceptable high levels of germination that shall be of interest to improve reliability of stand establishment.

3.3. Seed size

The large number of experiments are conducted on effect of seed size on crop performance. The yield differences which we get by sowing bold seeds are not because of their higher yielding potential but only because of their effects on plant stand establishment. Bolder the seed, better the crop performance is true in case of small seeded species only. Successful stand establishment of small seeded grasses, legumes and crops like mustard and sesamum is often a function of planting depth and soil tilth. Deeper the sowing and coarser the tilth, poor is the stand establishment. Thus, benefit of large seed size among the small seeded species becomes important. Increasing seed size within a species generally increases final emergence at normal depth of sowing. The greater benefits occur at maximum depth. In sunflower, Kausal *et al.* (1991) reported greater seed viability in bolder seeds as compared to smaller one.

Contrary to above proposition, in case of soybean and some bean species, the bolder seeds are more prone to mechanical damage. In such species, plant stand establishment is found to be always superior to planting small seeds of these species. Tiwari *et al.* (1976 & 1978) reported that bold seeded varieties in soybean were found to posses low germinability and emergence potential. The small seeds with 4.5-5.0 mm diameter and 140-150 mg weight/seed exhibited significantly higher emergence potential (85.88%) than bolder seeds (72.24%) with 6.0 to6.5 mm diameter and 150-160 mg weight/seed.

In sorghum, the findings of seed Technology Research Unit, Dr. Panjabrao Deshmukh, Krishi Vidyapeeth, Akola (1981) revealed that the seeds graded by using 8/64" screen was equally comparable to recommended 9/64" screen in respect of seed germination, vigour, field emergence and yield. Similarly, Chowdhari *et al.* (1984) reported non significant differences in grain yield from crops raised by using different seed size.

3.4. Seedling emergence strength

Emergence strength of individual seedling varies from a low of 0.15 newton for alfalfa to a high of 2.9 newtons for corn. Multiple seedlings in a group are able to rupture higher strength of soil crusts. For example, studies show that the maximum thrust of one, two and three cotton seedling is 3.8, 5.8 and 8.5 newtons respectively. Similarly, subterranean clover (*Trifolium subterraneum* L.) when seeded 2 cm deep in clumps of 1, 2, and 5 seeds has 28, 72, and 85% emergence (Card Well, 1984). Based on this principle, while sowing of precious seeds of hybrid cotton in India, it is advocated to sow few seeds of sunnhemp on every hill with 2 or 3 seeds of cotton to facilitate the ease in germination.

Large seeds exhibit greater emergence thrust but also encounter the more soil resistance due to greater surface area of their emerging seedling structures. Emergence data for many species and cultivars generally favour the larger seeded types when all other factors are equal. This is particularly true for the smaller seeded grasses and legumes where emergence from deeper planting is promoted by large seed size. Large seeded dicots occasionally fail to establish themselves as well as the smaller seeds of same species only because of differences in seed quality.

In cotton, hypocotyl elongation is more sensitive than radical to increasing soil impedance and decreasing soil moisture. Increased physical strength and decreased soil water reduce the depth from which seedling can potentially emerge from over 7.00 cm at 0.03 MPa soil moisture and 0.02 MPa soil strength to less than 2 cm at 1 MPa soil strength. Under this situation, incorporating clump of fast growing seeds of crops like sunnhemp improves the emergence.

The tall growing varieties of wheat recommended for dry sowing, could germinate even at the sowing depth of above 7.0 cm. However, the present dwarf varieties has a short plumule, and do not reach the soil surface if sown deep.

3.5. Seed health and quality

Several serious pathogens are carried as spores or chlamydospores, on surface of seeds and when harvesting is delayed, mycelia may penetrate to the inner structures such as cotyledons and kill embryo.

Pathogens inhibiting the seed or soil may reduce the stand establishment, where seedling emergence is delayed by low temperatures, high soil moisture and other conditions which are more favourable for pathogens than for germination and emergence of seed.

Fungi caused the largest number of plant diseases and occur more commonly in or on the seeds than bacteria or viruses. More than 8000 species of fungi have been identified as plant pathogen. Fungi associated with seeds consists of both saprophytic and pathogenic fungi. Saprophytic fungi are not specific to any particular host and may be found on seeds of various plants whereas, pathogenic fungi are usually confined to a limited host range. Both type may occur on the seed surface, in cracks or inside the seed coat but pathogenic fungi may also occur with seed embryo itself.

In the laboratory studies at Seed Technology Research unit, Akola; *Aspergillus flavus* and *niger, penicillium* spp., *Fusarium species, Alternaria species* are common fungi occurring on seed. In cotton *Fusarium moniliformae, Fusarium, oxysporium, Colletotricom dematium, Alternaria tenuis, Curvularia lunata, Rhizoctonia bataticola, Dreschslera* species and other saprophytes like *Rhizopus* species were identified inhabiting the cotton seed. In case of sorghum, more than 18 types of fungi were noticed on seed, the prominant occurrence was of *Aspergillus flavus, Aspergillus niger, Penicillium sp., Curvularia lunata, Fusarium moniliformae, Alternaria alternata, Alternaria padwicky* and *Dreschiera sacchari.* These fungi were also common on soybean and sunflower. In paddy, seed borne microflora consists of *F. moniliformae, D. oryzae, Periconia padwicky, P. oryzae* and *Rhizoctonia* sp.

Field invasion of seeds by fungi can occur when the seed is maturing, but requires 90-95% atmospheric relative humidity and least 22-25% seed moisture in cereal seed. Fungi genera commonly found in field infected seeds included *Alternaria, Aspergillus, Penicillium, Fusarium* and *Helminthosporium.* Alternaria does not reduce germination, although it does causes discoloration. All the other fungi may reduce germination.

The major storage fungi influencing seed viability include large number of *Aspergillus* species and *Penicillium* species. Optimum conditions for development of these species are 25-30°C temperature and 60-90% R.H. Safe seed moisture to prevent microbial deterioration of stored seed is less than 12.5% for cereals and below 8.5% for some oil seed crops.

Soil inhabiting microorganism influencing seedling establishment include those causing damping off disease (*Pythium, Rhizoctonia, Botrytis* and *Phytophthora*).

The symptoms may differ depending on the stage of germination or growth when infection occur i.e. (i) pre-emergence damping off associated with seed and seedling rot (ii) Post emergence damping off associated with stem rot near soil surface and (iii) root rot associated with plant stunting and eventual death. The optimum temperature range for these organisms is 20-30°C. The planting seeds, in general require temperatures of 25-30°C. If soil temperature goes near 20°C, it may be more favourable for pathogens than to the seeds. Cool season crops avoid these pathogens due to their faster growth rate at temperatures below 20°C.

4. AGRONOMIC FACTORS INFLUENCING SEEDLING EMERGENCE

4.1. Seeding depth

The sowing depth of many crops is supposed to be 4-5 times the average seed diameter. However, the rate of emergence is more dependent on soil temperature and moisture conditions than on planting depth. Corn emergence is delayed by 1-2 days with each 2-3 cm increases in depth (Card Well 1984). Very shallow plantings may fail due to rapid soil drying near the surface, since 1.25 cm of soil dries to below the permanent wilting point within 24 hours after wetting. The rate of soil drying varies with soil depth and atmospheric conditions i.e. temperature and humidity. Drying front advances linearing with time. Thus, slight increase in planting depth may make large difference in amount of moisture available to seed. A seed planted 2 cm deep requires 10 days to emerge and could dry out before emergence, whereas, planting 4 cm deep could solve the problem. Soil compaction can improve emergence from shallow planting depths because of better seed-soil moisture contacts and continued capillary movement from deeper region.

Seeding depths generally are greater on sandy than clay soils due to lower water holding capacity and faster drying rate at given depth. In regions where sowing occurs during periods of low soil moisture availability (rabi sowing) or low rainfall. Seeding depth may be doubled or more depth is commonly followed in more moist regions. The maximum depth for wheat planting is related to coleoptile ability to elongate and the mechanical resistance of soil.

4.2. Sowing methods

The seed sowing methods also influence the plant stand establishment. The seeding can be done by broadcasting, drilling or dibbling. Broadcasting is very crude method of seeding. Neither the sowing depth is controlled in this method nor uniformity in spacing is possible. During mixing the seeds, some seed goes deep beyond desired depth, while some remains on surface without soil cover. This method is followed to some extent in crops like paddy, wheat, linseed etc. Under conditions of low soil moisture, the seeds, falling deep to level of available moisture, shall emerge without fail, however, the seed on surface and at shallow depth becomes ill fated. Contrary to this under high moisture regimes, the seed going deep faces aeration problem and fails to emerge. Even under condition of high moisture content seed laying on surface without soil cover does not emerge. That way drilling method under dryland conditions as well as irrigated conditions and dibbling method under irrigated conditions are the better methods of seeding. Drilling with precise dibbler, with mechanism of adjusting depth of placement, number of seeds per hill, covering of seeds and compaction to bring seed in contact with soil is the best method of seeding for achieving better plant stand. Placement of seeds at desired depth and with desired number/hill helps in increasing seedling emergence.

4.3 Seed rate

Though lot of development took place in agriculture, the farmers adopting dry farming are not using Recommended seed rates. The seed rate per hectare to be adopted for sowing has a great relavance in contex to plant stand establishment and is dependent mainly on the planting distance (Plant to plant and row to row) and seed size. It is inversely proportionate to planting distance and directly proportional to seed size. The other factor which should be considered while deciding seed rate is seed quality. The low germination and poor vigoured seed lots always fail in plant stand establishment under normal rates of seeding. The acceleration of seed rate on prorata basis and sowing 2-3 seeds in clump may help in stand establishment. Thus acceleration of seed rate over normal and thinning after emergence if necessary shall be good practice for better stand establishment at least under unirrigated condition.

Plate 7.1. 1. Proper storage of harvested produce for better seed quality, 2. Raised wooden platform for air drying of harvested produce in field for good seed quality

Plate 7.2. 1. Drip irrigation system for improvement of crop stand, **2.** Virus affected plot of blackgram (TAU-1). Avoid collecting seed for sowing from affected plants. It helps spread diseases in next generation.

4.4 Seed Treatments

Seeds are uniquely equipped to survive as viable regenerative organism until the time and place are right for beginning of a new generation. Depending on such natural mechanisms available in different kinds of seeds, they may be short lived, long lived and very long lived seeds. However, as man has started farming, the problems like presence of seed dormancy, death of seeds and seedlings etc, were encountered. The crop seeds are fairly resistant to extreme external conditions provided they are in a state of desiccation. The length of time for which seeds remain viable is extremely variable and is determined genetically. Environmental factors and storage conditions had a decicive effect on the life span of any given seed. Some seeds do not germinate due to presence of dormancy. Therefore, efforts were made to increase life span of short lived seeds, to break the dormancy in the seed, to prevent the death of seed and seedlings, while in storage or when sown in field (prior and after germination). As a result of these efforts various types of seed treatments depending upon the type of problem are developed. They can be summerised as under.

5. DORMANCY BREAKING TREATMENTS

Some kinds of seeds have dormancy, may be of physical or chemical nature of varying range of period and proportion. Even after hundreds of years of development in agricultural crops, the dormancy for some degree exists in various, oil seeds, pulses and cereals as well as vegetable seeds. The scarification, hot water treatment, acid treatment etc. are recommended for breaking physical dormancy mostly, in pulses and other leguminous seeds. The chemical treatments like application of GA_3, IAA, KNO_3, Ethrel/ethepon, etc. are recommended for removing chemical dormancy or substitute for requirement of germination like light and temperature.

5.1. Seed/seedling protection treatments

Seed treatments with chemicals are cheap, easy for the grower to use and give useful protection from pests and diseases during storage and early stages of plant growth in field. There are various kinds of insects which damage the seed right from on the plant in field to storage in godowns and again in soil after sowing. Again in field, various kinds of fungi may invade the seed. Out of common genera, found in field large number of *Aspergillus* spp. and *Penicillium* spp. influence the seed viability in storage. Similarly, soil inhabiting microorganisms influencing seedling establishment include those causing damping off. (Pythium, Rhizoctonia, Botrytis and Phytophthora). Damping off due to *Pythium or Rhizoctonia* generally is greatest in warm season crops planted early or dry season crops sown in humid or wet season, (eg. coriander). The crops sown in their favourable season avoid these pathogens due to their faster growth rates than the fungi.

The seed or seedling protection treatments for control of insects and pathogens, promote good seedling establishment, minimise yield loss, maintain and improves seed quality and avoids the spread of harmful organisms. The treatments are planned to control seed borne organisms, and soil borne organisms. The way in which organisms are carried in the seed lot, influence the effectiveness of the particular treatment. Pathogens carried superficially on seed coat are controlled effectively by application of nonsystemic powder or liquid treatments.

However, internally seed borne mycelia may survive. Nonsystemic chemicals and systemic chemicals are more effective. The fungi like *Ustilago* spp. causing loose smut of wheat and barley, establish themselves directly in the embryonic tissues, and only systemic fungicide can penetrate to the embryo. Pathogenic fungi present in soil as spores or mycelium causing damage to seedling can be prevented by appropriate seed treatment.

5.2. Seed invigoration treatments

Seed invigoration in simple sense means any kind of seed treatment which will strengthen the seed or seedling to withstand against field stresses like water stress or insect/pathogen attack in the field and help in seedling establishment. It is to be understood that seed vigour is sum total of those properties of seed which determine the level of activity and performance of a non-dormant seed or seed lot during germination and seedling emergence (Pery, 1978). Therefore, any seed treatment which will supplement or strengthen one or more seed/seedling attributes manifesting vigour which will be beneficial for seedling establishment.

Preplant hardenity to hasten germination and precondition the seed to stress environment has been reported in the literature. The procedure consists of repeated imbition of seeds for 6 to 8 h. at 25°C and start the radical emergence followed by redrying. The cycle may be repeated 3-5 times. The process reportedly activates latent physiological processes. Using this techniques, the rate of emergence is observed to increase but not the per cent germination. The rapid water uptake by previously soaked and dried seeds, may explain the reported improved tolerance to seedling under stress conditions. Most of the work is reported, in wheat, rye, beet, carrot, sunflower etc.

The other type of invigoration is providing the fertilizers, growth regulators, and fungicides by way of pelleting, in close proximity to seed in precise amount which will bust the growth of newly emerged seedling and increase seedling establishment. Even the incorporation of some antagonistic fungi like *Tricoderma* in invigoration treatment help in checking the growth of fungi causing seedling decay, damping off or wilting in crops like safflower, gram, etc.

6. CROPWISE CONSIDERATIONS FOR BETTER STAND ESTABLISHMENT

6.1. Cotton

I. Seed rate:

Unirrigated cotton-Arborium cotton 15-20 kg/ha

Hirsutum cotton 10-12 kg/ha

Hybrid cotton 4 kg/ha

II. Seed Treatments :

1. The use of delinted seed is recommended.
2. Seed treatment of fungicides @1g carboxin + 3g thiram per killogram of seed.
3. For premonsoon cotton sowing, the linted seed should be hydrated for 8-10 h i.e. 4-5 h water soaking plus 4-5 h incubating in moist cloth or jute bag speed of germination and % emergence is improved.
4. If the number of seeds per hill is to be kept 1 or 2, then sowing of 4-5 sunnhemp seeds alongwith cotton seed is advocated.
5. To reduce the initial damage by sucking pests, the treatment of Emadochloropid. commonly called as Goucho @ 10 g/kg seed.

6.2. Green gram and black gram

I Seed rate : 12-15 kg/ha.

II Seed Treatments :

(1) Fungicidal seed treatment thiram 3 g/kg seed.

(2) Antagonistic fungus (Biocontrol) Trichoderma 4 g/kg seed

6.3. Gram - (Bengal gram)

I. Seed rate : 50-60kg/ha, for small seeded varieties

85-100 kg/ha, for bold seeded varieties

115-125 kg/ha for extra bold, varieties

II. Seed treatments :

(1) Fungicidal seed treatment thiram 3 g/kg seed.

(2) Antagonistic fungus (Biocontrol) Trichoderma 4 g/kg seed

6.4. Pigeonpea

I. Seed rate :

1. 40-50 kg/ha for determinant types
2. 25-30 kg/ha for indeterminant types
3. 5-7 kg/ha for Hybrids.

II. Seed treatment :

(1) Fungicidal seed treatment thiram 3 g/kg seed.

(2) Antagonistic fungus (Biocontrol) Trichoderma 4 g/kg seed

6.5. Groundnut

I. Seed rate :

100 kg kernels/ha for medium sized varieties eg. SBXI

125-135 kg kernels /ha for bold seeded varieties, TAG-24

150 kg kernels/ha for Extra bold seeded varieties.

II. Seed Treatment :

1. Thiram : Bavistin 2:1 of thiram 3g or captan 3g/kg
2. For cultivars having seed dormancy, seed treatment of 10 ppm etheral is recommended.
3. If crop is to be introduced in new area. Rhizobium inoculation @ 12 g/kg seed is recommended.

6.6. Sunflower

I. Seed rate - 10 kg seed/ha. for drilled crop

- 5 kg seed/ha. for dibbled crop.

II. Seed Treatment

1. Thiram, brassicol, captan, thiram : bavistin 2:1 any one above, @ 3g/kg seed.

6.7. Soybean

I. Seed rate - 75kg/ha

II. Seed treatment -*Rhizobium japonicum* @100 g/8kg seed.

6.8. Sesamum

I. Seed rate -1.5 - 2.0 kg/ha, kharif crop, 3 -4 kg/ha for summer crop,

II. Seed Teratment

1. Brassicol, thiram or captan 3g/kg seed.

6.9 Safflower

I. Seed rate - 10-15 kg/ha

II. Seed treatment

1. Brassicol, Thiram or Captan 3g/kg or carbendazim 2g/kg seed,
2. Tricoderma @ 4g/kg seed to check the wilting/root rot

6.10. Linseed

I. Seed rate -10-12 kg/ha

II. Seed treatment - 2g thiram/kg seed.

6.11. Rice

I. Seed rate - 50 kg/ha for coarse varieties, 35 kg/ha for fine varieties.

II. Seed treatment

1. The seed should be deeped in 3% solution of NaCl (Common salt). Remove the floating light seed. Drain the brackish water, retreat the remaining seed with plain water 2-5 times. Dry in the shade for 24h.
2. Before sowing dry-dress the seed with thiram @ 3g/kg.

6.12. Sorghum

I. Seed rate - 7.5 to 10 kg/ha

II. Seed treatment

1. Thiram 3g, or thiram 2g carbendazim 1g/kg dry-dressing.

6.13. Wheat

I. Seed rate -

75 kg/ha for dry wheat

100 kg/ha for irrigated wheat,

125 kg/ha for Bold seeded varieties

150 kg/ha for late sown irrigated wheat.

II. Seed treatment

1. Dry-dress the seed with thiram + Carbendazim (2:1) @ 3g/kg seed
2. Apply Azotabacter culture @ 25 g/20 kg seed, 2h prior to sowing. Dry in the shade.

Thus, the good quality seed is the key factor for better plant stand establishment. The soil factors and sowing technology is also equally important. The seed/seedling protecting treatments proved helpful for seed/seedling to overcome the soil biotic stresses and ultimately improve the plant stand in field under dryland conditions.

REFERENCES

Aksenova L.A; M.A. Bocharova; M.V. Dunaeva; E.A. Zak and N.L. Klyachko 1993. Effect of seed treatment by surfactants on germination of wheat, (*T. aestivum*) under water deficit. Seed Sci. and Technol. 21 : 483-485.

Agrawal, P.K. 1994. Principles of Seed Technology, ICAR publication New Delhi.

Anonymous, 1988. Indian Minimum Seed Certification Standards. The central Seed certification board, Department of Agriculture and Cooperation, Ministry of Agriculture, Govt. of India, New Delhi,

Anonymous, 1995. A bulletin on Seed Technology Research (1979-1994), Seed Technology Research Unit, Dr. Panjabrao Deshmukh Krishi Vidyapeeth, Akola

Anonymous, 1998. Progress report of Seed Technology Research Unit, Dr. PDKV, Akola, submitted to Quinquennial Review Team, 1 Nov, 1998.

Anonymous, 1999. *Krishi Margdarshika and Dainandini,* Directorate of Extension Education, Dr. Panjabrao Deshmukh Krishi Vidyapeeth, Akola.

Card Well, V.B. 1984 : Seed Germination and crop production : Physiological Basis of crop Growth and dvelopment, (ed.) M.B. Tesar, American Society of Agronomy, crop Science Society of America, Madison, Wisconsin : 53 -92.

Chowdhari, S.D.; V.K. Shinde and N.L. Bhale 1984. Effect of Seed Size on grain yield of Sorghum, Seed Res. 12(1) : 48 - 51.

Copeland, L.O. and M.B. McDonald 1985. Principles of Seed Science and Technology. Burgess Publishing company, Minneapolis, Minnesota, pp. 50 - 245.

Ellis, R. H., K. Osei-Bonsu and E.H. Robbert, 1982. Desiccation & germination of seed of cowpea (*Vigna unguiculata). Seed Sci. & Technol.* 10: 509-515.

Ellis, R.H. and E.H. Roberts 1982. Desiccation, rehydration, germination, imbibition injury and Longevity of pea seeds. (*Pisum sativum*) Seed Sci. & Technol. 10: 501 -508.

Feistritzer, W.P. 1975. The role of Seed Technology for Agricultural Development. Seed Sci. and Technol.; 3 : 415-420.

Jeffs, K.A. 1986 Seed Treatment. BCPC publications, 20 Bridport Road, Thornton heath, Surrey CR4, 7QG.

Kausal, R.T.,V.R. Zade, R.P. Chauke, and S.P. Changade 1991. Influence of Seed Size on seed quality of sunflower. Ann. Plant. Physiol. 5.(2): 288 - 290.

Matthews, S. 1980. Controlled deterioration; a new vigour test for crop seed in seed production (ed) P.D. Hebblethwaite, Butter worths, London P.P. 647 -660.

Matthews, S. 1993. Seed Vigour and crop performance. Seed Research, Special Vol., 1: 21 -27. Pery, D.A. 1978. Report of vigour test committee 1937 - 1977. Seed Science & Technol, 6: 159 - 181.

Powel, A.A.S. Matthews and De A. Oliveira 1984. Seed quality in grain legumes. Advances in Applied Biology. 10 : 217- 285.

Powel, A.A; S. Matthews 1984. Prediction of storage potential of Onion Seed under commercial storage condition. Seed Sci. & Technol, 12: 641 -647.

Rajput, R.P. and P.K. Gupta 1978. Effect of soil water potential on moisture absorption and seedling emergence of wheat, niger, and lentil. Seed Res. 6(1): 77 -84.

Robert, E. H. 1986. Quantifying Seed deterioration, Physiology of seed deterioration (eds) M.B. McDonald Jr. and C.J. Nelson C.S.S.A. special publication 11, Crop Sci. Society of America. Inc. Madison Wisconsin, pp. 101-103.

Tiwari, D.K.; J.P. Tiwari; and V.K. Agrawal 1978. Evaluation of soybeans for high germinability and field emergence. Seed Res, 6 (2): 125 - 128.

Tiwari J.P.; S.K. Shrivastava and U.R. Pandey 1976. Influence of Seed Size, Seeding rate and depth on Seedling emergence of *Glycine max* L. Merill, Seed Res, 4 (2): 174 -178.

Yadav, T.P. and R.P.S. Tomer 1988. High lights on seed production and Technology. Research, Published by H.A.U. Hissar

Chapter - 8

AGROMETEOROLOGICAL APPROACHES FOR SUSTAINABLE DEVELOPMENT IN DRYLAND AGRICULTURE

Surender S. Dhankar

Deptt. of Agrometeorology, Chaudhary Charan Singh Haryana Agricultural University, Hissar - 125 004 (India)

1. INTRODUCTION

The drylands are the areas of meagre or undependable rainfall, in which the average precipitation is deficient in relation to water requirements. These include the arid zones, in which arable crop production is not possible without irrigation, and also their semi-arid fringes, in which rainfall, though precarious, is sufficient techniques. It is an interesting fact of history that in most of the drylands in the Near East, India, North Africa, and Central Africa, very advanced agricultural civilizations developed, flourished, and eventually disappeared.

1.1 Arid Climates

Arid climates are those in which no arable crop can be grown, unless precipitation is supplemented by water harvesting and spreading or by irrigation. These are confined to livestock production under extensive conditions. The upper limit of 250 mm rainfall is generally used to characterize arid climates (Fuchs, 1973) (a higher limit, 400-500 mm, is justified in summer rainfall areas). Several years may sometimes elapse without rains. The occasional rains may be of high intensity and case floods. Solar intensities are very high, except when a high dust laden atmosphere reduces the intensity. The ability to survive under desert conditions is an essential characteristics of xerophytic natural vegetation of the arid regions. Typically, many of these plants have adopted themselves to arid conditions by a reduced transpirating surface resulting in dwarfed plant with very limited leaf surfaces. The traditional land use of world's deserts has been pastoralism, and its modern form is ranching (Arnon, 1992, Arnon and Gupta, 1995).

1.2 Semi-arid Climates

Actually, semi-arid climates are seasonal arid and are characterized by a completely dry season that alternates with more or less humid season, in which the amounts of precipitation may well exceed the amount typical of a humid climate. Low rainfall regions in India have an annual average precipitation of 500-900 mm (Venkateswarlu 1984), whilst the more humid semi- arid areas have upto 1250 mm (Willy *et al.,* 1989).

Atmospheric demand for water is very high; coupled with low and variable rainfall, periods of transient drought occur frequently. The reliability of rainfall at the beginning of season is generally low and consequently there is a high degree of risk of establishing crops because there are no soil reserves of moisture at this time (Kanemasu *et al.,* 1990). On the basis of temperature, the semi-arid tropics are warm throughout the year and suitable for the production of any crop that does not require a cold period in its life cycle. Unreliable moisture is the key limiting factor (Virmani *et al.,* 1980). Paradoxically, above average rainfall can have a more negative impact on yields than below average rainfall.

2. CAUSES OF ARIDITY

Rainfall generally occurs when moisture laden air rises, expands and cools. As a result, vapour condenses and precipitation occurs. The distribution of global rainfall is influenced by a great number of factors, of which the principal ones are the characteristics of the general circulation of the atmosphere, relief and the distribution of land and sea (Butzer, 1961). The most important factors which led to aridity (McDonald, 1959) are the low frequency of occurrence of an adiabatic rise of large bodies of air and the remoteness from an oceanic moisture source in the prevailing wind direction. The remoteness may be due either simply to distance. or to the so called 'hydrometeorological remoteness' caused by mountain barriers that block the inflow of moisture bearing winds (rain shadow deserts).

3. DRY REGIONS OF INDIAN SUBCONTINENT

Arid and semi-arid regions are over the whole of Pakistan and approximately one third surface area of India. Such regions are found in all the 11 states situated in the western half of country. The hot arid regions of India cover about 75 per cent of the 400 thousand Km^2 of arid lands, in the remaining area are the cold arid regions in the Ladakh area of Jammu and Kashmir State in North-West India (Das Gupta, 1977). Part of arid area of Pakistan and a contiguous area in India form the Thar Desert. The Indian sub-continent is primarily characterized by a tropical monsoon climate with four seasons (Dakshini, 1985).

North-east monsoon season	–	December -March
South-west monsoon season	–	June -October (Semi- arid region)
	–	July-September (Arid region)
Transitional hot weather	–	April - May
Transitional period of retreating SW monsoon	–	October -November

In the arid and semi-arid regions of sub continent, the same seasonal rhythm prevails, but variability in space and time is much greater.

3.1 Rainfall

Mean rainfall usually ranges from 0 to 200 mm in arid regions and 400 to 700 mm in semi-arid regions. The main rainy season is from June to September during which 90-95 per cent of the total annual precipitation occurs, under the influence of South-west monsoon. This is the hottest period of the year, thereby reducing the efficiency of the rains received. On an average, number of rainy days per year is 20 in the arid region, and varies from 20-50 in semi-arid region. In addition to the main rainy season, rainfall may occur for a few days during the cooler period of the retreating SW monsoon and occasionally during the rest of the year. Once in 10 year, the expected annual rainfall of the arid zone of North-west India may be less than 200 mm. In some areas it may be even less than 60 mm.

3.2 Air humidity

Average relative air humidity values are more or less similar in two regions. Mean annual evaporation is 200-300 mm in arid regions, and 200-350 mm in the semi-arid regions. Daily evaporation rate during summer exceeds 15 mm in the arid regions.

3.3 Temperature

Winters are cool with occasional frosts. The hot season begins in early April month. June is the hottest month, with day time temperatures reaching 50°C or even higher during heat waves. The diurnal variations in temperature in the arid regions may be as much as 11 to 17°C.

3.4 Winds

Hot winds are frequent during the summer months. Very high wind speed may be recorded throughout the year except during the post monsoon season. Occurrence of dust haze and dust storms is characteristic for the region. In the arid parts of Pakistan and some parts of India, dust clouds have been recorded at a height of 6000 to 9000 m.

4. DRYLAND FARMING

Also known as rainfed agriculture, is practised in regions where lack of moisture limits crop and pasture production to the part of year. This period is called the growing season and varies from a few months to about nine months in a normal year. Rainfall is often low and variable but the more successful dryland farming systems have been established where rainfall (however low) is reliable and the start and finish of growing season are fairly predictable (Squires, 1991).

5. AGROTECHNIQUES FOR SUSTAINABLE DEVELOPMENT IN DRYLAND FARMING

A major chunk (about 70 per cent) of arable land in India is currently rainfed and depends on natural precipitation for crop production. Concerned research efforts to improve productivity of drylands were initiated in 1970 with the establishment of All India Coordinated Research Project for Dryland Agriculture. These efforts were further strengthened with the setting of ICRISAT in 1972 and CRIDA in 1985. The following agro-techniques to mitigate problems of dryland farming have emerged.

5.1 Moisture conservation measures

It has been advocated for quite long that the provision of earthern structure like contour bunds and to minimize erosion and maximize moisture conservation. Such engineering measures, no doubt minimise the erosion instantly. In the recent past, financial outlay is computed for such bundings. Dry farming is practised in little over 60 per cent of area in India. Even with a modest cost of Rs 1000/ha for these bundings and only 50 per cent of the area is to be covered, an amount of Rs 7000 crores is required. Good crop management is the pre-requisite for most efficient use of water by the crop i.e. improved water use efficiency.

5.2 Timing and balanced fertilizer application

Drylands are improverished regarding fertility because most of the farmers seldom apply fertilizers. Dry farming is relegated to the background regarding nutrient management. Organic manures are mostly diverted to irrigated agriculture. With adequate fertilizer application, the yield could be increased substantially. Another area which requires attention is to promote 'less rainfall dependent crops. For example sunflower, samai and horsegram are promising rainfed crops than can be considered. Fertilizers especially nitrogen, may stimulate early growth and thereby exhaust the soil moisture supply before the period of maximum water requirement of the crop, as shown in the following example; Application of nitrogen and phosphorus at sowing to rainfed groundnut increased root growth, improved the root-shoot ratio, gave higher evapotranspiration values and increased total drymatter production but caused a considerable decrease in yields of pods, in India. The rapid increase in root development and leaf area apparently caused depletion of limited soil moisture supply early in the growing period (Bhan and Mishra, 1970). Nitrogen can also be effective under conditions of limited moisture. In Nebraska, wheat fertilized with nitrogen removed 25-50 mm more water from 2 m soil profile than non fertilized wheat. The additional soil moisture was removed at tensions greater than 15 atm., presumably as a result of the increased root growth stimulated by nitrogen (Ramig and Rhodes, 1963). Generally speaking, a balanced nutrient supply is beneficial even under conditions of limited rainfall, as it actually enables the crop to make more efficient use of the limited soil moisture available. By application of only phosphatic fertilizer at the time of sowing (or to the preceding crop) and with holding nitrogenous fertilizer in the cereal crop, by its appearance shows the need for a stimulant, but excessive initial growth can be avoided.

5.3 Concept of early seeding/planting time

In arid and semi-arid regions, the choice of an appropriate planting date may have a considerable effect on water use efficiency (in relation to economic yield) by ensuring that the pattern of growth of the crop is adjusted to the pattern of precipitation or to available soil moisture. In trials with hybrid sorghum grown entirely on stored soil moisture, Blum (1972) found that early planting under limited moisture supply increased grain yields due to more favourable seasonal water use distribution. During the period prior heading, early planed sorghum used roughly half as much water as later sown sorghum. Thus, more water was left for critical period from heading to seed formation.

5.4 Adjusting planting patterns to limited moisture supply

Plant population must be adjusted to available soil moisture levels, either within rows or between rows. Increasing the distance between sorghum plants, for example, within the row in order to adjust to low moisture levels generally defeats its own purpose : the young plants, with little or no intra-row competition, show excessive vegetative growth : soil moisture is rapidly depleted and the plants are unable to form satisfactory panicles or to mature their grain normally. The alternative is to space plants more closely within the row and to increase the distance between rows in order to compensate for low moisture levels. In the wide spaced rows, the soil moisture supply is not exhausted as rapidly as in narrow rows. In an experiment in rainfed groundnut, grown in dry region of India, the best water use efficiency was obtained from a planting distance of 60 x 25 cm in a poor rainfall season of 186.8 mm (plus an equivalent of 75.1 mm stored in soil upto a depth of 1 m before sowing). In the following year, with more favourable moisture conditions due to a rainfall of 357.3 mm (plus 67.2 mm of stored water), the best water use efficiency resulted from a spacing of 45 x 25 cm. In both seasons, the close spacing of 30 x 25 cm gave the lowest water use efficiency (Bhan and Mishra, 1970). Higher plant population of pearlmillet gave lower yields due to moisture every third row was found to be advantageous to mitigate moisture stress in sorghum crop at Bellary (Venkateswarlu and Singh, 1982).

It might be assumed that the wide row spacing, by exposing large areas of bare soil to radiation, would increase moisture losses due to evaporation, thereby defeating the purpose of the wider spacing. However, under dryland conditions, evaporation is influenced more by the moisture supply at the soil surface than by radiation. Therefore, once the upper soil layer has dried, further moisture losses by evaporation become negligible. At IARI, Kaushik and Gautam (1994) recorded higher yield of rainfed pearlmillet (Hybrid BK 560-230) by planting in furrows and maintaining a population of 20,000,00 plants per hectare.

5.5 Weed control

Controlling of weeds has been known to be one of means of increasing the amount of water available to crops and therefore, increases the water use efficiency.

5.6 Watershed management

Since water and soil fertility are two major constraints to be increased and stabilized in agricultural production in the rainfed farming areas in India, every effort must be made to conserve both water and soil resources. Rain water management and the implementation of soil conservation programmes hold the key to an ecologically balanced improvement in the quality of rainfed lands.

5.7 Alternate land-use system

Despite availability of an array of techniques for the development of dry farming, confining to arable cropping system in drylands is at best, risky as is witnessed every year with the failure of monsoon in India. Hence, suitable alternate land use system depending upon the land carrying capacity must be adopted.

5.8 Concept of dry seeding

Dry seeding has been recommended for situations where soil does not allow operation of seeding with the onset of monsoon. Dry seeding is done in anticipation of rain and it helps in the crop establishment at an early opportunity and gives advantage of better utilization of rainfall during crop growth.

5.9 Wind effects and their amelioration

Wind affects plants in several complex ways. It affects the exchange processes between plants and atmosphere, thereby influencing the heat and water balances (Monteith, 1973); it has mechanical effects leading to breakage or malfunction of plant parts; and it alters the rates of enzymatic reactions from wind induced water stress, mechanical perturbation, and changes in plant temperature (Baker, *et al.*, 1985).

Hence, microclimatic modification is needed by using wind-break and shelter-belts to mitigate the adverse effects of winds.

6. MONITORING OF DRYLAND FARMING SYSTEM

Agriculture is a man made eco-system. The fundamental feature of this eco-system is its continuing transformation. Since their evolution, agro-ecosystems, some of them have gained climax while others are still in transitional phase. These innovative changes have the bearing with the farmers who have adopted a wide array of management tools including tillage methods, choice of crops and their species, sowing time, changes in planting population, soil moisture regimes etc. perfecting thereby the art of management of fields within limits of knowledge. In this process, farmers advertently or inadvertently gained knowledge of plant interactions with the soil and aerial environment. The indigenous technical knowledge (ITK) of management of fields has improved with integration of knowledge in the discipline of plant ecology, plant nutrition and other sciences with the passage of time. The basic knowledge of ecological principles and related sciences is providing strong footing to exploit seed rate, methods of planting, plant population, rectangularity and sowing time for increasing the productivity of agro-ecosystems. These agro-techniques or cultural practices form basis of agronomic principles designed for the management of fields. In essence, agriculture involves harvesting the energy of sun to produce food and fibre for human populations. Growth of vegetation is necessary on farms to :

- Produce yields of crops and pastures and provides an income for farmers.
- Provides cover for controlling erosion. Grazing and tillage operations need to be carried out with care to maintain this cover.
- Provide organic matter from plant roots and residues, and thereby maintain the soil structure and microbial biomass.

The climate determines which crops and pastures can be grown and their potential yield and the soil properties and management skills of the farmer determine the degree of success in approaching the potential yields. The production of crops per hectare from the average farm has not increased very much during the last fifty to sixty years despite the investment of large sums of money into research (French, 1991).

7. ENVIRONMENTAL CONCERNS — AN EMERGING ISSUE

Farming has become much more mechanized and more diversified with greater regional and on farm specialization. By increasing its reliance on mechanization and technology, developing a heavy dependence on fossil fuels, exploiting the productivity of soil and enhancing crop yields with fertilizers and pesticides, agriculture have evolved to a state where short term benefits are being gained at the expense of long term costs. Such actions ignore the inter-dependence between maintaining productivity and achieving sustainability. The intensification of agricultural practices and the changing terms of trade have had an adverse effect on the environment, e.g. from pollution by pesticides and by land degradation through over cropping and overgrazing. In the cereal belt, land degradation is due to failure to apply appropriate conservation practices and to take into account processes which lead to salination, acidification etc. Agricultural inputs such as fertilizers, pesticides and feed additives have been responsible for many of the recent gains in agricultural productivity, but unfortunately also have or threaten to have, adverse side effects on the environment. Significant long term threats to agricultural production have only recently come under intense study. Among threats which are now perceived are :

- Increasing concentration of CO_2 which, in addition to being a factor in the 'greenhouse effect' of atmospheric heating, may also indirectly affect plant metabolism.
- Increase in ozone concentration at the earth's surface which appears to cause decline in crop yields.
- Decreases in ozone concentration in the upper atmosphere which allow greater amounts of UV radiation to reach the earth's surface thus undermining the plant growth.
- Pollution of the soil by persistent harmful chemicals which has led to a decline in food quality and increased cancer risk.
- Leaching of nitrate and phosphate into surface and underground water.
- Global climatic changes, including prolonged periods of unseasonal weather and apparent trend toward atmospheric warming which could have far reaching effects on growing seasons and precipitation. The climate change may lead to sea level rises, drier soils in some areas, raised groundwater tables in others, intensified rainfall and flash flooding.

Agriculture is far more sensitive to weather and weather changes than any other human activity both in relative and absolute terms. For agricultural systems, the prospects of change will prove challenging but are unlikely to prove fatal. Now developments in biotechnology and selection and breeding methods should allow relatively rapid adaptations to change in CO_2, temperature, rainfall and length of growing season. This will depend on the success of agricultural researcher and technologists to devise new farming systems for drylands.

8. DRYLAND MANAGEMENT IN INDIA

The importance of dryland agriculture in the Indian agricultural economy hardly needs any emphasis. In order to improve the overall productivity and to meet the growing needs to population, increasing attention has to be paid to dryland agriculture. The Government of India had identified priority areas and various measures to give a new direction to agricultural development.

The irrigated areas, which are already part of the Green Revolution, have limited scope for further development. Hence, priority will be given to dryland farming which covers 70 per cent of cultivated land at present. These areas grow nearly 90 per cent of the country's coarse grains and pulses, 75 per cent oilseeds and 70 per cent of cotton. Thus, dryland agriculture plays a vital role in our nation's economy. "It is on these lands alone that the battle for food self sufficiency will be won or lost" was remarked by eminent agricultural scientist, Dr. M.S. Swaminathan, while referring to our food requirements (210 million tonnes) by the turn of the century.

The problems of drylands are location specific and need research in that particular region to evolve suitable production techniques. This has been achieved to a great extent in India and we have research stations located in different agro-climatic conditions. The research output has provided varieties capable of performing better under stress environments, besides providing important cultivation techniques. To provide food security to our projected population of one billion by the turn of century, we need food availability of about 215 kg per capita. To ensure this food availability, contribution from dryland areas should increase to about 60 per cent. The major steps that count the most to manage drylands to sustain agriculture are as follows :

- Professional education and training of farmers in management skills are very important for increasing production in dryland. They should learn to adopt the full package of practices, rather than follow just one factor in crop production.

- Corporate efforts by plant breeders, agronomists and entomologists at the national and international levels should be intensified and the control of pests be achieved by combination of minimal pesticide input, rational, varietal and other agronomic practices, together with utilization of national control elements.
- In future planning, more funds should be invested for construction of bigger water harvesting ponds and dams for recycling water for irrigation purpose.
- It will be essential to extend agronomical practices to the entire rainfed areas.
- Suitable crop production measures such as taking two crops instead of one, blending variety of crop mixtures and relay cropping may be immediately taken in the rainfed areas.
- Under dryland agriculture, it was felt that inevitable run off water could be collected in farm ponds. The harvested water could be reused for irrigation. In good rainfall years, the stored water could be used for pre-sowing irrigation, thus making sequential cropping possible on drylands. It will also ensure timely sowing.
- The production capacity of dryland can be significantly enhanced if a mixed farming system is adopted. Fodder crops are required to be included in the regular cropping patterns of the farmers and grassland development should receive higher attention.
- Under the dryland situation, a conscious effort should be made, wherever possible, to take livestock farming and silvi pasture development as an important and integral part of dryland farming.
- Hardy fruits crops like ber, anola, guava, mulberry, bel, phalsa, pommegranate ec. can be successfully grown in dryland conditions. Hence efforts should be made to introduce these crops in the dry areas, wherever possible, so that farmers could get better economic returns from their holdings.
- In dryland, moisture being the main constraint, minimum time needs to be taken for land preparation and seedling. Hence, suitable implements for seed bed preparations, sowing operations and inter-cultural operation need to be devised for efficient results.
- The yield gaps between farmers technology and national demonstrations should be critically examined and identified with farmers having different levels of resources education and contacts with scientists and extension agencies.
- The most significant contribution of the recent dryland research lies in the formulation of contingency plans for all possible aberrament weather situations.
- Remote sensing and forecasting of weather, disease and pest epidemics should be strengthened, made more relevant and purposeful so as to guide the farmers to face any threat from the weather aberrations and disease epidemics.

However, enough time has lapsed and a number of alternative technologies are available for transfer to the farmers. Efforts should be made now to critically examine how effectively the existing technologies are working. research is a continuous process and further improvement and refinement in the technologies based on the field experiences will continue.

9. REMOTE SENSING TECHNIQUES FOR DROUGHT ASSESSMENT

The district level agricultural drought assessment and monitoring using Normalized Difference Vegetation Index (NDVI) generated from NOAA-AVHRR data in taking timely preventive and corrective measures for combating droughts is an area where remote sensing data has rised at operational level (Chari, 1988). Currently, the drought assessment is being done through visual interpretation of remote sensing data but needs to be improved through development of automated assessment procedures to improve objectively and time effectiveness. Comprehensive drought assessment would also involve the integration of information on cropping zones, soil types, coherent rainfall zones etc. For implementation of preventive and corrective measures for combating drought condition at Mandal/Block level with improved spatial accuracies is required. The availability of data with improved spatial resolution from IRS - IC LISS - III and future IRS P5 LISS-IV may enable region specific assessment of agricultural drought with improved spatial accuracies. Basically, the efforts need to develop agricultural drought prediction capabilities by integrating spectral information available from remote sensing data, *in situ* measurements/ observations and meteorological data. Decision support system in drought prone districts needs to be developed to extend drought warning into relief management activities.

10. AGROMETEOROLOGY FOR SUSTAINABLE AGRICULTURE IN DRYLANDS

A sustainable farming system needs management strategies with respect to varietal selection, soil fertility programs, and pest management approaches, in order to reduce the input cost and provide a sustained level of production and profit from farming. For planning an efficient agricultural production system, information on weather and climate is vital and also major resource for agricultural productivity and sustainability, especially in a stressed environments like drylands. Considering the complete agro-ecoclimate system it is necessary to develop farming systems that are sensitive to climate and weather, sustainable, and not susceptible to degradation on account of climate.

For an ideal farming system, a precise temporal and spatial weather information is needed so that the agriculturists can optimally plan their earlier described agro-techniques to achieve the above objectives. Hence, a method is presented for optimal use of information on climatic assessment and weather prediction as available for economically and environmentally viable, agricultural planning and management.

10.1 Weather and Climate

Weather of a place is generally defined in terms of the various weather elements, i.e., temperature, humidity, wind velocity (speed and direction), pressure, rainfall, sunshine duration and cloud cover. Climate is defined as the long term average conditions of weather at a particular place or region along with statistical variations. Weather relates to instantaneous atmospheric conditions at a place or region. Thus, variations in the weather are ever prevailing, and the changes or variations in climate are slow, within the various statistical extremes as defined in the long term data.

10.2 Weather forecasting and its scope in dryland management

The expected global population may exceed 850 crores by the year 2025 and at the same time the demands of the growing population for food and other agricultural commodities remain uncertain. Agriculture has to meet this challenge, mainly by increasing production in a suitable way in drylands. Agricultural production has increased in the last three decades mainly due to improved seed, fertilizer and better management in irrigated areas. It is vitally important that the measures taken to increase food production on short term basis be effectively integrated with long term policies. Any strategy to increase agricultural production should take explicit account of agrometeorology and its development.

World wide 615 crores hectares of land are under ramified agriculture. In India, crops on 67 per cent of the 14.2 crores hectares of arable land depends on rain. This contributes 44 per cent of the food and support the 40 per cent of Indian population. Drylands contribute more than 40 per cent of food grain (80 per cent of maize, 95 per cent of pearlmillet and sorghum). About 95 per cent of pulses and 75 per cent of oilseeds are also grown in these areas. Nearly 56 per cent of rice is grown in high rainfall areas without supplementary irrigation. Similarly, the country's 70 per cent of cotton and almost whole of jute and mesta are grown under rainfed conditions. Fortunately, India is endowed with high rainfall, on an average of 1250 mm as against world annual average of 850 mm. According to an estimate of ICRISAT, atleast for the next 100 years, rainfed agriculture would have to be necessarily relied upon. Our options to expand irrigation will generally narrow. The cost of creating irrigation facilities has already touched a lakh of rupees or more per hectare depending upon the situation. Even if the planned irrigation potential is realised, the country will still have more than 50 per cent of its cultivated area under rainfed conditions. This highlights the need for serious consideration to the strategies for improving the productivity and sustainability in rainfed areas.

10.3 Constraints and options

A basic constraint in rainfed areas is the uncertainty in rainfall. There is significant variability in the total annual rainfall as also in its seasonal distribution. Approximately 1/3 rd each of rainfed areas fall under high, medium and low rainfall regions. For stable agriculture, a departure from the annual and seasonal normal rainfall is extremely important in addition to total precipitation received. As the mean annual rainfall of region decreases, the availability in mean annual rainfall of region decreases, the availability in rainfall events increases leading to greater intensity of risk and a greater probability of drought occurrence. Thus, probability of occurrence of a drought in a low rainfall region, say less than 400 mm annually, may be 2 in every 5 years against one in 15 years in an area of rainfall of about 1000 mm. A better understanding of the climatic parameters and the quantification of relationship between these and crop growth parameters is the science of agrometeorology, which has made rapid strides in the past decade or so. This is an area where we will need to generate a lot of capability to be able to address the whole range of questions relating to variability in climate and response of crops, diseases, pests etc. for developing appropriate management strategies.

Weather forecasting has been done from the immemorial but the present day scientific weather forecasting practices are of recent origin, beginning with the invention of the barometer by Torricelli in 1643. This led to systematic observations of weather elements all over the world. To derive a three dimensional picture of atmospheric circulation, observations are made on the surface as well as in the upper atmosphere. The various elements observed include temperature, wind direction and speed, humidity, atmospheric pressure, cloud, and rain.

Weather forecasts, can with respect to time, be categorized into :

i. Short range weather forecasts valid for upto next 72 hours

ii. Medium range weather forecasts valid for 3 to 10 days in advance

iii. Long range weather forecasts valid beyond 10 days

iv. Assessment of climate changes in decades or greater time spans

10.4 Short and Medium range Weather Forecasting

Weather forecasting is basically an initial value problem. The techniques in use can be divided into three categories : (I) conventional or synoptic methods (II) statistical methods and (III) numerical weather prediction techniques (NWP) which are discussed in brief here.

(I) Synoptic methods

These involve detailed analysis of current weather reports from large area (a few hundred kilometers on either side of the area of interest). The current weather patterns are associated with the past analogues and forecasts are based on the assumption that a current situation will evolve on the lines of the past weather patterns. Often, selection of analogues is based on the experience and memory of forecaster. But with the advent of computers, picking up of analogues has become faster and more objective. Recent studies have shown that such methods can only provide useful forecasts for a short range (less than 3 days).

(II) Statistical methods

Statistical methods establish suitable regression equations or, in certain cases, more sophisticated statistical relationships between the predictant and the predictors. Normally, selections of the predictors is based on the possible physical relationship with the predictant. Such techniques have been useful for short and long range forecasting. Though tested for medium range forecasting too, the success has been limited. Statistical techniques have, however, become popular to improve output products from numerical weather prediction (NWP) models for predicting specific events such as hail or no hail, fog, minimum temperature, cloud cover, and rainfall.

(III) Numerical weather prediction

The NWP technique considers that the behaviour of the atmosphere can be represented and governed by physical laws which can be expressed by differential equations with their boundary conditions. These equations, when expressed as a closed system, are known as an atmospheric models. The basic equations which form the atmospheric model are :

- Equation of horizontal motion; two-dimensional Navier-Stoke equation
- Continuity equation for dry air
- Equation representing conservation of moist air
- Equation representing the First Law of Thermodynamics
- Hydrostatics equation
- Equation of state

For the numerical solution of these equations, it is necessary to discretize the atmosphere in horizontal and vertical planes. One day forecast would need 1200 steps and solving all equations at each grid point means about 10^{13} operations. Considering that the reasonable time needed for a 10 day forecast is four to five hours, only very powerful supercomputers, can meet the requirement. That is why the most powerful computers available at any time are put to solve meteorological forecasting problems.

10.5 Long range (seasonal) forecasting

For the preparation of long range (seasonal) forecast, say for the monsoon season, different antecedent meteorological parameters over the monsoon region and the rest of the globe have been considered and screened. On this basis important parameters have been identified which are used to forecast the monsoon rainfall in India. These parameters are physically connected with the monsoon circulation and are related to temperature, wind, and pressure. Snow cover over the Himalayas and Eurasia during winter also seems to have connection with the monsoon.

After elaborate research on the physical connections between monsoon flow and meteorological parameters, 16 parameters have been selected. The relationship of monsoon rainfall with individual predictors exhibits a nonlinear relationship. To take care of this nonlinearity, a curvilinear relationship with all the parameters has been combined and a power regression model has been developed. The forecasts from this power regression model (Gowariker *et al.,* 1989, 1991) has been found encouraging. This newly developed long range forecast model has given an excellent seasonal forecast during 1988 to 1998.

10.6 Climate studies for sustainability

Considering specifically, agriculture and livestock, with the present stage of research in biotechnology, it is possible to develop a genotype which produces maximum output in a given climate and soil, or what may be termed agroclimatic conditions. A country like India, has diverse climate and soil conditions. For efficient planning of the cropping strategy, it is necessary to divide into well defined agroclimatic zones. The India Meteorological Department, purely on the basis of rainfall homogenity and to some extent considering state boundaries divided the country into 35 sub-divisions. The planning commission (Khanna, 1989), however, divided India into 15 agroclimatic regions. Recently, The National Bureau of Soil Survey and Land Use Planning of ICAR, considering the agroclimatic resources and soil conditions, has divided the country into 21 zones (Sehgal *et al.,* 1990). Further sub- zoning has also been carried out on the basis of detailed climatic analysis along with soil characteristics.

Climate fluctuates naturally in various degrees and various time scale, over years, decades or even longer periods. The assessment of these climatic variations is very important for planning and ensuring sustainable agricultural production. It is more important for India, because here we want not only sustained but increased production to meet the need of a growing population.

10.7 Weather based long-term farm planning

For long term planning, the scenario development based on the Global Circulation Models (GCMs) becomes important. The prediction of the long term climate fluctuations in the present state of the art from precise, certain indicators are available that can help in long term planning of cropping strategies. As discussed elsewhere, all the work related to global warming and its projection at present should be taken for general guidance and not for any policy or economic decision making process. It is hoped that the ongoing research on more sophisticated coupled GCMs which would take into account the feedback processes of the complete climate system will help understand the scenario more accurately.

10.8 Medium-term planning

Medium term planning options include considerations of water and energy balance for an agroclimatic zone over a period of several months or a season in advance. For planning of the cropping strategy, a season in advance, the seasonal forecasts, like monsoon rainfall are of great use. The long range forecasts or the seasonal forecast primarily give information on departure of rainfall from normal for the country as a whole. This, therefore, is of limited use for planning at the micro level but could be of use for overall planning at central level.

10.9 Short-term operation decisions

Short term operational decisions relates to a period ranging from a few hours to a few days. These often involve decisions, based on the state of the crop and current and forecast weather (short and medium range), for farm operations such as weeding, irrigation, harvesting, chemical spraying, and fertilizer application.

The need for prior information on variability of pentad or weekly rainfall pattern (Biswas and Khambete, 1979; Sarkar *et al.,* 1978) and day to day precipitation forecast is of great importance for dry farming of the arid and semi-arid regions of the country. In addition to rainfall, a correct forecast of extreme temperature conditions, strong winds, dew, unseasonal rains, cloudiness, humidity and other conditions is of great importance for the increase in productivity and to save the costs of inputs (WMO, 1990). However, for this purpose, farmers have to be told in their language, sufficiently in advance, enabling them to take appropriate actions. The authorities must provide the necessary inputs so that the farmers can act on the advice.

Thus, realising the significance of agrometeorology, ICAR in 1983 launched the ALL India Coordinated Research Project on Agrometeorology with the headquarters at CRIDA, Hyderabad and 12 cooperating centres covering a range of agroecological conditions. The major objectives of the project are to study the crop weather relationship in order to eliminate the problems related to climatic parameters. Another sub project on agrometeorology was sanctioned from ICAR funding from USAID in June, 1988 to strengthen the research capability. Its main thrust was infrastructural development for agrometeorological research, training scientists in specialized areas of agrometeorology and providing US expert consultancy for improving agrometeorological research.

The Government of India recently launched the National Centre for Medium Range Weather Forecasting (NCMRWF) and established the agroadvisory services. This was primarily because Indian farmers need a longer reaction time to gear facilities for remedial actions based on weather based agroadvisories. This centre has started initial experiments based on its global spectral models through the state agricultural universities (SAUs). In March, 1991, issue of weather based agroadvisory bulletin was started from five state agricultural universities where a department of agrometeorology and its nucleus infrastructure were available. The NCMRWF supplies the weather information for the next three to five days, which the agrometeorologists of the universities interpret in term of agroadvisories with the advice of other area specific scientists and disseminate to identified progressive farmers to start improving the operational as well as Research and Development aspects of weather and crop weather relationship. The above said experiment is unique-nowhere in the world, a systematic way of integrating the weather and agriculture information is known to exist. This is a multi disciplinary effort of departments dealing with agriculture, meteorology, communication and information sciences which are involved in developing as well as implementing the project. It is expected in couple of years that the integration will take place at the zonal level. India has been divided into 127 agrometeorolgical zones (Ghosh, 1991). Each zone has a zonal agricultural research station of state agricultural universities. These zonal stations are supported by ICAR under the National Agricultural Research Program (NARP) to strengthen the regional faculties of the universities. It is proposed that agriculture expertise at these centres and meteorological informations from the NCMRWF be used so that area specific, crop specific advice can be issued through radio, television, and other media to the farming community. Through this process, it is expected to not only integrate the agroclimatic and soil data bases but also to considerably improve their use in optimizing the agriculture production system particularly in dryland areas.

11. FUTURE THRUST

There is an urgent need for more dense network of agrometeorological stations, so that location specific information can be oriented to clientele farmers. Satellite imageries of weather system, specially to the Advanced vehicular High Resolution of Radiometery (AVHRR) is required to obtain real time information for weather forecasting. Under dryland situations, effective growing based on moisture availability index has to be identified for each agro-ecological situation, so that right choice of crop would be determined. A detailed study between weather and yield reducers (pests, diseases and weeds) should be made on the various concentrations of CO_2 and increased air temperature. Integrated farming systems, no doubt, reduce the risk in agriculture. Combination of different sub-system requires different weathers and hence location specific integrated farming system should be identified. Exploitation of availability of power work stations and the geographic information system (GIS) with the use of remote sensing for precise delineation of agroclimatic zones and sub-zones need to be encouraged. The international, national and regional meteorological and dryland research centres can come together to deliver advice for farmers that will help them sustain their production and improve their lot. This program should serve as a model for technology transfer and should focus on important crops in stressed environments representing drylands.

REFERENCES

Arnon, I. 1992. Agriculture in drylands : Principles and practices. Elsevier Science Publishers, Amsterdam, The Netherlands, 979 pp.

Arnon, I and Gupta, U.S. 1995. Physiological principles of dry land crop production. In : U.S. Gupta (ed). Production and improvement of crops for dryland. Oxford & IBH publishing Co. New Delhi. 449 pp.

Baker, G.L. , Hatfield, J.L. and Wanjura, D.F. 1985. Cotton phenology parameters affected by wind. Field Crop Res. 12 : 33-47.

Bhan, S. and Mishra, D.K. 1970. Effects of variety, spacing and soil fertility on root development in groundnut under arid conditions. Indian J Agric. Sci. 40 : 10501055.

Biswas, B.C. and Khambete, N.N. 1979. Distribution of short period rainfall over dry farming tract of Maharashtra. Jour. Maharashtra Agril Univ. 4 : 145-156.

Blum, A. 1972. Effect of planting date on water use and its efficiency in dryland grain sorghum. Agron. J. 64 : 775-778.

Butzer, K.W. 1961. Climatic change in arid regions since the pleistocence. Arid Zone Res. 17 : 31-56.

Chari, S.T. 1988. National agricultural drought assessment and monitoring systems in India : The Satellite contribution, 39th IAF Congress, Special Publication on space and Drought Assessment, ISRO, Bangalore, India. 44 - 89 pp.

Dakshini, K.M.M. 1985. Indian Subcontinent. In : J.R. Goodin and D.K. Northington (eds). Plant Resources of Arid and Semiarid lands A - Global Perspective. Academic Press, NY, pp 69 -128.

Dasgupta, K. 1977. Ground water resources. In : H.S. Mann (ed), Desert ecosystem and its improvement, CAZRI, Jodhpur, pp 26 - 30.

French, R.J. 1991. Monitoring the functioning of dryland farming systems. In : V. Squires and P. Tow (eds), Dry land farming : A System approach, pp 223 - 238.

Fuchs, M. 1973. Climate and irrgiation. Ecological studies. 34 : 3-9.

Ghosh, S.P. 1991. Agroclimatic zone specific research : Indian persepctive under National Agricultural Research Project. New Delhi, India : Indian Council of Agricultural Research. 539 pp.

Gowariker, V, Thapliyal, V.; Sarkar, R.P., Mandal, G.S. and Sikka, D.R. 1989. Parametric regression models : New approach to long- range forecasting of monsoon rainfall in India. *Mausam* 40 : 115-122.

Gowariker, V, Thapliyal, V. Kulshrestha, S.M., Mandal, G.S. Sen Roy, N and Sikka D.R. 1991. A power regression model for long range forecast of south west monsoon rainfall over India. *Mausam* 42 : 125 - 130.

Kaushik, S.K. and Gautam, R.C. 1984. Drought evading studies in pearlmillet. India J. Agronomy. 24 : 254-256.

Kanemasu, E.T., Stewart, J.I., Van Donek, S.J. and Virmani, S.M. 1990. Agroclimatic approaches for improving agricultural productivity in semi-arid tropics. Adv Soil Sci. 13 : 273-309.

Khanna, S.S. 1989. The agroclimatic approach. The Hindu Survey of Indian Agriculture. Madras, India. The Hindu. 28 - 35 pp.

McDonald, J.E. 1959. Climatology of arid lands. Arid Lands Colloquia. Univ. Arizona, Temp, AZ, 1958-59 : 3 - 13.

Ramig, R.E. and Rhoades, H.F. 1963. Interrelationships of soil moisture level at planting time and nitrogen fertilization on winter wheat production. Agron J. 55 : 123-127.

Sarkar, R.P. Biswas, B.C. and Khambete, N.N. 1978. Probability analysis of short period rainfall in dry farming tract of India. Scientific Report 78/9. Pune, India : India Meteorological Department.

Sehgal, J.L. Mondal, D.K., Mondal, C. and Vadively, S. 1990. Agroecological zones of India. Technical Bulletin, NBSS Publication 24. Nagpur, India : National Bureau of Soil Survey and Land Use Planning. 73 pp.

Squires, V.R. 1991. A system approach to agriculture. In : V. Squires and P. Tow (eds), Drylands farming : A system approach, pp 3- 15.

Squires, V.R. Tow, P.G. and Boast, C.M. 1991. Future directions - Meeting the challenge. In : V. Squires and P. Tow (eds), Dryland farming : A system approach, pp 271-282.

Venkateswarly, J. 1984. Soil problems with vertisols with particular reference to surface soil conditions and water relations. In : Proc. Int Workshop on soils, Aust Centre for Int. Agril Res. Canberra, Australia, pp. 105-115.

Venkateswarlu, J and Singh, S.N. 1982. Crop management to mitigate moisture stress. In : 15 years of Agricultural Research under All India Coordinated Dryland Agriculture Research Project, Hyderabad. pp. 48-60.

Virmani, S.M., Sivakumar, M.V.K. and Reddy, S.J. 1980. Climatic classification of semi-arid tropics in relation to farming system research. In : Climatic classification : A consultants Meeting. ICRISAT, Patancheru, A.P., India, pp 27 - 44.

Willey, R.W. Singh, R.P. and Reddy, M.S. 1989. Cropping systems for Vertisols in different rainfall regimes in semi-arid tropics. In : J.R. Burford and K.L. Sahrawat (eds), Management of Vertisols for Improved Agricultural Production. ICRISAT, Patancheru, A.P., India. pp 119-131.

WMO (World Meteorological Organization) 1990. Economical and social benefits of meteorological and hydrological services. Proc. Tech Conf. 26-30, March, 1990. WMO No. 733. Geneva, Switzerland.

Chapter - 9

CLIMATE AS A TOOL TO DIAGNOSE SOIL MOISTURE STRESS AND DROUGHT IN RAINFED CONDITION

S.R. Ghadekar

Asso. Professor of Agrometeorology, Dr. Panjabrao Deshmukh Krishi Vidhyapeeth, College of Agriculture, Nagpur - 440012 (MS)

1. INTRODUCTION

Rainfed areas in India account for nearly 70 per cent of the total cultivated area but contributes only about 45 per cent to the total food grain production of the country. About 60 per cent of rice, 40 per cent of wheat and 75 per cent of pulses, oilseeds and cotton, normally grown in Kharif and Rabi seasons, account for dryland production. Despite better radiation and thermal regimes in these areas, the crop production is much unstainabilized with poor productivity. Among the various problems responsible for low productivity, weather born ones, mainly unassured rains with high variability, its inadequate spatial and temporal distribution stand first. It has been realized recently that weather induced variability in food production amounts to about 13 per cent or more, sometimes leading to total crop failure. The watershed technology for recycling runoff has created increased awareness in the recent years but the attempts to execute it are still much scarce. Even after complete exploitation of irrigation potential about 70 million hectares of cropped area would remain rainfed. To stabilize crop production, climatic techniques like crop monitoring through soil moisture scan, and its assesment enabling timely measures to be undertaken to mitigate water stress, can play significant role and have recently attracted attention.

2. CLIMATE AS A TOOL

Climate is an integrated effect produced by atmospheric thermal, humidic and dynamic conditions. The parameters such as air temperature, relative humidity, solar radiations, wind velocity, precipitation etc. can describe the climate well and are called as direct parameters while some synthetic parameters integrating more than one direct parameters are termed as derived parameters. Potential Evaporation (PE), Evapotranspiration (ET), Soil moisture (SM), Saruation Deficit (SD) are some of the derived climatic parameters and are used largely in conjunction with crop production.

Soil-plant-atmosphere continuum forms a closed and reciprocating system of supply, regulation and demand of water (Fig. 1). Soil functions as a reservoir and supplier of water while plant acts as a user and regulator and atmosphere decides demand of water. Water forms a communication link between these three distinct spheres (biosphere, lithosphere and atmosphere).

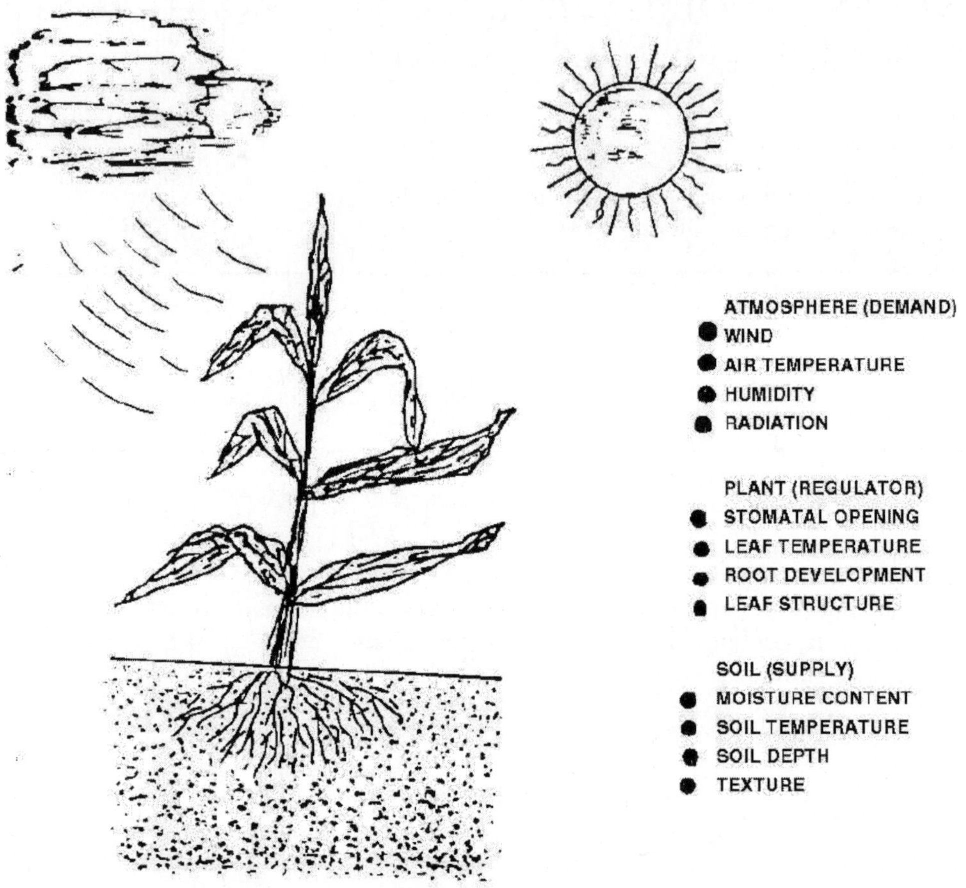

Fig. 1. Soil-water-plant atmosphere continuum; a system of supply, regulation and demand of water

Atmosphere, in the form of natural precipitation supplies the water at the first instance and the same through the soil medium is supplied to the plant which too lost through evaporation from the soil and transpiration (ET) through the plant. This loss is again decided by the atmospheric evaporative demand which in turn is controlled by climatic parameters viz., air temperature, humidity, wind velocity and solar radiations.

3. STRESS AND ITS TYPES

Stress has been defined as any environmental factor capable of inducing a potentially injurious strain in plants.

Plant stress is defined as a situation in which normal growth and development processes are compromised by extreme biotic pressures such that plants, tissues and cells enter into new and/or different physiological states.

Types :

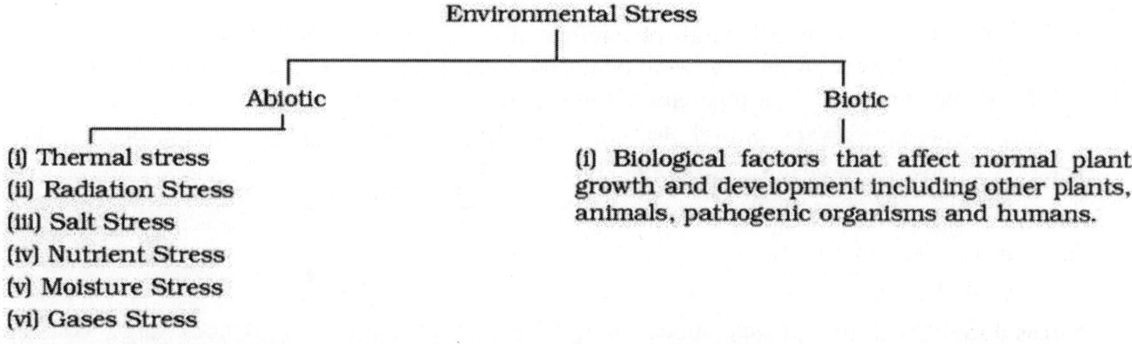

4. WATER BALANCE

Water balance concept conveys the water in balance after making suitable allowances for gains and losses of water to any fields, reservoirs and for any period say day, week, month, season, year etc. the water balance equation can be stated as follows for the soil,

$$P = ET + RO + D + SM$$

or $$SM = (P) - (ET + RO + D)$$

Where, P = precipitation

RO = Run off

SM = Soil Moisture Balance

D = Deep drainage

ET = Evapotranspiration

(All Units in terms of depth of water, mm)

Thus, day to day knowledge of soil moisture status and supply can be worked out by knowing climatic factors without any conventional measurement. The imbalance between the water demand of the plant and its (less) supply is nothing but the water stress. This forms the basis for determining moisture stress using climate as a tool or device.

5. SOIL MOISTURE STRESS : ITS EFFECTS ON CROP PRODUCTION

The plants need some specific amounts of water characteristics of the climate to have healthy and luxurious growth. This is water requirement (WR) of the crop. This amount exceeds the potential evapotranspiration (PET) which is defined by Penman (1948) as the water evaporated from short green grass uniformly covering the soil and under never short of moisture. The actual amount of water for the growth of the plant is however lessened when the moisture in the soil is depleted. This amount of water actually used by the plant is defined as actual evapotranspiration (AET). This difference between potential requirement (PET) and actual use (AET) i.e. (PET- AET) indicates water deficit causing moisture stress.

Moisture Stress : Water deficit that occurs when plants display measurable physiological or biochemical response, deviating from the normal is water stress. It is a condition of subnormal plant when transpiration exceeds water absorption. Atmospheric or climatic evaporative demand increases transpiration and water supply through roots not fulfilled. It is a imbalance between demand and supply and indicates water stress.

Effect of Water Stress: It affects all physiological morphological, anatomical and biochemical processes, all phenological growth, development processes, hormone relationships, nutrient uptake, metabolism, Photosynthesis, phytorespiration, turgor, etc.

Drought/stress and Yield : Growth development and finally yield are affected, following three stages are important in yield determination (Slatyer, 1969).

1. Floral initiation and inflorescence development.
2. Anthesis and fertilization.
3. Grain filling

Stress during these three stages reduces yield. Moisture stress does not effect all stages equally.

Stress syndromes : Plants in semi-arid regions always grow under fluctuating stresses of different degrees. Plants display stress syndrome when biotic/abiotic pressure exceeds some theoretical limit i.e. tolerance limit. Above this, plants can not tolerate pressure and normal functioning capitulates. Note that in a no stress situation, plants seldom function at full genetic potential or optimum conditions.

6. METHODS OF MEASUREMENT OF WATER STRESS

1. Soil moisture estimation (percentage or water potential)
2. Bark moisture content
3. By measuring leaf and soil water potentials using thermocouples.
4. Pressure bomb-measurements (measures non-osmotic pressures).
5. By IR- Thermometry from the differential between canopy temp (Tc) and air temp (Ta).

Data Required : Air temperature (t), wind velocity (V), Relative Humidity (RH) Solar Radiation (Rn), Precipitation (P), Pan Evaporation (Epan) and Pan Constant (K).

1. Climatic Data :
2. Soil Data : Field capacity (Fc), PWP, Soil depth, Bulk density.
3. Crop Data : Crop coeff, Rooting depth.

7. CLIMATOLOGICAL METHODS FOR MOISTURE AND STRESS DETERMINATION

Instrumental measurements of soil moisture have their limitations, merits and demerits and top and moisture meters have been described in length by Baver (1956). Scientists concerned with soil-moisture measurements are well acquainted with the problems. Such measurements are of localized nature and rarely represent tangible proporation of land.

Soil moisture determinations can be made by climatological methods based on water balance as follows :

i. Thornthwaite - Mather's (1955) book keeping method.
ii. Bair-Robertson's (1966) method of estimating simultaneous moisture withdrawals from several soil layers.

iii. Raman Venkataraman (1970) method based on effective rainfall (1975) and estimation of runoff.

iv. Smith method -using moisture with-drawls one by one from different layers.

THORNTHWAITE - MATHER'S METHOD

This method employs water balance principle in the root zone. In this method, on weekly basis, the account of precipitation (P), potential evapotranspiration (PET), moisture storage (STOR), potential water loss (P-PET) is worked out.

The potential evapotranspiration (PET) is determined by suitable method, usually by modified Penman method or PET = (0.70 - 0.80) x Pan Evaporation). There are about 32 methods to compute PET.

For crop growth, when soil is at field capacity (FC), the water is readily available but when it depletes, difficulty in moisture extraction is experienced. It is assumed that the moisture from the soil is released at exponential rate (Fig. 2) decided by depletion in the soil, and AWC.

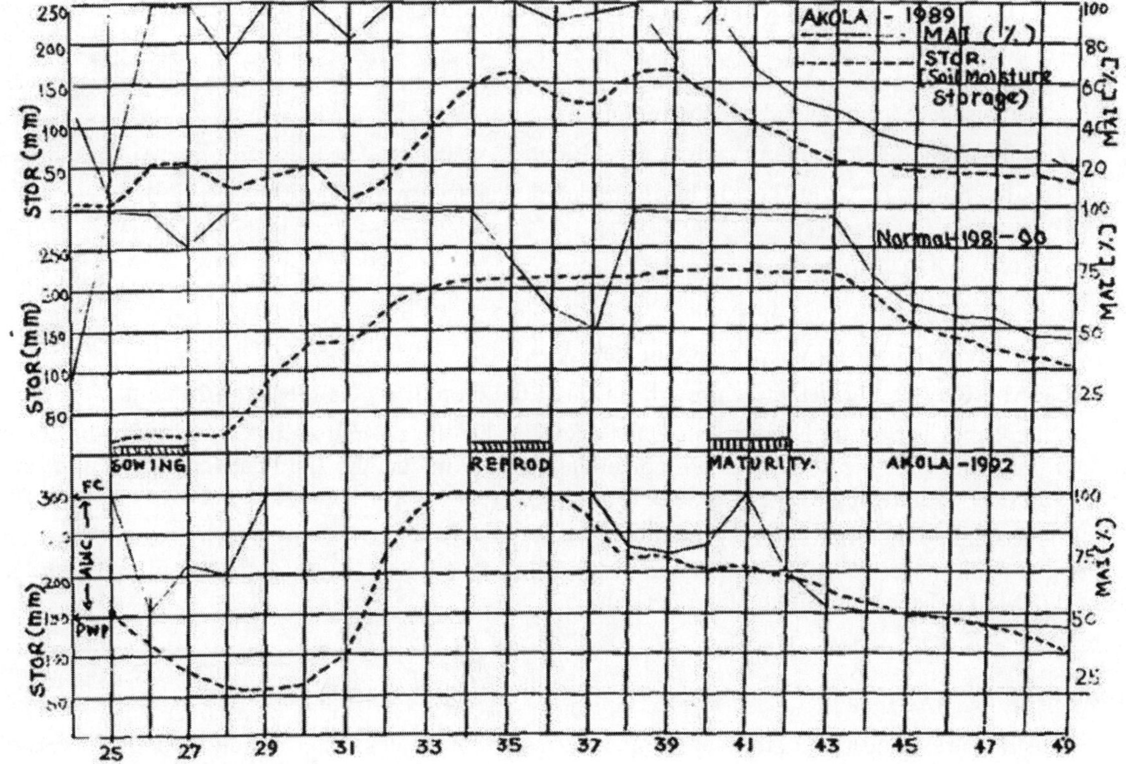

Fig. 2. Weekly soil moisture status (ASM) and moisture adequacy index at Akola for normal growing season (1981-90) and for Kharif seasons of 1989 and 1992 in Maharashtra state

Procedure

1. **PET** : Compute by suitable method for each week (Modified Penman)
2. **P-PET** : Compute the difference for each week between rainfall (P) and PET.

3. **STOR** : Computation of storage soil moisture PET for each week from the following formula.

$$STOR = FC\ e^{-\frac{ACC\ (P - PET)}{FC}}$$

Where, STOR = actual storage of moisture in the week

FC = moisture at field capacity.

- ACC (P-PET) = accumulated values of (P-PET)

(This value is always negative)

4. **STOR** : Change in soil moisture storage is difference in (STOR) of the previous week minus (STOR) in the current week.
5. **AET** : Actual evapotranspiration

   AET = P + STOR (When STOR is negative)

   AET = PET (When STOR is positive)
6. **DEF** : Deficit indicates short of moisture required to grow the plant and is the difference (PET-AET)
7. **SUR** : Surplus and takes place when the soil reaches field capacity. The excess water cause RO (runoff) and deep percolation.
8. **MAI** : Moisture Adequacy Index = AET/PET x 100

Classification of MAI on Weekly basis are decided as follows (A. Krishnan, 1979)

i. MAI = 75% - Excellent

ii. MAI = 50 to 74 - Good Adequate moisture

iii. MAI = 49 to 24 - Poor

iv. MAI = 24 - Very Poor Moisture stress

If MAI = 0 to 49, during active growth stages of the crop, it is considered as drought.

In the Table 1, Soil moisture balance for AKOLA for the normal values (19811990) has been worked out. The effective rooting depth considered is one meter, the field capacity (300 mm) and PWP (150mm) with available water AWG (150 mm).

It is assumed when there is no groundwater storage AET = R, During growing season, the soil is charged with water (SMR) from the rains, reaches maximum value and the soil moisture is utilized (SMU) subsequently in the later period.

8. MOISTURE STRESS

Moisture stress and water stress are similar terms and indicates water deficit inducting stress on the plants. When the full demand of the crop (PET) is not met through rains and water in the soil, the plant uses lower amounts and this constrained demand is indicated with reduction in actual evapotranspiration (AET). Thus, the difference (PET-AET) indicates water deficit. This deficit can be indicated by moisture Adequacy Index defined as

$$MAI = \frac{AET}{PET} \times 100$$

Cocheme and Franquin (1969) have suggested the limit of water availability as follows :

1. AET >PET or MAI >50% - No water stress but when MAI 50% the water stress is imposed. In the Fig. 3 the MAI alongwith soil moisture (STOR) is shown for the year 1989, 1992 years along with the normal data (1981-90). From the normal, it is clear that the season starts in 245 MW (11-17 June) and Soil begins to charge and reaches peak storage in the 35 MW, (27 Aug. - 21 Sept.) thereafter in 45 MW soil moisture begins to deplete. Year wise, this situation is likely to change. In 1992 season, water stress is experienced in the 26 MW and From 43 MW onwards till end of the season. In 1989 season, water stress was observed in the 25 MW and from 42 MW onward till the season ends. Soil moisture utilization (SMU) normally begins when the in the 39 MW (24 - 30 Sept.) monsoon weakens and withdraws. In 1989, thc soil moisture lowered down to PWP (150 mm) in the 40th weeks while in 1992, it did so in the 45th weeks. The weekly moisture balance with its different components viz., rainfall, PET, AET, water deficit, soil moisture, surplus, MAI for normal data, is given in Table 1.

Table 1. Weekly soil moisture balance for the normal data (1981-1990) for Akola (Vertisols) with different components

MW	PET	R	R-PET	STOR	AET	Surplus	Deficit	Runoff	MAI
23	56.8	28.2	–26.6 0	28.2	—	28.6	—	49.4	
24	45.3	42.3	–3.0 0	42.3	—	3.0	—	93.4	
25	38.9	57.1	18.2	18.2	38.9	—	0	—	100
26	31.7	38.0	6.3	24.5	31.7	—	0	—	100
27	37.4	30.8	6.6	24.5	30.8	—	0	—	82.4
28	33.5	34.4	0.9	25.4	33.5	—	6.6	—	100
29	26.6	97.6	71	96.4	26.6	—	6.6	—	100
30	25.5	58.5	33	120.4	25.5	—	0	—	100
31	35.9	37	1.1	130.5	35.9	—	0	—	100
32	26.9	75.4	48.5	179.0	26.9	—	0	—	100
33	26.9	52.8	25.9	204.9	26.9	—	0	—	100
34	29.9	39.9	10.0	214.9	29.9	—	0	—	100
35	32.9	26.6	–6.3	214.9	26.6	—	0	—	80.9
36	30.2	19.3	–10.9	214.9	19.3	—	10.0	—	63.9
37	30.9	15.6	–15.3	214.9	15.6	—	0	—	50.5
38	31.5	35.2	3.7	218.6	31.5	—	0	—	100
39	27	29.9	2.9	221.5	27	—	0	—	100
40	29.8	29.7	-0.1	221.5	29.7	—	0	—	99.7
41	26.8	18.9	-7.9	221.5	18.9	—	0	—	70.5
42	30.5	0.8	-29.7	191.0	30.3	—	0.1	—	99.3
43	30.5	6.0	-24.5	167.2	30.0	—	7.9	—	98.4
44	30.5	1.3	-29.2	146.6	21.9	—	0.8	—	71.8
45	31.6	0.9	-20.7	133.5	14.0	—	0.5	—	44.3

MW	PET	R	R-PET	STOR	AET	Surplus	Deficit	Runoff	MAI
46	21.6	1.5	-20.1	121.9	13.0	—	8.6	—	60.2
47	20.2	3.8	-16.4	113.2	12.4	—	7.8	—	61.4
48	20.2	0.0	-20.2	103.3	9.8	—	10.4	—	48.5
49	20.2	0.0	-20.1	94.4	8.9	—	11.3	—	44.1
50	18.8	2.4	-16.4	87.6	9.1	—	9.7	—	48.4
51	18.8	2.6	-16.2	81.5	8.7	—	10.7	—	46.3
52	19.2	6.4	-12.8	76.9	10.9	—	8.3	—	56.8

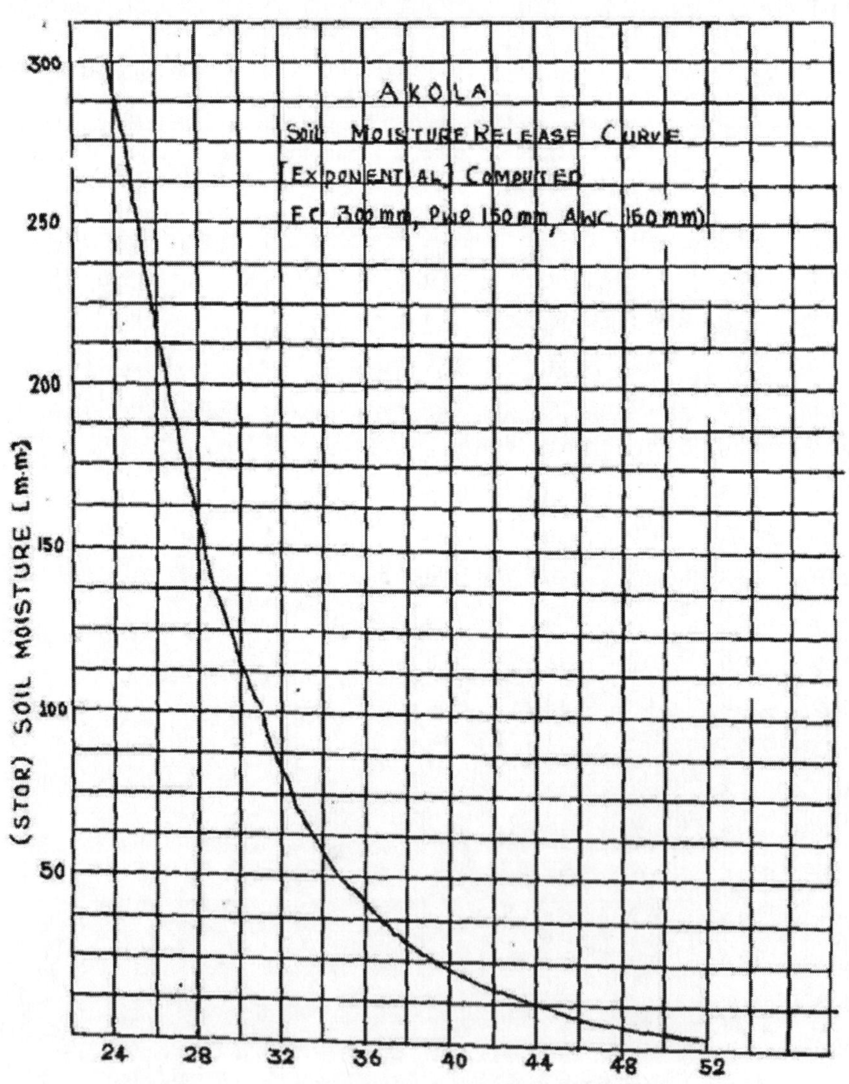

Fig. 3. The curve shows the release of Moisture in vertisols (Akola) when the soil is reached to field capacity in the 24th MW.

9. DROUGHT

There are about 150 definitions of drought in the literature rooting through the various sciences like climatology, hydrology, Agriculture, Physiology etc. (Ghadekar, 1991). Even though these definitions differ in their content, one thing agreed commonly is that it is a phenomena produced due to prolonged insufficient rains or no rains. Definition based on soil moisture and crop parameters can be constituted for agricultural drought as lack of available soil moisture to support plant growth, causing temporary or permanent wilt and resulting in crop yield reduction. Droughts develop due to lack of precipitation, increased saturation deficit and anomalies in planetary circulation. Under equal rainfall on the same soil, one crop may experience drought, the other may not while on different soil under the same rainfall the same crop may or may not experience drought.

Drought is a measure of the degree of water deficit and its duration, it effects growth and development and finally yields. Therefore, the process is much complicated and involve large number of parameters. Some of the climatological techniques use the following criteria for agricultural droughts.

1. Rainfall anomaly between the actual and long term mean or normal.
2. Palmer's drought indices (1965) : Technique uses the computation of CAFEC (Climatologically Appropriate for Existing Conditions). Precipitation and index is worked on difference between (P-Pcafec). The method is realistic but tedious.
3. Hilar and Clarke's (1971) stress day index (SDI) : The method involves calculation of stress day (SD) factor and crop susceptibility (CS) factor.

   $$SDI = SD \times CS$$

4. Yao's R - Index (1969) : $R = \dfrac{AET}{PET}$

   The Yao's R - Index and the MAI values are similar.
5. Drought recokning can also be made on the basis of anomaly between the highest water demand of crop (PET) and the actual amount utilized (AET). A period under rainfed farming can be rated with drought as follows :

   (i) Period with AET > 1/2 PET, Slight or no water stress

   (ii) Period with $\dfrac{PET}{2} > AET > \dfrac{PET}{4}$ moderate drought

   (iii) Period with $AET < \dfrac{PET}{4}$ Severe drought.
6. Drought Reckoning Model : Drought reckoning on the basis of the degree of wilt caused during the various growth stages viz., seedling, vegetative, reproductive and maturity can be done (Ghadekar, 1993). The back ground studies and the past experiences in the region (AKOLA, ASM 150 mm) make it clear that the crops during seedling, vegetative, reproductive and maturity stages can accomodate and withstand to the increasing amount of water deficit or dry spell duration, which for permanent wilt being 14, 28, 42 and 56 days respectively. Permanent wilt is irreversible while the temporary wilt is reversible process. This increase in capacity to accomodate larger reversible wilt with progressing stages is justifiable, firstly as the soil moisture recharge to support the crop increases with

time and secondly the inherent resistance due to established root system after achieving tangible growth, is developed into the crop.

Based on Boules (1971) principle, it is possible to find out relationship between the relative growth as below.

$$\frac{gs}{go} = \frac{AET}{PET}$$

Where,

go = growth of the crop in a given stage without stress at full PET

gs = growth under stress at given AET (AET< PET)

$$\text{or } 1 - \frac{gs}{go} = 1 - \frac{AET}{PET}$$

$$\text{or } \frac{go - gs}{go} = \frac{PET - AET}{PET}$$

$$\frac{\text{Reduction in growth}}{\text{Full growth}} = \frac{\text{Water deficit}}{\text{PET}}$$

Table 2. Progressive moisture status of the soil and water used from it for the soil charged to FC (300 mm) in the 24 MW with increasing water deficit

MW	Progressive Deficit (P-PET)	Progressive moisture status	Moisture used
24	50	300	—
25	100	254	46
26	150	215	39
27	200	182	33
28	250	154	28
29	300	130	24
30	350	110	20
31	400	93	17
32	450	79	14
33	500	57	22
34	550	56	01
35	600	48	08
36	650	41	07
37	700	34	06
38	750	29	05
39	800	25	04
40	850	21	04

MW	Progressive Deficit (P-PET)	Progressive moisture status	Moisture used
41	900	18	03
42	950	15	03
43	1000	13	02
44	1050	11	02
45	1100	9	02
46	1150	8	01
47	1200	6	01
48	1250	5	01
49	1300	4	01
50	1350	3	01
51	1400	28	0.2
52	1450	2.3	0.2

Thus, a linear relationship between the growth reduction and water deficit is possible. The best growth is possible at no or Zero stress or no drought while the worst growth or permanent wilt is the consequence of extreme stress or total drought. This leads to conclude the inverse relationship between growth and drought.

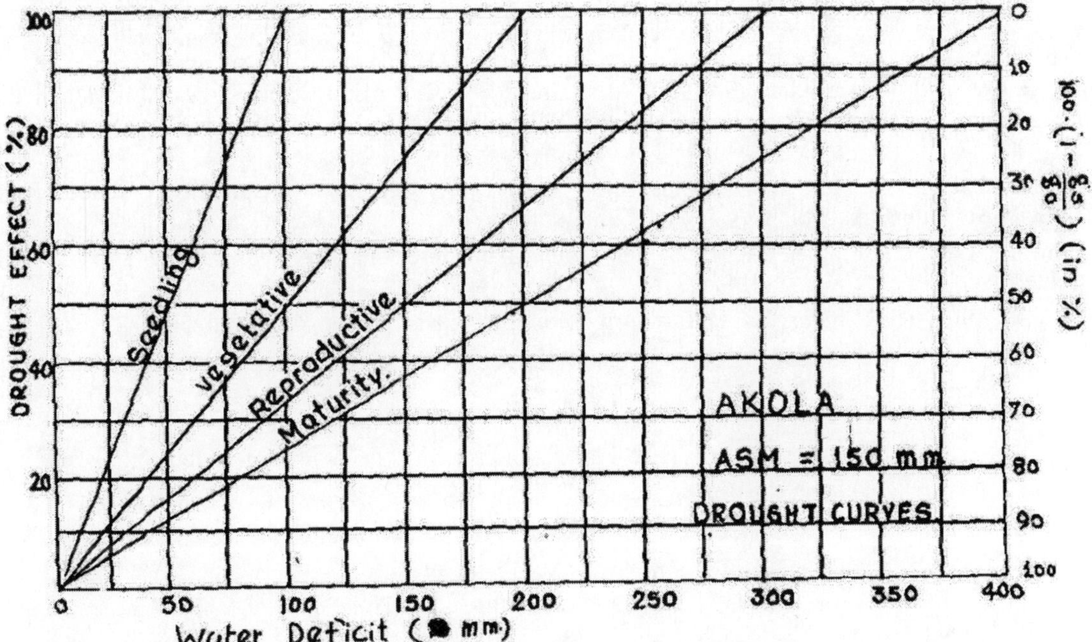

Fig. 4. Drought curves showing quantum of drought effect(%) produced on various growth stages by water deficit (mm) alongwith growth in %.

Drought curves, based on this Thus, the climatological methods based principle, are plotted for various stages in on measurement of few parameter and viz. the crop enabling quantification of drought air temperature, humidity, solar radiations (Fig. 4). and wind velocity, enable to compute

From the literature, it is evident, that moisture stress and drought, potential no absolute method for drought quanti-evapotranspiration, and crop water fication is complete in every respect. The requirements. The methods are much involvement of large number of parameters simpler and economic in nature and complicates the problem to resolve drought facilitates day to day scan and assessment quantification. for crop monitoring.

REFERENCES

Bair, W. and Robertson Geo, 1966: A new versatile soil moisture budget, Can. J. Plant Sci. 45 : 276-283.

Baver, L.D. (1956) : Soil physics, 290-298, John Wiley, New York.

Boule, S.N. 1963. Quoted by Jensen M.E. 'Water consumption by plants. In T.T. Kozlowski (Ed) Water deficits and plant growth, Vol. 2, pp-1-22, Acad. Press, New York.

Cocheme, J. and Franquin, P. 1967 : An agroclimatological survey of a semi Arid area in Africa, South of the Sahara, Technical Note No. 86, world Meteorological Organization.

Ghadekar, S.R. 1991. Meteorology, pp- Agromet Publ. Nagpur.

Ghadekar,S.R. 1993. A simple model to quantify drought. unpublished.

Hiler, E.A. and Clarke, R.N. 1971. Stress day index to characterize effects of water stress on yields. Trans. Am. Soc. Agril. Engrs. 14 : 751-61.

Krishnan, A. 1979. Key paper on definition of droughts and factors relevant to specification of agril. and hydrological droughts. Proc. I.H.P. International symposium, Dec. 1978, New Delhi, India

Palmer, W.C. 1965. Meteorological drought. Weather Bur. U.S. Deptt. of commerce, Washington, Res. Paper - 45.

Raman, C.R.V. and Venkatraman, S. 1970. Assessment of soil moisture storage from rainfall in dry farming areas of Mysore State. Pre-Publication Scientific Report, No. 131, India Met. Deptt. Pune - 5.

Smith, L.P. 1975. Development in Atmospheric Sciences 3, Methods in Agricultural Meteorology. Elsevier Sc. Publishing Co. : New York.

Thornthwaite, C.W. and Mather, J.R. 1955. The water balance published in : Climatology, 8 (1) : 1-104, Centerton.

Yao, A.Y.M. 1969. The R-index for plant water requirement. Agric. Meteorology. 6 : 259-273.

Chapter - 10

WATER REQUIREMENT AND IRRIGATION MANAGEMENT OF SEMI-ARID AND ARID CROPS

R.P. Samui

Director, Office of the DDGM Agrimet, Shivajinagar, Pune - 411 005

1. INTRODUCTION

Water requirement of a crop may he defined as the quantity of water required by a crop in a given period of time for its normal growth under field conditions at a place. It is the total loss of water from the soil and canopy of the crop. The evapotranspiration at a given place is controlled by the evaporative demand and availability of water from rain, irrigation and soil water storage. Though there is a considerable variation in water requirement and also in their yield response per unit of applied water yet the information on water use and water use efficiency (WUE) is very useful in efficient irrigation management and selection of proper crops in different agroclimatic zones. In India, dryland rainfed agriculture with life saving irrigation accounts for most of the food and fibre production in the semi-arid and arid regions comprising the state of Rajasthan, Gujarat, Parts of Punjab, Haryana, Madhya Pradesh, Maharashtra, Karnataka, Tamilnadu where mean annual rainfall ranges between 400 to 1000mm and meets the seasonal water requirement either fully or partially of crops such as wheat sorghum, millet, groundnut, cotton, pulses etc.

2. THEORY OF EVAPOTRANSPIRATION

Evapotranspiration (ET) is described as the total process of water transfer into the atmosphere from vegetated land surfaces. The process of evaporation of water from soil and plant is known as evapotranspiration. Both these processes occur simultaneously in nature. The rate of ET is controlled by net radiation received, availability of water and aerodynamic factor that acts as a mechanism for transport of vapour. This is a complex process and depends mainly on water availability at the level of water evaporation and energy available at the same level for partition in vaporisation and sensible heat flux. The process could be better understood through a schematic diagram shown in Fig. 1. This figure depicts the input data, regulation parameters and equilibrium variables which interact with each other near the earth surface.

The input data are : (i) radiative terms (global radiation, Rs; and downward longwave radiation from the atmosphere, Ra), (ii) air characteristics (air temperature, Ta; specific humidity or dew point temperature, Td; and wind speed, u) and Surface properties : (1) albedo, a (reflectance for Rs); emissivity, ε (ii) aerodynamic exchanges as a function of wind speed, u. (roughness parameter, Z_o;

zeroplane displacement, D; and two levels to define the exchange coefficient viz., reference levels, Z_r; where climatic data are measured, and the level Z_h at the top of the canopy). (iii) for heat and mass transfer coefficients between surfaces to the top of the canopy is expressed as resistance such as resistance inside the canopy, r_o; (diffusion through structure canopy) and stomatal resistance, r_s; (biological parameters).

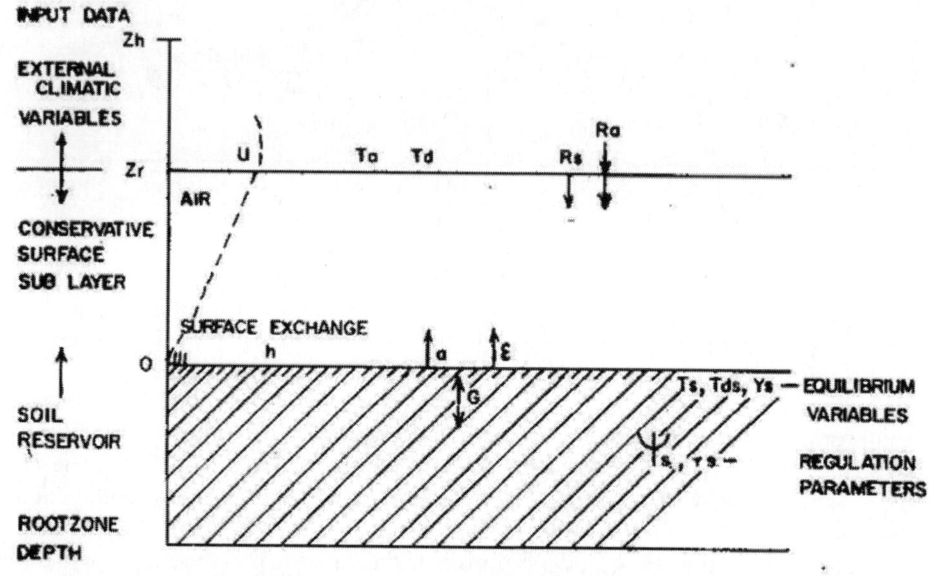

Fig. 1. Schematic diagram with input data, parameters (surface, soil and Air), Regulation parameters and equilibrium variables

Assuming conservative flux between the two chosen levels and admitting at the level Zh a given available energy (Rn+ G) where net radiation (Rn) is modified by soil heat flux (G), the energy balance equation becomes :

$$Rn + G + H = LET \tag{1}$$

Where H is sensible heat flux and LET is latent heat flux.

$$H = \rho Cp [h (u)_{Zr}^{Zh} \ (Ta - Ts) \tag{2}$$

$$LET = (ML/RT) \ [h (u)_{Zr}^{Zh} \ [P (Tds) - P (Td)] \tag{3}$$

Where M = Molar mass of water,

L = Latent heat of vapourization,

R = Gas constant,

T = Temperature, °K.

γ = Psychrometric constant = δ Cp(LM/RT)

δ = Air volume mass

Cp = Air specific heat.

For the surface (level, Z_h) Ts and Td are the equivalent value of air and dew point temperatures respectively. P(T) is the saturated vapour pressure function and P' is derivative function for T.

It is easy to express air pressure deficit at reference level, Da and at the surface level, Ds by the following expressions:

$$Da = P(Ta) - P(Td) = P'(Ta - Td) = P'Ya$$

$$Ds = P(Ts) - P(Tds) = P'(Ts - Tds) = P'Ys$$

Introducing in relation (3), the simple additive relation :

$$Tds - Td = (Tds - Ts) + (Ts - Ta) + (Ta - Td) = Ya - Ys + (Ts - Ta)$$

and replacing (Ts- Ta) by its value function of Rn, G and ET according to relations (1) & (2), the equation can be written as :

$$ET = P' [Rn + G + \rho\, Cph(u)\, (Ya - Ys)] / (P' + \gamma) \qquad (4)$$

Rn, G, h(u), Ya, P' (T) may be considered as factors dependent on climatic conditions, and Ys as the term of real availability of water at the surface which induces the real offer of water or the ET.

From this relation, it is possible to analyse several levels of evaporation or evapotranspiration :

(1) Ys = 0 or Ts = Tds = Tso i.e. saturation at the surface level and ET reaches the maximum value which is the potential evaporation, EP and represents the climatic demand. It leads to a minimum value of surface temperature, Tso and a maximum value of dew point at surface, Tds (Fig. 2).

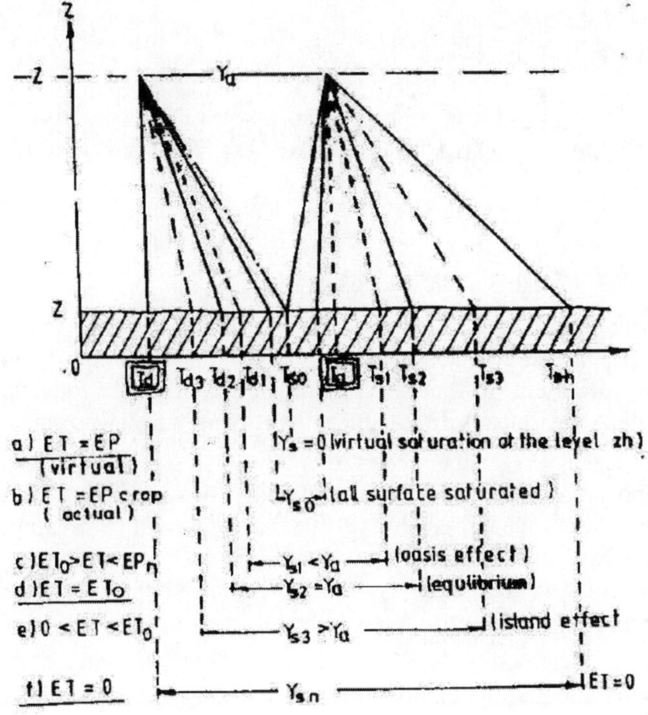

Fig. 2. Schematic profiles and values of equilibrium variables Ts, Tds OR Ys

(2) 0 < Ys < Ya; the water availability at the surface decreases and surface saturation deficit, Ds increases but remains smaller than the air saturation deficit (D < sa). Under such good water supply, ET is always less than potential evaporation, EP, but greater than equilibrium evapotranspiration ETo. In this case the surface temperature increases and

becomes similar to air temperature, sometimes less, sometimes more, and dew point at the surface decreases as function of ET (decreasing order).

(3) Ys = Ya, the water availability at the surface induces a surface deficit in equilibrium with air deficit, in such condition whatever may be wind speed, ET is equal to equilibrium evapotranspiration (ETo) :

$$ETo = (P' [Rn + G])/ (P' + \gamma)$$

In such a condition surface temperature is always greater than air temperature.

(4) Ya < Ys the water availability decreases strongly and the surface deficit becomes higher than air deficit and ET becomes lower (0 < ET < ETo). In such a condition surface temperature increases quickly as ET decreases and surface dew point temperature tends to reach air dew point temperature when ET = 0.

(5) Ys = Ymax, the water availability is so low that the surface deficit becomes maximum and its value is a function of climatic conditions. Under such extreme condition, no evaporation takes place and all the available energy is dissipated by convection in air, determining the value of maximum surface temperature so that

$$(Tsmax - Ta) = [(Rn + G) / \rho Cp [h (u)_{Zr}^{Zh}]$$

At the same time, the surface dew point temperature reaches the lowest value becoming equal to air dew point temperature.

$$Ymax = Tsmax - Td$$

3. STUDIES ON EVAPOTRANSPIRATION OF SEMI-ARID AND ARID REGION CROPS

WHEAT

Agrometeorological aspects of wheat production

Starting from germination, vegetative growth, flowering to grain formation, soil moisture plays the most important role for wheat production. Kulik (1965) observed that the germination of the seed was shown to be more dependent on the soil moisture content in the top 10cm of the soil than on temperature. More often the lack of sufficient soil moisture was the principal reason for nongermination.

Soil moisture supply from different soil especially deeper layers plays important role for growth and yield of winter wheat especially in regions where irrigation water is limited. Relationships developed by Ulanova and Rymar (1963) made it possible to determine the moisture supply in one layer from that of another layer. This information can be utilised in practical work in evaluating the existing and expected agrometeorological condition of growth of winter wheat in the fall.

Under conditions of adequate mineral nutrition the most appreciable influence on the grain yield of winter wheat brings productive moisture supply in the 0-50cm layer for 3, ten day periods before the ears form as well as for the duration of the renewal of the vegetation period (Kulik, 1995).

Entz and Fowler (1988) identified critical stress periods influencing the productivity of no-till winter wheat produced on the Canadian Prairies. Environmental stress between stages 21 and 65 days influences grain yield, aerial dry matter production, harvest index and grain protein yield. Among the agrometeorological parameters, evaporation, followed by root zone extractable soil water and precipitation had the largest influence on crop performance. Major (1988) observed increase in potential yield with latitude but rainfed wheat yields were similar at all sites confirming that moisture stress is the most limiting factor on the Great plain. There was a significant increase in yield under irrigated wheat.

Working with wheat crop under intensive irrigation in sub-humid environments, Laing and Fisher (1976) observed that grain growth rate was slower in the misted treatment apparently as a result of lower ear temperature. However irrigation treatments had no effect upon final grain size and grain yield. Atmospherically induced stress did not appear to limit grain yield under the meteorological conditions experienced in the north-west Mexico. Under irrigated and non-irrigated wheat, Ehrler (1978) observed that the difference of canopy-air temperature (ΔT) between irrigated and non- irrigated crops can be used to monitor stress conditions by remote sensing techniques. An approach to irrigation decisions for wheat under conditions of unstable rainfall is also presented by Yaron (1980) and a simulation model designed for tracing the variation of moisture in the soil profile during the growing season.

Water requirement of wheat

Seasonal ET - losses, Crop coefficient and ET - WUE and ET-yield relationship

When wheat was grown under irrigated condition at New Delhi, the total consumption of water was in the range of 250 to 320 mm. Under rainfed condition at Raipur ET varied in the range of 292 to 509 mm. Potential evapotranspiration (PET) also varied in the range of 334 to 517 mm. Rainfall received in the crop growing period was in the range of 31 to 107 mm (Table 1).

Table 1. Duration of crop, cumulative evapotranspiration, cumulative potential evapotranspiration, crop coefficient, rainfall, yield and water use efficiency of wheat crop at different stations

S.N.	Station	Year	Crop duration days	Cumulative ET (mm)	Cumulative PET (mm)	Ratio ET/PET	Rainfall (mm)	Yield (Q/ha)	WUE Q/ha mm	Irrigation No. & amount
1	Delhi	1978	154	250	342	0.73	144.4	26.8	10.7	3
		1982	154	291	361	0.8	199.6	34.6	11.9	4
		1987	133	302	420	0.71	79.8	37.3	12.4	3
		1984	140	308	475	0.64	19	31.2	10.1	4
		1988	156	313	513	0.61	93.6	37.2	11.9	3
		1979	147	316	347	0.91	40.2	31.9	10.1	3
		1986	140	431	474	0.9	114.2	31.5	7.3	3
2	Raipur	1984	119	292	437	0.66	58.8	10.2	3.5	–
		1981	119	329	334	0.99	86.4	9.2	2.79	–
		1983	126	351	427	0.82	106.3	10.4	2.97	–
		1982	126	502	476	1.05	31	21.8	4.33	–
		1980	133	509	513	0.09	95	25.4	4.98	–

Mean values of ET, PET and Crop coefficient (Kc) increased gradually with the advancement of crop growth stages in New Delhi. Kc (1.44) was maximum at 11th week after sowing. At Raipur Kc was more than 1.0 at jointing and heading stages (Table 2 and 3). Water use efficiency was maximum i.e. 12.5 kg $ha^{-1}mm^{-1}$ in New Delhi. It was minimum i.e. 2.81 kg $ha^{-1}mm^{-1}$ at Raipur (Samui et.al., 1999).

Table 2. Weekly mean evapotranspiration, potential evapotranspiration and crop coefficient (Kc) of wheat at Delhi, Ludhiana, Jodhpur and Banswara

Week after sowing	Delhi			Ludhiana			Jodhpur			Banswara		
	Mean ET	Mean PET	KC	Mean ET	Mean PET	KC	Mean ET	Mean PET	KC	Mean ET	Mean PET	KC
1	0.55	2.01	0.27	1.00	3.74	0.26	0.75	3.02	0.24	2.90	1.89	1.53
2	0.77	1.82	0.42	1.10	3.76	0.29	1.10	3.04	0.36	2.00	1.94	1.03
3	1.10	1.67	0.65	1.30	3.55	0.36	1.55	2.72	0.56	1.15	1.92	0.59
4	1.35	1.71	0.78	1.76	3.37	0.52	1.75	3.36	0.52	2.00	1.89	1.05
5	1.57	1.58	0.99	1.83	3.15	0.58	1.85	3.34	0.55	3.10	1.88	1.64
6	1.87	1.61	1.16	2.13	3.19	0.66	2.10	3.65	0.57	1.50	2.36	0.63
7	2.22	1.63	1.36	2.23	2.98	0.74	2.75	4.67	0.58	2.30	2.49	0.92
8	2.57	1.76	1.46	2.16	2.93	0.73	2.70	5.19	0.52	2.90	2.40	1.20
9	2.92	2.00	1.46	2.10	3.01	0.69	3.60	4.39	0.82	1.15	2.65	0.43
10	3.20	2.35	1.36	2.26	2.91	0.77	4.20	5.36	0.78	2.05	2.69	0.76
11	3.27	2.26	1.44	2.13	3.14	0.67	4.40	3.48	1.26	2.85	2.89	0.98
12	3.12	2.38	1.31	2.30	2.95	0.77	4.70	4.77	0.98	1.20	2.98	0.40
13	2.97	2.86	1.03	2.56	3.01	0.85	4.50	5.21	0.86	2.30	3.19	0.72
14	2.82	3.46	0.81	2.06	3.12	0.66	4.15	6.63	0.62	2.90	3.49	0.83
15	2.67	3.22	0.82	3.30	3.63	0.90	3.75	4.99	0.75	1.70	3.65	0.46
16	2.47	3.57	0.69	2.63	3.39	0.77	3.35	6.79	0.49	0.45	4.00	0.11
17	2.12	3.90	0.54	2.53	3.59	0.70	2.30	6.84	0.33	0.15	4.48	0.03
18	1.85	3.93	0.47	2.66	3.86	0.68				0.10	4.11	0.02
19	1.55	4.94	0.31	3.36	4.20	0.80				0.10	4.73	0.02
20	1.53	4.58	0.33	3.70	4.51	0.82				0.10	4.25	0.02
21	1.20	4.84	0.24	3.60	4.91	0.73						
22	1.05	7.01	0.14	1.73	5.53	0.31						
23				1.90	5.78	0.32						
24				1.20	5.54	0.21						
25				1.10	5.34	0.20						

Table 3. Weekly mean evapotranspiration, potential evapotranspiration and crop coefficient (Kc) of wheat at Raipur, Jabalpur, Dharwar and Akola

Week after sowing	Raipur			Jabalpur			Dharwar			Akola		
	Mean ET	Mean PET	Kc	Mean ET	Mean PET	Kc	Mean ET	Mean PET	Kc	Mean ET	Mean PET	Kc
1	1.03	3.63	0.28	1.40	3.35	0.41	2.35	3.61	0.65	0.82	3.60	0.22
2	1.10	2.81	0.39	1.60	3.18	0.50	1.80	3.44	0.52	1.77	3.20	0.55
3	1.33	3.41	0.39	1.30	2.70	0.48	3.05	3.23	0.94	2.82	3.15	0.89

4	1.50	2.55	0.58	1.50	2.53	0.59	4.00	3.21	1.24	2.90	3.13	0.92
5	2.03	2.77	0.73	2.40	2.25	1.06	3.25	3.29	0.98	3.55	3.41	1.04
6	2.80	2.73	1.02	2.70	2.03	1.33	5.10	3.27	1.55	3.75	3.67	1.02
7	3.53	2.88	1.22	3.00	2.53	1.18	3.25	2.97	1.09	4.22	3.55	1.18
8	3.70	3.14	1.17	3.40	2.20	1.54	4.40	2.93	1.50	4.85	3.58	1.35
9	4.70	3.39	1.38	3.70	2.43	1.52	4.35	2.83	1.53	5.05	3.83	1.31
10	4.30	3.53	1.21	3.90	2.67	1.46	3.05	2.94	1.03	5.02	4.08	1.23
11	4.93	3.94	1.25	3.70	2.56	1.44	3.40	2.84	1.19	5.00	4.32	1.15
12	5.56	3.85	1.44	3.60	2.56	1.40	3.25	3.08	1.05	4.65	4.82	0.96
13	3.86	4.43	0.87	3.50	3.11	1.12	4.05	3.19	1.26	4.05	5.18	0.78
14	3.43	4.96	0.69	3.30	3.23	1.02	2.35	3.29	0.71	3.03	5.59	0.54
15	3.13	5.42	0.57	2.80	3.99	0.70	3.50	3.48	1.00	2.82	5.87	0.48
16	3.80	5.67	0.67	2.60	4.29	0.60	1.95	3.33	0.58	2.30	7.10	0.32
17	3.80	6.22	0.61	2.40	4.04	0.59	1.70	3.61	0.47	1.63	5.58	0.29
18	2.30	6.45	0.35	2.00	4.90	0.40	0.90	3.97	0.22	1.90	6.40	0.29
19				1.40	5.04	0.27	0.50	3.69	0.13	1.70	5.18	0.32
20				0.70	5.03	0.13				1.20	7.61	0.15

An analysis of combined data set of New Delhi - Ludhiana - Jodhpur - Banswara and Raipur - Akola – Dharwar - Jabalpur which represents the semi-arid and arid climatic regions of the country showed that ET ranged between 194 to 431 mm and 292 to 509 mm respectively in these regions. Water use efficiency was maximum when the cumulative ET was in the range of 250 to 300 mm in the northern region. When cumulative ET increased beyond 300 mm water use efficiency decreased gradually. This can be seen from the graph of ET versus WUE (Figs. 3 & 4), where as in central part of India water use efficiency and yield was maximum when the water consumption of wheat was around 350 mm. Water use efficiency decreased with increase in the water consumption of wheat beyond 350 mm (Fig. 5 & 6). The curvilinear shape shows the importance of efficient and timely application of irrigation at critical growth stages of the crop. ET increased gradually with the advancement of the crop growth stages till the heading stage and then ET decreased gradually at dough and maturity stages. During jointing and heading stages Kc was more than 1.0 and it was maximum (Kc = 1.38) at heading stage.

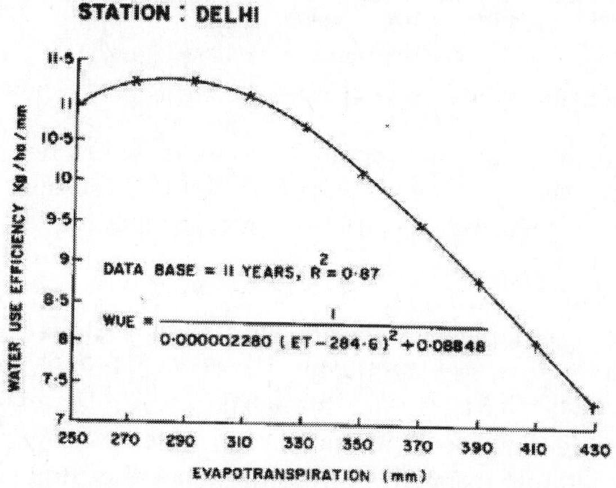

Fig. 3. Relationship of water use efficiency to seasonal evapotranspiration

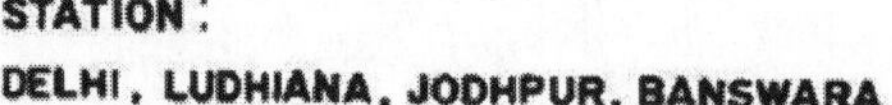

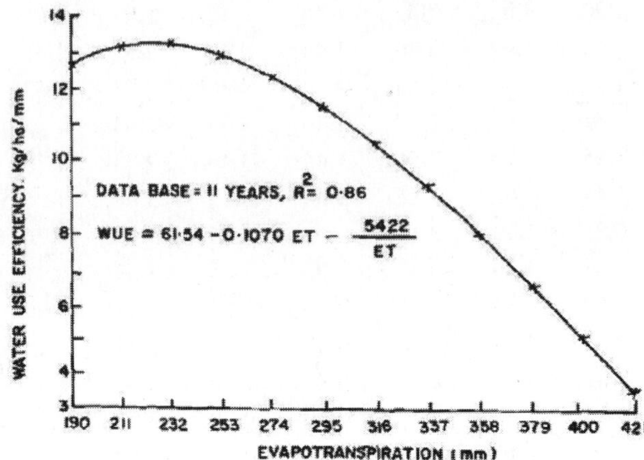

Fig. 4. Relationship of water use efficiency to seasonal evapotranspiration

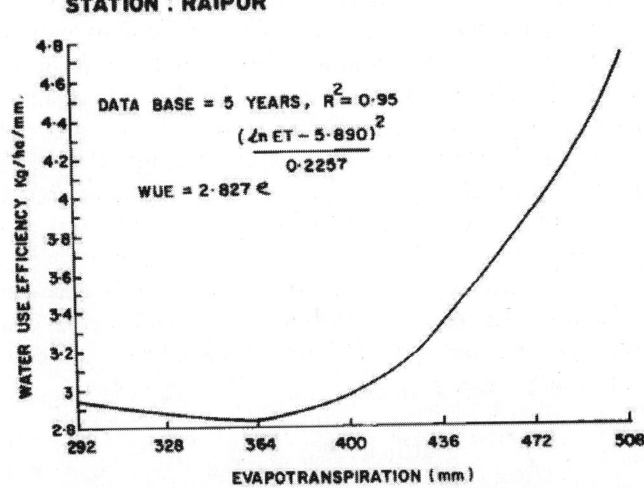

Fig. 5. Relationship of water use efficiency to seasonal evapotranspiration

The irrigated wheat in northern zone comprising some selected stations viz., New Delhi – Ludhiana – Jodhpur – Banswara and central zone comprising stations viz., Raipur – Akola – Dharwar performed differently because of variations in consumptive use at each of the locations. The relationship between seasonal ET and grain yield was curvilinear in nature in all cases except at Raipur (Fig. 7) where wheat was grown under rainfed condition. The relationship for combined data set of New Delhi – Ludhiana – Jodhpur – Banswara and Raipur – Akola – Dharwar were found to be curvilinear in nature and was statistically significant (R^2 = 0.812; P 0.05 for New Delhi – Ludhiana – Jodhpur – Banswara) and (R^2 = 0.81; P 0.05 for Raipur – Akola – Dharwar Fig. 8 & 9). The relationship indicated that yield increased with increase in water use of wheat till seasonal water use reached about 300 mm and 380 mm respectively for northern and central zones for India.

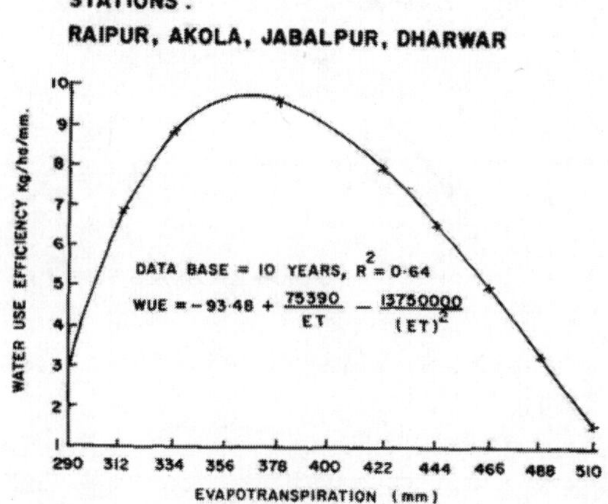

Fig. 6. Relationship of water use efficiency to seasonal evapotranspiration

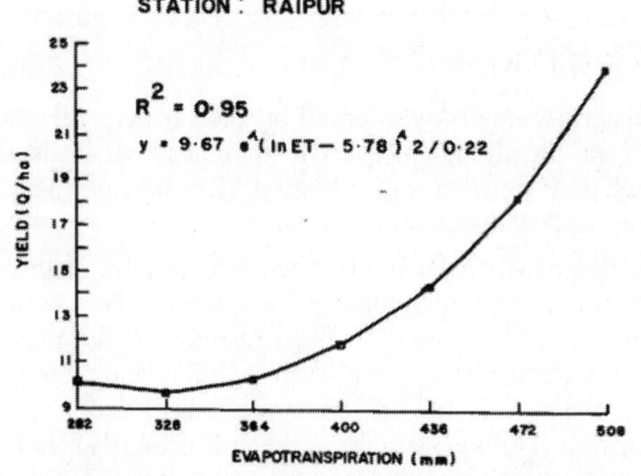

Fig. 7. Relationship of yield to seasonal evapotranspiration

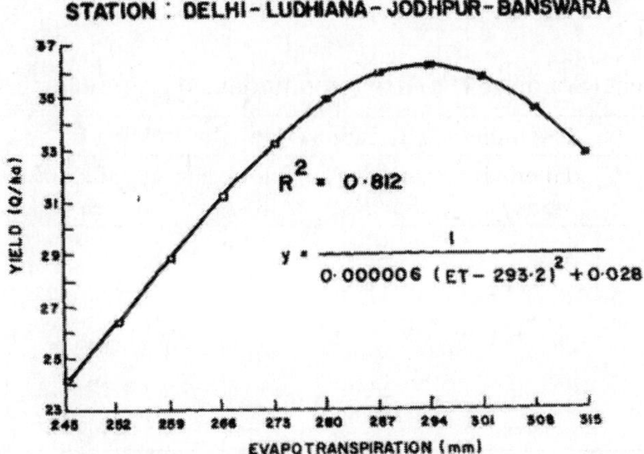

Fig. 8. Relationship of yield to seasonal evapotranspiration

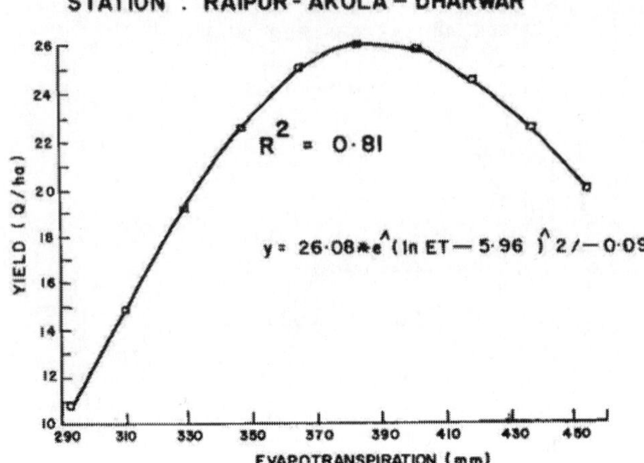

Fig. 9. Relationship of yield to seasonal evapotranspiration

4. SCHEDULING OF IRRIGATIONS

The most practical criteria commonly adopted by the farmers for scheduling of irrigations to wheat is the one based on the physiological growth stages critical in demand for water. In wheat, different growth stages such as (1) crown root initiation, (2) tillering, (3) jointing, (4) flowering, (5) milk and (6) dough could be well delineated.

In case of dwarf varieties of wheat, it is The second important stage was that of observed that irrigation at the crown root flowering which takes place at 80-85 days initiation (C.R.I) stage (20-25 days after sowing, jointing and milk ranked sowing) resulted in the maximum third, followed by tillering stage and then production per unit of water applied and, dough stage. The irrigation is applied at the therefore, this stage was considered as the initiation of the stage so that the moisture most critical stage for irrigation (Hukkeri supply in the soil is amply available and Chauhan, 1967). The delay in the first throughout stages. Gautam *et al.* (1968) irrigation reduced the yield considerably. observed that on the sandy loam soils of Delhi four irrigations applied at proper growth stages of crop yielded as much as an intensive irrigation practice with irrigation at all five growth stages (Table 4).

Table 4. Response of wheat (Sonora-64) to irrigation timings on sandy loam soils at Delhi, India

Total Irrigations (Frequency)	Timings of irrigation (Days after sowing)						Yield q/ha
	C.R.I. 25days	Tillering 45days	Jointing 65days	Flowering 85days	Milk 105 days	Dough 120 days	
6	+	+	+	+	+	+	51.4
4	+	—	—	+	+	—	51.3
4	+	+	+	+	—	—	48.5
4	—	+	+	+	+	—	45.2
4	—	—	—	+	+	+	44.0
4	—	+	+	+	+	+	37.0

+ Irrigation; - No Irrigation; (Source: Gautam et al. 1968)

An appropriate schedule of irrigation based on physiological stages of growth for early and late maturing varieties of wheat grown on light and heavy soil in northern India and irrigation intensity (after sowing) varying from 3-6 are in practice in India. In case the availability of water is still less, then it is recommended that irrigation may be applied at C.R.I. stage when water for only one irrigation is available and at CRI and flowering stages when water for two irrigations is available. Less number of irrigations spread over larger area helps increasing the overall production of the region.

Sources of irrigation water and methods of irrigation

Wheat being a low temperature crop is grown in winter season in most of the countries. Most of the wheat growing areas in the semi-arid and arid regions of Pakistan, Egypt and Sudan come under irrigation. It covers about 80%, 100% and 97% area under irrigation respectively in Pakistan, Egypt and Sudan on loam, clay loam and clay soils. In India, it covers more than 78% under irrigation on well drained clay loams, loams and sandy loam soils.

I.R. of Iran, Romania and Chile cover respectively 38%, 50% and 55% area of their land under irrigation. In Saudi Arabia, wheat crop is grown on 4,40,000 ha sandy-silty soil under irrigation. Bulgaria, South Africa, Lebanon and Czech Republic cover only about 2.5 to 25% land under irrigation (Samui, 1998).

The major sources of irrigation water for wheat crop are canal, dam, and lift irrigation from rivers and ponds. Surface, sprinkler and drip irrigation methods are used for applying water more effectively. In India, wheat is irrigated by applying surface irrigation from canal, dam and lift irrigation from rivers and ponds. In Pakistan 90%, 5%, 2% and 3% of wheat cropped area are irrigated respectively by applying water from canal, dam, lift irrigation and underground water. In Bulgaria more than 50% of its land is irrigated through lift irrigation from rivers and ponds by using drip irrigation methods and remaining 28 and 15% from dam and canal water respectively through sprinkler and surface irrigation methods. Only 4% cropped land is irrigated by using underground water. In Egypt, Sudan and Romania most of their wheat cropped area are irrigated through lift irrigation from rivers and ponds using sprinkler and surface irrigation methods. In Chile, 85% area is irrigated by using canal water through surface, sprinkler and drip irrigation methods (Samui, 1998).

In India, the best wheat producing region is the Indo-Gangetic Plain where cool and dry climate prevails during the growing season (rabi). In Central and Peninsular India, macaroni wheat is grown with 6-7 irrigations amounting each of 60 to 70 mm depth. Irrigation is given at an interval of 15-20 days at CRI, tillering, late jointing, flowering and milky dough stages. In Pakistan, Egypt and Saudi Arabia wheat is grown in winter season under irrigated condition. In Pakistan, 3 to 4 irrigations amounting 200-300 mm are applied during the critical stages of crop growth viz., vegetative to reproductive and maturity. In Egypt, 5-6 irrigation are required throughout the crop growing period. In Saudi Arabia, 1-2 days interval irrigations amounting total 7000 m^3 ha^{-1} season are required for wheat crop. In Bulgaria 2-3 irrigations each of 30-40 mm in earning to grain filling stages are applied. In I.R. of Iran where wheat crop is grown in the spring season, total 4-7 irrigations amounting each of 10 cm are sufficient for its good growth and higher yield. In Czech Republic, two irrigations in April and two irrigations in June, total amounting 600-800 mm are required respectively during panicle initiation and flowering stages. In Lebanon and Chile, 15-20 days interval irrigations are applied during flowering to grain formation stages. In Sudan, 7-8 irrigations are sufficient for crop growth and high yield.

5. STUDIES ON EVAPOTRANSPIRATION IN THE SORGHUM/MILLET GROWING AREAS OF INDIA

Seasonal ET losses

Sorghum is a deep rooted crop and has excellent drought tolerance and grown under stored soil water in deep soil applying one or two life saving irrigations at critical growth stages. In the drier region, rabi sorghum is also grown following kharif crop on stored soil water with limited irrigation. Regions benefitted from north east monsoon and having favourable thermal environment, grow rabi sorghum using available stored soil water. Occasional droughts and persistent drying trend over the regions point to the need for improved farm water management, if crop yield is to be sustained. This involves life saving irrigation at some of the critical growth stages of the crop. Thus the water requirement of sorghum at critical growth stages is important for optimum utilization of available water resources. The use of irrigation to overcome the water stress conditions normally increase the grain production. Several studies have reported Water Use Efficiency (WUE) to he higher during water stressed condition when irrigation is applied at critical stage of crop development.

ET observations of sorghum from 8 stations in central and southern part and of millet from 6 stations situated in arid and semi arid regions were compared with potential ET, normal rainfall and actual weekly rainfall. Some of the representative station's figures (2 for sorghum and 2 for millet) are presented. Depending on the stage of crop and the environmental evaporative demand, water requirement varies considerably at different growth stages. It is obvious that less water is required at early growth stages than in the active vegetative or reproductive stages. Similarly, water requirement of crop declines at maturity stage (Fig. 10 to 13). ET observations recorded from different ET stations, suggest an average water requirement for sorghum at different growth stages viz. seedling stage : 2.5 to 3 mm; flowering primordia stage : 4.0 mm; flowering stage : 4.5 mm and grain filling stage : 5.1 mm per day during kharif and 2.25, 4.2, 5.4 and 5.5 mm/day for above mentioned stages respectively during rabi season (Maske and Samui, 1982).

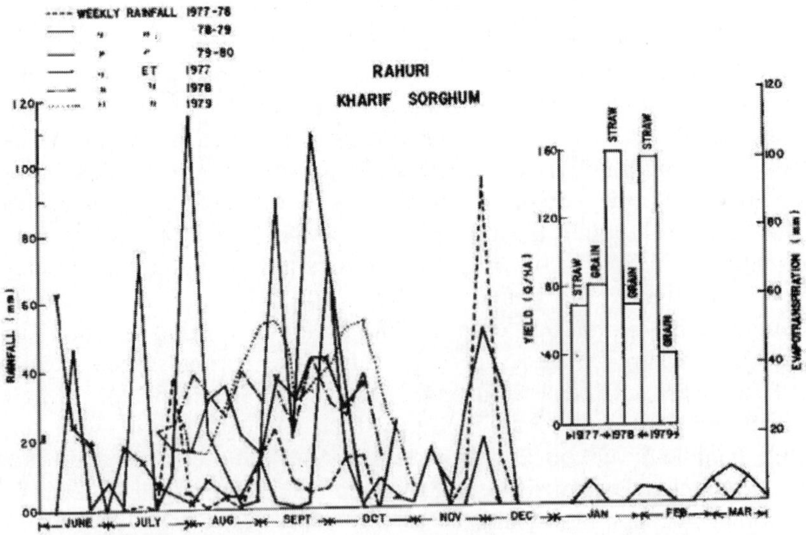

Fig. 10. Weekly rainfall during 1977-78, 78-79 and 79-80 and evapotranspiration during the years 1977-1979 of Kharif sorghum above the straw and grain yield of kharif during the years 1977-79.

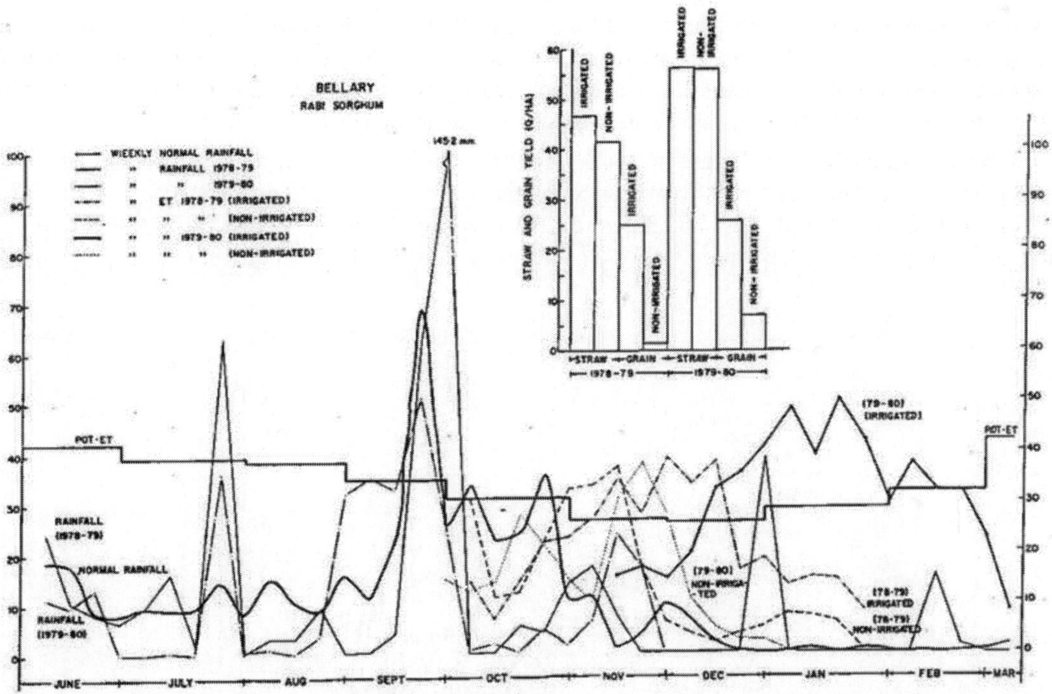

Fig. 11. Weekly normal rainfall, potential evapotranspiration Penman, Rainfall during 1978-79 & 79-80 and measured for irrigated and non-irrigated sorghum crop during 1978-79 and 79-80. Above the grain and straw yields of irrigated and non-irrigated crops are given.

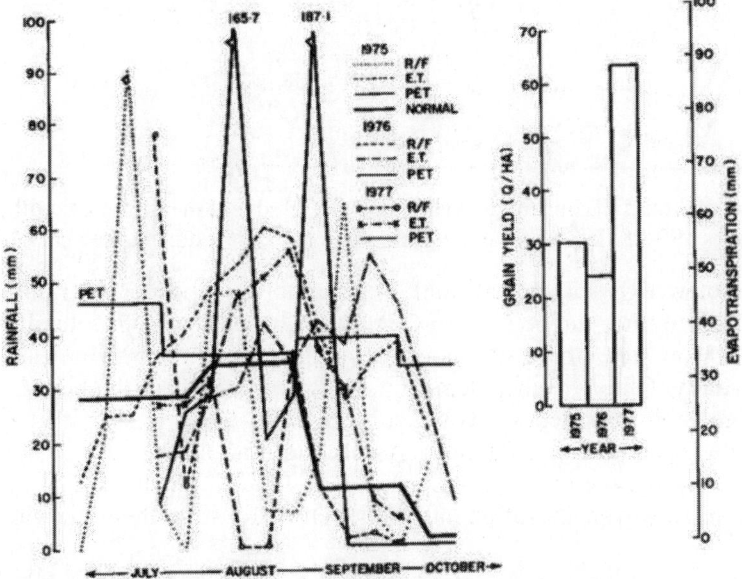

Fig. 12. Weekly normal rainfall, during the years 1975, 1976, & 1977 and measured evapotranspiration of millet during the years 1975, 1976 & 1977 above the grain yields of millet during the years 1975, 1976 and 1977.

Grain sorghum production was found to be satisfactory with approximately 350 to 450mm of water; maximum consumptive use approached 9.3 and 8.3 mm/day (irrigated crop) during kharif and rabi seasons respectively. These values are comparable with the estimated ET of sorghum (8.75 mm/day) by Jensen and Musick, 1960 and Musick and Grimes, 1961.

ET observations from different stations have revealed that the water requirement of millet at early vegetative, booting, earning-flowering and seed formation stages are 4.1, 7.2, 5.4 and 3.6 mm day^{-1} for irrigated crop and 3.2, 4.7, 3.1 and 2.0 mm/day for nonirrigated crop respectively. The nonirrigated crop was under water stress condition which resulted much reduction in grain yield (Fig. 13).

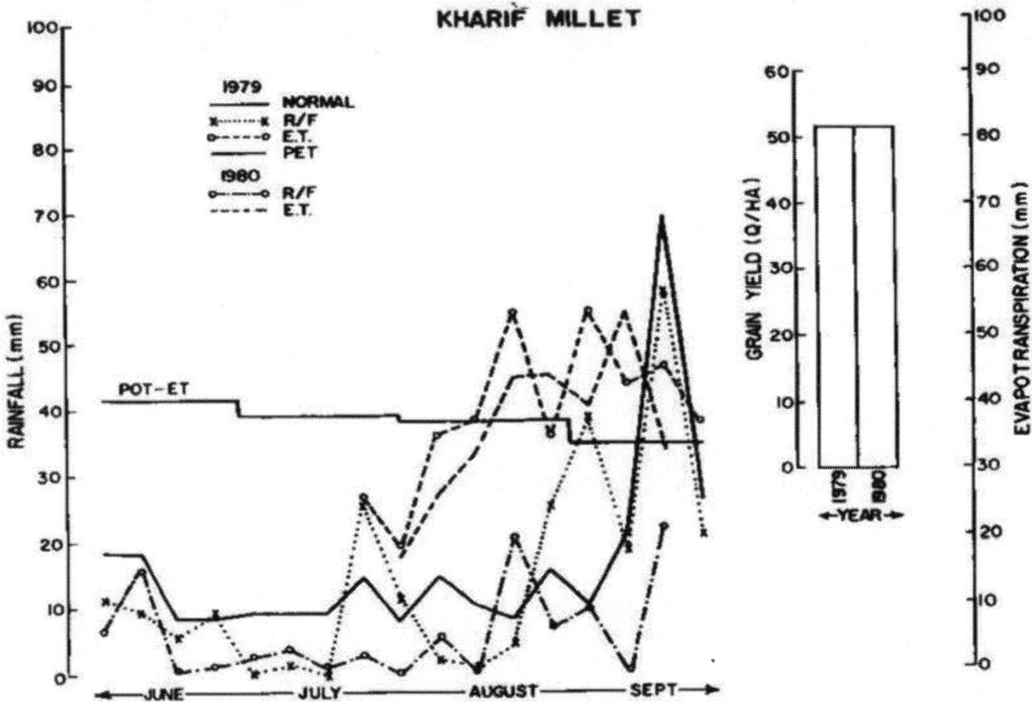

Fig. 13. Weekly normal rainfall, during the years 1979 & 1980 and measured evapotranspiration of millet during the years 1979 & 1980 above the grain yields of millet during the years 1979 and 1980.

Comparisons of weekly ET, actual rainfall, potential ET and normal rainfall (Figs. 10 to 13) show that water need during the active growing stages in rabi season is hardly fulfilled from the precipitation alone. However, stored soil moisture coupled with precipitation can somehow meet the water demand partially. Otherwise supplemental irrigation, atleast in the boot to dough stage, may be applied for high grain production. It is also seen from these figs. that kharif season has less evaporative demand (from potential ET line). Assured rainfall amount at 50% probability (Figs. 14 & 15) level is more steady, thereby, crops ET demand is fulfilled without irrigation in major part of the semi arid region. However, irrigation may be given to boost up the grain yield, where irrigation water is available.

For some selected years comprising good and deficit rainfall, graphical analysis was made to study the behaviour of water use of sorghum and crop coefficient Kc, (Figs. 16 to 18). Weekly total of ET, PET and rainfall during various stages of dryland sorghum were quite different than irrigated

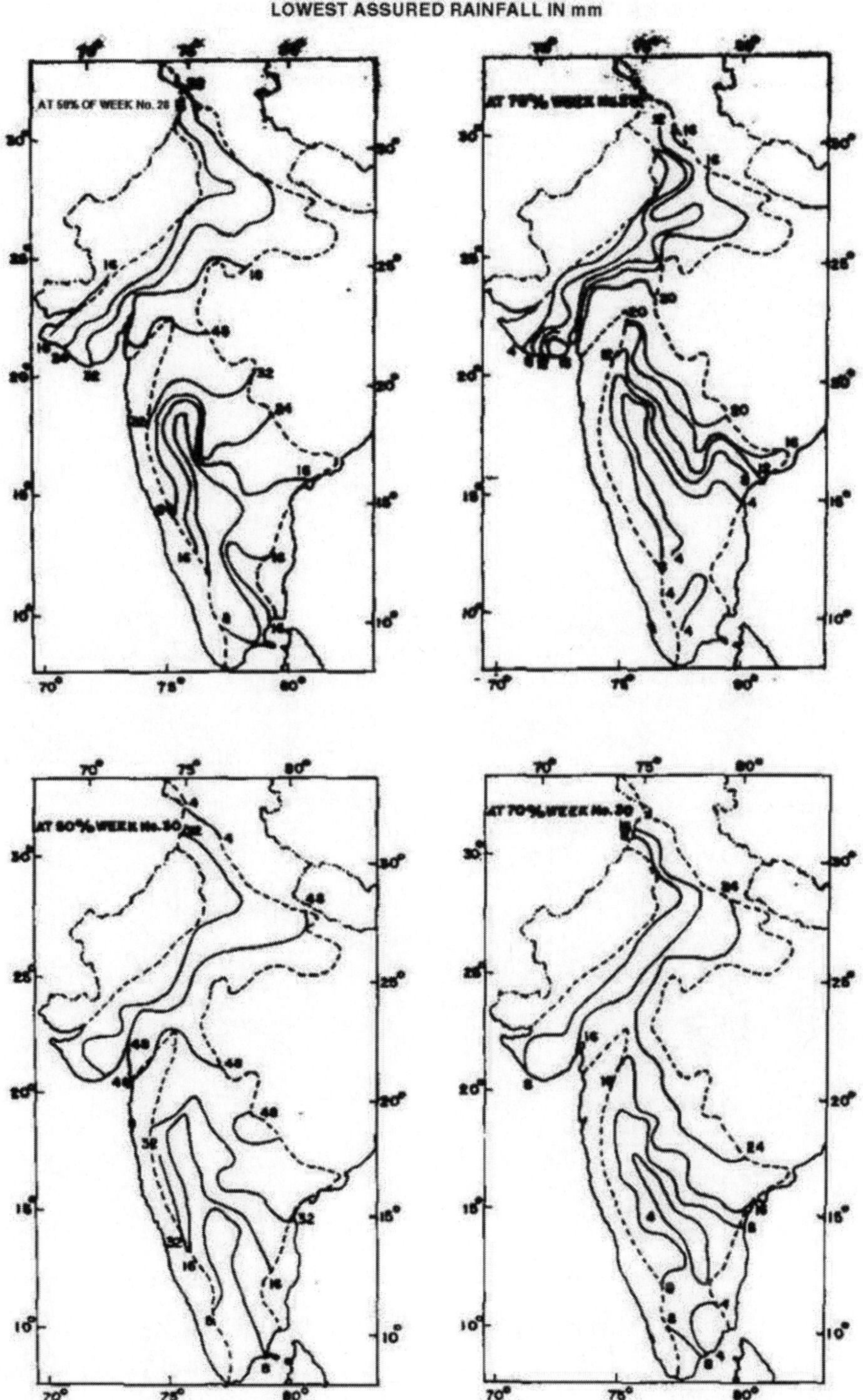

Fig. 14. Assured rainfall (mm) at 50 and 70% level, week Nos. 28 and 30. (Sarker *et al.,* 1978)

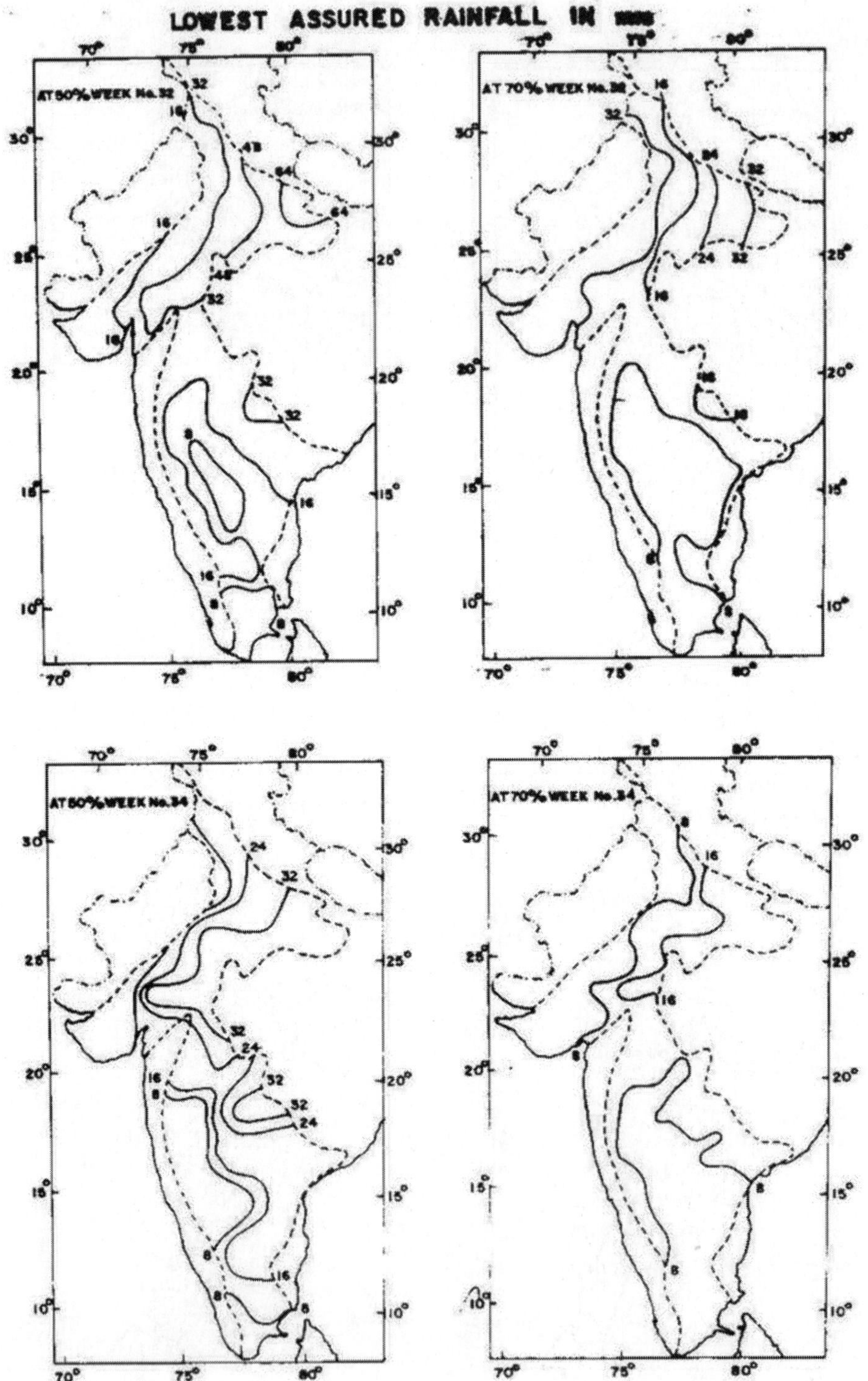

Fig. 15. Assured rainfall (mm) at 50 and 70% level, week Nos. 32 and 34. (Sarker *et al.,* 1978)

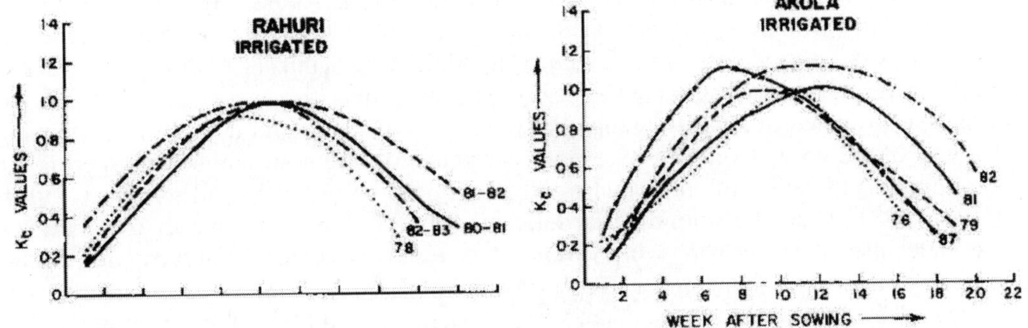

Fig. 16. Kc curves of sorghum

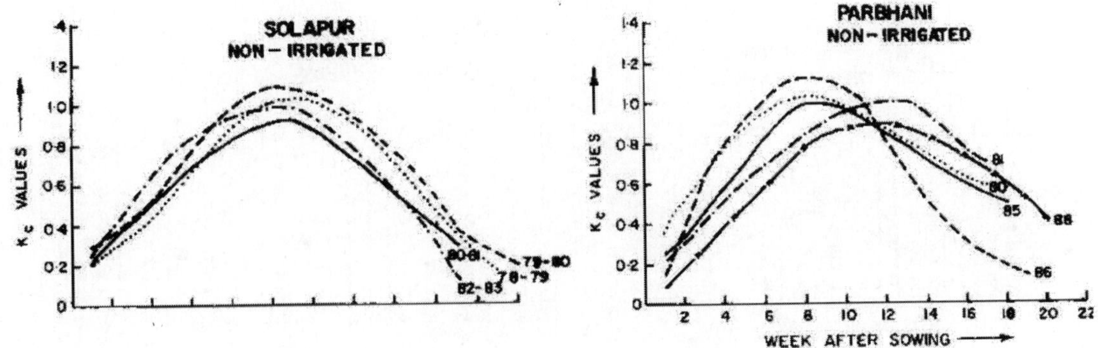

Fig. 17. Kc curves of sorghum

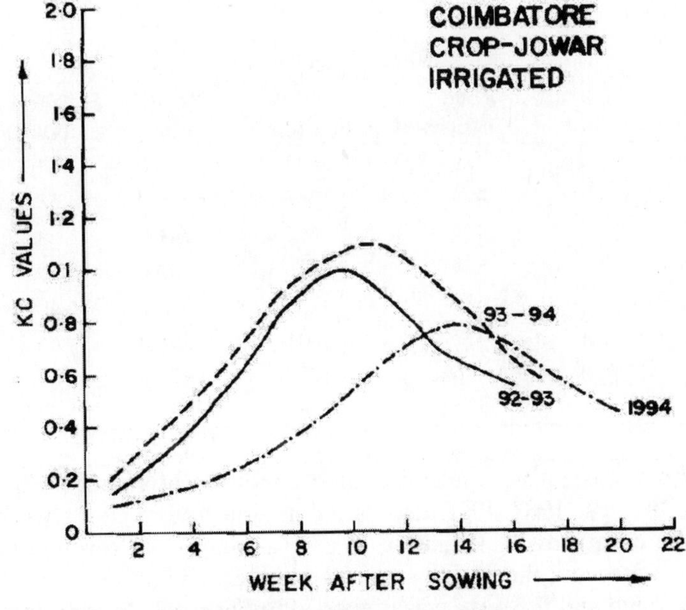

Fig. 18. Kc curves of sorghum

sorghum. When sorghum was grown under irrigated condition at Akola, the water use pattern in 1981, a good rainfall (rainfall = 741mm) year was as comparable to evaporative demand of the station. ET showed steady increase from 2 mm to maximum of 5 mm at 11th to 12th week after sowing i.e. at flowering stage. There was a steady decrease after 12th week. The PET during the crop growing season varied in a narrow range between 4.8 to 6.5 mm and it was all the time higher than the ET loss of the crop. Rainfall was evenly distributed throughout the growing period which resulted in optimum utilization of soil water and higher yield at 43.1 q ha^{-1} (Table 5). Contrary to this in the year 1987 when the rainfall was deficient (rainfall-426 mm) and its distribution was uneven, the water use of the crop was maximum at week no. 7 which coincides with the active vegetative stage of the crop. There was a steep increase in ET from 2 mm to 6.5 mm and maximum ET was preponed by 3-4 weeks. Subsequently due to deficient rainfall at the later stages of the growth period, ET dropped drastically from 5.5 to 3.0 mm (Fig. 19) whereas evaporative demand increased from 5 to 7 mm and it varied in a narrow range till harvest.

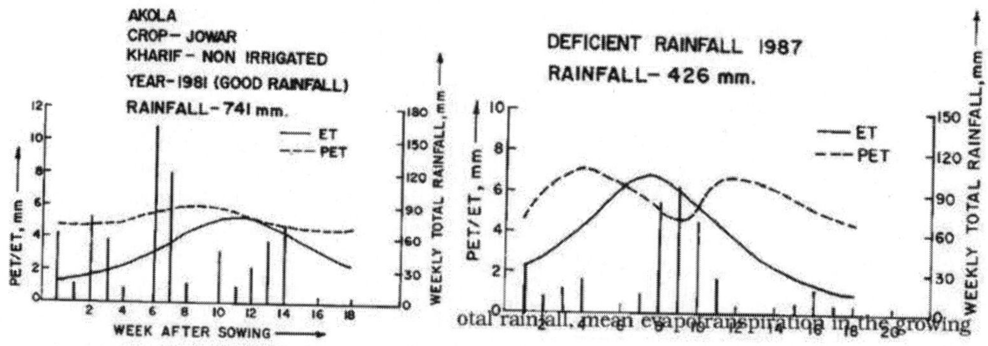

Fig. 19. Weekly total rainfall, mean evapotranspiration in the growing season.

Table 5. Evaporation, potential evapotranspiration, rainfall, yield and water use efficiency of Sorghum at *Akola*

Kharif-Irrigated

Year	Variety	Total ET (mm)	Total PET (mm)	Rainfall (mm)	Yield (Q/ha)	WUE Kg hec^{-1} mm^{-1}
1976	CSH-2	489	833	594.0	51.4	10.51
1979	CSH-1	476	738	606.0	58.3	12.24
1981	CSH-5	451	672	741.0	43.1	9.56
1982	CHS-5	554	710	392.0	40.3	7.27
1983	CSH-5	639	722	680.0	40.4	6.32
1984	CSH-9	412	501	398.0	46.2	11.21
1987	CSH-9	486	735	426.0	57.9	11.91
1988	CSH-9	522	666	452.0	57.9	11.09
MEAN		503	696	536.1	49.43	9.83

When rabi sorghum was grown under irrigated condition at Rahuri, ET varied in a narrow range of 2 to 3 mm in the year 1981. PET also varied in a narrow range between 3 to 5 mm during the entire period of the crop growth. Irrigation at critical stages of crop had resulted in optimum utilization of available water and the yield was on the higher side at 36.5 q ha^{-1} (Table 6). In 1982 ET varied from 2 to 6.5 mm and the peak was preponed to 6th week. During this period PET was in the narrow range of 3 to 5.5 mm. The water requirement of the crop was fully met by irrigation and stored soil water and the yield was at the higher side at 39.8 q/ha. At Rahuri in deficient rainfall year WUE of sorghum was considerably high at 8.63 kg ha^{-1} mm^{-1} and corresponding yield was 39.8 q/ha.

Table 6. Evaporation, potential evapotranspiration, rainfall, yield and water use efficiency of Sorghum at *Rahuri*

Kharif-Non Irrigated						
Year	Variety	Total ET (mm)	Total PET (mm)	Rainfall (mm)	Yield (Q/ha)	WUE Kg hec^{-1} mm^{-1}
1978	CSH-5	475	711	–	69.38	14.60
1979	CSH-5	576	618	–	41.25	7.16
MEAN		525.5	664.5	–	53.31	10.88
Rabi-Irrigated						
Year	Variety	Total ET (mm)	Total PET (mm)	Rainfall (mm)	Yield (Q/ha)	WUE Kg hec^{-1} mm^{-1}
1980	CSH-8R	371	573	36.1	38.5	10.38
1981	CSH-8R	429	562	82.8	36.5	8.51
1982	CSH-8R	461	528	60.6	39.8	8.63
MEAN		420	554	59.8	38.3	9.17

In a deficient rainfall (295 mm) year (1980), WUE of sorghum was relatively higher at 6.1 kg ha^{-1} mm^{-1} (Table 7) and corresponding yield was 28.81 q ha^{-1} at Parbhani. But in good rainfall year (1989) WUE and yield decreased. This may be due to excessive vegetative growth.

Table 7. Evaporation, potential evapotranspiration, rainfall, yield and water use efficiency of Sorghum at *Parbhani*

Kharif-Non Irrigated						
Year	Variety	Total ET (mm)	Total PET (mm)	Rainfall (mm)	Yield (Q/Ha)	WUE Kg hec^{-1} mm^{-1}
1980	CSH-6	470	719	295.2	28.81	6.13
1981	CSH-6	470	657	560.6	57.25	12.18
1985	CSH-6	522	767	347.5	2.28	0.44
1986	CHS-9	559	879	360.0	42.00	7.51
1988	CSH-9	413	706	440.2	28.31	6.85
1989	CSH-9	479	647	519.7	27.65	5.77
MEAN		485.5	729.2	420.5	36.80	7.69

At Solapur non-irrigated sorghum was raised with stored soil moisture accumulated during south-west monsoon season and rainfall received from north east monsoon. During 1979-80, the crop had experienced good and well distributed rainfall (rainfall = 458.8 mm) which resulted in variation of ET between 2 to 4 mm during the crop growth stages and peak ET was attained during 9 to 10th weeks after sowing (Fig. 20). In 1980-81, when there was deficient rainfall (rainfall = 82.5 mm) crop had experienced water stress condition and ET reduced drastically. This had resulted into early maturity of the crop and the crop period was shortened by 3 weeks. There was no significant change in WUE. It varied between 6.18 kg ha^{-1} mm^{-1} under good rainfall and 6.55 kg ha^{-1} mm^{-1} (Table 8) under deficient rainfall respectively.

Table 8. Evaporation, potential evapotranspiration, rainfall, yield and water use efficiency of Sorghum at *Solapur*

Rabi-Non Irrigated

Year	Variety	Total ET (mm)	Total PET (mm)	Rainfall (mm)	Yield (Q/Ha)	WUE Kg hec^{-1} mm^{-1}
1978	MALDANDI	495	669	310.4	33.30	6.73
1979	MALDANDI	419	653	458.8	25.90	6.18
1980	M-35-1	345	595	82.5	22.60	6.55
1982	M-35-1	314	434	266.5	14.50	4.62
1983	M-35-1	319	507	142.5	14.70	4.61
1984	M-35-1	419	620	336.8	21.60	5.16
1985	M-35-1	257	634	214.8	9.50	3.70
MEAN		366.8	587.4	258.9	20.30	5.36

When irrigated sorghum was raised at Coimbatore providing limited amount of irrigation at critical crop growth stages, ET ranged between 2 to 3.5 mm throughout the life period amounting average water use of 349 mm (Table 9). Life saving supplementary irrigations were applied to save the crop from the permanent wilting at the time of sowing and as well as at maturity. Yield in 1992 was considerably low (20 q ha^{-1}) compared to that in 1989 when grain-yield was 30 q/ha even under deficient but evenly distributed rainfall. During deficient rainfall year in 1989 WUE was 9.46 kg ha^{-1} mm^{-1} whereas during good rainfall year in 1992, WUE was 6.01 kg ha^{-1} mm^{-1} (Table 5).

Table 9. Evaporation, potential evapotranspiration, rainfall, yield and water use efficiency of Sorghum at Coimbatore

Rabi-Irrigated

Year	Variety	Total ET (mm)	Total PET (mm)	Rainfall (mm)	Yield (Q/Ha)	WUE Kg hec^{-1} mm^{-1}
1989	CO-26	318	899	372.0	30.10	9.46
1992	CO-25	333	463	555.0	20.00	6.01
1993	CO-25	416	519	452.0	39.45	9.48
1994	CO-25	331	470	366.1	18.14	5.48
MEAN		349	587	436.2	29.62	7.16

Crop Coefficient, Kc

For kharif sorghum at Akola Kc ranges between 0.22 to 1.35. It is below 0.5 for the first 3-4 weeks after sowing. It reaches maximum during elongation and vegetative stages at 10 to 12 weeks and then it decreases slowly at maturity stage. At Coimbatore for rabi sorghum Kc ranges between 0.12 to 1.30. Kc was observed between 0.22 to 1.16 at Parbhani when crop was raised in kharif sorghum. It was below 0.5 for first 3 to 5 weeks after sowing, then it increases and reaches to maximum of 1.33 during elongation and vegetation period at week Nos. 8 to 10. Kc values for 8th to 16th weeks were higher than 1.0 for kharif sorghum at Rahuri. Kc for rabi sorghum at Solapur varied between 0.11 to 1.27. Kc value was below 0.5 in weeks 4-5 after sowing. Maximum Kc reached in vegetative and flowering stages.

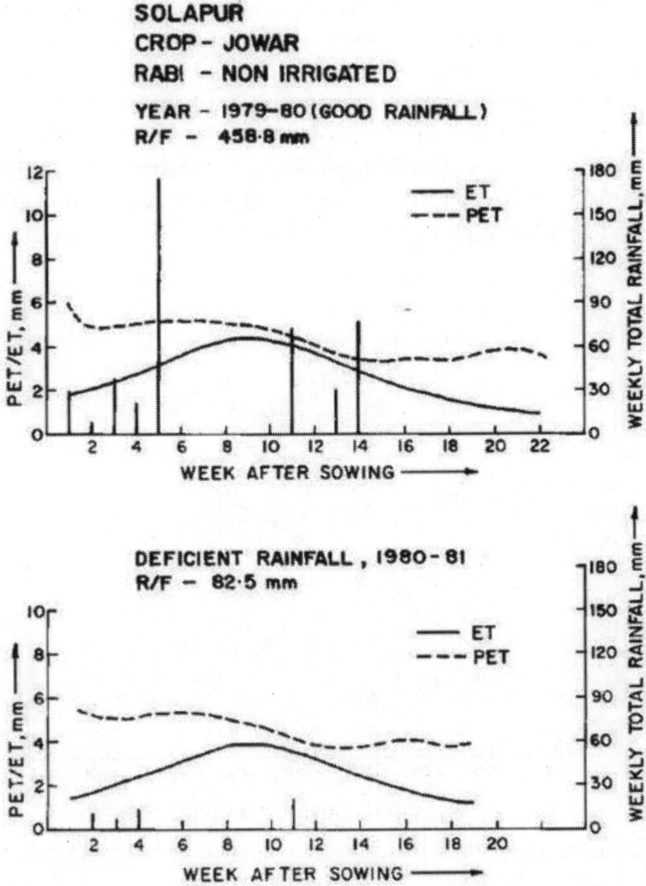

Fig. 20. Weekly total rainfall, mean evapotranspiration in the growing season.

6. SEASONAL ET-YIELD RELATION

The relationship between seasonal ET and grain yield is presented in Fig. 21 for the nonirrigated rabi sorghum grown at Solapur. Seasonal ET ranged from 250 to 500 mm and grain yield ranged from 9.5 to 33.3 q ha^{-1}. With the exception of four for rabi sorghum grown at Solapur is years, dryland sorghum yield ranged from presented in Fig. 22. The WUE ranged 9.5 to 14.7 q ha^{-1}. The linear regression between 3.7 to 6.7 kg ha^{-1} mm with a relationship was statistically significant (R^2 mean value of 5.4 kg ha^{-1} mm under = 0.90; P <0.001) with a slope of 9.46 kg of rainfed condition. The reciprocal hyperbolic grain yield per mm of seasonal ET. The relationship was statistically significant (R^2 relationship between seasonal ET and WUE = 0.73; P<0.01). With irrigated kharif sorghum, grown at Akola seasonal ET ranged from 412 to 639 mm and grain yield ranged from 40.3 to 58.3 q ha^{-1}. The relationship between seasonal ET and grain yield of irrigated kharif sorghum was 2nd order hyperbolic in nature and was statistically significant (R^2 = 0.44; P <0.03). The relationship indicated that yield increased with increase in water use of sorghum till seasonal water use reached about 500 mm. Further increase in water use beyond 500 mm had detrimental effect and yield decreased sharply beyond 550 mm of seasonal ET (Fig. 23). For irrigated kharif sorghum at Akola, the relationship between seasonal ET and WUE was "cauchy" type (Fig. 24). The relationship was statistically significant at 1 percent level ($R^2 = 0.79$; $P < 0.001$) (Samui et al., 1997). Excess vegetative growth under well watered condition which might have resulted in disproportionate change in the ratio of harvestible yield to above ground dry matter had caused such type of relationship. Similar converse yield-evapotranspiration function has been reported (Musick et al., 1963 : Grimes et al., 1969 : Kallsen et al., 1984) in studies with tall wheat, cotton and spring barley respectively. Under high water treatment and with tall wheat, Musick et al. (1963) observed lodging which had caused reduction in maximum grain yield. Lodging at high ET values had also been found to bring departure from linearity in four varieties of spring barley grown on two soils (Ali, 1979).

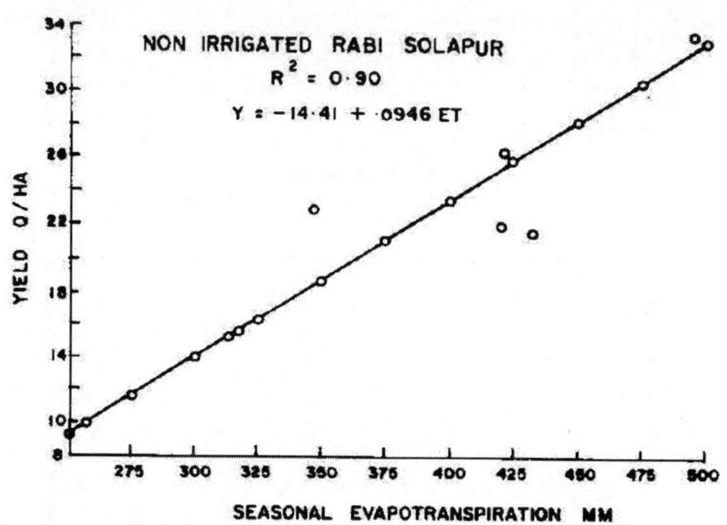

Fig. 21. Relationship of sorghum yield to seasonal evapotranspiration (Drawn not as per equation)

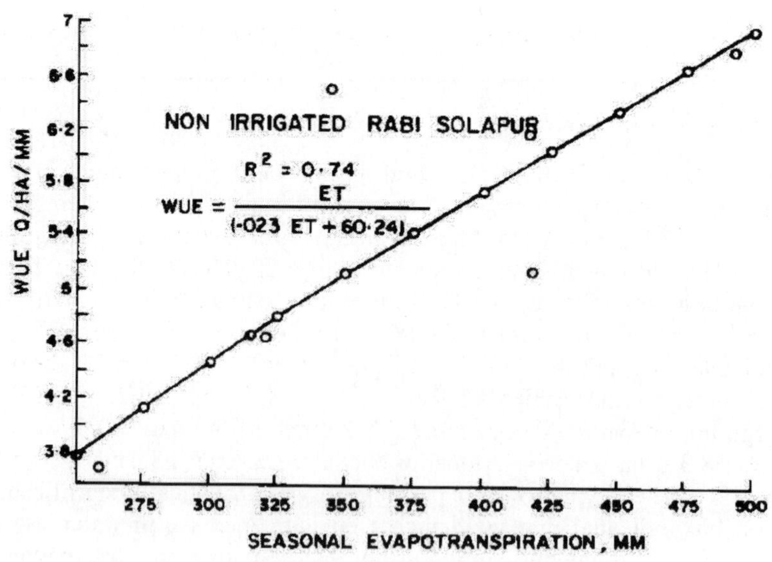

Fig. 22. Relationship of water use efficiency to seasonal evapotranspiration

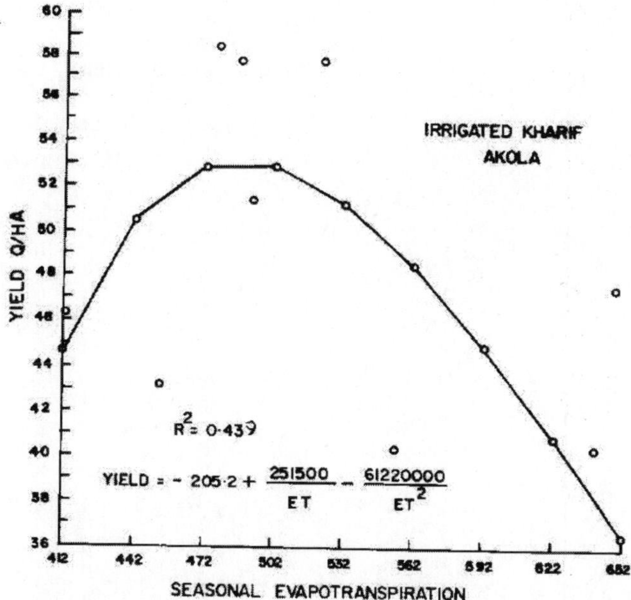

Fig. 23. Relationship of sorghum yield to seasonal evapotranspiration (Drawn not as per equation)

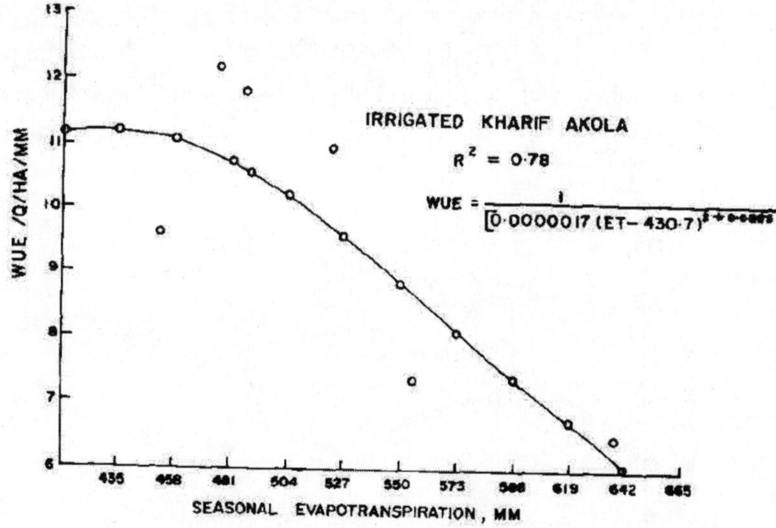

Fig. 24. Relationship of water use efficiency to seasonal evapotranspiration

When all the seasonal ET data generated in Kharif season both under irrigated and nonirrigated conditions were pooled together and best fit equations developed between seasonal ET Vs. grain yield and seasonal ET Vs. WUE respectively, the relationships were found to be 2nd order hyperbolic in nature (Fig. 25 and 26). In both the cases the relationships were not significant (R^2 = 0.15 for seasonal ET Vs. Yield and R^2 = 0.23 for seasonal ET Vs. WUE). However, with rabi sorghum grown at Solapur (nonirrigated), Coimbatore (irrigated) and Rahuri (irrigated) the relationship between seasonal ET and WUE seasonal ET and grain yield relationship was modulus

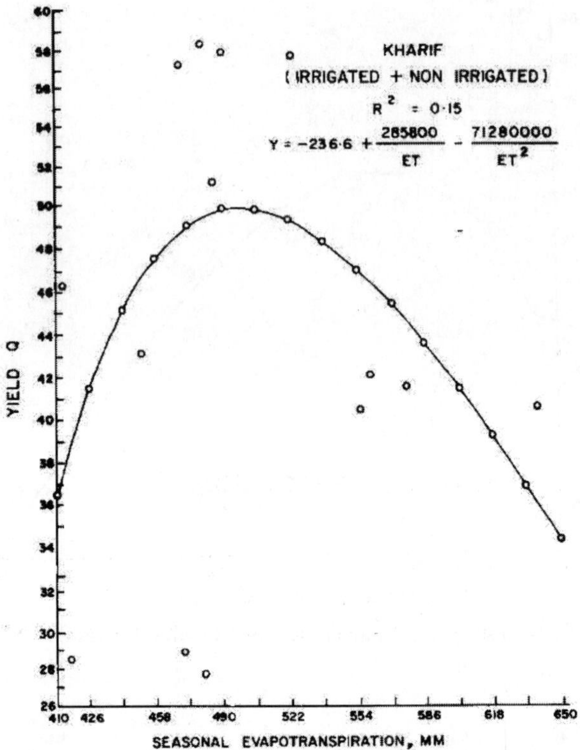

Fig. 25. Relationship of sorghum yield to seasonal evapotranspiration (Drawn not as per equation)

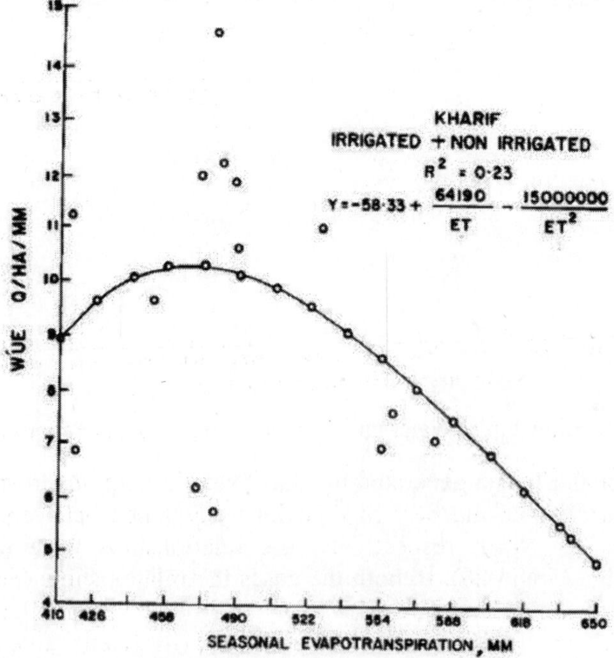

Fig. 26. Relationship of water use efficiency to seasonal evapotranspiration (Drawn not as per equation)

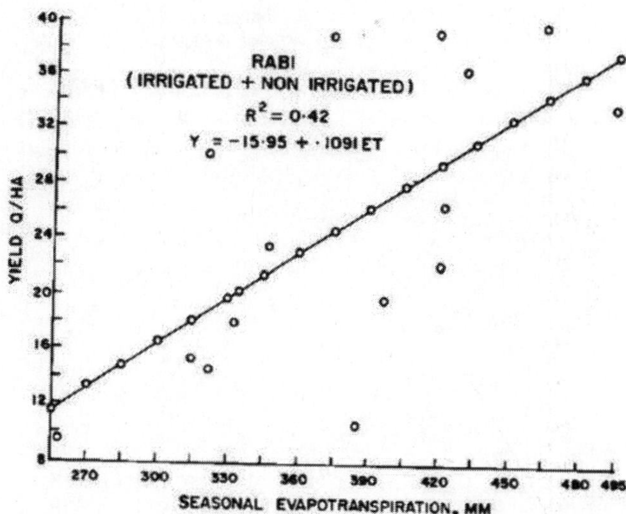

Fig. 27. Relationship of sorghum yield to seasonal evapotranspiration (Drawn not as per equation)

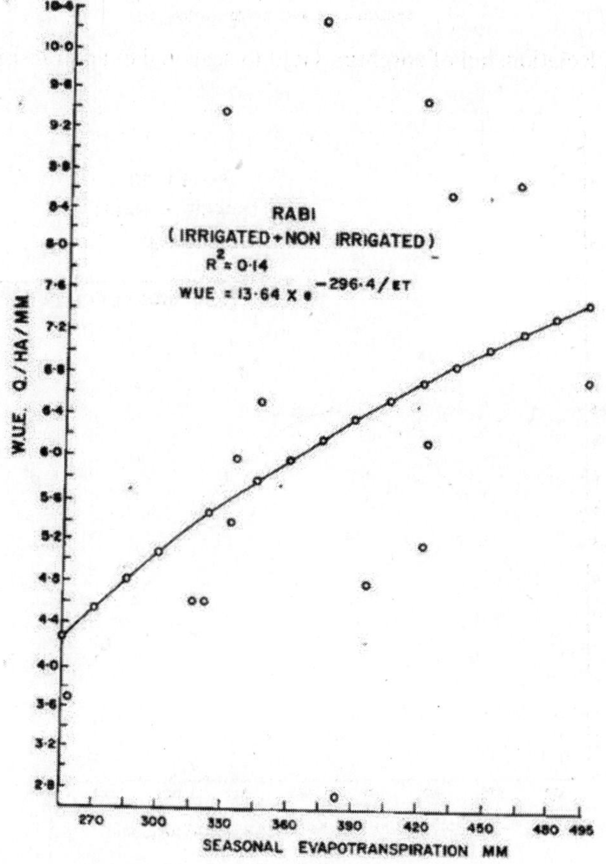

Fig. 28. Relationship of water use efficiency to seasonal evapotranspiration

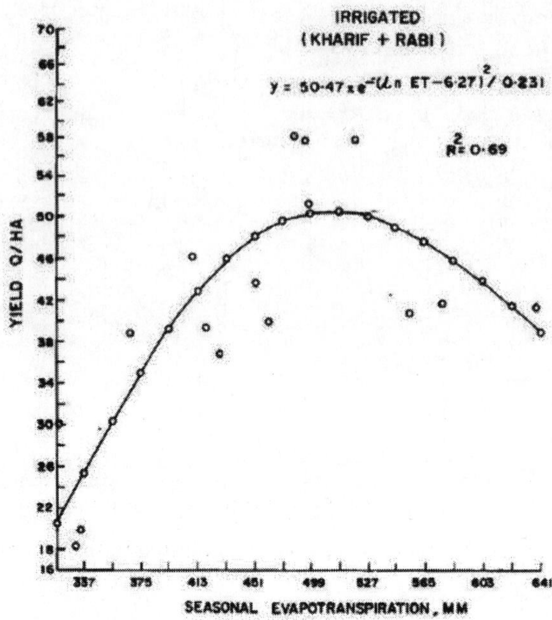

Fig. 29. Relationship of sorghum yield to seasonal evapotranspiration

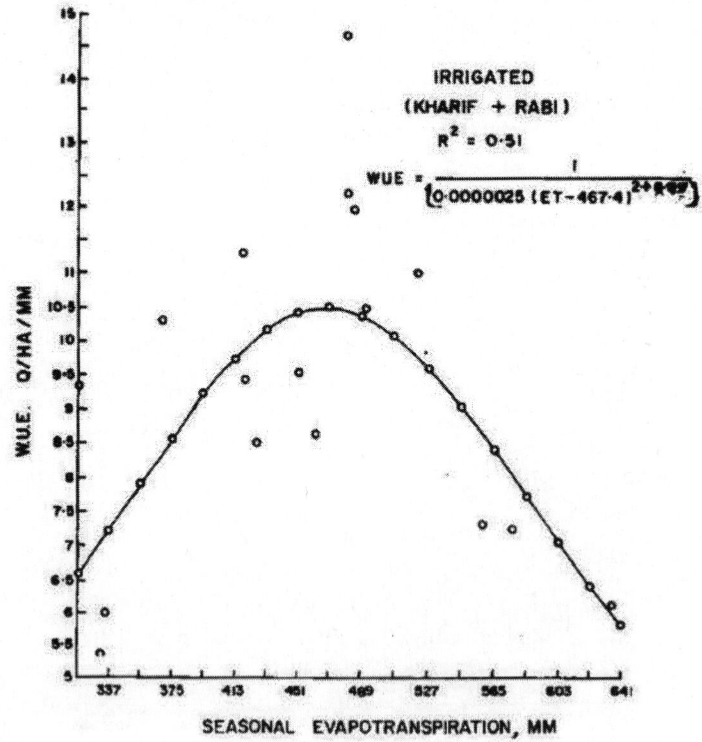

Fig. 30. Relationship of water use efficiency to seasonal evapotranspiration

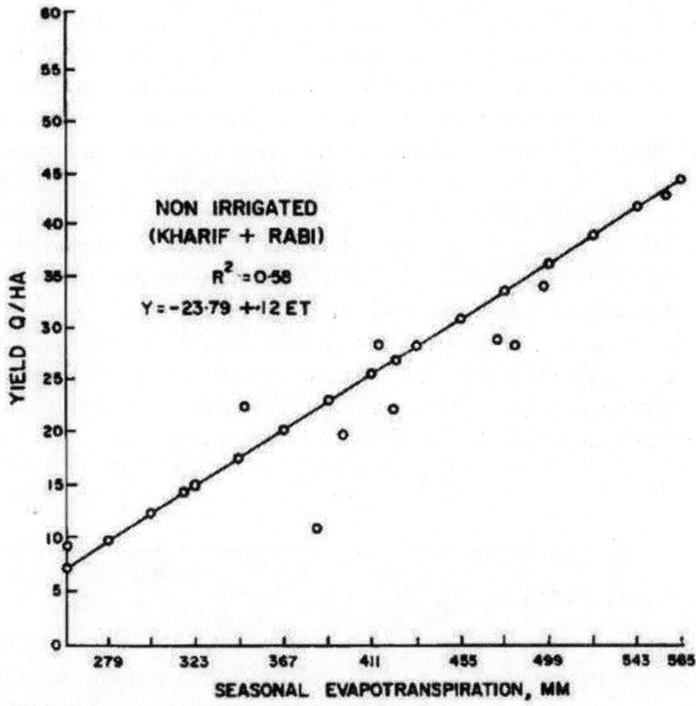

Fig. 31. Relationship of sorghum yield to seasonal evapotranspiration (Drawn not as per equation)

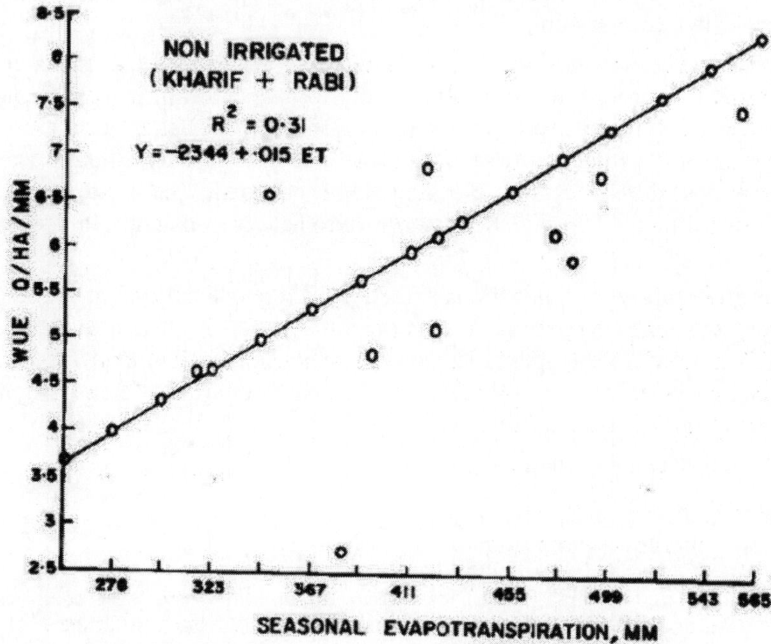

Fig. 32. Relationship of water use efficiency to seasonal evapotranspiration (Drawn not as per equation)

exponential in nature (Fig. was found to he linear (Fig. 27 and 28) and not statistically significant (R^2 = significant at 1 percent level (R^2= 0.42). The 0.14). When seasonal ET data of grain sorghum under irrigated condition both for kharif and rabi seasons were pooled together and best fit curves were developed for grain yield and WUE respectively. The relationships were found to be "log normal" and "cauchy" types (Fig. 29 to 30). Both the relationship were statistically significant (R^2 = 0.69; P) for seasonal ET Vs yield (R^2 = 0.51; P <0.001 for seasonal ET Vs WUE). From the nature of the relationships it may he interpreted that for grain sorghum crop, the lowest value of seasonal evapotranspiration found to produce maximum yield was 490 mm. Similarly, maximum WUE could be achieved with about 475 mm of seasonal ET. However, with nonirrigated sorghum grown at Parbhani in kharif season and at Solapur in rabi season the relationships were found to be linear in nature. The kharif season grain yield data combined with the rabi season grain yield data represent the full range of 9.5 to 57.25 q/ha yields for determining the seasonal ET grain yield relationship. The linear regression relationship was statistically significant (R^2=0.59; P <0.001) with a slope of 12 kg of grain yield per mm of seasonal ET (Fig. 31). The linear regression relationship between seasonal ET and WUE (Fig. 32) was also statistically significant (R^2 = 0.31) at 5 percent level (Samui, 1997).

7. IRRIGATION SCHEDULING

Although sorghum is considered drought tolerant and able to give yield under dry land conditions, it responds well to irrigation. Under ideal conditions sorghum has an extensive and deep root system. The depth of rooting in most varieties is about 2.0 m. To capitalize on these deep moisture extraction possibilities during later part of growing season, it is important to maintain surface moisture during the seedling stage, 3 to 4 weeks after planting to encourage development of a strong secondary crown root system.

The irrigation experiments indicate that the seasonal consumptive use of water by CSH - 1 was 331 mm and estimated irrigation requirements were 170 mm (90 mm in premonsoon period - 20th June to 10th July and 80 mm in the post monsoon period - September to early October). The rate of water use by sorghum is at its peak at flowering stage and rises rapidly 20 days after planting till 65 days and remains almost the same till late dough stage, but later falls rapidly with advancement in maturity. It is estimated that a 100 - 120 days crop may have a consumptive use of water of 450 to 500 mm.

Based on the irrigation experiments conducted at different locations, it is recommended that if limited irrigation is available, irrigations at vegetative stage (25-35 days after planting) and at boot-flowering stage (55 to 65 days) be given. These two irrigations both in kharif and rabi seasons have been found to be most beneficial compared to other treatment. If sufficient irrigation is available, irrigation at following growth stages be given.

1. Vegetative stage 25 - 30 days
2. Boot stage 50 - 55 days
3. Flowering and grain formation 70 - 75 days
4. Grain filling 90 - 95 days

The most critical stage is boot stage and if only one irrigation is available, it may be given at this stage. It has been observed that 35 to 40 percent losses in grain yield occur, if there is moisture stress at this stage.

The consumptive use of water by sorghum is maximum during 7th to 12th weeks after sowing which coincides with active vegetative and flowering stages of the crop. In order to get optimum yield and over come water stressed condition, the crop has to be irrigated as specified below.

Station	Period of life saving irrigation (weeks after sowing)
Akola	7 - 14th week
Parbhani	7 - 14th week
Solapur	7 - 14th week
Rahuri	5 - 12th week
Coimbatore	7 - 13th week

It is interesting to note that in most of years when nonirrigated rabi sorghum or nonirrigated kharif sorghum was grown, the seasonal water use was more than that of seasonal rainfall except at Coimbatore. The study revealed that during the growing season 100 to 200 mm of soil water was extracted from sub-soil layers.

8. CONSUMPTIVE USE OF WATER AND WATER USE EFFICIENCY (WUE) OF SOME MAJOR CROPS

Cotton : The average WUE of H-6 variety at Anand was 2.20 Kg ha^{-1} mm^{-1} and average value of K-9 and MCU-10 varieties grown at Kovilpatti was 3.03 Kg ha^{-1} mm^{-1}. At Coimbatore average WUE for MCU-5 and MCU-9 was 2.85 Kg ha^{-1} mm^{-1} as against 645 mm of cumulative ET. The average WUE and water use were comparatively low and found to be 1.17 Kg ha^{-1} mm^{-1} and 560 mm (Table 10) respectively at Banswara.

Table 10. Crop Cotton

	Station (No. of seasons)	Obser-vation	Crop durations ET (Days)	Cumula-tive ET (mm)	Cumulative EP (mm)	E.T./E.P. (mm)	Rainfall (mm)	Yield (kg/ha)	W.U.E. kg/ha/mm
1	Bansara 23°53'' N 74°27'' E	2	295	560	——	——	975	659	1.17
2	Anand 22°35''N 72°55''E	2	239	1123	909	1.23	844	2479	2.20
3	Coimbatore 11°00''N 77°00''E	7	197	645	953	0.67	644	1840	2.85
4	Kovilpatti 09°12''N 77°53''E	2	217	438	1047	0.41	573	1330	3.03

Table 11. Crop Mung

	Station (No. of seasons)	Obser-vation	Crop durations ET (Days)	Cumu-lative ET (mm)	Cumu-lative EP (mm)	E.T./E.P. (mm)	Rainfall (mm)	Yield (kg/ha)	W.U.E. kg/ha/mm
1	Banswara 23°53'' N 74°27'' E	2	70	261	——	——	——	860	3.29
2	Raipur 21°16''N 81°36''E	5	91	624	1008	0.61	222	1538	2.46
3	Jorhat 25°10''N 94°12''E	1	73	172	190	0.90	250	408	2.37

Mung (Green Gram) : Mung when grown as a rainfed kharif crop at Raipur (variety Pusa Baisakhi) used as much as 624 mm of water with WUE of 2.46 Kg ha^{-1} mm^{-1}. At Jorhat and Banswara the WUE was 2.37 and 3.29 Kg ha^{-1} mm respectively. The average cumulative ET and WUE were 352 mm and 2.71 Kg $ha^{-1}mm^{-1}$ (Table 11) during 78 days crop period.

Gram (Chickpea) : Gram being drought resistant crop with deep root system, its water use and WUE vary considerably depending upon water availability period and stored soil water in the root zone. The average WUE of gram was found to be 3.38 Kg ha^{-1} mm^{-1}. Variety Aunigeri at Dharwar and variety N-59 at Parbhani had cumulative ET 182 and 272 mm (Table 11) respectively.

Safflower : Safflower grown in the rabi season at Dharwar showed low water use efficiency (1.80 Kg ha^{-1} mm^{-1}). Cumulative ET and WUE of Annigeri variety ranged 325 to 369 mm and 1.25 to 2.44 Kg ha^{-1} mm respectively. This low WUE is due to low moisture availability as the crop was grown under residual soil moisture.

Groundnut : The overall average WUE and cumulative ET of groundnut grown at 4 stations in the semi-arid region of India was 2.87 Kg ha^{-1} mm^{-1} and 601 mm (Table 13) respectively. The average WUE and cumulative ET were respectively 2.72 Kg ha^{-1} mm^{-1} and 406 mm at Rajkot (average of three seasons); 2.92 Kg ha^{-1} mm^{-1} and 745 mm at Anand (average of two seasons); 3.53 Kg ha^{-1} mm^{-1} and 632 mm at Parbhani (average of three seasons) and 2.32 Kg ha^{-1} mm^{-1} and 620 mm at Rajendranagar (based on one season). Yadav (1975) observed WUE of 5.79 to 7.04 Kg ha^{-1} mm^{-1} at Hyderabad on red sandy loam soils in rabi season when it was irrigated at varying available soil moisture conditions. Subramanian et al. (1975) observed the least WUE (3.0 Kg ha^{-1} mm^{-1}) when irrigations were timed at 0.9 IW/CPE up to flowering and 0.6 IW/CPE thereafter; (IW = Irrigation Water; CPE = Cumulative Pan Evaporation). Total water use values reported in literature and summarized by Shivkumar and Sarma (1986) confirm the results.

Table 12. Crop Gram

	Station (No. of seasons)	Obser-vation	Crop durations ET (Days)	Cumulative ET (mm)	Cumulative EP (mm)	E.T./E.P. (mm)	Rainfall (mm)	Yield (kg/ha)	W.U.E. kg/ha/mm
1	Parbhani 19'16"N 76'47" E	2	94	272	550	0.49	033	549	2.01
2	Dharwar 15'26" N 75'07"E	1	104	182	471	0.38	003	868	4.76

Table 13. Crop Groundnut

	Station (No. of seasons)	Obser-vation	Crop durations ET (Days)	Cumulative ET (mm)	Cumulative EP (mm)	E.T./E.P. (mm)	Rainfall (mm)	Yield (kg/ha)	W.U.E. kg/ha/mm
1	Rajkot 22'17"N 70'48"E	3	109	406	570	0.71	884	1106	2.72
2	Anand 22°35''N 72°55''E	2	133	745	742	1.00	169	2178	2.92
3	Parbhani 19'16"N 76'47" E	3	128	632	1210	0.52	277	2233	3.53
4	Rajendranagar 17'14"N 78'33"E	126	620	630	0.98	017	1437	2.31	

Oat : There was considerable variation in water use and WUE of variety kent grown at IGFRI, Jhansi. Average water use and WUE were found to be 352 mm and 25.23 Kg ha^{-1} mm^{-1} respectively. WUE values ranged from 9.03 to 41.82 Kg ha^{-1} mm^{-1}.

Supplemental irrigation to increase production

In general, when planted in rabi season, rainfall usually provides half or less than half of the water requirement of the crops. Crops usually benefit from the plant available stored soil moisture and give almost similar vegetative growth in irrigated and non-irrigated plots. But our observations show that if the growing plant suffers from storage of water during the dough stage, even if recovery later appears to be complete, has a disproportionately severe effect upon subsequent yield (Fig. 18). The effect of water stress on yield (non-irrigated plot) and no water stress (irrigated) is demonstrated in Fig. 17 and 18 for sorghum and millet crops respectively. It is seen that both in 1978-79 and 1979-80, there were marked water deficit at Bellary throughout the growing period, with the results that the grain yields of sorghum in both the years were very low (1.5 and 7.0 q ha^{-1} respectively) in non-irrigated ET tanks, whereas grain yields in irrigated ET tanks were reasonably high (25.0 and 25.6q ha^{-1}).

9. IRRIGATION MANAGEMENT IN ARID LANDS

Irrigation water is the most important input limiting agricultural development in arid lands. This development is mostly achieved by importing water from other areas. This water has a very high economic value and many policy issues such as level of irrigation technology to be used, cropping pattern and area to be committed for irrigation are involved in its use for the agricultural development and for the overall economic growth of the region.

About 36% of the total land area of the world are arid. This area is water starved since time immemorial. However, current policies in water resources development have brought about transformation in this area. Water for agriculture is now available either in the form of precipitation or ground-water or surface water imported from other drainage basins. Good water harvesting techniques, which are usually capital intensive, are needed in rainfed agriculture. This helps to reduce the risk in dry land farming. The situation can be improved by developing better prediction methods of rainwater availability. Peak water availability period and critical growth stages when water requirement is high could then be matched so as to maximise production. However, policies relating to this dry farming irrigation involve issues of decisions under risk and under uncertainty as rainfall is scanty and erratic. Stations situated in semi-arid and arid regions receive much less rainfall compared to the evaporative demand of the place.

The fixed quantity of water available in the canal can be used to meet the full consumptive water requirements of the plants in a limited area thereby getting maximum crop yield per unit area or this could also be utilised for greater cropped area with water use by the plants per unit area less than their consumptive requirements and with crop yields less than the maximum. This policy question was looked into by determining the benefits and agricultural employment opportunities for the high agricultural technology for two cases (i) water supplied to the plants is 100% to their consumptive requirements and (ii) water supplied to the plants is only 60% of their consumptive requirements. The results indicate that when the water supplied to the plants was only 60% of their consumptive requirements, the total benefits of the Rajasthan Canal Project in India were reduced to about 40% while the cropped area of the project increased by about 50%. If the additional cost that will be required in developing on-farm works for the increased cropped area are taken into account, the total benefits of the project with water use of 60% of the consumptive requirements should be less than half to what these will be if full consumptive requirements of the plants are satisfied from the canal water. Water use policies in arid lands should be based on the principle of maximising benefits and / or agricultural production. High agricultural and irrigation technologies should be used to maximise benefits/production per unit of water and land area. Intensive agriculture at high technology level will result in higher economic efficiency as well as opportunities of more agricultural employment (Verma, 1983).

Bettering the conditions of arid and desert environments is a matter not only of the supply of adequate water but of permanent maintenance of fertility and quality of soil water through efficient irrigation, drainage, soil reclamation and optimally economic and sustainable land use pattern. The irrigation development program in arid regions should focus on human use systems and on problems of land and water resources management with minimum environmental deterioration (Bithu, 1983).

Due to high evaporation (EP) and scanty rainfall (R) it is quite evident that aridity is more intense in the western part of India as compared to other parts. For the purpose of crop water management, the EP and R values given for a location may be taken as valid for the adjoining areas having a radius of 50 km around the location. Due to erratic and low rainfall in kharif season and scanty winter rainfall, there exists a moisture deficit of 400 to 700 mm in either of the crop seasons, necessitating recourse to irrigation for successful crop production. Soils of the region are mostly loamy sand, sandy loam and sandy clay loam having low water holding capacity. Maximum water releases for crop growth from a typical loamy sand, sandy loam and sandy clay loam are 1.34, 1.56 and 1.68 mm/cm soil depth, respectively (Figs. 33 and 35) Besides, these soils due to their coarse texture have got low specific yield (drainable volume of water/volume of soil) values (0.10 to 0.14 cm^3 /cm^3 soil volume) indicating thereby the potential danger of over irrigation. In other words, deep-perlocational recharge to the water table of 10 cm to 14 cm of water will cause a rise of 1 m in the water table. Farmers should, therefore, be extremely cautious not to over irrigate their fields (Ray and De, 1983).

When to irrigate

Scarce water situation, less moisture retentivity of soil and dry climate of arid regions compels to irrigate crop with utmost care. Various approaches for scheduling irrigation and their applicability under different soil types and climatic conditions are highlighted.

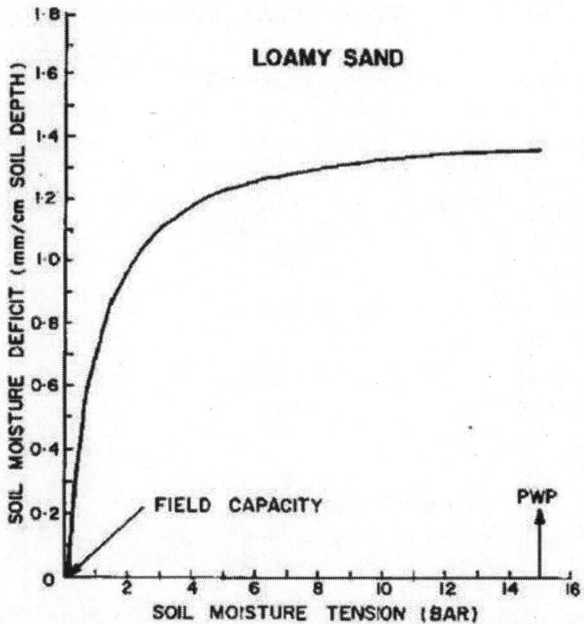

Fig. 33. Soil moisture release curve for a typical sandy loam soil (Source : After Roy and De, 1983)

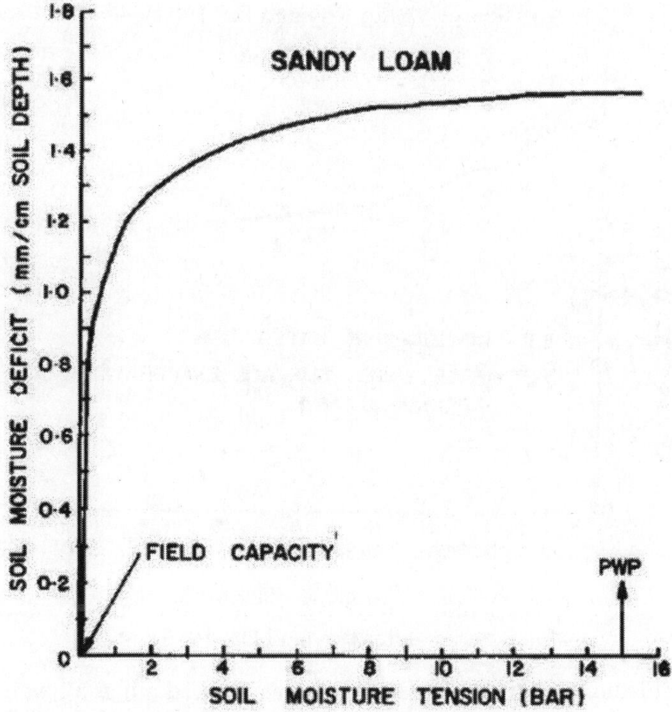

Fig. 34. Soil moisture release curve for a typical loamy sand soil (Source : After Roy and De, 1983)

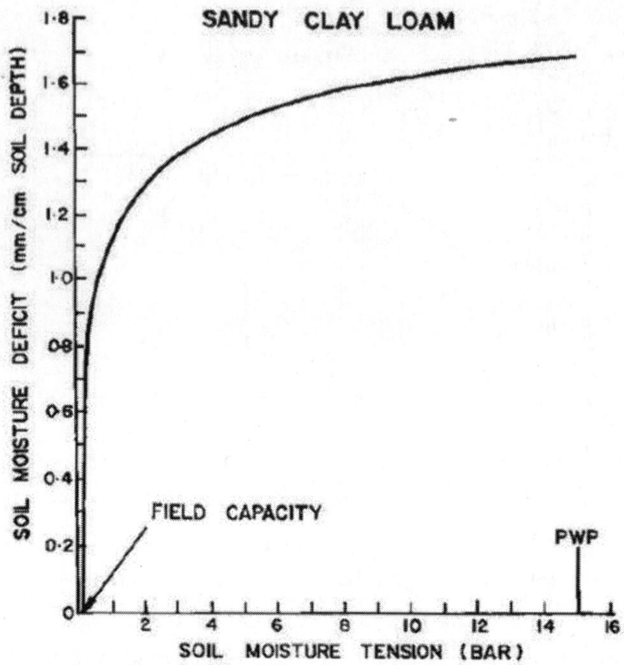

Fig. 35. Soil moisture release curve for a typical sandy clay loam soil (Source : After Roy and De, 1983)

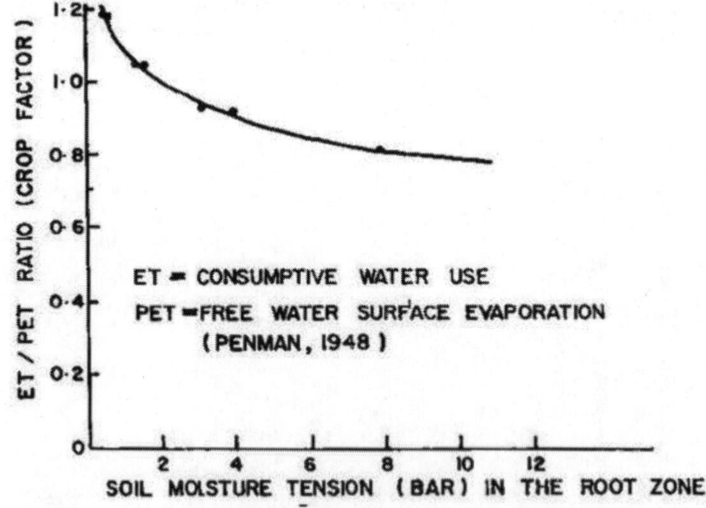

Fig. 36. Variation in ET/PET ratio at different soil moisture tension under arid climate (source : after Roy and De, 1983)

Optimum soil moisture tension to irrigate a crop, as worked out at a given location, can be used safely to irrigate the same crop at another place having similar soil type and climatic conditions. Irrigation interval and the depth of water to be applied per irrigation for the crop are determined depending on soil moisture release pattern and climatic dryness of the new location. Irrigation interval (I) is determined from soil moisture deficit (SMD) value in mm water/cm soil depth corresponding to the optimum tension from the SMD curves by using equation

$$I = \frac{SMD \times ERD}{ET}$$

Where,

ERD = Effective root zone depth of the crop in cm

ET = Consumptive water use of the crop for a given location and period (mm/day).

= PET (in mm/day) X ET/PET

ET / PET (crop factor) obtained from Fig. 36 against soil moisture tension.

Irrigation methods

Small amount of soil moisture deficit created in the root zone depth of light soils of arid zone at the time of irrigation poses a real problem for obtaining high application efficiency with satisfactory uniformity function when irrigation is applied by check -basin method. Usually in this method of irrigation a large portion of the applied water is lost as deep percolation. Such deep percolated water causes rapid rise in the saline ground water table and ultimately reduces the productivity of the land.

Further research requires to be done for standardising the shape, size and spacing of furrows and on their durability in light soils under very windy conditions of arid zone. Usefulness of seasonal wind breaks made by growing rows of sesbania plants against the prevalent wind direction in reducing crop ET as well as in increasing furrow life may also be studied. Sprinkler irrigation during less evaporative night time may be advocated for field crops grown on very light soils, while drip irrigation may prove advantageous for irrigating plantation crops grown on undulated land (Ray and De, 1983).

IRRIGATION UNDER LIMITED WATER RESOURCES

Special attention should be given when or at what growth stages should irrigation be withdrawn if availability of irrigation water is limited. This refers mainly to the irrigation of grain and other field crops, since optimal irrigation practice for high value crops such as flowers, fruits and vegetable crops will probably be less altered.

Early studies came to the conclusion that stress sensitivity is generally greatest during floral through pollination stages. Grain yield was claimed to be less sensitive to stress during early vegetative stage or grain filling stage (Salter and Goode, 1967). Irrigation frequency and quantities of water to be applied can thus be reduced during this growth stage if irrigation water is limiting (Samui, 1999).

These findings were verified and further interpreted by later studies (Stewart et al. 1975; Stegman 1982). Stress at an early growth stage may result in adjusting plant size and leaf area to low soil water availability and will thus decrease ET. When plant is subjected to water stress conditions at earlier growth stages and water is applied at flowering and part of it stored through grain filling, grain yield per unit water applied will be higher. Irrigation at vegetative stage only may result in insufficient water stored for grain filling so that yield per unit water applied will be lower. Irrigation at grain filling stage following dry early growth stages is not recommended because it will result in high transpiration rate late in the season, while plants might have shed part of their tillers, aborted flowers and lost assimilates during leaf drying.

The most favourable irrigation scheduling under limited availability of irrigation water will depend on the amount of water stored in the soil at planting, the size of the root system, the density of plant population and ET demands at different parts of the growing season (Samui and Jog, 1986).

In general, crops viz., corn, maize, sorghum produce more grain when the same quantity of irrigation water was applied over the season upto grain filling stage. Experimental results showed that irrigating corn during flowering and pollination stage which was expected to have highest grain yield, resulted in lower yield as compared to the same quantity of irrigation water spread over the season upto grain filling stage. Irrigation water during the vegetative growth stage prevented stress damages and retained much more assimilates within the vegetative organs, to be used later for grain filling stage.

10. WATER MANAGEMENT AND DEEP PERCOLATION LOSSES

Significance of the difference in soil management and crop production lies in the fact that soils with low moisture retention capacity will need to be irrigated more frequently. Besides, this will also lead to greater deep percolation losses. Under conditions of increasing scarcity of fresh waters and immense infrastructural cost involved in supply or irrigation water, permitting such heavy deep percolation losses amounts to a lavish use of a precious resource. What is more the infiltrated water will accelerate the process of perched water table formation and water logging under conditions of a slowly permeable substrata. Therefore, reduction of deep percolation losses is a highly desirable goal. In this regard a number of approaches can be attempted. One of these is the placement of 2-3 mm thick asphalt barrier at depth of 60-80 cm (Erickson et. al, 1968); Gupta and Aggarwal, 1980. Erickson et al (1968) has reported that barriered profile could provide 200% more water as compared to the no barrier treatment. Singh (1979, 1980) has tried bentonite in creating moisture barrier or in amending the soil. Compaction of soil is another means of reducing infiltration rate and increasing water retention in soil. In a sandy soil by making 20 passes of a 200kg roller the bulk density of 10-20 cm soil layer was increased from 1.56 to 1.67 g/cm^3 and infiltration rate decreased by 2.5% (Anon, 1977). A similar effect was seen by making four passes of a 1500 kg rollers. The amount of water required for each irrigation was reduced by 41% i.e. from 8.4 cm to 5.0 cm (Dhir, 1983).

Water Management and Erosion Control

Potential evapotranspiration demands and the heat stress on the growing crop are influenced to a large extent by wind regime. Krishnan (1978) has found that in May and June despite high radiating conditions, the value of aerodynamic term exceeded by 3 to 5 times the value of the energy balance term. The difference decreased during monsoon period and was high again during post-monsoon. Therefore, reduction of wind velocity is a highly desirable objective. On farming lands establishment of tree barriers across the wind direction is an appropriate means of achieving this. Skidmore and Hagen (1973) and Woodruff et al. (1959) reported that at a wind speed of 4 m/sec, such a barrier could reduce the potential evaporation demand by 50% upto a distance of 4 times the height from the barrier and by 20% at 12 times the height. The effectiveness of such a barrier increases with increasing wind velocity. On the other hand introduction of irrigation leads to a stabilisation of the surface of the otherwise erodible soils. This is achieved through the formation of a thin crust by the dissolved carbonates and the silt present in the irrigation water. Therefore, the problem of wind erosion in the irrigated field proper will be small indeed. But there is another aspect that needs to be looked into. In large parts of the commands yet to be developed, irrigated lands will not form big continuous tracts. More often there are small irrigable patches interspersed by high dunes thereby providing a potential source of fresh sand supply to cause choking of water distribution system and of encroachment on irrigated fields. An awareness of this problem as well as a programme of action to stabilise the non-command area already exists (Dhir, 1983).

ACKNOWLEDGEMENT

The author wishes to place on record his thanks and gratitude to the Director General of Meteorology, India Meteorological Department for giving permission to contribute this chapter. He also expresses his thanks to the Deputy Director General of Meteorology (Agrimet) for providing facilities. Thanks are also due to Mrs. P.S. Kulkarni, Miss S.R. Agale, Shri N.G.V aidya and J.P. Sable for their assistance.

REFERENCES

Anon., 1977. Annual Progress Report, A.I.C.R.P. Improvement of soil physical conditions, ICAR.

Baier, W. and Robertson George, W., 1967. Estimating yield components of wheat from calculated soil moisture. Canadian J. of Plant Sci. Ottawa, 46: 617-630.

Baier, W. and Robertson George, W., 1968. The performance of soil moisture estimates as compared with the direct use of climatological data for estimating crop yields. Agric. Meteorol. 5(1) : 17-31.

Dhir, R.P., 1983. Future technological needs of the Rajasthan Canal Command area with particular reference to stage II. Proceedings of the seminar on irrigation in arid lands. 1: 47-59.

Ehrler, W.L., 1978. Diurnal changes in plant water potential and canopy temperature of wheat as affected by drought. Agron. J. 70 (6): 999-1004.

Entz, M.H. and Fowler, D.B., 1988. Critical stress periods affecting productivity of no-till wheat in western Canada. Agron. J. 80(6): 987-992.

Erickson, A.E., Hansen C.M. and Smucker A.J.M., 1968. The influence of sub-surface asphalt barrier on water properties and productivity of sand soils. Trans 9th International Congress Soil Science 1: 331-337.

Gautam, D.P., Bharadwaj, R.B.L., Singh, Y. and Singh, R.P., 1968. Irrigation studies with tall and dwarf varieties of wheat. Proc. Symp., 1966. Water management monograph, Indian Soc. Agron., 110 -117.

Gupta et.al. (1979). "Effect of application of pond sediments on soil properties and yield of pearl millet and green gram in arid areas of Rajasthan. Indian J. Agri. Sci. 49 : 875-879.

Hukkeri, S.B. and Chauhan, D.S. 1967. Identifications of problems faced in field management of irrigation supplies and measures to be adopted for maximisation of production from available resources of water. Proc. Symp. Irrigation water management. 40th Annual meeting, Central Bd. Irrig. and Power, New Delhi.

Jensen, M.E. and Musick, J.T., 1960. The effects of irrigation treatment on evapotranspiration and production of sorghum and wheat in the southern great plains. Trans. 7th Int. Cong. Soil Sci. 1, 386-93.

Krishnan, A., 1978. Climatological Research at C.A.Z.R.I. Central Arid Zone Res. Inst., Monograph 10.

Kulik, M.S., 1965. Forecasting grain yield of wheat in the nonchemozem zone on the basis of new agrometeorological information. Meteorologiya Gigrologiya, Moscow No.5 : 100-105.

Laing, D.R. and Fischer, R.A., 1975. Response of wheat under subhumid irrigated conditions to atmospheric misting during grain filling. Agric. Meteorol. 15 (2) : 285 293.

Major, D.J., 1988. Winter wheat grain yield response to water and nitrogen on the north American Great Plains. Agric. For Meteorol. 44 (2) : 141-149.

Maske, S.J. and Samui, R.P., 1982. Some meteorological aspects of sorghum and millet production in India. Presented in the ICRISAT / WMO Symposium, Hyderabad, 15-19 Nov. 1982.

Musick, J.T. and Grimes, D.W., 1961. Water management and consumptive use by irrigated grain sorghum in western Kansas. Tech. Bull. Kansas Agric. Exp. Stn, 113-20.

Ray, S.B. and De, Rajat., 1983. On-farm water management with special reference to arid areas. Proceedings of the seminar on irrigation in arid lands. 1: 31-36.

Samui, R.P., Ghanekar M.G. and Agale S.R., 1999. Consumptive use, seasonal evapotranspiration and water use efficiency for wheat at some locations in India – Presented in Tropmet – 99 held at Chennai from 16 – 19 Feb. – 1999.

Samui, R.P., Ghanekar, M.G. and Ravindra, P.S., 1997. Consumptive use, potential evapotranspiration and water use efficiency of grain sorghum at some locations in semi-arid tract of India. Presented in Tropmet – 97 held at Bangalore, 10-14 Feb., 1997.

Samui, R.P. and Jog, A.L. 1986. Dryland farming under limited water resources. Mausam, 37(3) : 401-406.

Samui, R.P., 1999. Agricultural water management through Agrometeorological Advisory Services Units – Presented in Tropmet – 99 held at Chennai from 16-19 Feb. – 1999.

Samui, R.P., 1998. Rapporteur report on "Validation of information requirements for irrigated soils and crops." Submitted to WMO, Geneva (In press).

Sarkar, R.P., Biswas, B.C. and Khambete N.N., 1978. Probability analysis of short period rainfall in dry farming tract in India. Pre-published Scientific Report no. 78/9, I.M.D.

Singh, H.P., 1980. Improving the moisture storage in sandy desert soils by sub-surface moisture barrier. Arid zone research and development. Jodhpur. 245-252.

Skidmore, E.L. and Hagen, L.J. (1973). "Potential evaporation as influenced by Barrier Induced Micro-climate. In "Physical aspects of soil water and salts in ecosystems". Springer-verlag, Borila : 237-244.

Ulanova, E.S. and Rymar, A.L., 1963. Relationship of productive moisture supply in different layers of the soil in winter wheat fields in the fall. Tsentralnye Institute Prognozov Trudy No., 131 : 53-63.

Verma, R.D., 1983. Water use policy for irrigation in arid lands. Proceedings of the seminar on irrigation in arid lands. 1: 5-9.

Woodruff, N.P., Read, R.A. and Chopil, W.S., 1959. Influence of a field windbreak on summer wind movement and air temperature. Kansas Agriculture Exp. St. Bulletin 100: 1-24.

Yaron, D., 1980. Wheat response to soil moisture and the optimal irrigation policy under conditions of unstable rainfall. Cur. Sci. 42(8): 173-274.

Chapter - 11

DRY FARMING TECHNOLOGY IN RELATION TO EFFECTIVE RAIN WATER MANAGEMENT

***S.B. Atale*[1] *and G.M. Bharad*[2]**

1. Senior Research Scientist (Sorghum) & Head, Deptt. of Botany,
2. Ex Vice Chancellor, Dr. Panjabrao Deshmukh Krishi Vidyapeeth, Akola - 444 104 (MS)

1. INTRODUCTION

Natural resources includes water, soil, solar energy and air in the atmosphere. Water resource is at the prime place in resource management pathway. It is broadly agreed that management of rain water resource in dry lands follows management of other resources.

Rainfall pattern in most of the semi-arid tropics is uni-model and *kharif* cropping systems (June to October) are dominating. However, in some parts with assured winter rains with bi-modal rainfall pattern and or very high rainfall during *kharif* season, rabi cropping systems are predominant.

Rain water management to achieve sustainable production and productivity of vegetation over the entire toposequence through *in-situ* conservation of nonrenewable soil resource and harvesting of available solar energy and CO_2 is the management key in dry farming.

Monsoon rains are seasonal varying in duration, intensity and amount resulting into unpredictable dry and wet spells leading to adverse effects on vegetation growth and yields. Annual cropping systems are more prone to stress effects than the perennial systems. Existing cropping systems mainly include cotton and sorghum crops in combination with pulses like green gram, black gram and pigeonpea as well as oilseeds like groundnut and sunflower. Very recently, soybean as an oilseed crop is spreading fastly. Because of higher prices, some areas are being utilized for pulses and soybean as either an alternative to cotton and sorghum or as an intercrop. In spite of all these factors, cotton as a cash crop and sorghum as staple food and fodder crop are dominating the scenario mainly because of stability of these crops especially in Western Vidarbha region. However, seasonal variations in sorghum yields are noticed mainly due to erratic pattern of monsoon. Very high productivity is recorded in relatively dry seasons indicating the need for appropriate rainwater management layout for this crop. This paper mainly deals with the results related to *in-situ* rainwater management to reduce the yield fluctuations and conservation of natural resources.

2. EFFECT OF *IN-SITU* CONSERVATION MEASURES

Sorghum grain yield

Grain yield data indicated substantial increase in sorghum yield with continued cultivation along vetiver and leucaena live hedges over across the slope sowing and sowing along graded bunds on medium deep and deep soils (Table 1, 2 and 3).

Table 1. Effect of conservation measure on sorghum yield (q/ha) V.M.P.U. - Medium deep soils

Treatment		1988-89	1989-1990	Mean % increase Over T
T_1	Across the slope	29.3	38.2	33.8
T_2	Contour cultivation along Leucaena hedge	31.8	41.9	36.8
T_3	Contour cultivation along vetiver hedge	33.3	43.6	38.5
T_4	Sowing along graded bund	27.0	42.8	34.9

Table 2. Effect of conservation measures on sorghum yield (q/ha) on fara - Deep Soils

	Treatment	1991-1992	1992-1993	1993-1994	Mean
T1	Along the slope sowing	40.0	40.0	26.7	35.5
T2	Contour cultivation along vetiver hedge	45.8	48.8	37.5	44.0
	% Increase over T_1	14.6	21.9	40.6	23.8

Table 3. Effect of vegetative barriers in influencing sorghum yield (Cv. SPH -388) Dry Land Research Project, Akola (q/ha)

Treatments	Grain yield			Fodder yield		
	89-90	90-91	Mean	89-90	90-91	Mean
Khus keyline	45.0	28.0	36.5	88.4	87.3	87.9
Leucaena keyline	44.2	32.5	38.3	86.7	90.4	88.5
Graded bunds (0.36 m^2)	41.1	24.7	32.9	82.6	85.8	84.2
Sowing across the slope (control)	31.5	19.9	25.7	67.3	84.2	75.7

Table 4. Grain yield (q/ha) of sorghum as influenced by different treatments Dry Land Research Project, Akola

Treatments	Grain yield (q/ha)			
	1982-83	1983-84	1984-85	Mean
Sowing across the slope	28.3	35.4	17.2	26.9
Sowing on raised bed (150 cm bed)	29.8	40.3	19.0	30.0
Opening of furrow (at final)	– 39.8	19.6	29.6	
Verticle mulch	–	36.6	17.3	26.4
Live bund (green gram)	–	38.0	18.4	28.2

Sowing of sorghum on raised beds and furrow method recorded higher yield over across the slope sowing on flat lands in both low and high rainfall years (Table 4).

Similarly vertical mulch (opening a trench 30 x 30 m^2 across the slope and filling it with stubbles upto 15 cm and upper 15 cm with soil) was seen to be better than across the slope sowing, which was followed by opening of furrows in standing crop at final hoeing (Table 4).

2. *IN SITU* WATER AND SOIL CONSERVATION

Rain water management — A Review

In India, out of 328 m ha total land, 143 m ha is under agriculture uses (1993). Nearly half of total area is under stages of degradation. It is 175 m ha (Das and Mukherjee, 1980). Tejwani (1982) reported that 56% of agricultural land, 75% of fallow land, 86% of cultivable waste land, 95% of so called grazing and non-descript pastures and 33% of forest and tree land are suffering from soil erosion problems.

Inspite of 119.4 cm average rainfall in India (Michael, 1992) nearly 80% of the land area is semiarid and arid. Drought is a common feature and nearly, 259.7 m ha (70%) lands is the country are susceptible to droughts. Nearly 88% of land area is rain dependent in Maharashtra. This is characterised by limited seasonal rainfall having erratic pattern resulting dry and wet spells. Simultaneously, the evaporative demand is as high as 3000 mm in some parts. During rainy season on the same day we often see the vegetation suffering from water deficit on upper parts and due prolonged wet conditions on lower parts of toposequence of micro watershed (Bharad *et al.,* 1993). The adverse effect is visible in terms of severe erosion resulting in poor vegetative cover results in higher surface runoff and low infiltration. This situation is not good for long term scale. This has to be changed using scientific principles in line with nature. The soil, water, air and solar energy are main principle resource of nature. It is rotating in universe through hydrologic cycle and this cycle has to be monitored in desired direction for improvement in environment and social structure.

Tamhne (1967) defined watershed as drainage area on earth's surface from which runoff resulting from precipitation lowers fast into a single point as the area of land contained within a drainage dividing above a specified point of stream. Watershed sizes vary considerably and each watershed contains number of sub or micro watershed or in other way any watershed is part or a micro watershed of large one. Each field of a far is micro watershed for implementation of a planned programme. The watershed size should be such that the development components are completed within 2-3 seasons. Watershed comprises of arable and non arable lands and drainage lines which are utilized by permanent and transient populations (Doolette and Magrath, 1990). Considering utilization in semi-arid regions, lands in micro watershed can broadly be grouped in to:

i. Arable lands

ii. Non-arable lands

iii. Wet lands

iv. Drainage line area

Some areas are put to non agricultural uses and it is difficult to manage due to poor land use legislation. The report of selected pilot plans implemented at government level in different regions in India is explained hereunder for the use of readers.

The results of Manoli project in Akola district of Maharashtra showed dramatic changes in reducing runoff losses, increased cropping intensity and sustainability under different types of soils

(Bharad *et al.,* 1993). The use of contours bunding, live hedges on contours, runoff losses affected by treatments, *in situ* water conservation and continuous contour trenches have been studied for nine in five sub watersheds and 141 micro watersheds. The size of watershed varied from 2312 to 10255 ha lands. The slope varied from 1-3%. The soil type was shallow sandy loam with low water available capacity. The classification of soil was vertisol, inceptisol and entisol within the project area. The conclusions are given below (Bharad *et al.,* 1993):

1. Each field needs to be considered as micro watershed for developing rain water management.
2. Vegetative hedges on contour in the centre of the field and outlet point can be developed appropriately.
3. On arable lands, vegetative hedges on contour are found to be more effective than across the slope sowing and graded bund system for improving crop yield and *in situ* rain water and soil conservation.
4. Vetiver hedge was more effective than *leucaena* hedge.
5. Vegetative hedge on contour especially vetiver showed remarkable effect on daily rainfall and surface runoff on all types of soils. On arable crop lands with very shallow, medium deep and deep soil cover, contour cultivation along vetiver hedge at 1 m is recommended.

The studies on land use system on rainfall and surface runoff revealed following conclusions :

1. Development of very good perennial vegetation systems on degraded lands is possible with CCT layout.
2. In each field, one to two contour guide lines are necessary to lay down CCTs at 5 to 6 m surface interval.
3. Adoption of CCT layout at 5-6 m HI resulted in 96% *in situ* rain water and soil conservation uniformly over the entire toposequence.
4. Fruit species observed to grow well and early productive species are custard apple and zyzypus (Ber).
5. *Hardwickia binata* tree on trenches and *Stylo hemata* and *Siratro* bunds and strips were found to be a good combination for silvipasture system.
6. On farmer's fields, agroforestry or agro-silvi-horticulture system were better in terms of crop yield and tree growth.
7. Option for utilization of class IV lands needs to be oriented towards either development of perennial plantation systems on CCT layout or cropping with vetiver hedge on contour.

On very shallow soils, for opening of CCT with manual labour, it is recommended to excavate the trench in two stages, i. excavation of upper soil layer first and collection of same upstream side of trench and ii. excavation of lower parent material afterwards and stacking it on down stream side in form of bund with burm of 15 cm (Bharad, *et al.,* 1993).

Pande and Singh (1994) reported results of watershed at Vasad project in Gujarat. The project not only changed the land use pattern favourably but the land use intensity is also increased by 27% over the prevailing one. Monocropping system is changed to double cropping and cropping intensity is increased by 18%. The land productivity is increased by 52 to 130% in different crops. The B:C ratio was 2.65 which showed improved cropping system. In south India (Paiyur, Coimbatore) Kandasamy (1994) studied the impact of watershed farming system on marginal lands. He reported profitability of mixed farming system over pure cropping system. Dairy based farming system resulted in highest income and provided additional employment.

Dalvi et al. (1994) studied rain water management through contour cultivation at Akola (Maharashtra). The studies revealed 25% increase in productivity on contour cultivation along vetiver key line. The system was found very much suitable for *in situ* water harvesting in medium deep clay soils. The watershed management was studied by Bhati et al. (1994) in arid areas of Rajasthan in India. These areas have more evaporation, high wind velocity and non-existence of perennial rivers with scanty vegetation. Five year studies revealed that integration of one or two farming system coupled with appropriate soil could increase the agricultural productivity by 25-30% on sustainable basis. Such areas are found in Rajasthan (61%), Gujarat (20%), Punjab and Haryana (9%) and Andhra Pradesh and Karnataka (10%) contributing India's 12% of total area. These areas can be improved with watershed management systems and suitable rain water conservation measures.

Sarvanan and Kaleeswari (1994) studied soil erosion, run off and nutrient loss in Kodaikanal hills of Tamilnadu in India. They found relatively higher losses of nitrogen and potash than phosphorus. The differential soil and nutrient loss were attributed to the initial soil properties, soil conservation measures and topography of watersheds. In Madhya Pradesh, Sahu (1994) extensively studied micro catchment development through water harvesting system. Two dug out ponds each in two hectares microcatchment were constructed along with drainage and irrigation system to harness rain water for supplemental irrigation. The system was found to be very useful in converting barren land into double cropped area including fish rearing inside the ponds. Besides, it helped recharging ground water and raising the water table in micro catchment area. About, 72 to 92% rain water was conserved and it increased the cropping intensity by 75 to 148%. The C:B ratio increased from 1.10 to 1.66. The system was found very useful in dryland agriculture.

The rainfed micro watershed farming system were studied by Bharad (1994) for development of sustainable farming system in Maharashtra state. The results indicated that contour cultivation along vetiver hedge recorded 28.6 and 14.27% increase in yield of pearl millet, and hirsutum cotton on shallow soils over across/along slope sowing. The run off losses were reduced by 51.3 and 33.9% by vetiver and leucaena respectively. On very shallow soils, (Calss IV), the CCT layout was better option for rain water conservation which conserved rain water by 97% over the entire toposequence. He also reported that over 30% area can be brought under perennial irrigation with restricted lifting of conserved water. In sub Montane zone of Maharashtra, Chunale *et al.* (1994) reported that life saving irrigations can be had with watershed techniques. The yield of finger millet and chillies could be increased by 60 and 100% with one irrigation and with two irrigations, it increased by three folds. The seepage was good for paddy crop. It increased paddy yield by 28.9 to 53 q/ha by 84%.

Belgaumi and Radder (1994) reported that contour border strips wtih verticle interval of 0.3 m between two ridges were found equally effective to minimise runoff losses in vertisols. Ridges and furrows minimised the runoff by 66% and increased the yield of rabi sorghum by 50% over no land treatment. They reported 52.5% increase in yield with contour border strips in safflower in Dharwar area of Karnataka state in India. Pal *et al.,* (1994) studied rain water management for rainfed rice in Madhya Pradesh. They found that the use of green manure (*Sesbania rostrata*) or inclusion of food legumes may minimise over all fertilizer need of the cropping system. They found benefit of mulch for post rice crop. Pandey and Mishra (1994) further extended their studies to Chattisgarh area of M.P. and they have suggested the measures of strip cropping, stubble mulching, mixing cropping, proper tillage and crop geometry to be better for water and soil conservation to have sustainable farming.

In Konkan area of Maharashtra, 90% rain water is returned back to the sea as runoff. The studies on watershed based farming system in such hilly areas by Khanvilkar et al. (1994) revealed that the menace of soil erosion can be effectively controlled by adopting platform terraces, staggered trenches, farm ponds, bench terraces and planting of perennial fruit plantation on hill slopes for sustainable production. They found cropping system of rice, blackgram, niger, soybean, cowpea and vegetables most suitable for such terrace farming. On region of laterite soil in konkan, Badhe (1994) reported that staggered trenches (230ha) were found most effective for conservation of soil, water and nutrients in cashew plantation under hilly terrains and high rainfall conditions of south konkan.

In plateau region of Bihar, Singh (1994) reported his findings for Ranchi area. He did research on farmers field. Agroecosystem analysis revealed heavy erosion of soil in shallow sandy soils with acidic soils and poor fertility. The land was not found suitable for double cropping.

In Uttar Pradesh of India, Singh (1994) reported the benefits of watershed to the average farmers. He studied soil and water conservation during 1991-92 using contour bunding, submergence bunding, levelling, gully plugging and check dams, water harvesting bundhies, check dams, land smoothing, vegetative bunds, sodding of grasses on bunds, establishment of *Ipomoea cornia* on severe eroded land and inter bund treatment. He found enhanced productivity from 6.5 q/ha to 28 q/ha with watershed management with increased cropping intensity to 185%.

A comprehensive successful story of a village, Ralegansidhi in Ahmadnagar district of Maharashtra state where community watershed project was operated on 25043 ha area, has been reported by Bagchee and Bagchee (1994). Several projects are operated by the state government in different regions, but this project is unique because the farmers took part in maintenance, repairs and answering the social problems of affected poor farmers whose land was acquired under the programme. This has become a model example for others to follow for mutual benefits (may be poor or rich) under able and honest leadership of Shri Anna Hazare. Sharing the cost and benefits is the main achievement of this project. Co-operative movements in reality can bring dramatic revolution in life style of village people if it is adopted in right perspective. This region is prone to frequent drought (Rainfall 800 mm). Main crops are sorghum and pearl-millet in mono-cropping system. Guha (1985) reported that major projects were constructed on some rivers in this region. Traditional system of farming has been identified by Jodha (1980) and Gupta (1981). These included low technology low-input agriculture coupled with supplementary income. This was integrated approach of watershed management adopted in 1983. A strategy of restoring the environment of watershed has succeeded in rainfed farming system in this area. It reinforces the concept of water as common property resource, the water has to be managed by the community. A concensus has been built around the recognition of this principle (Bagchee and Bagchee, 1994). It can be concluded that the government project can be made successful and viable with community participation, local management and acceptable system of community participation. The philosophy lies again here that one should think in a manner of John F. Kennedy as *''Don't think what country can do for you but think what you can do for the country''*. This is the essence of success in management of watersheds in Ralegansidhi rural areas of the country.

Miscreants, bad social elements, greedy mentality of people may lower down the values of viable projects which should be taken on large scale instead having limited thinking of selfishness. The village, town, district and state should be in focus while operating the govt. projects in larger interest of rural folks. Recently, a *Sarpanch* who requested a farmer not to use the water from canal for crops when cattles are affected for paucity of water, was killed in MP in India (Nov. 15, 2000, News Paper flash). For good cause, honest people have to suffer and this is the tragedy of democracy. One should think upon this where we are in progress of rural folks. Unless this is changed, no one could come forward to implement good ideals in the society. There are people of determinations and unless they are supported by government, no revolution can be expected in days to come. Inspite of hurdles of democracy, some people with sincerity and commitments get involved in work without considering the deeds of miscreants. They need to be supported by the community and the police for good cause of society welfare in rainfed agriculture. Increase in yield levels (Pre and Post project) are given in following table :

Crop	Pre-project yield (Before 1983) (q/ha)	Post-project yield (1989-90) (q/ha)
Pearlmillet	3.5	7.0
Sorghum	4.5	15.0
Mungbean	3.0	5.6
Greengram	8.9	12-15
Wheat	10.0	16.0
Red gram	6.0	10.0

Source : Govt. Maharashtra, Deptt. of Agriculture ''A short note on comprehensive watershed programme at Village, Ralegansiddhi'' (Bagchee and Bagchee, 1994).

It is honestly suggested here that such projects should be taken up in every village of the country with co-operation of rural community in right perspective instead of spending huge public funds on big dams. Narmada dam contraversy is known to every one. Each field or a community fields can be taken as base units for developments of watersheds and miraculous results can be gained in five years. The fruits can be reaped by the next generation and can bring harmony, happiness in social structure. We talk much and do little. Can we not make reverse of this? What is needed today is action and not the reports. If benefits are in focus, no one would dare to reject the proposals. What we lack in us is honesty. Let us be with universal brotherhood as per Sant Vinoba Bhave.

Contour cultivation along vetiver hedge recorded substantial reduction in surface runoff of rain water and soil loss over across the slope sowing and graded bund land treatment. Contour cultivation along leucaena hedge also recorded reduction over these treatments. However, vetiver barrier was found to be more effective than leucaena in reducing runoff and arresting soil losses (Table 5).

Table 5. Effect of conservation measure on mean surface runoff and soil loss from sorghum crop, on Station medium deep soils 1988-89 to 1990-91 (1018.2 mm)

Treatment	Surface runoff %	Soil Loss t/ha
T_1 Across the slope	25.05	14.36
T_2 Contour cultivation along Leucaena hedge	19.34	7.55
T_3 Contour cultivation along vetiver hedge	15.39	3.88
T_4 Sowing along graded bund	16.03	11.25

3. SEEDING PERIOD

- There was marked decrease in sorghum yield when seeding was delayed from 29th meteo. week onwards.
- Sowing after normal onset of monsoon gave higher yield at all locations except at Achalpur. At Achalpur dry seedling was immediately followed by good rains (Table 6, 7).

- Sowing of sorghum with immediate onset of monsoon is recommended. Late seeding should be avoided.

Table 6. Grain yield (q/ha) of sorghum as influenced by sowing dates. Dry Land Research project, Akola.

Treatments	Grain yield (q/ha)			
	88-89	89-90	90-91	Pooled
23rd Meteo. week (Dry sowing)	41.2	54.6	37.6	44.4
25th Meteo. week (Normal sowing)	51.3	67.6	29.9	49.6
27th Meteo. Week (Late sowing)	28.5	58.4	27.5	38.1
CD P=0.05	7.9	–	–	–

Table 7. Grain yield (q/ha) of sorghum as influenced by seeding dates (Average of 1986-87, 87-88, 88-89)

Seedling dates	Grain yield (q/ha)		
	Buldana	Washim	Achalpur
23rd Meteo. week	34.2	42.0	46.5
25th Meteo. week	42.9	43.8	39.0
27th Meteo. week	33.1	25.9	27.4
29th Meteo. week	24.9	11.0	14.2
31st Meteo. week	17.0	9.5	12.1

Source: NARP Centres - Western Zone

4. NUTRIENT MANAGEMENT

Fertilizers

- Yield of sorghum was influenced by fertilizer levels and soil depth. It increased with increase in fertilizer dose and soil depth (Table 8).
- However, it was more controlled by fertilizer use rather than soil depth.
- Response to fertilizer (q/ha N) was higher on shallow and medium deep soils than deep soil.

Recommended dose of fertilizer (80: 40: 40 kg NPK/ha) is essential for sorghum on soil types to get higher productivity.

5. ORGANIC RECYCLING

Studies were conducted at Dryland Research Project, P.K.V., Akola on the addition of Leucaena loppings as mulch for improvement in Soil fertility, moisture content and finally sorghum yield.

Plate 11.01. 1. Unideal property of vertisols soil poor in organic matter. **2.** Sorghum stubble collection after harvest with laboures. Go for organic recycling to improve soil texture instead burning it as fuel or use as stubble mulch for rabi crops.

Leucaena loppings contain on an average 3.3% N on oven dry basis (Table 9). In this experiment 5.5 t/ha Leucaena loppings were added throughout the growth of sorghum crop. The addition of nitrogen due to loppings was 38.5 kg N/ha. Leucaena loppings has 7.3:1 CN ratio and increased the available N content of the soil by about 44 per cent over the control.

It was observed that addition of Leucaena loppings increased soil moisture content in 0-30 and 30-60 cm soil zone at various growth stages of sorghum as a result of mulching effect and an increase in organic matter content of the soil (Table 10).

Table 8. Mean sorghum grain yield (q/ha) as influenced by land capability classes and fertilizer doses over seasons and pooled means

Treatments	1988-89	1989-90	1990-91	Pooled
(A) Soil class				
II (60 cm)	24.6	53.4	25.4	34.5
III (35 cm)	21.3	37.8	21.3	26.8
IV (20 cm)	15.6	32.8	17.8	22.1
CD P = 0.05	6.27	6.86	3.15	6.49
(B) Fertilizer doses (MPK kg/ha)				
Control	5.6	26.8	10.0	14.1
40 : 20 : 20	22.6	43.1	23.3	29.9
80 : 40 : 40	33.5	53.5	31.2	39.4
CD P = 0.05	6.27	6.86	3.15	2.54

Only addition of Leucaena lopping @ 5.5 t/ha increased sorghum grain yield from 11.4 to 22.1 q/ha (Table 11). It was about 70% of the yield from full recommended fertilizer dose (80:40 kg NP/ha).

Table 9. Total N contents in Leucaena sample on oven dry basis Dry land Res. Project Akola

Cutting after days	N Content %
30	2.74
60	3.78
90	3.38
Average	3.3

Table 10. Total soil moisture (mm) in soil at different growth stages of sorghum as influenced by Leucaena loppings Dryland Res. Project Akola

Treatments	Dough		Harvest	
	0-30	30-60	0-30	30-60
No mulch	172	146	123	140
With mulch	139	167	131	165
Only fertilizer (80:40 kg/NP/ha)	125	143	126	149

6. CROPPING SYSTEMS

Kharif Cropping

Growing of green gram/black gram in mixture with local sorghum was an age old practice to minimize risk of crop failure in dryland agriculture (Table 12).

Table 11. Grain yield of sorghum as influenced by mulch x alley width during 1988-89, 89-90 and pooled means (q/ha).

Treatments	1988-89			1989-90			1990-91			Pooled means		
	6	9	Mean	6	9	Mean	6	9	Mean	6	9	Mean
No mulch (No fertilizer)	3.6	6.5	5.0	15.0	20.4	17.7	9.5	13.3	11.4	9.4	13.4	11.4
Mulch (4.5 t/ha)	13.0	16.7	14.8	31.4	31.8	31.6	17.4	21.9	19.9	20.8	23.5	22.1
Sole sorghum (80:40 kg/WP/ha)	–	14.3	–	–	44.4	–	–	34.4	–	–	31.2	–
Mean	8.3	11.6	–	23.2	26.1	–	13.7	17.6	–	15.1	18.4	–
CD P = 0.05		5.39			6.40	–		5.04			7.51	

With the introduction of sorghum hybrids in early sixties, systematic research was carried out on intercropping (sowing in row proportion) of green gram/ black/soybean and pigeonpea in sorghum.

Two lines of sorghum (CSH-1) and one line of pigeonpea (C-11); sown at 45 cm row spacing was found more stable intercropping. Likewise in case of green gram/ black gram it was 2:1 row proportion with 30 cm row spacing.

However, with the introduction of CSH-9 and SPH-388 high yielding sorghum hybrids, the row proportion of 2:1(sorghum two lines and intercrop one line) did not found congenial for the growth of an intercrop. Because of more plant height and spreading leaves; intercrop was adversely affected. With these varieties strip cropping (3:3 row proportion) is found more economical than sole cropping.

Table 12. Grain yield (q/ha), sorghum grain equivalent and gross monetary returns from sorghum sole and sorghum + pigeonpea intercropping (3:3) over locations Mean : 1988-89, 89-90.

Cropping system	Buldana			Washim			Achalpur		
	B+1	SE	GMR	B+1	SE	GMR	B+1	SE	GMR
Sorghum (CSH-9) Sole	35.0	35.0	15708	30.6	30.6	8788	34.7	34.7	10411
Sorghum	24.5+	24.5+	12096	18.5+	18.5+	16371	13.0+	10.3+	12991
Pigeonpea (C-11)	2.7	8.3		11.4	44.5		10.2	31.6	
		32.8			63.0			44.6	

Source: MARP station, Western Zone B+I - Base, Inter, SE - Sorghum equivalent; GMR - Gross monetary return

On medium deep soils (about 60 cm depth) 3 rows of sorghum and 3 rows of green gram/black gram are recommended with 30 cm row spacing, where as on deep soils (more than 60 cm) sorghum

+ pigeonpea 3 : 3 is recommended at 45 cm row spacing in areas receiving about 900 mm rainfall. Because of high price of pulses, this cropping system is found more paying than sole sorghum. In areas where soils are medium in depth and rainfall is around 800 mm, sole sorghum is more profitable than sorghum + pigeonpea cropping as pigeonpea yields are low in this situation.

Data reported in Table 12 reveals, that sorghum + pigeonpea intercropping was profitable than sole sorghum in deep and high rainfall areas of Washim and Achalpur, whereas it was not paying on medium soils and low rainfall of Buldana tract.

Table 13. Grain yield of main and intercrop (q/ha).

Treatments	1988-89			1989-90			199-91			Mean		
	Main	Inter	LER	Main	Inter	LER	Main	Inter	LER	Main	Inter	LER
Safflower + gram (6:3)	4.90	0.70	0.64	5.00	1.10	0.98	5.30	1.80	1.75	5.10	1.20	0.94
Gram + safflower	4.00	3.60	1.10	2.00	3.00	0.92	3.40	2.70	1.20	3.10	3.10	1.10
Gram + Sorghum	2.03	–	0.48	2.30	–	0.59	2.70	–	0.61	2.40	–	–
Sorghum + Gram	–	5.30	1.10	–	1.30	0.33	–	1.80	0.41	–	2.80	–
Safflower sole	9.90	–	1.00	7.20	–	1.00	6.30	–	1.00	7.80	–	–
Gram sole	4.80	–	1.00	3.90	–	1.00	4.40	–	1.00	4.40	–	–

Source : NARP/ARS - Buldana, LER = Land Equivalent Ratio.

Table 14. Grain yield of main crop and intercrop (q/ha) LER and GMR (Rs/ha) from different intercroppings

Treatments	Grain yield (q/ha)		LER	GMR
	Main	Inter		(Rs/ha)
Safflower + gram (6:3)	10.00	1.70	0.74	11390
Safflower + R. sorghum (6:3)	11.45	6.50	0.99	13941
Gram + Safflower (6:3)	7.40	7.60	0.94	11661
Gram + R. Sorghum (6:3)	7.10	7.00	0.94	7921
R. Sorghum + Safflower (6:3)	11.40	6.80	0.95	12270
R. Sorghum + gram (6:3)	12.50	2.80	0.84	7398
Safflower sole	16.80	–	1.00	16871
Gram sole	11.80	–	1.00	8645
Rabi sorghum sole	20.80	–	1.00	9028
CD P = 0.05	3.75	–	–	1906.9

Market rates (Rs/q) : Gram -675/-. Safflower 1005/- rabi Sorghum -180/- Fodder 60/-

Source : NARP/ARS -Tharsa, Nagpur

Sorghum + pigeonpea (3:3) is recommended on deep soils with good rainfall (above 900 mm), whereas sorghum + green gram/black (3:3) are recommended on medium soils with average rainfall of 800 mm.

7. RABI CROPPING

Trials on intercropping of gram and safflower with sorghum conducted at Buldana, Achalpur, Ekarjuna and Tharsa revealed that during rabi season sole cropping is more sustainable than intercropping on stored soil moisture. As compared to sorghum, safflower was more paying under rainfed situations. (Table 13, 14 and 15).

Intercropping during rabi season under dryland conditions should not be followed.

8. ALTERNATE LAND USE SYSTEMS

Based on land/soil resource located at different parts of toposequence, perennial plantation systems viz., Agri, horticulture, silvipasture, agro, horti silvi pasture on continuous trench layout are observed to be sustainable, interms of productivity and income (Refer Chapter 15).

Table 15. Grain yield (q/ha) and LER as influenced by the treatments

Cropping patterns	1988-89											
	Main crop	Inter crop	LER	Main crop	Inter crop	LER	Main crop	Inter crop	LER	Main crop	Inter crop	LER
Safflower sole	22.60	–	1.00	14.30	–	1.00	6.20	–	1.00	14.30	–	1.00
Gram sole	11.60	–	1.00	10.60	–	1.00	8.80	–	1.00	10.30	–	1.00
Rabi sorghum sole	4.20	–	1.00	14.40	–	1.00	14.00	–	1.00	10.90	–	1.00
Safflower + gram (6:3)	12.60	2.20	0.73	4.90	1.80	0.51	2.90	1.60	0.65	6.80	1.80	0.64
Safflower + R. sorghum (6:3)	7.10	1.60	0.69	6.40	2.30	0.61	4.50	7.10	1.23	6.00	3.66	0.75
Gram + safflower (6:3)	3.60	6.60	0.60	3.30	4.10	0.60	2.70	2.30	0.68	3.20	4.30	0.60
Gram + R. Sorghum (6:3)	3.90	1.30	0.65	3.10	7.20	0.79	2.60	6.30	0.75	3.20	4.90	0.75
R. Sorghum + Safflower (6:3)	3.10	7.70	1.08	5.10	3.80	0.62	5.90	2.60	0.84	4.70	4.70	0.75
R. Sorghum + gram (6:3)	2.70	3.00	0.97	5.60	1.80	0.58	6.00	1.80	0.63	4.80	2.20	0.77

Source : ARS -Ekarjuna (Chandrapur)

9. CONCLUSIONS

Grain yield was increased substantially in sorghum with contour cultivation along vetiver and leucaena live hedges. The verticle mulch across the slope with stubbles up to 15 cm with soil was better than across the slope sowing. Contour cultivation along vetiver hedge recorded considerable reduction in surface runoff of rain water and soil loss. Vetiver barrier was found more effective than leucaena in reducing runoff losses. Sowing of sorghum immediately after monsoon is recommended. Fertilizers should be applied for yield optimization. The studies on organic recycling revealed that an addition of leucaena loppings @ 5.5 t/ha increased soil moisture content and organic matter in 0-30 and 30-60 cm soil zone. Two lines of sorghum and one line of pigeonpea sown at 45 cm was found more stable intercropping. While using improved varieties (Sorghum SPH-388 and CSH-9), strip cropping (3:3 proportion) was better. The sorghum + pigeonpea is recommended in deep soils with good rainfall. The alternate land use system is recommended under different toposequence in terms of productivity and income. Thus, effective rain water management can be achieved with the recommendations given above in dryland farming.

REFERENCES

Anonymous, 1989. Report of sorghum Agronomy Research work, Dr. P.D.K.V., Akola (Maharashtra) submitted to RRC, meeting of Agronomy and soil Science.

Anonymous, 1990. Report of Dryland Research work, Dr. P.D.K.V., Akola, RRC meeting of Agronomy and soil science.

Anonymous, 1990. Report of NARP, Western Zone, Dr. P.D.K.V., Akola, RRC meeting of Agronomy and soil science.

Anonymous, 1991. Report of dryland research work, Dr. P.D.K.V., Akola, RRC meeting of Agronomy and Soil science.

Anonymous, 1994. Report of watershed management research work. Dr. P.D.K.V., Akola, RRC meeting of Agronomy and Soil science.

Anonymous, 1976. National Comm. on Agriculture. Govt. of India, Min. Agri. New Delhi.

Bagchee, A. and S. Baghcee 1994. Improving rainfed farming systems through watershed development - A case study of Ralegansiddhi. Proc. Nat. Sem. held at Nagpur (PKV) Dec. 29-30, 1994. pp. 132-142.

Badhe, V.T. 1994. Effect of watershed development of runoff, soil and nutrient loss in lateritic soils of south Konkan. Proc. Nat. Sem. held at Nagpur (PKV) Dec. 29-30, 1994. pp. 89-94.

Bharad, G.M.; B.G. Bathakal; M.B. Nagdeve; S.K. Kohale; S.N. Ingle; R.L. Gawande; V.K. Mahorkar and K.S. Kayande 1991. Watershed Approach in Rainfed Agriculture. Bull. PKV/WMR Pub. Director of Ext. PKV, Akola (MS) India.

Belgaumi, M.I. and G.D. Radder 1994. Land treatments for efficient soil and water conservation in vertisols. Proc. Nat. Sem. held at Nagpur (PKV) Dec. 29-30, 1994. pp. 49-53.

Bharad, G.M., M.B. Nagdeve; S.K. Kohale; S.M. Taley; A.N. Mankar; S.N. Ingle and S.N. Patil 1993. Watershed Management Research. PKV Res. Bull. PKV/WMR-2. Pub. Director of Ext. Edu. PKV, Akola. India.

Bharad, G.M. 1994. Rain water management for developing sustainable farming systems in rainfed micro watershed. Proc. Nat. Sem. held at Nagpur (PKV), Dec. 29-30, 1994, pp. 35-42.

Chunale, G.L., R.B. Patil and S.D. Kalke 1994. Rain water harvesting and its recycling for increasing crop yield in sub montane zone of Maharashtra. Proc. Nat. Sem. held at Nagpur (PKV) Dec. 29-30, 1994, pp. 43-48.

Dalvi, V.B., G.M. Bharad and A.J. Agrawal 1994. Rain water management through contour cultivation. Proc. Nat. Sem held at Nagpur (PKV) Dec. 29-30, 1994. pp. 12-16.

Dass, D.C. and B.K. Mukherjee 1980. National perspective on soil and moisture conservation. Agri. Eng. Today 4(3): 11-17.

Doolett, J.B. and W.B. Magrath 1990. Watershed development in Asia, Strategies and Technologies. World Bank Tech. Paper 127 : pp. 4-5.

Greenfield, J.C. and R. Grimshaw 1988. Vetiver grass. A method of vegetative soil and moisture conservation. 2 Ed. World Bank, New Delhi.

Gururajan, B., A.A. Dason and V. Natrajan, 1994. Integrated farming system for vertisol watershed areas of Kovilpatti (TN). Proc. Nat. Sem. held at Nagpur (PKV) Dec. 29-30, 1994. pp. 82-88.

Khanvilkar, S.A., J.R. Ramteke and U.V. Mohadkar 1994. Watershed based farming systems development in hilly areas of Konkan region - A case study. Proc. Nat. Sem. held at Nagpur (PKV, Akola) Dec. 29-30, 1994. pp. 75-81.

Kandasamy, O.S. 1994. Integrated farming system for sustaining the productivity of marginal farms. Proc. Nat. Sem. held at Nagpur (PKV, Akola) Dec. 29-30, 1994. pp. 6-11.

Magrath, W.B. and J.B. Doolett 1990. Strategic issues in watershed development. World Bank technical paper. 127 : 1-4.

Michael, A.M. 1992. Irrigation theory and practices. Vikas Publishing House, New Delhi.

Mishra, R.K.and V.K. Pandey 1994. Agro forestry - An essential component of watershed management in Northern hill zone of Chattisgadh, Proc. Nat. Sem. held at Nagpur (PKV), Dec. 29-30, 1994, pp. 54-57.

Pandey, V.C. and H.B. Singh 1994. Impact of integrated watershed management of the farming systems in semi arid tropics of Gujarat. V. Naik Memo. Proc. Nat. Sem. on Farming systems on watershed managment (PKV, Akola). Dec. 29-30, 1994 held at Nagpur, (MS). pp. 1-6.

Pandey, V.K. and R.K. Mishra 1994. Watershed management for sustainable farming with special reference to soil and water conservation measures for Chattisgarh region of Madhya Pradesh. Proc. Nat. Sem. held at Nagpur (PKV, Akola) Dec. 29-30, 1994. pp. 95-101.

Pal, A.R., A.L. Rathore and R.K. Sahu 1994. Land use soil and rain water management for rainfed rice and rice-based cropping systems. Proc. Nat. Sem. held at Nagpur (PKV, Akola) Dec. 29-30, 1994, pp. 64-74.

Sahu, R.K. 1994. Microcatchment development through water harvesting system. Proc. Nat. Semi. held at Nagpur (PKV, Akola), Dec. 29-30, 1994. pp. 29-34.

Saravanana, A. and R.K. Kaleeswari 1994. Studies on soil erosion, runoff and nutrient loss in selected watersheds of Kodaikanal hills, Tamilnadu. Proc. Nat. Sem. held at Nagpur (PKV, Akola), Dec. 29-30, 1994. pp. 23-28.

Singh, K.P. 1994. Farming system research and development based on watershed management in plateau region of Bihar. Proc. Nat. Sem. held at Nagpur (PKV, Akola) Dec. 29-30 Dec. 1994. pp. 102- 108.

Tamhane, R.V. 1967. The challenge of our watersheds. Proc. DEV centre on Watershed Management for Africa and far east. Soil Con. Soc. of India.

Tejwani, K.G. 1982. National policy for India. Paper for 12th Int. Cong. Soil Science, New Delhi 8-16 Feb. 1982.

Chapter - 12

ROLE OF PLANT GROWTH REGULATORS AND NUTRITION IN DRYLAND FARMING

A.M. Dhopte[1] and S.D. Ramteke[2]

1 Former Professor of Botany, College of Agriculture, Nagpur, 49, Vivekanand Nagar, Khamala Road, Nagpur – 440015 (MS)
2 Dr. S. D. Ramteke, Principal Scientist (Plant Physiology), National Research Centre for grapes, Manjri Farm, P. O. Box No. 3, Pune -Solapur Road, PUNE - 412 307

1. INTRODUCTION

Plant growth depends on the environment. The growth kinetics are subject to change due to weather parameters such as solar radiation, temperature, rainfall and relative humidity. Growth is promoted when conditions are ideal for development. Each crop or species has definite requirement of above parameters and this is exactly why the sowing dates are different for various crops. Some crops are adopted to hot weather and some requires cool environment for better growth. The crop species are, therefore classified on thermoperiodicity and photoperiodism as short day, long day and day neutral plants. Flowering is dependant on photoperiod and the classification of crops as kharif or rabi is mainly dependant on requirement of season for the particular crop. The dates of sowing of various crops are decided on its specific requirement of weather. Any deviation in sowing date for the crop therefore affects the growth and development. Sowing period is critical because growth is dependant on ideal weather requirements and by sowing at specified time we provide the required environment for the crop. For example, wheat grows well under cool climate and yields abandant to its maximum potential. Hot climate arrests the growth and yield is reduced. Day neutral crop like sunflower, cotton, tomato and chilli can be grown round the year because its flowering is not influenced by change in photoperiod. Different seasons are characterised by specific photoperiod, temperature and relative humidity. Ideal weather means the initiation of desired changes in hormonal levels that can promote the growth under normal nutritional and water requirements. Variation in soil water and temperature influence the endogenous hormonal levels and varies the physiological processes affecting the growth and yield (Fig. 1).

Dryland crops are subjected to variation in available soil water, changes in relative humidity and day and night temperature. Adverse soil and air conditions bring changes in endogenous hormonal levels thereby influencing the growth of a crop. The water stress promotes the level of growth inhibitors and suppresses the level of promoters. The balance of these two chemical substances determines the growth, either good or bad. High temperature also influences similar growth effects but in addition, it hits the adverse changes in reproductive growth. High temperature accelerates the physiological processes but it is not supplemented with increased inputs of water, nutrition and other unknown factors and therefore crop growth is highly affected. The tolerance of crop species depends on how it can adjust the balance of endogenous levels of promoters and inhibitors within a plant. Harsh environment promotes catabolic processes and finally it leads to termination of growth and senescence. Thus, it is essential to monitor the inputs for better growth. Weather, however can not be monitored but what can be managed is nutrition and water and this can be accomplished by a prudent farmer who understands the sign/symptoms of a plant. The damage to the crop can be minimised by hormonal and nutritional sprays.

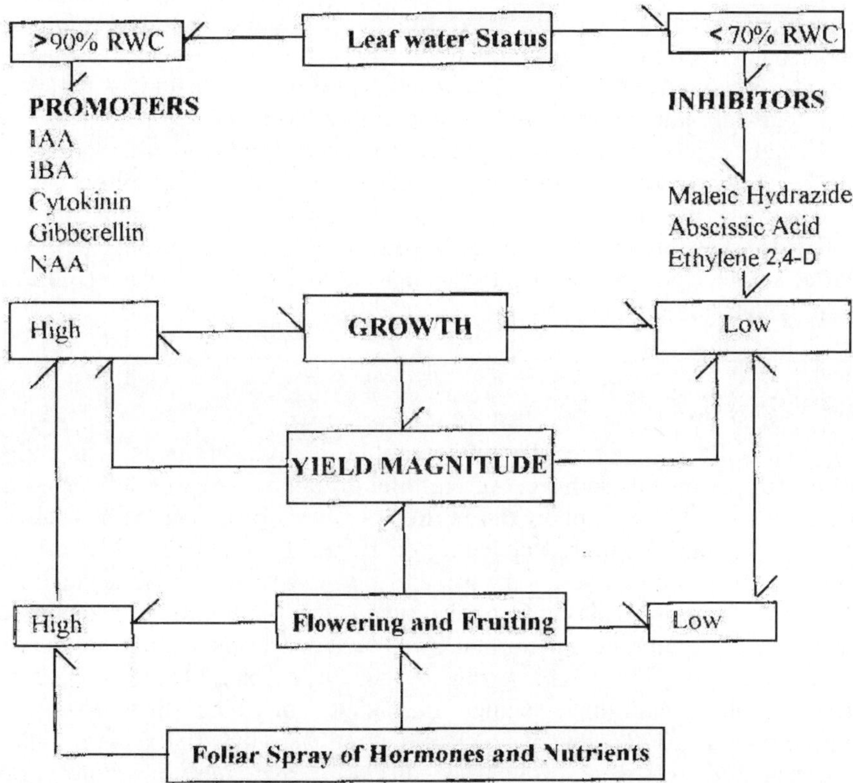

Fig. 1. Flow chart to demonstrate how correction of hormone and nutrient sprays help maintain yield levels.

Timely operations are needed. Late sprays may not be useful to remedy the bad effects. Thus, a farmer needs to have frequent visits to the field to speak with plant and know what it needs and arrange for deficit factors. The water stress/drought results in production of abscisic acid, ethylene phenoloc compounds and oxidants that deteriorates the growth kinetics. A ratio of growth promoter (GP) to inhibitor (GI) goes low in a plant and it remains stunted. It is the hormone level within the plant that decides the line of growth. Under ideal conditions of soil water, the level of promoters is more with negligible inhibitors and hence the crop growth is luxuriant. This happens only when light intensity, relative humidity and temperature are at optimum levels. The Blackman's law of limiting factor works here. The law states "The rate of the process is limited by the pace of slowest factor". Accelerated growth is observed only under optimum conditions of all factors. All the inputs should be balanced for ideal productivity. Such kind of situation is never found under dryland crops. Some time some factor is found deficit and thus growth and development of rainfed crops are always restricted and yields are poor as compared to well watered crop.

The endogenous levels of hormones and nutrients can be well monitored by exogenous sprays. Vigilance is however, necessary. Better crop stand and seedling vigour can be managed by presowing treatments with auxins and gibberellins. Seed pilleting is another beneficial attempt to maintain good crop stand (See Chapter 1). Age old practice was to treat the seed with cowdung and ash. This could help promote initial drifts in physiological processes at seedling stage. Treating the seed with rhizobium culture and few nutrient chemicals and auxins help promoting initial growth by providing the required inputs at the place where it is needed. Better initial growth at seedling stage inculcates good vigour to combat adverse changes. This philosophy worked well in old times and provided the food for limited population. The families survived with limited inputs in agriculture. With change in population explosion and continuous exploitation of land from centuries with negligible inputs, higher yields can not be expected under degraded land and eroded soils. Primarily, the soil should be enriched with organic matter for better nutrient absorption and water holding capacity. This is a key factor that needs to be improved for having optimum productivity. Deficit hormone and nutrient levels can be monitored with foliar sprays. The efficacy of plant growth regulators (PGRs) is either positive or negative under dryland farming. The erratic monsoon results in either drought or waterlogging.

2. WHAT IS DROUGHT ?

Any environmental factor potentially unfavourable for living organisms is called stress. It is elastic or plastic with permanent damage. Agricultural drought is usually defined as a period when insufficient water is available to support the normal activities of a crop over a fairly normal long period of time of a fortnight or more depending on stage of crop. Drought is distinguished from aridity and it may be expected that both very wet and very dry regions experience drought. All India Coordinated Research Project for Dryland Agriculture. (2003). All stresses move around the water stress. If it is optimum, every process is normal. Any deviation leads to deletorious effect on growth affecting crop productivity. The plastic strain results in irrepairable physical damage such as scorching, firing, abscission and this is evident only under long term stress. Elastic resistance is a measure of organism's ability to prevent reversible strain. (Levitt, 1980). If rainfall is 50% below normal, it is considered as severe drought condition. Dehydration of tissue results in drastic physiological changes and crop remains stunted. The central plains of geographical zone is classified as semiarid zone and it is oftenly subjected to intermittant drought. Long dry spell of 10- 15 days is harmful to the crop growth. Drought resistance of a crop is defined as the mechanism causing minimum loss in yield in a water deficit environment relative to maximum yield in water constraint free management of the crop (Sinha, 1986). It is the ability of a plant to grow under moisture stress without much loss in yield. Water use efficiency is a ratio of yield to evapotranspiration which is monitored by climate and soil moisture regime. The management practices involve plant character through breeding and agronomic practices. Increased soil moisture storage can be managed through reduced water losses either through plant or soil or both.

The water stress brings changes in physiological aspects, biochemical phenomena and morphological expression in a plant (Table 1).

Table 1. Role of growth regulators on physiological processes

SN.	Process	PGR
1.	Cell, division, enlargement, differentiation, Germination Flowering, Mitochondrial activity	Auxins, IAA, IBA, Kinetin, 2,4-D, AA
2.	Parthenocarpy, Fertilization, Ovule development, Germination, Pollen viability, Seed development, Sex expression, dwarfism, Dormancy	Gibberellins
3.	Plastid, chloroplast development, Deferment of senescence, Enzyme activity, Organogenesis	Cytokinins
4.	Leaf and fruit abscission, Defoliation, Embryogenesis Dormancy, Stomatal closure	Abscissic acid (ABA)
5.	Wilting, Fruit ripening and retension, Sex expression Enzyme activity, Breaking dormancy	Ethylene (CEPA)
6.	Dwarfing, Branching, Source-Sink relationship, Antigibberellins, inducing dormancy.	Growth Inhibitors (CCC, PCB, Resorcinol, Maleic hydrazide)

3. DROUGHT & PHYSIOLOGICAL EFFECTS

Physiological effects

1. Reduced water absorption and dehydration
2. Cesation of stem elongation and leaf elongation
3. Cesation of fruit development, cell division, elongation and cell differentiation due to desiccation of protoplasm.

Biochemical effects

1. Interuption of protein synthesis and proteolysis
2. Impairment in nucleic acid synthesis
3. Effect on enzyme synthesis
4. Reduction in levels of cytokinin
5. Accumulation of toxic products like ammonia-a main cause of drought injury
6. Inhibition of starch synthesis
7. ABA is increased in roots
8. Increase in organic acids in cotton

Malic/calcium action on permeability affected.

Huang *et al.* (2013), studied the effect of abiotic stress on the abundance of different vitamin B6 vitamins in tobacco plants according to them there are six different vitamin B6 (VB6) forms, pyridoxal (PL), pyridoxamine (PM), pyridoxine (PN), pyridoxal 50-phosphate (PLP), pyridoxamine 50-phosphate (PMP), and pyridoxine 50-phosphate (PNP), of which PLP is the active form. Although transcriptional regulation of the genes involved in the de novo and salvage pathways of PLP syntheses after stress treatments has been described for *Arabidopsis thaliana* and tobacco plants, it remains open as to whether this in turn affects VB6 levels. In their study, the effects of chilling, UV radiation, intensity of illumination, osmotic pressure, oxidative and drought stresses on the abundance of different B6 vitamins in tobacco plants were examined by using high performance liquid chromatography (HPLC). They found the abiotic stressors resulted in significant increase of PLP, and caused some corresponding changes with PL and PN. The highest increase of PLP was 2.5-fold compared to the control plants, followed by a continuous decline back to the control levels. These changes are presumably caused by the regulation and control mechanism on the VB6 metabolism in plants.

Kakumanu *et al.* (2012) reported that the drought stress affects cereals especially during the reproductive stage. The maize (*Zea mays*) drought transcriptome was studied using RNA-Seq analysis to compare drought treated and well watered fertilized ovary and basal leaf meristem tissue, more drought responsive genes responded in the ovary compared to the leaf meristem. Gene Ontology (GO) enrichment analysis revealed massive decrease in transcript abundance of cell division and cell cycle genes in the drought stressed ovary only. Among GO categories related to carbohydrate metabolism, changes in starch and sucrose metabolism-related genes occurred in the ovary, consistent with a decrease in starch levels, and in sucrose transporter function, with no comparable changes occurring in the leaf meristem. ABA-related processes responded positively, but only in the ovaries. Related responses suggested the operation of low glucose sensing in drought stressed ovaries.

Overall effects :

1. Reduction in plant stature and dry matter production
2. Inhibition of photosynthesis and intake of CO_2
3. Inhibition of chlorophyll activity and light reaction.
4. Decreased leaf area causing accelerated senescence and abscission
5. Abscission of flower and fruits.
6. Higher production of ethylene
7. Impairment in auxin distribution and failure of fertilization
8. Increase in respiration due to dehydration
9. Starvation of plant due to low energy

Hlavinka *et al.* (2009) deliberated the effect of drought on yield variability of key crops in Czech Republic, their study proved that a severe drought spell during the sensitive period of vegetative season does have a quantifiable negative effect, even within more humid regions, their results also demonstrate that, at least in some areas of the CR (and probably most of Central Europe), drought is one of the key causes of inter annual yield variability.

Effects on root system

1. Reduced mineral uptake and absorption ability
2. Increased suberization of root
3. Reduced N_2 fixation
4. Oxygen uptake is reduced
5. Root cell wall permeability is reduced
6. Root/top ratio is reduced
7. Deposition of toxic substances in root.

Gaul *et al.* (2008), calculated that the drought event resulted in an extra input of about 28 g C m_2 and 1.1 g N m^2 to the soil due to increased fine root mortality. They also concluded that even periods of mild drought significantly increase fine root mortality and the associated input of root-derived C to the soil organic matter pool in temperate Norway spruce forests.

Haluskova *et al.* (2009) investigated the activity of glutathione S-transferase (GST) and glutathione peroxidase (GPX) in barley root tip and their relation to root growth inhibition induced by different abiotic stresses. They reported that the GPX activity in barley root tips was not affected by cold, heat or drought treatment.

Thus, water stressed plants have increased levels of ABA and ethylene which results in ageing, senescence and absicission of leaf and floral parts. The levels of GA, cytokinin and IAA are low in stressed plants. Hence, drought tolerance is a ability of a plant to postpone dehydration, repair or rebuilt damaged mechanism and not to maintain it for long time. The exogenous sprays of auxins help maintain growth and realize the yield without much losses.

4. ABERRANT WEATHER AND CHANGE IN ENDOGENOUS HORMONES

The hormones or plant growth regulators (PGRs) are synthesied in plants according to phenological stages from germination to flowering and fruit/seed development. Ideal weather conditions favour formation of growth pormoters while any parameter above or below specific requirement can cause the formation of growth inhibitors. Broadly, the PGRs are classified on the basis of modification of processes that they bring in plant (Purohit, 1993). The formation of PGRs is dependant on temperature, relative humidity and soil moisture. The ratio of GP to GI can be monitored by exogenous application of hormones. When plant experiences the adverse condition, a signal is received for formation of particular hormone and physiological effects are observed. Harsh environment promotes the formation of GI and restricts the GP. The imbalance between these two drastically affects the plant growth. The team of scientist from Center for the Study of Carbon Dioxide and Global Change (2012), studied on Effects of Atmospheric CO_2 Enrichment on Plant Hormones. In that they pointed out that the atmospheric CO_2 enrichment magnifies the early growth and developmental impacts of plant hormones, which is truly a positive thing for a world that continues to struggle to feed its burgeoning population. On the contrary, Qaderi *et al.* (2012), revealed that water stress strongly, but higher temperature to a lesser extent, affected canola seedlings, leading to reduced growth and dry matter, and the negative effects increased with time. Patel and Franklin (2009) mentioned the regulation of plant architecture by temperature appears to involve the complex integration of multiple hormone signaling networks. Gibberellin (GA), Salicylic Acid (SA) and cytokinin have been implicated in the regulation of plant growth during chilling, whilst a predominant role for auxin was observed at high temperatures. Various plant growth regulators identified so far are :

Auxins : 1. IAA, 2. IBA, 3. NAA, 4. 2,4-D, 5. MCPA and precursors 6. Ethrel (CEPA) All promoters in low concentrations.

Gibberellins : 1. GA 1 to GA 84 : All promoters (GA_3 is widely used).

Cytokinins : 1. BAP, 2.6-Furfuryl amino purine, 3. Betain 4. Kinetin : All promoters

Growth : 1. ABA, 2. Cycocel (CCC), 3. TIBA, 4. Paclobutrazol, 5. Resorcinol

Inhibitors : All growth inhibitors acting as anti gibberellin compounds.

Ethylene 1. Ethrel, CEPA or ethylene producing compounds. Promoter as well as inhibitor depending upon concentration.

Hormones are specific organic substances that can bring physiological changes when applied in minute quantities to the plant. Physiological responses can be altered by exogenous application of hormones. The endogenous levels of auxins or hormones can be modified. The hormone levels that are observed in plant due to harsh environment (Mainly inhibitors) can be modified and expected responses can be had to a certain extent. If GI levels are increased, its ratio is modified with exogenous applications. The formation of GP or GI is basically dependant upon leaf water status or leaf water potential (Fig. 1). If tissue water level is below 15 bar, suddenchanges are seen in plant signaling through hormones and there is a formation of ABA or ethylene. Heat shock protein's (HSP) formation is also a result of response to prevalent environment. This aspect shall be dealt further. Thus, the ratio of GP to GI indicates the status of a plant experiencing the normal or adverse condition. The role of PGRs comes here to modify the growth as per desired goal. The expression of auxin/hormone is also dependant upon nutrient status of a plant. The interaction of these two can modify the growth. It is therefore extremely important to monitor both hormones as well as nutrients together to have better responses under harsh environment because promotion of growth needs additional inputs. The interaction of these two are important to have positive responses. This phenomena is called 'Biocheminization' The modification is possible to certain extent and it can not grossly be altered. This must be noted by the readers. For example, dwarfing can be altered by application of GA. The response that is mediated through a gene can be altered. Genes are responsible for synthesis of hormones and phenotypic expression of a plant is the result of genotypic and environmental constituents. The environmental component is dominant in most cases when phenotypic expressions are considered. The development of molecular biology has anserwered the riddles of these expressions in an organism.

Dwarfness is characterised by low GA levels while tallness is due to high GA levels (Saxena *et al.,* 1993). Thus, dwarf or tall character is dependant on GA synthesis of a species. A cauliflower can be grown tall with GA application. It happens in cool climate. Internodal elongation is increased and rosette type is changed to a tall plant streching its inflorescence like raceme. Temperature is the key factor in modifying the hormone levels in a plant. The tropical, temperate and cole crops are classified on the basis of their requirements of temperature. Any condition that deviates above or below normal (Optimum) brings about drastic changes in hormone levels affecting growth. Suitable approach of a farmer to modify the plant for optimization of yield under unideal climatic condition is the trump card to accomplish the targetted goal for management of dryland farming. Thus, PGRs play significant role in crop tailoring to the desired level. It does not mean that the apples can be grown in tropics with the use of hormones. The modification has some limitations. The PGRs are broadly classified in groups of promoters and inhibitors (Malik *et al.,* 1993).

Growth promoters 1. Auxins 2. Gibberellins 3. Cytokinins 4. Triacontanol 5. Phenolic compounds 6. Ascorbic acid 7. SA

Growth Inhibitors 1. Cycocel 2. Maleic hydrazide 3. Paclobutrazol 4. Resorcinol 5. 2,4-D, 6. MCPA 7. B-nine 8. TIBA

Auxins : Auxin is naturally occurring substance to be identified as IAA. The compounds which are structure similar to IAA are PAA and IBA. The IBA was thought to be an active synthetic auxin but it is shown to occur in maize leaves (Epstein *et al.,* 1989). Auxins are required for all growth processes.

Gibberellins : The gibberellins are a group of 84 very closely related compounds based on entgibberellin structure. GA_3 is the fungal product (*Gibberella fujikuroi*) and is available widely. Commercially, it is exploited for fruit set and development, increased size of grapes, increase maleness in gynoecious lines of cucumber, internodal elongation and flower retension. It may be a substitute for vernalization in some plants and breaking dormancy in seeds and buds. It induces flowering in many plants.

Cytokinins : These are derivatives of adenine capable of inducing cells division in tissue culture. It may be natural or synthetic (Kinetin, Benzyladenine). It stimulates cell division, cell enlargement, auxin activity, chlorophyll synthesis, deferment of senescence of leaves, cut flowers, fresh fruit and vegetables and transport of assimiates within plants.

Abscisic acid : It is natural growth inhibitors and known because of its role in abscission and dormancy. Exogenous application inhibits growth of a plant but ABA appears to act as promoter than as an inhibitor (Malik *et al.,* 1993). It increases cold hardyness, stomatal closure and reverse gibberellin effect on promotion of staminate flowers. It promotes transport of photosynthates to seeds.

Ethylene : It is volatile hormone which is naturally produced in plant tissues from the amino acid. in response to environmental stress and ripening processes. It can move from one plant to another through air. Its precursor is present in wide variety of plants. It promotes epinasty, abscission of leaves and flower petals, protein synthesis and respiration, regulates sex expression in plants, releases dormancy of seeds and buds and induces adventitious root formation. It is reviewed by Agarwal and Seervi (1995).

Obata and Fernie (2012) considered the use of metabolomics to dissect plant responses to abiotic stresses, they introduce the analytical methods used for plant metabolomics and describe their use in studies related to the metabolic response to water, temperature, light, nutrient limitation, ion and oxidative stresses. They also discussed stress compounds such as proline, c-amino butyrate and polyamines are reviewed, and the widespread importance of branched chain amino acid metabolism under stress condition as well as mechanistic insights into metabolic regulatory processes.

In general, different PGRs both natural and synthetic are described under five premier groups namely, four auxins, many gibberellins, several cytokinins, abscisic acid and ehtylene. Polyamines, brassinolides, long chain alcohols (Triacontanol) and phenolic compounds occur naturally and possess growth regulating properties. This is extensively described by Setia *et al.,* (1991).

5. WHAT IS ABERRANT WEATHER ?

Erratic rainfall and its distribution lead to changes in physiological activity. The rainfall less than normal (20%) leads to moderate drought. If deficit is 50%, it creates severe drought. Heavy rains during short period of growth leads to waterlogging and plant growth is affected due to oxygen stress and non availability of nutrients. This is the position when precipitation is above normal (20%). Flooding results in losses of nutrients and its availability is tremendously reduced. The average total rainfall may be the same but few phenological stages are only affected due to deviation from normalcy. If adverse condition happens to be the critical stage, the damage is more. Thus, even though the precipitation is normal, the crop can suffer drought or waterlogging during some phenological stage. If drought occurs at critical stages, the yield is declined drastically. The soil moisture availability is therefore important for better plant growth because it maintains the nutrient availability and avoid oxygen stress. The role of glutathione in plants under abiotic stress conditions is extensively reviewed by Kocsy (1999). The glutathion (GSH) is involved in enzymatic detoxification as active oxygen species as component of ascorbate - GSH pathway. Environmental stressors as low temperature, high temperature, drought or high salt concentration can induce oxydative stress (Prasad *et al*, 1994).

Temperature deviation from normalcy is common during phenological stages of crop growth. The day and night temperature influence the growth of crop through changes in hormonal and protein synthesis. The change in growth either good or bad is the result of prevalent temperature. Every crop has specific requirement for day and night temperature and called thermoperiodicity. If it is not experienced by the plant in normal limits, the adverse effects on growth become visible. The classification of crop species is based on this phenomena. The Kharif, rabi and summer climates are compatible to some of the species of crops. Some genetic modifications are being induced in some species and such varieties are developed for semi-kharif and semi-rabi environment. This is achieved with mutagens either physical or chemical and therefore there is no line of demarkation for crop to grow only in particular season.

The species have been adopted to varied climatic conditions depending on tolerance or susceptibility of a species. Tailoring a plant to the required conditions has been possible with biotechnological tools. It is important to study the basis of abiotic stress responses in order to manipulate plants and enable them to grow in unexploited areas. Basic studies on stress response (mainly on lea genes and proteins) are carried out on genetic models. The combination of methods used in agricultural biotechnology and the basic knowledge of biotic stress response will enable us in the future to improve stress tolerance in agriculturally useful plants (Pelah, 1999). The heat shock proteins (HSP) are formed during high temperature stress and it affects the reproductive growth namely the formation of flowers, fertilization and grain development. Three different size groups of hsp are present in higher plants including soybean, pea, tobacco, tomato, maize, wheat and rice. The formation of male and female gametophyte is delineated Anomalies occur during ontogenic development of an organ and profoundly affects the process of above flowering and reproduction (Dhopte and Eastin, 1999). For proper organogenesis to occur, appropriate temperature is essential. The formation of HSP limits the fertilization, increases the ovule abortion and brings the male sterility. Thus, it is the temperature that affects enzymatic process, hormone synthesis and leads to HSP in plants that limits the productivity of a crop (Fig. 2).

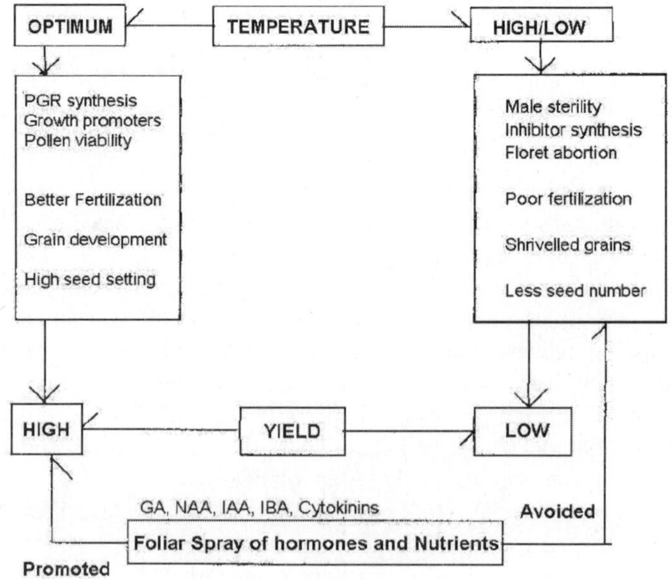

Fig. 2. Effect of change in temperature (Day/Night) on reproductive physiology of crops (After Dhopte and Eastin, 1988)

The influence of temperature stress on reproductive physiology of field crops is exhaustively reviewed by Dhopte and Eastin (1999) detailing the causes of limitations in crop productivity. Gupta *et al.* (2013) concluded that polyamines such as spermine, spermidine, and putrescine, in relation to the mechanism and adaptation to combat abiotic stress has been well established in plants. Because of their polycationic nature at physiological pH, PAs bind strongly to negative charges in cellular components such as nucleic acids, proteins, and phospholipids. Accretion of the three main PAs occurs under many types of abiotic stress, and modulation of their biosynthetic pathway confers tolerance to drought or salt stress. PAs can play important role in maintaining crop yield under adverse environmental conditions.

The stress proteins are the products of the expression of genetic programme in plants in stressful environment. Stress proteins are part of the primary cellular protective response from environmental stress induced by wide variety of stress factors, and are usually highly conserved in all organisms from bacteria to man. Because the plants are stand still, they have to make necessary metabolic and structural adjustments to cope up with the stress conditions. The pattern of protein synthesis is drastically altered when plants are subjected to sub-optimal environment. The induction or suppression of the production of specific proteins may constitute the integral defense mechanism of plants. This phenomena is precisely described by Gehlot and his associates (1999). They have concluded that over expression of single protein is not sufficient to alter stress response. Proteins work in cooperation with other substances like specific carbohydrates. Over expression of the superoxide dismutase gene has been found to enhance tolerance to oxydative stress in transgenic tobacco, alfalfa, potato and cotton; but it enhanced tolerance to freezing stress, chilling injury and water deficit. Expression of single sm HSP in carrot cells enhances thermotolerance. Recent technological advances promise further rapid progress in the study of cross tolerance in plants (Gehlot *et al.,* 1999). Besides heat shock, drought and salt stress, ABA, ethylene or auxin can also induce the synthesis of hsp (Pareek *et al.,* 1995).

6. ROLE OF PLANT GROWTH REGULATORS IN DRYLAND FARMING

The yields are plateaued since last two decades and there is no increase in maximum genetic yield. In united states, transgenic varieties and hybrids are introduced in crops like cotton, maize, oilseed rape, wheat and potatoes containing BT genes with tolerance to number of herbicides (Borlaug 1997). Plant tissue culture and genetic engineering shall have bright scope for evolving varieties tolerant to stresses through advancement in biotechnology. Until then, we have to depend upon the conventional practices in ameliorating the stresses through foliar application of hormones and nutrients. This is direct solution to the problem. If auxins/hormones and nutrients are deficient, they are provided through foliar spray. The plants shall be tailored to synthesize adequate auxins through the manipulation of genes using genetic engineering. We are supposed to resort to old techniques of supplementing the requirements of plants under stress until we get improved varieties through biotechnology.

Review of literature on foliar spray of hormones and nutrients indicates sufficient scope to modify the growth of plant. Every auxin or hormone has specific role to play at specific phenological stages of crop. The requirements vary according to the stage. Endogenous levels of auxins are almost found normal as long as the soil moisture levels are optimum. If this is affected, the synthesis of required auxins is also altered and the responses are negative. The trend of crop responses to foliar application of hormones and nutrients is subject to the condition of plant and the prevalent environment. Some times, the responses are positive but negative responses are also recorded by various workers. While reporting, the conditions are not specified in some literature and therefore it should not be considered that the responses are always grossly negative. We shall now consider the literature available on the role of foliar application of hormones and nutrients in field crops specially the use of NAA, Cytokinins gibberellins and ethylene in agriculture.

Abscission : Abscission is the process of shedding of flowers, leaves, fruits and it is regulated by various factors. Controlling abscission is extremely important in agriculture and horticulture to ensure the most effective crop growth. Leaves need to be retained healthy and green for long time for having average productivity. The flowers and fruit shedding are of usual occurrence in response to the adversities. The water stress or waterlogging results in fruit drop heavily and productivity is reduced. The auxin, ethylene and abscissic acid appear to be the hormones most directly involved in abscission. It is also indirectly coupled with other hormones and availability of nutrients. Apparently, the control of this phenomena does not reside in any one hormone or environmental factor but is regulated by complex interaction of environment, hormones and physiological status of the plant.

The role of NAA in boll shedding is extensively studied in cotton. The promotion of ABA synthesis regulates abscission and endogenous level of IAA plays the key role (Varma, 1978). The sprays of NAA (20 ppm) reduces the physiological boll shedding in cotton (Swamy, 1991; Khandagale *et al.,* 1992). The CCC sprays also reduced the shedding of bolls in cotton. Low concentration of NAA (20 ppm) reduced the boll shedding in MCU-5 cotton (Bhatt 1991). The time of application of auxin was shown to be important (Rubistin and Leopold, 1964). Application of auxin maintains the level of ethylene. The effects are brought out by enzymatic changes in plant after the application of auxins. The auxin inhibits the activity of cells wall degratating enzyme (Abels, 1967). Our studies under rainfed farming indicated that 20 ppm con. of NAA with 1% DAP reduced the boll shedding and resulted in increased seed cotton Yield (*Anon.*, 1994). The NAA or planofix or fruitofox is widely used by the farmers in cotton, fruit and vegetable crops, pigeonpea, soybean, gram and oilseed crops for increasing yield under dryland farming.

The literature on role of NAA is reviewed below for the benefit of readers. The flower and fruit number were found highest in plants sprayed with 50 ppm GA_3 and decreased fruit abscission by 10% (Castro *et al.,* 1985). When GA was sprayed at three days before anthesis in soybean, the fraction of associated node that set pods was reduced by 28-76% (Birnberg and Brenner (1987). Decrease in N-fixation was however reduced in GA treated plants (Guiamet *et al.,* 1987).

Pandey (1975) reported that planofix (NAA) reduced flower abscission and increased biomass production in pigeonpea and soybean. The NAA sprays (20-60 ppm) increased LAI, LAD and dry matter production and its distribution to pods in peanut (Reddy, 1978). The remarkable earliness in the time of budding, flowering and fruiting was found in plants treated with NAA (10-100 ppm). The uptake of N and P was found strongly increased with increased chlorophyll in sunflower. Sagare and Naphade (1987) reported that N and P uptake was increased in peanut plants treated with IAA and NAA. The 10 ppm sprays of NAA or 2,4-D at flowering increased seed yield by 38.2% in blackgram, 57.2% in greengram and 7.39% in soybean. The 10 ppm NAA and ethrel (500 ppm) on rapessed mustard significantly increased branches and seed weight (Singh and Kumar, 1991). There was increased in 100 seed wt. in plants treated with 20 ppm NAA and 100 ppm CCC at peak flowering in soybean (Ravikumar and Kulkarni, 1988). The spray of TIBA was however, not found much effective.

Exogenous application of NAA and IAA is found to reduce the fruit drop in apples. The studies on Kinno manadarin by Patil *et al.,* (1989) revealed that 10 ppm NAA was most effective in reducing fruit drop. They also studied the effect of 2,4-D and 2,4,5-T in various concentrations and above con. of NAA was ideal for controlling abscission. Lakshmanna (1988) reported reduced in fruit and flower drop in blackgram with 20 ppm NAA while in pigeonpea, the NAA with 1 and 2% urea was most effective in reducing pod drop. As a result, more pods were retained and the yield was increased. The sprays of NAA prevented abscission and reduced flower shedding in greengram (Subbian and Chamy, 1984). Foliar application of NAA (50-100 ppm) caused increase in number of branches, pods/ plants and increased seed yield (Setia *et al.,* 1993). The increase in oil content with increase in NAA concentration (50-100 ppm) was found in sunflower (Pain and Sarkar, 1980).

The spray of NAA increased yield of soybean (Table 2). Dhopte and Suradkar (1998) reported that, the spray of NAA+ $ZnSO_4$ maintained yield of soybean under low fertilizer dose (Table 3). Similarly, this combination was effective with half dose of fertilizer (Table 4, Armarkar, 1999). The seed setting efficiency was found increased by 25.9% in pilose type hirsutum cotton (Dhopte *et al.,* 1993, Table 6). The spray of NAA + KH_2PO_4 at flowering increased yield of hirsutum cottom (Anon. 1995, Table 7). Aslam *et al.* (2010) concluded that plant growth regulator (NAA) with 80 % Available soil moisture depletion level may be better to achieve higher seed yield and yield components in chickpea.

Gibberellins: The GA was less effective in controlling abscission than ABA and ethylene. It may be due to formation of ethylene formation. The promotion of abscission by GA can be reverted by auxin in citrus. The application of GA to intact plant prevents abscission of fruits and help development of young fruits. In ber (*Zizypus zuzuba*), the GA, sprays of 20 ppm reduced fruit drop by 70.93% (Singh *et al.,* 1987). In hybrid cotton (AHH-468), the sprays of GA, NAA and IAA (20 ppm each) controlled boll shedding by around 38% (Deshpande and Lakhdive, 1994).

Table 2. Effect of seed treatment with GA and NAA on soybean. JS 335. Maske, 1997.

Treatment	Pod/plant	Seed yield Kg/ha	Increase over control (%)
Control	48.1	3316	–
GA 50 ppm	55.0	3700	11.6
100 ppm	58.6	4003	20.7
NAA 25 ppm	57.2	3794	11.4
50 ppm	27.7	3776	12.7
SE	0.39	0.35	
CD 5%	1.13	1.01	

Table 3. Effect of foliar spray of hormones and nutrients on growth of soybean under suboptimal fertilizer dose (Half). Dhopte and Suradkar, (1998).

Treatment	Pod No./plant	Seed yield(q/ha)
Control (FRD)	37.3	22.2
Half Dose (HRD)	27.3	18.9
Half dose + NAA + ZnSO4	39.0	21.6
Half dose + GA + ZnSO4	40.3	22.1
SE	1.84	0.22
CD %	5.38	0.81

Note : Twelve treatments combinations were compared in this trial. Half dose of fertilizer can be saved with spray of NAA (20 ppm) and $ZnSO_4$ (0.5%).

Table 4. Response of blackgram to foliar sprays of NAA (50 ppm) and nutrients (0.5%) under half dose of fertilizers (Armarkar, Neeta, 1999).

Treatment	Seed yield (Kg/ha)	Increase over control (%)
Control (FRD)	4866	–
NAA + KH_2PO_4	3933	-19.0
NAA + KNO_3	5300*	+8.9
NAA + $ZnSO_4$	5477*	+12.6
NAA + $(NH_4)\ 2NO_3$	3933	-19.0
SE	5.0	–
CD (5%)	14.0	–

*Significant. Half dose of fertilizer can be saved with additional yield of 8.9 & 12.6% over full dose of fertilizer, FRD

Table 5. Effect of GA_3 on fibre of Ramie (g/plant) *Bohmeria nivea* Gaud (Dhopte and Kale, 1976) For first cut and third cut on oven dry basis

Variety	Control	25	50	100	125 ppm	
Diploid	0.97	2.32	1.86	4.09	3.4	First cut
Triploid	2.21	3.93	4.50	5.00	7.06	
Mean	1.59	3.12	3.23	4.54	5.23	
Diploid	10.50	13.75	14.30	16.90	19.60	Third cut
Triploid	10.08	10.42	10.33	14.50	16.40	
Mean	10.29	12.08	12.31	15.70	18.00	

CD (5%), Treatment First cut 2.37 and Third cut, 1.58

The 100 ppm GA increased number of pods and yield in chickpea (Khan and Rashid, 1983). The GA sprays increased pod number, seed size and yield in peanut (Mishra *et al.,* 1984) when sprayed at pegging stage. The sprays of GA (100 ppm) and IAA (15 ppm) improved pod yield in peanut and sprays were found more effective than seed soaking (Singh and Rathore, 1987).

Some types of seed dormancy can overcome by application of GA. The GA soaking improves the germination of seeds. Soaking seeds with GA (100 ppm) for 3 hours improved germination from 50 to 90% with increased seed vigour. Flowering percentage was improved from 2 to 6% by 100 ppm GA in bellary onion (Venangamudi *et al.,* 1988). The improvement in seed germination in *Vicia faba* and *Pisum sativum* was found with GA soaking (100 ppm, Sharma, 1982). The 50 ppm GA and 100 ppm NAA sprays could increase flowers, duration of flowers in Chiku plants (Delvadia *et al.,* 1994). The sex expression can be altered by GA (10 ppm) sprays in pumpkin (Arora and Partap, 1988). They further reported that the weight of fruit and fruit yield was increased with GA treatment (25 ppm) in pumpkin with more healthy seeds/ fruit. Thus, GA improves seed setting in fruits. It also increases chlorophyll content in cabbage (Kumar *et al.,* 1996). Saxena *et al.,* (1986) reported that soaking with GA (10 ppm) increased grain weight and number in wheat.

Beneficial responses of ramie (*Bohmeria nivea*) (Table 5), cotton and soybean to growth regulators are reported (Dhopte and Kale, 1976; Dhopte and Hudge, 1976; Dhopte and Zade, 1980;

Dhopte *et al.,* 1993, Maske 1997, Table 2). The foliar sprays of nutrients were shown effective in cotton (Dhopte and Zade, 1980; Dhopte and Wankhade, 1988, and Dhopte, 1987). The yield of brinjal was found increased with 50 ppm GA_3 by 38% (Chopde and Dhopte, 1971, Table 10).

Afrigan *et al.* (2013) carried out the study on the effect of plant hormone gibberellic acid on germination indices of *Secale montanum* in vitro and pot experiments under drought conditions, their results showed that pretreatment with Gibberellic acid to seed in drought stress conditions in vitro increased the germination percentage, germination rate, root length and shoot length and also in medium pot, seedling establishment percentage, seedling establishment rate, root length and shoot length as influenced by different hormonal treatments, with Gibberellic acid 50 ppm at 10°C and Gibberellic acid 100 ppm at 10°C and - 10 times surface in the experimental vitro and pot have had the most positive effect on germination and growth factors, respectively. The seed priming treatment at 10°C compared with 15°C displayed a better outcome.

Kaya *et al.* (2006) reported the combination effects of water stress and gibberellic acid (GA_3) on physiological attributes and nutritional status of maize (*Zea mays* L. cv., DK 647 F1) in pot experiment, Maize plants were grown in the control (well watered (WW)) and water stress (WS) subjected to treated both water stress and two concentration of gibberellic acid (GA_3 25 $mg{\cdot}L^{-1}$, 50 $mg{\cdot}L^{-1}$). WS was imposed by maintaining the moisture level equivalent to 50% pot capacity whereas the WW pots were maintained at full pot capacity, they pointed out exogenous application of GA_3 improved the water stress tolerance in maize plants by maintaining membrane permeability, enhancing chlorophyll concentration, leaf relative water content (LRWC) and some macro-nutrient concentrations in leaves.

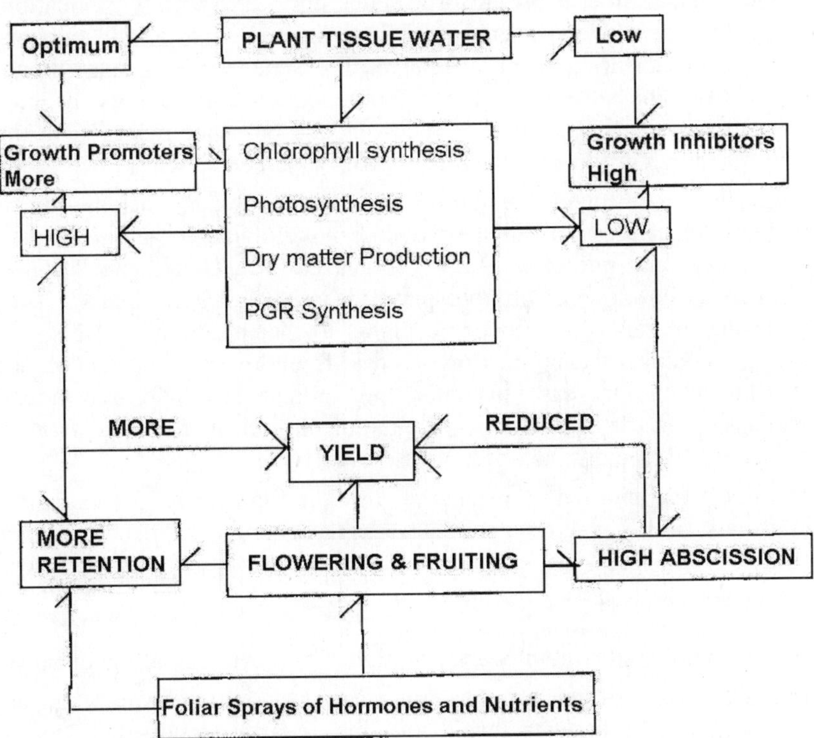

Fig. 3. Flow chart to show how leaf water status affects physiological processes and synthesis of plant growth regulators, flowering, fruiting, abscission and finally yield potential.

Role of cytokinin : The cytokinin is the hormone that is syntheised in roots and brings the changes in foliar parts. The water stress predominantly reduces the synthesis of cytokinins and it is shown that exogenous application of BAP helps prevent the adverse effect of drought in cereals. The cytokinin mobilizes the supply of nutrients and amino acids from which IAA is synthesised in plant and then it controls the abscission (Osborn and Moss, 1963). The response of auxin is always subject to nutrient status in plant. If they are sprayed with nutrient sprays, the response is additive/ synergistic. Deficiency of nutrient in a plant at the time of auxin spray may not have positive response. The effects are positive only when the nutrient status is moderate. Abscissed bolls contained more ABA than non abscissed ones (Bhatt, 1978, Madaghe, 1996). The endogenous auxin content may be playing key role in phenomena of abscission (Varma, 1978). The induction of abscission by ABA was mediated by ethylene production. Mature leaves were most sensitive to ABA than young leaves.

The role of cytokinin has been known in senescence. It defers the senescence in most of the species by maintaining the chlorophyll. Reduced productivity of dryland crops is mainly due to early chlorophyll degradation (Fig. 3). The leaf area duration is also reducced. This can be controlled by cytokinin. Cultivars of rice and soybean with early senescence show earlier decreases in foliar cytokinin. Treating isolated leaves of many species with cytokinin will delay their senescence. A dramatic affects are obtained when it is sprayed on intact plants. Cytkinins influence the movement of nutrients into the leaves from other parts of the plants.

Foliar application of cytokinins have shown to retard senescence in attached leaves of several species including common bean, oat, soybean, sunflower and wheat. While endogenous level of cytokinin is reduced in intact older leaves prior to senescence. Exogenous application of hormones is also effective in the deferment of senescence of foliar organs (Biswas and Choudhari, 1978). Genes, whose expression is induced during senescence are called senescence associated genes (SAGs). The functions of most SAGs have been deduced from sequence homology to known genes and confirmed by their enzyme activities. These genes encode enzymes involved in the conversion of remobilization of breakdown products (Oh *et al.,* 1996; O'Neil *et al.,* 1993).

Rivero *et al.* (2009) mentioned that the during water deficit conditions, the transgenic plants displayed an increase in catalase inside peroxisomes, maintained a physical association among chloroplasts, peroxisomes and mitochondria and increased the CO_2 compensation point, indicating the cytokinin-mediated occurrence of photorespiration in the transgenic plants. The contribution of photorespiration to the tolerance of transgenic plants to water deficit was also supported by the increase in transcripts coding for enzymes involved in the conversion of glycolate to ribulose-1,5-bisphosphate. Moreover, the increase in transcripts indicated a cytokinin-induced elevation in photorespiration, suggesting the contribution of photorespiration in the protection of photosynthetic processes and its beneficial role during water stress.

Merewitz *et al.* (2010), studied on effects of sag12-ipt and HSP.2-ipt expression on cytokinin production, root growth, and leaf senescence in creeping bentgrass exposed to drought stress. Their results demonstrated that expressing ipt resulted in the improvement of turf performance under drought stress in creeping bentgrass in some of the transgenic plants with SAG12-ipt or HSP18.2-ipt, which could be associated with the suppression of leaf senescence and promoting root growth relative to shoot growth due to the maintenance of higher CK level and a higher ratio of CK to ABA.

Cytokinin regulates root growth in a largely organ-autonomous fashion that is consistent with its dual role as a hormone with both paracrine and long-distance activities. Transgenic plants had a higher survival rate after severe drought treatment. The accumulation of several elements, including S, P, Mn, Mg, Zn, as well as Cd from a contaminated soil, was significantly increased in shoots.

Under conditions of sulfur or magnesium deficiency, leaf chlorophyll content was less affected in transgenic plants, demonstrating the physiological relevance of shoot element accumulation, Werner *et al*. (2010).

Kuppu *et al.* (2013) reported that the water-deficit induced expression of cytokinin biosynthesis gene (isopentenyl transferase gene) in cotton could significantly improve drought tolerance.

Role of Ethylene in Field crops : Abscission is increased when ethylene content of the tissue is increased. Ethylene production is increased under adverse environmental conditions like drought, waterlogging and high temperature stress. Thus, its production is entirely dependant on environmental conditions. The abscission due to ethylene can be inhibited by inhibitors of RNA and protein synthesis (Morgan, 1969). Ethylene biosynthesis inhibitor is Aminoethoxy Vinyl Glycine (AVG). The $AgNO_3$ absorbs the ethylene and CO_2 and inhibit the process of abscission. During ripening of fruit, ethylene production is increased (Durdon *et al.,* 1990). Ethylene concentration in mature fruit is significant. Ethylene is normally evolved during ripening of fruits and this hormone is known to have role in dormancy, germination and ripening. Artificial ripening can be induced by application of ethylene producing compounds. Sharp and LeNoble. (2002) suggested that the Abscisic Acid (ABA) plays important role to limit ethylene production. Bosch *et al.* (2004) demonstrated that the ethylene at concentrations similar to those found in polluted areas may reduce plant stress tolerance by altering, among other possible mechanisms, antioxidant defenses.

Sex Expression : Exogenous application of PGR can alter the sex expression in cucumbers. Hidayatullah *et al.* (2009), found that plant growth regulators affected the sex expression and the level of phytohormones in monoecious cucumber. Hidayatullah *et al.* (2012), studied the plant growth regulators affecting sex expression of Bottle Gourd (*Lagenaria Siceraria* Molina) plants. They applied ethrel exogenously with various levels (500, 1000, 1500 and 2000 $\mu mol\ l^{-1}$); Gibberrelic Acid (GA 15, 30, 45 and 60 µmol l) and Maleic hydrazide (MH; 200, 400, 600 and 800 $\mu mol\ ^{-1}$) to see its effect on sex expression of bottle gourd (*Lagenaria siceraria* Molina) plant. They found that the response of GA and MH was better. Exogenous application with 30 μmol^{-1} GA maximally increased the pistillate flower production as compared to control. The productivity of such crop is dependant on production of female flowers or ratio of male : female flowers. Initially the flowers produced are bisexual but at later node there is production of staminate or pistillate flowers. Ethephone is a substitute for ethylene. It is determined genetically and is modified by environmental factors. High nitrogen, short days, low light intensity and low night temperature are among the factors which favour the femaleness (Heslop and Harrison, 1957).

The ethephone is effective in promoting gynoecious condition in monoecious muskmelon, cucumber and squash (Christopher and Loy, 1982). The concept of sex expression in cucumber plants may be regulated by a balance between native auxin and GA_3 (Atsmon, 1963). Iwahori *et al.,* (1970) reported that ethrel caused increased femaleness when applied at 50 ppm at first to third leaf stage. Lower *et al.* (1970) observed higher percentage of female flowers on the main stem of cucumbers when plants were sprayed twice on fourth leaf stage with 120 ppm ethyphone. The CEPHA decreased staminate flowers (Rudich *et al.,* 1969) and increased number of pistillate flowers. Both sex expression and sex ratio can be modified by PGRs like GA, NAA, ethrel (100 ml/l) and BA (Elkholy and Hafez, 1982). They found ethrel to be more effective in increasing marketable fruits and its quality. The increase in femaleness from 8% to 84% was observed by EPCA, ethrel, 20 ppm (Alani, 1981). Thus, single spray of ethephone at 200-500 ppm increased significantly the number of femaleness. The chemical application was found cost effective.

Ascorbic acid

The ascorbic acid (AA) is known to act as plant growth regulator and it is reviewed by Tonzig and Marre (1961) and Chinoy (1977). The AA is most effective in increasing total plant nitrogen (Chinoy, 1962). The AA sprays (500 ppm) increased leaf respiration rate and nitrogen content and maintained micronutrient levels higher in hirsutum cotton. The leaf reddening was reduced by 30% and spray of AA increased yield of cotton by 32.1% (Dhopte and Lall, 1987).

Role of Cycocel

It is growth inhibitor and is used to reduce growth of plant, It increases harvest index by enhancment in enzymatic activity (Oben and Marcelle, 1975) It increases leaf thickness, pigment content and coefficient of assimilation in tomato (Bendarenko and Ledovskii, 1976). Dhopte and Lall (1987) observed increase in chlorophyll content and relative water content with sprays of 500 ppm CCC in cotton. Higher uptake of nutrients was also noticed with reduction in leaf reddening by 34.1%.

Role of Kinetin

Kinetin is known to defer senescence. Osborne (1962) reported that the ratio of RNA and DNA was maintained in xanthium leaves. She showed its critical action on abscission in bean explats. Kinetin (10 ppm) increased chlorophyll a/b ratio, reduced respiration rate and leaf reddening by 30% in hirsutum cotton and increased yield by 15.2% (Dhopte and Lall, 1987 Table 8 and 9). Jonas *et al.* (2012) concluded that the chlorophyll concentration index (CCI) was higher in all of phytohormones treated plants as compared to control.

Table 6. Effect of foliar spray of NAA (20 ppm) and DAP (1%) on seed setting efficiency (%) in hirsutum cotton (SRT-1) (Dhopte *et al.,* 1993).

Leaf type	Control	NAA + DAP
Broad Lobed Hairy	83.9	88.82
Pilose (Dense hairs)	65.5	91.37

Table 7. Effect of NAA (50 ppm) and KH2PO4 (1%) on hirsutum cotton (Anon. 1994-95), BLH = Broad Lobed Hairy and pilose = Dense hairy

			1994		1995	
Treatment	Bolls shed plant		Seed cotton yield (Kg/ha)			
	BLH	Pilose	BLH	Pilose	BLH	Pilose
Control	30.67	32.33	12.9	771	426	1065
NAA	19.00	27.67	1402	650	704	833
NAA + KH_2 PO_4	28.33	19.67	2431	456	699	833
SE	2.52		85.9		34.1	
CD (5%)	7.39		252.8		100.1	

Note : The yield increase is found only in normal type of variety (BLH). The pilose type did not respond to the treatment. Thus, the responses are always species specific.

Table 8 : Effect of plant growth regulators spray at flowering on yield (kg/ha) of hirsutum cotton, cv. SRT-1 (Dhopte and Lall, 1987).

Treatment (ppm)	1977	1978	1979	Mean
Control	739	1262	681	895
AA-500 ppm	635	1670*	758	1021
CCC-500 ppm	833	1268	581	894
Kinetin - 10 ppm	692	1458*	690	947
SE	43	59	34	
CD (5%)	126	172	101	

Note : During 1978, the yield increase in AA-500 and Kinetin 10 ppm was 32 and 15.2% over control under unfavourable season. The rainfall was low in this season. Thus, positive responses are evident under abnormal weather.

Role of Silicon

Jian Feng Ma (2004), studied the role of silicon (Si) in enhancing the resistance of plants to abiotic and biotic stresses. According to him, silicon exerts alleviative effects on various abiotic stresses including salt stress, metal toxicity, drought stress, radiation damage, nutrient imbalance, high temperature, freezing and so on. Gong *et al.* (2013) demonstrated that the leaves of stressed plants growing in pots with Si applied were thicker compared to those without Si applied. This may have a beneficial effect by reducing the transpirational loss of water and maintain high RWC and water potential. Therefore, application of Si may be one of the available pathways to improve growth of wheat crop and increase its production in arid or semi arid areas. Ullah *et al.* (2012), reported that the applied growth regulators maintained the water budget of canola plants, augmented the accumulation of osmolytes proline and protected photosynthetic pigments from adverse effects of drought stress. They found that SA (Salicilic acid) was effective to reduce the drought induced accumulation of glucosinolates and erucic acid in canola oil and both the growth regulators overcame the drought induced decrease in oleic acid (C18:1). It is inferred that SA is economical and environment friendly alternative and can be implicated to improve the plant growth and oil quality of canola in current scenario of drought and climate change.

Table 9: Effect of foliar spray of PGR and nutrients (micronutrient MN) on seed cotton yield (g/plant) of arboreum cotton (Dhopte and Zade, 1980).

Treatment	AK-235	AH-4
Control (WS)	18.6	16.0
CCC + PMA	17.1	23.0
CCC + PMA + MN	24.3	11.8

7. ROLE OF MINERAL ELEMENTS IN DRYLAND FARMING

Kumar *et al.* (1967) studied deficiency of mineral nutrients in Andrews cotton. They found deficiency of P and Cu in intermediate and unhealthy leaves grown in sandy alkaline soils. The application of boron @ 0.9 Kg/ha and one spray of 0.11 Kg/ha of B in 8 weekly applications was more effective for improvement in productivity (Keogh and Maple, 1969). Nageswarrao (1976) observed yellowing and reddening due to deficiency in MCU 5 cotton. He reported 18-23% increase in yield with 50 to 60% recovery in greenness by an application of 0.5% $MgSO_4$ as spray. Kadrekar and Kibe (1974) studied fixation of K in vertisol under wetting and drying cycles. Exogenous application of macro and micro elements increased physiological processes in horsutum cotton (Dhopte and Lall, 1987).

Kannan (1990) has advocated the foliar sprays of nutrients for correction of deficiency and to save the loss of fertilizers. The fertilizers are very costly and it is leached out in the soil if heavy rains occur. The P fixation is more than 70% and is not available to present crop. The K fixation is also the problem in vertisol soils. The influence of P is reviewed by Anuradha and Narayanan (1995) in field crops. Low organic carbon is found in cultivated soils and it affects the CEC and nutrient availability. It is always better to go for foliar nutrition under water stress and waterlogged conditions when nutrient availability to the crop is very low. This is one of the major cause of low productivity under dryland farming. The foliar sprays of KCl and KNO3 (1 x 10^{-2} M) increased pod size, number of seeds/pod and test weight in *Brassica campestris*. Similarly, the spray of urea and thiourea (2%) was found effective in improvement of above parameters (Asija, 2000).

This has been exhaustively reviewed by Kannan (1986). Sprays of nutrient solution (KNO_3, 0.5%, $CaNO_3$ 0.5% and Urea, 2%) on peanut improved LAI, CGR and pods/ plant (Patra *et al.*, 1995). The response of soybean to foliar sprays of nutrients (0.2% $ZnSO_4$) is found useful in improvement of growth and yield (Garic and Hanway, 1976; Thaloot *et al.*, 1989; Jat and Nepalia, 1994). Syverud et al. (1980) reported higher yield by weekly sprays of polyphosphates in maize and soybean. Beneficial responses are reported in cereals and vegetables to foliar application of ammonium nitrate and urea (Shapiro, *et al.*, 1984). The response of sulphur to soybean was better alongwith N, P and K (Shapiro, *et al.*, 1987). Highest seed yield was obtained when TSP was sprayed with 1% $ZnSO_4$ (Syaffruddin, *et al.*, 1991). Foliar application of K in KNO_3 for five times was better for yield improvement in Pecan (Diver *et al.*, 1985). Similarly, foliar application of 3% DAP at 70 DAS increased yield of pigeonpea (Soliappan and Ramiah, 1990). Fayyaz *et al.* (2012) carried out the research work to study the effect of sodium and potassium in *Atlas mastic* tree (*Pistacia atlantica* subsp. *mutica*) under drought stress condition. They found that the roots and stem of Atlas mastic tree seedlings exhibit high drought tolerance index for Na^+ and K^+. Raza *et al.* (2013) reported that the potassium applied under drought conditions improves physiological and nutrient uptake in wheat (*Triticum Aestivun* L.). They further reported that the maximum improvement in all the recorded nutrients uptake and physiological parameters were achieved when K was applied at grain filling stage of both cultivars.

8. COMBINED EFFECT OF HORMONES AND NUTRIENTS AND THEIR INTERACTION

The sprays of hormones alone were effective in yield increase in various crops but when hormone spray was practised with nutrients, an additive effect was observed by several workers in different crops. The response of crop to auxin/hormone alone is dependent upon availability of other inputs like nutrients. Better interaction was found when the sprays of auxins were coupled with nutrients. Foliar sprays of NAA and KNO_3 on cowpea showed better response in yield contributing parameters and increased yield by 33% (Shinde *et al.,* 1991). The response of soybean crop to NAA + Urea + K was good to improve yield of soybean (Sharma and Day, 1986). The spray of 3% P_2O_5 + 100 ppm NAA at 40 DAS on greengram improved physiological parameters like uptake of N and root nodulation (Kalita *et al.,* 1995).

Increase in grain yield of soybean with 15 ppm GA and 60 Kg P_2O_5 / ha was recorded by Dashora and Jain (1994). Similarly, increase in yield of cowpea with 45 ppm NAA is reported by Jain *et al.,* (1993) and Shinde and Jadhav (1995). The increased yield of hormones and nutrients was attributed by increase in flowers, pods/plant and better seed setting. The greenness of leaves remained for prolonged period and helped in grain filling. The endogenous levels of auxins in plants are reduced under water stress or waterlogged conditions and if it is supplemented with exogenous sprays, growth is improved and yields are not drastically affected under dryland farming. The spray of paclobutrazol (25 ppm) + $ZnSO_4$ increased yield of mustard by 21% (Table 14) and single spray of PCB (50 ppm) increased yield of pigeonpea (Anon. 1997).

In studies with foliar sprays of NAA and GA with nutrients under half dose of fertilizer in soybean, Dhopte and Suradkar (1998) reported that a spray of NAA + $ZnSO_4$ or KNO_3 (0.5%) recorded higher yield than full recommended dose with cost effectiveness of 1.52. The fertilizer economy can be achieved. Madghe (1996) reported higher yield of cotton with spray of NAA (50 ppm) with KH_2PO_4 (1%).

Maturity of a crop is found early under adverse conditions and it is due to early senescence. Senescence is a genetically programmed process and it is visible on greenness of leaves. It is an ordered series of cytological and biochemical events. Some organelles are destroyed and other remains active. Chloroplast is the first organelle to be degraded due to inactivation of stomatalen-zymes. This is regulated by specific genes. The genes regulate enzymatic activity and it controls synthesis of hormones. This phenomena can be altered to certain extent by providing the end product. If sprays are made, a reverse signal is transferred to the organelle and the biochemical messages at transpiration level are wrongly mis-interpreted thereby presuming no stress conditions by the organelle. The processes continue at subnormal rate (Not as faster as under normal ideal conditions) with available resources and plant completes life cycle assuring yields due to normal reproductive process in the plant. Ageing of a leaf and other plant parts is also a genetically programmed process. It can be reverted or modified by spraying required inputs a plant needs at a critical phenological stage. The accurate mechanism of chlorophyll degradation is not known due to lack of knowledge of the intermediates of pigment breakdown.

9. ABIOTIC STRESS AND ROLE OF HORMONES

The plant can be confused under certain adverse situation and its genetic programme is disturbed. During temperature stress, the heat shock proteins are formed and this chain of reactions can be altered giving break to further break down and expected results can be achieved successfully under adverse conditions. Catabolic reactions are arrested by exogenous application of hormones and nutrients. Rawson (1988) is of the view that the temperature stress is not the stress provided necessary inputs are made available to the crop. High temperature accelerates the physiological processes and if necessary inputs are not provided for accelerated reactions, growth is bound to affect. He further explains that as temperature increases, organs efficiency appears faster and complete their growth in less calendar time. Because, as the rate increases, the rate of availability of substrates (radiation, water and nutrients) must also increase in proportion. Any change in proportioning will result in change in size. He observed that high temperature *per se* has no effect on leaf area production/unit plant time and has only minimum effects on biomass production, which in his studies of wheat were linked with changed root/shoot allocation patterns.

During stress conditions, senescence results in increase in respiration. We breath faster when we are running. This is for higher O_2intake for more energy for cellular activity. Thus, faster reactions under stress conditions are demanding more inputs of energy and oxydizable sustrates. If it is supplied exogenously, the plants can grow normal under stress conditions. Many biological processes are controlled under stress by administering chemical drugs in human beings. The similar philosophy works here. Probably, respiration during senescence is higher due to low GA and cytokinin levels relative to ABA levels. It is usually agreed that plant hormone groups influence senescence. In most plants, cytokinins are anti-senescence compounds that can postpone plant death. Gibberellins have similar effects, but in fewer plant species, auxins influence senescence and abscission in very complex manners. The response depends upon age and receptivity of the tissue. When applied to leaf, auxin often retards senescence, but paradoxically, when auxin enters the leaf, it some times promotes the senescence.

Table 10 : Effect of GA (50 ppm) foliar sprays at flowering in brinjal (*Solanum melongena*) (Chopde and Dhopte, 1971).

Parameter	Treatment (ppm)				
	Control	50	100	SE	CD (5 %)
Seed number/fruit	2376	2488	2036 –	–	
Fruit yield/plant (g)	343.6	474*	368.3	37.5	118.2

Table 11 : Effect of paclobutrazol (Cultar) and $ZnSO_4$ spray on seed yield (kg/ha) of mustard *Brasica juncea* (Anon. 1997).

Treatment	Con.	1996	1997	Pooled Mean
Control	–	571	1097	834
ZnSO4	0.5%	656	1297	977
Cultar	25 ppm	662	1261	962
Cultar + $ZnSO_4$	–	681	1341	1011
SE	9.3	27.0	28.0	
CD (5%)	27.4	78.0	99.0	

Note : Based on pooled mean, the spray of cultivar (25) with $ZnSO_4$ (0.5%) increased yield of mustard by 21.2% over control.

Table 12: Effect of paclobutrazol (Cultar) on seed yield (Kg/ha) of pigeonpea (Anon., 1997)

Treatment	Con.	1996	1997	1997	Pooled
Location		Nagpur	Nagpur	Akola	
Control	–	1312	900	546	925
Cultar	50 ppm	1534	970	572	1025
Cultar	100 ppm	1494	1018	594	1035
Cultar	150 ppm	1450	1092	707	1083
SE		15.7	20.2	18.6	–
CD (5%)		46.4	60.4	55.8	–

Ethylene and ABA are the two important candidates for death hormones. Ethylene is closely involved in various senescence phenomena, especially flower wilting and fruit ripening. There is however little correlation between ABA concentration and plant death. The studies of Rawson

(1988) in wheat support the present view that the growth can be monitored with exogenous application of required auxins along with necessary nutrients. With available source of limited water under stress, the chemical reactions can process with moderate rate.

The role of PGRs is not always positive and this relates mostly to the interactions of many factors to achieve either positive or negative responses. The synthesis of PGR in plants is highly affected by edaphic or climatic factors. The interaction of adverse conditions and exogenous sprays are crucial for positive responses. The soil moisture affects porosity of soil and oxygen availability to roots is limited. The waterlogging depletes O_2 availability while water stress creates unfavourable water potential for optimum growth. High soil water potential brings reduction in kinetin levels and increases the levels of growth retardants such as ABA and ethylene (Fig. 3) which arrests the growth. Assimilation is affected and such impairment in metabolism is due to imbalance in soil water. High temperature under adequate water supply leads to formation of heat to changes in hormonal imbalance and shock proteins (HSPs) which affects the inadequate nutrients. How far these are reproductive growth. Meristematic growth effective is a matter of concern. Does branching, flowering, incompatibility in climate favours for correction? Does water fertilization, seed setting and seed scarcity interacts with exogenous sprays of development are influenced. This results in PGRs? Does frequent wetting and drying of reduction in crop production. The yield vertisol soils bring unfavourable changes in limitations under rainfed farming are due absorption of nutrients? Does exogenous sprays bring toxicity in metabolism? Such questions are not sitll answered while reviewing the negative responses under rainfed farming. The use of isotopes shall be better to know the mystery of negative responses under rainfed farming. The present concept of yield limitation can be refined after having the information on above proposed questions. The adaptation on a genotype to specific environment is comprised of genetic and environment components. There is specific need for breeding varieties for dryland conditions and this is not precisely attempted so far. The past breeding programmes were mostly centered for increasing yield with optimum resources. The transfer of gene through biotechnology for drought tolerance in a crop shall be more profitable under dryland conditions.

Pandey *et al.* (2003) studied the physiological effects of plant hormones in cotton under drought. They were examining the effects of plant hormones indole-3- acetic acid (IAA), gibberellic acid (GA_3), benzylaminopurine (BAP), abscisic acid (ABA) and ethrel (ETH) in 5 M concentration on gas exchange, ribulose-1,5-bisphosphate carboxylase/oxygenase (RuBPCO, EC 4.1.1.39) activity, pigment content and yield in cotton (*Gossypium hirsutum* L. cv. H-777) under drought. Their study suggested that the net photosynthetic rate (PN), stomatal conductance (gs), transpiration rate (E), carboxylation efficiency (CE), water use efficiency (WUE), RuBPCO activity, boll number per plant, seed number per plant and lint mass per plant were significantly decreased at drought while chlorophyll (Chl) b content and flower number per plant increased. ABA and ETH significantly reduced gas exchange parameters, Ch a and Ch b content. Detrimental drought effect on PN, gs, E, CE, RuBPCO and lint mass per plant was significantly alleviated by BAP and also its effect on seed number and lint mass per plant was significantly alleviated with the ABA treatment.

Zhang *et al.* (2006) put forth the hypothesis that ABA mediates so many stress responses, like the initial perception of dehydration and the subsequent changes in gene expression that lead to rapid ABA biosynthesis which constitute the most important stress signal transduction pathway among all the plant responses to stresses.

According to Agricultural Research Service (ARS) scientists, the application of naturally occurring plant hormones called cytokinins to cotton seeds or young cotton plants can increase yields 5 to 10 percent under drought conditions. USDA/Agricultural Research Service (2010)

Table 13 : Cost of plant growth regulators and approxmiate cost incurred/ha for single spray.

PGR	Cost Rs/g	Resposive Crops	Effective Con. (ppm)	Appro. cost incurred (Rs/ha)
IAA*	65/g	All crops	20-50	700-1000
IBA*	60/g	All	10-20	400-800
GA*	120/g	All	20-50	1300-3000
NAA	06/g	Legumes & Oilseeds	20-50	100-200
AA	03/g	-do-	50-100	100-200
BAP	112/g	All	5-10	300-600
2,4-D	02/g	Fruits & Vegetables	5-10	100-200
Succinic acid	28/g	Fibre crops	5-10	200-400
TIBA	160/g	–do–	50-100	200-400
CCC (Cycocel)	40/100 ml	Cotton and Pigeonpea	20-50	200-400
Paclobutrazol	3000 / litre	Fruits & vegetables	10-50	300-500
CEPA (Ethrel)*	210 / litre	–do–	50-100	200-400

* Used for seed soaking for improvement in germination and spray at flowering. For seed soaking, the expd. incurred is half the mentioned in the table.

Table 14 : Cost of chemicals used for spray (As on 2000) in foliar nutrition and Appro. cost (Rs./ha)

Chemical salt	Cost (Rs/500 g pack)	Effect Con (%)	Approx. Cost (Rs./ha)
KNO_3	95	0.5	500/-
KH_2PO_4	186	0.5	950/-
$Ca (NO_3)$	2 80	0.5	450/-
$(NH_4)_2 SO_4$	88	0.5	500/-
Urea (Com)	25	2.0	200/-
$ZnSO_4$	95	0.5	500/-
$MgSO_4$	80	0.5	450/-
$MnSO_4$	108	0.1	250/-
Thiourea	320	0.2	700/-
$Fe (NO_3)_2$	102	0.5	100/-

Foliar sprays are applied at flowering for positive response. Ready made formulations are also available in the market. It includes mixture of chemicals with chelating agents and sold under various trade names. It may also be used to correct the deficiencies of nutrients (Advocation is not for advertisement). The water to be used for spray should be clean. A surfactant (Teepol or Ethelene glycol (0.1%) must be added for effective response. Soap powder may also be used as surfactant (1 g/litre).

10. COST EFFECTIVENESS OF PLANT GROWTH REGULATORS AND FOLIAR NUTRIENT SPRAYS

The PGRs are used in very low concentrations ranging from 10-50 ppm and 100-500 ppm. They are successful in modification of plant growth under adverse conditions. Assuming the yield increase by at least 20% over unsprayed plots, the B:C ratio mostly ranges from 0.5 to 1.5. In some cases, it may be more than 2.0. The increase in yield is observed to be between half to 1.5 quintals/hectare. Some PGRs like, IAA,IBA and GA are used as soaking treatment before sowing and hence, its cost incurred is lower than given in the Table 13 & 14. The cost of PGRs as on Dec. 2000, its effective concentration for spray and cost incurred/ha are presented in Table 13 & 14 for the information of readers. The products of some companies are sold under different trade names like LIHOSIN for cycocel, FRUITOFIX and PLANOFIX for NAA, PLANTOVIT for hormones and nutrient sprays. Chloromquat is trade name for CCC used abroad. It is, therefore, advised to read the contents of product on the wrapper and know the crop it is recommended. The cropwise formulations are also available. The farmers may use it after their own experiences. It should be tried on small plot on 1 R with one R as unsprayed control to know the gain in yield. If gains are certain, it can be expended for more area under cultivation of crops.

11. CONCLUSIONS

The prevalent soil water and day temperature predominantly affects the synthesis of endogenous levels of hormones and it is mostly a signal that a plant receives under prevalent climatic conditions. How do we monitor these levels with exogenous sprays? According to soil/ plant analysis, the fertilizers are monitored. Can we monitor the levels of hormones and their ratios? What is the scale for manipulating these levels? We are helpless at this level because testing plant for hormone level is not feasible. Thus, indiscriminately, the sprays are advocated with assumption that it will have beneficial responses. This is just heating the problem without knowing the actual status of hormones. This is the missing link in advocation of hormone spray under dryland conditions. The probability of beneficial effects under adverse climatic conditions is hardly 60%. Failures are 40% in boosting crop productivity. The responses of hormone sprays are always not positive and this is the experience of farmers. Why then a farmer should go for such sprays? The strategies for such problems need to be found out. Who to do this? A committee should be formed by the state sector to formulate the strategic approaches regarding the use of PGRs in agriculture. Unfortunately, this has not been attempted at government level. On the basis of available literature, the research on missing links should be conducted and strategies should be charted out. The information gained should be published in book form as guide line for the workers at village level.

The adverse effects on flowering, seed setting and growth can be monitored to a certain extent and the probability of more damage can be minimised. Some yield can be assured instead having absolutely no yield out of the crop sown by a farmer with little expenses. This can be a consolation to the frustrated farmer. The use of NAA, GA, Kinetin, Ethrel, Cycocel, 2,4-D, TIBA, IAA, IBA and paclobutrazol is in practice by prudent farmers and they are reaping the advantage of PGRs. This practice is not adopted by poor farmers because he does not have paying capacity for hormones or he does not known the use of these chemicals in agriculture. The PGRs being used in very low concentrations are cost effective and B:C ratio works out to be between 1.5 to 2.0 in most of the cases of PGRs. Some PGRs are certainly costly and its B:C ratio is low. There is certainly no substitute for water but the magnitude of damage can be minimised and dryland farming can be made sustainable with precise use of PGRs. In some hormones like NAA, IAA, cytokinin, mix sprays are not advocated because of its incompatibility. It may not be mixed with sprays of insecticides and herbicides. The original auxin formulation may become ineffective. Some chemical recations occur and active ingredients looses its active sites of action. This must be remembered while undertaking the sprays. The sprayers used for 2, 4-D spray may be used for other purpose after through washing with soap and salt solutions to avoid ill effects on other crops e.g. use of 2, 4-D easter. Some anomalies may occur on meristematic tissue and size of leaf may change, epinasty is commonly observed phenomena in certain crops. The salts of nutrients can be mixed with auxin/hormone sprays because they are mostly neutral. When two chemicals are mixed, a third compound may be formed and it may be deleterious to the crop. Such precautions are, however necessary in order to have spray advantages. The negative responses reported in literature may be due to several factors. The sprays should be supplemented with surfactants that will allow the chemicals attached on surface of the foliage and it should not be washed out by rains. If after PGRs spray, rains follow, the spray must be repeated. Mostly the chemicals are absorbed by the tissue within one hour. The use of DAP and SSP (1 to 2%) should be practiced after its through soaking in water over night. The 'Teepol' surfactant (0.1%) can be replaced by soap powder which is neutral and has additive property. The sprays should normally be undertaken during morning hours or late evening hours when light intensity is low. The use of salts like KNO_3, $(NH_4)_2$ SO_4, $CaNO_3$, KH_2PO_4, $ZnSO_4$, $MgSO_4$, $(NH_4)_2$ NO_3 and $MnSO_4$ should be in appropriate concentration and normally it should be sprayed in very low concentration (0.5 to 1%) to avoid deleterious effects. The sprays of nutrients provide available element to correct the deficiency and promote growth. Two sprays of low concentration would be more appropriate instead one concentrated spray. The advantages of foliar sprays can be definit if precautions are rightly adapted.

REFERENCES

Agarwal, S. and S.K. Seervi (1995). The Plant hormone ethylene. In. Agro's Ann. Rev. Plant Physiolo. (Ed). Purohit *et al.,* AgroBios (India) Jodhpur.

Alani, A.M. (1981). Sex expression in cucumber plants cv. Bitalfa as affected by ethrel (CEPA) under heating green house. Uni. Bagdad, College of Agri. (Unpublished).

Anderegg, W. R., Anderegg, L. D., Sherman C and Karp DS. (2012). Effects of Widespread Drought-Induced Aspen Mortality of Understory Plants. *Conserv Biol.* 26**(6):1082**-90. doi: 10.1111/j.1523-1739.2012.01913.x

Anonymous (1994-95). Annual report of Plant Physiology and Biochemistry. Dr. PDKV, Akola.

Anonymous (1997). Annual report of Plant Physiology and Biochemistry. Dr. PDKV. Akola (MS).

Anonymous (2010). USDA/Agricultural Research Service. Plant Hormone Increases Cotton Yields in Drought Conditions. Science Daily, from *http://www.sciencedaily.com/releases/2010/03/100310142449.htm.*

Anonymous (2012). Center for the study of carbon dioxide and global change. Effects of atmospheric CO_2 enrichment on plant hormones. http://www.*co2science.org /subject/p/summaries/planthormones.php.*

Anonymous 2003 All India Coordinated Research Project for Dryland Agriculture. (2003). Guidelines on Drought Coping Plants for Rainfed Production Systems. *Central Research Institute for Dryland Agriculture*, Hyderabad.

Anuradha, PLakshmi and A. Narayanan (1995). Phosphorus nutrition in plants. In. Agros Ann. Rev. Plant Physiol. Vo III pp. 339-370.

Arora, S.K. and P.S. Partap (1988). Effect of plant growth regulators on vegetative growth, flowering and fruit yield of pumpkin. Haryana Agri. Uni. J. Res. 28 (4) : 282-290.

Asija, M.J. (2000). Ph.D. thesis in Botany. MDS. Uni. Ajmer, India.

Aslam, Muhammad., Ejaz, Ahmad., A.G. Saguu., Khalid Hussain., Muhammad Ayaz., Inayat Ullah., Amir Hussain and Himayatullah. (2010). Effect of plant growth regulator (NAA) and available soil moisture depletions on yield and yield components of chickpea. *Sarhad J. Agric.* Vol.**26, No.3,**

Atsman, D; and A. Lang and E.N. Light (1968). Contents and recovery of GA in monoecious cucum ber plants. Plant Physiol. 43 : 800-810.

Azadeh Afrigan1, Zahra Javdani2, Esfandiar Jahantab3, Ruhollah Jahanbin and Abdol Azim Bahari. (2013). The effect of plant hormone, gibberellic acid on germination indices *Secale montanum* in *vitro* and pot experiments under drought conditions. *Annals of Biological Research*, 4 **(6):1**-9.

Bangal, D.B. (1996). Studies on effect of growth regulators and urea on yield and yield attributes of pulses-Strategies for increasing pulses production in MS 27. MPKV, Rahuri, India.

Below, F.E.; R.J. Lambertz, R.H. Hageman (1984). Foliar application of nutrients on maize I. Yield and nitrogen content of grain and stover. Agro. J. 76 : 773-76.

Bendarrenko, G.L. and LYA Ledovski (1976). Doklay. Vses. Lenina. Akademic. No. 7 : 15-16 (Rulref)

Bhatt, J.G. (1978). Physiological losses of buds and bolls in cotton (Ed). ICAR. Pub. New Delhi

Bhatt, J.G. (1991). Auxin gradient and abscission of bolls in cotton. ISCI.J. 16 (1) : 36-38.

Biswas, A.K. and M.A. Chaudhari (1978). Regulation of leaf senescence in rice by hormone sprayed at different development stages and its effects on yield. Indian. J. Agri. Sci. 47 (1) :3.

Bondarda, B.R.; D.M. Oosterhuis; Norman, R.J. & W.H. Baker (1996). Canopy photosynthesis, growth, yield and boll N^{15} accumulation under nitrogen stress in cotton. Crop Sci. 36 (1) : 127-133.

Borlaug, N.E. (1997). Feeding a world of 10 billion people : the miracle ahead. Plant Tissue culture and Biotech. 3 : 119-127.

Bourely, J. and M. Brad (1989). Sulphur and cotton production quality. Tropicaleo. 44 (2) : 141-148.

Cengiz Kaya, A. Levent Tuna and Alfredo A. C. Alves (2006). Gibberellic acid improves water deficit tolerance in maize plants. *Acta Physiologiae Plantarum*, Vol. 28. No. **4. 331**-337.

Chima, S.S.; O.P. Malhotra and J.Singh (1975). Indian, J. Plant Physiol : 8 : 200-205.

Chinoy, J.J. (1977). Importance of coorelation in agro-physiology and physiogenetics. Indian J. Plant Physiol. 20 (1) : 1-36.

Chinoy, J.J. (1962). Formation and utilization of ascorbic acid in shoot apex of wheat as factors of growth and development. Indian J. Plant Physiol. 5 : 172-201.

Chopde, P.R. and A.M. Dhopte (1971). Effect of GA, IAA and MH on growth and yield of safflower. Agri. Coll. Mag. Parbhani. 10 : 60-68.

Chritopher, D.A. and J.B. Loy (1982). Influence of foliarly applied growth regulator on sex expression in watermelon. J. Amer. Soc. Hort. Sci. 107 (3) : 401 - 404.

Dashora, L.D. and P.M. Jain (1994). Effect of growth regulators and phosphorus levels on growth and yield of soybean. Madras Agri. J. 81 (5) : 235-137.

Delvadia, D.V; V.R. Chakarwar and V.G. Makne (1994). Effect of growth regulators on flowering and yield of chiku in rainy season. J. Maha. Agri. Uni. 19 (3) : 346-348.

Dhaval Patel and Keara A. Franklin. (2009). Temperature regulation of plant architecture. *Plant Signaling & Behavior* **4:7, 577**-579.

Dhopte, A.M. (1987). Influence of foliar application of macro and micro nutrients on leaf nutrition status and yield of hirsutum cotton in deep clay soils of Nagpur. Proc. Nat. Seminar on Micronutrients. May, 28-29, Banglore, India.

Dhopte, A.M. and A.T. Suradkar (1998). Physiological responses of soybean to foliar sprays of hormones and nutrients in relation to fertilizer economy. Ann. Plant Physiol. 12 (2) : 136-143.

Dhopte, A.M. and J.D. Eastin (1999). Influence of temperature stress on reproductive physiology of field crops. In : Agros Ann Rev. Plant Physiol. Vol. IV. Agrobios (India). Jodhpur.

Dhopte, A.M. and S.B. Lall (1987). Relative efficacy of antitranspirants, growth regulators and mineral nutrients in control of leaf reddening of horsutum cotton under dryland conditions. Ann. Plant Physiol. 1 (1) : 56-71.

Dhopte, A.M. and S.T. Wankhade (1988). J. Cotton Res. and Dev. 2 : 35-40.

Dhopte, A.M. and V.R. Kale (1976). Effect of gibberellic acid in growth and fibre yield of ramie (*Bohmeria nivea*, Gaud). Mah. J. Agri. Res. 1 : 18-20.

Dhopte, A.M. and V.R. Zade (1980). Combined effects of AA, CCC and PMA and nutrients on physiology of desi and American cottons. PKV. Res. J. 4 : 25-27.

Dhopte, A.M. and V.S. Hudge (1976). Effects of gibberellic acid on growth and yield of soybean. Oil seed J. 6 : 35-39.

Dhopte, A.M., S.L. Rahangdale and U.S. Satpute (1993). Role of leaf types in hormonal response and boll worm tolerance in isolins of hirsutum cotton. Ann. Plant Physiol. 7 : 270-173.

Dhopte, A.M.; V.R. Zade and B.N. Patil (1995). Effect of paclobutrazol spray on growth and yield of peanut and further anomalies in seed germination. Ann. Plant Physiol. 9 (2) : 99-100.

Dirk Gaul, Dietrich Hertel, Werner Borken, Egbert Matzner and Christoph Leuschner. (2008). Effects of experimental drought on the fine root system of mature Norway spruce. *Forest Ecology and Management* **256, 1151**–1159.

Diver, S.G.; M.W. Smith and R.W. McNew (1985). Foliar application of K_2SO_4, KNO_3, urea and NH45NP3 on pecan seedlings. Hort. Sci. 20 : 422-425.

Donato Chiatante , Antonino Di Iorio, Solena Sciandra, Gabriella Stefania Scippa and Stefano Mazzoleni. (2006). Effect of drought and fire on root development in *Quercus pubescens* Wild. and *Fraxinus ornus* L. seedlings. *Environmental and Experimental Botany* **56, 190**–197.

Durdon, J.H. and R. Sexton (1990). The role of ethylene in the shedding of red rasberry fruit. Ann. Bot. 66 (1) : 111-120.

El-Kholy, E. and H. Hafez (1982). Response of snake cucumber to some plant growth regulator. J. Agri. Sci. 99 (3) : 587-590.

Emily, B. Merewitz., Thomas Gianfagna and Bingru Huang. (2010). Effects of SAG12-ipt and HSP18.2-ipt expression on cytokinin production, root growth, and leaf senescence in creeping bentgrass exposed to drought stress. *J. AMER. SOC. HORT. SCI.* **135(3):**230–239.

Faizan Ullah, Asghari Bano and Asia Nosheen. (2012). Effects of Plant Growth Regulators on Growth and Oil quality of Canola (*Brassica Napus* L.) under Drought Stress. Pak. J. Bot., 44**(6): 1873**-1880.

Garcia, L.R. and J.J. Hanway (1976). Foliar fertilization of soybeans during seed filling period. Agron. J. 68 : 653-657.

Gehlot, H.S; S. Mali and S. Agarwal (1999). Stress proteins in plants. In. Agros Ann. Rev. Plant Physiol. Vol IV pp. 249-289. Purohit *et al.,* AgroBios (India) Pub., Jodhpur.

Goren, R.E. Dagan; O. Sogee; J. Rioxy and S.F. Yang (1993). Abscission in citrus leaf explant. : Role of ABA- induced ethylene. Acta. Hort. 329 : 43-50.

Gupta, Kamala., Abhijit Dey and Bhaskar Gupta. (2013). Plant polyamines in abiotic stress responses. *Acta Physiol Plant* **35:2015–2036**

Gupta, U.S. (1977). Foliar fertilization with macronutrients. In : Physiological aspects of dryland farming (Ed). U.S. Gupta. Oxford and IBH Pub. Co. New Delhi.

Haijun Gong., Kunming Chen., Guocang Chen., Suomin Wang and Chenglie Zhang. (2013). Effects of Silicon on Growth of Wheat Under Drought. *Journal of Plant Nutrition* Vol. 26, No. **5, 1055**–1063.

Heslop-Harrison, J. (1957). The experimental modification of sex expression in flowering plants. Biol. Rev. 32 : 38-39.

Hidayatullah, A. Bano and K. M. Khokhar. (2009). Sex Expression and level of phytohormones in monoecious cucumber as affected by plant growth regulators. *Sarhad J. Agric.* Vol.**25, No.2**.

Hidayatullah, T. Mahmood, M. Farooq, M. A. Khokhar and S.I. Hussain. (2012). Plant Growth Regulators Affecting Sex Expression of Bottle Gourd (*Lagenaria Siceraria Molina*) *Plants. Pakistan J. Agric*. Res. Vol. **25 No. 1**.

Hofmann, W.C. and P.T. Else (1989). The response of cotton to foliar spray application of a cytokinin containing compounds. Appl. Agri. Res. 4 : 25-29.

Huang, H.B. (1989). The water relation problem of fruit trees in Guangdong. J. South China Agri. Uni. 10 (1) : 33-39.

Iwahori, S; J.M. Lyons and O.E. Smith (1970). Sex expression in cucumber plants as affected by CEPA, Ethelene and growth regulators. Plant Physiol. 46 : 412-415.

Iwahori, S; J.M. Lyons and W.L. Sims (1969). Induced femaleness in cucumber by CEPA. Nature 222 : 271-272.

Jat, B.L. and V. Nepalia (1994). Effect of nitrogen and P application on NPK content of soybean plants at different stages. Indian Agri. 38 (2) : 410-411.

Jayalalita, K. and A. Narayanan. (1996). Growth and mineral composition of magnesium deficient cotton plants grown in solution culture. Ann. Plant Physiol. 10 (1) : 11-16.

Jian Feng Ma. (2011). Role of silicon in enhancing the resistance of plants to biotic and abiotic stresses. *Soil Science and Plant Nutrition*, 50:**1, 11**-18.

Kadrekar, S.B. and M.M. Kibe (1974). Res. J. Maha. Phule Agri. Uni. Rahuri (MS) 5 : 84-88.

Kakumanu, Akshay., Madana, M. R., Ambavaram, Curtis Klumas., Arjun Krishnan, Utlwang Batlang., Elijah Myers, Ruth Grene and Andy Pereira. (2012). Effects of Drought on Gene Expression in Maize meproductive and leaf meristem tissue revealed by RNA-Seq. *The American Society of Plant Biologists, Plant Physiology* Preview, **DOI:10.1104/pp.112.200444.**

Kalita, K; S.C. Dey and K. Chandra (1995). Influence of foliar application of P and NAA on nitrogen, dry matter accumulation and yield of greengram. Indian J. Plant Physiol. 38 (3) : 197- 214.

Kannan, S. (1986). Foliar absorption and transport of inorganic nutrients. CRC. Crit. Rev. Plant Sci. 4 : 341-375.

Kannan, S. (1990). Role of foliar fertilization on plant nutrition. In. Crops as enhancers of nutrient use. Acad. Press. Inc. New York.

Keogh, J.L. and R. Maple (1969). Bull. Ark. Agri. Rxpt. Sta. 740 : 19.

Khandagale, G.B; A.P. Suryawanshi; P.R. More and D.V. Wadkar (1992). Effect of cycocel, NAA and other nutrients on growth and yield of cotton under rainfed conditions. Ann. Plant Physiol. 6 (2) : 202-206.

Kocsy, G. (1999). Role of glutathione in plants under abiotic stress conditions. In. Agros Ann. Rev. Plant Physiol. Purohit *et al.,* Vol 4. pp. 215-225.

Kumar, D; K.D. Gujar and K. Singh (1996). Effect of GA and NAA on growth, chlorophyll content and maturity of cabbage. Crop Res. 12 (1) : 116-119.

Kumar, V; D.K. Mishra and P.C. Raheja (1967). Sci. & Cul. 33 (6) : 291-292.

Lakshamma, P. (1988). M.Sc. (Agri) Thesis. AP. Agri. Uni. Hyderabad, India.

Levitt, J. (1980). Responses of plants to environmental stresses (Ed. 2 nd). Vol. II Acad. Press Inc. New York.

Li, Y,C and G. Gupta (1993). Photosynthetic changes in soybean with and without nitrogen and increased CO_2. Plant Sci. 89 : 1-4.

Lower, R.L.; C.H. Miller; F.H. Baker and C.C. McCombs (1970). Effects of 2-chloroethyl phosphonic acid treatment at various stages of cucumber development. Hort. Sci. 5 : 433-434.

Lubica, Haluskova., Katarına Valentovicova., Jana Huttova , Igor Mistrık and Ladislav Tamas. (2009). Effect of abiotic stresses on glutathione peroxidase and glutathione S-transferase activity in barley root tips. *Plant Physiology and Biochemistry* **47, 1069**–1074.

M. Jonas, P. Salas and T. Baltazar. (2012). Effect of exogenously application Selected phytohormonal substances on the physiological and morphological indicators of philadelphus X hybrid in containers. Acta Universitatis Agriculturae Et Silviculturae Mendelianae Brunensis, **Volume LX**.

Madghe, P.W. (1996). Effect of growth regulators and nutrients on boll abscission and yield of cotton (*G. hirsutum* L.) M.Sc. Agri. Thesis, Dr. PDKV, Akola (MS), India.

Malik, C.P.; R.C. Setia and N. Setia (1993). Role of Plant growth regulators in increasing crop productivity and the Indian scenario. In : Hormonal Regulation of Plant Growth and Deve. (Ed). Purohit, S.S. AgroBotanical Pub. Bikaner pp. 161-197.

Maske, V.G. (1997). Physiological responses of soybean to growth regulators. M.Sc. (Agri) Thesis, Dr. PDKV, Akola (MS).

Masri, Al M. I. and R. Barakat. (2003). Role of the auxin, Naphthalene Acetic Acid (NAA) in the pathogenicity of *Sclerotinia sclerotiorum*, causative organism of white mold disease. *Hebron University Research Journal*, Volume **1/ No. 2.**

Mirwais, M., Qaderi, Leonid V. Kurepin and David M. Reid. (2012). Effects of temperature and watering regime on growth, gas exchange and abscisic acid content of canola (*Brassica napus*) seedlings. *Environmental and Experimental Botany* **75, 107**– 113.

Nageswarrao, P. (1976). Cotton Development. 6 (2) :11-12.

Open, G. and R. Marcele (1975). Lab. Plant Physiol. Belgium. 211- 216.

Osborne, D.J. (1962). Plant Physiol. 37 : 595-602.

Osborne, D.J. and S.E. Moss (1963). Nature 200 : 1299

Pain, S.K. and Sarkar (1986). Studies on growth and development of sunflower Plant cv. 68414. Effect of foliar spray of NAA at different concentrations. Indian. Agri. 24 (3) : 135-143.

Pandey, D.M., C.L. Goswami and B. Kumar (2003). Physiological effects of plant hormones in cotton under drought. *Biologia Plantarum* **47 (4):** 535-540

Pandey, S.N. (1975). Effect of planofix (NAA) on flower abscission and productivity of arhar (*Cajanus cajan*) and soybean (*Glycine max*). Indian Sci. Abst. 9(9) : 42-44.

Pareek, A; S.L. Singh and I. Grover (1995). Plant Mol. Biol. 29 : 293-301.

Patil, V.Z.; V.P. Keldar and P.K. Nagare (1989). Effect of foliar sprays of PGR and urea on premature and preharvest fruit drop in Kinno mandarine. PKV Res. J. 13 (2) : 161-163.

Patra, A.K.; S.K. Tripathi and R.C. Samu (1995). Effect of post flowering foliar application of nutrients on growth, yield and economics of rainfed groundnut. Indian. J. Plant Physiol. 38 (3) : 203-206.

Payam Fayyaz, Elham Etemadi, Nasim Julaiee-Manesh and Roghayeh Zolfaghari. (2012). Sodium and potassium allocation under drought stress in *Atlas mastic* tree (*Pistacia atlantica subsp. mutica*). *iForest – Biogeosciences and Forestry*, doi: **10.3832/ifor0856-006**.

Pelah, Dan (1999). Abiotic stress response in plants : Basic and Applied studies. In. Agros Ann. Rev. Plant Physiol. (Ed). Purohit *et al.,* AgroBios (India). Jodhpur.

Petr Hlavinka , Miroslav Trnka, Daniela Semera dova, Martin Dubrovsky, Zdeneˇk Zˇ alud and Martin Mozˇny.(2009). Effect of drought on yield variability of key crops in Czech Republic. *Agricultural and forest meteorology* **149, 431** – 442.

Prasad, P. V. V., S. A. Staggenborg and Z. Ristic (2008). Impacts of drought and/or heat stress on physiological, developmental, growth, and yield processes of crop plants. Response of crops to limited water: Understanding and modeling water stress effects on plant growth processes. *Advances in Agricultural Systems* Modeling Series 1.

Prasad, T.K; M.D. Anderson; B.A. Martin & C.R. Stewart (1994). Evidence for chilling-induced oxidative stress in maize seedlings and regulatory role of hydrogen peroxide. Plant Cell 6 : 65-74.

Purohit, S.S. (1993). Hormonal regulation of plant growth and development. (Ed). AgroBotanical Publ. (India), Bikaner. PP 175-178.

Ravikumar, G.H. and G.N. Kulkarni (1988). Effect of growth regulators on seed quality of soybean genotypes. Seeds and Farms. 14 (2) : 25-28.

Rawson, H.M. (1988). High temperature is not a stress. Proc. Int. Cong. Plant Physiol. New Delhi Vol. I. pp. 923-928.

Reddy, S.C.V. (1978). Growth and yield of groundnut varieties in relation to application of NAA Mysore. J. Agri. Sci. (Thesis); UAS, Bangalore

Robert E. Shapr and Mary E. LeNoble. (2002). ABA, ethylene and the control of shoot and root growth under water stress. *Journal of Experimental Biology*, Vol.53, No.**366, 36-37**

Rosa M. Rivero, Vladimir Shulaev, and Eduardo Blumwald. (2009). Cytokinin-dependent photorespiration and the protection of photosynthesis during water deficit. *Plant Physiology*, Vol. **150, 1530**–1540.

Rudich, J; A.H. Halevy and N. Kedar (1969). Increase in femaleness of three cucurbits by treatment with ethrel, an ethylene releasing compound. Planta. 86 : 69-76.

Sagare, B.N. and K.T. Naphade (1987). Effect of hormones on yield, economics and nutrient uptake by groundnut. PKV Rev. J. 11 (1) : 19-22.

Sahay, R.K. and R. Pandarikakshudu (1987). Growth regulating properties of synthetic pyrethroides. Ann. Plant Physiol. 1 (2) : 136 - 142.

Samar Raza1., M. A., M. Farrukh Saleem., G. Mustafa Shah., M. Jamil1 and I. Haider Khan. (2013). Potassium applied under drought improves physiological and nutrient uptake performances of wheat (*Triticum Aestivum* L.). Journal of Soil Science and Plant Nutrition, 13**(1), 175**-185.

Saxena, O.P. and G. Singh (1993). Hormonal regulation of dwarfism. In. Hormonal Regulation of plant growth and development (Ed). Purohit. Vol. VI pp. 147-160. Agro Botanical Pub., Bikaner, India.

Sergi Munne Bosch., Josep Pen˜uelas., Dolores Asensio and Joan Llusia (2004). Airborne ethylene may alter antioxidant protection and reduce tolerance of holm oak to heat and drought Stress. *Plant Physiology*, Vol. **136, 2937**–2947.

Setia, N; R.C. Sangeeta; R. Setia and C.P. Malik (1993). Alterations in growth and yield components of lentil in response to foliar application of NAA, Indian, J. Plant Physiol. 36 (1) : 47-52.

Setia, R.C; N. Setia and C.P. Malik (1991). Plant growth regulators overview and role in crop physiology In : Recent advances in plant biology (Ed). C.P. Malik and Y.P. Yabrol. Narendra Pub. House, New Delhi pp. 47-77.

Shapiro, C.A; T.A. Peterson and A.D. Flowerdax (1987). Irrigation method, composted manure, variety and foliar fertilizer effect on soybean. J. Fert. Issues. 4 : 122-129.

Sharma, C. (1982). Effect of hormonal treatments during seed development on the vigour of subsequently formed seeds of *Pisum sativum* and *Vicia faba*. Indian J. Plant Physiol. 24 (4) : 377- 381.

Sharma, S.C. and S.C. Dey (1986). Response of soybean to foliar application of some growth regulators and in combination of urea and potash. Soybean Gene. News Letter 13 : 71-74.

Shinde, A.K. and B.B. Jadhv (1995). Influence of NAA, Ethrel and KNO_3 on leaf physiology and yield of soybean. Ann. Plant Physiol. 9 (1) : 43-46.

Shinde, A.K.; B.M. Jamdagni and S.P. Birari (1991). Effect of foliar spray of growth regulators and KNO3 on growth and yield of cowpea. cv. VCM-8. Indian J. Plant Physiol 34 (4) : 392-395.275

Shuo Hao Huang., JianYun Zhang, Ling Hong Wang., Long Quan Huang. (2013). Effect of abiotic stress on the abundance of different vitamin B6 vitamins in tobacco plants. *Plant Physiology and Biochemistry* **66,** 63-67

Singh, R.P. and A.V. Kumar (1991). Effect of phytohormones on yield attributes and seed yield of mustard. Indian J. Agron. 36 (3) : 379-381.

Sinha, S.K. (1986). Drought resistance in crop plants. A physiological and Biochemical analysis. In. Approaches for incorporating drought resistance in crop plants (Ed) V.L. Chopra and R.S. Paroda. pp. 56-86. Oxford and IBH Pub. Co. Pvt. Ltd. New Delhi.

Soliappan, V. and S. Ramiah (1990). Effect of seed treatment, soil and foliar fertilization on N and P and yield and yield attributes of pigeonpea growth under rainfed conditions. Indian. J. Agron. 35 (3) : 234-237.

Subbian, P and A. Chamy (1984). Effect of zinc on growth of soybean. Indian J. Plant physiol. 33 (3) : 239.

Sundaram Kuppu., Neelam Mishra., Rongbin Hu., Li Sun., Xunlu Zhu., Guoxin Shen., eduardo blumwald., paxton payton and hong zhang. (2013). Water-deficit inducible expression of a cytokinin biosynthetic gene IPT improves drought tolerance in cotton. *PLoS ONE 8(5): e64190. doi:10.1371/journal*.**pone.0064190**.

Swamy, N.R. (1991). Assesment of boll shedding and its control in cotton under rainfed conditions. Ann. Plant Physiol. 3 (2) : 266- 269.

Syaffruddin, M.R.; Saenong and Djamaluddin (1991). Response of soybeans to P and Zn application in calcareous alluvial soils. Bull. Penelition Pertanion Maros. 6 (1) : 1-6.

Syvrud, T.D.; L.D. Walsh; F.S. Oplinger and K.A. Kelling (1980). Foliar fertilization of soybean (*G. max*). Comm. Soil Sci Plant Analysis. 11 : 637-651.

Tamoi, M. (1993). Phosphorus and zinc interaction in shoot of cotton plant. Angewandle Botanik. 67 : 138-140.

Thaloot, A.T.; M.A. Elseessy. (1989). Response of chemical composition and yield of soybean to foliar spray of zinc and different phosphatic fertilizers. Ann. Agri. Sci. (Cairo) 34 (2) : 925-937.

Tomas Werner., Erika Nehnevajova., Ireen Kollmer., Ondrej Novak., Miroslav Strnad., Ute Kramer and Thomas Schmulling. (2010). Root-specific reduction of cytokinin causes enhanced root growth, drought tolerance, and leaf mineral enrichment in arabidopsis and tobacco. *The Plant Cell,* Vol. **22: 3905**–3920.

Tonzig, S. and E. Maree (1961). Ascorbic acid as a growth hormone in plant growth regulation. Proc. Int. Conf.Plant Growth regulation. 725-734.

Toshihiro Obata and Alisdair R. Fernie. (2012). The use of metabolomics to dissect plant responses to abiotic stresses. *Cell. Mol. Life Sci.* **69:3225**–3243

Turner, A.D. and H.C. Wien (1994). Dry matter assimilation and partitioning in pepper cultivars differing in susceptibility to stress-induced buds and flower abscission. Ann. Bot. 73 (6) : 617-622.

Varma, S.K. (1978). Effect of NAA and ascorbic acid on flower bud and boll abscission in cotton. Indian J. Plant Physiol. 21(3) : 235-241.

Varma, S.K. (1990). Ascorbic acid contents in cotton bolls and its possible role in abscission. Ann. Rev. Plant Physiol. pp. 55-57.

Venangamudi, K; D. Kalavathi and K. Ramamoorthy (1988). Effect of gibberellic acid on dormancy, flowering, bulb and seed production in Bellary onion. Indian J. Plant Physiol. 31 (2) : 190-192.

Walter, S.S. (1970). Effect of CEPA and GA on sex expression and growth of pumpkins. Physiol. Planta. 23 : 762-768.

Zhang, Jianhua., Wensuo Jia., Jianchang Yang and Abdelbagi M. Ismail. (2006). Role of ABA in integrating plant responses to drought and salt stresses. Field Crops Research **97,** 111–119.

Zlatko Zlatev1 and Fernando Cebola Lidon. (2012). An overview on drought induced changes in plant growth, water relations and photosynthesis. *Emir. J. Food Agric.* 2012. 24 **(1): 57**-72.

Chapter - 13

AGRO-TECHNIQUES FOR SUSTAINED PRODUCTION IN DRY LAND AREAS

M.N. Sheelavantar and B.M. Chittapur

Department of Agronomy, University of Agricultural Sciences, Dharwad - 580 005 (Karnataka)

1. SCOPE OF DRY LAND AGRICULTURE

Dry land agriculture or dry farming simply means raising of crops with rainwater inadequate for crop production. World wide, 6150 million ha of land area is reported to be under rainfed situation. In India, 67 per cent of 143 million ha arable land depends on rain, contributing 44 per cent to national food basket and supports 40 per cent of the population. Further, it is estimated that 450 million cattle heads are thriving on rainfed farming. Almost 80 per cent of maize, 95 per cent of pearl millet and sorghum, about 75 per cent of pulses, 75 per cent of oil seeds, nearly 70 per cent of cotton (fibre) and almost whole of jute and mestha and to a little extent rice are under rainfed situation. Productivity of crops in dry lands is low due to inadequate supply of moisture and other inherent constraints. Since production gains from irrigated lands can not meet the growing demands from the ever-increasing population, emphasis needs to be paid for dry land agriculture. It is estimated that to meet the projected targets of 260 million tonnes for 2020 AD for a projected population of 1400 million the productivity of dry lands has to be raised from the current level of less than 0.8 t per ha to 2.0 t per ha as 50 per cent of land in the country will continue to depend on rains even after harnessing all the irrigation potential (113 million ha). Available technological information, nevertheless suggests the tremendous possibility of increasing productivity of dry lands. It requires concerted efforts to modify and adopt technologies to local situation with a great urgency. Increasing production and sustaining it calls for the successful management of resources while maintaining and enhancing the quality of environment and conserving natural resources.

2. CONSTRAINTS

Dry land agriculture suffers from twin problems of low productivity and sustainability. The basic constraint is low annual rainfall characterised by erratic distribution from season to season (with 30-50% cv.). Coupled with high solar incidence (18.9-21.0 $MJ/m^2/day$) and high wind velocity (20 km/h) these areas suffer from high potential evopotranspiration (PET 6 mm/day) and exhibit high mean aridity index. The peninsular India South of 15°N gets less than 500 mm rainfall (average annual rainfall of the country is 1170 mm). The lowest rainfall received in the country during the season is in the extreme south-east peninsula (Kanniyakumari and adjoining areas). In the western and northwestern India, the 500 mm isoline covers the region from Saurashtra and Kutch to Delhi and then to Punjab, Haryana and Himachal Pradesh. The rainfall decreases rapidly to less than 100 mm in western Rajasthan. As the mean annual rainfall of a region decreases the occurrence of rainfall events decreases (<30 rainy days in a rainy period of 120-140 days) leading to greater intensity of risk and a greater probability of drought occurrence. The probability of occurrence of drought in a low rainfall region, say less than 400 mm annually may be 2 in every five years as against one in 15 years in areas with a rainfall of about 1000 mm. Global warming with a estimated temperature rise of 1.5 to 4.0 by2020 AD may further aggravate the situation.

Secondly, lands often being unprotected are exposed to severe erosion. It is estimated that 5000-6000 million tonnes of soils are lost annually due to erosion in India. The less of nutrients is estimated to be in the order of 2.5 million tones of nitrogen, 3.3 million tonnes of phosphorus and 2.6 million tonnes of potash annually in India. About 60 per cent of cultivated land suffers from soil erosion, water logging and salinity problems. Consequently lands are getting degraded and progressively become unfit for crop production. Wind blown soil with high sand content (850 g/kg soil) and low organic matter (1.0 to 4.5 g/kg soil) retain less moisture (50-280 g/kg) and have very high infiltration rate (>90 mm/hr). Contrastingly medium to deep black soils in the central and south India suffer from low infiltration rate (<8 mm/hr), salinity and alkalinity problems, multiple nutrient deficiencies and certain elemental toxicity (B) in many instances. Red soils and acid soils have the tendency to form surface crust after sowing followed by light showers, which impair seedling emergence. In general, soils are poor in fertility and critically deficient in nitrogen, phosphorus and potassium (red soils) and micronutrients viz., Fe, Zn and S. Insufficient carry over of residual moisture from season to season, scarce ground water resources and deep water table (40-50 m) coupled with brackish ground water EC 1.25 to 10.0 dS/m) at many places aggravate moisture supply situation.

Further, the limited choice of crops, low cropping intensity, low yields of conventional crops and varied socio-economic situation also account for low productivity in rainfed areas. With changing environmental situation (as feared due to global warming) crop production in dry land areas remains to be more challenging unless it is practiced on scientific lines while keeping in mind conservation, development and optimum use of resources.

3. SOIL AND WATER CONSERVATION

The vary nature and characteristic of soils particularly those of black soils combined with erratic and intense rains causes severe runoff and soil erosion and need intensive soil and moisture conservation measures for sustained crop production under rainfed agriculture. Estimates suggest a runoff of about 10-40 per cent of annual rainfall in a black soil watershed and a erosion of 12 to 43 tonnes per ha per annum. Unless the erosion is checked and rainfall is conserved, the cultivation is no longer going to be sustainable. A three-tier approach for effective soil and water conservation involving, (a) terrace level practices, (b)inter-terrace land management, and (c) runoff harvesting and recycling, therefore, has to be considered.

3.1 Terrace level practices

Different terrace level practices vis-a-vis the mechanical structures are in practice in different parts of the country. The suitability and adoption of these structures mainly depend on rainfall, soil type, its depth, topographic features, socioeconomic conditions and the extent of soil erosion.

3.1.1 Bunds

Bunds are small earthen barrier either of impounding type (contour bunds) or of drainage type (graded bunds) executed on engineering principles on agricultural lands with slopes ranging from 1 to 6.0 per cent. These structures control effective length of slope and thereby reduce the gain in velocity of runoff to avoid rill and gully formation. While contour bunds are constructed in a relatively low rainfall areas having annual rainfall of less than 600 mm particularly in light textured soils or on soils with moderate to high infiltration, the graded bunds are constructed in medium to high rainfall areas having annual rainfall exceeding 600 mm and in soil with poor permeability (mm per hour).

Research/demonstrations conducted under NWDPRA at UAS, Dharwad revealed that contour bunds of 1.0 to 1.2 m^2 cross section at a vertical interval of 1.0 to 1.5 m with deviations 15 cm on the ridge and 30 cm in the valley in shallow black soils are found suitable (Table 1 and 2). On such bunds land shaping towards the upstream side of the bund is advised to correct the bund to contour at the deviated points. Waste weirs are to be located with crest level 15 cm above the contour. Graded bunds with 0.2 per cent grade and cross section of 0.5 to 0.75 m^2 at a vertical interval of 1.0 to 1.5 m (depending on land slope) with strip levelling towards upstream side of the bund and a protected water way are found effective (Table 1 and 2).

Table 1. Yields of crops at different locations as influenced by different terrace level practices

Structure	Crop	Yield (q/ha)		
		Treated	Traditional (control)	Per cent increase in the yield over control
A. Contour bunds				
Bijapur	Pearl millet	4.10	2.63	55.9
Gulbarga	Sunflower	10.45	9.63	8.5
Hanumanmatti	Groundnut	10.20	8.30	22.9
	Rabi sorghum	5.00	4.50	13.6
B. Graded bunds				
Bidar	Blackgram	11.50	10.25	12.1
Gulbarga	Sunflower	8.75	7.50	16.7
Hanumanmatti	Dry chilli	4.50	3.90	15.4
C. Zingg conservation				
Bidar	Blackgram	10.50	9.50	10.5
Gulbarga	Sunflower	15.20	12.50	21.6
Raichur	*Rabi* sorghum	5.00	3.50	42.8
	Chickpea	13.37	8.75	52.8
Bijapur	Wheat	4.30	3.00	43.3
D. Contour border strips				
Bidar	Blackgram	11.25	9.50	18.4
Bijapur	Sunflower	3.75	2.70	38.0

	Rabi sorghum	5.00	3.53	41.6
E. Graded border strips				
Raichur	*Rabi* sorghum	8.00	3.50	12.8
	Chickpea	11.75	8.75	34.3
Gulbarga	Sunflower	13.80	10.80	27.8
Bidar	Blackgram	12.25	10.25	21.9
Bijapur	Chickpea	2.16	1.14	89.5
	Sunflower	1.61	1.25	28.8

Source : Itnal *et al.*, (1998)

Table 2. Evaluation of soil and water conservation structures in medium to deep black soils of Bijapur (Av. of 5 years)

Structure	Run-off (% of rainfall)	Soil loss (t/ha/yr)	Grain yield		Per cent increase over control	
			Rabi sorghum	Safflower	*Rabi* sorghum	Safflower
Control bunds	5.2	0.74	9.08	4.89	17.4	11.6
Graded bunds	10.8	1.43	9.64	5.06	24.7	15.5
Broad based terrace	6.5	0.82	10.00	5.01	30.4	14.4
Graded border strips (0.2% grade)	7.2	0.91	11.00	5.70	42.3	30.1
Contour border strips	5.2	0.72	11.79	6.24	52.5	42.5
Zingg conservation terrace (1:3)	5.1	0.76	10.75	5.74	39.1	31.0
Conservation ditch	13.3	14.32*	9.50	4.95	39.1	13.0
Control	15.1	3.05	7.73	4.38	–	–
Rainfall (mm) during the evaluation period						
Year	1983	1984	1985	1986	1987	
Rainfall	529.8	516.5	484.1	444.8	639.8	

* Silting, Belgaumi (1998)

3.1.2 *Zingg conservation terrace*

Zingg conservation terracing is adopted for undertaking cropping programme in areas having rolling topography and on soils relatively heavy in nature ranging from sandy loam to clay loam with moderate depths. In medium deep soils having infiltration rate of more than 8 mm per hour, Zingg conservation terrace of 1:3 levelled to unlevelled portion with a ridge section of 1.2 to 1.5 m^2 at a vertical interval of 1.0 to 1.5 m have been found suitable (Table 1 and 2).

3.1.3 *Contour border strips*

Contour border strips are to be laid out in medium to deep black soils having infiltration rate more than 8 mm per hour. The strips are constructed along the contour key lines at a vertical interval of 0.3 m with a suitable drop structure at the end of each strip in the water way with ridge section of 0.24 m^2 and stabilization of vegetation (Table 1 and 2).

3.1.4 *Graded border strips*

Graded border strips are laid out in medium to deep black soils of more than 45 cm depth with permeability of less than 8 mm per hour in areas of medium to low rainfall. Graded border strips of 0.2 per cent grade laid out at a vertical interval of 0.3 m with a ridge section of 0.24 m^2 and stabilized through vegetation and by providing suitable drop structures at the end of each strip on the water way are also found suitable in different places.

3.2 Inter terrace management practices

3.2.1 Tillage practices

Cultivation of soil helps to increase pore space for storage and also keeps the soil loose so that high rate of infiltration is maintained toward of runoff. Musgrave and Free (1936) found that cultivation of the surface greatly enhanced water intake of soil particularly in the beginning of storms. In the absence of cultivation, the highly crusting red soils produce as much or even more runoff than the low permeability Vertisols under similar rainfall situations. Larson (1962) stated that pulling a tillage implement through soil results in the total porosity and thickness of the tilled area being greatly increased temporarily. Surface roughness and microdepressions, therefore, play great role in higher retention of water (Unger and Stewart, 1983).

Off season tillage or early tillage after the harvest of standing crop keeps the land in rugged condition and helps to avoid the runoff apart from having other benefits such as reduction in pests and weeds, burial of previous crop residues and reduction in lapse of time between first rain and actual seeding time. Beneficial effects are much pronounced during the year of low rainfall (Table 3).

Table 3. Effect of off-season tillage on yield of sorghum in red soils of Hyderabad

System	Yield (q/ha)			
	1977	1978	1979	Mean
No off season tillage	19.50	9.34	10.52	13.12
Off season tillage	24.30	13.36	19.65	19.10
Percent increase due to off season tillage	24	43	31	46
Rainfall in growing season (mm)	595	391	508	

Source : Sanghi and Korwar (1987)

Table 4. Runoff and soil loss as influenced by tillage

Treatment	Area	Runoff		Soil loss t/ha	Peak flow cum/sec
		mm	% of rainfall		
Minimum tractor tillage	0.20	135.3	21.4	3.08	0.0240
Deep tillage	0.24	44.7	7.1	1.18	0.0094
Farmers method	0.22	173.0	27.4	4.74	0.0529

CRIDA (1993)

Shallow tillage soon after rain is beneficial for retaining water in the soil profile particularly in the fine textured soils. At Dry Farming Centre, Bijapur on medium to deep black soil, deep tillage reported to increase soil moisture content and infiltration rate and decrease bulk density and consequently improved grain yield of sorghum (1877 kg/ha) by 22 and 45 per cent over medium and shallow tillage, respectively (Patil, 1988). On coarse textured soils deep tillage could be more desirable. Deep ploughing with inversion plough besides creating surface inequalities forms an infinite number of small depressions between/among the clods as storage spaces for impounding and later leisurely absorbing rain water received from intense summer storms. Due to increase opportunity time for infiltration and reduced peak runoff rate, runoff and soil loss are low (Table 4 and 5).

Table 5. Penetration of moisture as influenced by ploughing before the onset of rains

Depth of soil (cm)	Per cent of moisture after a total of 81 mm rain in May	
	Ploughed area	Unploughed area
0-15	10.74	3.59
15-30	13.22	7.13
30-60	13.27	8.59
60-90	13.44	Dry

Dry Land Agriculture Project, Bangalore (Anon, 1983)

3.2.2 *Contour farming*

Contour farming or contouring involves application of all tillage operations including planting on the contour. Since each ridge acts as a barrier to the flow of water, contouring ensures less runoff and erosion. The effectiveness of contour farming in erosion control varies with slope, crop cover and soil and rainfall characteristics. Maximum effectiveness is on medium slopes and deep permeable soils. Effectiveness decreases as the land slope becomes vary flat or step (Smith and Wishmeier, 1962). On 4 to 6 per cent slopes with proper rotation erosion is controlled upto a length of 100 meters. The maximum length on which contouring is effective is 60 m on a 8 per cent slope, 30 m on a 10 per cent slope, 25 m on 12 per cent slope land and 20 m on steeper slope. In India, contour cultivation has been effective on all major soil groups with an yield increase of 10 per cent and a reduction in soil loss by 50 per cent on 2-4 per cent slope.

For laying out contours, generally bunds are used as guidelines. On gentle slopes (0.5-2.0%) where bunding is not essential, contour guidelines need be marked in the field. On uniform slopes such guidelines marked at 50 m interval will be sufficient. Mark the first line at 1.5 m vertical interval down the slope from the top of the hill. First line is generally laid out with 0.5 per cent slope towards water ways. A deviation not more than 3 per cent from the contour is allowed. If deviations exceed this limit, mark a new line 1.5 m vertically down the hill.

3.2.3 *Rubble checks*

Arranging the rubbles on representative contour key line at a vertical interval of 0.3 m forms rubble checks. These can be constructed in shallow, medium or deep black soils receiving medium to high rainfall and in areas where rubbles are available in plenty. Bunds section of 0.75 m^2, base width of 0.5 m and height of 0.3 m were found to act as very good inter terrace land management practice. Rubble checks in addition to reducing the soil erosion, increase the crop yields appreciably over other vegetative barriers (Table 6).

3.2.4 *Furrowing*

Plough furrows opened at frequent distances intercept runoff and as a result more and more rain water is conserved in between furrows. Furrows may be at a distance ranging between 0.45 and 10.0 m. In rolling lands the spacing may vary from 2.5 m or 4 to 7 crop rows. The furrows should be laid out on contour or across the major slope. The depth of furrows could be 20-30 cm. Increase in crop yields due to furrowing have been reported from Dry Farming Centre, Bangalore (red soils) and Bijapur (black soils). Furrowing needs to be done prior to sowing of crops. The furrows could be tied at certain distances by earthen checks constructed at 3-5 m interval (Furrow dyking). The series of earthen dams hold rain water *in situ* and provide more opportunity time for it to infiltrate into the soil. On alfisols, the tied ridges reported to reduce runoff by 39 per cent and solid loss by 28 per cent (Pathak and Laryea, 1995). Furrow damming is a proven and acceptable conservation practice used in Great Plains in the USA for row crops. At Dry Farming Centre, Bijapur ridges and furrows during monsoon resulted in an yield increase by 26 per cent in *rabi* sorghum due to increased soil moisture (Patil 1998).

3.2.5 *Broad bed and furrows*

The International Crop Research Institute for Semi-arid Tropics, Hyderabad has developed technology of broad bed and furrow System (BBF). Here the land is converted into series of alternate beds and furrows. Beds are used for cultivation and furrows for conservation as well as safe disposal of excess water (runoff). Furrow (0.15 m) and bed (0.9 to 1.5 m wide depending on row spacing of crops and number of rows to be sown in each bed) width and length depends on peak rate of runoff and infiltration rate of soil. The beds are laid out on a grade (0.2 to 0.4%). The furrow holds rain water in years of low rainfall while the same furrow serves as a disposal system in years of heavy rainfall. A pre-requisite appears to be necessary for opening broad beds and furrows. The land needs to be graded before laying of BBF. This system is most suitable for high rainfall vertisols areas with kharif cropping. In very clayey soils the BBF reduces hazards of waterlogging.

3.2.6 *Broad furrows and ridges*

At Dry Farming Centre, Bijapur this system has been proved effective on medium to deep black soils having a slope of 2 per cent. Here the land is laid out into broad furrows and ridges across the slopes. Furrows hold more moisture and a more moisture requiring crop may be grown in furrows. Ridges may be uncropped or planted with tall components like pigeonpea. In heavy rainfall areas in upland situation such a system helps to grow paddy in the furrows and arable cotton on the ridges to make best use of available rains.

3.2.7 *Scooping*

Scooping is a low cost *in situ* conservation measure. The scoops or small depressions created with the help of a wooden hoe with plank attached to prongs or basin lister (scooper) hold water and allow it infiltrate. This measure is best suited for deep black soils (vertisols) where post-rainy season cropping on stored soils moisture is predominant. Yield increases by 19 to 20 per cent due to scoops have been reported from Dry Farming Centre, Hagari (Bellary). As compared to flat beds scoops found to reduce runoff and solid loss by 53 per cent on alfisols (Pathak and Laryea, 1995).

With increased infiltration as a result of reduction in runoff scoops (40 mm storage capacity) resulted in a significant yield increase in pearl millet (2.42 t/ha) over flat seed bed (1.79 t/ha).

3.2.8 *Compartment bunding*

In lands sloping upto two per cent, compartment bunding is a best low costing *in situ* water conservation measure. This is recommended for medium and deep black soils where *rabi* cropping is taken up. Earlier investigations at Bijapur Dry Farming Centre, suggested Compartment bunds of 10 m x 10 m formed with the help of bund former are ideal. However, the refinement of the earlier recommendations reveals that compartment bunds could be practiced even on steeper slopes. In lands sloping between 1 to 2 per cent, compartment bunds of 6.0 m x 6.0 m could be formed, while the size of the compartment may be 4.5 m x 4.5 m for lands sloping between 2 to 3 per cent, and the same way be 3.0 m x 3.0 m for areas having slope of more than 3 per cent. About 24 to 56 per cent increase in grain yield of *rabi* sorghum was obtained by forming compartments of 4.5 m x 4.5 x and 3 m x 3 m, respectively in black soils of Dry Farming Centre, Bijapur over no compartment bunding (Itnal *et al.,* 1998). The data further reveal that runoff could be reduced by 40-50 per cent besides checking the soil erosion to a greater extent. Patil (1998) observed a yield increase by 23 per cent in sorghum due to compartment bunding.

3.2.9 *Live bunding*

On contour or across slope planting of one or two lines of vegetation at close intervals of 10-20 m in medium to deep black soils have following advantages, i. They build natural terraces because they filter the soil and retain it, ii. Increase area under cultivation by bringing rolling lands under cultivation, iii. Provide stable barrier on Vertisols, iv. Provide more opportunity time for water to soak into the soil, and v. These are simple and easy methods of conservation for common man to follow, *Leucaena leucacephala, Desmanthus* sp., *Sesbania* sp., *Glyricidia* sp. *Calliandra* sp. and variety of grasses could be used for live bunding. The live bunds have to be pruned frequently leaving 25-30 cm of stumps above the ground so that they do not become weedy. The research conducted at Dry Farming Centre, Bijapur (black soil) and Bangalore (red soils) reveals that *Eragrastic kuruala* and *Pennisetum ohinikere*. are also suitable as live bunds. The leguminous species and other grasses besides helping to conserve rain water *in situ* provide fodder for the animals. The harvested material could be even used as mulch in between the crop rows in dry lands. In fact live bunds are much more stable than mechanical measures and allow more water to infiltrate. Yield responses are also encouraging (Table 6).

While selecting species, its effectiveness as barrier, survival per cent, etc. have to be considered. General experience favours planting of hedge species in twin lines with a intra row spacing of 10 cm with a small section bund of 30 cm width and 15 cm height in the middle at a vertical interval of 0.3 m (Chittapur, 1998). Two to three seeds or slips can be planted per hill by opening a furrow of 15 to 20 cm deep. In extreme dry areas of 300 mm rainfall, it is required to establish the hedge species in the centre of two wide and about 30 cm deep 'V' ditch with 160° inner angle. Gap filling increases effectiveness of bunds.

In some hedge species border effect to a distance of 30 to 50 cm on either side of hedge row with a consequent yield reduction of field crops are observed. However, there would not be overall yield reductions due to improved performance of interior rows. Gaddanakeri (1995) observed a border effect upto 2.7 m in subabul hedges on sunflower crop. He suggested proper cutting and root pruning practices to make live bund economical. A cutting height of 30 cm and opening trenches on either side of the hedge of subabul were found advantageous over other management systems on Vertisols of Bijapur.

Table 6. Effect of different barriers an run-off less soil loss, sorghum crop yield economics

Barrier	Runoff (% of rainfall)	Soil loss (t/ha)	Yield (q/ha)			B:C ratio
			Grain	Fodder	Green fodder	
Control	43.60	6.10	8.33	27.17	–	0.93
Rubbles	21.10 (53.9)	2.90 (52.5)	12.50	41.70	–	2.25
Khas grass	25.60 (41.4)	3.13 (48.7)	11.45	37.40	7.30	2.07
Weeping love grass	26.10 (40.2)	3.34 (45.2)	9.50	35.70	6.30	2.02
Subabul	30.70 (29.7)	4.33 (29.0)	9.37	33.30	15.60	1.80

Patil *et al.,* (1995)

3.2.10 *Vertical mulching or slot mulch*

Vertical mulch is a practice of embedding crop residues in trenches across the slope and allowing the residues to project 30-40 cm above the ground level. Series of vertical mulches in dry land areas provide continuous barrier for the runoff water resulting in more conservation of water *in situ*. Increases in crop yields of *rabi* sorghum have been reported from Dry Farming Centres, at Bellary, Bijapur and Solapur. Since residues get decomposed in the soil, it needs to be replaced. Availability of residue may be another constraint in many dry farming areas. Results indicate superior effects of vertical mulching spaced at a interval of 4.5 m particularly on deep vertisols with a very low infiltration rate. A modified form of vertical mulching termed "Slot mulch" is being tested in Pacific North West of USA to control runoff on steeply sloping lands. The approach is to compact loose residues and chaff of a green crop into a narrow continuous slot approximately 7 to 13 cm wide and 20 cm or more deep formed on the hill side contour. The straw packed in the slot is left well exposed at the soil surface. The slot mulch is highly effective in increasing water infiltration in soil, especially where there are restrictions such as frozen surface.

3.2.11 *Residue mulching*

Residue management practices reduce run-off and erosion loss, improve water content, aggregate formation, porosity, hydraulic conductivity and modify soil temperature. Surface straw mulch facilitates infiltration by slowing runoff and absorbing rain crop impact while reduces soil ruddling and crusting. Mulches also reduce evaporation loss by reflecting radiation and thermally insulating the soil surface. At Bijapur on Vertisols incorporation of sorghum stubbles and safflower stalks improved infiltration significantly (Badanur *et al.,* 1990). Residues and subabul incorporation also enhanced field capacity. Bond and Willis (1969) showed that the crop residues on the soil surface slow down rates of first and second stage evaporation of water from soil surface resulting in additional water storage. At, Cinzana Experiment Station, Mali (South Africa) in an experiment on a sandy soil with rainfall intensity of 93 mm/ hr, a final infiltration rate of 40 mm/hr was noticed after one hour under bare soil while it was double this rate under residue covered soil.

The actual amount of moisture conserved by residue mulch depends on a number of factors including type, amount and placement of residues, precipitation and other climatic characters, length of fallow, tillage practiced and soil types. The long term field experiments showed that increases in soil water from the mulches were significant and roughly proportional to the amounts of surface residues over the ranges of 2.2 to 6.6 t/ha (Greb, 1983). Crop residue retention in the SAT is sometimes impractical because of their value for other uses. Unger *et al.* (1991) suggested some possible alternative strategies such as limited or selective residue removal, substitution of high value forages for residues, alley cropping, utilization of waste lands for forage and wood production and the control of livestock number.

3.2.12 *Sand mulching*

Sand mulching has been practiced by farmers in some pockets of North Karnataka. Experiments conducted at Dry Farming Centre, Bijapur and Main Research Station, Dharwad indicated distinct advantages with sand mulching (Guled, 1999 and Sudha, 1999). The benefits were directly proportional to the quantity of sand applied or mulch thickness (Table 7). Guled (1999) attributed the beneficial effects to runoff control, increased wetting front and reduced runoff. Haung (1983) attributed improved crop yield to increased soil temperature, conservation of rain water, reduced evaporation and controlled wind and water erosion which inturn increased water content at any time under sand mulch compared to non-mulched soil. The utility of sand mulch, therefore, needs intensive study in comparison with non-mulched soil.

Table 7. Effect of sand mulch on soil moisture (cm) and pod yield of groundnut

	Soil depth (cm)	30 DAS	45 DAS	60 DAS	90 DAS	Pod yield
No mulch		0-15	3.56	3.32	3.43	4.99
	16-30	3.60	3.37	3.77	5.14	9.60
	31-60	7.56	6.98	8.55	11.29	
Sand Mulch (5.0 cm)		0-15	4.87	3.67	4.01	5.66
	16-30	5.38	4.85	4.83	5.80	13.76
	31-60	11.13	9.98	10.37	12.74	
Sand mulch (7.5 cm)		0-15	4.95	3.71	4.02	5.67
	16-30	5.42	4.87	4.84	5.86	12.76
	31-60	11.72	10.12	10.45	12.83	
CD. 0.05 P		0-15	0.87	0.22	0.35	0.29
	16-30	1.48	0.27	0.47	0.62	2.19
	31-60	2.75	0.48	2.42	NS	

Sudha (1999) DAS - Days after sowing

3.3 Water harvesting and recycling

Water harvesting is a broad term covering the various techniques to collect rain water from natural terrains or modified areas and concentrating it for use on smaller sites or cultivated fields to assure economic crop yields. Some of the *in situ* water harvesting measures found to increase the yield upto 15 per cent area at any given level of productivity (Table 8). The additional benefit of some water harvesting systems is "nutrient harvesting" through runoff water. Runoff farming is basically a water harvesting system specially designed to provide supplemental or life saving irrigation to crop or crops, especially during the periods of moisture stress. While designing a water harvesting system, factors like precipitation patterns, equipment, labour and acceptability of water harvesting concept by the water users need to be considered (Fraiser, 1997).

Table 8. *In situ* water harvesting measures for red and black soil areas

Soil type	Rainfall	*In situ* water harvesting measures
Red soils	Low	Dead furrows at 3-6 m intervals
	Medium	Sowing on flat and ridging later with eventual cultivation
	High	Graded border strips
Black soils	Low	Contour cultivation
	Medium	Dead furrows at 3-6 m intervals
	High	Graded open furrows (0.2 -0.3 m_2) at 10 m interval across the slopes

(ICRISAT, 1984)

Catchment area treatments include, topography modifications, soil modifications and impermeable covering to get maximum runoff for storage. Water collected from a catchment area are stored in the soil profile or monolith and tanks or ponds. In designing soil monolith storage, factors like the depth of profile, water holding capacity and the infiltration rate need to be considered.

Contour strip rainfall harvesting consists of parallel and alternating runoff areas covered with polyethylene sheets and cultivated strips. In an experiment, Zaongo *et al.,* (1988) used runoff areas and cultivated strips of 100 cm and 80 cm wide in millet and sorghum plot. In both the crops contour strip rainfall harvesting proved advantageous (Table 9).

Table 9. Effect of contour strip rainfall harvesting (CSRH) on crop yields (Kg/ha) at different places

Crop	Methods	Kalapali	N'Dounga	Chikal
Millet	Traditional	464a	923a	401a
	CSRH	1019b	1667b	644b
Sorghum	Traditional	321a	1140a	288a
	CSRH	501b	1814b	312b
Rainfall (mm)	464	354	218	

Zonal water harvesting may be tried in soil with 150 mm of available water storage capacity (All India Coordinated Research Project for Dryland Agriculture, ICAR, Bellary). Such a practice has been successful in stabilising yields of millet crop (Randhawa and Venkateshwarlu, 1987). A watershed based water harvesting system that has been developed at ICRISAT in Hyderabad may prove instrumental in stabilizing dry land crop production. A part, about 30 to 50 per cent of the watershed, can be left bare or uncultivated, which may be treated to encourage runoff which could be stored in a reservoir. The stored water could be applied to the cultivated area of the watershed (Kanemasu *et al.,* 1990).

In the watershed the water harvesting structures such as percolation tank, nala bunding, checkdam and form pond etc. provide water for irrigation. The technology of farm pond is not new to India. From the ancient times, the excess rain water flowing from the surfaces of cultivated non-arable areas is being collected in excavated ponds for human and animal use. Runoff is inevitable in the tropical and subtropical climates of our country. It varies from 10-40 per cent of the total rainfall. Of this, at least 30 per cent could be harvested into dugouts and water so harvested could be recycled to the donar area. The payoff is very high and works out 2q/cm/ha of critical irrigation. In the low rainfall areas of Karnataka such excavated ponds are noticed in black soils. These ponds can serve also as a loci for aquaculture, multiplication of biological fertilizers like Azolla and also help greatly in sediment collection. It has been observed that a form pond of 150 cubic meters capacity per hectare of catchment is found to be optimum for vertic group of soils.

Rain water harvesting has been found possible in areas with as little as 50-80 mm rainfall but collection of water in ponds from cultivated areas is feasible only with annual rain above 550 mm.

4. EFFICIENT CROP STAND, CROPPING SYSTEMS

In most of the dry land areas crops and varieties that are being grown are not efficient in moisture use and are subsistence oriented. Crop cultivars grown are photosensitive, tall with a low harvest index and are of long duration than the water availability period. Hence, these cultivars suffer from moisture stress and yield low often resulting in total failure. For example, the traditional varieties of groundnut with 110 days duration usually suffered from drought stress at the later stages of crop growth in Saurashtra region. Introduction of a relatively short duration cultivar JL-24 (100 days duration) found favour with this environment in groundnut growing areas. The selected crop's growth rhythm should, therefore, follow the water availability pattern.

In rainfed areas, the length of growing season is determined mainly by the moisture availability period, that in turn depends on rainfall amount and distribution, soil moisture storage properties and potential ET demands of the atmosphere. Singh (1978) computed the length of growing season based on the ratio of actual evapotranspiration (AE) and potential evapotranspiration (PE). He considered the week that has a value of AE/PE above 0.5 as dependable and identified continuous number of such weeks for different regions of India (Table 10). The depth and water holding capacities of soils may have a significant role in deciding the length of growing season. For instance, Virmani *et al.* (1978) pointed out that the length of growing season could vary from 15 to 23 weeks at Hyderabad depending on the soil type where annual rainfall is 760 mm.

Table 10. Length of effective growing season at different locations in India

Location	Soil type	Duration (weeks)
Bellary	Vertisol	8
Hisar	Aridisol	13
Kovilpatti	Vertisol	21
Solapur	Vertisol	23
Akola	Vertisol	27
Varanasi	Alluviums*	32

Indore	Vertisol	36
Bhubaneshwar	Alfisol	45
Jodhpur	Aridisol	11
Anatapur	Alfisol	13
Hyderabad	Alfisol	22
Agra	Alluviums*	24
Ranchi	Alfisol	31
Bangalore	Alfisol	32
Samba	Submontane soils**	44
Dehra Dun	Submontane soils**	51

*Inceptisol, alfisol, entisol; **Mollisol, inceptisol, entisol; (Singh, 1978)

4.1 Selection of suitable crops and varieties

Crops grown in a specific area are determined by number of factors. Even with adequate precipitation and sunlight, optimum temperature and fertile soils, it is quite possible that other factors such as economic concerns (commodity prices, transport costs, marketing facility etc.) or even political reasons (marketing boards, price controls, price stability) determine the crop choices that farmer makes. In the past, selection of crops and varieties was governed by the needs of a farm family rather than the crops suitability for a given environment. The choice of crops infact should be dictated by the land use capability which inturn is mainly decided by soil water availability. Incase of rainy season crops, it is the rainfall pattern and length of the effective growing season that should decide the choice of crops and varieties. But in case of post-rainy season crops which are grown on conserved soil moisture, it is the available moisture in the profile at sowing time which dictates choice of crops (Table 11, 12 and 13).

Table 11. Broad soil groups and their moisture storage capacities in rainfed areas of India

Broad soil group	Sub group (based on soil depth)	Moisture storage capacity (mm)
Vertisols and related soils	Shallow to medium (upto 45 cm)	135-145/45 cm
	Medium to deep (45 to 90 cm)	145-270/90cm
	Deep (>90cm)	300/m
Alfisols and related soils	Shallow to medium	40-70/45 cm (sandy loam)
		70-100/45 cm (loam)
	Deep	180-200/90 cm
Aridisols	Medium to deep	80-90/90 cm
Inceptisols	Deep	90-100/m (loamy sand)
		110/140/m (sandy loam)
		140-180/m (sandy loam)
Entisols	Deep	110-140/m (sandy loam)
		140-180/m (loam)

(Randhawa and Singh, 1983)

Table 12. Suggested cropping strategy for conserved soil moisture storage

Soil/ location	Conserved soil moisture storage (mm)	Suggested crops after *Kharif*
Aridsoils/Hissar	>300	Wheat, pea, Chickpea
	200-300	Wheat (local), barley
	150-200	Chickpea, barley, raya (*Brassica juncea*), Sarson (*B. compenstris*)
	50-75	Taramina
Vertisols/Bellary	>500	Sorghum (M-35-1, SPV 86)
		Safflower (A 300, 7-13-3)
	375-500	Chickpea (A-1)
		Sorghum (CSH-2, CSH-3)
	250-375	Safflower, Chickpea
	<250	Safflower, chickpea, sorghum as fodder *Dolichos,* lima bean, *Phoseolus vulgaris*

(Singh, 1987)

Table 13. Relative yields of traditional and efficient crops on different soils

Place/Soils	Traditional crop	Yields (t/ha)	Efficient crop	Yields (t/ha)
Bellary/Vertisols	Cotton	0.20	Sorghum	2.67
Bijapur/Vertisols	Wheat	0.94	Safflower	1.85
Varanashi/Entisols	Wheat	0.86	Chickpea	2.85
Ranchi/Alfisols	Upland rice	2.88	Maize	3.36
Indore/Vertisols	Greengram(k)	1.18	Soybean	3.33
	Wheat (R)	1.12	Safflower	2.42
Agra/Entisols	Wheat	1.03	Mustard	2.04
Hissar/Aridisols	Wheat	0.32	*Ernea sativa*	1.61
Udaipur/Verti/Inseptisols	Maize	1.80	Hy-sorghum	2.90
Rewa/vertisols	Kalitur	0.40	Soybean (yellow)	1.20

K -Kharif, R - Rabi ; (Singh, 1987)

The crops and their genotypes differ in their yield potential and water use efficiency, some being more efficient than others in the same environment. For example in the shallow to medium deep Vertisols of Solapur, safflower proved more efficient in water use than groundnut and pigenopea crops (Umrani *et al.,* 1981). Though the WUE of all the varieties of sunflower and groundnut was almost equal, pigeonpea variety No. 148 has greater WUE than S-5 and Prabhat (Table 15). At the same centre, in post-rainy season, the WUE of sorghum, chickpea and safflower was practically the same, but chickpea and safflower may be advocated to farmers due to higher net returns from these crops. The varietal differences in chickpea and safflower were not appreciable, but sorghum CSH-8 had greater WUE compared to M-35-1. In deep vertisols, safflower was more efficient and remunerative than cotton (Table 14).

Table 14. Yield and moisture use efficiency of different crops at Solapur

Crop	Variety	Yield (q/ha)	CU (mm)	MUE (kg/ha)
Pearl millet	Local	7.4	268	2.8
	Local	10.5	264	4.0
	HB-3	18.3	261	7.0
Sunflower	EC 68414	10.9	239	4.5
	Morden	9.6	241	4.0
	EC69874	11.7	238	4.9
Groundnut	SB XI	10.5	295	3.8
	TMV 10	10.2	274	3.7
	K 4-11	10.3	316	3.3
Pigeonpea	Prabhat	7.5	295	2.5
	5-5	9.2	324	2.8
	NC, 148	12.2	325	3.7
Sorghum	M 35-1	17.5	247	7.0
	CSH 8R	23.2	220	10.5
Chickpea	N59	13.8	199	7.8
	Chafa	13.6	212	6.6
Safflower	N 62-8	14.4	212	6.7
		15.4	227	7.3

Similarly, unconventional crops have been identified in various region as more profitable and efficient (Table 14).

Table 15. Unconventional crops for different dry land regions

Region	Crops
Vidarbha, plains of Bihar	Redgram as late *Kharif* crop
Red soils of south Karnataka	Redgram
Deccan *rabi* region	Field beans, cluster beans, castor
Sierozems of NW India	Castor
Southern black soils of Tamil Nadu	Castor
Deccan medium black soils	Setaria, castor
Bihar plateau	Groundnut
Red soils of Rayalseema	Redgram and castor
Submontane region of north west UP	Ragi
Medium black soils of Karnataka, AP, Maharashtra and MP	Sunflower

Venkateswarlu (1985)

Selection of crops and varieties also depends on the time of sowing. The following preference was found ideal for Bangalore region (Alfisols).

Table 16. Choice of crops depending on sowing time at Bangalore

Month	Choice of the crops
May	Cowpea, pigeonpea, sesamum
June	Groundnut, pigeonpea
July	Maize, long duration finger millet
August	Medium duration finger millet, soybean, cowpea
September	Cowpea, horsegram, short duration niger, transplanting short duration finger millet

Varieties that are suitable for normal sowing time are not necessarily suitable or efficient when grown late in the season (Table 17).

Table 17. Suitable crop varieties for normal and late sowing conditions

Locations		Varieties for different sowing time	
		Normal	Late
Anantapur	Groundnut	Kadiri-1	Kadiri-3
			TMV 2
Hyderabad	Sorghum	CSH-5	CSH-2
Bangalore	Fingermillet	Indaf-8	Indaf-9

Thus, choice of genotypes/varieties assumes a great importance based on time of sowing. The desired characters of varieties for dry lands include faster rate of growth in the initial period, short duration to escape late season drought, high grain number and strong and deep penetrating root system to exploit moisture and nutrients from lower depths to withstand adverse situation. Sequential cropping has showed that *kharif* crop should be a shorter duration one followed by a longer duration *rabi* which will facilitate good establishment of *rabi* crop.

4.2 Potential cropping systems

Traditional cropping systems that are in vogue in arid and semi- arid regions are often inefficient in terms of resource utilization. In arid regions single cropping system, involving a long fallow period (October to June) is the rule rather than exception. Deep vertisols in the semi-arid regions are left fallow during rainy season and cropping is done on stored soil moisture during post-rainy season. Traditionally mixed cropping is used mainly to provide subsistence or as a means of insurance rather as a efficient resource use system of high yielding potential. Under utilization of cropping system, period is also common in assured, medium to high rainfall areas (800 mm or more). Therefore, it is necessary to put to use the available technology to generate more.

A cropping system is the intensification of crops in space and time dimensions. It presents farmer with a range of decisions that need to be made such as what combinations of species should be planted and which varieties be selected from within these species, what planting arrangements should be followed and how many plants each components should be sown with and what should be the input levels for the component crops?

The length of growing period is the basis that decides the make up of a cropping system. In *kharif* season the rainfall both in terms of quantum and distribution decides the effective cropping season and it becomes critical in selecting cropping system for a given area (Table 18).

Table 18. Potential cropping systems in relation to rainfall and soil type

Rainfall (mm) Soil type	Effective growing season weeks	Suggested cropping system
350-600 Alfisols and shallow Vertisols	20	Single rainy season cropping
Deep Aridisols and Entisols	20	Single cropping during rainy or post-rainy season
Deep Vertisols	20	Single-post-rainy season cropping
Alfisols, Vertisols and Entisols	20-30	Intercropping
Entisols, deep Vertisols deep Alfisols and Inceptisols	30	Double cropping with monitoring
Entisols, deep Vertisols, deep Alfisols and Inceptisols	>30	Double cropping assured

Rain dependent lands are generally monocropped either during rainy season or on conserved soil moisture (October-March) and the productivity is stable during rainy season while it is low and unstable during post-rainy season. In these situation intercropping systems ensure efficient utilization of resources as well as stability in production. Subsistence farmers who practice low input farming are particularly dependent upon this form of crop production. The system also ensures reduced damage due to pests and diseases. Double cropping is possible where rainfall is more than 650 mm per annum and the moisture storage capacity of the soil is about 30 cm per meter of soil depth.

In India, water availability period varies from less than 110 days to more than 250 days in different parts. Depending on soil characters, inter or double cropping can be advantageously practiced under these situations (Table 19). Legumes have special utility in these systems. As a component or understorey crops, legumes establish rapidly, reduce soil erosion, reduce soil temperature and lower weed pressure beside giving yield additions. Yield advantages vary based on time of planting that is simultaneous or staggered, population levels used and planting geometry followed. Production technologies have been standardised for many locations. Some methods have special advantages too. For instance, strip planting of legumes and cereals (eg. 3:3 sorghum + pigeonpea) ease weeding, spraying and reduce influence of competition in comparison to alternate row planting. In certain situations due to border effect' compensation in yield also occurs.

In drier regions having light textured soils (arid soils) pearl millet and cluster beans are the predominant crops which are intercropped in 2:1 row ratio with either cowpea or greengram. The LER (Land equivalent-ratio-is a index of intercropping productivity which is the sum of fractions of yields relative to respective sole yields of component crops) of such a system ranges between 1.00 to 1.39. On heavier soils, groundnut is intercropped with pigeonpea (LER 1.51 to 1.87). Pearl millet or sorghum or clusterbean go well with greengram, mothbean or pigeonpea in regions having water availability period of 110-170 days (LER is around 1.40). In areas where moisture availability period is more than 175 days, there is wider choice of principal crops for intercropping with pigeonpea as the first choice and blackgram sesamum as second choice (LER = 1.16 to 1.93). At solapur with soils being more productive during *kharif*, intercropping not only minimise risk, being productive has better rainfall use efficiency, a concept that needs to be considered seriously in dry lands (Umrani, 1995) (Table 20). In addition, pearl millet + pigeonpea improve soil organic carbon and soil structure compared to a pure cereal cropping or keeping the land fallow. Umrani (1995) observes that such a system fulfills the concept of kharif-rabi continuum instead of compartmentalising kharif and rabi season cropping, which prevails in the hard core of Deccan rabi dry farming tract.

Table 19. Efficient cropping systems for different rain dependent regions of India

Water availa-bility period (days)	Inter cropping system		Double cropping system	
	Base crop	Inter crop	Rainy season crop	Post rainy season crop
Upto 110	Pearl millet/cluster bean	Cowpea / Greengram / Blackgram		
	Groundnut / pearlmillet	Pigeonpea		
110-150	Pearlmillet / Sorghum / Clusterbean	Greengram / Pigeonpea	Cowpea / Blackgram Soybean Green gram / Pearlmillet	Safflower / Chickpea Mustard / Safflower Mustard Chickpea / barley
150-175	Pearlmillet / Chickpea	Pigeonpea mothbean Barley / linseed / mustard / safflower	Greengram	Sorghum
175-200	Pearl millet / groundnut Sorghum	Pigeonpea	Greengram Cowpea Greengram	Sorghum Sorghum, Safflower
	Chickpea	Safflower	Blackgram	Barley / mustard
	Barley	Mustard	Pearlmillet Maize Rice Sesame Chickpea	Chickpea Wheat/ wheat + Chickpea/ Mustard
200-250	Sorghum /Pearlmillet / Male	Pigeonpea	Sorghum / groundnut Maize Soybean	Chickpea/Safflower Wheat
	Soybean Chickpea	Pigeonpea Mustard	Soybean + Maize	Safflower / Chickpea
>250	Rice/fingermillet / Soybean maize	Pigeonpea	Rice/ Maze/ Fingermllet / groundnut	Wheat / Chickpea / lnseed / lentil / Horsegram / barley
	Wheat/ barley Chickpea	Chickpea / mustard Mustard	Soybean Pearlmillet	Fingermillet Wheat

Source : Singh and Das (1995) Ann. Rep. AICRPDA, Solapur, 1983.

Table 20. Yield, moisture use, MUE, rainfall use efficiency and soil properties in different cropping systems at Solapur

System	Yield (q/ha)	CUM (mm)	MUE (kg/mm)	RUE %	Organic carbon % (0-15 cm)	Structural stability index (MWD in mm)
Pearl millet	24.6	190	13.0	30.0	0.35	0.341
Groundnut	7.6	240	3.2	38.2	0.44	0.457
Pigeonpea	6.71	25	1.9	51.2	0.45	0.444
Pearl millet + Pigeonpea	20.5+3.9	340	17.2	53.5	0.42	0.387
Sorghum (M 35-1)	10.9	172	6.28	26.	0.35	0.351
Safflower	5.9	180	3.3	28.0	0.39	0.387
Chickpea	4.7	154	4.1	18.9	0.35	0.359
Fallow	–	–	–	–	0.33	0.329

Double cropping systems are more common in regions having moisture availability period of 175-200 days or more. A wide variety of crops are grown including a short season (70 days) pulse crops (greengram, blackgram, cowpea), cereals (pearl millet, maize, rice) maturing in 95-100 days, and sesame (110 days). Short duration pulses are followed by *rabi* sorghum, safflower, barley, mustard etc. while cereal/oilseed crops are followed by chickpea, mustard or wheat. Sorghum/ safflower double cropped system is common in better moisture availability endowed regions (200-250 days). Soybean in rainy season followed by safflower or wheat in post-rainy season is also common in some areas. In wetter areas (moisture availability of 250 days) rice is more common, although, groundnut and soybean also find a place in double cropping systems.

Table 21. Evaluation parameters for some cropping systems grown on a deep vertisol at KRIS, Hyderabad (mean of four years)

	Cropping system			
	Intercrop	Sequential	Sequential	Sequential
Base/ first crop Inter/second crop	Sorghum + Pigeonpea	Mung- sorghum	Maize- Chickpea	Sorghum - Safflower
Cropping period (days)	180	150	190	190
Yields (t/ha) 1st	2.88	0.65	2.44	3.34
Yields (t/ha) 2nd	0.94	2.07	1.31	0.79
Nutrient value				
M. cals	11032	7956	10120	10765
Kg protein	508	371	494	453
Kg oil	71	48	157	271
Biomass (t/ha)	9510	6776	9363	9023
Energy (M cals/ha)				
Seed yield	17351	12407	17250	19464

Biomass	41567	29821	41094	44287
Gross returns (Rs/ha)	6727	4770	7490	6900
Costs (Rs/ha)	1600	1750	2200	2450
Net returns (Rs/ha)	5727	3020	5290	4450
Rate of return (%)	320	173	240	182

Source : Willey (1987)

Another important aspect that needs consideration in selection of cropping system for a dry farming situation is the risk of failure. One of the problems in trying to ensure full use of the potential growing period is that as the required growing period for a given system increases, the probability of end-season water stress also increases. For instance, on Vertisols with 990 mm rainfall, after a 91 day rainy season sorghum crop, there is still quite high probability (73%) of having sufficient water stored in the soil profile for a sequential second crop (assumed to be 100 days) (Willey *et al.,* 1982) (Table 22). After taking into account those years when there is sufficient water, but surface soil conditions are too dry to establish the crop, this probability drops to 60 per cent. If the rainy season sorghum has 105 day growing period, the probability of wet harvest conditions is slightly reduced, but so too is the probability of getting good sowing conditions or sufficient stored water for the second crop; the overall probability of success, therefore, drops substantially to only 27 per cent. On the other hand, a relay system of planting based on the 105-day first crop and a 14 day over lap does of course give same probability of success as a sequential system based on 91-day crop (provided there is no injury to second crop at harvest of first crop). The advantage of the sorghum + pigeonpea system in which pigeonpea acts as an already established second crop is very striking; there is extremely high probability of success and the 105-day first crop can be accommodated as easily as the 91-day crop.

Table 22. Probability of success (%) of different cropping systems at Indore, India calculated from 37 years of rainfall data using a water balance model

Cropping systems	First crop		Second crop		
	Sufficient moisture for grwoth	Wet week harvest (>50 mm)	Sufficient moisture for grain (>200 mm)	Insufficient rain for establishment (>200 mm) even though sufficient for growth	Overall probability of success
Sequential system (1st crop 105-days soghum)	100	24	51	24	27
Sequential system (1st crop 91-day sorghum)	100	24	73	13	60
Relay system (1st crop 105-day sorghum)	100	24	73	13	60
Sorghum + pigeonpea (105 day - 180 days)	100	24	97	–	97

Source : Willey et al. (1982)

5. AGROTECHNIQUES

Plant growth and yield is the function of potential of the environment and its management. Once the potential of the environmental factors is understood, through management techniques efforts have to be made to provide best possible growing conditions to achieve potential crop yields. Various agro-techniques or cultural practices developed on ecological principles help to manipulate agroecosystem to ensure i. minimum intra-specific and interspecific competition for the resources, ii. maximum utilization of available resource pool, iii. near optimum climatic parameters at each phenological stage, and iv. favourable conditions for germination, root growth and development of storage organs (Ahlawat and Rana, 1998). Various agro- techniques, therefore, aim at regulating physiological functioning of the crop plants by improving the nutrients, water, light and other inputs use efficiency by providing optimum conditions to each plant to express genetic potential within the constraint of environment. Following are the important agrotechniques, which play key-role in improving and sustaining crop productivity in dry land areas.

5.1 Land preparation

Land preparation including postharvest cultivation and preparatory tillage influences intake rate of water, reduces runoff and soil erosion and permits timely seeding of crops. In dry land areas primarily tillage is done by the country plough in the *Kharif* areas during non-crop season. This off-season tillage, however, is always not practical in many areas more so in the red soil regions of the country. Research results indicate yield advantages with off-season tillage alone by 7-9% over the control particularly during sub-normal years. Intensive primary tillage is generally found necessary for creating a favourable root proliferation zone and for enhancing rainfall acceptance. Deep tillage with inversion plough is particularly beneficial on Alfisols (Vijayalaxmi, 1987 and Hedge, 1987) (Table 23).

Table 23. Effects of tillage on crop yields

Centre	Crop	Yield (t/ha)	
		Shallow tillage	Deep tillage
Anantapur	Castor	1.01	1.30
	Pigeonpea Pearlmillet	0.71 0.68	1.04 1.00
Bangalore	Groundnut	0.86	1.17
	Maize	3.20	4.51
	Pigeonpea	1.23	1.90

(AICRPDA, 1982)

The turnover of soil materials to the surface in deeper soils with distinct textural profiles as in case of Royalseema and southern Karnataka region found to improve yields of sorghum (in normal rainfall years), castor (in both in normal and above normal years) and other crops (Table 24). At Dry Farming Centre, Bangalore, Hedge *et al.* , (1987) observed a yield improvement to the tune of 59 per cent in finger millet with one additional ploughing with a mould board plough to a depth of 15 cm over farmers practice of 2-3 ploughings with wooden plough (Table 24). Sorghum was found to respond to a ploughing depth of 25 cm at Hyderabad.

Table 24. Effect of mould ploughing on the yield

Crop/centre	System	Grain yield (Kg/ha)	Reference
Finger millet/ Bangalore	Local practice (2-3 times wooden ploughing)	1130	Hedge *et al.*, (1987)
	One additional mould board ploughing in May	1635	
Sorghum/ICRISAT, Hyderabad	Normal tillage (mould board ploughing at 12 cm deep)	2160	Pathak and Laryea (1995)
	Deep tillage (cross chiseling 25 cm deep)	2720	

Deep ploughing improves the physical characteristics in black soils but it does not always reflect in yields. Experiments at Dry Farming Centre, Bijapur indicated advantage with deep ploughing on Vertisols.

5.2 Timely sowing

Early sowing has several advantages both general and specific (Venkateswarlu, 1985). The general advantages are i. improved water use efficiency, ii. good early seedling vigour, and iii. longer growing season. The specific advantages are i. avoidance of pests and diseases (eg. shoot fly in sorghum, gallfly in rice, downy mildew on bajra, leaf sugar malady in *rabi* sorghum), ii. provide opportune time for second crop in double cropping systems, and iii. avoidance of cooler weather at flowering for *kharif* crops in north India. Hence, success of crops in dry lands largely depends on time of seeding especially with improved cultivation. Delayed sowing leads to reduced length of growing period and exposure of crop to terminal stress. The very first showers, therefore, should be utilized for sowing by keeping the land ready prior to probable rainfall date.

Agro-meterological analysis being made available now it would be possible to define available effective rainy period (number of consecutive week in the 14 week moving average of rainfall (R)/potential evopotranspiration (PE) is greater than or equal to 0.75 with a condition that initial week of growing period is preceded by R/PE ratio of 0.5, sowing rain (the rain needed to fill top 30 cm of the soil for better seed generation, i.e. weekly R/PE value around 0.5 preceding commencement of effective rain period (R/PE>0.75), pre-sowing cultivation (earlier to sowing rains when R/PE for that week is 0.25), wet spell, dry spell, etc. for all the stations which would enable precision and timeliness in agricultural operations.

Seeding prior to potential week helps to increase crop yields. For instance seeding in 24th meteorological week at Solapur, in dry soil proved beneficial in 5 out of 7 years and on remaining years it was at par with seeding at break of monsoon (Annual Report, 1984). In traditional Deccan *rabi* tract it is a practice to sow sorghum and safflower by the end of September to early October. An advance by 20 days has substantial effects on yields of sorghum at Solapur, Bijapur and Bellary with improved cultivar CSH-8R while the traditional M-35-1 produced more vegetative growth at the cost of grains (Umrani, 1995). A progressive delay of 7, 14 or 21 days in sorghum result is found to decrease yield to the extent of 319 to 370 kg/ha/day (Umrani and Patil, 1985). The reduction in grain yield is attributed to the decreased in grain number per panicle and seed weight. Late sown crops also attract more pests.

At Udaipur a delay in sowing by 20 days brought down the yield of CSH-5 by 1/3 (Table 25). This emphasized that this initial operation has to be done more timely. Reductions in yields of other crops are also evident (Patel *et al.*, 1981, Krishnamurthy *et al.*, 1977, Lomte *et al.*, 1994 and Pothiraj, 1982) (Table 26, 27 and 28). Singh and Das (1984) reported a relation between reduction in yield and delay in sowing with mathematical fractions for sorghum, rice and finger millet (Sorghum -Y = 39.19 (1-4.24/100)^x, R2=0.97; upland rice Y= 44.22 (1-3.17/100)^x, R2 = 0=0.86; finger millet = y = 25.82 (1.2.9/100)^x, R=0.995). Of all the crops studied, the per day per cent decrease in yield for sorghum was the highest (4.24) followed by upland rice (3.17) and ragi (2.94). At Coimbatore, Pothiraj (1982) observed reduction in yield due to delayed sowing in cotton beyond 38th week (Table 29). Now, optimum dates of seeding have been established for most of the crops in different locations in India.

Table 25. Effect of sowing dates on yield of *kharif* sorghum (CSH 5) at Udaipur

Sowing date	Rainfall in cropping period (mm)	Grain yield (q/ha)	% reduction in yield	Reduction in yield (kg/ha/day)
July 3	744	38.7	–	–
July 13	488	25.0	35.5	137.5
July 23	388	13.7	64.5	125.0
August 3	345	10.6	72.6	90.7

Source : Singh and Das, 1984.

Table 26. Effect of sowing date on yield (q/ha) of safflower at Solapur

Sowing time	Meteorological weeks	1973-74	1974-75	1975-76	Mean
Aug. 27 2nd Sept.	35	10.27	16.02	8.01	11.43
Sept. 10-16th	37	8.26	15.99	13.87	12.71
Sept. 24-30th	39	5.45	5.62	16.37	9.15
Oct. 8-14th	41	–	5.19	8.97	7.08
Oct. 22-28th	43	5.48	2.93	5.44	4.62

Source : Patil *et al.,* (1981)

Table 27. Effect of sowing date on yield (q/ha) of rabi crops at solapur

Sowing date	Sorghum	Safflower	Chickpea	Wheat
September 8	8.8	12.4	8.4	4.7
September 20	5.4	9.9	7.4	4.6
October 7*	4.2	7.2	4.4	2.1
October 23	1.5	4.0	4.1	2.0

*Normal date of sowing (Krishnamurthy *et al.,* 1977)

Table 28. Effect of sowing time on the yields (Kg/ha) of sorghum

Treatments	1986-87	1987-88	1988-89	1989-90	1990-91	Pooled mean
Advance sowing (dry sowing)	3437 (19th June)	2744 (10th June)	2418 (14th June)	4055 (9th June)	2601 (8th June)	3051
Normal sowing	2578 (19th July)	2007 (25th June)	2036 (28th June)	3662 (1st July)	1945 (25th June)	2445
S.Em ±	103	81	91	89	67	40
CD (0.05)	298	235	262	247	193	111
Rainfall (mm) (June-October)	533	707	1525	1177	1563	
	39	48	61	42	57	

Source : Lomte *et al.* (1994)

Table 29. Effect of seeding time in cotton

Seedling time	Seed cotton yield (kg/ha)	
	1980-1981	1981-1982
Pre-monsoon sowing on 17th Sept. (38th std. week)	1092	1583
Pre-monsoon sowing on 24th Sept. (39th std. week)	807	1127
Early monsoon sowing on 1st October (40th std. week)	739	677
Late monsoon sowing on 8th October (41st std. week)	670	588

Source : Pothiraj, 1982

The ideal time for sowing is found to be early part of the sowing season. It appears most crucial and loss in productivity as a result of delayed sowing can not be compensated by any other improved agronomic practice. Some crops like castor, clusterbean and some varieties of maize and sorghum however, grow tall and become too vegetative if sown early. There is a operational problem of how to sow all the crops as early as possible. Widening plant rows of several crops is found not to decrease yields, on the other hand improves coverage of land at a given time. It also saves energy, for instance, in case of sorghum and pearl millet yield is invariant from 30 to 75 cm rows spacing provided the population is maintained constant. Similarly, in case of castor and pigeonpea, the range is much wider from 45 to 135 cm.

A word of caution, however, is essential as many fields sown with pre-monsoon showers may have poor plant stand in red soils. This happens primarily because sowings are continued even upto fourth day after the receipt of these showers in red soils. Before sowing it is necessary to ensure that 20 cm soil is wet. And, on red soils seeding operation should be carried only upto two days after the pre-monsoon showers. Weather forecast should be followed closely to avoid risk.

5.3 Seed hardening

Pre-sowing hardening of seeds is one of the methods of modifying the physiological and biochemical nature of seeds so as the get the characters that are favourable for drought tolerance. Drought tolerance of many plants can be increased by subjecting them for several days to moderate dehydration, low temperatures or treatment with growth retarding substances (May and Milthrope, 1962). Seed hardening is, therefore, done to overcome the adversity of moisture stress that is not uncommon in dry farming areas. Although, it varies from crop to crop, the principle remains same. When dry seeds are soaked in water/chemical solution (Table 30), the quiescent cells get hydrated and germination initiated. It also results in enhanced mitochonidral activity leading to the formation of high energy compounds and vital biomolecules. The latent embryo gets enlarged. When the imbibed seeds are dried again, the triggered germination is halted. When such seeds are sown, reimbibition begins and the germination event continues from where it stopped previously. As a consequence there will be multitudes of benefits, which make the plants prepared to resist the adverse weather if any. Beneficial effects of seed hardening include, i. accelerated rapid generation and growth rate of seedling, ii. hardened plants recover much more quickly from wilting than those from untreated plants, iii. induces resistance to salinity as well as to drought condition, iv. seeds withstand higher temperature for prolonged period without loss of viability, v. flowering is slightly accelerated in treated plants, vi. early emergence enable seedling to compete more effectively with weeds, and vii. hardened seeds perform better and result in more yield.

Table 30. Chemicals recommended for good hardening in different crops.

Crop	Chemicals used
Sorghum	KH_2PO_4 (2%)* or $CaCl_2$ (2%)
Paddy	KCl (1%)
Maize	CCC (250 ppm)*
Pearl millet and finger millet	NaCl (0.5%)*
Millets and cotton	KCl (2%) for a period of 12-16 hrs soaking
Sunflower	GA_3(500 ppm) or Soaking in $MgSO_4$ or $FeSO_4$ (0.2-0.5% solution)
Groundnut	$CaCl_2$ (0.5%)
Redgram	Zinc sulphate (100 ppm)
Greengram	Manganese sulphate (100 ppm)
Chickpea	KH_2PO_4 (1%)
Gingelly	$MgSO_4$ or $FeSO_4$ (0.2%)

* Dharmalingam and Jagathambal (1996) Balasubramanian (1996)

Table 31. Effect of seed hardening and seed pellating with organics

Places/ treatment	Germination	Field emergence (%)	Ear head weight (g)	Seed yield (kg/ha)
Coimbatore				
Control	79	68	24	810
Treated	86	76	35.5	1080
CD 0.05	4.3	2.7	2.3	36
Arappakatti				
Control	79	70	28.2	1050
Treated	86	80	39.6	1550
CD 0.05	4.3	1.4	2.7	125

Dharmalingam and Jegathamabal (1996).

Seed hardening involves seed soaking in solution/water for a specified period (16 hr) at room temperature followed by shade drying (24 hr). Seed soaking in wheat (24 hr), groundnut (8 hr), greengram (4 hr), bengalgram and sorghum has been found to result in higher yields. At Kovilpatii, water soaking 4 hrs with Rhizobium found to increase the blackgram and greengram yield by 92 and 107 per cent and 27 and 94 per cent, respectively during two years (Arunachalam, 1996). Some of the chemicals recommended for different crops are given in Table 30. Dharmlingam and

Jegathambal (1996) tried some organic substances (leaf extracts of Prosofis and Pungam leaf in 100 ml water) for 16 hrs using 1 : 0.6 w/v (seed : solution) ratio [The air dried seeds were then coated with a fine leaf powder of Pungam @ 300 g/kg seed using 5% mida gruel as adhesive. Seed coating was done by mixing the seed with maida gruel (300 ml/kg seed)]. Organic farmers are coming out with many innovative materials and methods.

5.4 Seeding technique

Seeding operation is an important practice which helps in obtaining good stand establishment, early vigour and better yields of crops. Line sowing, dry seeding and depth of seeding are different cultural operations of considerable importance.

Line sowing : Finger millet is often broadcast with heavy seed rate (25 kg/ha) and plant stand is thinned down later by cross cultivation with fine hoes. This method, however, results in non-uniform stand or too much population. Common broadcast method, therefore, has to be discouraged on account of uneven stand establishment. Moreover, late emerged seedlings are also weak and fail to bear ears. Seed drill sowing is more useful in view of precision in the depth of sowing, uniformity in spacing and above all rapid coverage that it ensures. At the same time it makes post-sowing operations such as weeding, interculturing etc. more convenient and effective. Use of seed-cum fertilizer drill helps to attend two operations simultaneously while ensuring optimum placement of seed and fertilizer in moist zone thus enabling higher fertilizer use efficiency. Of late, tractor drawn seed drills have been locally manufactured which have become popular among farmers (Table 32). These are useful for sowing when surface conditions are dry and difficult for using bullock drawn seed drills.

Table 32. Effect of sowing method on rainfed sorghum and pearl millet at Kovilpatti

Methods	Grain yield (t/ha)			
	Sorghum	% increase over broadcast	Pearl millet	% increase over broadcast
Tractor drawn seed dril	2.08	23.0	2.04	10.9
Bullock drawn seed drill	1.92	13.6	1.95	6.0
Fertilizer-cum seed drill	1.97	16.5	1.99	8.2
Behind country plough	1.74	3.0	1.87	1.6
Hand dibbling	1.82	7.7	1.92	4.3
Broadcasting	1.69	–	1.84	–

Source : Anon., 1977

Dry seeding : Sowing of crops immediately after the receipt of soaking rains is the traditional practice of rainfed farmers which may be called monsoon or wet sowing. But in vertisols where dependable rains (P>60%) followed by reliable monsoon rains exist, sowing bold seeded crops before the onset of pre-monsoon rains in dry soil is advantageous and confers multiple benefits. Since here even the first drop of pre-monsoon rain itself is used for crop germination, growing period is lengthened as against taking up wet sowing which can be done only with considerable loss of moisture in the event of first rains happen to be heavy and incessant making the soil too sticky thus postponing the seeding unduly. The sowing period is, therefore, shortened. Dry seeding in anticipation of rains helps to establish a crop at an early opportunity and gives an advantage of better utilization of rainfall during crop growth. Dry seeding is found beneficial in sorghum, castor, cotton, redgram etc. (Table 33).

Experiments conducted at TNAU, Coimbatore in rainfed cotton clearly proved the advantage of pre-monsoon sowing. The reduction in yield due to sowing after 38th standard week was 2.9, 2.5 and 4.2 q/ha fore one, two and three weeks delay respectively, in one year and in the next year, the corresponding reductions were 4.5, 9.1 and 10.2 q/ha, respectively (Pothiraj, 1982). At Parbhani during monsoon, dry sowing (24th meteorological week) of sorghum was found more stable and productive compared to normal sowing (25th meteorological week) and gave 24.8 per cent higher grain yield over normal sowing (Table 28, Lomte *et al.*, 1994).

Table 33. Grain yield (kg/ha) of sorghum as influenced by time of sowing during *Kharif* at Dharwad

Varieties	Dry sowing		Normal sowing	
	100 kg N/ha	50 kg N/ha	100 kg N/ha	50 kg N/ha
CSH-5	4746	3385	2778	2278
SPV-351	4853	4345	4162	3698
SB-905	4011	4093	1238	1369
Mean	4537	3924	2726	2448
	4230		2587	

Source : Hosmani *et al.,* 1987

Soil necessarily should be dry for dry sowing otherwise seeds get damaged under semi wet condition leading to poor stand establishment. There may be also the problem of insects particularly of ants and other pathogens. To avoid damages to seed 'seed pellating' with protective chemicals could be followed. For instance in case of sorghum, seeds are made, wet by sprinkling water and stirring. Add a little fine soil or chalk powder and stir the contents till a uniform fine coating is obtained. To the mixer while stirring also add carbofuran (@ 1% by weight) and organo mercurials (0.1% by weight). Addition of gum or teepol helps in retaining coating. After uniform coating, dry the seeds in shade. Open or sundrying results in cracking and removal of coating. Treated seeds remain in soil unaffected for 20-25 days or more. Sowing is to be done based on rainfall probability. Alternatively, double seed rate can also be followed to make up any seed loss.

Deep sowing : Depth of seeding has been observed to be critical for the establishment of satisfactory plant stand. During rainy season a sowing depth of 2-3 cm is optimum in case of pearl millet, finger millet, other millets, sesamum and rapeseed mustard. For maize, sorghum, wheat, barley, pulse crops and most of the oil seeds, 5-6 cm seeding depth has been found optimum (Singh, 1995).

Deep sowing particularly in post-rainy season cropping or cropping or stored soil moisture is necessary for crops having long coleoptile like durum wheat, sorghum etc. and in bengalgram. This practice helps to place seed in moist zone for successful establishment and which in later stages of growth can withstand water stress to a greater degree due to well established roots deep into the soil. Frequent intercultivations that are taken up during vegetative stages in such system help to create dust mulch which wards of weed problem as well as loss of moisture through evaporation. Realizing the importance of deep sowing, farmers in the dry Vertisols of Karnataka increase the weight of the body of seed drill by pasting it with mud or keeping some weight or even by standing on the drill. Such heavy seed drills in dry soils are pulled by engaging to pairs of bullocks.

5.5 Optimum plant densities

Use of optimum plant population is a pre-requisite for higher yields. Higher plant population than the recommended creates more interspecific competition resulting in reduced yields. In sorghum, maize and other crop plants even become barren. Number of plants per unit area largely depends on soil moisture. For example it could be as low as 5 plants per m^2 for post-rainy sorghum. Whereas under better moisture situation (rainy season) it could be between 5 to 10 plants per m^2 (Virmani, 1995). Interestingly, it was observed that local varieties of sorghum posses a high flexibility in population where 0.75 lakhs to

1.75 lakhs population produce the some level of yield (Singh and Das, 1984). In tillering crops like pearl millet, plant density of more than 10 plants per m^2 does not show any advantage. However, in the years of low rainfall, high plant density (22.5 plants/m^2) results in lower yields (Umrani *et al.*, 1983). In branching crops like safflower, plants show considerable adjustments to plant density and spacing by regulating branching. At low density plants produce higher number of secondary and tertiary branches compensating the low density. On the contrary high plant density produces less secondary branching. Thus over a range of population yields remain at the same level (Umrani and Bhai, 1984). The traditional practice of planting with maximum population pressure in dry lands does not help. In fact it may work negatively by producing barren plants. The recommendation should be therefore, based on the study of the local situation.

Table 34. Optimum range of population for rainfed cropping in different crops

Crops	Optimum	Range (No./ha)	Reference
Pearl millet	1,00,000	1,41,000	
Maize	45,000	75,000	
Fingermillet	4,00,000	6,00,000	
Pigeonpea	50,000	75,000	
Groundnut	1,66,000	2,66,000	
Soybean	60,000	80,000	
Sunflower	56,000	1,00,000	
	60,000	80,000	Venkateswarlu (1985)
	50,000	70,000	Umrani and Bhoi (1985)
Cotton	25,000	75,000	
Safflower	60,000	90,000	
Sorghum	1,20,000	2,20,000	Venkateswarlu (1988)
	1,80,000	(*Kharif*)	Randhawa and Venkateswarlu (1979)
	1,50,000	(*Rabi*)	

Traditional varieties which are of long duration and taller in nature exert a high degree of competitiveness among plants hence perform better with relatively low density. These crops exhibit reduced width and height of the yield plateau. By contrast the high yielding varieties are bred for performance in a population and exhibit what is known as 'co-operative competition', hence it is necessary to increase the population density. Moreover, the yield plateau is so broad that the population can be cut down to adjust a intercrop much easier than traditional varieties. Knowledge on plasticity of crops with respect to population level, therefore, is necessary to make best use of the situation. Optimum population levels have been identified for different places (Table 34).

Use of optimum stand is something like having a good foundation to a building, opined by Venkateswarlu (1985). This ensures healthier and vigorous plant stands. Early vigour is particularly important to withstand the problem of intermittent moisture stress, weeds and even to some extent pests.

5.6 Planting pattern/crop geometry

Arrangement of plants in the field is equally important as that of optimum population. Normal geometry can be square, rectangular, elongated rectangle or irregular. The arrangements for a given population can be adjusted either by changing the inter or intra-row distances. Such adjustments have special effects on aspects like utilization of net radiation, root growth, nutrient uptake and water use. Under adequate moisture situation planting pattern does not show any advantage. Whereas under limited soil moisture planting pattern also helps to improve yield by increasing efficiency of individual plants. Square planting has reduced intra specific competition due to well distributed planting. Rectangular planting, however, has its own advantages particularly in crops grown on conserved soil moisture. There are different arrangement methods of planting.

5.6.1 *Wider row spacing*

Plants have palsticity within a certain range of row spacing (Table 35). Experiments of All India Coordinated Sorghum Improvement Project have shown no reduction in yield when row spacing of *kharif* sorghum was increased from 30 to 60 cm and that of *rabi* sorghum from 45 to 90 cm. Similarly in maize, row spacing varies from 45 to 75 cm, in finger millet from 30 to 45 cm, in chickpea from 30 to 60, in groundnut from 30 to 45 cm, in sunflower from 30 to 67.5 cm and in safflower from 45 to 90 cm without much variation in yield. Wide rows apart from the advantages of rapid land coverage, efficient weeding, prolonged intercultivation and dust mulching are distinctly advantageous during low rainfall years. Arnon (1972) postulated that when soil moisture is limited, there would be an advantage by increasing the inter-row spacings and decreasing the intra-row spacings. Intra-row competition would result in smaller plants and interrow reservoir of soil moisture would be maintained over a longer period of growing season.

Table 35. Interrelation of row spacings and profile moisture in chickpea at Hissar

Row spacing	Grain yield (q/ha)		
	Profile moisture (mm)		
	110	150	190
30	14.7	19.1	29.4
60	25.1	22.8	27.4
90	13.4	23.5	23.9

Balasubramanian and Venkateswarlu (1985)

The results of research under *rabi* situations at Solapur, Bijapur and Bellary in India support the above view (Hegde, 1995). The advantages are also due to deeper and more proliferous root system in wider row planting (Umrani and Bhoi, 1977). Water extraction in the initial stages is usually higher with closer row spacing leaving smaller amounts for future use (Hegde *et al.,* 1976). In wider rows usually inter row contains more moisture and rate of evaporation also comes down due to surface drying in a short time. Brown and Shrader (1959) and Karchi and Rudich (1966) observed that when soil moisture was not severely limited, highest grain yields of sorghum were obtained with narrow row spacings (50 cm or less) and high population, whereas in dry climates wider row spacings were desirable. Similar, observation was made in chickpea at Hissar (Table 36).

Table 36. Effect of row spacing on the yield of *rabi* sorghum (Av. of three years)

Variety	Spacing	Grain Yield (kg/ha)
CSV7R/8K	45	2620
	90	2860
M 35-1	45	2340
	90	2450

Annual Report, Bellary Soil Conservation Centre.

It is difficult to predict the season especially under *kharif* rainfall situations. It may be, therefore, advisable to start the season with permissible optimum spacing and optimum population and later on thin out the alternate or every third row to provide wider spacing or by removing unhealthy seedlings within a row itself if the season is not favourable. The experience of Bellary indicated that in subnormal years every third row of sorghum should be removed to obtain higher yields (Venkateswarlu, 1983).

Table 37. Effect of planting pattern on yield of *rabi* sorghum

Planting Pattern	Grain yield			
	1973-74	1974-75	1979-80	1980-81
	Soil depth 60 cm		Soil depth 150 cm	
Uniform planting in 45 cm rows	8.5	9.6	32.3	29.5
Paired planting at 30-30-60 cm	11.3	12.2	34.5	31.8
Paired planting at 45-45-90 cm	–	–	35.9	33.2

5.6.2 *Paired row planting*

As discussed in the previous paragraph, reduction in plant density by removing plants (alternate lines/plants in a line) though has scientific base, removal of plants or plant rows may not help to conserve the moisture which is already exhausted by plants. As an alternative to this changing of planting pattern has yielded good results in case of post-rainy sorghum. Planting crops in pairs of close rows alternated by a large cultivated patch results in initial intense intra specific competition (due to closed rows) leading to restricted vegetative growth while there would be improved root growth laterally as well as in deeper soil layers to meet the water requirement. This extensively established root system helps the plant during later stages. In this system there is border effect of root training and convenience in inter culture operations which adds to the efficacy of individual plants to provide more dry matter and finally yield. In case of sorghum, paired planting pattern (30-30-60 or 45-45-90 cm) found to increase the yield over normal planting (Umrani, 1985) (Table 37). Similar observations were made in other crops.

Yields during normal years under paired system of planting are comparable if not better than normal planted crop (Table 38). During years of scarcity, however, yields are higher with paired system of planting. In pearl millet 30-30-60 cm and in castor 60- 60-120 cm indicated similar response. Under good moisture supply situations, the large space provided between the pairs could be used for introducing a second crop as a intercrop. Under situations of late *kharif* sowings this space or the space created in wide row planting could be used for relay planting of second crop in succession.

5.6.3 *Skip row planting*

Skipping a row after every two to there lines has similar yield improvement as that of normal method of planting in dry lands but it is distinctively advantageous over normal planting during scarcity years. The advantage is due to border effect occurring in roots as well as in above ground growth. In an experiment at Dry Farming Centre, Bijapur skip row planting technique was comparable to normal planting method in sorghum and chickpea (Table 39). The averages indicated that skipping one line after every three lines found more beneficial compared to skipping after every two lines or no-skipping at all. In safflower, however, skipping one line after every two lines was advantageous.

Table 38. Effect of geometry of planting on yield (kg/ha) of *rabi* crops (Bijapur)

Crop	Spacing	1974	1975	1976	Average
Cotton	60 cm uniform rows	1123	882	816	940
	90/30 cm (pair)	1492	1230	1011	1264
Sorghum	45 cm uniform rows	593	2225	1083	1951
	60/30 cm (pair)	542	2239	1235	2003

Table 39. Effect of border method of planting on the yields (Kg/ha) of *rabi* sorghum, safflower and bengalgram

	Sorghum				Safflower				Chickpea			
	83-84	84-85	85-86	Mean	83-84	84-85	85-86	Mean	83-84	84-85	85-86	Mean
Planting 3 rows skipping 1 row	2066	1486	1462	1671	870	1088	689	882	1366	1522	827	1238
Planting 2 rows skipping 1 row	1906	1148	1536	1530	963	1120	718	938	1238	1278	760	1091
Normal planting	1595	1128	1441	1405	903	921	721	848	1087	1193	778	1019
Cd 0.05	NS	NS	NS	–	NS	34	NS	–	–	8	NS	–

5.6.4 *Hill planting*

Hill planting is a old practice where 2-3 seeds or seedlings are planted at a spot (concentrated spot planting). This practice seems to have promise in dry lands as concentrated planting increases competition among the plants planted at a spot. This would reduce the excess vegetative growth that otherwise would have occurred and wasted the available moisture. At the same time roots are forced to grow laterally and deep. This well developed roots system later on can make the plants yield better under scarcity situation. At Bijapur sowing of sorghum in 90 cm rows with three seeds per hill at 30 cm within the row was observed to give similar yields as that of normal spaced crop during normal years with little reduction in dry fodder yields but it out yielded the normal planted crop during scarcity years. The problem is that it requires manual seeding as seeding with drills in hills is difficult.

5.6.5 *Set row cultivation*

This is a traditional practice in parts of Saurashtra region. This is a improvement over placement technique. The farmers mainly grow pearl millet and groundnut in rotation in the same rather widely spaced permanent rows applying the little manure and fertilizer available with them. Areas between the rows are untilled and used for water harvesting. Such a system leads to a good root environment and it was found that in the root zone there is increase in the available nutrients and better microbial activity and improved physical characters. The system is advantageous under scarcity areas with resource poor farmers.

5.7 Nutrient management

The results of All India Co-ordinated Research Project for Dryland Agriculture and other crop improvement projects have clearly indicated that fertilizers can be profitably used in dryland agriculture. Response however, depends on crop varieties and situations (Table 40). Farmers are now convinced about the utility of fertilizers especially with improved varieties since soils are inherently poor in fertility. However, cost constraints, poor economic situation, environmental considerations, sustainability etc. call for rational decisions in fertilizer use.

5.7.1 *Crop response to applied nutrient's*

Moisture is the prime factor which decides the level of fertilizer use. Crop responses mainly depend on moisture availability interms of profile depth or soil water holding capacity and quantity of rainfall. For example the response to applied nitrogen in post-rainy season sorghum varied from 6.5 kg/kg nitrogen at Bellary (rainfall 100 mm), 9.7 kg grains at Bijapur (rainfall 680 mm), 19 kg at Solapur (rainfall 722 mm) to 27.7 kg at Kovilpatti (730 mm rainfall) depending on rainfall received (Randhawa and Singh, 1983). Generally, responses are higher with normal or above normal rainfall compared to below normal rainfall (Table 41). Responses also depend on depth and nature of soil. For example Randhawa and Singh (1983) observed response up to 25 kg N per ha in sorghum and a shallow soil with 50 mm stored moisture which increased to 50 kg N per ha on deep soil with 110 mm stored soil moisture, (Table 42). Similarly responses are generally of higher magnitude on sandy loam soils as compared to loamy sand soils (Table 43). Yield variations, thus, depend on available soil moisture (Table 44). It is the soil moisture in the profile available at seeding and the precipitation there to decide the crop performance to applied nutrients.

Response to fertilizers also depends on use of other supplemental organic sources, management (mulching), cropping system and treatment to crops in rotation. Mulching is useful for moisture conservation for post-rainy season crops. It has also been found to increase response of applied fertilizers (Table 45). Use of sheep manure, farm yard manure, green loppings, green manuring and crop residues are found useful as supplemental sources of nutrients in dry lands crops (Singh *et al.,* 1981, Hadimani *et al.,* 1982, Narkhede and Ghagre, 1987). These materials enable a reduction of 20-25 kg fertilizer N. In post rainy season crop at Bijapur application of sorghum stubbles, sunnhemp or glyricidia (5 t/ha) has resulted in maximum response with only 50 per cent of the recommended dose of fertilizers (Table 46).

Biological routes of improving soil fertility for optimum crop production are vital components of integrated nutrient supply systems. Microbial inoculants are biologically active products containing active strains of specific bacteria, algae, fungi alone or in combination which help in increasing crop productivity by way of helping in the biological nitrogen fixation, solubilization of insoluble nutrient sources stimulating plant growth or decomposing of plant residues. A number of biofertilizers are now commercially available. Research on biofertilizers in dry land agriculture is, however, scarce. Results with rhizobium inoculation are not consistent. However, there is no harm in treating legume seeds with specific cultures especially in new areas and new introductions. Non-symbiotic microbial inoculant *Azospirillum* seed or soil treatment at different locations increased the grain yield in sorghum equivalent to the application of 20-40 kg/ha minus N. Similar was the effect due to *Azotobacter*. Information on phosphate solubilizers, VAM or vermicompost is not forthcoming.

Table 40. Response (kg grains/kg nutrient) of rainy and post- rainy season crops to nitrogen and phosphorus under dry land condition

Crop	Rainy season		Crop	Post rainy season	
	Nitrogen	Phosphorus		Nitrogen	Phosphorus
Sorghum	3.4-43.4	2.4-59.0	Wheat	4.9-52.3	2.6-41.3
Pearl millet	2.1-24.8	1.7-14.3	Barley	4.0-31.6	1.9-9.2
Finger millet	5.0-42.4	6.4-38.9	Oats	8.3-20.0	–
Maize	4.1-67.4	6.8-80.0	Chickpea	1.9-23.5	4.5-38.2
Rice	5.9-17.9	5.1-31.8	Peas	–	3.7-38.2
Setaria		–	Lentil	–	3.2-6.3
Greengram		1.5-11.6			
Blackgram		1.8-6.7			
Pigeonpea		3.1-8.3			

Source : Rao & Das (1982)

Table 41. Response of crops to fertilizers as influenced by rainfall

Soil	Crop	Rainfall distribution		Yield without fertilizer (t/ha)	Response (kg grains/kg nutrient)	
					At 20-40 kg/ha	At 60-80 kg/ha
			Nitrogen			
Aridisols	Pearlmillet	Above normal	(3)	1.09	20.8	–
		Normal	(3)	0.75	13.7	–
		Below normal	(1)	0.25	6.3	–
Alfisols	Rice	Above normal	(2)	1.35	18.1	16.0
		Below normal	(3)	1.57	22.1	18.3
	Barley	Above normal	(1)	0.62	6.3	5.3

		Below normal	(2)	0.44	9.2	6.9
Inceptisols	Pearlmillet	Above normal	(2)	2.31	7.3	–
		Below normal	(2)	0.77	6.4	4.9
Entisols	Barley	Normal	(1)	2.50	10.0	11.8
		Below normal	(2)	1.29	14.8	13.7
	Wheat	Above normal	(1)	0.87	15.3	11.9
		Normal	(1)	0.72	3.2	1.5
Vertic	Sorghum	Above normal	(1)	1.84	36.1	29.6
inceptisols		Below normal	(1)	0.28	13.5	7.4
Entisols	Barley	Above normal	(1)	3.15	14.8	14.6
		Below normal	(1)	1.38	29.0	27.0
			Phosphorus			
Artisols	Greengram	Above normal	(1)	1.32	2.7	1.8
		Normal	(1)	0.30	1.0	1.3
		Below normal	(1)	0.12	2.7	1.8
Vertic	Sorghum	Above normal	(1)	2.87	13.0	–
inceptisols		Below normal	(2)	1.33	4.5	–
Inceptisols	Maize	Above normal	(1)	2.90	40.5	11.3
		Normal	(1)	1.93	7.2	0.10

Figures in parenthesis indicate number of seasons Source : Vittal (1987)

Table 42. Response of sorghum to nitrogen at different soil moisture availability on a Vertic inceptisols

Soil depth (mm)	Available moisture storage (mm)	Grain yield (t/ha) N (kg/ha)				S.E.	C.D.
		0	25	50	75		
0-30	50	0.66	1.23	1.30	1.30	0.11	0.33
0-60	110	1.66	1.78	2.13	2.33	0.12	0.33

Source : Randhawa and Singh (1983)

Table 43. Effect of N application on grain yield (t/ha) of wheat and maize in different soil types

N rate (kg/ha)	Wheat		Maize	
	Loamy sand	Sandy loam	Loamy sand	Sandy loam
0	1.27	2.11	1.26	2.07
40	1.79	3.93	2.17	2.89
80	1.98	3.45	2.00	3.41
120	2.05	–	1.97	3.12

Source : Randhawa and Singh (1983)

Table 44. Response of chickpea to fertilizer application under varying soil moisture conditions

Available Moisture	Control	Grain yield (t/ha)		Mean	CD
		50% RDF	RDF		
25%	1.04	1.32	1.77	1.38	Moisture
50%	1.27	1.57	2.33	1.73	0.35
75%	1.43	2.46	2.56	2.15	Fertilizer
Mean	1.25	1.78	2.22		0.38

RDF : Recommended dose of fertilizer - 20 kg N and 28 kg P/ha
Source : Singh & Das (1983)

Table 45. Yield and water use of dry land wheat followed by maize as affected by mulching the previous crop and N rates

		No mulch			Mulch		
Soil	Average of years	0	40	80	0	40	80
Yield (t/ha)							
Loamy sand	4	1.90	2.57	2.96	2.60	3.18	3.48
Sandy loam	3	2.52	3.56	3.99	3.69	4.49	4.72
Water use (mm)*							
Loamy sand	4	221	246	259	247	270	272
Sandy loam	3	239	254	262	262	275	283

*Sum of water depletion from 180 cm profile plus seasonal rainfall
Source : Dinesh (1982)

To sustain crop yields and land productivity, judicious combination of organic manures and biofertilizers with inorganic fertilizers is recognised as a prime factor. Combined use of organic and inorganic manures improve physical, chemical and biological properties of soil (Shashidhar *et al.*, 1995). Incubation of chemical fertilizers particularly phosphorus helps to provide the supply of applied nutrients on sustained basis. At Paiyur (Tamil Nadu) on red calcareous sandy loams, enriched FYM (125 kg SSP + 750 kg well decomposed and powdered FYM incubated for a month) along with *Azospirillum* inoculation resulted in higher grain yield of sorghum (Table 47). In post-rainy season crop at Dryland Project, Bijapur, application of sorghum stubbles, sunnhemp and glyricidia (5 t/ha) has resulted in maximum response with only 50% of the recommended dose of fertilizer (Badanur, 1993). Sorghum stubbles in combination with 20 kg N per ha or treating with cellulolytic organism also significantly increased sorghum yields over incorporation of stubbles alone (Badanur, 1995). Residues (sorghum stubbles or cotton stalks (@ 5 t/ha) with inorganic fertilizers also found to increase water use efficiency (Badanur, 1993).

In India with ecologically disadvantaged dryland areas dominated by resource poor farmers forging a strategy relying on integrated nutrient use is vital. Integrated use of organics and inorganics has been found promising in maintaining productivity, and bringing stability in crop production. The experiment on alfisols at Bangalore indicated higher and stable yields with FYM + RDF (Table 48).

Table 46. Effect of organic residues in conjunction with fertilizer on yield (kg/ha) of *rabi* sorghum

Organic residues	Fertilizer levels					
	0		50% RDF		RDF	
Sorghum stubbles @ 5 t/ha	1975	(101.2)*	2391	(114.1)	2485	(122.0)
Cotton stalks @ 5 t/ha	1905	(92.3)	2083	(88.1)	2484	(120.3)
Redgram stalks @ 5 t/ha	1929	(72.4)	1774	(78.1)	2026	(88.2)
Subabul loppings @ 5 t/ha	2253	(106.4)	2289	(115.4)	2134	(104.1)
Glyricidia loppings @ 5 t/ha	2201	(94.3)	2224	(90.8)	2047	(88.3)
Sunnhemp *in situ* @ 5 t/ha	1620	(72.2)	2186	(112.0)	2186	(109.7)
CD (0.05)	355					

* Figures in parenthesis are values of water use efficiency (*kg/ha cm) Source : Badanur, 1993

Table 47. Effect of nutrient management on sorghum yield

Nutrient management	Grain yield (kg/ha)	Straw yield (t/ha)
FYM @ 12.5 t/ha	346	8.6
FYM + *Azospirillum*	346	9.5
30:20 NP kg/ha + FYM	496	11.0
30:20 NP kg/ha+ enriched FYM	565	11.5
30:20 NP kg/ha + enriched FYM + *Azospirillum*	665	11.8
CD (0.05)	28	NS

Source : Veerbadran and Rajendran, 1993

Nutrient management has to be always based on the cropping systems adopted on the lands as previous crop treatments have lot of influence on the performance of current season crop and its nutrient management. Growing of same crop on the same piece of land may have adverse effect even under good soil fertility conditions due to nutrient depletion from the same layers of soil apart from many other crop associated factors. Mann and Singh (1977) observed 62 per cent reduction in pearl millet yield in pearl millet - pearl millet rotation in contrast to greengram-pearlmillet. Taking a legume crop in rotation or sequence, therefore, is advantageous (Table 49 and 50).

Table 48. Effect of continuous application of FYM and fertilizers on the productivity of dry land finger millet over a period of 18 years on alfisols of Bangalore

Treatment (N:P:K)	Stability Index	No. of years when yield (t/ha) was observed				
		<1	1.-2	2-3	3-4	>4
Control (no fertilizer)	0.12	6	11	1	0	0
FYM (10 t/ha)	0.42	0	3	12	3	0
FYM (10 t/ha) + 25 : 25 : 12.5	0.52	0	1	10	7	0
FYM (10 t /ha) + 50 : 50 : 25	0.63	0	0	5	10	3
RDF (50 : 50 : 50)	0.37	1	3	9	5	0

Source : Anon., 1998

Table 49. Effect of previous crop in rotation on grain and stover yield and N uptake of sorghum

Previous crop	Average yield (t/ha)		Total N uptake (kg/ha)
	Grain	Stover	
Groundnut	2.55	4.47	69.6
Greengram	2.62	4.09	79.5
Cowpea	2.73	4.63	75.4
Pigeonpea	2.03	3.95	64.6
Sorghum + pigeonpea	2.08	3.74	61.9

Source : Reddy *et al.,* 1993

Table 50. Effect of previous crop in rotation on the yield of sorghum in alfisols of Hyderabad

Previous crop	Yield of sorghum (t/ha) at different levels of N (kg/ha)				
	0	20	40	60	Average
Groundnut	1.70	2.87	2.44	3.21	2.55
Mungbean	2.26	2.86	2.63	2.74	2.62
Cowpea	2.52	2.48	2.90	3.03	2.73
Pigenopea	1.84	1.92	2.13	2.22	2.03
Sorghum + pigeonpea	1.73	1.93	2.44	2.22	2.08
Castor	2.03	2.09	2.44	2.13	2.10

Source : Singh and Das (1984)

Such rotation helps to save fertilizer N to the tune of 20 kg N/ha. Legumes, however, differ in their residual effect depending on crop variety used in the rotation. For instance, Singh *et al.* (1985) observed that rotation of pearl millet with greengram or clusterbean was better than its rotation with moth bean. The efficient legumes are pigeonpea, blackgram, cowpea and peanut (Singh and Das, 1984 and Reddy *et al.,* 1993). Better results with black gram were obtained when the crop was green-manured after the first picking (Singh and Das, 1983). On the basis of long term experimental results, Singh *et al.* (1985) reported that in pearl millet cluster bean rotation pearl millet yield was 11 per cent higher in comparison to continuous growing of pearl millet. Similar results were also obtained by Oswal *et al.* (1989).

Table 51. Changes in N status of soil after cultivation of legumes for three years at three levels of P application

Crop	P_2O_5 (kg/ha)	N_2 fixed (kg/ha)	Increase in total N over initial status (Kg)	NO_3N+NH_4-N (ppm)	Hydrolysable NH_4-N (Kg/ha)
Mungbean	0	131.6	9.5	32.0	39.2
	40	273.0	34.0	35.7	85.8
	60	299.4	40.4	35.8	35.3
	80	215.3	24.1	30.1	47.0

Mothbean	0	84.6	4.7	35.8	43.1
	40	247.6	34.0	44.8	95.8
	60	264.5	36.3	47.0	107.6
	80	231.3	31.3	42.5	74.6
Clusterbean	0	75.6	8.3	47.6	81.4
	40	145.7	19.4	48.2	105.3
	60	195.7	27.8	38.0	117.4
	80	209.2	34.0	45.9	125.4

Source : Kathju *et al.,* 1987

The beneficial effects of legume cultivation may be attributed to improved soil fertility. Singh *et al.* (1985) in a long term study found an increase in soil organic carbon by 12 per cent and available phosphorus by 25 per cent. Similarly, Das and Rao (1986) and Oswal *et al.* (1989) reported an increase in NO_3-N and available N, P, K due to inclusion of legume in crop rotation. Phosphorus application to these legumes found to increase the N fixing capacity of legumes, particularly the hydrolysable NH_4-N rather than total N (Table 51).

Legumes also find place in traditional intercropping systems. In spite of forming 50 to 75 per cent additional plant population in a intercropping system, these do not reduce the main crop yields. Performance of legumes in a intercropping system may vary depending on situations. For instance, Misra (1971) found that growing pearl millet in association with cluster bean was better in normal rainfall years while in sub normal years cowpea was a better associate. Some intercropping systems such as sorghum + cowpea/ soybean/greengram/sword bean (all as forages) had favourable influence on succeeding wheat crop compared to preceding sorghum (Balasubramaniam *et al.,* 1986). Studies under AICRP on sorghum on nutrient management in sorghum based cropping systems indicated higher yields and gross returns with recommended dose of fertilizer (RDF) (Anon., 1996). However, in some systems 50% RDF and FYM did maintain productivity and returns close to that of RDF when continued on the same site for many years due to fertility build up.

Response to added fertilizers also depends on moisture availability. Not much information is available on nutrient management in a intercropping system. Nevertheless, it may be noted that in a cereal + legume intercropping system, care should be taken so that cereal does not suffer for want of nitrogen and legume for phosphorus. Top dressing of nitrogen has to be localised to cereal component only.

5.7.2 *Recommendations of fertilizer management*

Singh and Das (1995) summarized the fertilizer recommendations for rainfed crop of rainy and post rainy seasons based on average annual rainfall, moisture storage capacity and native fertility status of the soil (Table 52). In case of rainy season cereals (Sorghum, pearl millet, maize, finger millet, upland rice), the recommended levels of N and P range from 40-80 kg and 8-22 kg/ha, respectively. For maize and finger millet 10-25 kg K/ha is also recommended. For pulse crops, 20 kg N and 13 kg P/ha are recommended.

Table 52. Fertilizer recommendations for principal rainfed rainy and post-rainy season crops

Crop	Annual rainfall (mm)	Moisture storage (mm)	Native fertility status		Fertilizer recommendations (kg/ha)			Management practices
			N	P	N	P	K	
Rainy season								
Pearl Millet	300-400	80-90/90cm	Low	Medium	40	8	0	Nitrogen to be applied in splits, half as basal and half at heading, P and K all as basal
	600-700	80-90/90cm	Low	Medium	75	11	0	
		135-145/45 cm	Low	Medium	50	13	20	
Sorghum	600-800	40-70/45cm	Low	Low	40	13	0	Nitrogen to be applied in splits half as basal, half at 3 o be applied in weeks, P and K all as basal
		135-145/45 cm	Low	Medium	90	13	0	
	800-1000	140-270/90 cm	Low	Medium	80	17	0	
Finger Millet	800-1000	180-200/90 cm	Low	Low	50	22	25	Nitrogen to be applied in splits, half as basal, half at tillering
	1400-1600	180-200/90 cm	Low	Medium	40	8	20	
Maize	800-1000	180-200/90 cm	Low	Low	75	22	25	N in 2 splits, half at sowing, half at panicle initiation stage
	1000-1200	90-100 m	Low	Medium	40	9	10	
		110-140 m	Low	Medium	80	18	20	
Upland rice	1000-1200	148-170/90 cm	Low	Medium	40	18	0	N in 3 splits, 1/4 at sowing/ transplanting, 1/2 three weeks thereafter and 1/4th at PI stage
	1400-1600	180-200/90 cm	Low	Medium	60	18	0	
Clusterbean / moth bean	300-400	80-90/90 cm	Low	Medium	0	13	0	All as basal
Horsegram/ blackgram/ cowpea pigeonpea	400-600	145-270/90 cm	Low	Medium	20	13	0	All as basal
Post-rainy season								
Sorghum	150-200	135-145/45 cm	Low	Medium	30-50	13	0	
Wheat	40-80	110-140/m	Low	Medium	30-40	13-6	0	
	100-150	90-100 /m	Low	Medium	40	9	0	
	150-200	110-180 / m	Low	Medium	60-80	18	0	All to be drilled below seed at sowing
	200-250	140-180/m	Low	Medium	60	13	0	
Barley	50-100	100-140/m	Low	Medium	60	13	0	
	100-250	140-180/m	Low	Medium	40-60	9	10	
Chickpea	60-70	110-140/m	Low	Medium	10	11	0	
	100-150	145-270/90 cm	Low	Medium	20	17	0	

Source : Singh & Das (1995)

For wheat, barley, *rabi* sorghum grown on conserved/residual soil moisture, 30-80 kg N and 6-18 kg P/ha are considered adequate to obtain fairly good productivity. In case of chickpea, 10-20 kg

N and 11-17 kg P/ha are recommended. However, the extra amount of nutrients to be added will depend on rainfall, moisture storage and soil test values.

Venkateswarlu (1985) stresses the importance of considering cordial points of law of diminishing returns and law of minimum. Under moisture constraint since the pay-of would be maximum with the first increment of fertilizers, lower fertilizer level need to be applied due to unpredictable risk factor. Secondly, yields of crop are often limited by a particular constraint like lack of good plant stand, timely control of pests, diseases and weeds, deficiency of particular nutrient element etc. By correcting these constraints the yield responses could often be synergistic. On the other hand, if a given deficiency is not corrected, the response to other applied nutrients would be limiting. For instance, in high levels of production Zn was found limiting in maize and sorghum in the southern Karnataka red soils and Rayalseema black soils in Andhra Pradesh, respectively. Calcium and sulphur deficiencies were found in case of groundnut grown in subhumid red soils region of Chotanagapur. Therefore, to get maximum output for applied nutrient it is necessary to take care of the law of minimum. Generally, this is achieved through balanced fertilizer use.

5.7.3 Increasing fertilizer efficiency

Fertilizers are costly and also add to the problems of environmental pollution when escaped in plant use. High fertilizer use efficiency, therefore, need to be considered. Use efficiency of fertilizer can be realized through proper placement of fertilizer and time of application.

Table 53. Method of application of fertilizer nitrogen on yield (q/ha) of *rabi* sorghum

Centre	Method of Application		Reference
	Drilled	1/2 drill + 1/2 foliar	
Solapur	8.91	7.29	Umarani *et al.,* (1978)
Bijapur	21.50	20.64	Havanagi (1979)

Broadcasting of fertilizers is inefficient. They need to be placed in or near the plant rows to increase their use efficiencies (Table 53). Drilling fertilizer is advantageous (Havanagi, 1979, Umrani *et al.,* 1978). In post-rainy season crops, deep placement has been found to be useful for increasing the fertilizer use efficiency (Singh and Das, 1984). While placing fertilizers within the row, there could be salt injury due to the soluble fertilizers when the seeds are about to germinate. When small quantities are used, seed-cum fertilizer drill could be advantageously used. It also helps for reducing labour needs both of human and bullock.

In dry lands the rainfall is unpredictable and hence, it would be prudent to adopt split application of nitrogen in *kharif* situations so that under aberrant weather a part of nitrogen could be saved. Under high/assured rainfall situations split application of nitrogen increased the use efficacy of applied nitrogen (Table 54). Deep placement of fertilizer P is very useful. In *rabi* season, where crops are grown on receding soil moisture, placement of P in advance of actual seeding by 10-20 days was found to be a useful practice. A detailed study on advanced application of P to *rabi* sorghum at Solapur resulted not only in higher production of both grain and fodder but also increased uptake of P due to advanced placement when there were early rains and soil was deficient in P (Table 55).

Table 54. Effect of time of nitrogen application on grain yield of sorghum on Vertic inceptisols

Treatments (kg N/ha) applied at Sowing	Split		Grain yield (t/ha)			
	25 DAS	45 DAS	1972	1974	1975	Mean
80	40	40	1.26	0.77	2.50	1.51
60	40	20	1.66	1.09	2.84	1.56
40	20	20	1.57	1.09	2.89	1.85
Control (without N)			0.70	0.69	0.81	0.58
CD 0.05			0.30	0.15	0.21	

Source : Randhawa and Singh (1983)

Table 55. Effect of advanced application of P on *rabi* sorghum in Vertisols of Solapur

Time of application	Yield (q/ha)		P uptake (kg/ha)	P concentration in index leaf (%)
	Grain	Stower		
Without P	12.0	31.6	11.44	0.24
P applied at seedling	14.4	36.7	15.41	0.28
P applied 20 days in advance of seedling	15.7	40.5	19.23	0.31
CD 0.05	1.10	3.76		

Ann. Rep. AICRPDA, Solapur, 1989

Nutrient use efficiency also depends on pre-planting tillage, use of improved crop varieties and integrated nutrient management involving major and minor elements and organics.

Integrated plant nutrient system which is perceived to have production as well as ecological considerations has following steps. 1. Identification of suitable and potential cropping systems for an area, 2. Assessment of nutrient removal by the cropping system for the desired yield levels. 3. Assessment of soil fertility status and its contribution towards nutrient supply. 4. Identification of local and on farm manurial resources and their potential in the region, 5. Possible level of processing and utilization, and availability of nutrient from these resources. 6. Establishing fertilizer rates and methods on the basis of total nutrient requirements for the cropping systems and the availability of nutrients through rotations soil, organic, biological and non-conventional sources, and considering also production levels and other input use/supply levels.

5.8 Weed management

In dry lands, it is estimated that weeds cause about 37 to 79% loss in yields of different crops (Table 56). Weeds emerge along with crops and during early stages of crop growth compete for nutrients. Later, competition for light and moisture becomes critical as weeds produce abundant foliage. In fact, weeds absorb nutrient from soil more rapidly and relatively in larger amount than does the crop (Rao and Agarwal, 1984). For producing equal amount of dry matter weeds transpire more water than do most of crop plants. Therefore, in dry lands the actual evapotranspiration from a weedy crop field, is much more than evapotranspiration from a weed free crop field (Kanitkar *et al.*, 1960). Transpiration coefficient of weeds ranged from 221 (*Tribulus terrestris*) to 1402 (*Tridax procumbens*) while that of maize and sorghum was 352 and 394, respectively (Kanitkar *et al.*, (1960). That apart, weeds also limit crop growth through allelopathic inhibition and by harbouring crop pests, hence need a control.

Table 56. Crop losses caused by weeds under different situations in dry lands

Crop	Centre	Crop yield (q/ha)		% yield loss
		Weeds controlled	Weeds not controlled	
Upland rice	Dehra Dun	33.7	7.1	79
	Varanasi	27.0	17.0	37
	Rewa	27.0	16.6	39
Maize	Dehra Dun	46.0	17.6	62
Ragi	Bangalore	27.2	15.0	45
Pearl millet	Agra	15.2	6.8	55
	Solapur	11.9	3.6	70
Groundnut	Bangalore	19.1	4.2	78
Soybean	Dehra Dun	18.4	9.2	50

Source : Sanghi and Korwar (1987)

Weeds, when not controlled can take away as much as 30% moisture and nutrients. Competition for moisture becomes critical with increasing soil moisture stress as found in arid and semi-arid areas. Kasasian and Seeyave (1969) proposed that all crops need a weed free period for one fourths to one thirds of their growing season. In general the initial 4 to 6 weeks period is more critical from the point of view in almost all crops (Table 57). However, the duration of critical period vary with crop, crop cultivation, duration, plant population, planting geometry, crop growth as well as the weed species and their characteristics. Yield losses are reported to be higher in drought year (15-67%) compared to losses during favourable season (21-35%) (Carson, 1987). Weeds and crop seeds germinate almost simultaneously and compete with each other particularly between 20-30 days due to steep rise in their growth. Incessant rain during initial stage not only hinder conventional weed control measure but also provides ideal condition for weed emergence over an extended period of time. The moisture in the surface soil is critical for weed emergence. The relative growth of crops, its population, ground cover etc. would decide the establishment and competitive ability of weed seedlings that have emerged. Hence, there will be variation in crop losses due to weeds from year to year and from place to place.

Table 57. Critical periods of competition and percentage yield reduction

Crop	Critical period	% yield reduction	Remarks
Rice upland	16-40 DAS*	60-80	Approx. per cent loss by seeds. However, the extent of reduction in yield mainly depends upon the type of weeds present and other factors
Maize	15-40 DAS	30-60	
Sorghum	15-35 DAS	30-40	
Cotton	30-60 DAS	45-85	
Groundnut	25-55 DAS	55-70	Erect

Sunflower & Safflower	30-60 DAS	20-25 30-50	Spreading
Sesamum	15-30 DAS	45-50	
Castor	30-60 DAS	35-40	
Soybean	20-45 DAS	25-35	
Mungbean	25-45 DAS	50-55	
Pigeonpea	20-60 DAS	40-60	
Blackgram	25-35 DAS	45-60	

* Day after sowing Source : AICRPWC, Comprehensive report for the period 1978-84. (Singh and Sarswat, 1986).

5.8.1 Integrated weed management

The semi-arid and arid environments being ecologically fragile endowed with poor resource base of farmers weed control options should be seeking to develop effective and economically viable weed management practices that will minimize drudgery and be environmentally neutral or favourable. Various weed management practices (Preventive, manual, mechanical, cultural and chemical) developed aim to minimise the weed growth and to maximize crop growth and yield. Although complete weed control through adoption of any single method is highly doubtful, combination of suitable techniques normally have complementary effects.

Preventive methods : Prevention of the spread of specific weed species embodies all measures to deny the entry and establishment of new weeds in an area. They include, use of weed free seeds, keeping field bunds and irrigation channels free of weeds and checking spread of perennial weeds that reproduce vegetatively.

Direct methods : Off-season tillage helps to minimise weed problem in *kharif* crops on light soils. During cropping, time of weeding is important. First weeding has to be necessarily completed within 20-25 days, for crops and varieties of about 100 days duration. On the other hand for about 150 day duration crops and varieties the period of first weeding can be extended upto 35 days. Manual method is the oldest and simplest method of removing weeds within rows and hills of crops, where weeds can not be controlled by mechanical means.

The efficiency of manual weeding could be increased using various hand tools such as hoes, krupi, spade and double bladed knives depending upon the crop and spacing and density of weeds. For all mechanical weed control methods, straight row planting is essential. Competition against weeds in field crops is similar whether the crop is broadcast or row seeded. However, seed drills can be used for row seeding. Wide row planting with recommended population facilitates weeding operations. Sweeps instead of blades are found to be useful in intercultivating as they cut the soil deep and uproot weeds more efficiently. In the process the plant rows would have some earthing up operation as well.

5.8.2 Indirect methods

Cultural methods of weed management can be regarded as an indirect method as their primary objectives are to provide optimum environment for crop growth. The type of tillage implements used (bullock/ power tiller/tractor drawn) and depth of ploughing influence the degree of weed control. Tillage directly affects the weed seed bank by physical mixing with soil and inturn reducing the weed number. Under chisel tillage weed seeds are buried deep (30 cm) compared to shallow burial (10 cm) in conventional tillage. Weed seeds buried deep, germinate but fail to emerge and die due to thick soil layer length above it.

Stale seed bed technique involves primary cultivation followed by a period in which seeds are induced to germinate and they are destroyed by further cultivations or herbicides use before sowing the crops. This technique is of great importance in rainfed cropping where moisture regime dictates weeding during calving cropping period.

Increasing competitive ability of the crops is one of the cheapest and most useful methods of weed control. The time of planting, cultivar grown, plant density, the nature and time of fertilizer application are to be manipulated to optimize crop growth and to minimise weed growth.

Crop selection influences the crop weed balance. Selections of competitive crops like horsegram (*Macroptylima uniflora*), cowpea (*Vigna unguiculuta*) and soybean (*Glycine max*) shift the crop weed balance in favour of crops. The high yielding cultivars with their short stature and erect leaves allow more light to penetrate through the canopy and are less competitive against weeds and to mitigate this, such cultivars need be planted closely.

In cropping systems, each crop has its own characteristic weeds and therefore, monoculture tends to increase crop associated weeds. By planning suitable crop rotation, the possibility of build up of such weeds can be greatly reduced.

Crop rotation plays a significant role in the management of certain host specific parasitic weeds. Cotton, sunflower, cowpea, groundnut and castor are important trap crops of *Striga* sp., a parasitic weed in sorghum. Similarly, *Orobanche* sp. a parasitic weed in tobacco, tomato and brinjal could be trapped via crop rotation with sunnhemp, sorghum and sesamum.

Ecological suppressions of weeds is also possible by intercropping system. This phenomenon is particularly important for 'smother crops' or living mulch system in which low growing weed suppressive species are grown between tall growing crops. Some of the useful combinations are maize and soybean or cowpea, sorghum and redgram or fodder cowpea or soybean or blackgram and cotton and blackgram and coriander.

Weed growth is usually stimulated by application of nitrogen and phosphorus. Early to mid season application of nitrogen and basal application of phosphorus are often beneficial to crop for effective canopy development and competition with weeds.

Herbicides use in dry land agriculture is limited as mixed and intercropping is common in dry lands. The selecting of crops to herbicides vary very much. Another problem is uncertainty of soil moisture, which influences the herbicide efficacy as well as seed germination. Further herbicides are not used at present for dry land crops on account of cost considerations. The selective herbicides for different crops are atrazine for sorghum, maize and pearlmillet, fluchloralin, pendimetalin and metolachlur for pulses and oilseed crops (Table 58 and 59). The effectiveness of the herbicides could be increased by proper method of application, precise time of application, use of surfactants, safners and slow release formulations.

Another approach for weed suppression through allelopathy provides us with a new alternative in weed management. The field use of allelopathy could be mulching of allelopathic residues, planting of strips of crops toxic to weeds but not to the crops and integration of allelopathic strategies with other methods of weed control.

The subsistence dry land farmer who is slow to accept weed control technology and herbicide use can be made to accept these technologies if losses caused by weeds, timeliness of weed control and cost-benefit ratio of various weed control practices are convinced to him effectively.

5.9 Pest management

Pest is also an important constraint in dry land production. Precision and timeliness of operation and selection of suitable durationed and pest tolerant crop cultivars may reduce the severity problem and protect farmer from total failure. Early sowing leads to vigours seedlings which can easily withstand the knockout problems due to wholly triple alliance of pests, diseases and weeds. The important pests of crops are weather bound hence can be managed to some extent through production practices. There are some low cost and no cost inputs also. For instance seed treatment with Apron 35 SD at 6 g/kg seed can control the downy mildew or green ear disease of pearl millet (Muthuswamy and Narayanswamy, 1981). Phytosanitory measures like destruction of crop residues and long term crop rotation also reduce the primary inoculum.

Table 58. Selective herbicides for weed control in major dry land crops

Crops	Herbicides combination	Rate (kg/ha)	Time of application	Remarks
Rice (upland)	Butachlor	1-1.5	PE (6-10 DAS)	Annual grasses and some broad based weeds
	Thiobencarb	1-1.5	PE (6-10 DAS)	
	Oxadiazon	0.5-1.0	PE	
	2,4-D	1-1.5	Post E	Broad leaved weeds, annual grasses
	Oxadiazon fb. one HW	0.5	PE	
Maize	Atrazine	0.5-1.0	PE	Broad spectrum control of broad leaved weeds and annual weeds
	Pendimethalin	1.0-1.5	PE	
	Simazine	0.75-1.0	PE	
	Alachlor	1.0-2.0	PE	
	Atrazine+Pendimethalin	0.5 + 0.5	PE	Many annuals including *Rottboellia* spp.
	Fluchloralin	0.75-1.0	PE	
Sorghum	Atrazine fb one HW	0.25-0.5	PE	Broad spectrum control of annual grasses and broad leaved weeds
	Simazine	0.5-1.0	PE	
	Alachlor	1-1.5	PE	
	Isoproturon	0.5-0.6	PE	
		0.2-0.5	PE	
		0.5-1.0	PE	
Millets	Atrazine	0.2-0.5	PE	Many grassy weeds
	2,4-D	0.75-1.0	Post E	Top control *Striga* sp.
Cotton	Fluchloralin	0.75-1.0	PPI	Broad spectrum control of annuals
	Pendimethalin	1.0-1.5	PE	
	Diuron	0.5	PE	
	Alachlor	1.0-1.5	PE	
	Oxadiazon	0.5-0.75	PE	
Groundnut	Alachlor	1.0	PE	
	Fluchloralin	1.0-1.25	PPI	

	Pendimenhalin	1.0-1.5	PE	
	Butachlor	1.0-1.5	PE	
Sunflower	Alachlor	1.0	PE	Broad spectrum control of animal weeds
Safflower	Fluchloralin	0.75-1.0	PPI	
	Pendimethalin	1.0	PE	
Sesamum	Pendimethalin	1.0	PE	Control of grassy and broad leaved weeds
	Fluchloralin	0.75-1.0	PPI	
	Alachlor	1.0	PE	
Castor	Alachlor	1.0-1.5	PE	Broad spectrum control of annual and grassy weeds
	Fluchloralin	0.75-1.0	PEI	
	Alachlor	1.0	PE	
	Fluchloralin	1.0	PE	
	Pendimetnalin	1.0-1.5	PE	
Soybean	Fluchloralin	0.75-1.0	PPI	Annual grassy and broad leaved weeds
	Alachlor	1.0-1.5	PE	
	Pendimethalin	0.75-1.0	PE	
	Metri buzin	0.25-0.85	PE	
Pigeonpea	Alachlor	1.0	PE	Broad spectrum control of annuals
	Pendimethalin	1.0-1.8	PE	
	Fluchloralin	0.75-1.0	PPI	
Greengram	Alachlor	0.75-1.0	PE	Broad spectrum control of annuals
	Pendimethalin	0.5-1.0	PE	
	Fluchloralin	0.75-1.0	PPI	
Blackgram	Pendimethalin	0.5-1.0	PE	Broad spectrum control of annuals
	Fluchloralin	0.5-1.0	PPI	
	Alachlor	0.75-1.0	PE	

PE- Pre-emergence, Post E- Post emergence, PPI -Pre plant incorporation, HW - Hand weeding, DAS - days after sowing, fb - followed by Ramakrishna and Tripathi, 1995

Table 59. Promising herbicides identified for managing in intercropping systems

Intercropping system	Promising herbicide	Reference
Maize/Mungbean	Butachlor	Bantilan *et al.* (1974)
Maize/Cowpea	Butachlor	Pamplona and Imlan (1976)
Maize/Cowpea	Alachlor	Damodaran and Sankaran (1974)
Maize/Lablab		
Sorghum/Cowpea		
Sorghum/Lablab		
Maize/groundnut	Trifluralin	Damodaran and Sankaran (1974)
Sorghum/blackgram	Alachlor	Boopathi Babu (1978)
Sorghum/Cowpea		

Sorghum/Lablab	Dinitamine	
Maize/Mungbean	Alachlor and Linuron	
Sorghum/pigeonpea	Ametryn Prometryn Terbutryn Pendimethalin Fluchloralin	ICRISAT (1977) Vijaykumar *et al.,* (1988)

Butachlor Alachlor

Ergot affected seeds while sowing can be separated in pearl millet before sowing by immersing seeds in 10% NaCl solution in which the affected reds float. Deep ploughing can also reduce the incidence. Smut is more prevalent in Tamil Nadu, Andhra Pradesh and Maharashtra. Removal and destruction of diseased ears early in the season and treatment of seeds with 3 g captan/tiram/kg of seed can reduce the disease.

Now resistant plant varieties for important diseases are available. In endemic areas use of these strains is beneficial. In case of sorghum shoofly incidence is more under delayed sowing conditions. To avoid the severity of the pest, sow the crop with the onset of monsoon. Enmass sowing of the crop in an area helps to avoid incidence. Extended sowing situation is congenial for establishment of the pest. Sorghum ear head bug is a problem if the crop comes to milk stage by late August. If the crop is sown in early June, the crop can as well escape this pest. If the sowing is delayed, it is necessary to change over to a shorter duration variety.

Many of the seed borne diseases can be controlled though seed treatments with appropriate chemicals e.g. smuts of cereals. Biological agents are also useful in natural conditions. For instance, *Trichoderma harzianum* can control charcoal rot of sorghum in field conditions (Elad *et al.,* 1986). Certain rots can be reduced through deep ploughing (e.g. stem rot-*Sclerotium rolfsii* of groundnut). Pigeonpea wilt is found to be less severe in a intercropping situation with sorghum.

Castor semilooper can be managed through adjusting time of planting. The pest completes its life cycle in about 21 days. Egg laying usually starts with the on set of monsoon. A peak build up of semilooper has been usually noticed during first fortnight of August. There are two natural larval parasites available in nature which are abundant during second fortnight of August when decline of the population of semilooper is usually noticed.

5.10 Water use efficiency and protective irrigation

Water being the most scarce and critical input that is not always within the control of grower, efforts need to be made to enhance its use efficiency, preferably use coefficient a ratio of yield (Y) per unit are a to the water used (Cu or E) to produce it, through management practices. Management practices through their influence an yield directly or through moisture conservation and increased supply of water to plants thereof help enhance water use efficiency (Table 60). Practices like seed hardening, selection of water efficient crops/varieties and use of antitranspirants are also associated with WUE.

Early and large canopy can help to reduce evaporation loss from the soil surface. This is possible through fertilizers or by improving soil fertility through any means. As far fertilizers are concerned one has to be careful since in semiarid regions with summer rainfall rapid early growth associated with excess early water use may result in insufficient water being available at grain filling at many instances leading to poor harvests compared to unfertilized condition. Fertilizer use, therefore, remains to be a trickly issue related with available moisture which itself is unpredictable.

Table 60. Some agronomic practices used to improve water use efficiency.

Reducing evaporation from the soil surface	Increasing the water supply to plants
Mulch	Rain harvesting
Modifying plant population and spacing	Supplementary irrigation
Selecting varieties with rapid early growth	Cultivation to improve infiltration and reduce runoff
Early sowing	Weed control
Application of fertilizers	Fallowing
Post-sowing tillage	Application of fertilizers, Multiple/relay cropping selection of varieties with deeper roots

The other alternative is to collect and recycle runoff water which is common place in dry lands due to intense rain storms. Runoff occurs when rainfall exceeds water storage capacity of soil or rainfall intensity exceeds the infiltrability of the soil. Traditionally, runoff used to be collected in the farm ponds and reservoirs in India. With recent technologies a refinement can be brought about in moisture consumption and excess water can be effectively collected. Once this is achieved, the question that needs to be considered is its sufficient use. It comprises of i. conveyance of water from pond to the field, ii. time of irrigation, iii. method of irrigation, iv. quantity of water for irrigation, and v. selection of crops and cropping systems, including agro-horti, and agro-forestry system that respond to limited water supply.

5.10.1 Response of crops to supplemental irrigation

The results of application of supplemental irrigation at different locations were summarised by Singh (1983) (Table 61). Application of 5 cm of water at critical stage of growth found to increase crop yields considerably. Similar observation was also made by Gajri *et al.,* (1981). On medium deep black soils of Bijapur, the average responses over a period of 4-5 years indicated yield increase by 33 to 92 per cent with one life saving irrigation (Table 62).

Table 61. Effect supplemental irrigation (5 cm) on crop yields at different locations

Centre	Crop	Yield (q/ha)		Per cent increase over control
		Control	Irrigated	
Agra	Wheat	21.9	27.4	25
Bellary	Sorghum	4.3	13.5	213
Chandigarh	Wheat	6.7	20.0	198
Dehra Dun	Wheat	21.4	35.5	66
Hyderabad	Sorghum	13.9	19.9	43
Jodhpur	Pearl millet	19.8	36.0	82
Varanasi	Beedi tobacco	14.9	17.5	17

Singh (1983)

Table 62. Response of crops of supplemental irrigation in medium deep black soil

Crop	Grain yield (kg/ha)		Per cent increase over control
	Irrigated	Unirrigated	
Rabi sorghum	2574	1845	40
Safflower	1702	1196	42
Bengal gram	887	569	33
Sunflower	1429	873	64
Pigeonpea	381	198	92
Hybrid cotton	1514	1094	38

Radder *et al.,* (1995)

Results at different places also indicated improved response due to increased, irrigation levels alone or incombination with increased nitrogen rate. For instance the grain yield of *rabi* sorghum increased with an increase in the level of nitrogen under supplemental irrigation (Radder, *et al.,* 1995). Such irrigation when provided to a high value crop like hybrid cotton or fruit crops or vegetables can enhance income levels too (Vijayalakshmi and Vittal, 1981) (Table 63 and 64). Responses among crop however vary.

Table 63. Evaluation of cropping system under farm pond water utilization

Cropping systems	Gross income (Rs/ha) with irrigation			
	Control	1	2	3
Hybrid cotton (DCH-32)	7348	10968	14904	19228
Bengalgram + safflower (4:2)	4227	5321	6117	6646
Rabi sorghum + Bengalgram (2:1)	6315	6413	6920	7129
Sunflower – Bengalgram (sequence)	7664	8222	10205	11182

Source : Radder *et al.,* (1995)

Table 64. Influence of supplemental irrigation on different horticultural crops (Bijapur)

Crop	Fruit yield (kg/ha)		Gross income (Rs/ha)	
	Irrigated	Control	Irrigated	Control
Ber	5236	3265	26,180	16325
Fig	1341	946	8046	5676
Guava	2772	1245	5544	2490

Source : Radder *et al.,* 1995

5.10.2 Time and depth of irrigation

It is an extremely important aspect so as to derive the maximum benefit of the water collected. In the *kharif* regions, the supplemental irrigation may be provided to save the crop from cyclic stress that will generally occur in July and August in the Deccan Plateau. In potentially double-cropped areas, the supplemental irrigation may be used to start of the second crop as the surface soil is dry enough to adversely affect the crop stand after *Kharif* harvest.

Under *rabi* conditions the stress is continuous as the crops are grown on stored soil moisture. In areas which do not receive rain during early growth of *rabi* crops, a small early irrigation with harvested water help to stabilize yields in light textured soils (Brihar and Gajri, 1980 and Gajri *et al.*, 1981). Pre-sowing irrigation in post rainy season crops is much beneficial. During subsequent growth depending on critical stages as pre-flowering and pod development in pulses or at flag leaf and grain formation stages irrigation may be scheduled. Since the terminal stress is common in *rabi* crops scheduling at flag leaf stage in sorghum is advantageous.

The depth of irrigation has direct bearing on the yield of crop depending on time of irrigation, available soil moisture in the root zone. At Hyderabad, it is noticed that water to a depth of two centimeters may be effective in shallow rooted crops and at least 3 to 5 cm of wetting is essential for deep rooted crops (Table 65). Thus, response to extensive irrigations are appreciable and this also provides an opportunity of large area irrigation with available water compared to intensive irrigation.

Table 65. Response of long duration *kharif* crops to supplemental irrigation

Crop	Year	Level of irrigation (cm)	Yield (q/ha)		Per cent increase
			Irrigated	Unirrigated	
Castor	1977	1	9.8	7.4	32
	1978	2	9.6	5.3	81
	1979	3	13.2	10.1	31
Redgram	1980	4	1.7	0.5	240
	1980	5	3.5	0.5	600

Source : Vijaylakshmi & Vittal, 1981

Application methods to increase the efficiency of water and cut down on unproductive losses were tried over years and it has been clearly shown that for shallow rooted narrow spread crops like sorghum combined middle alternative furrow has an advantage over furrow irrigation (Anon., 1988). At Bijapur drip irrigation using pond water saved 50 per cent of water compared to modified sprinkler and surface methods of application (Anon., 1991). Yield resources were also higher with drip method (Table 66).

Table 66. Response of fruit crops to different methods of water application in black soils at Bijapur

Method of irrigation	Yield increase over control (%)		
	Ber	Fig	Pomegranate
Drip	252	228	133
Modified sprinkler (jet method)	189	128	52
Surface	149	82	60

At Solapur, sprinkler method of irrigation increased grain yield of *rabi* sorghum compared to contour furrow irrigation (Anon., 1988). However, drip or sprinkler units may be beyond the purchasing power of dry land farmer.

5.11. Plemental irrigation

A farming system approach aimed at realizing higher returns using farm pond water selectively. At Bijapur, horticulture components proved superior to arable crops. Ber, curry leaf, hybrid cotton and mulberry gave substantially higher income than pastures or bengalgram + safflower (4:2) sequence.

5.12 Antitranspirants

Water stress during reproductive stage limits crop yields substantially in dry land areas due to early senescence or dropping of fruiting parts. Arnon (1975) reported that there is a need to reduce transpiration to a minimum to get high water use efficiency. Therefore, the use of antitranspirant materials when applied to transpiring plant surface reduce water loss from the plant hence conserve water, was introduced to solve stress problems. Several chemicals were identified to have transpiration suppressant property. These chemicals work by closing stomata, or by forming a film over the stomata or by reflecting the incident solar radiation.

Slatyer and Bierhuizen (1964) reported that phenylmercuric acetate (PMA) was useful in increasing water use efficiency in cotton. It is a stomata closing type of material, which prevents complete opening of guard cells by decreasing the vapour loss from the leaf. Growth regulators such as ABA/alkanyl; succinic acid and retardants like CCC, mepiquat chloride and herbicides like atrazine and alachlor at lower concentrations found to have similar effect. Das and Santkumari (1975) observed that the amide form of alachlor and butachlor led to rapid closure of stomata within 24 hrs of application. Shantkumari (1977) further reported that alachlor at 20 mg/l on 2 week old maize remarkably reduced transpiration upto 20 days after spraying. Likewise, Atrazine reduced transpiration of maize upto 50 per cent (Smith and Buchlotz, 1962), increased moisture in the plants after 12 to 24 hrs (Wills *et al.,* 1963) and increased grain yields under limited water condition (Fuehring, 1978).

Narang *et al.,* (1988) reported that alachlor and atrazine increased grain yield significantly over control when applied at concentrations of 20 mg/l water, sprayed twice at monthly interval using 500 1 spray solution per ha (Table 67). Mixtalol- a long chain fatty alcohol – on the other hand failed to bring about similar effects. They noticed conservation of water with antitranspirants which enabled plants to maintain a high leaf water potential and photosynthetic activity.

Table 67. Effect of antitranspirants on grain yield of maize

Treatment	Pooled mean (t/ha)
Control	3.89
Mixtalol (1 mg/l)	4.14
Alachlor (20 mg/l)	4.37
Atrazine (20 mg/l)	4.47
C.D. (0.05)	0.38

Source : Narang *et al.* (1988).

Reflecting materials when sprayed on leaves reduce solar energy absorbed and reduce leaf temperature and thus transpiration. Kaoline a radiation reflectant type had beneficial effect on cotton (Gidda and Morey, 1981; Dhopte *et al.,* 1987), groundnut (Khan and Morey, 1980) and rapeseed (De and Giri, 1978).

In an experiment with Cotton, Dhopte and Jamdar (1998) used Kaoline (1.25%), Hico 110-R along with Ascorbic acid (500 ppm) + PMA (10 ppm) by spraying once at peak flowering after termination of rainfall (September). He noticed a yield increase by 23 per cent due to spray of Atonik + Kaoline (1.25%) under rainfed conditions. A yield increase of 9.4 per cent was also evident with Hico 110-R (Table 68). Hence, they suggested the use of kaoline and Hico 110-R to improve cotton yields under dry farming areas.

The forgoing discussion, claudicates the possibility of use of anti-transpirants under scarcity situation.

6. CONTINGENT CROP PLANNING FOR ABERRANT WEATHER

In dry lands, crop production depends on monsoon rainfall and the trend of monsoon is unpredictable. Factors of unpredictable nature are the starting of monsoon, its distribution, and closure of the rainy season. Since these factors are uncontrollable, modifications need to be made in cropping practices to over come the undesirable effects of rainfall vagaries. In other words, the technology of production developed for normal years/ with normal distribution of rainfall may not hold good under aberrant weather conditions. A plan known as contingency plan, therefore, has to be considered to overcome the uncertainty.

Contingency may arise in four situations

I. Timely onset of monsoon followed by prolonged dry spells

II. Delayed onset of monsoon

III. Early withdrawal of monsoon and

IV. Extended monsoon

Early commencement of rain is rather rare. Under such situation normal crops can be planted as early as 15 days without any adverse effect. The aberrations listed above have serious impact on normal production processes and hence, warrant certain contingency measures detailed here.

Table 68. Effect of antitranspirant (Hico -110 R) on growth and yield of hybrid cotton AHH-458

Particulars	Treatments		Deviation over control (%)
	Control	Treated*	
Stomatal conductance (cm/s)	0.26	0.22	-15.4
Transpiration rate (μg/cm^2/s)	8.32	7.24	-12.9
Leaf temp. difference ΔT°C	-0.48	-0.60	
Height (cm)	56.2	61.0	+8.5
Seed cotton yield (g/plant)	30.22	29.78	–
Boll No./plant	11.8	13.4	+13.6
Seed cotton yield (kg/ha)	950	1048.3	9.4

CD for yield at 5%-73.8 kg/ha *Three sprays at peak flowering @1 ml/L

Source : Dhopte and Jamdar, 1995

6.1 Timely onset of monsoon followed prolonged dry spells

There may be variation in the onset of monsoon for 5-10 days. This situation can not for the practical purposes, be treated as a serious aberration requiring corrective or remedial measures but it is those spells extending over 3-4 weeks and at such times with severe effects on yields need strategy to reduce their adverse effects. Effects of breaks also depend on soil storage capacity and nature of crops cultivated. For example, a break of 15 days may not affect adversely on soils medium to deep but a break of a week's period can cause wilting of plants on shallow light soils. Similarly, crops response also differ. Crops like pigeonpea or castor modify their growth pattern. These crops produce several flushes of flowers. In other words, crops with indeterminate growth habit are less affected and hence are desired under dry land conditions. Different practices of practical value are as follows.

6.1.1 Dry spell immediately after sowing

During some of the years a dry spell may occur immediately after sowing of the crop. This may result in poor germination due to soil crusting, withering of seedling or poor establishment of crop stand due to inadequacy of soil moisture at sowing.

Resowing : It is necessary to maintain proper plant stand to ensure better yields. When there is a severe dry spell and the population is too poor to expect any reasonable yield, resowing of the entire field using same or new crop using subsequent rains is necessary. At Solapur, resowing in case of pearl millet gives a good illustration of the advantage gained by resowing (Table 69 and 70). In crops like sorghum reseeding with the same crop is discouraged. Resowing with new suitable crop may be considered under such condition. For instance, it is a practice in post rainy season sorghum to reseed with Wheat crop in the vertisols of Deccan Plateau.

Table 69. Performance of pearl millet (var : MBH - 110) in relation to occurrence of ear season drought

Date of sowing	Plant production/ha		Grain yield (q/ha)	Fodder yield (q/ha)
	Initial	At harvest		
28th June	1,50,000	21,900	3.7	10.9
15th August	1,50,000	1,11,806	6.2	20.1

Source : Singh, 1995

Table 70. Effect of reduced plant density on production during dry spells of 197677 on sunflower

Plant density/ ha	% reduction	Yield (q/ha)	Additional grain yield (%)
74,000	Normal	7.12	–
56,000	24	9.87	39
49,000	34	9.12	28

Source : Singh, 1995

Gap filling : If the dry spell following sowing results in suboptimum population, with the subsequent showers gap filling may be considered advantageously. If the sow up crop is suitable for transplanting and the stage of the seeding is adequate for transplanting, transplant in the gaps using excess seeding growing at some places in the field with subsequent rains eg. finger millet, pearl millet. If the break in rains is rather long and do not permit transplantation of already sown crop with existing seedlings due to age, other crops grown in the region can also be used for transplanting eg. pearl millet, finger millet, chilli, tomato etc.

Reseeding can also be considered in the gaps using crops suited for late planting. At Dry Farming Centre, Bangalore it is suggested to fill the fields of groundnut and finger millet with the seeds of cowpea horsegram, niger or sunflower in the event the gaps occupy more than 25 per cent of the area. At Dharwad in crop production plots during one of the years the crop stand was not satisfactory in a groundnut + pigeonpea (3:1) intercropping. After three weeks of dry spell, hybrid maize was hand dibbled in the gaps with the fresh showers. This practice helped to obtain reasonable income during that year.

6.2 Delayed onset of monsoon

Rains may be delayed by about 20-40 days as observed in some places. Such a late onset shortens the length of growing season, restricts the choice of crops and varieties, limits use of inputs, results in low yields of *Kharif* crops due to delayed sowing, and leaves less moisture for post-rainy season crops grown in succession. Change of crop may also be required due to pest and diseases problems. For example in Maharashtra, Pearl millet is likely to suffer from ergot (*Claviceps microcephala*) disease under delayed sowing condition. The suggested measures for delayed onset of monsoon are as under.

6.2.1 Sowing under dry condition

Whenever onset of rains is delayed it is preferable to go for dry seeding with necessary seed treatment if the seed bed is timely prepared. This practice helps to avoid further delay in sowing that could happen once showers of high intensity are received in medium to deep soils. Dry seeding needs to be followed based on probability of rains. Under dry seeding basal fertilizer application may be delayed till the establishment of crop.

6.2.2 Raising nursery and transplanting

In endemic areas community nurseries using farm pond or tank water needs to be raised and seedlings could be transplanted with the receipt of rains. For instance, extensive surveys in Bangalore (Hegde and Seetaran, 1985) indicated that medium to late varieties of fingermillet should be transplanted for maintaining yield levels whenever sowing is delayed. Fingermillet transplanted at Bangalore during the month of September yielded as high as 25.6 q/ha as against a yield of 7.5 q/ha with directly sown crop. Tomar *et al.,* (1974) also reported the advantages of transplanted pearl millet over seeded crop under delayed sowing situation. The limitation with this technology is the availability of water for raising nurseries and amenable (pearl millet, finger millet) crops.

6.2.3 Changing plant population and planting geometry

In northern dry zone of Karnataka increased population level has been recommended in case of pigeonpea under delayed sowing condition to get near normal yield. Since the per plant productivity is reduced under delayed sowing, use of one and half and recommended population has been suggested on light soils.

Another practice to remedy the yield is to use wider or skip row planting and obtain benefit of border effects. In double cropped area delayed onset may reduce length of growing season and availability of moisture for the second crop grown in succession. If adjustment with short duration

Kharif season crops is not possible, the regular *Kharif* crop may be sown in wider rows (preferably those followed for the crop in succession) than normally followed and the second crop could be planted at normal time in the space available between the rows. This has been a practice in some parts of northern dry zone in Karnataka where the *Kharif* groundnut is sown in wide rows (45 cm instead of the usual 30 cm rows) and *rabi* sorghum is relay planted between the rows behind the plough during October.

6.2.4 Selection of right crops and species, depending on the onset of monsoon

Certain crops are adopted to particular environmental conditions and will not perform better when sown late while some crops are relatively less sensitive or intensive to day length variation and perform fairly well under extended period of sowing (Table 71). Therefore, the first pre requisite is to select suitable crop depending upon the onset of monsoon. For instance following is the suggested crop at Dry Farming Centre, Bijapur (Venkateswarlu, 1983).

Table 71. Yield of legumes (q/ha) as influenced by normal late sown conditions (Pooled mean of 2 years)

Legumes	Sowing period		Mean
	Annual	Late	
Cluster bean	5.57	5.68	
Green gram	5.52	2.25	
Kidney bean	7.25	3.50	
Cowpea	6.34	1.63	
C.D. 0.05	2.18		

Rainfall situation	Suggested crop in black soil areas
Normal rains in June	Hybrid pearl millet/pigeonpea/ greengram/groundnut/setaria, *Kharif*/sorghum or pearl millet + pigeonpea (2:1)
Failure of rains in June but rains in July	Hybrid pearl millet/groundnut (spreading)/sunflower on medium to deep soils, sesamum /pigeonpea/cowpea horsegram on light soils
Failure of rains in July but good rains in August	Pulses/sunflower/setaria/pigeonpea/herbaceum cotton
No good rains in *Kharif* but in early September	Setaria/safflower/kidney bean
Normal *rabi* rains (mid Sept.)	*Rabi* sorghum/safflower/cotton, bengalgram
Normal rains in middle October	Bengalgram/safflower/sorghum or sorghum + safflower (3:3)

Castor (cv. Aruna) found to yield better than pigeonpea under late sown conditions on red soils in Telangana region. In Western Rajasthan, short duration crops like greengram and cowpea were found to perform better than pearl millet under late sown conditions (Singh, 1995).

All dry land research centres are equipped with alternate crop strategy as contingent crop planning. One such crop strategy for Solapur region and central dry region of Karnataka are presented in Table 72, 73 and 74 respectively.

Table 72. Contingent crop planning for late onset of monsoon in Solapur region

Rainfall situation	Suggest crops	Remark
Rains during 1st fortnight of July (normal onset of monsoon)	Pearl millet, setaria, sorghum (CSH 6) Groundnut (M-13), Castor (Aruna) Redgram (No. 148) Horsegram (K42) Blackgram (9), Sunflower	Adopt intercropping of pearlmillet + redgram (2:1)
Rains during 2nd fortnight of July (Rains delayed by 15 days)	Setaria (Arjuna), Sunflower Castor Redgram, Horsegram (D-40-1)	Intercropping redgram + Setaria (2:2)
Rains during 2nd fortnight of August (Rains delayed by 40 days)	Castor, Sunflower, Redgram	Intercropping sunflower + redgram (2:1)

Table 73. Suggested crops for central dry zone of Karnataka

Rainfall situation	Suggested crop/cropping system
Normal onset in June	*Kharif* sorghum, groundnut pearl millet, groundnut and redgram (4:1) Chinamung/cowpea/blackgram in double cropped areas with rains in May
Normal rains in July	Long to medium duration finger miller (Indaf-1, Indaf-3, Indaf-5 or PR 202), Sunflower, castor, finger millet nursery
Normal rains in August	Finger millet (PR 202, Indaf 5) transplant finger millet/setaria/cotton/ sunflower (Morden)
Normal rains in September/ October	Transplant finger millet, Setaria, cowpea, horsegram safflower/ bengalgram/ linseed/sunflower/*rabi* sorghum

Source : Venkateswarlu, (1983)

Table 74. Average yields (q/ha) of various crops under and season correction trial sown at fortnightly intervals during 1974-80

	June II	July I	July II	Aug I	Aug II	Sept. I	
Cereals	Pearlmillet	17.5 (2)	19.1(5)	9.5(5)	5.5(5)	4.0(5)	0.4(5)
	Sativa	19.9 (2)	22.7(5)	12.2(5)	7.4(5)	4.4(5)	5.8(5)
	CSH-5	17.5 (2)	23.6(5)	15.1(5)	5.4(5)	2.7(5)	3.3(1)
Oilseeds	Groundnut	10.4 (2)	7.1(5)	5.9(5)	3.1(5)	2.5(5)	4.6(4)
	Sunflower	9.3 (2)	10.3(5)	8.9(5)	4.5(5)	4.2(5)	2.7(4)
	Castor	17.0 (2)	20.1(5)	14.8(5)	10.2(5)	10.8(5)	6.0(4)
Pulks	Pigeonpea	5.4 (2)	5.6(5)	9.4(5)	6.5(5)	5.4(5)	3.2(4)
	Horsegram	8.5 (2)	11.9(2)	13.0(2)	6.2(2)	4.0(2)	4.4(2)
	Greengram	6.0 (2)	2.2(1)	1.8(3)	0.5(5)	Failed (2)	Failed(2)

Source : Venkateswarlu *et al.*, (1991)

Table 75. Choice of crop varieties for normal and late sown conditions

Centre	Crop	Choice of crop varieties for different sowing times	
		Normal	Late
Akola	Sorghum	CSH-5/CSH-9	CSH-1
Anantapur	Groundnut	Kadiri 1	TMV-2 or Kadiri 3

Bangalore	Indaf 1 or Indaf 3	PR 202 (Godavari)	
Bellary	Sorghum	SPV 86 or CSH 7R or CSH 8R	M-35-1
Bhubaneswhar	Rice	Ratna or Kumar	BR-92
Dantiwada	Sorghum	CSH-5 or CSH-6 for grain	S 1049 or Malwa for fodder
Dehra Dun	Maize	C.5 or C.2 or Vikram	Ageti
Hissar	Greengram	Varsha	S-9
Hosiarpur	Wheat	WL410	C306
Jhansi	Sorghum	CSH-5	CSH-6, SPV 224
Jodhpur	Greengram	S-8	S-9
	Cowpea	FS-68	K-11
Udaipur	Maize	Ganga 5	"Bassi selected" Satha
	Sorghum	CSH-5	CSH-6
Varanasi	Greengram	Varsha or K-851	T-44

Source : Venkateswarlu *et al.,* (1991)

Certain crop species are more efficient either due to shorter duration or capacity to produce better yields even under relatively unfavourable moisture regimes when sown late in the season. Therefore, varieties have to be chosen depending upon the date of occurrence of sowing rain (Table 75). Selection of varieties however, is a continuous.

6.3 Early withdrawal of monsoon

If monsoon terminates earlier than usual, it cuts the length of growing season. Normally crops in their maturity and those growing on shallow soils suffer heavily. Sowing of *rabi* crops may also gets affected. It is therefore advisable to use short duration and early maturing varieties to face the situation. Suggested measures to tide over the situation are as follows.

6.3.1 Reduce the plant density

Remove alternate plants in a row or blade out alternate rows or every third or fourth row. At Bangalore it is suggested to harvest crops like maize for fodder leaving the intercrop component pigeonpea for full maturity. In northern dry region of Karnataka, it is suggested to thin *rabi* sorghum or safflower based on moisture availability within 40 to 45 days of sowing and in *rabi* sorghum+ chickpea and in safflower + chickpea, uprooting sorghum and chickpea component, respectively is suggested. The practice of removing top portion of plant (above the cobs) after milky stage in maize is also prevalent with farmers in some areas.

6.3.2 Dust mulching

Carry out frequent and deep intercultivation in post monsoon areas to create dust mulching to burry the soil cracks.

6.3.3 Supplemental irrigation

If possible, protective irrigation may be given. Usually protective irrigation is proposed to be given at 55-60 days of growth. Due to early withdrawal of monsoon, the same may be applied at 35 to 40 days growth (Table 76).

Table 76. Effect of protective irrigation at 40 days stage on yield of *rabi* sorghum (1976-77)

Treatments	Yield of grains (q/ha)	Additional grain (q/ha)	Grain/cm water
Without irrigation	8.22	–	–
2 cm irrigation in alternate furrow	12.89	4.67	2.33
4 cm irrigation in all furrows	15.29	7.37	1.84

Singh (1995)

6.3.4 Harvest the crop early

Crops may be harvested at physiological maturity in case of *kharif* crops in a double cropping system or they can also be harvested for fodder.

6.4 Extended Monsoon

Such aberration is rarely experienced. This is a welcome situation as double cropping would be possible in medium to deep soils. After the harvesting *kharif* crop take a short duration crop preferably a pulse or fodder crop. In crops like *kharif* sorghum ratooning can also be practiced and additional grains could be obtained.

7. MID SEASON CORRECTIONS
(Break in monsoon-mid or late season drought)

In dry lands moisture is insufficient for crop production quite often. Farmers take up sowing when sufficient moisture is there in the soil. When the soil profile is wet for 30-40 cm it is sufficient to meet initial crop requirement of moisture. Plants tolerate stress upto 30-35 days and later on start wilting. During this stage there is need to introduce or super impose mid season correction to save the standing crop otherwise whole crop may be lost. Important remedial measures suggested are discussed here.

7.1 Thinning of the stands

If there is failure of rain immediately after sowing that is at about 30-35 days of sowing or a bit later it is required to thin down the crop. At this stage the crop is about to enter grand growth stage with increased leaf area which starts to deplete soil moisture at rapid pace and if there is no replenishment of water, the crop suffers at reproductive stage resulting in poor harvest or no harvest at all. At this stage if thinning is attended the competition for moisture is reduced, the rate of depletion is reduced and the remaining stand can make best use of available moisture and could yield near normal.

Thinning, therefore, has been suggested by removing alternate lines or every line after two to three lines in crops like pearl millet, sorghum, groundnut etc. This can be attended by blading out the row at inter-cultivation around 30 to 35 days. A spacing of 90 cm, thus, will be created from a 45 cm row spacing. This practice increases water availability to individual plants. In a intercropping system the most sensitive/exhaustive crop (such as maize) may be removed.

At Solapur, thinning of population in sunflower increased the yield over normal population by 39 and 28 per cent due to 24 and 34 per cent reduction in population (Table 70).

At Bijapur, thinning of population had differential effects on yields depending on rainfall received (Table 77). During scarcity years the thinning proved advantageous. In years of good rainfall such practice was not comparable with normal population. Therefore, thinning needs to be attended based on rainfall situation.

Table 77. Influence of seasonal correction in plant population as life saving on rainfed *rabi* sorghum yield (kg/ha)

Population corrections	1975	1976
No thinning (original)	1550	2420
Alternate plant removed	2150	NT
Alternate row removed	2070	1450
Alternate row + alternate plant in the existing row removed	1500	1370
3rd row removed	2110	2020
3rd plant removed	2160	2190
3rd row and 3rd plant in the existing row removed	1530	NT
CD 0.05	58	60
Total rainfall	224.9	447.6

7.2 Removal of week seedlings

In most of the dry land crops there may not be uniform emergence because of soil variation, differential depth of sowing or inadequacy of soil moisture at sowing. In depressions, there may be quick germination and on ridges there may be slow germination. In most of the cereals the late germinated seedlings will not be vigorously growing and may not be competitive to give any yield. In sorghum 20-25% of weaklings are observed. These have to be removed within 30-45 days because at this stage flower primordial initiation starts and any correction after this stage is not going to help even in irrigated soil. In the sorghum belt of northern Karnataka, farmers usually remove barren plants during grain formation stage and use them to feed the cattle. But since this practice is introduced late in the season the competition that has occurred by then would, limit the ear size. The improvement would therefore, will be only the form of hold seeds.

7.3 Repeated and deep intercultivation

Early breaks in rainfall brings about soil cracking early in the season which enhances evaporation loss of moisture through the cavities. To prevent soil cracking and to break the capillary loss, early and frequent intercultivation are required. Instead of the normal 3 hoeings, six hoeings can create a good dust mulch with good impact on existing crop. Generally entire blade hoes or slit blade hoes are employed which penetrate hardly to a depth of 5 to 7.5 cm. Shovel type of interculturing hoes with iron prongs or counters need to be used which due to sharp cutting edge can tear off the soil and create a dust much to a depth of 10-12 cm.

7.4 Mulching

If the break in monsoon is very brief, soil mulching with residues may be helpful to extend the period of storage of water in soil profile by way of reducing heat lead on soil and by avoiding direct contact of dry desiccating wind with soil surface. In groundnut (cv : S206), beneficial effects of mulching after germination of crop were found equivalent to the application of one protective irrigation.

Availability of mulch material may be a problem. There is also added cost due to procurement and spreading. The locally available material like unwanted vegetation on bunds or in waste lands may be made use for this purpose. A minimum of 5t/ha residue is required to have any cognizable impact of mulching.

7.5 Stripping of older leaves

Older leaves are still active and are not dead. As they are shaded they transpire more than helping in photosynthesis. So to avoid moisture loss, stripping of older leaves to direct moisture and nutrients to actively growing parts is suggested. In sorghum 12-15 leaves are produced. When the plants are in flag leaf stage strip of the bottom 3-4 leaves and use them for mulching or for feeding the livestock. Singh (1995) opined that such a practice might control moisture loss temporarily and is not useful under prolonged condition as observed at Solapur. In maize growing areas it is a practice to cut the top portion above the cob after milky stage. This practice may save moisture to some extent.

7.6 Returning ratooning

In case of *Kharif* season in all five dry zones of Karnataka, intermittent dry spells ranging from 4-6 weeks are common adversely affecting crop yields. Finger millet is a drought resistant crop and it rejuvenates with occurrence of rains. In case of pearl millet when crop is exposed to dry spells of 30-35 days during peak growth, yields are drastically reduced. Under such condition cut/harvest the crop for fodder purpose and allow the ratoon for seeding. Such practice should be done with the onsets of rains (after experiencing drought). For quick rejuvenation supplemental addition of fertilizer is suggested. Top dressing through foliar spray may also be useful to activate the plant.

7.7 Application of nutrients to rejuvenate growth

Sometimes to rejuvenate the growth activity immediately after receiving rains after a long break supplementation of nutrients through urea, DAP, $ZnSO_4$ or $FeSO_4$ is suggested. For instance, in the northern dry zone of Karnataka, foliar spray of urea (2%) and $FeSO_4$ (0.25%) immediately after rains in groundnut crop is recommended for quick revival of the plant.

7.8 Life saving irrigation

If there are farm ponds/seasonal streams, life saving irrigation could be given to save the crop. Runoff is inevitable in the tropical and sub tropical climates of our country. It varies from 10-40 per cent of the total rainfall; of this at least 30 per cent could be harvested and recycled for crop use.

7.9 Use of antitranspirants

As already discussed, there are chemicals which are found to reduce water loss from the plant surface on application. Such of the chemicals may be used whenever condition warrants crop failure due to severe moisture loss.

7.10 Harvest the crop for fodder

If there is complete failure of rains and crop is not in a position to produce any seed yield especially the *kharif* crop in a double cropped area, such crop may be harvested for fodder e.g. sorghum, groundnut etc. and the land could be kept ready to make use of shower for *rabi* season cropping.

8. ALTERNATE LAND USE SYSTEMS

In our efforts to produce more, more and more forest areas, unsuitable marginal lands and range lands are brought under cultivation. No doubt this has led to acute shortage of fuel, timber ward and fodders in the country but also rendered cultivable land progressively unsustainable. These problems could not have achieved if this alarming proportion had these lands been treated according to their capabilities. Alternate land use system, a land capability biased production system which plays important role is stabilizing natural ecosystem has been emphasized in the ecologically sensitive dry land areas (Table 78). There are other advantages too. (1) Trees and pastures will be more remunerative and impart stability to dry lands. (2) Perennial vegetation makes use of off season rains and also checks soil erosion, and (3) Farm plan involving use of fodder, fuel, timber and cash crops meets the energy requirement of rural family.

Table 78. Farming systems options for different agro-climatic conditions

Annual rainfall (mm)	Soil type	Farming systems	Suitable tree/grass/legume species
<500	Shallow (0-30 cm)	Tree farming	*Prosopis cinearia, P. jubiflora, Acacia aneura, A. nelotica, A. tortilis, Pitheoellabium dulce*
	Medium (0-45 cm)	Pasture management	*Lasiurus sindicus* (Light textured soils) *Cenchras setigerus, Sehima rervoium, Stylvanthes scaboh, Clitoria ternatea*
500-7 50	Shallow (0-30 cm)	Silvi-pastoral system	*Acacia nilotica, Celophospermum mopane, Dlbergia sissoo, Hardwickia binata, Cassia sturti, Albizia-mora, Leucaena sp. Cenelirus ciliaris, C. setigerus, Dicanthium annulatum, Panicum antidotale, Stylosanthes homata, Macroptilium antropurpurelum*
	Medium (0-45 cm)	Horti-pastoral system	*Custard apple, Ber Jamun, Anola, Tamarind, Wood apple, Bacl, Cenchrus ciliaris, Panicum antidotate, Urochloa mosauficensis, Shylosanthes hornata, Macrptilium antropurpureum, Clitoria ternatea*
>750	Shallow (0-30 cm)	Ley farming silvipastoral system	3 years stylosanthes homata and 4th year arable crop (sorghum on heavies soil, pearlmillet on lighter sorts) silvi-pastoral system as above
	Medium (0-45 cm)	Ley farming horti pastoral system	Ley farming as above, Mango, Sapota, Guava, *Anola, Stylosanthes homata, Macroptilium autropurpureum*

Source : Singh and Osman (1995)

There are two approaches in this direction. One is to put land belonging to class IV and above under suitable grasses, fodder crops, fruit and forest species integrated with animal production systems. Alternately, the same land or a portion of it could be simultaneously or sequentially put to food, fodder, fuel, fruit or forest trees production to serve immediate benefits as well as to meet long term environmental concerns. Ley farming, alley- farming, agri-horticulture, silviculture, tree farming agri-silviculture are some of the alternate uses of lands.

8.1 Ley Farming

The word "ley" comes lie from the old English term "lea" and denotes former arable land allowed to fallow or sown to grass land. Ley farming is thus nothing more than the rotation of annual or perennial field crops with temporary postures for the purpose of regenerating the fertility of the soil and improving the supply of forage. In this context ley farming is useful under Indian conditions where under dryland situation farmer is reluctant to apply fertilizer.

Ley farming (cereal/pasture) with annual self regenerating legumes is becoming more and more popular in areas with mediterranean climate in Australia where extensive cereal cultivation is practiced. The advantages of ley farming include i. Pasture legume constitute high quality pasture feed, ii. The pasture needs to be sown only once, iii. Cereal yield due to increased nitrogen accretion in the soil. iv. Favourable effects on soil protection and soil improvement, and v. This agro-pastoral form of land utilization leads to increased and stabilized income for farmers.

In the cereal growing sheep wool production zone in the semi-arid medeterranean (rainy winter) climate of Australia with 250-650 mm rainfall, introduction of ley farming in 1940's has increased the cereal yields by 50 per cent and doubled the sheep population and wool production by 1980's. The important ley species are *Medicago* spp and *Trifolium subterraneum* (Subterraneum clover).

Studies at CRIDA, India led to the identification of *Dolichos, Macroptilium, Leucacena, Clitoria, Desmodium, Detrocema, Glycine, Calapogoniuam, Vigna, Tephrosia, Alysicarpus, Indigo-fera, Cassia* and *Stylosanthes*. Though some species perform better, their persistance is poor eg. *Macroptilium Clitoria, Dolichos* etc. Ideal legume among these is *Stylosathes* which is high yielding, persistent, nutritive and palatable. *S. hamata* is as remunerative as grain sorghum *S. hamata* and *S. scabra* could also be grown as intercrops with dry land cereals such as sorghum, pearl millet, fingermillet or along with range land grasses like *Dicanthium, Urochloa, Lasiureas, Sehima, Cenchrus*.

Substantial amount of nitrogen to the soil was found to be added with *Stylosanthes hamata*. At Hyderabad yield of sorghum and pearl millet were higher in plots where *S. hamata* was grown previously for two years (Reddy and Hampaiah, 1980) (Table 79).

Table 79. Influence of stylopasture on yield of sorghum and pearl millet

Crop	Previous crop	Grain yield (q/ha)		
		Nitrogen level kg/ha		
		0	50	100
Sorghum	Castor	2.1	11.6	16.6
	Stylo	10.8	12.1	17.3
Pearl millet	Castor	6.3	17.0	22.8
	Stylo	24.2	25.4	27.2

Source : Reddy & Hampaiah (1980)

Veerashankar (1984) observed that soil organic carbon increased (0.56%) with three years of stylosanthes pastures leys compared to castor, sorghum or fallow rotations (Table 80). Similar trend was noticed with available nitrogen. There was a build up of 35 kg N/ha with three years stylo pasture. At least two years are required to buildup soil nitrogen level.

Table 80. Soil fertility as influenced by duration of stylosanthes leys

Treatment	Before fourth year sorghum		After fourth year
	OC %	Soluble N (kg/ha)	Sorghum available N (kg/ha)
C-S-C-S	0.22	110	82
C-S-Sh-S	0.31	113	97
S-Sh-Sh-S	0.35	141	107
Sh-Sh-Sh-S	0.56	145	106
C-F-F-S	0.15	88	63

C-Castor, S-Sorghum, Sh-*Stylosanthes hamata*, F-Fallow
Source : Veerashankar (1984)

Niranjan, *et al.* (1985) observed no significant differencies in grain and stover yield in one year ley while four years leys had significant effect (Table 81) due to buildup of soil fertility.

In ley farming, the animal component is necessary to have economic evaluation apart from improved soil productivity.

Table 81. Grain yield of sorghum and green fodder yield of grasses in difference ley farming systems

Treatment	Grain yield and green forage yield (t/ha)				
	1988	1989	1990	1991	1992
S-S-S-S-S	1.15	1.70	1.09	0.96	1.51
G-G-G-G-S	(3.2)*	(13.75)	(11.66)	(7.25)	1.42
G+Sh-G+Sh-G+Sh-G+Sh-S	(3.6)	(16.60)	(20.66)	(11.80)	1.75
Sh-Sh-Sh-Sh-Sh-S	(2.10)	(7.35)	(13.75)	(9.00)	1.19
S-Sh-S-Sh-S	1.31	10.80	1.0	8.70	1.88
CD (0.05)					0.56

S-Sorghum, G-Grass, Sh-Stylosanthes hamata,
*Figure in parenthesis indicate the forage yield Naranjan *et al.,* (1985)

Harvesting interval ranging from 30 to 60 days have been studied; looking to the growth of the components a narrow cutting interval may be preferred.

Crop management : All recommended improved agronomic practices of sole crop have to be followed in case of alley crops and necessary crop rotations recommended in the region may also be adopted to maintain yield levels. Sorghum, pearl millet, groundnut, pigeonpea, sunflower castor etc.

have been made use as alley crops. Use of loppings for mulching between the crop rows have enhanced the yield performance due to multiple benefits.

8.2 Agro-horticulture

Most dry land farmers do not provide surplus crop for sale. Fruit crops will bring cash income to dry land farmers hence it is attractive. Most fruit tree grown in drylands take 5-8 years to cover the inner space. Further, fruit trees like ber need pruning every year so interspace is available. In such cases intercrops can be grown successfully and profitability between the fruit plants in a system of agro-horticulture of food-cum- fruit system. This system not only keeps the interspace weed free but also provides higher returns per unit area. In addition there could be improved soil fertility and availability of leafy fodder as well as fuel wood from the pruning.

The system works out best in medium to deep black soils with good water holding capacity. Individual farm ponds and supplemental watering in the off-season will make the system viable and successful. Jamum (*Syzigium cuinii*), custard apple, guava, mango, sapota ber etc. are the suitable species and these can be established on contours.

The crops grown in the interspace should be normally low stattured and of short duration. So that they do not compete with the fruit trees for light, moisture and nutrients, short duration legumes vegetable etc. are more suitable. Horsegram and clusterbean along with ber proved superior to cowpea and greengram (Table 82). Ber pruning in the system also provide nutritions leaf (12% CP) and fuel word from the annual pruning.

8.3 Alley cropping

Alley cropping is essentially an agro forestry system in which food crops are grown in alleys formed by hedge rows of trees or shrubs. The main objective is to get green and palatable/fodder from hedge rows in dry season and produce reasonable quantum of grain and stover from the crops grown in the alleys during rainy/cropping season. This necessarily calls for management of hedge rows as well as alley crops.

Table 82. Effect of cutting height on crop and *luecaena* yield

System	Grain yield (q/ha)		Luecaena yield (DM q/ha)	
	Cutting height (cm)		Cutting height (cm)	
	60	15	60	15
Pearl millet				
Sole	18.0	21.1	–	–
7.8 m mulch	11.9 (66)	23.7 (108)	18.4	30.5
7.8 m fodder	11.7 (65)	17.3 (86)	26.9	36.2
Sorghum				
Sole	15.2	18.5	–	–
7.8 m mulch	8.4 (55)	19.1 (103)	16.0	36.6
7.8 m fodder	7.3 (48)	16.7 (90)	23.1	46.5

Figures in parentheses are grain yield per cent of sole crop.
Source : Korwar (1995)

Table 83. Performance of agro-horti system at Hyderabad

System	Yield (kg/ha)			
	Ber		Intercrop	
	1986-87	1987-88	1986-87	1987-88
Ber + cowpea/cluster bean	919	1828	250	1918
Ber + greengram/ horse gram	919	1812	245	1025
Ber sole	920	1830	–	–

Source : Singh & Osmam (1995)

There are three version of alley cropping system. 1. Forage alley cropping, 2. Forage-cum-mulch alley cropping and 3. Forage cum pole alley cropping. Thus, there are two essential components, the hedge row, and the crop. To optimize production management of both becomes extremely important.

Hedge rows : Selection of hedge species, width of alleys, cutting height and frequency are the important management factors. *Luecaena leucocephala* proved to be a good hedge species. *Glyricidia, Calliandra Sesbania* are other species which have shown promise. The species selected should withstand frequent cutting and should have good coppicability.

Varying alley widths have been studied ranging from 2 m to 10 m or more at various places. Moisture being the main limiting factor in dry land situation, wider rows of widths 10, 15 or even 20 m could be tried with advantage. Preferably, the hedge rows should be set up on contours.

Studies on cutting height indicated reduced competition as the alley crop with increased green loppings yield. Too lower cutting height however, would reduce regeneration capacity. A cutting height of less than a meter may be ideal from the point of shading and competition (Table 82).

8.4 Horti-pastoral system

In a horti-pastoral system, along with perennial fruit trees, pasture crops could be grown in the interspace. This avoids weed growth, conserves moisture and provide fodder as well as mulching material. In the initial stages, care should be taken so that the pasture does not overtake the fruit trees. Stylosathes proved superior to *Cenchrus ciliaris* (Singh and Osman, 1995).

8.5 Silvi- pastoral system

Integrated, silvi-pasture system is a low input technology for utilization of marginal and submarginal lands. Through this forage and leafy fodder is obtained besides fuel wood and timber.

Forage productivity under silvi-pastoral system entails biomass production from understorey by taking advantage of multicanopy layers on the same piece of land.

Combination of woody and herbaceous (Table 84, 85 and 86) biomass producing plants when allowed to grow in a defined geometrical pattern (taking advantage of their canopy or manipulating the canopy) may allow radiation infiltration at a lower level. In areas where forage production is priority this system is suitable.

Selected tree species should be a fast growing multipurpose tree with leaves of high digestibility and palatability. It should also have a good coppicing ability.

Silvi pasture system (cut and carry feed) involving Leucaena trees and *Stylosanthes hamata* or *Cenchrus ciliaris* demonstrated high productivity and superiority at Hyderabad. *Hardwickia binata* was promising at Jhansi.

Table 84. Suitable pasture species for drought prone areas

Pasture species	Minimum rainfall (mm)	Soil type	Seed rate (kg/ha)	Dry forage yield (t/ha)
Grass		Light	3-5	3.5
Sewan *(Lasiurus sindicus)*	250	Versatile	5-6	4.0
Buffel *(Lenchrus ciliaris)*	300	Versatile	8-10	3.0
Birdwood *(Cenohrus setigerus)*	250	Versatile	2-4	3.0
Bluepanic *(Panicum antidotale)*	500	Medium to heavy	4-5	2.5
Marvel *(Dicanthium annulatum)*	375	Mixed red and black	7-8	2.5
Sum *(Sehima nervosum)*	300	Versatile	8-9	3.0
Dinanath *(Pennisetum pedicellatum)*	400	Light	11-17	3.5
Bahia *(Paspalum notatum)*	450	Versatile	2-4	3.0
Sabi *(Urochloa mesambicensis)*	450	Medium to heavy	12500*	4.0
Napier *(Pennisetum purpureum)*	750	Sodic	8-10	2.5
Rhodes *(Chlorvis gayana)*	600	Light to medium	5-7	3.5
Legumes				
Stylo *(Stylosanthes hamata)*	450	Light to medium	4-6	2.5
S. scabra	325			
Sirato *(Macroptilium atropurpureum)*	500	Light to medium	7-8	2.8
Banklthi *(Atylosia scarabacoides)*	300	Light to medium	10-12	2.0
Aparajit *(Clitoria ternatea)*	300	Light to medium	15-20	3.0

* Cuttings

Among grasses, guinea grass in found most suitable. Legumes like *Stylosanthes hamata, Clitoria tematea* and *Macroptilium autropurpureum* are suitable.

8.6 Tree farming

Dry lands not fit for cultivation (non-arable) and when fodder production is not a priority, can be put under tree farming as block plantation (farm wood lot). The uncultivated wastelands which lack soil of any kind can not support any vegetation. These could be utilized and improved by planting a sparce tree cover in selected microsites or soil pockets. These can flourish and yield abundantly in many places where arable crops would fail to grow.

Leucaena (in short felling cycles), *Posofis cineraria*, Zizyphys and *Albezia lebbeck* (higher felling cycle) are suitable for tree farming *Acacia tortilis* (8 years, felling cycle, 12% crude protein). *Albezia lebbeck* (15 year, 29% crude protein), *Prosopii jubiflora* (6 year), *Prosofis cineraria* (24 year, 14% crude protein), *Zizyphus* (15/25 year, 12% crude protein) and *Leucaena* (8 yr, 34% crude protein) are more productive.

Tree farming resulting in other intangible benefits like water and soil conservation and moderating influence on the extreme variation in climate, a common feature in many dry lands.

Trees are also farm components of arable farming. They can be used as shelter belts or for planting as along the boundaries or on bunds along with arable crops. Under such situation there could be adverse border effect due to root interference, shading, leaf dropping or alleolopathic effects (Table 87) which needs to be considered while developing such systems.

Tabe 85. Suitable top feed species

Common name	Botanical name
Babool	*Acacia lencophloea, Acacia nilotica*
Gum arabic	*A. senegal*
Khairwal	*Bauhinea purpurea*
Mopane	*Calophospermum mopane*
Shishem	*Delbergia sisso*
Rosewood	*Delbergia latifolia*
Gumhar	*Gmelina arborea*
Dhanman	*Gtrewia optiva*
Anjan	*Hardwikia binata*
Subabul	*Leucaena leucocephala*
Khejri	*Prospis cineraria*
Safeda, Tras	*Populus* sp.
Ageti	*Sesbania grandiflora*
Farash	*Tamrix aphylla*

Table 86. Tree species suitable for areas having different annual rainfall

<500 mm	500-750 mm	750 mm
Acacia nilotica	*Acacia nilotica*	*Acacia nilotica*
Acacia aneura	*Acacia ferruginea*	*Albezia lebbeck*
Acacia tortilis	*Albezia lebbeck*	*Azadirachta indica*
Acacia albida	*Albizia amara*	*Casturina equisetifolia*
Colophospermum mopane	*Azadirachta indica*	*Dalbergia sisso*
Prosopis cineraria	*Casuarina equisetifolia*	*Dalbergia latifolia*
Prosopis juliflora	*Cassia sturti*	*Gamlina arborea*
Pitheoellobium dulce	*Dalbgia sisso*	*Grewia optiva*
	D. latifolia	*Grevillearo busta*
	Leucaena leucocephola	*Hardwickia binata*
	Tamarindus indica	*Melia azadchta*
	Terminalia sp.	*Poputus* sp., *Seshania* sp., *Tumnalia* sp.

Plate 13.1. 1. Model of contour bunding across the slope (5%) for rain water conservation, 2. Contour bunding with vetiver hedge for best results (Model)

Plate 13.2. 1. Suppressed growth of sorghum due to heavy infestation of weeds. Crop shows wilting symptoms, **2**. An example of agri-horticulture system. Planting of Awla trees (*Phyllanthus emblica*) at 5 m distance and utilization of land in between for short duration crops during kharif & rabi season. Multi-storied cropping system for better harvest of solar energy and best use of available water under dryland.

Neem is one plant whose medicinal properties stand out of its many uses. Its leaves, fruits, seed oil, oil cake, timber etc. are all valuable.

Timber-cum-fibre system has been developed at Dry Farming Centre, Bijapur, Intercropping and subabul with agave was found more remunerative than arable crops on marginal shallow soils. Fibre yield increased with a increase in plant population without any adverse effect on subabul growth (Table 87). Sisal (Agave) fibre is strong and superior. The potential of this plant needs to be exploited in the drier areas.

Table 87. Border effects on field crop and runoff collection as influenced by strip planting of tree species

Tree species	Border effect (m)	Rainfall collected in ponds (m^3) Av. 3 years
Delbergia sissoo	9.6	139.4
Tectona grandis	6.8	14.4
Casurina equisetifolia	7.6	15.0
Acacia auriculiformis	12.3	100.1
Albizia lebbeck	15.9	113.8
Ecalyptus spp.	17.9	107.3
Acacia catechu	15.1	110.4
Acacia nilotica	19.5	91.5
Leucaena leucocephala	6.8	27.1
Control	–	141.3

Table 88. Permanent intercropping of subabul and agave at Bijapur

Plant population/ha		Mean fibre yield (kg/ha) after 3 years
Agave	Subabul	
10000	6944	516
6666	3888	334
5000	3472	311
3333	2083	236
2966	1889	199
2500	1250	189

9. CONCLUSIONS

With traditional agricultural practices, the productivity of crops grown under dry land conditions is not only low but almost stagnant. In these practices, the use of chemicals is avoided mainly through the adoption of time tested local cultivars and marginal use of FYM to rejuvenate soil fertility. Adopting appropriate crop rotations which incldue legumes is yet another means to maintain soil fertility under traditional dry land agriculture. For resource -poor farmers, these practices are sustainable and guarantee a harvest and are borne out of traditional wisdom. However, increasing demand for food grains with increasing population and alternate uses of food grains for industrial uses, make it incumbent to double the productivity of food grain crops grown in drylands. Experiences gained at AICRPDA and SAU's have shown that higher production targets are achievable. The main components of the technology that ensure higher targets are, soil and water conservation, use of input responsive cultivars and cropping system, use of efficient production practices, moderate use of fertilizers, pest management, planning for aberrant weather and capability biased land use systems. Obviously the resource poor dry land farmers are constrained in adopting the recommended technology mainly due to the uncertainties of the returns from the investment.

The sustainability analysis (estimation of sustainable yield index SYD = (Y - o)/Y max, where Y is estimated average yield of a practice over years, o is estimated standard deviation and Y max is the observed maximum yield) attempted by Singh *et al.* (199) has made some interesting revealations. In case of plant density, the average productivity over varying plant densities is positively correlated with sustainability in most of the situations. Sustainability in yield due to plant density is interacting with crop, variety, season, rainfall, soil type and fertilizer.

The average sustainability in terms of gross returns increased with the annual rainfall both for intercropping and sequence cropping systems. In many cases, intercropping systems paid more sustainable returns than sole cropping. The sustainability of intercropping is dependent on the crop and crop growing season. Intercropping in the post-rainy season was found to be more sustainable. In case of sequence cropping systems, short duration crops grown during the rainy season, and deep rooted crops taken during the post-rainy season resulted in more sustainable income.

In most of the situations sustainability of yield was higher when the recommended dose of fertilizer was applied. Available nitrogen, organic carbon and phosphorus content in soil was increased with organic fertilizer application. Application of crop residues in combination with chemical fertilizer resulted in higher sustainable yield and maintained higher levels of nitrogen, phosphorus and organic carbon. Green leaf manuring proved promising in increasing sustainability in yield and improving organic carbon, infiltration rate and hydraulic conductivity of the soil.

Inter terrace land treatment like, BBF and ridging resulted in higher sustainable yield. Mulching increased sustainability both in rainy and post rainy seasons. In case of supplemental irrigation, higher sustainability in yield was obtained when the limited water was applied to post rainy season crops.

The practice which is productive is not necessarily sustainable. Further, sustainability of a given practice varies from location to location, depending on the agro-climatic conditions.

REFERENCES

Ahalawat, I.P.S. and Rana, D.S., 1998. Agro-techniques. In : Fifty Years of Agronomic Research in India (Eds.) R.L. Yadav, Punjab Singh, Rajendra Prasad and IPS Ahlawat. Indian Society of Agronomy, New Delhi, pp. 33-50.

Ahlawat, I.P.S. and Rana, D.S., 1998. Agro-techniques In : *Fifty years of Agronomics Research in India.* Ed. R.L. Yadav, *et al.,* Indian Society of Agronomy, New Delhi, India, pp. 33-50.

AICRPDA, 1982. A Decade of Dryland Agricultural Research in India 1971-80. All India Co-ordinated Research Project for Dryland Agriculture, Hyderabad, Andhra Pradesh, India.

AICRPDA, 1988. *Annual report*, 1987-88. All India Co-ordinated Research Project for dryland Agriculture (ICAR), Main Centre, Bijapur, Univ. Agril. Sciences, Dharwad, India, pp. 59-69.

AICRPDA, 1989. *Annual Report*, 1998 Chief Scientist, AICRPDA Centre Solapur.

Annual Report 1984. The chief scientist, All India Co-ordinated Research Project Dryland Agriculture, Solapur, pp. 19-21.

Anonymous, 1977. *Dryland Agricultural Research Progress Report*, 1970-75, Indian Council of Agricultural Research, New Delhi, India. pp. 12.

Anonymous, 1983. Highlights of dryland research in red soil of Karnataka 1971-83. AICRP on dryland Agriculture, Bangalore, Centre, India (mimeographed) pp. 98.

Anonymous, 1983. Improved Agronomic Practices for Dryland Crops in India, All India co-ordinated Research Project on Dryland Agriculture, Hyderabad.

Anonymous, 1988. *Annual Progress Report 1987-88*. AICRPDLA, Solapur.

Anonymous, 1991. *Annual Progress Report, 1990-91*. AICRPDLA, Bijapur.

Anonymous, 1998. Role of organic manures in soil health and productivity Documentation of field experiences, UAS, Tech. Info. Series, Ed. T.K.T. Krishne Gowda, M. Chandrappa, M.A. Singlachar and B.R. Hegde, Univ. of Agri. Sci. Bangalore, 1-52.

Arnon, I. 1975. Physiological principles of dryland crop production. In *Physiological Aspects of Dryland Farming*, Ed. U.S. Gupta, Oxford and IBH, publishing Co. 3 - 145.

Arnon, I. 1972. *Crop Production in Dry Regions*. Barnes and Noble Books New York, pp. 650.

Arunachalam, N., 1976. Input management in dryland agriculture. In : *Recent Advances in Weather Forecasting and Dryland Management* (Ed. N. Balasubramaniam). Short training course, 10-23 April, 1996. Centre for advanced studies in agronomy. TNAU, Coimbatore, pp. 25-28.

Badanur, V.P., 1993. AICRPDA Annual Report of Bijapur Centre, pp. 83-85.

Badanur, V.P., Poleshi, C.M. and Balachandra Naik, 1990. Effect of organic matter on crop yield and physical and chemical properties of vertisols. *Journal Society of Soil Science*, 38 : 426 -429.

Badnur, V.P., 1995. Crop residue management for sustainable agriculture under dryland conditions. *Abstracts Seminar on Conservation of Natural Resources for sustainable Production*, 16-17 Nov. 1995. Univ. Agri. Sciences, Dharwad (India), 5.

Balasubramani A.N.V. and Venkateswarlu. J. 1985. Chickpea In : *Efficient Management of Dry land Crops*. Ed. V. Balasubramanian and J. Venkateswarlu, CRIDA, Hyderabad pp. 209- 219.

Balasubramaniam, N., 1996. Sowing time and technique in dryland agriculture (seed hardening, seed rate and planting pattern). In : *Recent Advances in weather forecasting* and dryland management (Ed. N. Balasubramaniam). Short training Course 10-23 April 1996. Centre for advanced Studies in Agronomy, TNAU, Coimbatore, pp. 41-43.

Balasubramanian, A., Arunachalam, L. and Palaniswamy, S., 1986. *Sorghum Newsletter*, 25 : 29-39.

Bantilan, R.T., Palada, M.C. and Harwood, R.R. 1974. Integrated weed management, 1. Key factors affecting crop weed balance. *Philip. Weed Sci. Bull.*, 1 : 14-36.

Belgaumi, M.I., 1998. Soil and water conservation structures for low rainfall black soil areas In : *Integrated Dryland Farming Practices for Sustainable Production*. Tech. Series 7 (Eds.) C.J. Itnal *et al.,* Univ. Agri. Sci. Dharwad, India pp. 45- 48.

Bhoopati Babu, B., 1978. Studies on weed control in sorghum based intercropping systems. M.Sc. (Agri.) Thesis, TNAU, Coimbatore.

Bond, J.J. and Wills, W.U. 1969. Soil water evaporation in surface residue rate of placement effects. *Soil science of America Proceedings*, 33 : 445-448.

Brown, P.L., and Shradler, W.D., 1959. Grain yields, evapotranspiration and water use efficiency of grain sorghum under different cultural practices. *Agronomy Journal*, 17 : 339-343.

Carson, A.G., 1987. Tropical Pest Management. 33 (4) : 359-365.

Chittapur, B.M., 1998. Vegetative measures for erosion control In : *Integrated Dryland Farming Practices for Sustainable Production* (Ed. C.J. Ithanl *et al.*) Tech Series 7. University of Agril. Sci. Dharwad, India, pp. 75-84.

CIAT, 1976. Bean production systems. *Annual Report for 1975*. Cali. Columbia, p. 58.

CRIDA, 1993. *Annual Report* 1001-02 Central Research Institute for Dry land Agriculture, Hyderabad pp. 46-47.

Damodaran, A. and Sankaran, S., 1974. A note on evaluation of herbicides in cereal legume mixtures. *Madras agricultural Journal*, 61 : 924-925.

Das, S.K. and Rao, A.C.S., 1986. Effect of preceeding crops on nitrogen management for rainfed sorghum in black soil. *Journal of Indian Society of soil Science*, 34 : 510-513.

Das, V.R.S. and Shantkumari, M. 1975. Stomatal behaviour towards four classes of herbicides as basis of selectivity to certain weeds and crop plants. *Proceedings Indian Academic Sciences*, 82 : 385-390.

De, R. and Giri, G., 1978. Effect of mulching and kaoline foliar spray on mung (*Vigna radiata*), rapessed (*Brassica compestris*) double cropping system under rainfed conditions. Journal of Agricultural Sciences, Cambridge, 91 : 191-197.

Dharmalingam, C. and Jegathambal, R. 1996. Seed hardening technology for rainfed crops. A case study with botanicals. In : *Recent Advances in weather Forecasting and Dryland Management*. (Ed. N. Balasubramaniam) Short training course, 10-23 April 1996. Centre for Advanced studies in Agronomy, TNAU, Coimbatore, pp. 41-63.

Dhopte, A.M., Akarte, M.M. And Jamdar, S.L. 1987. Effect of foliar sprays on new aromatic nitro compound, Atonik and antitranspirant, kaolin applied singly and combinations on seedlings growth, physiology and yield of hirsutum cotton under semi-arid environment. *Journal of cotton Research and Development*, 3 : 52-63.

Dhopte, A.M. and Jamdar, S.L., 1995. Effect of antitranspirants and growth regulators on productivity of upland cotton. *Journal of cotton Research and Development*, 9 (2) : 136 - 141.

Dinesh Kumar, 1982. *Ph.D. Thesis*, submitted to Punjab Agricultural University.

Elad, Y., Zvieli, Y. and Chet, T., 1986. *Crop Protection*, 5 (4) : 288 - 292.

Fraiser, G.W., 1987. Water harvesting for collecting and conserving water supplies. *Alfisols in the semi arid Tropics*. Proceedings of the consultants workshop on the state of art and management. Alternatives for optimizing the productivity of SAT Alfisols and related soils. 1-3 Dec. 1983. ICRISAT Centre, Patanchery, AP. pp. 67-77.

Gaddanakeri, S.A. 1991. Investigations on the effect of *Leucaena leucocephala* (Lam) De wit live bund on growth and yield of companion *rabi* crops under rainfed conditions. *Ph.D. Thesis* submitted to Univ. Agril. Sciences, Dharwad, India.

Gajri, P.R., Verma, H.N. and Prihar, S.S. 1981. Rain water harvesting and its recycling for maximization of crop production. *Indian Journal of soil conservation*, 9 (2&3) : 116 - 123.

Gidda, V.R. and Morey, D.K., 1981. Effect of tillage practices and antitranspirants on relative water contact, leaf water potential and yield of rainfed cotton (SRT -1). Journal of Maharashtra Agricultural Universities, 6 : 219-20.

Greb, B.W. 1983. Water conservation : Central great plains In : *Dryland Agriculture, Agronomy*, No. 23 (Ed. H.E. Dregue and W.O. Wille, American Society of Agronomy, Madison Wisconsin, pp. 57-72.

Gregory, P.J. 1988. Plant and management factors affecting water use efficiency of dryland crops. *Challenges in Dryland Agriculture. A Global Perspective*. Proc. Int. Conf. of Dryland Farming 15-19 August 1988. Amarillo, Bushland and Texas, USA, pp. 171-175.

Hadimani, A.S., Hegde, B.R. and Satyanarayana, T., 1982. Management of red soils In : *Reviews of soil Research in India* Part II. Transactions. 12th international congress of soil science, New Delhi, 8-16 February, pp 689-700.

Havangi, G.V. 1979. Fertilizer use for dryland crops in Karnataka Proc. FAI, Group discussion on fertilizer use in drylands. Fertilizer Association of India, New Delhi, pp. 39-54.

Hegde, B.R., 1995. Improving water use efficiency under dryland conditions. In : *Sustainable Development of Dry land Agriculture in India* (Ed. R.P. Singh), Scientific Publishers, Jodhpur pp. 177-193 India.

Hegde, B.R., Channappa, T.C. and Anandram, B.K., 1987. Management alternatives for increased productivity of red soils. Experience in Karnataka In : Proceedings of the consultant workshop at the state of art and management alternatives for optimizing the productivity of SAT alfisols and related soils. ICRISAT Centre, Patancheru, India pp. 137-143.

Hegde, B.R., Major A.J., Wilson, D.B. and Krogman, K.K., 1976. Effects of row spacing and population density on grain sorghum production in Southern Atlanta. *Canadian Journal of plant science*, 56 : 31-37.

Hegde, B.R. and Sitharam A., 1985. Fingermillet and minor millets. In : *Efficient management of Dryland Crops*, Ed. V. Balasubramanian and J. Venkateswarlu, CRIDA, Hyderabad, India, pp. 79- 100.

Hosmani, S.A., Patil, R.V. and Gulaganji, G.G., 1987. Effect of dry sowing and nitrogen levels on some sorghum genotypes. *Sorghum Newsletter*, 30 : 33.

ICRISAT, 1977. *Annual Report 1996-97*. Agronomy and weed science, Hyderabad, India, p. 288.

Itnal, C.J., Warad, S.M. and Hundekar, S.T. 1998. Three-tier approach of resource conservation, development and utilization in rainfed areas. In : *Integrated Dryland Farming Practices for Sustainable Production*. Tech Series 7 (Eds.) C.J. Itnal *et al.,* Univ. Agri. Sci. Dharwad, India, pp. 38-44.

Kanemasu, E.T., Stewart, J.I. Van Donk, S.J. and Virmani, S.M., 1990. Agroclimatic approaches for improving agricultural productivity semi-arid tropics. Dryland Agriculture: Strategies for sustainability. *Advances in soil Science*, 13 : 273- 308.

Kanitkar, N.V., Sirur, S.S. and Gokhale, D.H. 1960. *Dry Farming in India*, ICAR, Publication, 2nd ed. New Delhi.

Karchi, Z. and Rudhich, Y. 1966. Effects of row with and seedling spacing on yield and its components in grain sorghum under dryland conditions. *Agronomy Journal*, 58 : 602-604.

Kasasian, L. and Seeyave, J. 1969. *Pans*, 15 : 208.

Kathju, S., Aggarwal, P.K. and Lahiri, A.N. 1987. Evaluation of diverse effects of phosphate application on legumes of arid area. *Tropical Agriculture*, 64 : 91-96.

Khan, G.Z., and Morey, D.K. 1980. Influence of Kaoline spray on transpiration and water use efficiency of groundnut (*Arachis hypogaea*) under different soil moisture regimes. *Journal of Maharashtra Agricultural Universities*, 5 : 131-134.

Korwar, G.R., 1995. pp. 380. Unpublished data quoted in R.P. Singh and Osman 1995. Alternate land use systems. (Drylands) In : *Sustainable Development of Dryland agriculture*, Ed. R.P. Singh, Scientific Publishers, Jodhpur, India, pp. 375-398.

Krishnamurthy, C.H. Chouwdhury, S.I. Anderson, D.T. and Deyclen, R.D., 1977. *Crop Management for Semi-Arid Farming*. Second FAO/SIDA Seminar on fied food crops in Africa and the near East at Lahore, Pakistan from Sept. 18 October, 1977, pp. 1 -29.

Larson, W.E., 1962. Tillage requirements for corn. *Journal of soil and Water conservation*, 17 : 3-7.

Lomte, M.H., Kawarkhe, R.K. and Ateeque, M., 1994. Dry sowing for improving the stability and productivity of rainfed sorghum in deep Vertisol. *Journal of Maharashtra Agricultural Universities*, 19(2) : 195-197.

Mann, H.S., and Singh, R.P., 1977. Crop production in Indian arid zone. In : *Desertification and its control*, ICAR, New Delhi, pp. 215-234.

May, L.H., and Mithorpe, F.L., 1962. Drought resistance of crop pants. *Field Crop Abstr.*, 15 : 171-180.

Mishra, D.K. 1971. Agronomic investigations in arid zone *Proc. Symp. Problems of Indian Arid Zone*, pp. 165-169.

Musgrave, G.W. and Free, G.R. 1936. Some factors which modify the rate and total amount of infiltration of field soils. *Journal of American Society of Agronomy*., 28 : 727-739.

Muthuswami, M. and Narayanswamy, P., 1981. *Indian Phytopathol*, 34 : 418- 420.

Narang, R.S., Brar, R.S. and Nathu Singh, 1988. Antitranspirants for conservative use of water in fall planted corn grown under limiting water regime. In : *Challenges in dryland Agriculture- A Global Perspective*. Proceedings of the International Conference on dryland farming. August 15-19, 1988, Amarillo/Bushland, Texas USA, : 227-229.

Narkhede, P.L., And Ghugare, R.V., 1987. Organic recycling in drylands II. Effect of organic recycling of subabul on yield, nutrient uptake and moisture utilization by winter wheat. *J. Indian Soc. Soil Science*, 35 : 417-420.

Niranjan, K.P. Arya, R.L. Yadava, R.B. and Gangawar, K.S., 1995. Performance of different dry farming systems under rainfed conditions. *Range Management and Agroforestry*, 16 : 93-99.

Oswal, M.C., Bakshi, R.K., Kumar, V. and Kumar, S., 1989. Response on dryland pearl millet to fertilizer under two cropping sequences. *Journal of Indian Soc. Soil Science*, 37 : 337- 342.

Pamplona, P.P. and J.S., Imlan, 1976. MIT, Res. J. 6 : 1-9.

Pathak, P.A. and Laryea, K.B. 1995. Soil and water conservation in the Indian SAT : Principles and Improved practices. In: *Sustainable Development of Dryland Agriculture in India*, (Ed. R.P. Singh) Scientific Publisher, Jodhpur, pp. 83-94.

Patil, N.D., Umrani, N.K., Shinde, S.A. Manke, B.J., Kale, S.P. And Shigte, A.K. 1981. Improved crop production technology for drought prone areas of Maharashtra. Mahatma Phule Agricultural University Rahuri, India.

Patil, S.L. 1998. Response of *rabi* sorghum to tillage, moisture conservation practices, organics and nitrogen in vertisols of semiarid tropics. Ph.D. Thesis submitted to Univ. Agril. Sciences, Dharwad, India.

Patil, Y.M., Belgaumi, M.I. Maurya, N.L. Kubsad, V.S., Mansur, C.P. and Patil, S.L., 1995. Economic evaluation of different barriers on growth and yield of *rabi* sorghum. *India Journal of Soil Conservation*, 23 : 56-57.

Pothiraj, P., 1982. Seeding techniques for pre, early and late monsoon sowings of rainfed cotton in vertisols. *Ph.D. Thesis*, TNAU, Coimbatore.

Prahar, S.S. and Gajri, P.R., 1980. Water harvesting - a management practice in dryland farming. *Indian farming*, 28 (9) : 22.

Radder, G.D., Itnal, C.J. and Belgaumi, M.I., 1995. Protective irrigation (life saving) : Principles and practices. In : *Sustainable Development of Dryland Agriculture in India*, Ed. R.P. Singh, Scientific Publishers, Jodhpur, India, pp. 207 - 215.

Ramakrishna, A. and Tripathi, B., 1995. Changing scenario of weeds and their management in dryland crops and cropping systems. In : *Sustainable Development of Dryland Agriculture in India*, Ed. R.P. Singh, Scientific Publishers, Jodhpur, India, pp. 425-444.

Randhawa, N.S. and Singh, S.P., 1983. Fertilizer management in varied areas available technologies and future needs. *Fertilizer news*, 28 (9) : 17 -32.

Randhawa, N.S. and Venkateswarlu, J. 1979. India experiences drought in semi-arid tropics, prospects and retrospects. In : *Development and Transfer of Technology*. Proceedings International symposium or intercropping. 10-13, January, 1979, held at International Crops Research Institute For Semi-arid Tropics, Hyderabad, pp. 306-317.

Randhawa, N.W. and Venkateswarlu, J., 1987. Farming systems researchers : Indian experience In : *Proceedings of International Agriculture Research Centre Workshop on Farming Systems Research*, 17-21 Feb, 1986. ICRISAT. Centre, India, Patancheru, AP. JU 2324, PP. 110-112.

Rao, A.C.S., and Das, S.T. 1982. Soil fertility management and fertilizer use in dryland crops In : *A Decade of Dryland Agricultural Research In India 1971-80 AICRPDA, Hyderabad, (cited by Prihar et al.,* 1985) pp. 120-139.

Rao, A.S. And Agarwal, J.P., 1984. Seed production root/shoot ratio, stomata number and nutrient uptake by efficient crop and weed species. *Indian Journal of weed sciences*, 16 : 250- 256.

Reddy, G.S., Das, S.K. and Singh, R.P., 1993. Prospects of green manuring on alternative to fertilizer nitrogen in dryland farming. In : *Recent advances in Dryland Agriculture* (Ed. LL Somani), Scientific Publishers.

Reddy, N.V. and Hamapaiah, R. 1982. Alternate land use in dryland agriculture. In : *A decade of Dryland Agricultural Research in India*, 1971-80, AICRPDA, Hyderabad.

Saghi, N.K. and Korwar, G.R., 1987. Integrated pest management In : *Technological Advances in Dryland Agricultures*, (Ed) S.p. Singh *et al.,* Central Research Institute for Dryland Agriculture, Hyderabad, pp. 101-119.

Sanghi, N.K. And Korwar, G.R., 1987. Integrated pest management. In : *Technological Advances in Dryland Agriculture*, Ed. S.P. Singh *et al.,* Central Research Institute for Dryland Agriculture, Hyderabad, pp. 101-119.

Shantakumari, M., Reddy, C.S. And Ramdas, V.S., 1977. Alachlor a new potent antitranspirant on maize plants. *Proceedings Indian Academic Sciences*, 86 : 143-150.

Shashidhar, B.C., Kachapur, M.D. Chittapur, B.M. and Ravi Hunje, 1995. Effect of crop residues on physical, chemical and biological properties of soil in sorghum based cropping system.

Singh, G.B., and Sarswat, V.N., 1986. Paper presented at the joint Indo-Pakistan weed control workshop and 7th USDA/ICAR AICRPDA workshop at Dhurgapur, Agric. Research Station, Jaipur 22-24 March, 1986, India.

Singh, R.D., 1983. *Farm ponds*, Project Bulletin No. 6 All India co-ordinated Research Project for dryland Agriculture, Hyderabad.

Singh, R.P., 1995. Efficient management of dryland crops - principles and practices. In : *Sustainable Development of Dryland Agriculture in India*, Ed. S.P. Singh, Scientific Publishers, Jodhpur, India, pp. 285-298.

Singh, R.P., DAS, S.K. Bhaskara Rao and Narayanareddy, M. 1990. *Towards sustainable Dry land Agricultural Practices*. Central research Institute for Dryland Agriculture, Hyderabad, 1990, pp. 1 - 106.

Singh, R.P., Singh, H.P., Daulay, H.S. and Singh, K.C., 1981. Effect of periodical application of nitrogen in organic and inorganic form on the yield of rainfed pearl millet. *Indian Journal of Agricultural Sciences*, 51 (6) : 409-416.

Singh, R.P. 1995. Crop management options for aberrant weather situation and agricultural drought. In : *Sustainable Development of Dryland Agriculture in India*, Ed. R.P. Singh, Scientific Publishers, Jodhpur, India, pp. 299 - 312.

Singh, R.P. and Das, S.K., 1983. Dryland agriculture -a glimpse of recent research in India. *Advances in Indian agriculture*, 1 : 147-173.

Singh, R.P. and Das, S.K., 1984. Nitrogen management in cropping systems with particular reference to rainfed lands of India In : Nutrient management in drylands with special reference to cropping systems and semiarid red soils. Project bull No. 8, AICRPDA, Hyderabad, p. 87.

Singh, R.P. and Das, S.K., 1984. Nitrogen management in cropping systems. India, part A of Project bulletin No. 8. AICRPDA, Hyderabad.

Singh, R.P. and Das, S.K. 1995. Agronomic aspects of plant nutrient management in rain dependent food crop production systems in India In : *Sustainable Development of Dryland Agriculture in India* (Ed. R.P. Singh), Scientific Publishers Jodhpur, pp. 117-138.

Singh, R.P. and Osman, M., 1995. Alternate land use systems for drylands. In : *Sustainable Development of Dryland Agriculture in India*, Ed. R.P. Singh, Scientific Publishers, Jodhpur, India, pp. 375-398.

Singh, S.D. Bhandari, R.C. and Aggarwal, R.K. 1985. Long term effects of phosphate fertilizers or soil fertility and yield of pearl millet grain in rotation with grain legume. *Indian Journal of Agricultural Sciences*, 55: 274-278.

Singh, S.P., 1987. Efficient crops and cropping systems In : *Technological Advances in Dryland Agriculture* Ed. S.P. Singh *et al.,* Central Research Institute for Dryland Agriculture, Hyderabad, pp. 47 -66.

Singh, S.P., 1988. Efficient cropping systems for dryland areas in India In : *Challenges in Dryland Agriculture - A Global Perspective*, Proc. of Int. Conf. on Dryland Agric. Amarillo -Bushland Texas USA - Aug, 15-19, 1988, pp. 770-775.

Singh, S.P. and Das, S.K. 1995. Agronomic aspects of plant nutrient management in rainfed dependent food crop production systems in India. In : *Sustainable Development of Dryland Agriculture in India*, Ed. S.P. Singh, Scientific Publishers, Jodhpur, India, pp. 117-138.

Slatyer, R.O. And Bierhuizen, J.F., 1964. The influence of several transpiration suppressants on transpiration photosynthesis and water use efficiency of cotton leaves. *Australian Journal of Biological Science*, 17 : 131- 146.

Smith, D. And Buchlotz, K.D., 1962. Transpiration rate reduction in plants with atrazine. Science, 136 : 263-264.

Sudha, K.N., 1999. Resource of rainfed groundnut (*Arachis hypogaea* L.) to sand mulching and organics in vertic inaceptisols. *M.Sc. (Agri.) Thesis* submitted to Univ. Agril. Sciences, Dharwad, India.

Tomar, B.S., Singh, R.S., and Sainy, S.L. 1974. Effect of direct seeding and transplanting on the growth and yield of hybrid Bajra 3 pearl millet grown at different nitrogen levels under late sown conditions. *Indian Journal of Agricultural Sciences*, 44 : 317-328.

Umrani, M.K., Daftardar, S.Y., Patil, C.P. and Walujkar, R.B., 1983. Effect of inter row spacing, plant density and nitrogen application on pearl millet, BJ 104 under rainfall conditions. *Annals of Arid Zone Research*, 22 (2) : 173-175.

Umrani, N.K., 1995. Crop production technology for dryland areas. In : Sustainable development of dryland agriculture in India Ed. R.P. Singh, Scientific Publishers, Jodhpur. India, pp. 241-270.

Umrani, N.K., Kale, S.P., Diam, I.S. and Pharande, K.S., 1978. Response of M-35-1 winter sorghum to levels and methods of application of nitrogen. *Indian Journal of Agricultural Sciences*, 48 : 559-560.

Umrani, N.K., Kale, S.P. and Bhoi, P.G., 1981. Studies on crop planting for drought prone area of Maharashtra. *Journal of Maharashtra Agricultural Universities*, 6(2) : 110-113.

Umrani, N.K. 1985. Sorghum (Post-rainy season) In : *Efficient Management of Dryland Crops*. Ed. V. Balasubramaniam and J. Venkateshwarlu, CRIDA, Hyderabad, pp. 113-128.

Umrani, N.K. and Bhoi, P.G., 1984. Effect of plant density on growth and yield of Safflower under two rainfall situations. *Indian Journal of Agronomy*, 29 (3) : 282-286.

Umrani, N.K. and Bhoi, P.G., 1985. Performance of safflower as influenced by plant density under drylands. *Journal of Maharashtra Agricultural Universities*, 10(1) : 34-35.

Umrani, N.K. and Bhoi, P.G. 1977. Root development in relation to inter row spacing of sorghum. Sorghum News Letter 20 : 49.

Umrani, N.K. and Patil, C.B. 1985. Improved technology in bajra production under different environmental situations. *Journal Maharashtra Agricultural Universities*, 10 (1) : 31-33.

Unger, P.W., Stewart, B.A. Parr, J.F. and Singh, R.P., 1991. Crop residue management and tillage methods for conserving soil and water in semi-arid regions. *Soil Tillage Research*, 20 : 219-240.

Unger, P.W. and Stewart, B.A., 1983. Soil management for efficient water use : An overview. In : *Limitations of Efficient Water use on Crop Production* (Eds.) Taylor *et al.,* Madison Wisc. Am. Soc. Agron. pp. 419-460.

Veerabhadran, V. and Rajendran, P., 1993. Integrated nutrient management for rainfed sorghum in red soils of northwestern zone of Tamil Nadu. *Madras Agricultural Journal*, 80(5) : 255- 258.

Veerashankar, P., 1984. Ley farming. *M.Sc. (Agri.) Thesis*, Submitted to APAU, Hyderabad, India.

Venkateswarlu, J., Vishnumurthy, T. and Padmanabhan, M.V., 1991. *Contingent crop production strategy in rainfed areas under different weather conditions*, project Bulletin No. 5. Central Research Institute for Dryland Agriculture, Hyderabad. pp. 76.

Venkateswarlu, J. 1985. Dry Farming extension bulletin No. 1. Central Research Institute for Dryland Agriculture, Hyderabad, India, pp. 1-57.

Ventakeswarlu, J., 1983. Contingent crop production strategy in rainfed areas under different weather conditions. Project Bulletin No. 5 AICRPD. Hyderabad.

Vijayakumar, B., Shivshankar and Rao, A.N., 1988. Weed control in pigeonpea + groundnut - intercropping system. *Indian Journal of weed science*, 26 : 96-97.

Vijayalakshmi, K., 1987. Rain water management in drylands. In : *Technological Advances in Drylands Agric.* (Ed. S.P. Singh *et al.*,) Central Research Institute for Dryland Agriculture, Hyderabad, pp. 19-65.

Vijayalakshmi, K. and Vittal, K.P.R., 1981. Rainwater harvesting and recycling in semi-arid red and sierozem soils. *Indian Journal of soil Conservation*, 9 (2 & 3) : 87-89.

Virmani, N.K., 1995. Crop management technology for drylands. In: *Sustainable Development of dryland agriculture in India*, Ed. R.P. Singh, Scientific Publishers, Jodhpur, pp. 241-270.

Vittal, K.P.R. 1987. Soil fertility management. In : *Technological Advances in Dryland Agriculture*, CRIDA, Hyderabad, pp. 83-99.

Willey, R.W., 1988. Cropping systems for Drylands. In : *Challenges in Dryland Agriculture - As global Perspective. Proc. of Int. conf.* on dry land agric. Amarillo/Bushland, Texas, USA, August 15-19, P. 77-774.

Willey, R.W., Rao, M.R. Reddy, M.S. and Natarajan, M. 1982. Cropping systems with sorghum p. 477-490. In : Proceedings of Sorghum in the eighties. Hyderabad, India : ICRISAT.

Wills, G.U., Davis, P.E. and Funderburk, H.M. 1963. The effect of atrazine on transpiration in corn, cotton and soybean. *Weeds*, 11 : 253-255.

Zaongo, C.G., Wendt. C.W. and Hossner, L.R., 1988. Contour strip rainfall harvesting for cereal grains production in the sahel. In : *Challenges in dryland agriculture - A Global Perspective*. Proceedings of the international conferences on dryland farming. August 15-19, 1988, Amarillo, Bushland, Texas, USA : 242-244.

Chapter - 14

CROP LIFE SAVING MEASURES IN DRYLAND AGRICULTURE

J.D. Patil

Ex-Chief Scientist, Agricultural Research Station, Solapur
Head, Dept. of Agricultural Chemistry and Soil Science, Mahatma Phule Krishi Vidyapeeth, Rahuri - 413 722 (MS)

1. INTRODUCTION

Out of 178 million hectares of cultivable land in India, 119 million hectares is under rainfed/dryland agriculture. The country is focusing major attention on irrigation for the development of agriculture and for increasing the foodgrains, but almost half of the country's arable land will continue to be rainfed/dryland even after all the surface ground water resources are fully utilised. Soil erosion is one of the major problems limiting crop production in dryland areas. Majority of the rainfall (about 80%) is received in period of about three to four months (June to September). Quite a sizeable portion (15 to 40%) of this is lost as run-off, causing erosion of top fertile soil. These areas are most vulnerable to aberrent weather conditions. Generally, the rainfall in this region is scanty (<750 mm), erratic and ill-distributed coupled with hot weather, high wind velocity and prolonged dryspells. These situations lead to partial or complete failure of crops.

Scientists have made efforts to diagnose and rectify the limps and shrinks of drylands. During the last two decades, the primary emphasis was given on generation of technologies for better management of soil, rain water and cropping system to optimise crop production and this was achieved by and large through use of improved seed, moderate levels of fertilizers and management. Considerable efforts have been made to identify ways to meet the challenges of weather aberrations by evolving alternate crop strategies. The scientists are now involved in optimising use of the environment for bringing prosperity and economic stability to the dryland farmer through watershed development and identifying efficient crops based land use systems. Dryland technology is more site specific and warrants participation of farmers in developing the technology for its adoption. Timely judgement is very important for the success of dryland agriculture which calls for need-based training both at trainers and trainee level. This is being achieved through establishment of ORPs and Pilot Projects.

The dryland agriculture in Maharashtra is represented by 87 tahsils of 14 districts. The average rainfall of this region varies from 500 to 720 mm and has bimodal distribution. The first peak is usually experienced during June and second during September. Rainfall during September is more assured (150-200 mm). Even though monsoon sets in by the end of June, July and August are characterized by dry spells of varying duration (2 to 8 weeks at stretch) and frequencies (1 to 5).

Usually dry spells of more than four weeks or 3 dry spells of two weeks duration result in failure of Kharif crops.

Majority of the soils (30%) are shallow (<22.5 cm) having low moisture storage capacity and low fertility, approximately 45 percent soils are medium deep black (22.5-90 cm) and 25 percent soils are deep black (>90 cm). These soils exhibit adverse physical characters due to high clay content (35 to 65%) of smectite type clay minerals. They have high volume expansion when moist and shrink when dry, producing deep cracks, which accelerates the process of soil moisture loss. Soil moisture is the most important single factor in dryland agriculture which is governed by amount and distribution of rainfall, clay content in soil and soil depth. Usually soils with limited depth (45 cm) are cropped during *kharif.* Soils with better depth (<45 cm) are put under *rabi* crops or double cropping (sequence cropping) depending upon rainfall situations.

Soil and water conservation is the pre-requisite for successful dry farming which can be achieved through adoption of contour bunding in shallow and medium deep soils and graded bunds, in deep soils with surplusing arrangements. Dryland crop production can be increased and stabilized substantially through choice of efficient crop/crops variety/cropping system depending upon soil depth, timely operations, use of fertilizer, good management practices and plant protection measures whenever needed.

2. CROP LIFE SAVING MEASURES

The yield of dryland crops can be increased and or stabilized by adopting following crop life saving measures, viz.

(1) Early sowing of rabi crops.

(2) Dust mulching and or mulching of organic materials.

(3) Timely control of weeds.

(4) Adjustment in plant density.

(5) Recycling of runoff water as a protective irrigation.

(6) Crop planning for aberrant weather condition.

(7) Plant protection measures.

(1) Early sowing of rabi crops

Rabi crops are grown on receding soil moisture. This puts limitation on growth and full expression of plant vigour. As a measure to correct it, early sowing of *rabi* crops has been recommended. This helps to make more soil moisture available during plant growth period. It has been found that sowing of sorghum in the first fortnight of September improved the grain and stover yield of sorghum by 30 and 93 percent, respectively over traditional sowing. Similarly the yield of safflower was found to be increased by 78 per cent over traditional sowing.

(2) Dust mulching of organic materials/ crop residues

(a) Dust mulching through interculturing

Timely interculturing operations with suitable implements are equally important in proper conservation of moisture in the soil and making it available to the crops for their satisfactory growth. The experiment conducted on medium black soils at Solapur for six years proved that interculturing with hoe at third, fifth and eighth week from sowing of *rabi* sorghum (M. 35-1), conserved more moisture and increased the grain yield of sorghum to a reasonable degree (Table 1). Out of the three interculturings, the first interculturing should be given by an entire blade hoe and the later two by toothed hoe. The first hoeing separates out the surface layer from underlying soil profile and thereby acts as a dust mulch in the first instance and breaks the capillary movement of the soil moisture from top in the second instance. This results into more storage of receding soil moisture in the soil profile, which in turn is useful for boosting up the crop production. The second and third hoeings subsidise the micro cracks developed in the soil and thereby reduce the moisture loss through the micro-cracks.

Table 1. Grain yield of sorghum as influenced by interculturing (Average of six years)

S.N.	Treatment	Grain yield (q/ha)
1	One interculturing	6.36
2	Two interculturings	6.50
3	Three interculturings	6.99
4	Control	5.91
	C.D. at 5%	0.66

(b) Organic/crop residue mulching

Interculturing the field during *rabi* season to create dust mulch is a useful practice. Better practice than this is, however, the use of crop residue of non economic value as surface mulch between the crop rows. Crop residue of less economic value such as sorghum stubbles, dry grass, wheat straw or pigeonpea stalks can be utilized as surface mulch @ 5 t/ha. Before application to the field, the material should be cut to about 20- 25 cm length. This type of mulching is specifically recommended for moisture conservation to *rabi* sorghum. However, the mulches were not found effective in increasing the yields of safflower, gram and *rabi* cotton. It should be applied immediately, after the emergence of seedlings or upto 15 days from emergence. It should be spread between two rows of crop. It prevents the loss of soil moisture through thermal insulation and prolongs the moisture retention period. Thus, the moisture level is better during the flowering and grain formation stages of crop. The results of the experiment conducted on medium deep black soil at Dry Farming Research Station, Solapur indicated that application of organic mulch of crop residues @ 5.0 t/ha increased the grain and stover yield of sorghum substantially over control (Table 2).

Table 2. Effect of different mulches on grain and stover yield of sorghum (Average of six years)

S.N.	Treatment	Yield of sorghum (q/ha)	
		Grain	Stovar
1	Control (without mulch)	7.28	34.3
2	Sorghum stubbles	10.35	47.2
3	Wheat straw	11.20	51.9
4	Pigeonpea stalks	11.82	56.8
5	Dry grass	12.65	62.4

(3) Timely control of weeds

Weeds reduce the yield of crops by competing with crop for moisture, nutrient and light. The problem of weeds is severe in *Kharif* season than *rabi* season. The extent of damage by weeds to *Kharif* crops varies from 35 to 97 percent, depending upon the intensity of weeds, rainfall and pretillage operations. In shallow *Kharif* soils weeds like cockscomb (*Pelasia argenta*) and commelina (*Commelina forskalli*) cause huge damage. Rains during *Kharif* season cause fast growth of weeds. Therefore, timely control of weeds became very important. The study conducted at Dry Farming Research Station, Solapur in pearlmillet crop indicated that weeds should be controlled by interculturing or by weeding before 20-30 days of the crop growth. Every 10 days delay causes considerable damage. It was estimated that about 2.5 q/ha grain is lost due to every 10 days delay (Table 3).

Table 3. Yield of pearlmillet as affected by time of weed control

S.N.	Time of removal of weeds from sowing of crop	Grain yield(q/ha)
1	Weed free till maturing of crop	11.02
2	Weeds removed after 10 days	8.54
3	Weeds removed after 20 days	7.32
4	Weeds removed after 30 days	4.45
5	Weeds not removed	2.02

Timely control of weeds not only saves the expenditure on weeding but also minimises the weed troubles in the subsequent years. Under certain situations like shortage of labourer/bullocks, inaccessibility of field due to continuous rains, control of weeds by chemical method becomes inevitable. It was found that spraying of atrazne @ 0.5 kg a.i./ha as pre emergence spray helps to control the weeds satisfactorily in *Kharif* season.

(4) Optimum plant population

In *rabi* sorghum amount of stored soil moisture becomes a key factor for production. Adjusting the density and row spacing is desirable which will effectively utilize the stored moisture for grain production. The results of the experiment conducted at Solapur revealed that for low moisture storage (50 mm), it is better to keep 5 plants/m^2 and spacing of 135 cm between rows (Table 4). For better moisture availability (100 mm), plant population should be 5-10 plants/m^2 with 90 cm spacing between the rows. A linear increase in MUE was noticed as the spacing increased.

Table 4. Grain and stover yield of rabi sorghum as affected by plant density, row spacing and moisture storage

S.N.	Plant density / Row spacing	Yield (q/ha)			
		50 mm moisture		100 mm moisture	
		Grain	Stover	Grain	Stover
1. Plant density (Plants/m^2)					
	5	8.39	11.47	16.73	19.57
	10	7.80	13.33	16.36	21.40
	15	7.52	12.47	15.61	22.05

2. Row spacing (cm)

5	7.36	13.01	15.88	24.54
10	8.11	13.84	17.18	18.96
15	8.26	12.76	15.35	17.78

In case of safflower and sunflower low plant density of 50,000 to 1 lakh/ha would be optimum for maximum production.

(5) Recycling of runoff water as a protective irrigation

Field runoff from the mechanical soil conservation structures is stored in dug out farm pond which could be utilized as life saving irrigation whenever needed. If the farmer is having well, he can utilize the water from existing well for protective irrigation. For many field crops three stages namely germination, tillering/branching/ elongation (30-55 days) are the critical for moisture stress. Normally, in dryland, the care of first stage can be taken of by good management. At the later stages, soil moisture becomes limiting for proper growth which affects crop yield adversely. If water is made available just before these stages, moisture sufficiency would ensure good crop production. If the water in the pond is not expected to last for longer period, the critical irrigation should be applied at early stage.

The life saving irrigation experiment conducted at Solapur, indicated the possibilities of minimum (4 cm) use of water for extensive irrigation. It was concluded that one protective irrigation at the flowering stage either by surface or alternate furrow method produced 65 and 15 percent more grain and stover yield of sorghum (M. 35-1) over control, respectively (Table 5). However, providing protective irrigation through alternate furrow method saved 50 per cent quantum of water thus more area is covered by this method.

Table 5. Grain and stover yield of sorghum as influenced by protective irrigation

S.N.	Method of Irrigation	Yield (q/ha)	
		Grain	Stover
1.	Control	8.84	18.62
2.	Surface irrigation	14.77	21.38
3.	Alternate furrow irrigation	14.37	21.38
4.	% increase over control	64.82	14.82

The influence of 5 cm protective irrigation at the time of flowering of bengal gram and sorghum was studied at Solapur for two years. It was revealed that the grain yield of bengal gram and sorghum was increased by 57 and 39 percent by applying one protective irrigation at flowering (Table 6).

Table 6. Yield of Bengal gram and sorghum as influenced by one protective irrigation (Average of two years)

S.N.	Crop	Yield (q/ha)	Unirrigated	Irrigated	% increase
1	Bengal gram (Chafa)	Grain	4.17	5.80	39.1
2	Sorghum	Grain	17.61	27.67	57.12
		Stover	40.20	52.16	29.75

Plate 14.1.1. Well maintained canal of Pench project in Maharashtra leading to increased crop intensity in the area, **2.** Ridges and furrows for irrigation in rabi season.

Further, experiments were conducted at Solapur for three years on use of protective irrigation at critical stages of safflower and bengal gram. It was observed that one protective irrigation of 4 cm depth by border method was optimum for boosting the yield of safflower by 34 percent, when applied at 55-60 days growth stage. One irrigation at 30-35 days growth stage, increased the yield by 26 percent. Applying two irrigations was not found much beneficial. For Bengal gram applying one irrigation either at 30- 35 days or at 55-60 days growth stage increased the yield by 32 per cent. Applying two irrigations at both stages increased the grain yield by 50 percent (Table 7).

Table 7. Effect of protective irrigation on grain yield of safflower and bengal gram

S.N.	Stage of crop for applying irrigation	Safflower		Gram	
		Yield (q/ha)	Per cent increase	Yield (q/ha)	Per cent increase
1	Control (No irrigation)	7.69	–	7.95	–
2	One irrigation at 30-35 days	9.66	25	10.48	32
3	One irrigation at 55-60 days	10.30	34	10.38	32
4	Two irrigations one each at 30-35 days and 55-60 days	10.54	36	11.81	51

(6) Crop planning for aberrant weather situations

The crop production in dryland suffers from instability due to aberration in weather. The common weather aberrations are delayed onset of monsoon, early withdrawal of monsoon, dry spells during monsoon and extended monsoon. Under such circumstances, crop planning according to aberrant weather situation becomes very important to stabilize crop production. The specific recommendations evolved for Scarcity Zone to tackle aberrant weather situations are given below.

(a) Delayed onset of monsoon

For delayed onset of monsoon, alternate cropping plans (contingent crop planning) have been suggested for kharif soils (45 cm) which help in stabilizing crop production (Table 8).

Table 8. Contingent crop planning

S.N.	Rainfall situation	Suggested crops/ cropping systems
1	Rains during July but seedling by end of July or early August	Setaria, Red gram, Castor, horse gram, sunflower, intercropping of sunflower + red gram
2	Seedling during August	Red gram, sunflower, horse gram, castor, setaria, intercropping or sunflower + red gram
3	Rains during late August and seedling in late August or early September	Red gram, sunflower, castor, sorghum for fodder, sorghum for grain (Selection-3 TL).

(b) Good start of monsoon followed by dry spells

This situation is also common in dryland agriculture. If the dry spell exceeds two weeks period, corrective measures are required to be taken with the hope that the dry spells would be broken after about three to four weeks. Such measures are usually aimed at conserving available soil moisture through timely weed control and increase in interculturing operations. For efficient utilization of conserved moisture, plant population is reduced. A usual plant population of sunflower (74000 plants/ha) should be reduced to 38,000 plants per hectare either by removing alternate plants in a row or by removing alternate row.

(c) Early withdrawal of monsoon

Under such situation early sowing of *rabi* crops is advocated. The stored moisture should be efficiently utilized through reducing plant density, increasing frequency of interculturing operations/ applying surface mulch. If available one or two protective irrigation may be given at critical growth stages of the crop.

(d) Extended monsoon

In case of extended monsoon, postpone the sowing of *rabi* crops and replace sorghum by safflower, or gram or sunflower. The risk due to disease damage should be reduced by adopting proper seed treatment and following zonal system of planting.

(7) Plant protection measures

Proper schedule of integrated disease pest control measures should be followed to prevent the loss of crop due to pest incidence whenever required.

The yield of dryland crops can be increased and stabilized by adopting various crop and soil management practices, use of protective irrigation and adopting integrated disease pest control measures. The above measures should be adopted on watershed basis which is aimed at optimizing the use of land, water and vegetation and thereby alleviating droughts, moderating flows and preventing soil erosion as well as ensuring increased food, fuel and fodder production on sustained basis.

REFERENCES

Dryland Agriculture in India, State of Art of Research in India (1992). Edt. L.L. Somani, Scientific Publishers, Jodhpur.

Umrani, N.K. (1983). Dryland Research Project, Solapur - A Profile. Technical Bulletin No. 2, AICRP for DLA, Solapur - 2.

Umrani, N.K. (1984). Management Practices for Black Soils in Drought Prone Area of Maharashtra. Technical Bulletin No. 4, AICRP for DLA, Solapur-2.

Chapter - 15

AGRO – HYDROLOGICAL MODELLING FOR DRYLAND WATER MANAGEMENT

B. Bapuji Rao and M. Raghu Babu*

Department of Agronomy, Agricultural College, BAPATLA – 522 101 (Andhra Pradesh)
*Saline Water Scheme, Agricultural College Farm, BAPATLA – 522 101

INTRODUCTION

Dryland agriculture still remains the backbone of Indian agriculture, as large areas of cultivated land are rainfed, which contribute about 42 per cent to country's food basket. Characterization and understanding of the environment is imperative to dryland agricultural research. To formulate judicious soil and crop management practices for varied dryland conditions, the crop growth processes under stress conditions are to be properly understood. The information on moisture and nutrient uptake pattern under moisture stress condition is presently incomplete. The studies on these lines will unravel many of the unanswered questions that would lead to a better understanding of nutrient and moisture interactions. Moreover, information on moisture deficits in different periods is very helpful in crop planning. The efforts are on for some years to develop models that would predict the crop water use pattern. An attempt is made here to present a comprehensive review of the currently available agro-hydrological models and to illustrate the basics behind the development of such models.

Some of the currently available agro-hydrological/soil water balance models are listed in Table 1. They differ in complexity, operation and purpose. The models also differ in the use of theoretical or empirical descriptions of the process in the model. These can be broadly classified as:

(a) Agroclimatic models - These are usually single layer models used for regional characterization of environments for water availability.

(b) Management models - In these models the soil profile is divided into two or three layers. Information generated on soil moisture availability is used for soil and crop management.

(c) Physical process models - The soil profile is divided into many layers for studying the flow processes more precisely.

Table 1. List of some agro-hydrological models

Name of the model	Reference
Versatile Soil Moisture Budget (VSMB)	Baier *et al.* (1979)
WATER	Burt *et al.* (1980)

Unnamed	Belmans *et al.* (1983)
Unnamed	Brisson *et al.* (1992)
Unnamed	Cordery and Graham (1989)
SMEP	Edey (1980)
Unnamed	Greacen and Hignett (1984)
SWATRE	Feddes *et al.* (1976)
Unnamed	Jagtap and Jones (1989)
SWACRO	Feddes *et al.* (1984)
Unnamed	Hansen (1975)
Unnamed	Holst and Madsen (1984)
Unnamed	Jones and Smajstrla (1980)
Unnamed	Lascano and van Bavel (1986)
Unnamed	Norman and Campbell (1983)
Unnamed	Place and Brown (1987)
Unnamed	Rama Prasad (1984)
PLANTGRO	Retta and Hanks (1980)
Unnamed	Robinson and Hubbard (1990)
SPAW	Saxton *et al.* (1974)
Unnamed	Seliorio and Brown (1979)
Unnamed	Stockle and Campbell (1985)
SIMBAL	Stuff *et al.* (1975)
EMWATBAL	van Bavel and Lascano (1987)
Unnamed	Victor *et al.* (1988)
Unnamed	Visser (1974)
Unnamed	Vossen (1990)
Unnamed	Wright *et al.* (1994)

1.1 Model components

The basic components of all the soil water balance models are presented in Fig. 1. The components are not independent but they are interrelated. For instance, amount of runoff depends partially on the rainfall intensity, hydraulic properties of the soil, and surface water content.

Precipitation or irrigation is usually measured and the other components of water balance are estimated, in the same units.

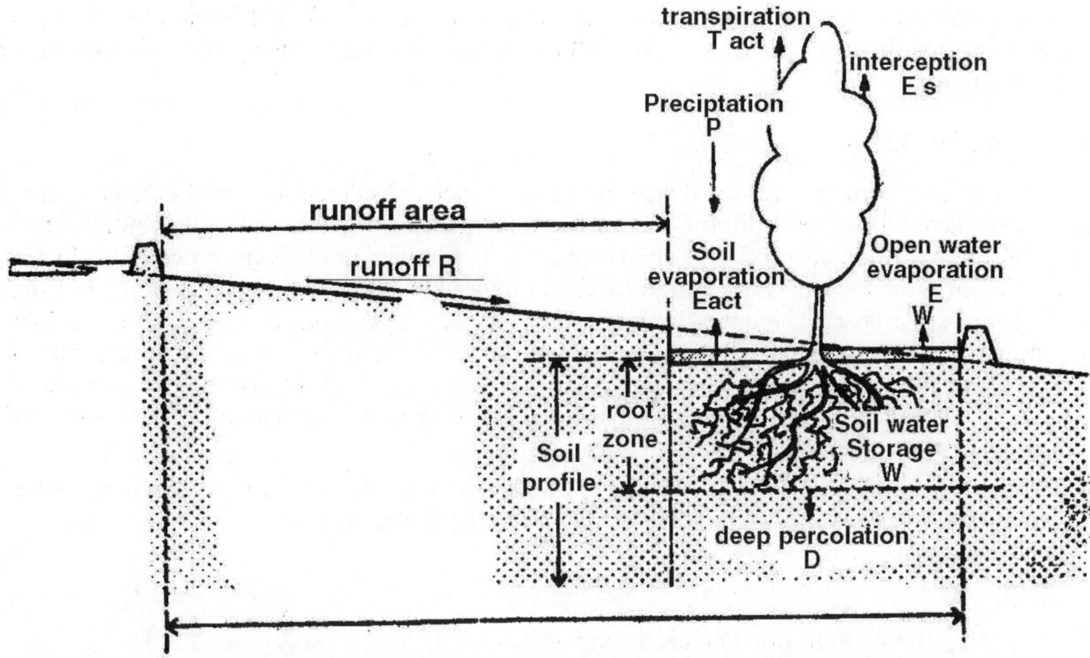

Fig. 1. Components of the water balance in a typical basin area

1.2 Surface runoff and infiltration

Whenever rainfall occurs, some amount of water often runs off and becomes unavailable to the crop. Therefore, to estimate the recharge of soil profile it is important to estimate the amount of runoff that occurs with each rainstorm. Various approaches have been made to estimate this component in water balance models. Some models do not consider runoff and deep drainage separately. Any amount of water input into the soil after the soil profile is full to its maximum storage capacity (field capacity) is considered as water loss (runoff + deep drainage). However, in some models the runoff or infiltration is estimated to calculate profile recharge. Kanemasu *et al.* (1978) calculated effective precipitation (Pe) and runoff as follows :

$$Pe = R^{0.75} \text{ when } R > 25.4 \text{ mm} ; Pe = R \text{ when } R \leq 25.4 \text{ mm}$$

Therefore, runoff = R – Pe. (1.1)

Baier *et al.* (1978) used a simplified relationship between moisture content in the topsoil zone and daily total precipitation to estimate infiltration into the soil. On days with rainfall 25.4 mm, the total amount of rainfall is considered to infiltrate into the soil. On days with rainfall 25.4 mm, the amount of infiltration (Infl) into the soil is less than the daily rainfall, because it is limited by run-off as a function of rainfall and the moisture already in the top zone of soil, and is computed as

$$Infl = 0.9177+1.811 \log RR_i - 0.97\ [S_{j(i-1)}C_j] \log RR_I \quad (1.2)$$

where,

RR_i = rainfall on day i

$S_{j(i-1)}$ = soil moisture in the jth zone on day i-1

C_j = available water capacity of the jth zone j = 1.

The remainder of the daily rainfall is assumed to be lost as runoff. Dale *et al.* (1982) employed the same relation in their SIMBAL model for cropland drainage effects on soil moisture and evapotranspiration.

1.3 Soil water recharge and drainage

The soil profile in the water balance models is either considered as a single layer or divided into discrete layers of either uniform or variable thickness. Infiltration and redistribution of water throughout the layered soil profile is often treated in two rather in different procedures. In simple models, the infiltrated water is freely transmitted to lower layers by gravity or out of the profile if it was the lower most layer. The upper limit of water for each layer is set at field capacity. When the antecedent water content plus inflow of water exceeds field capacity of that layer then the excess water is allocated to next lower layer. This process is repeated for all layers and excess water from the lowest layer is considered as deep drainage. In this method upward movement or redistribution of water is not allowed unless it is an added feature.

In the Versatile Soil Moisture Budget (VSMB) of Baier *et al.* (1979), the partitioning of infiltrated water to each zone is simulated by the following function.

$$Infl_{ij} = \{ 1-[S_{j(i-1)}/C_j]\ b\} \{Infl_i - Infl_{in}\} \qquad (1.3)$$

where,

$Infl_{ij}$ = new infiltration into each zone of soil

$S_{j(i-1)}$ = soil moisture in the jth zone on day i-1

C_j = available water capacity of the jth zone

b = percolation coefficient ranging from 0 to 1.

This equation is applied only when the ratio of soil moisture to capacity in any zone is less than 0.9. The amount of water that can infiltrate into and remain in each zone can not exceed the deficit (DEF) for that zone (j) and day (i). The DEF is given by

$$DEF_{ji} = C_j - S_{j(i-1)} + AE_{ji} \qquad (1.4)$$

In addition, the VSMB assumes that the amount of water that can be budgeted to the jth zone can never exceed what remains from the total water infiltrated after water has been budgeted to zones 1 to j-1. If, after all zones have been recharged, there is still infiltration water remaining, then this water is allocated to subsurface drainage. This means that the infiltration is distributed over the zones as a function of the amount of infiltration, the relative moisture content in each zone, and the percolation coefficient (b). The percolation coefficient is the fraction of water infiltrating to the next zone. For b=0, the water content of each zone must reach field capacity before the remaining water infiltrates into the next zone. For b=1, a fraction of the infiltration water percolates to the next zone before field capacity is reached, depending on the moisture content in the upper zone. This feature of the infiltration equation was found to be useful in heavy textured soils.

In WTGROWS - a wheat growth model of Aggarwal *et al.* (1994), the amount of water available either by rainfall or irrigation after deducting runoff is allocated to various soil layers starting from the surface layer. In this model, inter layer fluxes of water are considered only at the time of rainfall or irrigation and at all other times the fluxes are considered negligible. Depending upon the amount of water available, the layers are charged to field capacity. Water in excess of field capacity of a layer, if available, is immediately drained to the next layer. The amount of water above field capacity of the bottom layer is drained out of the profile and is not available for crop use. Similar procedure was adopted for profile recharge in COTTAM model of Jackson *et al.* (1990).

In the other method of profile recharge, it is customary to use Darcy's unsaturated flow equation, in which each layer is assumed to be uniform in moisture content, capillary pressure, and unsaturated conductivity. Mathematical solutions vary from the simple finite difference with large time steps to finite element with near analytical results. There are several models considering soil water movement in response to pressure-head gradients in accordance to the Darcy and continuity equations, for example, SWACROP of Feddes *et al.* (1978), SPAW of Saxton *et al.* (1974) and Rama Prasad (1984). This treatment of water flow can be used to represent nearly all situations including upward or downward flow between layers, widely varying characteristics within the profile, time distribution of infiltration and redistribution among layers, water tables and plant water withdrawal. But this approach requires specifications of the soil water-retention and hydraulic-conduc-tivity curves, upper boundary conditions of precipitation and potential evapotranspiration and a lower boundary condition appropriate for the site under consideration. The choice of which soil water movement calculation to employ depends upon the accuracy required. For readily drained soils where withdrawal of water by the plant dominates the water profile development and casual accuracy is required, the free flow procedure would suffice.

1.4 Potential evapotranspiration

The estimation of potential evapotranspiration (PET) is essential to know the evaporating power of the environment so that crop evapotranspiration (ET) could be estimated. Soil moisture models use different methods for estimating PET. In IBSNAT crop models, PET is calculated using an equilibrium evaporation concept developed from the Priestley and Taylor (1972) model. The equation calculates the approximate daytime net radiation and equilibrium evaporation, assuming that stomata are closed at night and no ET occurs then. Equilibrium evaporation, E_{eq} is computed in CERES - Maize model of Jones *et al.* (1986) as

$$E_{eq} = R_s(4.88 * 10^{-3} - 4.37 * 10^{-3} * \alpha)(T + 29) \quad (1.5)$$

where,

R_s = solar radiation, MJ m^{-2} day^{-1}

α = albedo of crop soil surface

T = average daily temperature (°C), estimated as

$$T = 0.6 * T_{max} + 0.4 * T_{min} \quad (1.6)$$

T_{max} = daily maximum temperature (°C)

T_{min} = daily minimum temperature (°C).

The PET is then computed by

$$PET = 1.1 * E_{eq} \text{ if } 5 < T_{max} \leq 35°C \quad (1.7)$$

$$PET = E_{eq} [1.1 + 0.05 (T_{max} - 35) \text{ if } T_{max} > 35°C \quad (1.8)$$

$$PET = E_{eq} * 0.01\, e^{[0.18 (Tmax + 20)]}$$

$$\text{if} \quad T_{max} > 5°C \quad (1.9)$$

The PET is calculated as the equilibrium evaporation times 1.1 to account for the effects of unsaturated air. The multiplier is increased above 1.1 to allow for advection effect when the maximum temperature is greater than 35°C, and reduced for temperatures below 0°C to account for the influence of cold temperatures on stomatal closure.

The versatile soil moisture budget (VSMB) of Baier *et al.* (1979) uses regression formulae for estimating PET from various combinations of available meteorological data. This is because of the development of these relations, for most of the Canadian sites where the VSMB was tested, from the historical weather data. For the sites where these regression equations are not available, the VSMB calculates PET from the daily maximum (T_{max}) and minimum (T_{min}) temperatures and radiation at the top of the atmosphere (Q_a) as

$$PET = 0.0034 [(T_{max} * 0.928) + 0.933 (T_{max} - T_{min}) + 0.0486 Q_a - 87.03] \quad (1.10)$$

The Penman's formula has been used to compute PET in several models like WTGROWS of Aggarwal *et al.* (1994) and WATER of Burt *et al.* (1980). In determining the PET, primarily the choice of method should be based on the type of meteorological data available.

1.5 Actual evapotranspiration

The actual evapotranspiration (AET) is generally estimated by two approaches in water balance models as

(a) using crop coefficients and soil dryness curves

(b) separating evapotranspiration into evaporation and transpiration.

1.5.1 Crop coefficients

Crop coefficients are generally empirical ratios of crop ET (ET_c) to some reference ET (PET) that have been derived from experimental data according to the relationship

$$K_c = ET_c/PET \quad (1.11)$$

where,

K_c = crop coefficient for a particular crop for a given growth phase and soil moisture condition - dimensionless

ET_c = daily crop ET (mm)

PET = daily reference ET (mm).

The reference ET characterizes the evaporative demand determined by meteorological conditions and a standard crop surface and K_c indicates the relative ability of a specific crop - soil surface to meet that demand. Since PET is affected by many variables, it can not be simplified for all climate and crop situations. This is because of the effects of relative leaf area and the morphological and physiological characteristics of the reference crop canopy on the energy exchange and aerodynamic diffusion processes within the atmosphere over a field. Different methods have been proposed for estimating PET for either grass or alfalfa with corresponding crop coefficients (Doorenbos and Pruitt, 1979; Wright, 1981).

Normally, the crop coefficient includes the effects of evaporation from both plant and soil surfaces and is dependent upon available soil water within the root zone and the wetness of the exposed soil surface. Soil water depletion data obtained by gravimetric or neutron probe methods and lysimetric data can be used to obtain K_c values.

The crop coefficient values for different crops are presented in Table 2. A generalized crop coefficient curve proposed by Doorenbos and Pruitt (1979) is presented as Fig 2. The Kc values for growth stages 1 and 2 for the crops in Table 2 can be obtained by interpolation from Fig 2. A crop coefficient curve represents the seasonal variation of the empirically derived K_c values. Once this curve has been developed for a given location, daily crop ET can be estimated using Eq. 1.11. The variation in time of season of the crop coefficient in Eq. (1.11) has not proven to be generalizable because it is often management, site and weather specific as evident from the values furnished in Table 2. Values of K_c may be management specific as a result of planting date, plant population and row spacing. The values may be site specific because of large-scale soil spatial variability and may not be reproducible from one year to the next for a given location because weather sequences are usually not reproducible from year to year. Crop coefficients are dependent on weather because air temperature, radiation and frequency of rainfall effect E_s and E_p directly and temperatures influence the rate of crop development. Hanks (1985) compared crop coefficients measured at Logan and Davis for the same crop. The crop coefficient curve differed markedly for the two locations, especially early in the season. He attributed this site specificity to the dependence of E_s on the rainfall frequency and amount or irrigation regime when plant cover is low and suggested that the crop coefficients vary from year to year for the same reason. Wright and Jensen (1978) recognized this limitation of the crop coefficient procedure and developed a crop coefficient curve that was based only on conditions where the soil surface is dry. Their model accounts for increased evaporation when the soil surface is wet and efficiently reduced the estimation of ET of snap bean during the leaf area development. Wright (1982) defined these modified crop coefficients, designed to represent conditions where the soil surface is dry but water is readily available, as "basal ET crop coefficients (K_{cb})". Estimates of an adjusted crop coefficient in terms of K_{cb} were accomplished through use of the following equation:

$$K_a = K_{cb} + (1 - K_{cb})\ [1 - (t/t_d)^{0.5}]\ f(w) \tag{1.12}$$

Where,

K_a = adjusted crop coefficient

t = number of days after major rain or irrigation

t_d = usual number of days for the soil surface to dry

f(w) = relative proportion of the soil surface originally wetted.

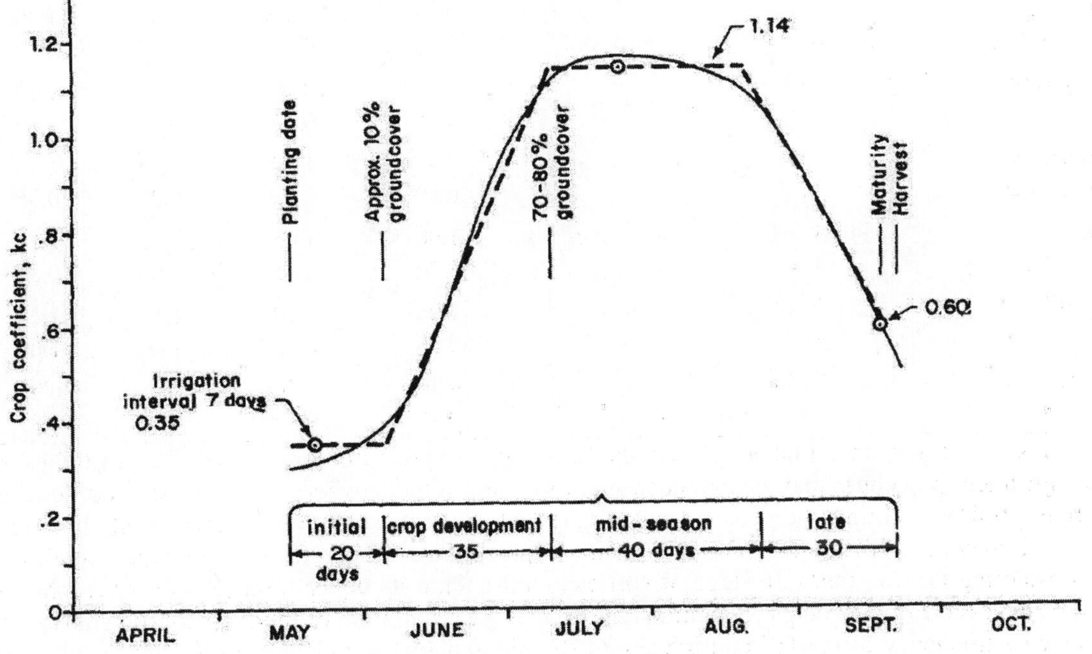

Fig. 2. A typical crop coefficient curve

Table 2. Crop coefficient (Kc) values for different crops and weather conditions (after Doorenbos and Pruitt, 1979)

Crop	Crop stage	Humidity wind speed (m/sec)	> 70%		<20%	
			0 – 5	5 – 8	0 – 5	5 – 8
Wheat	3		1.05	1.10	1.15	1.20
	4		0.55	0.25	0.20	0.20
Maize	3		1.05	1.10	1.15	1.20
	4		0.65	0.55	0.60	0.60
Cotton	3		1.00	1.15	1.20	1.25
	4		0.30	0.65	0.65	0.70
Millet	3		0.95	1.05	1.10	1.15
	4		0.55	0.30	0.25	0.25
Groundnut	3		1.00	1.00	1.05	1.10
	4		0.50	0.55	0.60	0.60
Sorghum	3		1.00	1.05	1.10	1.15
	4		0.45	0.50	0.55	0.55
Soybeans	3		1.05	1.05	1.10	1.15
	4		0.40	0.45	0.45	0.45
Sunflower	3		1.05	1.10	1.15	1.20
	4		0.70	0.40	0.35	0.35
Potato	3		0.95	1.10	1.15	1.20
	4		0.75	0.70	0.75	0.75
Onion (dry)	3		1.05	0.95	1.05	1.10
	4		0.25	0.75	0.80	0.85
Cabbage &	3		0.95	1.00	1.05	1.10
Cauliflower	4		0.80	0.85	0.90	0.95
Carrot	3		1.00	1.05	1.10	1.15
	4		0.70	0.75	0.80	0.85
Sugarbeet	3		1.05	1.10	1.15	1.20
	4		0.90	0.95	1.00	1.00
Radish	3		0.80	0.80	0.85	0.90
	4		0.75	0.75	0.80	0.85

Most of the adjustment takes place in the first few days after wetting the soil. The usefulness of this adjustment procedure may be extended mainly to arid regions where evaporative conditions are relatively uniform during the season and t_d may indeed be constant for a given soil. In humid regions, it may be necessary to accommodate the unpredictable temperature and radiation conditions by considering the constant rate stage of soil evaporation and its upper limit (Ritchie and Johnson, 1990). However, Eq. 1.12 helps to diminish the year-to-year variation of the crop coefficients caused by varying frequency of rainfall and irrigation. Ritchie and Johnson (1990) proposed the following relationship which incorporates the effects of temperature and leaf appearance and expansion for the influence of varying plant population on ET as:

$$K_{s+p} = (E_s + E_p)/PET \quad (1.13)$$

where,

K_{s+p} = a daily crop coefficient based on separate calculation of E_s and E_p.

The calculations of E_s and E_p require prediction of daily values of LAI and a temperature-based relation for leaf area development can accomplish this. Several authors have developed relations to evolve K_c values as a function of thermal time or days after emergence. Sammis *et al.* (1986) applied the following polynomial for estimating K_c values of winter wheat and spring barley as

$$K_c = B_o + B_1 \sum G + B_2 \sum G^2 + B_3 \sum G^3 \quad (1.14)$$

where,

$\sum G$ = accumulated growing degree days starting at planting

B_i = regression coefficients.

The CALEX/COTTON irrigation module program has been evaluated by Plant *et al.* (1992) with the following relation for K_c as

$$K_c = 0.038 + 21.158\ x - 56.579\ x^2 - 263.86\ x^3 \quad (1.15)$$

where,

x = DD/10000

DD = total elapsed growing degree days 15°C.

The K_c values for spanish peanuts were determined from the third-order polynomial (Elliott *et al.*, 1988) as a function of the fraction of the peanut growing season as

$$K_c = -1.644 + 12.05\ F - 17.155\ F^2 + 7.499\ F^3 \quad (1.16)$$

where,

F = fraction of the growing season (ratio of days since planting to days between planting and harvesting).

In this case, a growing season length of 140 days was assumed when calculations of F were made. Since this polynomial relationship gives a negative result for small values of F, a minimum value of 0.4 for K_c is suggested.

A second order polynomial equation has been employed by Idike *et al.* (1982) for estimating K_c of corn as function of days after emergence as

$$K_c = 0.152 + 0.0164\ D - 0.00012\ D^2 \quad (1.17)$$

where,

D = days after emergence of the crop.

A third degree polynomial has been fitted for spring wheat, barley, canola, sugarbeet and potatoes by Foroud *et al.* (1992) with a lower and upper limit of 0.1 and 1.2 for K_c, respectively.

1.5.2 Soil dryness curves

Depending on the energy available in the atmosphere, the soil is able to provide all or part of the water requirement of the plant for transpiration. This depends on soil characteristics, crop type, crop stage and the magnitude of PET itself. As the soil dries, its hydraulic conductivity decreases, and water moves more slowly towards the roots. Only in certain circumstances AET becomes equal to the PET, and normally the former is a fraction of the later. When the moisture tension becomes too high, AET becomes less than PET. This break - off tension is higher for lower values of PET and vice -versa. It is high for some crops like horsegram (drought -resistant) and low for sugarcane. For a given crop it is higher during the initial and late stages than during mid-season stage (Rama Prasad, 1984). In the range of plant available water i.e., amount of water from field capacity (FC) to permanent wilting point (PWP), the amount of available soil water gradually decreases to become zero when PWP is reached. Contradictory view points exist on the availability of soil moisture and the versatile soil moisture budget of Baier *et al.* (1979) who have used 8 types of relationships between the available water for plants and the ratio AET/PET. From these, an adjustment factor 'Z' was proposed for different types of soil dryness curves as:

$$Z = (AET/PET)/(AW/AWC) \quad (1.18)$$

where,

AW = available water in the zone concerned on a given day

AWC = available water capacity of that zone.

The AET on a given day can thus be computed as

$$AET = (AW/AWC) * Z * K_C * PET \quad (1.19)$$

1.5.3 Separating PET into evaporation and transpiration

Several types of models are available for calculating E_s and E_p. Because of their differences in their purposes and organization, they can be broadly categorized according to Addiscott and Wagenet (1985) into (a) deterministic or stochastic (b) mechanistic or functional and (c) rate or capacity types.

The deterministic models produce a unique outcome for a given set of events. However, due to the spatial variability of the mediating processes there will be a certain degree of uncertainty associated with the results. Stochastic models have been developed to accommodate this spatial variability and to quantify the degree of uncertainty. Stochastic models produce an uncertain outcome because they include one or more parameters that are random variables with an associated probability distribution. But stochastic models have been applied little in modelling ET (Ritchie and Johnson, 1990).

Most models used for estimating E_s and E_p are deterministic and can be further categorized as mechanistic or functional. The mechanistic models are based on dynamic rate concepts and incorporate basic mechanisms of processes such as Darcy's law or Fourier's law and the appropriate continuity equations for water and heat fluxes, respectively. Functional models are usually based on capacity factors and treat processes in a more simplified manner, reducing the amount of input required. Mechanistic models are useful primarily as research tools for better understanding of an integrated system, and are usually not used by non-authors due to their complexity. On the other hand, the functional models have modest input requirements making them useful for management purposes. Because of their simplicity, functional models are more widely used and independently validated than mechanistic models.

The most important causes of unproductive loss of water is direct evaporation from the soil surface and especially so under arid and semi-arid conditions where deep drainage can generally be ignored. Thus, the ratio of soil evaporation to transpiration is of decisive importance for overall water use efficiency. Evapotranspiration can be divided into two parts as follows

(a) firstly, under fallow conditions evaporation proceeds at rate depending upon soil type, frequency of wetting and evaporative demand; and

(b) secondly, under a cropped situation evaporation and transpiration proceed and in turn depend on the soil type, evaporative demand, available water in the root zone and the type of crop cover at different stages of crop growth.

1.5.2.1 Soil evaporation (E_s)

There are several mechanistic models available on E_s in which general equation of water flow is used. Some of these models predict evaporative losses of water from a bare soil (Gardner and Gardner, 1969; van Bavel and Hillel, 1976; Hillel and Talpaz, 1977; Lascano and van Bavel, 1986). Some models facilitate the separate calculation of E_s and E_p in the presence of a crop (Feddes *et al.*, 1976; Norman and Campbell, 1983; Huck and Hillel, 1983; Lascano *et al.*, 1987). Functional models are less evident in the literature and some models that have been used to calculate E_s and E_p separately are the models of Ritchie (1972), Hanks (1974), Kanemasu *et al.* (1976) and Tanner and Jury (1976). Because of the complexity in applying mechanistic models, as explained in section 1.5.2, the review presented here mostly deals with functional models.

The rate of evaporation from the soil can be grouped into several stages. During the first stage, which may be lost for only one to three days in mid-summer, the rate of evaporation is controlled by heat energy input and is about 90 percent of PET (Jensen *et al.*, 1990). The duration of first stage is influenced by the rate of evaporation, soil depth and hydraulic properties of the soil (Gardner and Hillel, 1962). By noting the changes in albedo, the transition from first to second stage of drying can be identified (Jackson *et al.*, 1976). Immediately after wetting, the evaporation from a wet bare soil is approximately same as that from a free water surface, the duration of which is again dependent on soil type and evaporative demand of the atmosphere. The period is shortened under coarse textured soils. This relates to the amount of water retained in the top 10 cm of soil layer (Reddy, 1993).

During the second or falling stage, the surface has begun to dry and evaporation is occurring below the soil surface. Water vapor reaches the surface by molecular diffusion and mass flow caused due to the fluctuations in soil air pressure. The dry surface soil greatly influences the effective internal resistance. After the mulch has been formed then the rate of evaporation is less than PET and the rate is controlled by soil characteristics like hydraulic conductivity but not by the meteorological conditions. During the second stage the cumulative evaporation tends to increase with the square root of time for a given soil and evaporation potential as

$$E_s\,dt = K\,(t - t_1)^{0.5} \text{ if } t > t_1 \,. \qquad (1.20)$$

where,

K = empirical constant for a given soil that depends on the soil characteristics and water content (Black *et al.*, 1969).

t_1 = time at which the falling stage begins.

The value of K can be determined experimentally from cumulative evaporation data for a single drying cycle of a given soil. Several direct measurements of the coefficient K in a diversity of soils have consistently resulted in values of about 3.5 mm day$^{-0.5}$. Mason and Smith (1981) used a value of 5 in their model. Thus, the cumulative loss of water after an irrigation or rain can be approximated by

$$E_c = 0.9\ \text{PET}\ dt \text{ for } t\ t_1 \qquad (1.21)$$

and

$$E_c = 0.9\ \text{PET}\ dt + K\,(t - t_1)^{0.5} \text{ for } t\ t_1 \qquad (1.22)$$

where,

E_c= cumulative evaporation

t = time since evaporation began.

Ritchie (1972) summarized E_c at t_1 and K value for several soils and the values are reproduced in Table 3.

Table 3. Typical coefficients for second- stage evaporation

Soil	Ec (mm)	K (mm $day^{-0.5}$)
Adelanto clay loam	12	5.08
Yolo loam	9	4.04
Houston black clay	6	3.50
Plain field sand	6	3.34

PLANTGRO model of Retta and Hanks (1977) computes E_s on the assumption that E_s occurs only from the top layer, usually of 20 to 30 cm in thickness, as

$$E_s = PET\ t^{-0.5} \quad (1.23)$$

where,

t = time in days since last irrigation or rain.

It also allows the top layer to loose moisture below wilting point to the air-drying moisture content. The amount of water that may be lost by evaporation after the soil has reached its permanent wilting water content is taken as equal to the amount of water lost when the first 10 cm of the top layer is dried to air-drying. Thus during the constant rate stage, the evaporation occurs at the potential rate until the upper limit of stage 1 evaporation (U) is reached (Ritchie and Johnson, 1990). This U is reached more rapidly under condition of high PET than under low PET. Ritchie and Johnson (1990) reported value of U to be about 5 mm in sands and heavy shrinking clays to about 14 mm in clay loams. Mason and Smith (1981) assumed a value of 7 mm.

However, contradictory view point exists on the second-stage evaporation rates. Arora *et al.* (1987) used the following empirical relation for sandy loam soils

$$E_s = PET\ t^{-0.30} \quad (1.24)$$

It was also assumed that the top 30 cm soil layer contributed towards soil evaporation. Hill *et al.* (1979) used the following relation to estimate E_s as

$$E_s = E_p / (2^{t-1}) \quad (1.25)$$

and $$E_p = K_s * PET \quad (1.26)$$

where,

E_p = potential soil evaporation

K_s = soil evaporation factor which depends on the value of K_c.

t = time in days since the last soil surface wetting.

It is subject to the constraint that the surface soil can not be drier than air-dry. The top 10 cm soil is assumed to be dried by evaporation and transpiration to the wilting point and then by evaporation only to air dry.

However, Ritchie's (1972) model is the most frequently used model for E_s. This has been used by Cull *et al.* (1981), Mason and Smith (1981), Sharpley and Williams(1990) and Jain and Murthy (1985).

The calculation of E_s in the Ritchie's (1972) model requires prediction of the net radiation through the canopy to the soil surface. Ritchie and Burnett (1971) quantified the influence of partial cover on ET and found that LAI of sorghum and cotton were more generally related to plant evaporation (ET) as fractions of PET than ground cover or plant dry weight. The empiricism used to estimate ET from crops with an adequate supply of water in the root zone usually make ET a function of PET, LAI or plant cover.

In the presence of canopy, the fraction of energy (R_{so}) supplied to the soil surface depends on crop cover or leaf area index (LAI) and is given by

$$R_{so} = R_{ns}/R_n = \exp(-0.4\ \text{LAI}) \text{ for LAI} < 2.7 \quad (1.27)$$

where,

R_{ns} & R_n = 24-hour net radiation at soil surface and above the crop canopy, respectively.

Thus, the soil evaporation during the constant rate stage (stage 1) is calculated using a Priestley - Taylor type equation (Ritchie, 1974):

$$E_{sp} = \frac{\Delta}{\Delta + \gamma} \theta_n R_{so} \quad (1.28)$$

where,

$$\theta_n = 0.92 + 0.4 (R_{ns}/R_n) \text{ for LAI} < 2.7 \quad (1.29)$$

$$\theta_n = 1 \text{ for LAI} > 2.7 \quad (1.30)$$

Some of the functional relations reported in the literature are presented here. CERES-Maize model of Jones et al. (1986) computes potential soil evaporation as

$$E_{sp} = \text{PET}[1.0 - 0.43(\text{LAI})] \text{ for LAI} < 1 \quad (1.31)$$

and $$E_{sp} = (\text{PET}/1.1)\, e^{-0.4(\text{LAI})} \text{ for LAI} < 1 \quad (1.32)$$

In this case, a modified Preistley-Taylor equation is used to calculate daily values of PET although other equations could be used. The potential rates for PET and E_{sp} are equivalent in this example when applied to bare soil conditions (i.e., LAI = 0).

For wheat crop of North-Western part of India, Jain and Murthy (1985) computed E_{sp} as

$$E_{sp} = 0.1 \frac{\Delta}{\Delta + \gamma} R_n \exp(-0.428\ \text{LAI}) \quad (1.33)$$

Sammis *et al.* (1986) used the following relation for winter wheat and spring barley to compute daily soil evaporation as

$$E_s = \text{PET} * e^{-0.623\ \text{LAI}} \quad (1.34)$$

In a water balance model that was used to calculate dry matter yield of wheat, Hanks and Puckridge (1980) computed potential soil evaporation as a function of LAI and dry matter production (DM) in g m^{-2} as

$$E_{sp} = (\text{PET} - T_p)(1 - \text{DM}/2000) \quad (1.35)$$

where,

$$T_p = 0.9\ \text{PET if LAI} > 3 \quad (1.36)$$

and $$T_p = 0.9\ \text{PET}(\text{LAI}/3) \text{ if LAI} < 3 \quad (1.37)$$

In the Soil Water Leaf Extension of Winter Wheat and Wheat Growth (SWLEWW-WTGRO) model of Farshi *et al.* (1987) for the values of ground cover (Gc) smaller than 0.45, it was assumed that PET of winter wheat was equal to ET_o, the PET of a full grown grass cover. The PET was separated into potential transpiration and evaporation on the basis of predicted values of leaf area index as

$$Gc = 1 - \exp(-0.6\ LAI) \tag{1.38}$$

$$E_p = PET \text{ if } Gc = 0 \tag{1.39}$$

$$E_p = 0.9\ (1 - Gc)\ PET \text{ if } Gc > 0 \tag{1.40}$$

The equation (1.32) was used for LAI up to 3.5. For LAI >3.5, it was assumed that the crop covers the ground surface completely, consequently Gc was put equal to 1.

In an attempt to simulate water content under barley using simple empirical approach, Al Khafaf *et al.* (1989) estimated bare soil evaporation using the expression

$$E_s = a\ (t)^b - a\ (t-1)^b \tag{1.41}$$

where,

a,b = empirical coefficients depending on soil type

t = time after irrigation (days).

The potential soil evaporation was calculated as

$$E_{sp} = PET \quad \text{if } Fs < 0.08 \tag{1.42}$$

that is, if PET is not higher than the value of 'a' in Eq.(1.41), otherwise E_{sp} = 7.24 mm day^{-1}. Further, they used

$$E_{sp} = (1.21 - 2.343\ Fs)\ PET \quad \text{if } 0.08 < Fs < 0.48 \tag{1.43}$$

$$E_{sp} = 0.08\ PET \text{ if } 0.48 < Fs \tag{1.44}$$

where, Fs is fraction of degree of shading which is the ratio of LAI at any given time to the maximum value of LAI during the season under non-limiting water conditions.

1.5.2.2 Plant transpiration (E_t)

As the case with modeling E_s, there are also mechanistic and functional models for estimating plant transpiration. Two classical examples of mechanistic models are Penman-Monteith model (Monteith, 1973) and the EMWATBAL model (Lascano *et al.,* 1987). Though they differ in their input requirement and detail but were stated to have good theoretical bases, the former being simpler than the later and intended to calculate E_t or AET. Both these models operate on 1-hour time step and Penman-Monteith has been used extensively to estimate AET in recent years. This equation requires hourly input or estimates from empirical functions for humidity, temperature, net radiation, soil heat flux, heat storage rate in the canopy air column, transport resistance from the leaf surface to instrument height (r_a), and the resistance of the crop (r_c). However, use of this relation to predict AET in applications such as irrigation scheduling is a problem because the required meteorological inputs are difficult to obtain. The value of the surface resistance is a complex function of many climatological and biological factors (Monteith, 1985). All the meteorological variables that must be known are dependent to some extent on the properties of the vegetation. Thus, they must be estimated on the basis of previous experience or calculated from models of the exchange processes (McNaughton and Jarvis, 1984).

The other mechanistic model, EMWATBAL calculates the water and energy balance for both the soil surface and crop canopy. This model calculates water evaporation and transpiration fluxes.

At each time step, the model calculates and updates values of water content and temperature for each soil layer. From these values the inter-layer fluxes of water and heat below the soil surface are calculated. At the soil surface the net radiation, latent, sensible, and soil heat fluxes are calculated from an energy balance equation. Instead of assuming that evaporation at the soil surface at the equilibrium rate corresponding to the net radiation flux above that surface (Ritchie, 1972), the EMWATBAL calculates evaporation at the soil surface by finding the surface temperature that satisfies the energy and water balance at the soil surface by the method of van Bavel and Hillel (1976). The model requires information on soil water retention curve and the unsaturated hydraulic conductivity for each soil horizon, and the number and thickness of the soil layers that determine the geometry of the soil system. Plant inputs are the relation between leaf conductance and leaf water potential, the root distribution as a function of soil depth and time, and the LAI as a function of time. Weather inputs are daily total solar radiation, daily maximum and minimum air and dew point temperatures, daily wind speed, and the quantity of rain or irrigation as a function of time of day. In addition, initial values of the water and temperature profiles must be specified at the start of the simulation period. The daily weather data will be disaggregated into hourly values using empirical relations to produce 24 estimated values from one or two measured values.

Though EMWATBAL is a mechanistic model, it uses empirical relations also. For example, several polynomials were used to compute short wave absorptance of the crop (ABSC), of the soil (ABSS), and the view factor from soil to sky (FTSR), all as a function of LAI, as

$$ABSC = 0.5809\ LAI - 0.2231\ LAI^2 + 0.04640\ LAI^3 - 0.004759\ LAI^4 + 0.0001875\ LAI^5 \quad (1.45)$$

$$ABSS = 0.825 - 0.6447 LAI + 0.2646 LAI^2 - 0.05695\ LAI^3 + 0.005937\ LAI^4 - 0.0002355\ LAI^5 \quad (1.46)$$

$$FTSR = 1.0 - 0.6780\ LAI + 0.2052\ LAI^2 - 0.02799\ LAI3 + 0.001383\ LAI^4 \quad (1.47)$$

Further, the LAI is estimated with third-order polynomials as a function of calendar day number (CD). For the irrigated cotton crop, this relationship was given as

$$LAI = 105.6713 - 1.61088\ CD + 0.008035084\ CD^2\ 0.0000130257\ CD^3 \quad (1.48)$$

and for the dryland cotton crop, the relation was given as

$$LAI = 70.93205 - 1.13014\ CD + 0.005848976\ CD^2 - 0.0000097607\ CD^3 \quad (1.49)$$

Ritchie and Johnson (1990) have identified several problems that may preclude the use of a mechanistic model for prediction purposes and for supporting farm decisions such as irrigation scheduling. Mechanistic models may provide the information needed to derive some of the empiricism upon which functional models are based.

The second type of models for estimating E_t are functional models. These are quite widely used in many crop growth models because they require few inputs, most of which are readily obtainable.

In an earlier attempt to estimate the evaporation rates from developing cotton and grain sorghum canopies under water non-limiting conditions, Ritchie and Burnett (1971) estimated the potential transpiration (T_P) as

$$T_p = PET\ (-0.21 + 0.70\ LAI^{0.5}),\ 0.1 \leq LAI \leq 2.7 \quad (1.50)$$

The non-linearity of the relation between T_p and LAI is stated to be the result of two interacting factors

(a) less competition for radiation per unit of leaf area during initial stages of plant growth and

(b) the partitioning of a large fraction of net radiation at the dry soil surface between plant rows to sensible heat flux causing increased canopy temperature and consequently increased T_p (Ritchie and Burnett, 1971).

Upper limit of 2.7 of LAI represents minimum requirement of LAI necessary for full cover of canopy. For crop canopies with LAI > 2.7, T_p = PET. When LAI ≤ 0.1, T_p is considered negligible.

The T_p is computed from actual (Ea) and potential soil evaporation (E_{sp}) and PET by Brisson *et al.* (1992) using the following relation

$$T = (PET - E_p)\ [\beta + (1 - \beta)\ E_s/E_{sp}] \quad (1.51)$$

where, the value of β was 1.1, which simulates a 10% increase in T_p/PET between wet soil conditions (E_s=E_p) and dry soil conditions (E_s= 0).

The potential transpiration for a wheat crop was estimated by Hanks and Puckridge (1980) using the relations already explained in section 1.5.2.1 as Eq. (1.53) to (1.54), which are reproduced here, as T_p = 0.9 PET if LAI is 3

$$T_p = 0.9\ PET\ (LAI/3)\ \text{if LAI is 3.}$$

Assuming that T_p is largely controlled by evaporative demand and degree of shading under non-limiting water conditions, Al-Khafaf *et al.* (1989) estimated T_p using the following functional relations

$$T_p = 0\ \text{if}\ Fs < 0.08 \quad (1.52)$$

$$T_p = [1 - (1.21 - 2.343\ Fs)]\ PET\ \text{if}\ 0.08 < Fs < 0.48 \quad (1.53)$$

$$T_p = 0.92\ PET\ \text{if}\ 0.48 < Fs \quad (1.54)$$

where,

Fs = fraction of degree of shading .

In most of the soil water balance models like that of Farshi *et al.* (1987), Murthy *et al.* (1992) and Sammis *et al.* (1986), the T_p is calculated as

$$T_p = PET - E_s \quad (1.55)$$

where E_s is calculated by one or the other method discussed in section 1.5.2.1.

In CERES-Maize, the calculation of E_t is through a functional model on the lines of Ritchie (1972). The functional model for estimation of E_s has been described earlier in section 1.5.2.1. Where the soil water is non-limiting, the functional model calculates E_t using the relationships

$$E_t = PET\ (1.0 - e^{-LAI})\ \text{if LAI} \leq 3 \quad (1.56)$$

$$E_t = PET\ \text{if LAI} > 3 \quad (1.57)$$

and if $E_s + E_t$> PET, then

$$E_t = PET - E_s \quad (1.58)$$

The conditional Eq (1.58) was felt necessary because values for E_s and E_t are calculated independently and their sum can exceed the potential rate on a given day estimating E_t when the soil surface is dry.

Once the potential transpiration (T_p) is computed, most of the models estimate actual transpiration (E_t) on the basis of soil water availability. As the soil dries, the conductivity of soil to water flow decreases, thereby decreasing the uptake of water by the root system. Actual transpiration (E_t) by the crop falls below the potential transpiration demand (T_p). There are essentially two approaches to estimate E_t. In the first approach T_p is decreased in proportion to the water deficit in the rooting zone. The transpiration from sorghum or corn as observed by Ritchie (1973) is not affected by soil water deficit until the available water (θ_A) in the root zone is less than 0.3 of the maximum available moisture content (θ_{max}). Thus, when the available water content in the root zone is in between 1 and 0.3 of the maximum, Et is considered equal to Tp. When available water content is less than 0.3 of the maximum then

$$E_t = T_p \, \theta_A/0.3 \, \theta_{max} \tag{1.59}$$

The concept was used by Cull *et al.* (1981) for cotton crop. However, Hanks because Eq (1.56) and (1.57) are for (1974), Sammis *et al.* (1986) and Abdul Jabbar *et al.* (1983) assumed a value of 0.5 of θ_{max}. Singh and Wolkewitz (1988) adopted the critical value of 0.65 to 0.84 for different growth stages of wheat. In the second approach, potential water supply (Pw) by the root system is considered in relation to the potential demand by the crop (T_p). If water supply is greater than demand, then demand is the actual transpiration. If water supply falls below the demand due to water deficits in the soil, then supply is the actual transpiration.

REFERENCES

Addiscott, T.M. and R. J. Wagenet. 1985. Concepts of solute leaching in soils: A review of modeling approaches. *J. Soil Sci.*, 36: 411-424.

Aggarwal, P.K., N. Kalra, A. K. Singh, S.K. Sinha and A. Kumar. 1994. WTGROWS – a model to simulate growth, development, evapotranspiration, nitrogen uptake and productivity of spring wheat in semi-arid tropical and sub-tropical environments. P 13-52. (In) P.K. Aggarwal and N. Kalra (eds). *Simulating the effect of climatic factors, genotype and Management on productivity of wheat in India.* Division of environmental Sciences, IARI, New Delhi.

Al-Khafaf, S., F.A. Sharhan; P.J. Wierenga and A. D. Iyada. 1989. Some empirical relations for the prediction of soil evaporation, transpiration and root water uptake under field conditions. *Agric. Water Manage.*, 16 : 323-335.

Arora, V.K., S. S. Prihar and P.R. Gajri. 1987. Synthesis of a simplified water use simulation model for predicting wheat yields. *Water Resour. Res.*, 23 : 903-910.

Baier, W., J. A. Dyer and W. R. Sharp. 1979. The versatile soil moisture budget. *Tech. Bull. 87*, Agrometeorology Section, Land Resources Research Institute, Agriculture Canada, Ottawa, Ontario, P 52.

Belmans C., J.G. Wesseling and R.A. Feddes. 1983. Simulation model of the water balance of a cropped soil. *J. Hydrol.*, 63 : 271-286.

Black, T. A., W. R. Gardner and G. W. Thurtell. 1969. The prediction of evaporation, drainage and soil water storage for a bare soil. *Soil Sci. Soc. Am. Proc.*, 33: 655-660.

Brisson, N., B. Seguin and P. Gertuzzi. 1992. Agrometeorological soil water balance for crop simulation models. *Agric. for. Meteorol.*, 59 : 267-287.

Burt, J. F., J. T. Hayes, P. A. O'Rourke, W. H. Terjung and P. F. Todhunter. 1980. WATER: A model of water requirements of irrigated and rainfed agriculture. *Pb. Climatol.* 33(2). Lab. Climatol. Elmer, New Jersey, P. 119.

Cordery, I. And A.G. Graham. 1989. Forecasting wheat yields using a water budgeting model. *Aust. J. Agric. Res.*, 40 : 715 - 728.

Cull, P. O., R. C. G. Smith and K. Mc Caffery. 1981. Irrigation scheduling of cotton in a climate with uncertain rainfall. II. Development and application of a model for irrigation scheduling. *Irrig. Sci.*, 2 : 141-154.

Dale, R. F., W.L. Nelson; K. L. Scheeringa; R. G. Stuff and H. F. Reetz. 1982. Generalization and testing of a soil moisture budget for different drainage condition. *J. Appl. Meteorol.*, 21 : 1417-1426.

Doorenbos, J. and W. O. Pruitt. 1979. Guidelines for predicting crop water requirements. *FAO Irrig. and Drain. Paper No. 24*, 2nd ed., FAO. Rome, Italy, P 156.

Edey, S.N. 1980. SMEP- An approach to aerial assessment of soil moisture reserves in the prairie provinces. *Canada Agriculture. Spring issue*, Ottawa, Canada, p.26 – 30.

Elliott, R. L., S. L. Harp; G. D. Grosz and M. A. Kizer. 1988. Crop coefficients for peanut evapotranspiration. *Agric. Water Manage.*, 15 : 155-164.

Farshi, A. A., J. Feyen, C. Belmans and K. De Wijngaert. 1987. Modelling of yield of winter wheat as a function of soil water availability. *Agric. Water Manage.*, 12 : 323-339.

Feddes, R. A., P. J. Kowalik, K. Kolinska – Malinka and H. Zaradny. 1976. Simulation of fields water uptake by plants using a soil water dependent root extraction function. *J. Hydrol.*, 31 : 13-26.

Feddes, R. A., P. J. Kowalik and H. Zaradny. 1978. Simulation of field water use and crop yields. *Simul. Monogr.* PUDOC, Wageningen, The Netherlands, P 189.

Feddes, R. A., J.G. Wesseling and Wiebing. 1984. Simulation of transpiration and yield of potatoes with the SWACRO-model. *9th Tri-annual conference of the European Association of Potato Researchers (EAPR)*, Interlaken, Switzerland, July 2-6, 1984.

Field, M. 1983. The meteorological office rainfall and evaporation calculation system-MORCES. *Agric. Water Manage.*, 6 : 297 – 306.

Foroud, N.E., H. Hobbs; R. Riewe and T. Entz. 1992. Field verification of a micro-computer irrigation model. *Agric. Water Manage.*, 21 : 215-230.

Gardner, H. R. and W. R. Gardner, 1969. Relation of water application of evaporation and storage of soil water. *Soil Sci. Soc. Am. Proc.*, 33 : 192-196.

Gardner, W. R. and D. I. Hillel. 1962. The relation of external evaporative conditions to the drying of soils. *J. Geophys. Res.*, 67 : 4319-4325.

Greacen, E.L. and C.T. Hignett. 1984. Water balance under wheat modeled with limited soil data. *Agric. Water Manage.*, 8 : 291 – 304.

Hanks, R. J. 1974. Model for predicting plant yield as influenced by water use. *Agron. J.*, 66 : 660-65.

Hanks, R. J. 1985. Crop coefficients for transpiration. P 431-438 (In) *Advances in evapotranspiration*. Proc. Natl. Conf. on Advances in Evapotranspiration, Chicago. 16-17 Dec. ASAE, St. Joseph, MI, USA.

Hanks, R. J. and D. W. Puckridge. 1980. Prediction of the influences of water, sowing date and planting density on dry matter production of wheat. *Aust. J. Agric. Res.*, 31 : 1-11.

Hansen, G.K. 1975. A dynamic simulation model of water state and transpiration in the soil-plant-atmosphere system. I. The model and its sensitivity. *Acta Agric. Scand.*, 25 : 129 –149.

Hansen, G. K. and H. B. Madsen. 1984. Soil evaluation based on simulation by a model. *Acta Agric. Scand.*, 34 : 9 – 16.

Hayhoe, H.N. and R. De Jong. 1988. Comparison of two soil water models for soybeans. *Canada Agric. Engg.*, 30 : 5 –11.

Hill, R. W; D. R. Johnson and K. H. Ryan. 1979. A model for predicting soybean yields from climatic data. *Agron. J.*, 71 : 251-256.

Hillel, D. and H. Talpaz. 1977. Simulation of soil water dynamics in layered soils. *Soil Sci.*, 123 : 54-62.

Holst. K. A. and H. B. Madsen. 1988. Modelling the irrigation need. The influence of climate and soil water content on the irrigation need for barley and grass. *Acta Agric. Scand.*, 38 : 261 – 269.

Huck, M. G. and D. Hillel. 1983. A model of root growth and water uptake accounting for photo-synthesis, respiration, transpiration and soil hydraulics. P. 273-333. (In) D. Hillel (ed.) *Advances in irrigation. Vol. 2*. Academic Press, New York.

Idike, F. I., C. L. Larson and D. C. Slack. 1982. Modeling soil moisture and effects of basin tillage. *Trans ASAE*, 25 : 1262-1267.

Jackson, B. S., G. F. Arkin and A. B. Hearn. 1990. COTTAM : A cotton plant simulation model for an IBM PC micro-computer. *Texas Agril. Expt. Sta. Texas. MP-1685*, P 242.

Jackson, R. D., S. B. Idso and R. J. Reginato. 1976. Calculation of evaporation rates during the transition from energy-limiting to soil-limiting phases using albedo data. *Water Resour. Res.*, 12 : 23-26.

Jagtap, S.S. and J.W. Jones. 1989. Evapotranspiration model for developing crops. *Trans ASAE*, 32 : 1342 – 1350.

Jain, A. K. and V. V. N. Murty. 1985. Simulation of soil moisture profiles for scheduling of irrigations. *Agric. Water Manage.*, 10 : 175-181

Jensen, M. E., R. D. Burman and R. G. Allen. 1990. Evapotranspiration and irrigation water requirements. *ASCE manuals and Reports on Engg. Practice No. 70*, ASCE, New York, P. 327.

Jones, C. A., J. T. Ritchie, J. R. Kiniry and D. C. Godwin. 1986. Subroutine structure. P. 49-112. (In) C. A. Jones and J. R. Kiniry (eds.) *CERES-Maize: A simulation model of maize growth and development*. Texas A & M Univ. Press, College Station, Texas.

Jones, J.W. and A.G. Smajstrla. 1980. Application of modeling to irrigation management of soybeans. p.49 – 111 (In) F.T. Corbin (Ed.) *Proceedings of the World Soybean Research Conference II*. Westview Press, Boulder, Colorado, USA, pp. 49 – 111.

Kanemasu, E. T., L. R. Stone and W. L. Powers. 1976. Evapotranspiration model tested for soybean and sorghum. *Agron. J.*, 68 : 569-572.

Kanemasu, E. T., V. P. Rasmussen and J. Bagley. 1978. Estimating water requirements for corn with a pocket calculator. *Kansas State Univ. Agric. Expt. Sta., Bulletin 615*, Manhattan, USA.

Lascano, R. J. and C. H. M. van Bavel; 1986. Simulation and measurement of evaporation from a bare soil. *Soil Sci. Soc. Am. J.*, 50 : 1127-132.

Lascano, R. J., C. H. M. van Bavel, J. L. Hatfield and D. R. Upchurch. 1987. Energy and water balance of a sparse crop : simulated and measured soil and crop evaporation. *Soil Sci. Soc. Am. J.*, 51 : 1113-1121.

Mason, W. K. and R. C. G. Smith. 1981. Irrigation for crops in a sub-humid environment. III. An irrigation scheduling model for predicting soybean water use and crop yield. *Irrig. Sci.*, 2 : 89-101.

McNaughton, K. G. and P. G. Jarvis. 1984. Using the Penman-Monteith equation predictively. *Agric. Water Manage.*, 8 : 263-278.

Monteith, J. L. 1973. Principles of Environmental Physics. Edward Arnold, London.

Murthy, V. V. N., A. H. Azar; A. Sarwar and K. Sudsaisin. 1992. Simulation of tertiary unit efficiencies in large irrigation systems. *Agric. Water Manage.*, 21 : 13-22.

Norman, J. M. and G. C. Campbell. 1983. Application of a plant environment model to problems in irrigation. P 155-188. (In) D. Hillel (ed.). *Advances in irrigation*. Vol. 2. Academic Press, New York.

Place, G. and D.M. Brown. 1987. Modelling corn yields from soil moisture estimation: Description, sensitivity analysis and validation. *Agric. For. Meteorol.*, 41 : 31-56.

Plant, R. E., R. D. Horrocks; D. W. Grimes and L. J. Zelinski. 1992. CALEX / COTTON : an integrated expert system application of irrigation scheduling. *Trans ASAE*, 35 : 1833-1838.

Priestley, C. H. B. and R. J. Taylor. 1972. On the assessment of surface flux and evaporation using large scale parameters. *Mon. Weather Rev.*, 100 : 81-92.

Rama Prasad. 1984. Influence of time step in the simulation modeling of evapotranspiration. *Sadhana*, 7 : 91-118.

Reddy, S. J. 1993. Agroclimatic / Agrometeorological techniques as applicable to dry-land agriculture in developing countries. Jeevan Charitable Trust, Secunderabad, India, P 205.

Retta, A. and R. J. Hanks. 1981. Manual for using model PLANTGRO. *Utah Agric. Expt. Sta. Res. Rep. 48*. Logan, Utah, USA : Utah State University.

Ritchie, J. T. and E. Burnett. 1971. Dryland evaporative flux in a subhumid climate : II. Plant influences. *Agron. J.*, 63 : 56-62.

Ritchie, J. T. 1972. Model for predicting evaporation from a row crop with incomplete cover. *Water Resour. Res.*, 8 : 1204-1213.

Ritchie, J. T. 1973. Influence of soil water status and meteorological conditions on evaporation from a corn canopy. *Agron. J.*, 65 : 893-897.

Ritchie, J. T. 1974. Evaluating irrigation needs for south-eastern U.S.A. P 262-273. *Proc., Irrig. And Drain, Spec. Conf.*, ASCE.

Ritchie, J. T. and B. S. Johnson. 1990. Soil and plant factors affecting evaporation. P 363-390 (In) B. A. Stewart *et al.* (ed.) *Irrigation of agricultural crops*. ASA, CSSA and SSSA, Madison, WI, USA.

Robinson, J.M. and K.G. Hubbard. 1990. Soil water assessment model for several crops in the high plains. *Agron. J.*, 82 : 1141-1148.

Sammis, T. W.; S. Williams, D. Smeal and C. E. Kallsen. 1986. Effect of soil moisture stress on leaf area index, evapotranspiration and modeled soil evaporation and transpiration. *Trans ASAE*, 29 : 956-961.

Saxton, K. E., H. P. Johnson and R. H. Shaw. 1974. Modelling evapotranspiration and soil moisture. *Trans ASAE*, 17 : 673-677.

Seliorio, I.S. and D.M. Brown. 1979. Soil-moisture based simulation of forage yield. *Agric. Meteorol.*, 20 : 99 – 114.

Sharpley, A. N. and J. R. Williams (eds). 1990. EPIC. Erosion/Productivity Impact Calculator : 1. Model documentation. *U. S. Dept. Agric., Tech. Bull. No. 1768*, P 235.

Singh, P. and H. Wolkewitz. 1988. Evapotranspiration, pan evaporation and soil water relationship for wheat (*Triticum aestivum*). *Agric. Water manage.*, 13 : 65-81.

Stockle, C. and G. Campbell. 1985. A simulation model for predicting effect of water stress on yield : an example using corn. Adv. Irrig., 3 : 284 – 311.

Stuff, R.G., K.L. Scheeringa and R. F. Dale. 1975. Simulation of soil moisture balance in FORTRAN IV. SIMBAL-2. Agronomy Dept., Purdue Univ., West Lafayette, USA.

Tanner, C. B. and W. A. Jury. 1976. Estimating evaporation and transpiration from a row crop during incomplete cover. *Agron. J.*, 68 : 239-243.

van Bavel, C. H. M. and D. Hillel. 1976. Calculating potential and actual evaporation from a bare soil surface by simulation of concurrent flow of water and heat. *Agric. Meteorol.*, 17 : 453-476.

van Bavel, C. H. M. and R.J. Lascano. 1987. EMWATBAL : A numerical method to compute the water loss from a crop by transpiration and evaporation. Texas Agric. Expt. Sta., Texas A & M Univ., Texas, USA, pp 73.

Victor, U. S., N. N. Srivastava and B. V. Ramana Rao. 1988. Quantification of crop yields under rainfed conditions using a simple soil water balance model. *Theor. Appl. Climatol.*, 39 : 73 – 80.

Visser, W.C. 1974. The model of the soil water balance and nutrient uptake as a basis for hydrological, agrohydrological and other projects. *Mathematical models in hydrology. Misc. Rep. 172*, Inst. For Land and Water Management Research. The Netherlands, pp 580 – 597.

Vossen, P. 1990. Comparative statistical validation of two ten-day water-use models and of three yield reduction hypotheses for yield assessment in Botswana. *Agric. For. Meteorol.*, 51 : 177 – 195.

Wright, J. L. 1981. Crop coefficients for estimates of daily crop evapotranspiration. *Irrig. Scheduling for Water and Energy Conserv. in the 80's*, Am. Soc. of Agric. Engrs., Dec. 18-26.

Wright, J. L. 1982. New Evapotranspiration crop coefficients. *Am. Soc. Civil Eng. J. Irrig. Drain. Div.*, 108 (IA 1) : 57-74.

Wright, E., M.K.V. Carr and P.J.C. Hamer. 1994. Crop production and water-use. I. A model for estimating crop water use with limited data. *J. Agric. Sci., (Camb.)* 123 : 9 – 13.

Wright, J. L. and M. E. Jensen. 1978. Development and evaluation of evapotranspiration models for irrigation scheduling. *Trans ASAE*, 21 : 88-96.

Chapter - 16

TARGETTED YIELD APPROACH FOR FERTILIZER RECOMMENDATIONS TO RAINFED CROPS

K.R. Sonar*

Retd. Professor and Head, Deptt. of Agricultural Chemistry and Soil Science,
Mahatma Phule Krishi Vidyapeeth, Rahuri (MS);
*4, Swastic Apt. Savedi, Ahmadnagar - 414 001 (MS).

1. INTRODUCTION

Rainfed farming is declined as the production of crops in sub-humid or semi-arid regions without irrigation, while dryland farming is the practice of crop production in low rainfall areas by conserving natural precipitation. Supplemental irrigation may be given from harvested run-off water. The term rainfed and drylands are used interchangeably by many people. It can be said that while all drylands are rainfed, the reverse is not always the case.

Rainfed farming is the dominant sector of Indian agriculture. It accounts for over 100 m ha (million hectare) or 70% of the net sown area. A large portion of this area falls under dryland agriculture due to the fact that inadequate and/or poorly distributed rainfall results in frequent periods of water stress for the crops. The drylands in India is grown more than 90% of the coarse grains and pulses, 75% of oilseeds, 70% cotton and several other crops. Crop yields in general are low (Tandon, 1991). These are several reasons for low yields per unit area. One of these is anadequate application of fertilizers to dryland crops. During the last 25 years, results of research have convincingly and repeatedly proven that fertilizer use is a profitable proposition and infact one of the key steps for improving dryland agriculture. It is crucial to improve the productivity of dryland areas to increase their contribution in overall agricultural production and to improve the incomes of the farmers (Tandon, 1991). Weather being unpredictable, timeliness in farm operations assumes greater importance. The rainfall in the dryland areas varies from less than 400 mm to more than 2000 mm. Broadly speaking, about one third area of the dryland is under 400 to 700 mm (low), 700 to 1100 (Medium) and more than 1100 mm (High) rainfall. Generally rains occur with high intensity leading to 10 to 40% run off and considerable erosion and soil loss. Further, the rainy season is short and interspersed with drought (Venkateswarly and Ranjodh Singh, 1982).

Dryland agriculture has vast potentials when it is managed scientifically. Results in dryland agriculture are not quick due to poor adoption of packages. Concentrated multidisciplinary, multi-agency approach is necessary (Umrani, 1995).

Instability in dryland agriculture arising from shifts in long term rainfall norms can be moderated by maximum conservation of available precipitation. Infact, insufficient attention on rain water harvesting and its recycling undervalue the significance of nutrient application and other technologies in crop improvement and management.

2. SOILS OF DRYLAND

Arid and semi-arid areas, often called drylands are predominant in India where crops are either rainfed or under limited irrigation. Out of the total geographical area of 329 million hectare (m ha) in India, the net cultivated area is 1453 m ha which may shrink to 120 m ha by the end of this century due to biotic pressure that has set in motion the process of land degradation (Kanwar, 1994; Sehgal and Sharma, 1994). It seems we have reached to physical limits with respect to area, which could remain under agriculture because of the increasing demands for land by industry, urbanization, roads, dams, canals, parks, airports, railways etc. The human population growth rate is making the man to land ratio more unfavourable (Table 1). As against 0.48 ha agriculture land per capita in 1951 which may be reduced to 0.15 ha in the year 2000 (Kanwar, 1994a). The rainfed areas constituting about 70% of net cultivated area and contributing to a significant fraction of the total foodgrains (40%), oilseeds (70%), pulses (90%), cotton (70%) etc. are spread over in the black and red soil zones of peninsular India, the eastern states, coastal ecosystem and the mountainous regions of the country. These areas produce much below their capability.

Table 1. Land use classification of India

Particular	Area (m ha)	% of total
Total geographical area	329	–
Reporting area	304	–
Forest	67.0	20.3
Fallow	10.6	3.2
Other uncultivable area	30.2	9.1
Area not available for cultivation	41.0	12.4
Net sown area	143.0	43.5
Gross sown area	181.1	55.0
Problem area (wind and water erosion, salinity, alkanity, acidity etc.)	174.97	53.2*
Land available per capita		
1951	0.48 ha	
1981	0.20 ha	
2000	0.15 ha	

*Needs to be improved on top priority in 12th V year plan.

Soils of rainfed region belong to several orders and their representation can be delineated on climatic bias. Among the three climatic variables-rainfall, temperature and sunlight play an important role in the sustainability of semi-arid tropical rainfed crops. Quantity and duration of rainfall along with water holding capacity of soil determine the effective span of cropping. Soils occurring in semi-arid tropical climate (Mean annual rainfall 500-1000 mm) of peninsular and central India (Maharashtra, Madhya Pradesh, Andhra Pradesh, Tamil Nadu, Karnataka and Parts of

Gujarat and Uttar Pradesh) comprise mainly Vertisols (Chromusterts, Pellusterts), Alfisois (haplustalfs, Rhodustalfs), Inceptisols (Ustropepts, Ustochrepts) and Entisols (Ustifluvents, Ustorthents). Those occurring in arid zone (mean annual rainfall 500 mm) of Rajasthan and parts of Haryana, Punjab and Gujarat belong mainly to Aridisols (Camborthids, Calciorthids), whereas those occurring in western Himalayas belong to Entisols (Cryorthents) and Aridisols (Cryorthids) (Sehgal and Sharma, 1994; Venkateshwarlu, 1987, Randhawa and Singh, 1983).

Vertisols (black soils) vary in depth (15-240 cm) and clay content (30-70%). The predominant clay minerals are smectite, montmorillonite, nontronite and beidellite. They hold more water, alkaline in reaction, low in organic matter and crack on drying. Affisols (red soils) are low in clay content and predominant clays are kaolinite and lillite. These soils are deficient in N and P and sometimes in K. Vertisols are more productive than Alfisols. Aridisols (sierozems) are extremely light - sandy, loamy sand, rich in $CaCO_3$, and low in water holding capacity. Inceptisols and Entisols (alluviums/submontane soils) are shallow to deep, predominant in lillite, smectite, and chlorite clay minerals. They are neutral to alkaline in reaction, and water holding capacity low (Sehgal and Sharma, 1994; Venkateshwarlu, 1987).

3. NUTRIENTS REMOVAL BY CROPS

Crops remove large quantities of nutrients from the soils which must be replenished to maintain the soil fertility at a higher level. For each tonne of grains produced dryland crops remove no less nutrients than the irrigated crops. This and the fact that nutrient deficiencies are wide spread in the drylands has led to the often phrase that "the drylands are not only thirsty, but they are also hungry" (Abrol, 1991). The data presented in Table showed the nutrient removal by important crops from the soil indicating the deficit of N P K nutrients (Negative balance) from 30 to 168 kg ha^{-1} in various foodgrain ad oilseed crops (Sonar, 1996; Somawanshi *et al.,* 1997). Because of the intensive agriculture and use of high yielding, fertilizer responsive varieties of crops, the demand for nutrient requirements has been also increased. Although there is a direct relationship between the yield and nutrient uptake by the crop, adequate quantities of fertilizer nutrients are not added. This has resulted in depletion of the soils in respect of available N, P and K. The data in Table 3 give the fertility status of the soils in six states with a total cropped area of 98 m ha, 80% of which is rainfed (Tandon, 1991). The most important nutrient deficiencies in dryland soils are those of nitrogen and phosphorus. Nitrogen deficiency is universal since these soils contain less than 0.1% total N. Phosphorus deficiency is widespread and is a limiting factor for producing high yields. Soils in 44% districts are low in available P and in 50% districts the average P status is medium (Table 3).

On all India basis, about 95% districts are low to medium in available N and P while 65% districts are low to medium in respect of available K (Table 4), (Anonymous, 1996).

Table 2. Nutrient removal by rainfed/dryland crops

Crop		Yield q/ha^{-1}	Nutrient removal kg ha^{-1}				Nutrient applied				Nutri. deficit
			N	P_2O_5	K_2O	Total	N	P_2O_5	K_2O	Total	
Sorghum (kh)	(R)	50	145	60	140	345	80	50	50	180	168
	(D)	25	70	25	110	205	50	25	0	75	130
Sorghum (Rabi)	(R)	30	120	40	140	300	80	50	50	180	120
	(D)	25	55	30	105	190	50	25	0	75	115
Pearl millet	(R)	25	65	30	120	215	60	30	0	90	125

	(D)	15	45	25	100	170	50	25	0	75	95
Cotton	(R)	15	100	65	80	245	50	25	25	100	145
Pigeonpea	(R)	15	50	75	65	190	20	40	0	75	115
	(D)	12	40	60	60	160	13	25	0	60	100
Green gram	(R)	10	45	45	55	145	20	40	0	75	70
	(D)	7	35	40	52	127	13	25	0 120	67	
Sunflower	(R)	15	80	55	50	185	60	30	30	75	65
	(D)	12	70	45	40	155	50	25	0	60	80
Groundnut	(R)	12	80	55	65	200	20	40	0	38	140
	(D)	8	65	45	55	165	13	25	0	60	127
Black gram	(R)	10	75	50	60	185	20	40	0	60	125
	(D)	7	60	35	45	140	13	25	0	38	102
Soybean	(R)	16	90	65	60	215	20	40	0	60	155
	(D)	10	80	55	55	190	13	25	0	38	152
Chickpea	(R)	20	70	55	70	195	20	40	0	60	135
	(D)	12	55	50	60	165	13	25	0	38	127
Safflower	(R)	10	50	20	35	105	50	25	0	75	30
	(D)	6	35	15	25	75	30	15	0	45	30

R = Rainfed; D = Dryland

Table 3. Generalised P and K fertility status of some dryland areas of India

State	Available P			Available K		
	No. of districts			No. of districts		
	Low	Medium	High	Low	Medium	High
Maharashtra	15	10	0	0	13	12
Andhra Pradesh	17	4	0	2	13	6
Gujarat	5	13	1	0	0	19
Madhya Pradesh	15	30	0	3	9	30
Rajasthan	1	22	3	0	5	14
Karnataka	15	4	0	3	9	6

Table 4. Nutrient status of soils of India

Soil nutrient	Soil fertility status		
	Low	Medium	High
	No. of districts		
Nitrogen	228	118	18
Phosphorus	170	184	17
Potassium	47	192	122

Since our major concern under dryland is about the productivity of agricultural systems, it will be necessary to study the effect of fertilizer, which is the key input and which has contributed to major increase in production and productivity next to irrigation water.

Though the fertilizer consumption in India has increased to 72.4 kg ha^{-1}, it is not enough to balance the nutrient withdrawal. Fertilizer is a king pin of important technology contributing to 50 to 60 per cent increase in productivity of food grains in India. Whether it is irrigated or nonirrigated crops, without use of fertilizers increased production is not possible (Kanwar, 1994 b). Fertilizer consumption has increased from 60000 t in 1950-51 to 11.6 M t in 1989-90 and is expected to rise in 20 M t by the end of this century to produce about 240 M t foodgrains besides other cash crops. Low productivity of our soils is due to low available nutrient status and low use of fertilizers. Hence for increasing the crop yields by 2 or 3 times, the fertilizer nutrient application must increase in higher proportion. With higher yields, there occurs proportionately higher withdrawal of nutrients from the soil reservoir and the gap between inputs and outputs goes on widening and fertilizer use efficiency is declining.

4. BALANCED NUTRITION

Fertilizers are the costliest inputs in world agriculture today which in turn necessitates their efficient and economic utilization. The cost of fertilizers has been increased recently.

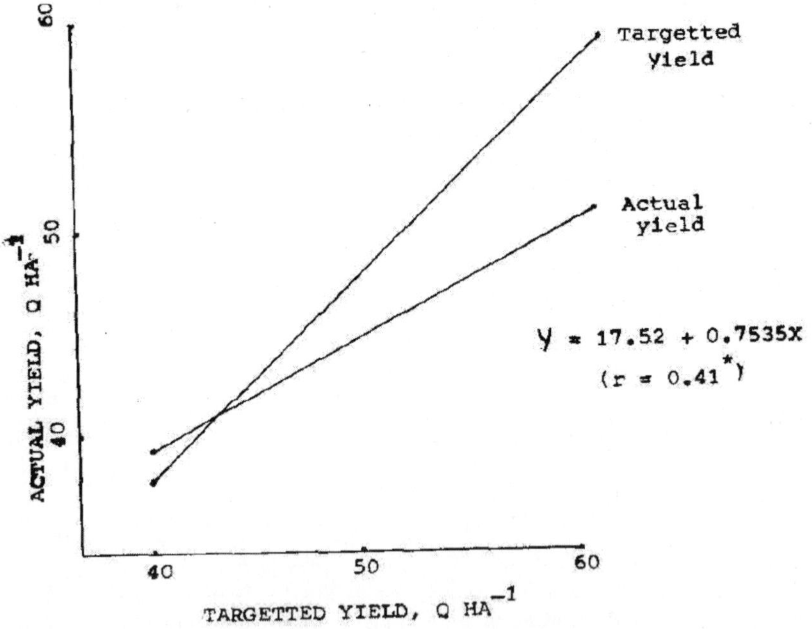

Fig. 1. Relationship between targeted and actual yields of rainfed sorghum

The data in Table 5 showed that due to withdrawal of subsidy by the Govt. on Aug. 25, 1992, the cost of fertilizer (P and K) has been increased by three times while fertilizer N cost is increased further by about 20% since Feb. 1, 1999. The cost of fertilizer is likely to be further increased in near future. For this purpose, fertilizer nutrients should be applied in balanced proportion.

Table 5. Cost of fertilizer nutrients before and after decontrol (Aug. 25, 1992; Feb. 1, 1999).

Fertilizer nutrient	Cost before decontrol	Cost after decontrol	Present cost
	Rs. Kg-1		+
N	6.35	7.20	8.70
P_2O_5	5.62	13.70	18.00
K_2O	2.10	6.70	7.00

Balanced nutrition means that nutrients in the soil should be in adequate quantities and in required proportion for the plants to produce maximum yields (Reddy, 1992). Most soils are only able to supply a fraction of the nutrients needed to achieve high yields. Even available nutrients are rarely present in a balanced condition to meet the nutrient requirements of all the crops. Balanced fertilization is must for getting higher efficiency and economy in fertilizer use. The most conventional and practical way of predicting balanced fertilization is through soil testing.

5. SOIL TESTING AND FERTILIZER USE

Soil testing is a rapid chemical analysis to assess available nutrient status of the soil. Soil testing can measure available status of the soil as well as acidity or alkalinity, salt concentration and organic carbon content. Thus, in general we know the health of the soil from soil testing.

A sound soil testing programme requires an enormous amount of background research on a continuing basis. The background research should determine among other things, significant chemical forms of the available nutrients in the soils of the area, the extractants most suitable for accurately measuring the available nutrient forms, the relative productive capacity of various crops, the differential response of the various rates and methods of fertilizer application for the different crops (Reddy, 1992).

Several extractants/methods have been in use for determining available N P K status of the soil. It is therefore, necessary to use suitable methods for our soils/ regions for determining the fertility status of the soil. Among the various methods/ extractants, alkaline $KMnO_4$ method for available N, Olsen's method for available P and ammonium acetate extractant for available K were found to be suitable for swell-shrink soils of Maharashtra as seen from the higher correlation coefficient values with crop yields (Table 6) (Nazirkar and Sonar, 1991; Palwe and Sonar, 1989; Akolkar and Sonar, 1994).

Table 6. Evaluation of soil test methods and critical levels of nutrients in soil.

Soil/ location	Test crop	Extraction/ method	"r" with % yield	Critical level
Available N				
Vertisols	Kh.	Alkaline	0.97**	280
(Rahuri)	Sorghum	$KMnO_4$		
Available P				
Vertisols	Wheat	Olsen	0.91**	29
(Rahuri)				
Available K				
Vertisols	Kh.	NH_4OAc	0.94**	750
(Rahuri)	sorghum			

6. FERTILIZER RECOMMENDATION

At present fertilizer recommendations under rainfed conditions are made on the basis of soil testing in different states of the country which is known as soil analysis correlation approach. Since there are certain problems in the use of this method, another method viz., targetted yield approach has been developed to overcome the problems (Ramamoorthy, *et al.,* 1967). In this paper these two methods of fertilizer recommendations have been compared.

Dryland crops are known to remove from the soil no less nutrients than the irrigated crops. The nutritional deficiencies of the drylands are also more pronounced. Hence, there is a need for a scientific approach to fertilizer for dryland crops.

(a) Soil analysis - correlation approach

In this method surface soil samples from the fields are analysed for available nutrients (N P K) and classified into six categories such as vary low, low, moderate, moderately high, high and very high (Table 7). On the basis of soil test values, the general recommended fertilizers doses (RD) for the crops are modified for making fertilizer recommendations to the farmers.

Table 7. Classification chart for soil test data

Category	Organic Carbon (%)	Available nutrients, kg ha^{-1}		
		N	P	K
Very low	<0.20	<141	<7	<100
Low	0.21-0.40	141-280	7-14	100-150
Moderate	0.41-0.60	281-420	14-21	151-200
Moderately high	0.61-0.80	421-560	21-28	201-250
High	0.81-1.00	561-700	28-35	251-300
Very high	>1.00	>700	>35	>300

If the level of nutrients in soil is very low, the RD for a crop is increased by 50%, if low by 25% and for moderate to moderately high, the RD is recommended as it is. When the level of nutrients is high, the RD is reduced by 25% and very high by 50% (Sonar *et al.,* 1984).

In this method only the available nutrient status of the soil as per six tier system is considered and not the actual soil test values. Similarly, other soil properties viz., depth, physical and chemical properties, soil types etc. can not be considered and hence the fertilizer recommendations are of general nature and not precise. It is also not possible to consider financial condition of the farmer while making fertilizer recommendations (Sonar, 1999). The following example will explain the problems in making fertilizer recommendations based on soil testing alone.

Calculate the fertilizer dose for rainfed sorghum based on soil testing for the following two soils.

	Soil 1	Soil 2
Available N (kg ha^{-1})	150	270
Available P (kg ha^{-1})	7.5	13.5
Available K (Kg ha^{-1})	310	550
Soil depth (cm)	50	150
RD of sorghum	80:50:50	

As per the norms presently followed (Table 7), in both the soils available N and P are low and K very high. So the fertilizer dose recommended for sorghum on both the soils is same as 100 : 63 : 25.

Although there is variation in available nutrient status and soil depth, same fertilizer dose is recommended for both the soils for sorghum. This method dose not consider financial condition of the farmer while making fertilizer recommendation.

(b) Targetted yield approach

To overcome the problems in fertilizer recommendations based on soil analysis correlation approach, Ramamoorthy *et al.* (1967) used targetted yield approach. This approach is based on available nutrient status and first developed by Truog (1960) and Ramamoorthy *et al.* (1967) established theoretical basis and experimental proof. They observed linear relationship between yield and nutrient uptake of the crop. For obtaining given yield, definite quantity of nutrient must be taken-up by plant. Once this requirement is known for a given yield, fertilizer needed can be estimated considering efficiency of soil and fertilizer nutrients. This forms the basis of fertilizer recommendations for targetting of crop yields.

Thus, targetted yield approach considers nutrient requirement of the crops, available nutrients present in the soil and efficiency present in the soil and those applied through fertilizers.

(i) Rainfed Kharif sorghum

Standard field experiments were conducted on sorghum (c.v. CSH 9) at Agril. Research Station, Digraj, dist. Sangli, Maharashtra during the year 1981 based on Colwell's approach at 12 different locations having varying physio-chemical properties(Sonar, 1984; Somawanshi *et al.*, 1994). Digraj comes under plain zone of Maharashtra where the soils are Vertilsols with assured rainfall. There were 11 selected treatment combinations (Table 8) out of 4 levels of N (0,50, 100 and 150 kg ha^{-1}), 4 levels of P_2O_5 (0, 30, 60 and 90 kg ha^{-1}) and 3 levels of K_2O (0, 30, and 60 kg ha^{-1}).

Table 8. Treatment combinations for rainfed sorghum field experiment

Sr. No.	Treatment	Fertilizer nutrient applied, kg ha^{-1}		
		N	P_2O_5	K_2O
1.	NoPoKo	0	0	0
2.	NoP60K60	0	60	60
3.	N50P60K60	50	60	60
4.	N100P60K60	100	60	60
5.	N150P60K60	150	60	60
6.	N100PoK60	100	0	60
7.	N100P30K60	100	30	60
8.	N100P90K60	100	90	60
9.	N100P60Ko	100	60	0
10.	N100P60K30	100	60	30
11.	N100P60K30	100	60	30

These set of treatments were replicated at 12 different locations. Sorghum crop was grown with recommended package of practices. At harvest crop yields (grain and stover) were recorded. Grain and stover samples were obtained at harvest, analysed for N, P and K concentrations and their uptake calculated. The data on sorghum yields, N P K uptake by sorghum, soil available nutrients and fertilizer doses applied were used for calculating nutrients requirements (NR), contribution from soil nutrients (CS) and contribution from fertilizer nutrients applied (CF) were calculated separately for N, P and K from which the fertilizer prescription equations were derived (Ramamoorthy *et al.*, 1967; Sonar, 1984).

$$\text{N R (kg q}^{-1}\text{)} = \frac{\text{Nutrient uptake kg ha}^{-1}}{\text{Grain yield, q ha}^{-1}}$$

Nutrient uptake, kg ha $^{-1}$ control plot

CS (%)= × 100 Soil test value control plots kg ha $^{-1}$

From these parameters, fertilizer prescription equations were derived separately for N_1 P_2 O_5 and K_2O (Sonar and Tamboli, 1997).

$$\text{FD} = \frac{\text{NR}}{\text{CF}} \times 100\ \text{T} - \frac{\text{CS}}{\text{CF}} \times \text{STV}$$

NR = Nutrient requirement for N_1 P_2 O_5 or K_2O_1 kg ha $^{-1}$

CS = Contribution from soil available N or P_2 O_5 or K_2O_3 %

CF = Contribution from fertilizer N or P_2O_5 or K_2O, %

T = Yield target, q ha $^{-1}$

STV = Soil test values for available N or P or K_1 kg ha $^{-1}$

Using the above method, fertilizer, prescription equations for rainfed kharif sorghum were calculated which are given as below.

FN = 4.58 T - 0.96 SN

FP_2O_5= 2.21 T - 6.94 SP

FK_2O = 3.34 T - 0.22 SK

Where FN_1 FP_2O_5, FK_2O = fertilizer N, P_2O_5 and K_2O in kg ha $^{-1}$; T = yield target in q ha $^{-1}$; SN, SP, SK = Soil available N, P and K in kg ha $^{-1}$, respectively.

The soils, were silty clay, pH 7.8 - 8.2, EC 0.12-0.30 dSm $^{-1}$, available N 145 - 210 kg ha $^{-1}$, P 18.6 - 42.0 kg ha $^{-1}$, K 615-920 kg ha $^{-1}$. Sorghum grain yields obtained were 12.6 - 61 q ha $^{-1}$, and stover yields 30-140 q ha $^{-1}$.

The above fertilizer prescription equations have been transformed into ready reckoners for fertilizer N, P_2O_5 and K_2O requirements of rainfed sorghum for different yield targets with soils having varying soil test values (Table 9).

These equations were tested for their validity by conducting follow - up trials on rainfed sorghum (cv. CSH -9) on the University farms as well as on the farmers fields. The results of 23 trials conducted during 1982 to 1992 showed that yield targets of 40, 50 and 60 q ha $^{-1}$ of rainfed sorghum were achieved within ± 10-15% variation by applying fertilizers based on targetted yields approach. All the targetted yield treatments registered higher returns per rupee invested on fertilizers than that of general recommended dose and as soil test treatments (Table 10). The results further showed that the differences between yields targetted and the actual yields obtained were non-significant indicating that the yields could be achieved by applying fertilizers based on these equations (Table 11). The coefficient of correlation value (r = 0.41*) also indicated significant relationship targetted and actual yields of rainfed kharif sorghum and importance of targetted yield approach for getting desired yield targets (Fig. 1). (Somawashi *et al.,* 1994). In this method, it is possible to make fertilizer recommendations for crops considering the financial condition of the farmers.

Very limited work has been done on targetted yield approach for rainfed crops. Fertilizer prescription equations developed for various rainfed crops at different locations (Reddy, 1992) are given in Table 12.

In rainfed agriculture, the first limiting factor for crop production is the availability of soil water. The law of minimum operates and the yield is governed by this limiting factor. It is therefore imperative to target the crop yield suited to soil water availability and then only tailor the fertilizer prescription to achieve the yield targetted (Reddy, 1992).

7. FERTILIZER USE EFFICIENCY

Integrated nutrient management (organic and inorganic nutrients), time and method of application besides balanced fertilizer use are directly related to fertilizer use efficiency irrespective of the agroclimatic region, the soil or the crops grown. Phosphorus and potassium are generally required to be applied as a basal dose. Secondary and micronutrients wherever possible have to be supplemented for improving the efficiency of applied nutrients (Prihar and Ranjodh Singh, 1983; Katyal, 1994; Rao, 1992).

Table 9. Fertilizer nutrient requirements of rainfed kharif sorghum for different yield targets and soil test values

Available nutrient in soil kg ha^{-1}		Sorghum yield target, q ha^{-1}						
		30	36	40	45	50	55	60
Nitrogen	100	41	64	87	110	133	156	179
	120	22	45	68	91	114	137	160
	140	3	26	49	72	95	118	140
	160	0	7	30	52	75	98	121
	180		0	10	33	56	79	102
	200			0	14	37	60	83
	220				0	18	41	64
	240					0	22	44
	260						2	25
	280						0	0
Phosphorus	6	25	36	47	58	79	80	91
	8	11	22	33	44	55	66	77
	10	0	8	19	30	41	52	63
	12		0	5	16	27	38	49
	14			0	2	13	24	35

Potassium	16				0	0	11	22
	200	56	73	90	106	123	140	156
	250	45	62	79	95	112	129	145
	300	34	51	68	83	101	118	134
	350	23	40	57	73	90	107	123
	400	13	29	46	62	79	96	112
	450	1	18	35	51	68	85	101
	500	0	7	24	40	57	74	90
	550		0	13	29	46	63	79
	600			2	18	35	52	68

Table 10. Yield targetting in rainfed sorghum (cv. CSH 9) (Mean of 23 trials)

Treatment	Fert. nutrient applied			Sorghum yield	Cost of fertilizer	Monetary returns	Increase over control	Returns per Rs. spent on ferti.
	N	P_2O_5	K_2O					
		Kg ha^{-1}		q ha^{-1}		Rs ha^{-1}		
Control	0	0	0	29.94	0	14970	—	—
Recom. dose	80	50	50	47.98	1825	23990	9020	4.94
As per soil test	100	62	30	49.01	2045	24505	9535	4.66
40 q ha-1 target	20	00	02	40.92	160	20460	5990	34.31
50 q ha-1 target	66	16	35	45.70	1010	22850	7880	7.80
60 q ha-1 target	112	38	68	52.42	1965	26210	11240	5.72
C.D. at 5%		2.08		2.08				

Soil = Available N 170, P 13.6, K 601, kg ha^{-1}, medium - deep black
Rates = Fert. N 7.20, P_2O_5 18.00, K_2O 7.00, Rs. kg^{-1}, sorghum grain Rs. 500 q^{-1}

Table 11. Test of significance between targetted and actual yields of rainfed kharif sorghum

Targetted yield of sorghum q ha^{-1}	Actual yield of sorghum q/ha^{-1}	"t" calculated
40	40.46 (N=12)	0.22 NS
50	46.19 (N=23)	-1.85 NS
60	51.93 (N=12)	-2.00 NS

N = number of trials, NS = non-significant

Table 12. Soil test based fertilizer prescription for targetted yields of dryland crops

Sr. No.	Crop/variety	Location/soil Type	Targetted yield equation		
1	Pearlmillet	New Delhi	FN	=	10.25 T - 0.67 SN
	HB1	Alluvial	F P_2O_5	=	4.82 T -5.19 SP

			F K_2O	=	3.87 T -0.24 SK
2	Maize	Chindwara	FN	=	3.46 T -0.18 SN
	CMS2	Black	F P_2O_5	=	2.30 T -1.15 SP
			F K_2O	=	2.07 T -0.08 SK
3	Rice	Raipur	FN	=	2.78 T -0.27 SN
	IR8	Red and Yellow	F P_2O_5	=	5.45 T -11.03 SP
			F K_2O	=	1.88 T -0.23 SK
4	Soybean	Jabalpur	FN	=	3.89 T -0.42 SN
	Bragg	Black	F P_2O_5	=	7.67 T -7.63 SP
			F K_2O	=	4.24 T -0.14 SK
5	Pigeonpea	Jabalpur	FN	=	4.00 T -0.29 SN
	JA 3	Black	F P_2O_5	=	4.11 T -3.44 SP
			F K_2O	=	4.55 T -0.19 SK
6	Groundnut	Dholi	FN	=	14.75 T - 0.77 SN
	M 13	Alluvial	F P_2O_5	=	7.11 T -6.66 SP
			F K_2O	=	5.92 T -0.82 SK
7	Sorghum rabi	Solapur	FN	=	9.80 T -0.82 SN
	SPV 86	Black	F P_2O_5	=	2.28 T -2.99 SP
			F K_2O	=	3.52 T -0.08 SK

FN, F P_2O_5, F K_2O are fertilizer N_1, P_2O_5 and K_2O in kg ha^{-1}; SN, SP, SK are soil available N, P and K in kg ha^{-1}; T is yield target in q ha^{-1}

Fertilizer use efficiency can be increased by efficient management of inputs. Further, rain water being the main source of water in drylands, all efforts to improve intake of rain water would also improve fertilizer use efficiency. Similarly, at a given fertilizer level, soil depth in Vertisols and soil texture in the alluvium and submontane soils would affect the efficiency of fertilizer use.

Increased availability of moisture, early sowing, optimum plant stand, right choice of crops and varieties and timely weed control, all help in increasing the fertilizer use efficiency. Infact under dryland situation timely, weed control (within 3 weeks after sowing) should receive priority as weeds rob away substantial amounts of moisture and nutrient from the soil.

(a) Integrated use of fertilizers and manures

Integrated nutrient management aims to make the best possible use of diverse sources of nutrients such as fertilizers organic manuers, green manuers and contributions through biological route. It is well established that integrated use of fertilizer and FYM is essential for sustaining moderate to high crop yields in the drylands. The possible effect of FYM is through several ways. It adds nutrients to soil, improves soil physical properties enabling retention of more water and better root growth, and it can improve the availability of soil nutrients as well. The long term soil fertility experiments for over 25 years all over India have clearly demonstrated that fertilizer NPK (100%) when supplemented with FYM could sustain high yields and maintain soil fertility.

It has been shown that the integrated use of fertilizers and manuers not only increases yields, but also helps to sustain the system at a high level of productivity. The data in Table 13 showed that yield and N uptake by rabi sorghum (cv. SPV 86) were increased due to application of inorganic N plus organic N treatments in Vertisols at Solapur over control. Application of 50 kg N ha^{-1} to rabi sorghum (1/2 through sorghum residue + 1/2 through leucaena loppings) recorded the highest yield (12.54 q ha^{-1}) and sustainability yield index (0.64) followed by application 1/2 N through FYM + 1/2 N through urea (11.68 q ha^{-1}) with a sustainability yield index of 0.54 (Anonymous, 1995). Organic carbon contents were also higher in these treatments (0.51 and 0.54 per cent) at the end of 8 years of experimentation. Increase in organic matter content and simultaneous rise in water retention capacity of the soil profile were the additional benefits arising from leucaena lopping treatment (Narkhede *et al.,* 1984).

Table 13. Integrated use of fertilizer and organic N for rabi sorghum (pooled means of 8 years)

Treatment	Yield q ha^{-1}		N uptake (kg ha^{-1})	Org. C at harvest (%)	S.Y.I.
	Grain	Fodder			
Control	7.40	21.8	24.3	0.28	0.39
25 kg N ha^{-1} through urea	10.25	29.6	26.9	0.33	0.52
50 kg N ha^{-1} through urea	11.60	31.6	27.6	0.35	0.54
25 kg N ha^{-1} through sorghum stover	8.08	21.3	18.7	0.43	0.41
25 kg N ha^{-1} through FYM	10.88	30.1	35.1	0.43	0.54
5 kg N through sorghum stover + 25 kg N ha^{-1} through urea	10.90	29.9	25.6	0.46	0.54
25 kg N through FYM + 25 kg N ha -1 through urea	11.68	32.8	36.5	0.54	0.55
25 kg through sorghum stover + 25 kg N ha^{-1} through leucaena loppings	12.54	31.9	39.1	0.51	0.64
C.D. at 5%	0.64	1.4			

SYI = Sustainable yield index

(b) Sources of fertilizers

It is well known that the forms of fertilizers and their transformation products respond differently to varying moisture regimes. For instance, nitrate containing fertilizer will readily denitrify under soil saturation, as anaerobiosis sets in. On the other hand, with soil saturation urea will not nitrify and denitrify. Owing to poor water holding capacity, arising from depth situation, shallow soils are subjected to extreme moisture fluctuations more often than deep soils (Katyal *et al.,* 1997).

Shallow depth soil encouraged higher crop response to P particularly for pulse and oilseed crops. On account of neutral to alkaline soils supporting largely to short duration crops, water soluble P sources have been preferred.

(c) Time of fertilizer application

This aspect is important only for nitrogen management which can be given in 1, 2 or 3 splits during the crop season. The general guideline is to apply N in splits during the rainy season but all N as basal in the dry season.

Split application has the same significance in nitrogen management as drilling or placement has for phosphorus. Crop yields and fertilizer use efficiency is usually higher with split application than with a single application (Tandon, 1991). The exact number and timing of splits has to be decided on the basis of crop duration, rainfall pattern, total N to be applied and whether there is any possibility of providing an irrigation out of the stored rain water.

Placement of fertilizer N adds to the efficiency of split application because it curbs losses of N through surface run off and ammonia volatilization. Nitrogen loss could be reduced from 26 per cent with surface incorporation to 6 per cent with band placement.

Application of N in splits, half at sowing and the rest two in equal installments proved to be better than single N application at sowing. Application of N in 2 to 3 splits has two distinct advantages besides increase in yield (1) basal dressing including N application at sowing induces early seedling vigour (2) economics on fertilizer use as split application could be withheld, if the season turns out to be unfavourable.

In dryland the rainfall is unpredictable; hence it would be prudent to adopt split application of N in all situation during the monsoon period so that under aberrant weather a part of N could be saved.

(d) Methods of fertilizer application

The standard recommendation in the drylands is to drill or place the basal dose deep in the root zone. In other words, broadcasting is not recommended. The basal dose in Kharif usually consists of part of the N dose plus all the other nutrients to be applied. In rabi, in most cases, the entire dose for the crop season is recommended as basal. This is particularly so where little or no rainfall is expected during the crop season.

The benefits from the fertilizer placement have been demonstrated in several trials (Katyal, 1994; Venkateswarlu, 1987). Yield gains varying from 340 to 1500 kg grain ha^{-1} have been reported from placement over broadcast application. Sowing of seeds and fertilizer application at a time using a two bowin seed cum fertilizer drill has been found to be quite satisfactory (Tandon, 1991; Sharma and Das, 1992).

Table 14. Effect of placement of fertilizer on crop yields

Centre	Crop	Average yield, q ha^{-1}	
		Broadcast and mixing	Placement below or aside seed row
Hyderabad	Sorghum	16.1	21.0
Varanasi	Rice	29.3	29.4
Hoshiarpur	Wheat	33.9	37.8
Dehra Dun	Wheat	26.2	28.2
Ranchi	Chickpea	7.0	7.8
Agra	Black gram	8.6	9.4
Indore	Soybean	15.7	16.7

For maximum utilization, fertilizer should be placed in the moist zone below the seed rows, Drilling the fertilizer, 10-15 cm deep in the seed furrows was superior to other methods. Deep placement of fertilizer N and P helped in increasing grain yield of sorghum by 11 per cent over 24.7 q ha^{-1} obtained by broadcasting. Deep placement of fertilizers is important for post rainy season crops (Katyal, 1994).

Method of fertilizer application is also dependent on (i) type of soil, (ii) nature of fertilizer, (iii) nutrient requirement of crop and (iv) nature of field crop.

Broadcasting of fertilizer is inefficient. They need to be placed in or near the plant rows to increase their use efficiency. Deeper placement of fertilizers for the post monsoon period is essential. The data in Table 14 show that cereals respond well to placement. In food legume and oilseed crops, the response is not conspicuous because the general yield levels are low (Venkateswarly and Ranjodh Singh 1982).

Organic manuers (eg. FYM) should be preserved under shade, if necessary, spread over soil just before ploughing so as to mix well with soil. It should not be applied to the soil and expose to sunheat. These should be applied 3-4 weeks before sowing of crops.

8. CONCLUSIONS

The productivity of rainfed/dryland crops is low due to deficiency of N and P in soils, inadequate and poorly distributed rainfall and use of low amount of fertilizers. There is a vast potential for increasing crop yield through fertilization of rainfed/dryland crops. However, fertilizers need to be used economically, efficiently and scientifically because of their increased cost. In this regard, fertilizer application based on targetted yield approach was found to be superior to the present method based on soil testing along. By targetted yield approach, it is also possible to make fertilizer recommendation considering financial conditions of the farmers. The targetted yield approach for fertilizer application works well for getting desired yields of rainfed dryland crops. However, there is a need to work on rainfed/ dryland crops (foodgrains, oilseeds, pulses etc.) as very limited work has been done so far.

REFERENCES

Abro I.P. 1991. Foreword, Fertilizer management in dryland agriculture - a guidebook, Tandon, H.L.S. (Ed.), FDCO Publ. New Delhi.

Akolkar, R.D. and Sonar K.R. 1994. Calibration of soil test methods for available potassium in vertisols for sorghum. J. Potassium Res. 10 (1) : 63 -67.

Anonymous, 1995. Research review report. Dept. of Agril. Chemistry and Soil Sci. MPKV Rahuri.

Anonymous, 1996. A few facts about fertilizers in the western region, FAI, Mumbai pp. 132.

Kanwar, J.S. 1994a. Management of soil and water resources for sustainable agriculture and environment. In "Management of land and water resources for sustainable agriculture and environment", Indian Soc. Soil Publ., pp. 1 - 10.

Kanwar, J.S. 1994b. Relevance of soil management in sustainable agriculture. In. "Soil management for sustainable agriculture in dryland areas". Bull. 16, Indian Soc. Soil Sci., pp. 1-11.

Katyal, J.C. 1994. Nutrient management in soils of arid and semi-arid regions. In. "Soil management for sustainable agriculture in dryland areas". Bull. 16, Indian Soc. Soil Sci. pp. 78 - 91.

Katyal, J.C., Sharma K.L. and Srinivas K. 1997. Nutrient management practices in dryland agriculture and cropping systems. In. "Nutrient management practices in crops and cropping systems," Ghonsikar C.P. and Shinde V.S. (Eds.) Scientific Publishers (India), Jodhpur, - pp 191-211.

Nazirkar, R.K. and Sonar K.R. 1991. Calibration of soil test methods for available nitrogen in black soils (Inceptisols) for sorghum. J. Indian Soc. Soil Sci. 39 : 383-385.

Narkhede, P.L., Umrani N.K. and Ghugare R.V. 1984. Organic recycling of subabul for winter sorghum cultivated on drylands. Indian J. agric. Sci. 54 (5) : 444 - 446.

Palwe, C.R. and Sonar K.R. 1989. Evaluation of soil test methods for phosphorus in black soils of Maharashtra for wheat. J. Indian Soc. Soil Sci. 37 : 186-189.

Prihar S.S. and Ranjodh Singh 1983. Advances in fertilizer management for rainfed wheat. Fert. News. 28 (9) : 49 - 56.

Ramamoorthy, B.; Narasimhan R.L. and Dinesh R.S. 1967. Fertilizer application for specific yield targets of Sonora -64. Indian Fmg. 27 (5) : 43 -44.

Randhawa N.S. and Singh R.P. 1983. Fertilizer management in rainfed areas - available technologies and future needs. Fert. News. 28 (9) : 17 - 32.

Rao K.P.C. 1992. Soil moisture and nutrient management. In "Dryland agriculture -state of art of research in India, Somani L.L. (Ed. in Chief), Scientific Publishers (India), Jodhpur, pp. 374-394.

Reddy, K.C.K. 1992. Soil test calibration research for dryland soils. In "Dryland agriculture - State of art of research in India." Somani L.L. (Ed. in Chief), Scientific Publishers (India), Jodhpur, pp. 283.

Sehgal J.L. and Sharma J.P. 1994. Soil and climatic resource characterization of dryland ecosystems in India for sustainable agriculture. In. "Soil management for sustainable agriculture in dryland areas", Bull. 16, Indian Soc. Soil Sci., pp. 12-25.

Sharma K.L. and Das S.K. 1992. Nitrogen and phosphorus management in dryland crops and cropping systems. In. "Dryland agriculture -state of art of research in India,"

Somani L.L. (Ed. in Chief), Scientific Publishers (India), Jodhpur, pp. 306-350.

Sonar, K.R. 1984. Highlights of research (1968-1984) - fertilizer recommendations based on soil test crop response. Dept. of Agril. Chem. and Soil Sci. MPKV, Rahuri Publ. pp. 1-39.

Sonar K.R., Patil N.D., Kadam J.R., Dhage A.R. and Bagul. K.M. 1984. Soil testing - key to efficient fertilizer use and higher profits. Dept. of Agril. Chem. and Soil Sci., MPKV Rahuri, Publ. pp. 1-28.

Sonar, K.R. 1996. Soil testing for ensuring balanced use of fertilizers. In "Balanced fertilizer use for increasing foodgrains production in western states. Dev, G. (Ed.), PPIC, Gurgaon Publ. pp. 63 -80.

Sonar, K.R. 1999. Soil testing for increasing crop production. In "Advances in soil science research in Maharashtra" State level seminar at MAU Parbhani, Jan. 21-22, Parbhani Chapt. Indian Soc. Soil Sci. Publ. pp. 54-69.

Sonar K.R. and Tamboli B.D. 1997. Targetted yield concept. In "Nutrient management practices in crops and cropping systems. Ghonsikar C.P. and Shinde V.S. (Eds.) Scientific Publishers (India), Jodhpur, pp. 251-265.

Somawanshi, R.B., Tamboli B.D., Patil Y.M., Bhakare, B.D., Kadu, P.P. 1994. Twenty five years of research on soil test crop response correlation studies in Maharashtra (1968-1993). MPKV, Rahuri, Bull. 14, pp. 1-142.

Somawanshi R.B., Tamboli B.D., Kadu P.P., Patil Y.M., Bhakare B.D., and Sonar K.R. 1997. Technology for getting desired crop yields. Dept. of Chem. and Soil Sci., MPKV. Rahuri Publ. pp. 1 - 45 (in Marathi).

Tandon H.L.S. 1991. Fertilizer management information for dryland agriculture. In "Fertilizer management in dryland agriculture - a guidebook" Tandon H.L.S. (Ed.), FDCO New Delhi Publ. pp. 50-73.

Truog E. 1960. Fifty years of soil testing. 7th Int. Congr. Soil Sci., Madison, Wi. USA, III & IV, pp. 36-45.

Umrani, N.K. 1995. Crop management technology for drylands. In " Substainable development of dryland agriculture in India, Singh R.P. (Ed.) Scientific Publishers (India), Jodhpur, pp. 241-270.

Venkateswarlu, J. 1987. Efficient resource management systems for dryland of India. Adv. Soil Sci. 7 : 165-221.

Venkateswarlu, J. and Ranjodh Singh. 1982. Crop responses to applied nutrients under limited water conditions. In. Review of Soil research in India. Part 1, 12th Int. Congr. Soil Sci. New Delhi, pp. 442-452.

Chapter - 17

PRECAUTIONS TO AVOID ABIOTIC STRESSES

A.M. Dhopte[1] *and S.D. Ramteke*[2]

1 Former Professor of Botany, College of Agriculture, Nagpur, 49, Vivekanand Nagar, Khamala Road, Nagpur – 440015 (MS)

2 Dr. S. D. Ramteke, Principal Scientist (Plant Physiology), National Research Centre for grapes, Manjri Farm, P. O. Box No. 3, Pune -Solapur Road, Pune - 412 307

1. INTRODUCTION

Water is necessary for passive existence of protoplasm, essential for maintaining the turgidity of cells which is the basic requirement for carrying normal metabolic processes. If a spell of drought occurs constantly, a protoplasm is subjected to dehydration and desiccation injury and in turn the yield of crop is adversely affected. Moreover, lack of water in soil in later stages of crop growth results in low yield. Under such conditions, crop either becomes complete failure or produces low yield depending upon the severity of moisture stress. Therefore, soil moisture penetration, its conservation and efficient utilization is an important line in tackling the problems of crop production under dryland conditions (Plate 17.1).

Satisfactory completion of vegetative stage of crop with the help of few inches of rains does not ensure yield as the drought condition at later stages is detrimental. Production is adversely affected if the moisture stress is experienced at the critical stage of reproductive phase of the crop. An effort has been made to discuss some of the important crop life saving devices in this paper.

2. REDUCE EVAPOTRANSPIRATION

(a) Reducing the soil evaporation

Water harvesting is one of the practice followed for collecting and storing water from rainfall especially in arable areas having insufficient rainfall. In semi-arid areas, crop growth is greatly limited by water. Amount of available water in soil can be increased by surface mulching and other soil management practices.Wang et al. (2009), studied the effects of rainfall harvesting system and mulching technologies and found that the integrated use of Rainfall harvesting system (RFRHS) results in improvement in rainwater availability for high sustainable crop yield in China. Andersson et al. (2011), explored the potential impacts of two strategies, namely in situ water harvesting (in situ WH) and fertilisation with stored human urine (Ecosan), to increase the water and nutrient availability in rain-fed smallholder agriculture in South Africa's Thukela River basin (29,000 km^2). In situ WH does not generally affect the river flow regimes. Occasionally, significant regime changes occur due to enhanced lateral and shallow aquifer return flows. This leads to higher risks of flooding in some areas, but also to enhanced low flows, which help sustain aquatic ecosystems in the basin. Subrat Kumar Mishra. (2012), conceded study on the issues and concerns of drought prone areas related to watershed management, he mentioned in that watershed management progammes are implemented in drought prone areas to attempt the special problems faced by those areas habitually affected by severe drought conditions. The main objective of the watershed approach is to diminish adverse effects of drought on the production of crops, livestock and productivity of land. The key problem in water harvesting is the run-off inducement in the catchment area which is achieved by several techniques, such as mechanical and chemical treatments of the surface, as well as surface covering by means of plastic or rubber sheets. In heavy soils where crevices are developed due to its shrinking nature, various types of mulches can be adopted to check the evaporational losses from the soil. Different types of mulches that are in vogue now a days are listed below:

(i) Stubble mulch (ii) Plastic mulch (Plate 17.1.3)

Yaghi et al. (2013) showed that Traspirant mulched Drip Irrigation (DI + TM) treatment gave the highest soil temperature and moisture during both of the seasons in comparison to black mulched drip irrigation (DI + BM). This treatment also enhanced its vegetative growth and almost doubled its productivity compared to the surface furrow irrigation (SI) treatment. Diaz et al. (2005), studied the influence of the thickness and grain size of tephra mulch on soil water evaporation, according to them the thick layers of mulch reduced maximum evaporation. McMillan. (2013), conducted experiments on the effect of mulch type and thickness on the soil surface evaporation rate, he conducted research to determine which of readily available mulching materials would be best at conserving soil moisture, and at which thicknesses, 5, 10, and 15 cm. Soil water content was monitored every three days for a duration of three weeks, when no discernable differences were measured. Within the first 3 days, a mulch layer of at least 5 cm reduced surface evaporation to 40% compared to the water losses from bare soil, and all mulch types were equally effective. While there were no differences between the mulch types, the mulching rate did have a significant effect on water loss. Doubling the mulching rate from 5cm to 10 cm maintained soil moisture 10% higher throughout most of the experiment. However, increasing the rate further to 15 cm had no discernible effect. This experiment clearly demonstrates that what might otherwise be agricultural waste can be used to significantly conserve soil moisture, providing more resources for crops and reducing overall costs of production. Xie *et al.* (2005), according to them evaporation was decreased little when the soil moisture was reduced from 27% to 8% with gravel mulch. Evaporation and the ratio of evaporation to evapotranspiration (E/ET) increased with the diameter of sand and gravel. The E/ET ratio was 40.7% in the growing period for the control, and it was only 17.8–25.0% for treatments mulched with sand and gravel. Soil evaporation with non-mulching was reduced by 78.0–93.7 mm when plastic film was mulched on the gravel surface and by 16.9–26.3 mm with gravel mulching only. The yield was not obviously different between treatments with gravel of various sizes, but water use efficiency was significantly higher for mulching with 2–5 mm diameter gravel than for mulching with 20–60 mm diameter gravel. In addition, the soluble carbohydrate content decreased for small-diameter gravel. Moratiel and Martı´nez-Cob. (2010), deliberated on the evapotranspiration (ETc) of a table grape vineyard (Vitis vinifera, cv. Red Globe) trained to a gable trellis under netting and black plastic mulching was determined under semiarid conditions in the central Ebro River Valley during 2007 and 2008.

(iii) Rubber sheets (iv) Scooping etc.

All above practices cannot be followed on larger area, but some of which certainly serves as a tool in decreasing the soil evaporation for crop life saving under moisture stress. However, mechanical and chemical treatment and stubble mulch as described below may to some extent, help in conservation of moisture.

Plate 17.1. 1. View of watershed in field. 2. Undeveloped peanut pegs due to water stress (TAG-24), 3. New polyculture system for peanut cultivation between vetiver live bunds, A case of polythene mulch

3. APPLY MECHANICAL TREATMENTS

It includes levelling, contour ditching scooping and surface compaction. Land forming consists of providing a systematic slope to the catchment area so that depressions and crevices are obliterated and run off is captured without appreciable loss by evaporation from depression storage and by percolation through crevices. Compaction of surface soil with roller enhances the run off yield by decreasing the water intake rate. This would be particularly useful in case of coarse textured soil. Hill side run off, which is otherwise lost, can be collected by building contour ditches.

4. APPLY CHEMICAL TREATMENTS

It includes spraying of salt solution, water repellant chemicals and asphalt compounds on the soil surface. Spraying of NaCl solution causes clay in the soil to disperse and form sodic crust. Hillol and Rawitz (1968) obtained more than 70 per cent run off by treating lossial sicrozom of fine sandy loam texture with NaCl at the rate of 820 lbs./acre. Water repellant compounds offer possibilities for producing high rate of precipitation run off without obtaining the hazard of erosion. Encouraging results have been obtained by the use of compounds like sodium rosinate, sodium methyl silanolate and dialkyl quarternary ammonium chloride (Myers, 1967). Run off as high as 94 per cent has been obtained. The effect of water repellant compounds get decreased due to oxidation of the compound or its removal by erosion. Sprayed asphalt offers considerable promise for building low cost impermeable catchment. It can be dissolved in kerosene and sprayed on the surface. As the solution soaks in, it deposits the asphalt which binds the soil particles in to a reasonably strong but porous pavement. The pavement surface is then sealed water tight by spraying with the homogenised mixture of asphalt and water (Mechael, 1969). Application of KNO_3 (0.5%) can ameliorate water stress (Plate 17.2 & 17.3) when strayed at pre flowering. Youssef *et al.* (2013), reported that the applications of salt solutions before and after harvest affect the quality and incidence of postharvest gray mold of 'Italia' table grapes. They mentioned the salts applied just before harvest may be an effective way to minimize gray mold during storage.

5. USE OF STUBBLE MULCH

Stubble mulching refers to the maintenance of crop residues on the soil surface. The managing of plant residues on a year round basis in which harvesting, tilling, planting and cultivation operations are performed in such a way as to keep protective amounts of vegetative material on the surface of the soil until the time of seeding the next crop. Stubble mulching aids in moisture conservation by increasing moisture intake and reducing evaporation. When stubble mulch is used on the surface of the soil, the impact of rain drops is reduced and infiltrations is maintained at a higher rate for a longer period of time. Stubble mulch also acts like a dam to slow the movement of run off water and increases the opportunity for infiltration and surface storage (Meyer *et al.,* 1970). With vegetation mulch, the constant rate of evaporation is decreased and the reduction is about 0.1 cm. per day per 560 Kg/ha of residue. (Bond and Willis, 1969). Under condition of limiting residues, residue management system has to be devised considering the relationships between residue rates, evaporation rates and precipitation events (Bon and Willis, 1970). It would be desirable that during rainy season, for greatest evaporation reduction, the residue available for mulching might best be utilized over larger surface area (Plate 17.5). Zhang *et al.* (2011), studied on the effects of stubble retention and tillage practices on surface soil structure and hydraulic conductivity of a loose soil, they concluded that the improvement in soil structure and stability was confirmed by higher soil hydraulic conductivity measurements. Consistently higher Ksat was detected in the no tillage with stubble retained soil compared to other treatments. Therefore, no-tillage with stubble retention practice is an effective management technique for improving physical quality of this fragile soil in the long term. Moradi. (2012), carried his study on the Role of the Stubble Much on control of dry Farming lands soil temperature and its economic efficiency, he concluded that the stubble mulch which appears because of residue remaining on soil level in contrast with soil free of mulch could decrease soil temperature and also alleviates the of avoidance temperature fluctuations as well as could avoid remarkable fluctuation of circadian and seasonal temperature of soil, more maintenance of water and decreased permeability of soil, deliberation of weeds, avoiding of erosion and decreased soil salinity and finally decreased evaporation, surely, decreased evaporation causes to increased humidity storage in soil and consequently leads to increased yield which is important economically.

Plate 17.2. 1. Leaf rolling in sorghum due to water stress, 2. Amelioration of drought with potash application (Sprays, KNO_3, 0.5% and soil applied 30 Kg MP/ha)

Plate 17.3. 1A. Green cotton leaf, B. leaf reddening, 2. A,B & C. Types of leaf reddening due to nutrient deficiency. An example of abiotic stress and can be corrected with timely foliar sprays with nutrients (Urea 2%, KNO_3 0.5%) and auxins (NAA, CCC, 25 ppm)

Plate 17.4. 1. Poor growth of arbareum cotton due to no weeding and interculture, 2. Parthenium weed growing along farm roads. Make compost or uproot and burn to prevent rapid coverage. This creates menace and can not be controlled easily. Isolated area can be sprayed with 2,4-D (0.1%) before flowering. If should not be tried in a crop.

Plate 17.5. 1. Glyrsidia plantation along bunds for green manuring during monsoon to improve organic matter; 2. Preparation of organic mulch from field organic waste. It should be incorporated in field during active rainfall for decomposition.

6. THINNING THE PLANT POPULATION

Simonin *et al.* (2007), conducted a study on the influence of thinning on components of stand water balance in a ponderosa pine forest stand during and after extreme drought. To understand the effect of restoration thinning on the water balance of upland semi-arid ponderosa pine (Pinus ponderosa) forests of the southwestern US, they compared the components of forest water balance between an unthinned plot and a thinned plot using a paired water balance approach. Forest overstory transpiration (EO) was estimated from tree sapflow scaled to the plot level. Understory evapotranspiration (EU) was estimated from the difference between through fall precipitation and changes in soil water content measured in trenched plots that excluded tree roots. The thinning treatment in 2001 reduced plot basal area by 82% and leaf area index by 45%. Difference in stand-level evapotranspiration (E) between the thinned and unthinned plots, and partitioning of E between EU and EO during the first post-treatment summer and spring, varied between drought and non-drought periods. The importance of EU in stand-level E was greater in thinned compared with unthinned plots and increased during extreme drought when EO was low due to stomatal closure. The results highlight the importance of drought and climate as factors determining the impact of thinning on water balance in southwestern ponderosa pine forests. Tremendous quantity of water from plant is lost through the transpiration process. The cell survives in low water up to 10 per cent of the water absorbed by the root, less than one per cent is retained in plant. It, therefore, becomes indispensable to reduce the transpiring surface of the plant by thinning the plant population to a optimum before the commencement of temporary wilting or even before it by proper judgement. This stage of the crop is to be judged by visual observation so that transpirational loss of water is minimised and the moisture is utilized by the limited plant population, which in turn, will give some yield where nothing would have been produced due to competition of water in heavier plant population.

7. CHECKING THE UNWANTED VEGETATION (WEED CONTROL)

Parker (2003), according to him weeds, like other plants, consume large quantities of water, and most of it is lost by transpiration to the atmosphere. Plants with deep roots have an advantage under conditions of moisture stress. Growth of common lambsquarters, kochia, and Russian thistle is less affected by a shortage of moisture than is growth of many crops. A few showers of rains are sufficient to create a favourable condition for weed seed germination and it is observed that after a commencement of rains, within a week soils get covered with the green blanket of weedy vegetation which starts consuming soil water, mostly from the surfaces, at such a faster rate that within a fortnight if no rains occur, soils show the symptoms of drying. The transpiration efficiency of weeds is very high and indeed it grows faster suppressing the growth of economic crop plants. Weeds exorbitantly utilize the soil moisture as indicated by the following data on T.C. (Table 1).

Table 1. Transpiration co-efficient (T.C.) of weeds

Sr. No.	Local name	Scientific name	T.C.
1	Unhali	*Tephrosia purpurea*	1108
2	Kunda	*Ischamum pilosum*	556
3	Hariyali *Cynodon dectylon*	813	
4	Jowar	*Andropogan sorghum*	430

(Data obtained from Hand Book of Agriculture, I.C.A.R. Publication pp. 667).

Considering the voracious consumption of water by weeds from the soil, it is highly essential to check the growth of weeds for soil moisture conservation. Unfortunately, the experience is that the practice of clean cultivation is not followed by majority of farmers. It is, therefore, highly desirable to check the weed growth either by mechanical means or by hormonal weedicides. Various types of herbicides are available in the market and their selectivity is also reported by different workers. List of herbicides commonly available for use in controlling the weeds is given below (Plate 17.4):

(i) 2, 4-D
(ii) Dalapon
(iii) MCPA
(iv) 2, 4, 5-T
(v) MCPB
(vi) 2, 4, D-B etc.

Among all the above weedicides 2, 4-D is commonly and cheaply available in the market. Pre-emergence and post emergence spray of 2, 4-D wherever possible will help in checking the moisture losses significantly (Refer Chapter 18 for details).

8. USE OF ANTITRANSPIRANTS

An effective and efficient device of crop life saving in later stages of crop growth when depletion of soil moisture is acute, is the use of antitranspirants which reduces the rate of transpiration and brings economy in water utility by crop plants. Most important aspects of dry land farming is the maximum utilization of limited rain water for crop use, rather than its wastage. The amount of water transpired from plant leaves may be significantly reduced by preventing the opening of leaf stomata.

Some chemicals called "Antitranspirants" have been introduced, on the basis of valid ficid experimentation to reduce the loss of water through stomata (Gale and Hagen, 1966). Two types of chemical compounds have been used as anti- transpirants namely : (i) those that cover the foliage with a thin film which ideally is impervious to water but pervious to CO_2 and O_2 and acts as reflectant of light to save the leaves from rise in temperature and (2) Chemicals which control the closing and opening of stomata, by affecting the biochemical reactions, which influence the turgidity of stomatal guard cells.

Success has been obtained by the chemicals of later category. Some chemicals which have shown considerable effects on crop plants are —

1. Adol, 52 (A formulation of acetyl alcohol)
2. OED green (A combination of decanol and octadecanol with ethylene oxide)
3. S-600 (A plastic transplanting spray)
4. 8-hydroxyquinoline
5. Sodium-L-hydroxydecanosulfonate
6. Alkenyl succinic acid
7. Monomethyl ester of decanyl succinic acid
8. Fatty alcohols as hexadecanol and octadecanol.
9. Phenyl mercuric acetate (PMA)
10. Abscisic acid.
11. CCC
12. AA
13. Kaolin

Amongst all, PMA and alkenyl succinic acid have preference over others. Since stomatal movement is brought about by changes in the turgidity of the guard cell, the mode 'of action of both these two compounds depends on prevention of complete turgidity. Thus, PMA is believed to react with sulphydryl groups in guard cell membrane, thereby altering their permeability. The important caution in using antitranspirants is that decrease in the rate of transpiration should not be on the cost of reduction in photosynthesis.

Adequate available water during the periods from tillering to filling of grains in cereal is necessary for the production of high yield in arid and semi-arid regions. Storage of enough water in the root zone during this phase is essential. Antitranspirants may be used to retard the rate of soil water depletion and thus can extend the period over which precipitation may be received without serious decline in yield. The persistance of PMA spray and its reduction in transpiration on various crops are presented alongwith the optimum concentrations for favourable results in Table 2.

Table 2. Persistance of P.M.A. Spray in various crops.

Crop	PMA spray (Concen.)	Reduction in transpiration per cent	Persistance in days	Reference
Wheat	3.5×10^{-5} M	20-30	15	Brengle (1968)
Tomato	100 ppm	Reduced	–	Cotter (1971)
Cotton	10^{-3} M	40-60	30	Slatyer & Bierhuizon (1964)
Chrysanthemum	5×10^{-5} M	30	4	Cocking and Tukey (1970)
Sunflower	5 ppm	48	15	Aher (1973)

Pasternak and Wilson (1971) reported the variation in effectiveness of PMA as a stomatal closing antitranspirants. In a comparison of crop including sorghum, maize, cotton and groundnut, stomatal apertures were affected to different extent by PMA spray. In pot culture, PMA spray conserved soil water. Field and glass house trials with sorghum showed that the successive application may be necessary for persistant effect. The result obtained by Majernik (1971), however suggests that the concentration of PMA ($10^{-3.5}$M) normally applied as an antitranspirant is unfavourable for healthy growth of maize and barley. Therefore, PMA should not be used on maize and barley. Antitranspirants are expected to curtail growth via. reduced photosynthesis. However, by increasing plant water potential, they can enhance the growth of shoots fruits (Davenport *et al* (1971).

Mishra and Pradhan (1972) reported no significant effect on yield in Tomato but PMA reduced 80 per cent stomatal opening and wilting was found to be delayed by 8 days when water was withheld 59 days after final spray application. PMA reduces the transpiration rate by 20-50 per cent and can persist on plant for its effect for 20 to 25 days.

Recently in field investigations carried out at Marathwada Krishi Vidyapeeth, Parbhani, at Post-graduate Research Scheme, Aher (1973) (unpublished) obtained significant increase in seed yield of sunflower treated with 1 and 5 ppm. PMA when sprayed at 2 stages of plant growth viz. seedling and preanthasis stage along-with 50-70 per cent stomatal closure. There was 45 cent reduction in transpiration with considerable increase in carbon assimilation under field trials. The effect of the sprays persisted for 20 days, even when the water was withheld for 70 days after final spray application. Chen and Murata (2008), concluded that the Glycinebetaine (GB) increased tolerance to abiotic stress have led to considerable progress in the characterization of the roles of GB in stress tolerance in plants. In particular, the reproductive organs of GB-accumulating transgenic plants exhibit enhanced tolerance to abiotic stress. Due to accumulation of GB results in increased yield potentials under non-stress conditions.

Plate 17.6. 1. Sorghum hybrid (CSH-9) showing ununiform growth due to slope. Go for contour bunding, 2. Critical stage of maize (Cob development, 50 DAS), cv. Ganga-5. 3. Poor seed setting due to poor fertilization. Apply potash spray KNO_3 (0.5%) at Cob development.

Zhang *et al.* (2006), suggested the plant hormone abscisic acid (ABA) plays a major role in plant responses to stress. Although rapid production of ABA in response to drought and salt stresses is essential to define ABA as a stress hormone, an equally rapid catabolism of ABA when such stresses are relieved is also essential in that role. Since ABA mediates so many stress responses, the initial perception of dehydration and the subsequent changes in gene expression that lead to rapid

ABA biosynthesis constitute the most important stress signal transduction pathway among all the plant responses to stresses. Identification of the genes involved and understanding their roles during stress perception and physiological regulation has become an important and exciting research field in recent years. This review covers mainly our understanding of this aspect. ABA-induced changes in gene expression and their roles in physiological regulation are dealt with in less detail.

Jian Feng Ma (2004), studied on role of silicon in enhancing the resistance of plants to abiotic and biotic stresses, according to him silicon exerts alleviative effects on various abiotic stresses including salt stress, metal toxicity, drought stress, radiation damage, nutrient imbalance, high temperature, freezing and so on.

In a third annual report of Haryana Agricultural University, Hissar, (1971-72) it is reported that the use of antitransparants (Kaoline 1.2%) with dust mulch yielded bajra (H.B-1) 14.2 q/ha as against 12.31 q/ha when no mulch and no antitranspirant was used. Field applicability of the antitranspirants for increasing moisture use efficiency of crops, should not be restricted so much as by the cost of the material, because of the extremely low concentrations used in any easy process of solution spraying. However, the economics of PMA spray application has been worked out on the basis of data obtained by Aher (1973) on sunflower in Table 3 (Refer Chapter 19 for details).

Table 3. Economics of PMA for use as Antitranspirant

1.	Optimum spray for economic turnover	5 ppm
2.	Recommended spacing	60 x 23 cm
3.	Total number of plants/hectare	7246
4.	Volume of PMA solution required for two sprays/plant.	25 ml
5.	Total volume of solution required for spraying total plant population/ hectare	181.15 litre
6.	Grams of PMA required/hectare	906 mg/~ 1g
7.	Cost of PMA required/hectare	Rs. 25.0 (cost Rs. 250/- per 10 gm.)

*Estimations based on the experimental data obtained by Aher (1978) on Sunflower crop under field trial conducted at Botany Section, College of Agriculture, Parbhani.

9. SEED TREATMENT FOR DROUGHT INDUCTION

In an areas where the moisture stress or drought is of common occurrence, presowing seed treatment may be advantageous. As a precaution by measure, the seeds of the varieties can be pre-treated and then sown in the field with a view to induce drought tolerance in successive periods. Parija and Pillay (1945) found that plants of summer paddy, raised from seeds soaked in water (for 24 hours) and dried before sowing survived better after wilting, transpired less and required less water than plants from untreated seeds. The grain yields were, however, similar under these two treatments. Chinoy (1962-67) investigated the problems of drought resistances in wheat and found that alternate soaking and drying of seeds before sowing resulted an average grain yield of 948 1b/acre from treated seeds as against 866 1b/acre from untreated seed, on dry land over a period of three years. Seed treatment trials on farmers field conducted at 1971-72 (Bapat and Umale, 1973) on various phosphate salts revealed that there is a significant increase in yield of crops as shown in Table 4 & 5.

Plate 17.7. 1. Know physiological maturity of soybean (JS-80-21). Yellow pod wall, 4th from left., 2. Proper stage of harvest in sunflower (Leaves green but heads are mature), 3. Soybean crop waiting for harvest (JS-335). Avoid & prevent shattering losses

Kursanow (1956) and Henkel (1961) stated, that by a presowing treatment of grain, protoplasmic qualities favouring drought resistance could be conferred on subsequent crop plants. A similar presowing treatment with a solution of NaCl is recommended from crops of saline land (Refer Chapter 1 for details).

Unfortunately very little work on this aspect is reported in India and hence planned programmes in this direction may be worth while trying.

Table 4. Effect of Seed treatment trials on farmers' field during 1971-72

Crop	Variety	Treatment 3 g salt/ 100 g of seed in equal proportion	Yield in kg/plot	Percentage increase over control
Cotton	MCU-5	$KH_2PO_4 + NH_4NO_3$	190	26.6
Cotton	Buri-147	$NH_4NO_3 + KH_2PO_4$	110	22.2
Moong	Kopargaon	$(NH_4)_2HPO_4$	310	19.2
Urid	Local	–do–	130	30.0
Wheat (irrigated)	Hira	KH_2PO_4	1250	22.5
Wheat	Local	–do–	130	30.0

* Bapat and Umale (1973)

Table 5. Effect of seed soaking with brassinosteroid (4 hours) on germination under normal and saline condition in tomato

Treatment (mg/l)	Germination (%)		Root length (cm)		Vigour index	
	N	S	N	S	N	S
Control	50.0	30.0	1.06	1.00	800	510
0.1	53.3	43.3	1.53	1.7	1242	909
0.2	60.0	50.0	1.10	0.96	996	1100
0.8	53.7	36.7	1.36	1.83	2028	1173
0.9	53.3	36.7	2.23	1.46	4783	1136
1.0	52.0	33.3	1.80	1.42	3640	699
CD 5%	31.3	23.8	0.91	0.92	865	568

Source : Prakash *et al.*, (1999) N = Normal, S = Saline conditions.

REFERENCES

Abdollah Moradi. (2012). The Role of the Stubble Much on Control of Dry Farming Lands Soil Temperature and Its Economical Efficiency. *J. Basic. Appl. Sci. Res.,* **2(5)**5203-5205.

Aher S.S. 1973. Effect of Antitranspirants on Water-turnover, physiological process, growth and yield of sunflower (unpublished. Thesis submitted to M.K.V., College of Agriculture, Parbhani).

Anonymous 1971-72. The Annual Report, Dept. of Agronomy, College of Agriculture, Haryana Agricultural University, Hissar.

Bapat, M.V. and Umale, S.R. 1973. Seed treatment studies with phosphates (1) Wheat. *P.K.V.J.* **1** (2) : 214-220.

Bond, J.J. and Willis, W.O. 1969. Soil water evaporation - surface residue rates and placement effects. *Soil Sci. Soc. Am. proc.* **33** : 445-448

Bond, J.J. and Willis, W.O. 1970. Soil water evaporation. First stage drying as influenced by surface residue and evaporation potential *Soil Sci. Soc. Am. Proc.* **34:** 928.

Brengle, K.G. 1968. Effect of PMA on growth and water use of spring wheat. *Agro. J.* **60** (2) : 246-247.

Chinoy, J.J. 1962. Physiology of drought resistance in wheat. V. Drought co-efficient in relation to drought intensity. (Quoted from crop Physiology in India, Asana and Sarin, 1968 pp 27).

Cocking, W.D. and Tukey H.B. 1970. Reduction of water loss from *Crysanthemum* spp. following foliar application of phenyl mercuric acetate. *J. Amer. Soci. Hort. Sci.* **95** (3) : 382384.

Cotter, D.T. 1971. Phenyl mercuric acetate effects on water loss of the tomato. *J. Amer. Soc. Hort. Sci.* **95(6)** : 774-776.

Davenport, D.C., M.A. Fisher and R.M. Hagan 1972. Some counteractive effects of antitranspirants. *Plant Physiol.* **49:** 722-724.

F. Diaz, C.C. Jimenez, M. Tejedor. (2005). Influence of the thickness and grain size of tephra mulch on soil water evaporation. Agricultural Water Management 74, 47–55.

Gale, J & R.M. Hagen 1966. Plant antitranspirants *An. Rev. Pl. Physiol.* **17** : 269.

Guosheng Zhang, K.Y. Chan, G.D. Li, Gaobao Huang. (2011). The effects of stubble retention and tillage practices on surface soil structure and hydraulic conductivity of a loess soil. *Acta Ecologica Sinica* **31, 298**–302.

Henkel, P.A. 1961. Drought resistance in plants: Methods of recognition and of intensification (Quoted from water deficits and plant growth Vol. II. T.T. Kozlowsky, 1968 pp. 157).

Hillel, D. and Ramitz, E. 1968. A preliminary field study of surface treatments for run off inducement in Neger of Israel. *9th Intern. Soil Sci. Conq. Adelaide. Aust.* ***1*** **:** 303-311.

Jafet C.M. Andersson, Alexander J.B. Zehnder, Johan Rockstrom, Hong Yang. (2011). Potential impacts of water harvesting and ecological sanitation on crop yield, evaporation and river flow regimes in the Thukela River basin, South Africa. *Agricultural Water Management* **98, 1113**–1124.

Jian Feng Ma. (2011). Role of silicon in enhancing the resistance of plants to biotic and abiotic stresses. *Soil Science and Plant Nutrition*, **50:1**, 11-18.

Jianhua Zhang, Wensuo Jia, Jianchang Yang, Abdelbagi M. Ismail. (2006). Role of ABA in integrating plant responses to drought and salt stresses. *Field Crops Research* **97, 111**–119.

Khamis Youssef, Sergio Ruffo Roberto. (2013). Applications of salt solutions before and after harvest affect thequality and incidence of postharvest gray mold of 'Italia' table grapes. *Postharvest Biology and Technology* **87, 95**–102.

Kursanov, A.L. 1956. Recent advances in plant physiology in USSR. *Ann. Rev. Plant Physiol. 7 : 401*

Majernik, O. 1971. Response of stomata of barley and maize to phenyl morcuric acetate. Biol. Plant (Quoted from Bioiogical abstract, 1971 No. 51534).

Mayor, L.D., Wischmeier, W.H., and Fostor, G.R. 1970. Mulch rates required for erosion control on steep slopes. *Soil Sci. Soc. Am. Proc.* **34:** 928-931.

Michael McMillen. (2013). The Effect of Mulch type and Thickness on the Soil Surface Evaporation rate. Horticulture and Crop Science Department California Polytechnic State University San Luis Obispo.

Michael, A.M. 1969. New water supplies from rainfall harvesting. Presented at symposium on planning of drought areas under the auspices of NISI, New Delhi, May 4-6.

Mishra, D. and Pradhan, G.C. 1972. Effect of transpiration reducing chemicals on growth, flowering and stomatal opening of tomato plants. *Plant Physiol.* **50**: 271-274.

Myers, L.E. 1967. Recent advances in water harvesting *J. Soil* and *Water Cons*. **22**: 95-97.

Parija, P and Pillay, K.P. 1945. Effect of presowing treatment on the drought resistance in rice. *Proc. Nat. Acad. Sci. India.* **15**: 6-14.

Pastornak, D and Wilsen, G. D. 1971. Some factors responsible for varying effectiveness of stomatal closing antitranspirants. Australian *J. Expt. Agri. An. Hub.* **11 :** 48-52.

Prakash, M; K. Kannan; J.S. Kumar; B. Bharathi Veermani & J. Ganesham 1999. Effect of brassino-stcroids on germination and seedling behaviour of tomato. Ann. Plant Physiol. **13(2):** 178-180.

R. Moratiel and A. Martı´nez-Cob. (2012). Evapotranspiration of grapevine trained to a gable trellis system under netting and black plastic mulching. *Irrig Sci* **30:167**–178

Robert Parker. (2003). Water Conservation, Weed Control Go Hand in Hand. Drought Advisory, Washington State University Cooperative Extension and the U.S. Department of Agriculture (EM4856).

Simonin, K., T.E. Kolb, M. Montes-Helu, G.W. Koch. (2007). The influence of thinning on components of stand water balance in a ponderosa pine forest stand during and after extreme drought. *Agricultural and Forest Meteorology* **143, 266**–276.

Slatyer, R.O. and J.F. Bierhuizon 1964. The effect of several foliar sprays on transpiration and water use efficiency of cotton plants. *Agri. Meteorol.* **1:** 42-53.

Subrat Kumar Mishra. (2012). Watershed Management: Issues and Concerns of Drought Prone Areas. *Odisha Review*, **April**.

T. Yaghi, A. Arslan, F. Naoum. (2013). Cucumber (*Cucumis sativus* L.) water use efficiency (WUE) under plastic mulch and drip irrigation. *Agricultural Water Management* **128, 149**– 157.

Tony H.H. Chen and Norio Murata. Glycinebetaine: an effective protectant against abiotic stress in plants. *Trends in Plant Science* **Vol.13** No.9.

Yajun Wang, Zhongkui Xie, Sukhdev S. Malhi, Cecil L. Vera, Yubao Zhang, Jinniu Wang. (2009). Effects of rainfall harvesting and mulching technologies on water use efficiency and crop yield in the semi-arid Loess Plateau, China. *Agricultural Water Management* **96, 374** – 382.

Zhongkui Xie, Yajun Wang, Wenlan Jiang, Xinghu Wei. (2005). Evaporation and evapotranspiration in a watermelon field mulched with gravel of different sizes in northwest China. *Agricultural Water Management* **81, 173**–184.

Chapter - 18

WEED MANAGEMENT IN DRYLAND FARMING

A. Muthusankarnarayanan & B.J. Pandian

Professor of Agronomy, Agriculture College and Research Institute, Killikulum, Vallanad (TN).

I. INTRODUCTION

Weeds are unwanted and undesirable plants that interfere with the utilization of land and water resource and thus adversely affect crop production and human welfare. India is loosing Rs. 6000/- crores worth of Agricultural production due to pests, of which weeds alone cause one third of loss amounting to Rs. 1980 crores (Gambhir, 1987). Weeds, however, have always been rather casually accepted as an nuisance both for dryland farming community and policy makers. Weeds compete with crop plants for nutrients, soil moisture, space and sunlight. Generally an increase in one kilogram of weed growth leads to reduction in one kilogram of crop growth. Weeds remove plant nutrient more efficiently than crop plants.

Most of the weed management research and applications have been conducted under assured water supply areas. Immense potentiality exists for increasing productivity in dryland areas by adopting proper weed management technology and developing economically viable technology for managing troublesome weeds in dryland crops and cropping systems. In dryland farming, adoption of monoculture rarely exists and the cropping system approach, i.e, either inter cropping or mixed cropping is in vogue in many parts of India. Most of research works were carried out on the weed management for individual crop cultivation in dry land areas but the weed management under cropping system approach needs to be intensified.

The cropping in dryland areas will be successful when the agricultural operations are attempted in time by utilising the rainfall. Like wise, weeding in dry land crops also requires sufficient soil moisture. Human labour weeding in dryland areas is restricted as the available labourers is quite inadequate to meet the demand. Further efficiency of mechanical or human weeding are reduced due to unpredicted rainfall that leads to excessive soil moisture. Hence, the option left for dryland farmers is either to adopt chemical weeding or any cultural methods for effective weed control. In this chapter, available weed management technology for dryland farming studies is discussed.

II. WEED FLORA IN DRYLANDS

Weed problems in dryland crops and cropping system vary from one vegetation zone to another. Even within the same zones, variation in weeds as well as intensity of weed infestation occurs because of differences in cropping systems, intensity of land use, rainfall, soil fertility and methods of weed control adopted. The weed plant growth, architecture, morphology, growth cycle and canopy development varied with the environment and due to competition in crops and cropping system.

Some perennial weeds (Eg. *Cyperus rotundus* L., *Cynodon dactylon* (L.) Pears; *Sorghum halepens* (L.) Pers. pose a serious problem and due to that they are very competitive and difficult to control when regular crop husbandary practices are employed. The broad leaved perennials eg. *Convolvulus arvensis* L., and *Solanum elaegnifolium* Cav. *Moninda tinctoria* (a shrub) are also some of the most pernicious weeds which are difficult to control. The parasitic weeds like *Orbanche* spp. are widely spread in semi-arid regions damanging the tobacco crop and some legume vegetables. *Striga* spp. a root parasite are restricted to hotter arid regions affecting cereals viz., sorghum and maize. The stem parasitic weed *Cuscuta* spp., is also becoming a major problem in crops like greengram and blackgram that reduces the growth posing a serious problem in certain vegetation. In pearl millet and in cotton *Asphodelum tenuifolius* (L.) are found in large scale in northern belts of the country. Incidence of *Ageratum conzoides* (L.). is also reported to be a major problematic weed in

maize and also in other major crops in dry farming. *Trianthema portulagastrum* L. and *Boerhaavia diffusa, Celosia argentea* (L.)., *Phillanthus madaraspatensis, Cyanotis cuculliata* Kunth. Alysicarpus vaginalis (L.), Dc. A. rugosus (Will) Dc. *Aristolochic brachata* (L.) Retz. *Panicum* spp. *Elusine indica* (L.) Gaertn, *Euphorbia* spp. *Sorghum haleapense* (L.) Pers. are the major weeds under blacksoil dry farming areas of southern tracts. *Leucas utricaefolia, Merremia emerginata, Physalis minima, Tephrosia spinoses, T. purponea panicum* spp., *Eleusine indica, Chloris barbata* are the major weeds in red soils of southern tract of the country. The weeds like *Boerbaavia diffusa, Phyllanthus madaraspetensis, Digera arvensis, Cyonotis cuculliata, Elcusine indica, Panicum* spp., *Euphorbia hirta* are also found in drylands.

III. WEED COMPETITION IN DRY FARMING REGION

(a) Competition for Nutrients

In dry land areas, during early stages of crop growth, weeds compete for nutrients. A weed produces abundant foliage in the early stages, they compete more for light and moisture, Weeds are used to absorb nutrients from soil more rapidly and relatively in larger amount than does the crop. The nutrients taken up by maize and its associated weeds viz., *Amaranthus* spp. *Portulaca* spp., *Digitaria* spp. showed that weeds contain high concentration of nutrients (Table 1). The weeds biomass may account for 50 per cent of the total biomass in maize field causing around 64 to 83 per cent of nutrients lowered by weeds (Sankaran and Subbaiah Mudaliar, 1993). In groundnut (Table 3), the nutrients absorbed by weeds are considerable, especially phosphorus and potassium which are more than 80 percent of total nutrient absorbed by the crop and its associated weeds.

Table 1. Chemical composition of Maize and associated weeds.

Plants	Nutrient Content (%)				
	N	P	K	Ca	Mg
Maize	1.20	0.21	1.19	0.16	0.15
Amaranthus	2.61	0.40	3.86	1.63	0.44
Portulaca	2.39	0.30	7.31	1.51	0.64

Table 2. Dry matter and nutrient uptake of crop and associated weed in Maize.

Plants	Dry matters	Nutrient uptake Kg/ha				
		N	P	K	Ca	Mg
Maize	6195	30.2	5.49	62.8	14.06	5.98
Weeds	6171	55.0	9.69	297.5	43.45	39.94
Total nutrients	—	85.2	15.18	36.2	57.51	45.92
Percentage to total nutrient taken by weeds	—	64	64	83	76	83

Table 3. Nutrient uptake by Groundnut and its associated weeds.

Plant	Nutrient uptake Kg/ha		
	N	P_2O_5	K_2O
Groundnut	48.27	4.41	8.06
Weeds	21.23	28.33	38.86
Total	69.50	32.74	46.92
Percentage to total of nutrient taken by weeds	30.50	86.53	82.80

NPK content of common weeds and crops are compiled (Table 4)

(b) Competition for Light

Generally, weeds grow faster and covers the crop plants if not checked. Even in the case of shorter weeds, the lower leaves of crops are shaded by them. Weeds deplete photosynthetically active radiation resulting in reduction of the photosynthesis and shortening the life of lower leaves. There are instances where *Amaranthus* spp. reduced the photon flux density reaching the cotton canopy by 90 per cent at mid day. The light intensity reaching maize leaves may be only around 56 to 63 per cent in cloudy weather and 50 to 57 per cent in sunny weather due to the interference of tall plant of *Sorghum helpense* (L.) Pers. Due to shading of crop plants, they produce thin pale green leaves with reduced number of vascular bundles and meristem cells.

Table 4. Nutrient composition in different crops and weed spread

Plant Species	Nutrient on dry weight basis %		
	N	P_2O_5	K_2O
Crops :			
Paddy	1.13	0.33	1.15
Wheat	1.34	0.66	1.50
Maize	1.21	0.20	1.18
Sorghum	0.41	0.24	2.16
Pearl millet	0.64	0.74	2.50
Blackgram	1.35	0.52	1.22

Greengram	1.24	0.49	1.21
Mustard	0.76	0.30	0.62
Gram	1.18	0.43	1.10
Weeds :			
Echinochloa coloum (L.) Link.	2.95	0.65	2.50
Digeria arvensis Forsk	3.25	1.65	3.15
Chenopodum album L.	3.50	1.40	3.45
Phalaris minor Retz	1.78	0.84	1.90
Cynodon dectylon (L.) Pers.	2.05	1.45	1.25
Cyperus rotundus L.	1.65	1.50	1.24

(Adopted from Rao and Agarwal, 1984).

(c) Competition for Moisture

Research has established that competition for moisture usually occurs along with other forms of competition. It becomes critical with increasing soil moisture stress as found in arid and semiarid areas. In general, for producing equal amounts of dry matter, weeds transpire more water than do most of the crop plants. Therefore in drylands, the actual evpotranspiration from weedy crop field is much more than that from weed-free crop field (Kanitkar *et al.,* 1960). Transpiration coefficient of weeds ranged from 221 (*Tribulus terrestris*) to 1402 (*Tridax procumbense*) while that of maize and sorghum was 352 and 394 respectively (Kanitkar *et al.,* 1960). The transpiration ratio of *Cynodon dactylon* (L.) Pers is around 813 compared to 450 for pearl millet and 430 for sorghum. Weeds consume water luxuriently and exhaust soil moisture quickly under dryland condition. The leaf water potential of cotton falls by 5.3 bars if *Ameranthus* spp. occurs at a population density of 2.5 plants / m^2. The *Ameranthus* spp. maintains higher leaf water potential and targur potential than cotton throughout the growth period by depleting moisture from deeper layers of soils. Weeds also consume valuable stored soil moisture from fallow fields. Weeds growing in fallow fields consume as much as 7 to 12 ha cm of soil moisture which is sufficient to produce 1.5 to 2.0 t/ha of sorghum grain. Another study at Rohtak revealed that the amount of water remaining in the soil is 5 ha cm when it was infested with weeds in the previous season compared to weed-free fallow which has 10 ha cm of water.

(d) Competition for Carbon-dioxide

Competition for carbon-di-oxide may occur when the weed infestation is high. Since, most of the weeds are C4 plants, they can deplete CO_2 to a very low level compared to C_3 crops. When CO_2 content is increased artificially from 350 ppm to 700 ppm, the competitive ability of C_3 crop like soybean increased compared to its associated C_4 weed *Sorghum halepense* (L.) Pers.

IV. CRITICAL PERIOD OF CROP-WEED COMPETITION

Critical period of crop weed competition can be defined as time span before and after which weed competition does not reduce the crop yield. Critical periods vary from crop to crop depending upon competitiveness of weeds, environment, location, plant density, soil fertility and crop variety etc. In general 4-6 week period is more crucial from the competitive point of view in almost all the major crops. Kasassian and Seeyave (1969) proposed that all crops need a weed free period of one fourth to one third of their growing season. The critical period of important dryland crops is given below :

Table 5. Critical period of weed competition and percentage of yield reduction

Crop	Critical period DAS	% of yield reduction
Rice (upland)	10-40	60-80
Maize	15-40	40-60
Sorghum and millets	15-35	30-40
Cotton	30-60	45-85
Groundnut	25-35	55-70 (erect)
Groundnut	25-35	20-25 (spreading)
Sunflower	15-35	30-50
Sesamum	15-30	45-50
Soybean	20-45	25-35
Greengram	20-45	50-55
Blackgram	25-35	45-50

Source : AICRPWC comprehensive report for the period for the period of 1978-84 and Singh and Saraswat, 1986.

(a) Factors Influencing the Critical Period of Weed Competition in Dryland

The critical period of Crop-weed competition starts with the beginning of interference from weeds and ends when the crop covers at least 80% of the soil. The length of critical period of weed competition depends on the nature of crop, its competitive ability, variety and growing condition.

(b) Nature of Crop/Variety

Certain crops like sorghum, maize and sunflower grow faster and cover the field quickly. Hence, the length of critical period of weed growth is short. In other crops like cotton whose initial growth is slow and as they are widely spaced, they take longer time to cover the soil. In such situation the critical period is longer.

Tall growing traditional varieties with profuse tillering and more foliage cover the ground earlier and suppress the weeds effectively. Erect growing short statured plant allow more light to penetrate initially favouring emergence of weed for longer periods. Critical period of weed competition is therefore longer for dwarf varieties.

In drylands, the critical period of crop weed competition is also based on the availability moisture required for the crop. If moisture is adequate in the early stage of the sorghum crop, the crop growth will be at higher rate and suppress the weed growth and hence the competition period will be shorter. If the moisture availability is lesser during the early stage, the canopy coverage will be lesser but the growth of drought resistant weeds will be more which leads to compete crops growth and at this condition crop weed competition will be longer for the crop.

V. WEED PROBLEM AND ITS EFFECT ON CROP

Weeds deplete the crop's share of resources like nutrients moisture and light and thereby reduce the economic yield considerably. In addition weeds interfere in agricultural operations and also reduce the quality of produce.

(a) Yield loss due to Weed

The crop yield loss due to weeds depends on intensity of infestation, time of occurrence and type of weeds when winter wheat was infested heavily (435 weeds/m^2) with wild oats (*Avena fatua*), the yield loss occurred mainly due to reduction in tiller number and grain No. per panicle. Seed yield of soybean decreased linearly with duration of weed competition from 45 to 90 per cent in wet season. The yield losses in upland rice due to weeds infestation was higher and ranged from 66 to 9.28 per cent compared to transplanted rice which was lower (45.5 to 63.9 per cent). Rainfed chilli was most sensitive crop to weed competition in early stages and the yield loss is higher, if not weeded. The average losses due to weeds in different crops were 30 to 40 per cent in maize, soybean, vegetables and fodder.

(b) Interference

Weeds not only reduce the yield of crops, but also interfere with agricultural operations. Sowing operation in drylands with seed drill is not possible, if the weed infestation is at high density as they block free flow of seeds. The preparatory cultivation will be costlier when the weed infestation was heavy in previous crop. In black soil dryland areas, in the southern districts of Tamil Nadu, the *Morinda dinctoria* growth was heavy with deep roots during summer which interfered with the off season/seasonal tillage operations and preparatory cultivation. Some times the plough share may be broken at ploughing due to this morinda roots.

VI. WEED CONTROL METHODS IN DRY FARMING

Weed control methods are grouped into cultural, physical, chemical and biological. Every method of weed control has its own advantages and disadvantages. No single method is successful under all weed situations. Many time, a combination of these methods gives effective and economic control than single method.

(a) Cultural Methods of Weed Control

Cultural operations like tillage, planting method, after cultivations fertilizer applications etc., are employed for creating favourable condition for the crop for getting higher yield. These cultural methods alone will not help to control weeds, but help to reduce the weed population. In addition to cultural methods to control weeds, selection of variety, cropping system, time of sowing and field sanitation are also important in checking the weed growth.

(i) Preparation for Cultivations

Field preparation in drylands, has to be done for the crop requirement, soil moisture conservation and also control the weeds. During the summer season deep ploughing exposes the underground parts likes rhizomes, tubers of perennial and obnoxious weeds to scorching summer sun and kill them. It is one of the effective cultural methods to check the growth of perennial weed populations. Initially before cropping, one should encourage clod formation. These clods, which have the weed propogules get desiccated upon drying. Subsequent tillage operations will break the clods into small units to further expose to hot sun.

The off season tillage is being practiced in many parts of the drylands in India. This operation brings the soil free from weeds and helps to conserve the moisture. Deep ploughing with disc and leaving the fields in a cloudy condition will help better absorption of rain water. This off season tillage eradicates the perennial weeds like *Cynodon dactylon*. The preparation of field after the summer or off season ploughing also provides better seed bed.

In the alfisols of semi-arid Telengana, off season ploughing followed by bladding was found to reduce the weed density. The effect was more pronounced in sub-normal rainfall years. Such system also permits early sowing, and also the operation weakens weed plants through injury or root and stem prunning, reducing their competitiveness or regenerating capacity. This tillage operations also burried the weeds. Tillage operations at the sowing includes ploughing, discing, harrowing and leveling which promote the germination of weeds through soil turnover and exposure of seeds to sunlight which can be destroyed effectively later. In case of perennials, both top and under ground growth is injured and destroyed by tillage. Tillage during the dry season keeps the soil free of weeds and helps for the conservation of soil moisture, Krantz (1979).

(ii) Minimum and zero tillage

Frequent cultivation of the field is not possible because of the time factor and minimum and zero tillage techniques; former reduction is turned around time, thereby result in early crop establishment. Rangaiah (1989) observed a small increase in weed dry weight under minimum tillage (the land was tilled only once) compared to conventional tillage in a cotton-sorghum ragi systems. Weeds were effectively checked by pre-emergence herbicide followed by a late weedling. This technique reduced the turn around time, cost of land preparation and labour requirement. In zero tillage system, the rotation from the first crop will act as weed and may offer severe competition to the second crop (ICRISAT; 1980-81). Herbicides kill weeds in the planted crop and at the same time also kill the previous ratooning crop. Further research is needed on the role of herbicides in minimum and zero tillage; and the physiological basis for yield increase due to reduction in turns around time and reduced tillage needs to be further understood.

A *stale seed* bed is one where initial one or two flushes of weeds are destroyed before the planting/sowing of a crop. In well prepared summer or off season ploughed fields, on receipt of rain, the weeds will germinate. At this stage, shallow tillage may be given or non-residual herbicide like paraquat may be used to destroy the dense flush of growing weed seedlings. This technique allows the crop to germinate in almost weed free environment. This method of filed preparation is being adopted for raising post rainy season crops i.e. Rabi crops in northern states, on the cessation of south west monsoon for sorghum, sesamum, mustard and safflower, etc., and for late Rabi crops in southern States after the cessation of north east monsoon for coriander, bengal gram and horse gram. Here the fields will be ploughed with the receipt of every rain during the monsoon and destroy the sprouted weed seedlings.

(iii) Seeds and Sowing

Using weed free clean seed bed can reduce the weed intensity. It is a preventive method against the introduction of weeds. In dry lands, sowing operations are taken up one to three days after the receipt of rain depending upon soil type and quantum of rainfall received. If the sowing is delayed, the weed seeds will germinate and control them by working with blade harrow by uprooting the germinated weed seedlings. In addition, due to loosening of surface soil, weeds will dry up quickly and does not allow weed seeds to germinate until subsequent rains.

(iv) Blind Tillage

The tillage of the soil after sowing a crop before the crop plants emerge is known a blind tillage. It is extensively employed to minimise weed intensity in drill sowing crops where emergence of crop seedling is hindered by soil crust formed on receipt of rain or irrigation immediately after sowing. In upland rice cultivation in Southern States, the fields are prepared well in advance before the monsoon rains and make the soil more friable. The seeds of rice are sown either by drilling or sown behind the country plough. After the third day onwards wooden planks are worked on the field for levelling and compaction. This wooden planks are also being worked on 5th, 7th and 9th day of sowing. By this operation, the germinated weeds by compaction are controlled and the paddy seedlings emerged will not be affected and uniform / germination is obtained because of compaction of soil and levelling.

(v) Crop Density

It is one of the cultural method of weed control. Closer planting/sowing suppresses the weed growth. For example in Southern districts of Tamil Nadu, during north east monsoon season, fodder sorghum (Irungu types) are used to sow with high seed rate of 150 kg ha^{-1}, so as to get thin stemmed fodder sorghum for cattle feeding. It was observed that dense population of fodder sorghum will be faster in growth and early stages will suppress the weed growth. It was observed even in the nut grass and growth found arrested in other germinated weed species.

Increasing the seed rate from 40 to 100 for direct seeded rice decreased weed weight to 18 g/m^2 from 52 g/m^2 and increased the yield in rice. Further increase in seed rate from 120 to 160 kg ha^{-1} was not effective in further reducing the weed weight and increasing the yield. In vegetable pea, the weed number and weed DMP were reduced when close row spacing of 30 cm was adopted. In finger millet 10 x 10 cm planted crop suppressed the weed growth at higher rate compared to 15 x 10 cm under rainfed condition.

(vi) Crop Sequence

The possibilities of a certain weed species or group of species occurring in greater intensity are more if same crop is grown year after year. In many instances, the crop rotation can eliminate or atleast reduce the weed problem. The intensive cultivation results in shift in weed flora towards more persistence. Perennials such as *Sorghum halepens, Cynodon dactylon, Cyperus rotundus, Panicum maximum* and *Paspalam* sp. have become threat to crop production in drylands. A shift in broad leaved weeds to perennial sedges and grasses was noticed by Rangiah (1989). The rotation of crops having dissimilar life cycle or cultural condition is an effective method of weed control. By crop rotation, the possibility of build up of certain weed species is usually reduced.

The sequential cropping aims at growing two or more crops in sequence on the same field in a year. The problems of weed management in a sequential cropping are different from those in intercropping. Continuous presence of crop cover, residual toxicity of herbicides applied to the previous crop and change in weed flora need a different approach in weed management. Apart from optimum weed control, agronomic considerations which would determine the success of sequential cropping include early crop establishment, time and method of tillage, date and method of planting and harvesting, crop rotation and length of the turn around time (Moody, 1976). The obnoxious weeds like *Cyperus rotundus* can be effectively controlled by crop rotation growing sorghum with thick population using high seed rate (100 kg ha^{-1}). in rainfed vertisol of Tamilnadu. The parasitic weed striga in rainfed sorghum can be effectively controlled by growing sunflower followed by sorghum (Muthusankaranarayanan, 1983). When an upland crop is rotated with a low-land-crop, the total number of weeds is lowered (Moody, 1977). Gill and Brar (1972) observed *Phalaris minor* weed, is a relatively minor weed when tall traditional wheat was grown as a single crop and is increasing at alarming rates in an intensive rice wheat rotation system. In the sequential cropping system, if the weeds were allowed to grow later in the season of the first crop it may cause a serious threat to the second crop apart from difficulty, they impose during the harvest of the first crop and land preparation for the second crop.

(vii) Growing Intercrops

Inter cropping suppresses weeds better than sole cropping and thus provides an opportunity to utilize crop themselves as tools of weed management. Many short durations pulses viz., green gram, cowpea and soybean effectively smother the weeds without causing reduction in the yield of main-crop.

When the intensive cropping/ intercropping is followed in rainfed situation, mechanical weeding may be difficult or even become impossible in certain spatial arrangements or row spacing of the component crops which are too close to each other. Comparing the ability of intercrops and sole crops in competing with weeds, it has been reported that intercrops are better than either or both of the component crops as sole crops in weed suppression. Experiments conducted at ICRISAT, Hyderabad, indicated 50-70% reductions in weed infestation by intercropping (Shetty and Rao, 1977). Although intercropping may reduce weed infestation and growth, there is still a need for some weeding in most cases. But the labour input for weeding is reduced because of limited weed growth. The results obtained by Rao and Shetty (1976) in sorghum/pegionpea intercropping and by Moody (1980) and Shetty and Rao (1981) in sorghum/mungbean intercropping systems indicated that one weeding would be sufficient to get yields equal to weed free control.

Inter crop of slow growing crop with fast growing crop is an effective cropping system for reducing weed competition for cotton based intercropping system. (Chhokar and Balyan, 1996), cotton intercropped with greengram recorded lesser weed population and dry matter (AICRPDA 1982 and 1983). Here the intercrop suppressed the weeds in cotton + green-gram intercropping systems and gave additional yields. Solaiappan and Chelliah (1998) found in cotton, the dry matter production of weeds in cotton + blackgram intercropping system significantly reduced the weed density compared to sole cotton and cotton + cluster beans in dry land vertisols. The weed DMP reduction was 12.5 percent and they considered that this weed intensity reduction might be due to early growth and shading effect of blackgram. They also found that the cotton + cluster beans system reduced the weed DMP at the harvest stage than in cotton + blackgram. The intercropping with legume crop is also effective in controlling the weeds (Chatterjee and Mandal, 1992, Thakur, 1994). Weed infestation was considerably lesser under intercrops than in sole crops. Thakur (1994) reported that suppression effect of weed infestation was higher in maize + groundnut intercropping system and in maize + horse gram than in maize + Black gram and Maize + soybean system. In the pigeonpea + sesamum intercropping system, pigeonpea + 100% of sesamum population had the maximum weed smothering efficiency of 22.5% than with 50% sesame population (8.3%) (Singh and Singh, 1993).

Crop differs in relation to growth rates, spreading habit, height, canopy structure and inherent competitive character and accordingly differ in their weed suppressing ability. Pearlmillet and maize showed, high initial weed smothering ability followed by cowpea and groundnut. Pigeonpea and castor have poor competitions with weeds. Weed growth in compact pigeonpea genotypes was higher than that observed in spreading type. In sole cropping systems, Rao and Shetty (1976) recorded decreased weed dry weight with increased crop population. The spread of higher yielding wheat varieties is contributing towards dissemination of *Phalaris minor* Retz. in wheat growing states, causing major weed problems in wheat and other *Rabi* crops. The tall wheat genotypes exerted a strong suppressing effect on weed intensity and growth of *P. minor* and *Chenopodum album* L. (Gill and Mehra, 1981).

Inter cropping was proved to be superior to component crops in weed suppressing ability (Shetty and Rao, 1977 and 1981) and thus it provided an opportunity to utilize crops themselves as tools of weed management. The weed suppressing ability of intercrops is dependent on many factors such as the component crops selected, the genotypes used, plant density adopted, proportion of component crop, their special arrangement and fertility and moisture status of soil (Moody, 1980, and Moody and Shetty, 1981). The combined density of sole and intercrops if increased, there is likelyhood of reduction in weed infestation. Rao and Shetty (1976) reported that the weight of weeds growing in association with intercrops such as maize/mungbean, maize/ soybean, sorghum/ pegeonpea, sorghum/ cowpea and soybean/mungbean is as low or lower than that growing in association with the sole crops. Light interception patterns and leaf area index (LAI) observations revealed that inclusion of smother crops viz., cowpea and mungbean resulted in quicker and earlier attenuation of maximum LAI and maximum percentage of light interception by component crops. The growth habit of the genotypes used in intercropping has great influence on weed suppression. Early canopy coverage seems to be more important than rapid increase in height. Mungbean Cv. MG 50 intercropped in maize competed with weeds better than CES 14, Since, MG 50 has a better early lateral growth (Bantilan *et al.,* 1974). Spreading genotype of pigeonpea; ST 1 was found to be more competitive than a compact genotypes HY 3C against weeds under sole cropping. However, when intercropped with sorghum the genotype differences in weed suppressing ability were not significant (Rao and Shetty, 1976).

Weed competition usually increases with increase in soil fertility. Bantilan *et al.,* (1974) observed that intercrops were less effective in controlling weeds at high nitrogen levels than at low levels. Similarly, in a naturally fertile vertisol more weed growth was noticed than in a comparatively low fertilie alfisol (ICRISAT, 1977). Tripathi and Singh (1983) studied the density of weeds in soybean intercropping with maize and found that across fertilizer levels, the total weight of crop seeds produced and competitive suppression of weeds were positively correlated. Sandhu and Gill (1973) showed that weedy-maize yield decreased as the competition period increased at high fertility, but not at low fertility.

Besides the direct competition, some of the weeds exert chemical stress on crops by these *allelo-chemicals* phyto-toxic root excudate and plant leachates secreted or excreted into the soil. Lall and Sevongdy (1981) reported that root exudates of plants and leachates of plant residues of purple nutsedge caused significant reduction in plant height and dry matter productions of pearlmillet. Roots of *Lolium temulentum* and roots of *Phalari's minor* and *Avena fatua* had inhibitory effect on growth, and growth characteristics of wheat (Bansel and Singh, 1986).

Mulching is a protective covering of material maintained on soil surface. Mulches when applied on soil surface don't allow weeds to germinate or to grow the germinated weeds as light does not reach the soil. Weed control through organic mulch on other hand is ecofriendly in nature and leaves no residual toxicity to the environment. Mulching is very active against some perennial and annual weeds like bermuda grass (*Cynodon dactylon*) and *Sorghum halepense*. Mulches also provide barrier against emerging weeds and thus post emergence weeds control ability. Rice straw mulch at 16 t/ha reduced the grass weeds to a maximum of 89% broad leaved weeds (71%) and sedge weed (91%).

Ghorai and Bera (1998) found that there was a reduction in weed population by 23% as against unweeded control and the grain yield of soybean also increased by 93%. Use of mulches in dryland has an additional advantages of moisture conservation besides suppressing weed biomass (Chhokar and Balyan, 1996). Both subabul and bajra straw mulching had suppressed the weed density and weed DMP compared to no mulch throughout the growth stage of rainfed castor at Kovilpatti. (Table 6). The reduction in weed DMP at harvest stage with bajra straw mulching was 56.3 and 35.7 percent as compared to unmulched control. The subabul loppings and bajra straw had identical effect on controlling the weed growth both at initial and at harvest stages (Solaiappan and Chelliah, 1998). The lesser weed growth in mulched plot was also recorded by Shairah *et al.* (1994) under rainfed situation.

Table 6. Density and DMP of weeds as influenced by mulch in rainfed cotton.

Treatments	Weed No/m^2		Weed DMP kg ha^{-1}	
	45 DAS	Harvest	45 DAS	Harvest
No mulch	57.5	28.9	271	210
Subabul mulch	41.0	19.4	191	141
Bajra straw mulch	39.9	18.6	182	135
CD (P = 0.05)	4.6	2.2	16.1	18.8

(viii) Solarisation

This is another method of utilization of solar energy for desiccation of weeds. In this method, soil temperature is further raised by 5 to 10ºC by having a pre-soaked fallow field with thin plastic sheet. The plastic sheet checks the large wave length back radiation from the soil and also prevent loss of energy by hindering moisture evaporation.

(ix) Fertilizer Application

Plants differ in their capacity to respond to fertilizer application. Crops like sorghum, maize, pearlmillet and rice grow at faster rate, when nitrogenous fertilizers are applied and cover the soil earlier. Weeds like *Cynodon dactylon and Cyperus rotundus* do not respond to nitrogen application and they are suppressed by fast growing crops. If the weeds present are responsive, the applied nitrogen did not influence weed population or weed weight in maize at Bangalore.

(b) Mechanical Weed Control

Mechanical and physical method of weed control are being employed ever since man began to grow crops. The mechanical methods include hoeing, hand weeding, digging, sickling, moving, burning, flooding, etc.

(i) Dryland Weeders (manually Operated)

There are different types of weeders developed for weeding and hoeing in row crops. They are :

Peg Type

Dryland weeder with long handle is suitable for weeding in row crops. It can be easily operated by a man or woman. This weeder is useful and economic under ideal soil conditions i.e., when the soil moisture content is about 10 per cent. It performs well in plain fields where the weeds are shallow rooted and the land is not too thickly infested with weeds. The blade can be adjusted to the desired angle and depth. The peg teeth permit the movement of the roller in clay soil without getting clogged.

Gogoi (1997) evaluated the different dryland weeders comparing with grubber and twin wheel hoe for wheat and found that the grubber recorded the maximum performance index than the peg type and blade type weeders.

Star Type

The function and operation are similar to the peg type weeder excepting that the star type roller facilitates easier operation of the weeder in loamy and sandy soils. Both tools can cover an area of 0.2 ha per day and the cost of operation per ha is around Rs. 36. The cost of tool is Rs. 100/-each approximately.

Abhijit Sarma and Gogoi (1995) have studied the ability of five manually operated weeders in rainfed rice and compared with butachlor 1.5 kg ha^{-1} + hand weeding of 35 DAE weed free and weedy check in rainfed rice and reported that pre-emergence application of butachlor 1.5 kg ha^{-1} + hand weeding at 35 DAE name considerably reduce population dry weight and, growth rate of weeds.

Table 7. Total weed count/m^2 as influenced by different weed control treatments

		Treatments	1993		
			15 DAE	30 DAE	45 DAE
T1	–	Weedy	13.9	19.0	20.0
T2	–	Weed free	(19.1)	(361)	(401)
T3	–	DLW (peg type) at 20 DAE	3.2	4.1	4.9
T4	–	DLW (peg type) at 20 and 35 DAE	(9)	(16)	(23)
T5	–	DLW (Blade type) at 20 DAE	13.5	14.9	16.6
T6	–	DLW (Blade type) at 20 and 35 DAE	13.1	16.7	14.2
T7	–	TWH at 20 DAE	13.7	16.2	17.4
T8	–	TWH at 20 and 35 DAE	13.5	16.4	13.5
T9	–	IITWAM-82 at 20 DAE	13.8	15.7	16.6
T10	–	IITWAM-82 at 20 and 35 DAE	13.6	15.50	12.9
T11	–	Modified IITWAM 82 at 20 and 35 DAE	13.4	17.2	18.3
T12	–	Modified IITWAM-82 at 20 and 35 DAE	13.9	17.1	14.6
T13	–	Butachlor 1.5 kg ha^{-1} H.W. at 35 DAE	7.1	15.7	10.7
	–	CD at 5%	1.0	1.0	1.6

Source : Abhijit Sarma and Gogoi, (1995)

Among the weeder dry land weeder peg type. (20 and 35 DAE) and 11-TWAM - 82 (20 and 35 DAE) showed better weed control ability in respect of reducing weed population, weed dry weight and weed growth rate.

(ii) The Guntaka

It is the blade harrow that is commonly used for primary tillage in the ceded districts, instead of the plough. It is very simple in construction. A horizontal beam of about 15 cm diameter is the main frame, to which are fixed the handle, the shaft pole and the harrow blade. The blade is fixed to the beam near the ends, at a distance of about 25 cm by means of two standards. The blade is 1.0 m long, about 7.5 cm broad and 1.25 cm thick, with a cutting edge in front. Big size guntakas, 1.8 m wide, called 'bara guntakas' are also in use. Small guntakas, called 'danties' are largely used for intercultivating line sown crops. The danties are 15 to 33 cm wide and are suitable for crops and spaced 28 to 46 cm apart. Five or six danties are attached to a common yoke. The ploughman guides one of them and rest are guided by women, boys and girls. These implements cover about 0.4 hectare a day of 8 hours, for each 0.3 metre width of strip is covered by the implement.

(iii) Harrows

Some of the common types of intercultivating blade harrows are in use in ceded districts. The depth of penetration of the blade harrows into the soil is regulated by the distance between the blade and the yoke. When it is increased, the angle made by the blade with the land is increased and the blade tends to go deep into the soil. Conversely, when the blade is nearer, the yoke takes less depth. There are some improved models of blade harrows, whose depth of working is adjusted by modifying the angle of the shaft pole. Their high cost is against their general adoption by the ordinary farmers.

(c) Chemical Weeding in Drylands

Selective herbicides are available for sole crops and effective weed control can be obtained by using herbicides. But the residual effects of these herbicides have to be carefully evaluated before using them in the crop sequence. There are reports on the hazards of herbicides in the succeeding crop especially if legumes are planted after cereals. Atrazine applied at 0.5 kg ha^{-1} as pre-emergence gave effective weed control in sorghum but the establishment of legumes crops such as mungbean or groundnut which followed sorghum was poor but following cotton was not affected (Palaniappan and Ramasamy, 1976). However atrazone, diuron and amiben applied to maize had no harmful residual effect on the succeeding crop of soybean (Hugar and Hosmani, 1977). Atrazine applied at 1, 2 and 4 kg ha^{-1} and alachlor at 1, 2 and 4 kg as pre -emergence spray to maize gave good control of weeds in maize and left no harmful residues that control effect emergence, growth or yield of peas, chickpea and lentil (Singh and Mani, 1981).

In sequential cropping system, the weeds allowed to grow later in the season in the first crop may cause a serious threat to the second crop, apart from the difficulty they impose during the harvest of the first crop and land preparation for the second crop.

(d) Integrated Weed Management

Among the weed control methods namely chemical, mechanical, biological and cultural methods, no one method is superior to the others under all situations. Too much reliance on any one method of control at the exclusion of other method will not be successful and economical. The mechanical methods such as hoeing and weeding should follow herbicides to the extent necessary to remove persistent weeds inspite of herbicide application. Too much of reliance in herbicide at the cost of other mechanical measures will not yield desired results. The chemical weeding use in dryland will be effective when there is optimum moisture in the soil at the time of application. The herbicides like fluchloralin which gets photo decomposed, when applied as incorporation will have more effective weed control. For the incorporation it has to be done before sowing to avoid photo-oxidation which can reduce the activity of herbicide. The integeration of cultural, mechanical chemical and biological methods of weed control will have better effect on weed control and is eco-friendly approach for weed control.

VII. WEED CONTROL/MANAGEMENT PRACTICES FOR DRYLAND CROPS

(a) Rice (upland)

Direct seeded rice under rainfed upland situation is prone to severe weed infestation due to aerotic soil conditions with optimum temperatures. Besides weed competition is severe in during and slow growing rice high yielding varieties. Very high yield losses (upto 97%) due to unchecked weed growth were reported (Saxena *et al.,* 1990). By establishing higher crop stand and crop coverage by the crop, it is possible to minimise the weed growth to a considerable extent. Use of higher seed rates may increase the crop stand and improve canopy. The major constraints of growing rice in upland situation include severe weed competition, nutrient loss due to coarse textured soil, inadequate crop stand, poor growth due to low input use and occasional drought. In this ecosystem unchecked weed compete with rice plants for light, nutrient and moisture resulting the grain yield loss upto 80% (Behera and Jha, 1992). Weeding the upland rice field by hand, though effective, is not feasible as it coincides with transplanting of rice in low land and expensive. The resource poor upland farmers are unable to go for herbicide application because of higher cost and non-availability in nearby area. Traditional weeding is still a common practice to remove weeds in upland rice. Hoeing with narrow bladed tools is commonly employed. Frequent weeding (3-4 times) is often required to control weeds. Use of herbicides offers the most practical and economical solution.

In rainfed rice on upland during Rabi season, Ramamoorthy (1991) have studied the effect of integrated weed management and reported that pre-emergence application of pendimethalin 1.25 kg ha^{-1} appreciably brought down the nutrient removed by weeds possibly due to reduced weed intensity and dry matter accumulation. Pre-emergence application of pendimethalin 1.25 kg ha^{-1} registered a remarkable improvement in uptake of nutrients by upland rice. Singh (1992) also reported that minimum weed dry matter was recorded under pendimethalin + one late hand weeding in rainfed rice in which maximum panicle number and yield was recorded; Kamalam Joseph and Bridget (1993) also found that pendimethalin followed by one hand weeding effectively controlled the weeds. Patankar and Sorte (1992) found that pre-mergence application of ronstar 25% EC @ 1 kg ai ha^{-1} + post emergence application of 2, 4-D 10% WP @ 1.5 kg ha^{-1} oil + 5% jaggary on 30 DAS had reduced the weed population to minimum level in rainfed rice. Post emergence spraying of proponil 35% EC @ 2.5 kg ha^{-1} on 20 DAS + one hoeing and one weeding on 40 DAS has reduced the number of perennial weeds to the minimum.

Ajay Pandey and Tiwari (1994) have reported that total dry weeds, number of grass weed, sedge weed and broad leaf weeds were minimum in the treatment with pendimethalin 30 EC followed by 2, 4-D Na at 25-30 DAS to rainfed rice. The grain yield and net income were also higher in this rainfed rice.

(b) Wheat

In wheat the most dominated weed is *Phalaris minor*. The critical period of crop weed competition is between 15 to 30 DAS. Adopting higher seed rate of 150 kg ha^{-1} with the spacing 20 x 20 cm along with the pre-emergence application of isoproturon 1.00 kgha^{-1} is controlling the weeds in wheat. There is also one method to reduce the weed population by inducing to germination of weeds prior to sowing of wheat crop with early showers and sow the crops later. By this practice, weeds can be effectively controlled in the early stages of crop. Integrated weed management practices increased the grain yield by reducing the weed density. According to Singh (1997) fluchloralin + isoproturon mixture enhanced grain yield and was comparable with two hand weedings by checking the weed population of annual weeds like *Polygonum plobeium, Malilotus indica and Chenopodium album.*

For wheat + blackgram and wheat + Mustard and wheat + linseed, application of isoproturon at 1.00 kg ha^{-1} proved quite effective in controlling weeds.

(c) Maize

Maize is very sensitive to weed competition during the first 3 to 6 weeks after sowing (Sandhu and Gill, 1973; Thakur and Singh 1990). Uncontrolled weed growth causes yield loss of 40-60 per cent. Some weeds like *Rottboellia cochinchinensis* and *Striga* sp. can cause total crop loss. The extent of infestation of different weeds varies from one zone to another.

In Maize-based cropped system, the weed intensity of different weed flora was at higher rate, in both the *Kharif* and *Rabi* seasons under rainfed condition. The intercropping of groundnut and soybean suppressed the weed infestation considerably (Moody and Shetty, 1981). Suppression of weed growth was at higher rate when intercropped with soybean (Angiras and Singh, 1989). The weeds took lesser nutrient uptake in maize + soybean system as compared to other intercropping system (Chennabasappa and Nanjappa, 1992). Thakur (1994) reported that the horsegram also suppressed the weed growth in maize when it was raised at 1 : 1 ratio with maize, than soybean and blackgram. Intercropping of cowpea in maize resulted substantial increase (26%) in grain yield of maize. Intercropping cowpea supplemented with one hand weeding produced the same grain yield as was obtained under two manual weedings in sole maize cropping.

Integrated method of weed control with the inclusion of chemical weed control revealed that pre-emergence application of atrazine, simazin 0.50 to 1.05 kg ha^{-1} followed by one late hand weeding was found effective in controlling the weeds in maize at its critcal period of weed crop competition and resulted higher level in growth, yield attributes and yield of rainfed maize (Mittal *et al.,* 1989). Atrazine at 0.5 kg ha^{-1} pre-emergence followed by post-emergence application of 2, 4-D on 30 DAS was also found effective in controlling weeds. The pre-emergence application of mixture of chemicals alachlor 1.00 kg ha^{-1} + atrazine 0.50 kg ha^{-1} (Sangwan *et al.,* 1991) and primextra (metalachlor + atrazine) by sand mix at 1.88 kg ha^{-1} gave better control of weeds (Shad *et al.,* 1993 and Johnson *et al.,* 1997).

(d) Sorghum

Sorghum crop is very much susceptible to weed competition as the seedling starts off weak under dry condition. As the annual and perennial weeds are common problems in sorghum, the *Striga lutea, striga arabica and Striga hermonthica*, the parasitic weeds were found damaging the crop heavily under dryland condition. Two hand weedlings are normally required, the first can be carried at thinning and other one at 35-50 DAS. Sorghum roots are easily damaged by inter-cultivation and hence careful hoeing is essential. Animal drawn blade harrow is normally used by farmers to control weeds in row planted sorghum.

Intercropping of cowpea and lab-lab suppressed the weed population to some extent in rainfed sorghum. The *Striga* can be effectively controlled in vertisols by growing sorghum followed by sunflower (Muthusankaranarayanan, 1986) and it can also be controlled by the application of 12.5 t/ha of FYM to sorghum crop.

Integrated weed management for rainfed sorghum showed that atrazine herbicide at 0.25 kg ha^{-1} as pre-emergence followed by one late hand weeding resulted in better weed control in rainfed sorghum. The application method of herbicide was standardised by Muthusankaranarayanan (1991) that the sand-mix application (50 kg sand per ha) of atrazine herbicide was found effective in control of weeds. The time of application of this herbicide as preemergence at the time of pre-monsoon sowing (15 days period to North-East Monsoon) also found effective and the chemical degradation was not noticed in pre-monsoon sowing (Muthusankaranarayanan 1986). The post emergence application of 2, 4-D at 1 kg ha^{-1} followed by pre-emergence applications of atrazine @ 0.25 kg ha^{-1} on 30 DAS found to control all the weeds and also late grown weeds (Korwar and Friesen, 1984). Gangwar (1992) found that three intercultivations reduced the weed population and was comparable with one chemical weeding by pre-emergence application of fluchloralin @ 0.5 kg ha^{-1}.

(e) Pearl Millet

Pearl millet is very sensitive to early weed infestation. Delayed weeding could reduce the yield of bajra by 44% (Uppal *et al.,* 1969). Significant yield increase was noticed by timely weeding in moisture deficit condition than when soil moisture was adequate (Misra and Kumar, 1962). In drylands, working with country plough at 20-35 days is a normal practice in broadcasted sowing areas during which the thinning operation and removal of weeds will be done simultaneously. For this ploughing, optimum moisture is required. By this method, the bajra will produce more number of tillers and also it provides better aeration and better absorption of rain water received afterwards.

The intercropping is not much familiar in pearl millet. The integrated weed management was found effective to control weeds. The application of pre-emergence herbicide atrazine @ 0.20 kg ha^{-1} was found effective to control the weeds. The post emergence application of herbicide, 2, 4-D at 25 DAS to 30 DAS also controlled the weeds formed in the later season. The crop rotation with cotton, sunflower and chilli was effective in control of striga.

(f) Pulses

The rainfed pigeonpea will be affected due to severe weed competition in early 25 to 30 days of sowing. When the crop is developed in the early stages, it will not be affected and the weeds were well suppressed. (Kasasian and Seeyave, 1969). Dwarf cultivars of pigeonpea are more susceptible to weed competition than tall cultivars.

The mungbean, blackgram and cowpea require two weedings upto 35 days of sowing to control weeds effectively. These crops are mostly grown under rainfed crop as mixture/inter crops. These pulse crops reduce the population of weeds when they are grown in inter cropping system to the extent possible.

The herbicides viz., pendimethalin, alachlor, fluchloralin are effective preemergence application followed by one hand weeding controlled the weeds at its critical weed competition stage. Fluchloralin @ 1.25 kg ha^{-1} with one hand weeding on 35 DAS (Yadhav *et al.,* 1997) controlled weeds in pure crop of blackgram under rainfed condition.

(g) Oil Seeds

(i) Groundnut

It is very sensitive to weed competition because of its initial slow growth, short status and prostrate growth habit. Weeds also interfere at the time of pegging and harvesting. Hand weeding and inter row cultivation are generally done after the emergence of weeds and before peg formation. Pre-emergence application of fluchloralin at 1.00 kg ha^{-1} followed by one hand weeding on 30- 45 days after sowing resulted better weed control in groundnut.

(ii) Sunflower

Weed competition is more severe and damaging the crop during the early stage of crop growth. Singh and Singh (1978) recorded highest yield when weed free condition was maintained for first 60 days of sowing. Muthusankaranarayanan *et al.* (1988) reported that the weed competition from 15 to 35 days of sowing reduced the yield of sunflower under rainfed condition.

Performance of cultural, chemical and integrated weed management method in sunflower in vertisols is found better by Giri *et al.,* (1998). The pre-emergence application of fluchloralin at 1.0 kg ha^{-1} (or) oxyflourfen 0.10 kg + one hand weeding and hoeing on 6th week of sowing controlled weeds in rainfed sunflower and is found economical than cultural method of weeding ie., weeding and hoeing on 11th and 35th day.

(h) Cotton

Being a slow growing and wide spaced crop it suffers from heavy weed competition just after emergence. Higher inputs used in this crop have intensified the weed problem and is one of the main reason for low yield in seed cotton. The weeds which emerged along with crop caused a heavy drain of fertilizer applied and loss of moisture conserved before sowing and during crop season in drylands.

Major weed flora associated with cotton in drylands are *Cynodon dactylon, Cyperus rotundus, Panicum* sp., *Eleusine indica, Bocrhaavia diffusa, Dactyloctenium aegyptium, Phyllanthus madrapetensis, Alysicarpus rugosus, Euphorbia hirta, Cyanotis cuculiata, Celosia argentia, Amaranthus spinosus, Ageratum consoides, Phaseolus minima, Biophytum sensitivum,* and *Eclipta alba.*

Identified critical period of weed competition helped for dryland cotton crop to provide potential for eradication of weeds to get higher yield. In cotton, studies conducted in black soil tract revealed that 15 to 45 days found to be the critical period to cotton crop (Muthusankaranarayana and Dasan, 1977). The yield loss during this period was higher when the weeds were not controlled. If chemical means are to be adopted for control of weeds, the herbicides should have residual effect lasting atleast upto the end of this critical period.

The cotton + blackgram is a predominant intercropping system followed in South India in dryland vertisols and the chemical weed control for this system developed at Kovilpatti revealed that the pre-emergence application of herbicide butachlor at 1.00 kg ha^{-1} sand-mix application (50 kg/ ha^{-1}) for the pre-monsoon sown crop was also found effective in controlling weeds in dryland (Muthusankaranarayanan, 1988). Thus, herbicides are found effective for cotton + coriander/ greengram/cowpea intercropping system followed by one hand weeding/ hoeing.

Higher plant population (2 plants and hill) had greater stress on weed growth and decreased the weed biomass significantly and decreased the seed cotton yield over weedy check (Hneidi and

Mehta, 1991). As light is a major factor for weed growth, cutting out the sunlight on soil surface by cotton crop canopy after 35 days provides excellent weed suppression.

Cotton crop is infested with both grassy and broad leaved weeds. Efficiency of soil applied herbicide depends on moisture status of soil. At the time of cotton sowing, temperature remains very high in drylands which results in rapid depletion of moisture and hence, performance of soil applied herbicide efficiency decreases. Further, the available herbicides will loose their virulence before 30-40 days and that is the pre-requisite for any herbicide and hence a pre-emergence herbicide application followed by a manual or mechanical hand weeding/ hoeing is essential to keep the crop free from weed competition at the critical stage of the cotton crop. The preemergence herbicide application of fluchloralin at 1.00 kg ha^{-1} by incorporation effectively controlled the weeds in rainfed cotton. Studies conducted in rainfed vertisols, the sand mix application with 50 kg ha^{-1} of sand was found equally effective as that of spraying (Muthusankaranarayanan, 1988).

Pre-plant application of trifluralin, oxadiazon, fluchloralin and pendimethalin gave satisfactory control of weeds (Balyan *et al.,* 1983, Brar and Gill 1985) and Pre-emergence herbicides pendimethalin (0.75 to 2.0 kg ha^{-1}) diuron 0.75 to 1.5 kg ha^{-1}) and alachlor (1.00 to 2.00 kg ha^{-1}) can control the annual weeds in *Kharif* cotton. Broad spectrum weed control can be obtained by tank mixing of different herbicides viz., pendimethalin at 1.5 kg ha^{-1} + diuron 1.00 kg ha^{-1} (Brar and Brar 1992 and Jain *et al.,* 1985). New flushes of weeds which germinate either at onset of monsoon or after first irrigation can be effectively controlled by post emergence herbicides. Directed spray of paraquat at 0.7 kg ha^{-1} in mid August gave complete control of weeds and produced equivalent yield to its sequential application after diuron and pendimethalin (Brar and Walia, 1992). Panwar and Malik (1991) found that post emergence application haloxyfopmethyl and fomestifer effectively controlled barnyard grass (*E. colonum*) and carpet weed (*Prianthema portulacastum*) respectively.

Sequential application of herbicides ie., pre-emergence of diuron at 1.0, pendimethalin 1.5 kg ha^{-1} followed by directed post emergence of paraquat at 0.2 kg ha^{-1} gave season long control of weeds (Singh and Brar, 1990). By using adjuvent, the efficiency of herbicides can be enhanced. These adjuvants can increase the foliar penetration of post emergence herbicides. Panwar and Malik (1991) reported that use of surfactant (Selwat) increased the efficiency of haloxyfop on banyard grass.

VIII. EFFICIENCY OF HERBICIDES IN DRYLAND CROPS

Time and method of application of herbicide determine the efficacy of applied herbicides. The herbicide efficacy can be improved by selecting suitable formulation and mixing with suitable adjuvants. Rethinam and Sankaran (1976) reported that in cotton EC formulation of alachlor was having better selectivity than the grannular form. The pH of spray solution, the addition of detergents and sucrose were found to affect the penetration and translocation of foliar applied, 2, 4-D in purple nutsedge (Veerabhadriah *et al.,* 1980) and *Solanam elaegnifolium* Cav. (Devendra *et al.,* 1986). The rate of symplast translocated herbicide was enhanced by addition like sucrose (Shastry *et al.,* 1973).

The scope of reducing the dose of isoproturon and metoxuron by 20 per cent was reported by Walia and Gill (1987) with 0.1% sandovit without any loss in the efficacy to control *Phalaris minor* in wheat.

The application of herbicide like fluchloralin and pendimethalin was found effective when they were applied by pre-plant incorporation in cotton-based pulses intercropping system in vertisols. During North-East monsoon season the sowing of crops prior to monsoon (ie. 15 days prior to onset of monsoon) is a practice in southern districts of Tamil Nadu for cotton and sorghum. The studies in pre-monsoon sown cotton and sorghum based intercropping in vertisols revealed that application of herbicides by sand mix with 50 kg of sand on the day of sowing after sowing with butachlor at 1.0 kg ha^{-1} was found effective in controlling the weeds in cotton and for sorghum, 0.25 kg ha^{-1} of atrazine was effective. The herbicide effectiveness was comparable with the monsoon sown crop. The herbicide effect was not degraded if rain was delayed by 10-20 days. The herbicide has controlled effectively the weeds found in monsoon and pre-monsoon sown crop of cotton + blackgram intercropping system and sorghum. The effectiveness of herbicide was noticed on receipt of rains only (Muthusankaranarayan, 1988).

IX. HERBICIDE TOLERANCE

Differential tolerance to herbicides among crop varieties was reported (Sahi, *et al.,* 1978, Lall and Chaudhary, 1974, ICRISAT, 1977). Selectively of herbicides depends on factors such as genetic, morphological charactersitics, and mobilising capacity of a crop variety. Tolerant crop culture would permit the use of higher rates to reduce the selection for tolerance within some susceptible weed species. It would also permit the use of herbicides which are previously nonselective. This is a need for intensifying bio-technology research efforts to improve herbicide tolerance in crop cultivars and development of herbicides tolerance in crop varieties would be more efficient and economical way to control unmanageable weeds than consistently screening new herbicides.

X. CONTROL OF PROBLEMATIC WEEDS IN DRYLANDS

(a) Control of *Cyperus rotundus*

The nutsedge are problematic weeds in all type of dry land soil. Eradication of this weed will be more difficult. Deep ploughing once in two to three years with disc or mould board plough exposes the rhizomes of this weeds to hot sun for 3-7 days and makes the rhizome to dry. Sowing of thick seed rate fodder sorghum (60 kg ha^{-1}) reduced the density of sedges. However, the biological control of this weed is possible by allowing pigs to feed on the rhizome. Regarding herbicides application of 2, 4-D @ 2.00 kg ha^{-1} or MSMA @ 4.5 kg ha^{-1} twice at interval of 10 days on standing nutsedge reduced the incidence in next season. Application of herbicide glyphosate @ 0.8 to 1.5 kg ha^{-1} on standing sedge or directed spray to sedges controlled the *Cyperus* spp. (Table 8).

(b) Control of *Cynodon dactylon*

Effective method of controlling this weed in dryland areas is by digging this weeds infested area with crowbar and expose roots (stolen) to hot sun during summer which will reduce the incidence. The spraying of herbicide glyphosate @ 1.00 - 4.00 kg ha^{-1} spraying on 15 to 20 cm high *Cynodon daclylan* will control this perennial grass.

(c) Control of Morinda dinctorius

It is worst problematic shrub weed in vertisols. The growth rate will be more after harvest of the *Rabi* crops in drylands. The infestation of this weed will be more and it will putforth more branches if cut at one end. It also obstructs the preparatory cultivation during the ploughing operation; the plant roots may cut the plough share. Morinda was effectively controlled by peeling the bark of 2.5 cm at the junction of root and shoot merge portion and the paste of 2, 4-D (Sodium) herbicide be applied on the peeled portion. It should be kept covered with cotton and kept moist for 2 to 3 days. After 3 to 5 days onwards the plant will start drying and it will take 15 days to dry completely.

XI. CONCLUSIONS

Very little work has been done in developing weed management strategies for cropping systems in dryland as a whole and on a year round basis. The role that weeds play in the determination of cropping patterns is not well documented. Little information is available on the implications of weeds on cropping systems and vice-versa. Weed control in cropping systems needs to be considered on a different basis from that in monoculture. Weed management strategies for the small farms should involve combinations of crop production practices and specific weed control technologies aimed at reducing weed competition, reducing the cost of weeding and should focus on the entire cropping system with emphasis on year round and multi-year management of weed populations.

Table 8. Herbicides recommended for dryland crops and cropping system

S. No.	Crop	Herbicide	Dose kg ha^{-1}	Time of Application
1	Rice (upland)	Thiobencarb	1.50	Pre-emergence - 8 DAS
	Rice (upland)	2,4-D EE	1.00	Pre-emergence - 25 DAS
2	Sorghum/Maize cumbu	Atrazine	0.25	Pre-emergence - 3 DAS
3	Finger Millet	Pendimethalin	1.00	Pre-emergence - 3 DAS
		Butachlor	1.25	Pre-emergence - 3 DAS
4	Cotton	Flucloralin	1.00	Pre-emergence - 3 DAS
		Pendimethalin	1.00	Pre-emergence - 3 DAS
5	Groundnut/Sunflower	Flucloralin	1.00	Pre-emergence - 3 DAS
		Pendimethalin	1.00	Pre-emergence - 3 DAS
6	Pulses	Pendimethalin	0.75	Pre-emergence - 3 DAS
		Flucloralin	0.75	Pre-emergence - 3 DAS
7	Groundnut + Pigeon Pea	Pendimethalin Fluazifop-p	1.00 –	Pre-emergence - 3 DAS Pre-emergence - 3 DAS
8	Groundnut + Sorghum	Alachlor	1.00	Pre-emergence - 3 DAS
9	Cotton + Greengram or blackgram	Alachlor	1.00	Pre-emergence - 3 DAS
10	Wheat + Mustard	Pendimethalin	–	Pre-emergence - 3 DAS
		Isoproturon	–	Pre-emergence - 3 DAS
11	Maize + Pigeonpea	Metolachlor	–	Pre-emergence - 3 DAS

Plate 18.1. Lake infested with water hycienth. Affects more evapotranspiration and hence needs to be cleaned. This biomass can be used for FYM processing. Note half surface is cleaned and half still with weeds.

Herbicide Recommended before planting/sowing

To minimise the cost of labour on tillage, herbicides can be used to control the weeds in fallow field one month prior to sowing. The herbicide glyphosate @ 0.5-1.0% conc. can be applied as post-emergence for the control of perennial weeds like *Cyperus rotundus* and *Cynodon dactylon*. This herbicide is a translocated and non-selective herbicide, it kills the rhizomes and stolens of the said perennial weeds. Due to the drying up of these weeds, the soil gets organic matter and also reduces the number of preparatory tillage operations.

REFERENCES

Abhijit Sarma, and A.K. Gogoi, 1995. Mechanical weed control in upland rice. World Weeds, 2 : 137 - 143.

Ajay Pandey and K.L. Tiwari, 1994. Integrated weed management in rainfed direct seeded upland rice. World weeds, 101-104.

Angiras, N.N. and C.M. Singh. 1989. Effect of weed control methods, cropping systems, plant population and soil fertility levels on nutrient removal and soil fertility status of maize field. Indian J. Weed Sci. 21(1-2) : 69-74.

Annual Report (AICRPDA), 1982-83. All India Co-ordinated Dryland Agricultural Research Project, Agric-Res. Station, Kovilpatti, India.

Balyan, R.S., V.M. Bhan, and S.P. Singh 1983. Chemical and cultural weed control studies in cotton. Trop. Pest. Manage, 29: 56-59.

Bansel, G.H. and C.M. Singh, 1986. Indian J. Weed Sic. 18, 108- 110.

Bantilan, R.T., M.C. Palada and R.R. Harwood, 1974. Philippines Weed Sci. Bull. 1, 14-36.

Behera, V.K. and K.P. Jha, 1992. Technology for improving and stabilizing rice yields in drought prone regions of Kalahandi, Indian Farming, 42 (4) : 9-13.

Bran, H.S. and H.S. Gill. 1985. Study on pre and post emergence herbicides for weed control in cotton. J. Res. Punjab, Agri. Univ., 22 : 437-442.

Brar, A.S. and L.S. Brar. 1992. Bio-efficacy of herbicides of weed control in American cotton. J. Cotton Res. and Dev., 6 : 143-150.

Brar, L.S. and U.S. Walia 1992. Bio-efficiency of pre and post emergence herbicides for controlling weeds in American Cotton. J. Res. Punjab agric. Univ. 29 : 23-28.

Chatterjee, B.N. and B.K., Mandal, 1992. Present trends in research on intercropping. Indian J. agri. Sci. 62 : 507 - 518.

Chennabasappa, K.S. and H.V. Nanjappa, 1992. Nutrient uptake by crop and weeds as influenced by crop weed competition under maize + Soybean intercropping system in rainfed condition. Mysore J. Agric. Sci. 26 : 225-229.

Chhokar, R. and R.S. Balyan, 1996. Integrated weed management in cotton - A Review. J. Cotton Res. Dev. 10 (1) : 44-50.

Devendra, R., T.G. Prasad, S. Ramarao; U. Udaya Kumar and K. Krishnamoorthy, 1986. Indian, J. Weed Sci., 18 : 153-158.

Gambhir, O.P., 1987. Pesticides information, 13, July - September, 36-38.

Gangwar, K.S., 1992. Weed control in Sorghum + Pigeon pea inter cropping system under dryland conditions. Indian J. Agron., 37 (4) : 91-92.

Ghorai, A.K. and P.S. Bera 1998. Weed control through organic mulching in pointed gourd. Indian J. Weed Sci. 30 (142) : 14-17.

Gill, H.S., and L.S. Brar, 1972. Punjab Agricultural University Research, J. 287-294.

Gill, H.S., and S.P. Mehra, 1981. Growth and Development of *Phalaris minor, Chenopodium album* and *Melitotus indica* P. 175-179 In. Proc. 8th APWSS Conf., Bangalore, India.

Giri, A.N., R.H. Bhosle and O.G. Lokhande, 1998. Performance of cultural, chemical and integrated weed - control methods in sunflower (*Helianthus annus*). Indian J. Agron. 43 (1) : 143-148.

Gogai, A.K., 1997. Evaluation of dryland weeder in rainfed wheat. Indian J. Weed Sci., 29 (142) : 75-76.

Hneidi - M. Haythem and H.M. Mehta, 1991. Study on integrated weed management in hybrid cotton. Indian J. Weed Sci., 24 (1 & 2) : 78 - 80.

Hugar, P.V. and M.M. Hosmani, 1977. Chemical and cultural methods of weed control in sorghum. Mysore J. Agri. Sci. 11 : 165 - 167.

ICRISAT (International Crop Research institute for semi-arid Tropics) 1977. Annual Report 1976-77 : Agronomy and Weed Science, Hyderabad, India, 239P.

ICRISAT (Internation Crop Research Institute for Semi-Arid Tropics) 1980. Annual Report 1978-79 : Agronomy and Weed Science, Hyderabad, India, 288.

ICRISAT, (International Crop Research Institute for Semi-Arid Tropics), 1981. Annual Report 1979-80 : Agronomy and Weed Science, Hyderabad, India, 304 P.

Jain, H.C., S.C. Jaris and C.Deshmukh 1985. Cultural and chemical weed control in cotton in nuimar region of Madhya Pradesh. Indian J. Weed Sci. 17, 1-8.

Jain, V.K., et al. 1997. Effect of cultural method of weed management on weeds and yield of blackgram. Indian J. Agron. 42 (4) : 661-664.

Johnson, W.G., M.S. Defelice and C.S. Holman, 1997. Application timing affects weeds control with metalachlor plus atrazine in no-till corn. Weed Technol., 11 : 207-211.

Kamalam Joseph and J.K. Bridget, 1993. Effect of chemical and integrated weed management in rainfed rice. J. Tropical Agri. 31 : 77 - 80.

Kanitkar, N.V., Sivan, SS., and Gokhale, D.H. 1960. Dry farming in India I.C.A.R. Pub., 2nd ed, New Delhi.

Kasasian, L. and J. Seeyave, 1969. PANS, 15, 208.

Korwar, G.R. and G.H. Frieson, 1984. Tropical Pest Management 30, 14-17.

Krantz, B.A., 1979. Intercropping on an operational scale in an improved farming system PP 319-329. Proc. International workshop on intercropping. Hyderabad, India.

Lall, M. and Savondy, 1981. Allelopathic effect of purple nutsedge on the growth of pearlmillet. 8th Asian Pacific Weed Science Soc. Conf. Bangalore, India.

Lall, M., and P.N. Chaudhary, 1974. Indian J. Pl. Protection 1, 53-60.

Misra, D.K. and V. Kumar, 1962. Indian J. Agron. 6, 260-269.

Mittal, S.P., Pratap Singh, S.K.N. Dyal., A.D. Sud and Madanlal. 1989. Weed control measures in rainfed maize in shiwalik foot hill region. Indian J. Agric. Sci. 59 (4) : 251-254.

Moody, K. and S.V.R. Shetty 1981. Weed management in intercropping system PP 229-237. In Proc. International workshop on intercropping. 10-13 Jan. 1979.

Moody, K. 1980. Weed control in inter cropping in tropical Asian. P. 101-108. In Akobudu, I.O. (Ed.) Weeds and their control in the humid and sub-humid tropics. Int. Inst. trop. Agric. Proc. Ser. no. 3 Ibadab, Nigeria.

Moody, K. 1977. Philli. Weed Sci. Bull. 4, 27-38.

Moody, K. 1976. Weed control in Asian soybeans p. 69-73. In RM. Goodman Expanding the use of soybeans IMTSOY Ser. No. 10.

Muthusankaranarayanan, A., 1983. Annual Report (AICRPDA), All India Co-ordinated Dryland Agricultural Research Project, Agrl. Res. Station, Kovilpatti, India.

Muthusankaranarayanan, A. 1986. Studies in the control of striga in rainfed sorghum, Annual report 1985-86 Agrl. Res. Station, Kovilpatti, India.

Muthusankaranarayanan, A., 1977. Critical period of weed competitions for rainfed cotton. Annual report AICRPDA, Kovilpatti, India.

Muthusankaranarayanan, A., 1984. Annual report Agricultural research Station Kovilpatti India.

Muthusankaranarayanan, A., 1988. Effect of herbicides on the weed control in rainfed cotton/blackgram intercropping system. Indian J. Weed Sci. 20 (3) : 91-93.

Muthusankaranarayanan, A., 1991. Method of application of herbicide for rainfed sorghum. Annual Report, 1990-1991, Agrl, Res. Station, Kovilpatti, India.

Muthusankaranarayanan, A., V. Chellamuthu, A. Rajamanickam and J. Sundarsingh Rajapandian, 1988. Effect of herbicides on the weed control in the rainfed cotton/blackgram intercropping system. Indian J. Weed Sci., 20 (3) : 91-93.

Palaniappan, S.P., and R. Ramasamy, 1976. Residual effect of altrazine applied to sorghum on the succeeding crops. Madras Agri. J. 63 : 196-197.

Panwar, R.S. and R.K. Malik, 1991. Competition and control of weeds in cotton. Haryana Agri. Univ. J. Res. 21 : 226-234.

Patnakar, M.N. and N.V. Sorte, 1992. Effect of herbicides on annual and perennial weeds growing in rainfed paddy. J. Soils and Crops 2 (1) : 72-75.

Ramamoorthy, K., 1991. Effect of integrated weed management on nutrient uptake by rainfed rice and associated weeds. Indian J. Agron. 36 (2) : 213-217.

Rangiah, P.K. 1989. Studies on weed control in a cropping system of irrigated ragi, cotton and sorghum. Ph.D. Thesis, TNAU, Coimbatore, India. Rao, A.N. and Agarwal, 1984. Indian J. Weed Sci. : 16 250-254.

Rao, M.R., and S.V.R. Shetty, 1976. Indian J. Weed Sci. 8, 32-43. Rathinam, P and S. Sankaran, 1976. Indian J. Weed Sci. 18, 74-76.

Sahi, B.K., Lall, M. and Chaudhary, P.N., 1978. Differential tolerance of sugarcane varieties to herbicides. Indian J. Weed Sci., 10 : 54-62.

Sandhu, K.S., and G.S. Gill. 1973. Indian J. Weed Sci. 5, 1-5.

Sangwan, N., R.S. Panwar and R.K. Malik, 1991. Effect of various herbicides on the nutrient uptake by weeds in winter maize. Pestology. XV (6) : 12-15.

Sankaran, S. and V.T. Subbaiah Mudaliar, 1993. Principles of Agronomy. The Bangalore printing and publishing Co. Ltd., Bangalore. pp. 143-174.

Saxena, A.; R.D. Vaishya and V.C. Upadhyay, 1990. Weed management in upland drilled rice. Extented summary Proc. Int. Symp. on Rice research, New Frontiers held at Hyderabad : 31 P.

Shad, R.A., M.Q. Chatha and H. Nawaz, 1993. Weed management studies in maize. Pakistan, J. Agric. Res. 14 (1) : 44-50.

Shaikh, A.A., Jadhav, A.S., Koll, B.D. and Wattamwar, M.J., 1994. Effect of moisture conservation technique and fertilization on weed growth and yield of rainfed pearlmillet (*Pennisetum glaucum*). Indian J. Agron. 39 : 307-208.

Shastry, K.S.K., S. Ramarao; T.R. Muniappan and J.R. Viswnath, 1973. Studies on the use of weedicide in the control of nutsedge and preventive regeneration of tubers. Final Technical Report. U.S.A., Bangalore.

Shetty, S.V.R., and M.R. Rao. 1977. Weed management studies in pigeonpea based intercropping. p 655-672 in proc. 6th Asian pacific weed Sci. Soc. Conf., Jakarta, Indonesia.

Shetty, S.V.R., and A.N. Rao, 1981. Weed management studies in Sorghum/Pigeonpea and Pearlmillet/ground nut intercropping system - some observation P 238 - 248 proc. International workshop on intercropping, ICRISAT, Hyderabad, India.

Singh *et al.,* 1982. Studies on weed control in bean. Abstract papers. Annual conference of Indian Society of weed sci., held at Haryana Agri. Univ. Hissar. During 27-30 Dec. 1982.

Singh Thakur and L.S. Brar, 1990. Efficacy of different herbicides for weed control in American cotton. J. Res. Punjab. Agri. Univ. 17 : 164-169.

Singh, A.K. and R.A. Singh, 1993. Weed management in rainfed Pigeon pea- Sesamum intercropping system Indian J. Agron. 38 (4) : 541-546.

Singh, G.B. and V. Singh 1978. Indian J. Weed Sci. 19, 79-82.

Singh, G.B. and R.N. Saraswat, 1986. Paper presented at the joint India-Pakistan Weed control workshop and 7th USDA/ICAR-AICRPWC workshop at Durgapur, Agricultural Research Station, Jaipur held on 22-24 March, 1986.

Singh, S.P., 1992. Weed control in rainfed upland rice. Indian J. Dryland Agri. Res. Dev. 7 (2) : 94-97.

Singh, R. and V.S. Mani, 1981. Effect of atrazine and alachlor in maize, pulses (peas, gram and lentil) rotation. Pp 127-130. In Proc. 8th Asian Pacific Weed Science Soc. conf. Bangalore, India.

Solaiappan, V. and N. Chelliah, 1998. Effect of sowing Time and mulching on weed control in cotton - based intercropping system in rainfed vertisol. Madras Agri. J. 85 (2) : 117 -120.

Thakur, D.R. and K.K. Singh, 1990. Effect of weed and fertilizer management on nutrient uptake by weed and maize under rainfed condition. Indian J. Agron., 35 (4) : 357-360.

Thakur, D.R., 1994. Weed management in intercropping system based on maize (*Zea maize*) under rainfed midhill condition. Indian J. Agron. 39 : 203- 206.

Tripathi, B., and C.M. Singh 1983. Trop : Pest management 29, 267-270.

Uppal, H.S., S.K. Agarwal, and M.K. Moolani 1969. J. Res. Punjab, Agri. Univ. 6, 18-24.

Vaishya, A.R.D. and V.C. Upadhyay, 1990. Weed management in upland drilled rice extended summary. Proc. Indian Rice Research new frontiers held at Hyderabad. 31P.

Veerabhadraiah, G.P., S. Ramarao; M. Uday Kumar, and K.S. Krishna Sastry, 1980. Indian J. Agric. Sci. 50, 342-356.

Walia, U.S. and H.S. Gill, 1987. Indian J. Weed Sci. 19, 37-46.

Yadav, R.P., K.S. Yadav and V.K. Shrivastava, 1997. Integrated weed management in blackgram. Indian, J. Agron., 42 (1), 124-126.

Chapter - 19

ROLE OF ANTITRANSPIRANTS IN DRYLAND AGRICULTURE

***A.M. Dhopte*[1] *and S.D. Ramteke*[2]**

1 Former Professor of Botany, College of Agriculture, Nagpur, 49, Vivekanand Nagar, Khamala Road, Nagpur – 440015 (MS)

2 Dr. S. D. Ramteke, Principal Scientist (Plant Physiology), National Research Centre for grapes, Manjri Farm, P. O. Box No. 3, Pune -Solapur Road, PUNE - 412 307

1. INTRODUCTION

Globally 2/3 of earth surface is occupied by water but paradoxically only 1% of ground water is vital for sustenance of life on earth. Crops grown exclusively on rain water are classified as dryland crops. Out of total land on earth, 1/4 of the surface (11%) is cultivable, other land is either drought prone, waterlogged or poor in available nutrients. Twenty eight percent of land is rainshade area, 22% is without soil microbes, 10% is waterlogged, 23% of soil is without nutrients and 6% land is covered with ice. Land area under cultivation at global level is given in Table 2.

In south Asia, the land is over exploited and degradation is fast. Many of the land pockets are deficient in essential nutrients and productivity is very low compared to world average. In Africa, only 6% of total cultivable land is under cultivation, India, in particular, is the country where cover of 90% of land is under rainfed farming and it is very essential to manage every drop of water received through rains. Various measures are adopted to conserve the rain water but rainfed farming is either subjected to drought or waterlogged conditions. As per world bank report, 2/3 of global population shall face shortage of water by 2025 AD and it will lead to social turmoil. Today, 2.8 billion people of the world is poor. In such situation, proper rain water management is a dire need of the day.

Agro climatic zones are categorically identified at national level. In central India, following zones are denoted on the basis of rainfall.

Table 1. Climatic zones

Climatic zone		Avg. Rainfall (mm)	Number of rainy days
1	Assuared rainfall zone	650-950	42-52
2	Medium rainfall zone	950-1250	42-62
3	High rainfall zone	1250-1750	59-75

Table 2 : Land area (%) under cultivation in various continents

Continent	Cultivable	Uncultivable	Remarks
North and Central America	22	13	Converted in roads
South America	21	10	Nutrient deficiency seen
Europe	33	31	Nutrient deficiency seen
South Asia	20	24	Nutrient deficiency seen
South east Asia	17	14	Nutrient deficiency seen
North and Central Asia	50	10	Cool temperature
Africa	16	06	44% wasteland
Australia	16	06	55% drought prone

Table 3 : Drought incidence and their intensity in Maharashtra (1875-1989), India.

Region	Moderate drought (Year)	Severe drought (Year)
Konkan and Goa	1877, 1899, 1905, 1918, 1920, 1941, 1968, 1972, 1986	Nil
Madhya Maharashtra	1877, 1904, 1905, 1911, 1972, 1985, 1987	1899, 1918
Marathwada	1877, 1885, 1899, 1905, 1907, 1912, 1918, 1925, 1929, 1939, 1941, 1972, 1974, 1984, 1985	1920
Vidarbha	1877, 1902, 1904, 1918, 1920,1950, 1952, 1965, 1971, 1972, 1974, 1982, 1985, 1987	1899

Source : Ann Report of DDG (Agrimet), Pune (MS)
Moderate drought - Rainfall deficiency 26% to 50% of the normal
Severe drought - Rainfall deficiency >50% of the normal.

The dought incidence and their intensity for Maharashtra is mentioned in Table 3.

Based upon 100 years of study, the frequency of moderate drought is more in *Vidarbha* and *Marathwada* as compared to Konkan and Goa and Madhya Maharashtra. The occurrence is also erratic and unpredictable. The rain water management is therefore highly essential for sustainable agriculture. Wide variation in rainfall and land topography are special characters of Maharashtra. The degree of aridity depends upon rainy days. Central penninsular zone has over 90% area under rainfed farming. In central plains, monsoon is expected from June, 6th to 10th every year but in some years it is either delayed till June 20 or after receipt of pre-monsoon showers for couple of days, it is terminated for 10- 15 days (Fig. 1). Later situation is quite critical because crops are sown assuming normal onset of monsoon. Resowing is necessary in such case due to failure of first attempt. The seeds get germinated and dried due to lack of enough water. High VPD due to high solar intensity is highly fatal to the seedlings. Poor farmers are unable to resow the crop for paucity of funds. This is the major constraints for poor productivity of rainfed crops. Therefore, it has become risky preposition. Farmers have to pay very heavily for wrong decisions. Late sowing results in very poor productivity. Timely sowing is the key factor for maximum productivity. Erratic behaviour of monsoon does not allow the farmer to adopt right sowing period. Timely and early sowing preferably before June, 20th are certainly remunerative in central plains.

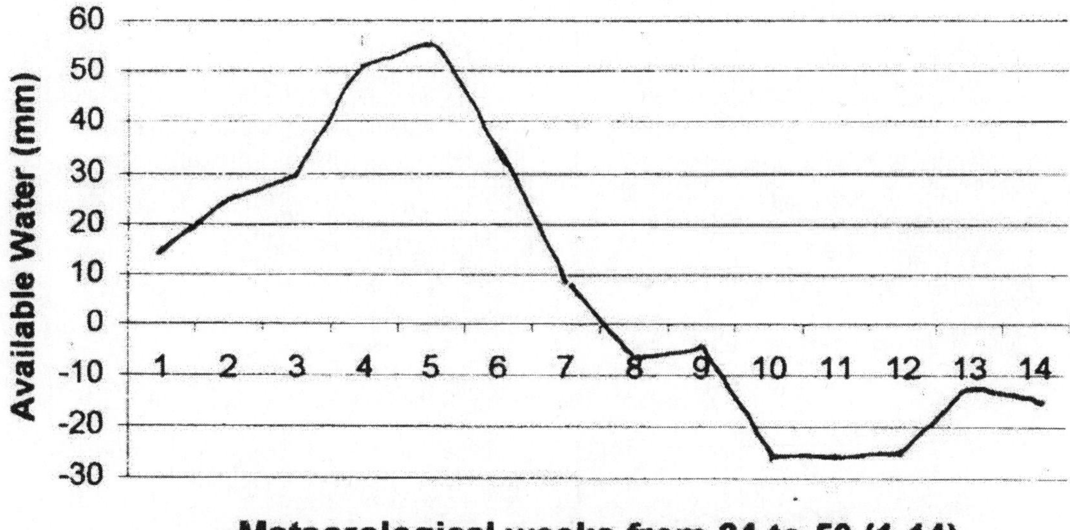

Fig. 1. Weekly available water (mm) in Vidarbha (1970-1989)

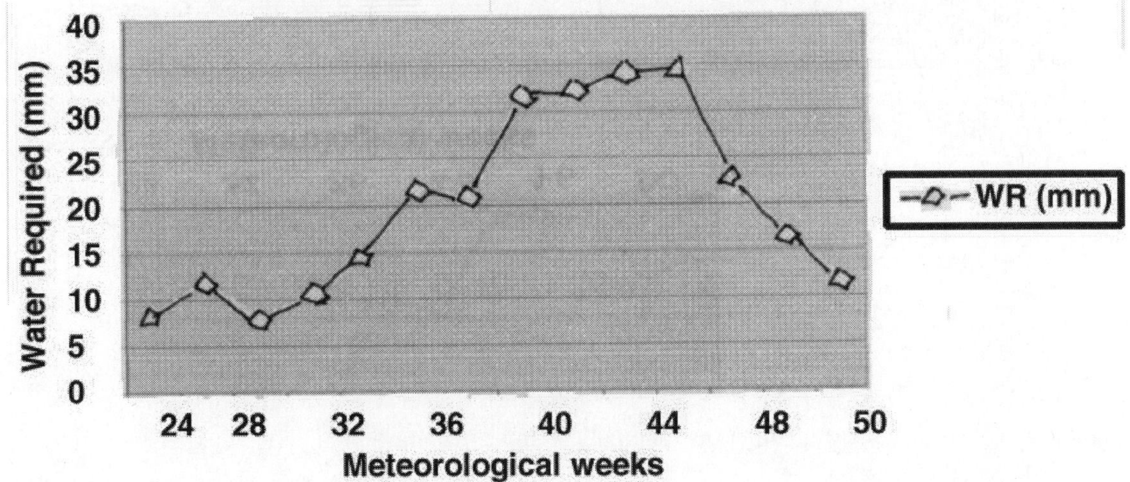

Fig. 1A. Weekly water requirement (mm) for cotton in Vidarbha (1970-1989)

2. MONSOON PATTERN AND STRATEGIC APPROACH

Late onset of monsoon is equally detrimental because ideal climatic conditions and period of growth are altered and productivity is affected. This situation is beyond the control of farmer and drought fetches poor yield or no yield at all and famine is likely in some parts of plains. In India, over 90% of land is rainfed. Technology is however developed for ideal conditions of water and other inputs. The aridity map of Maharashtra state is given in Fig. 2.

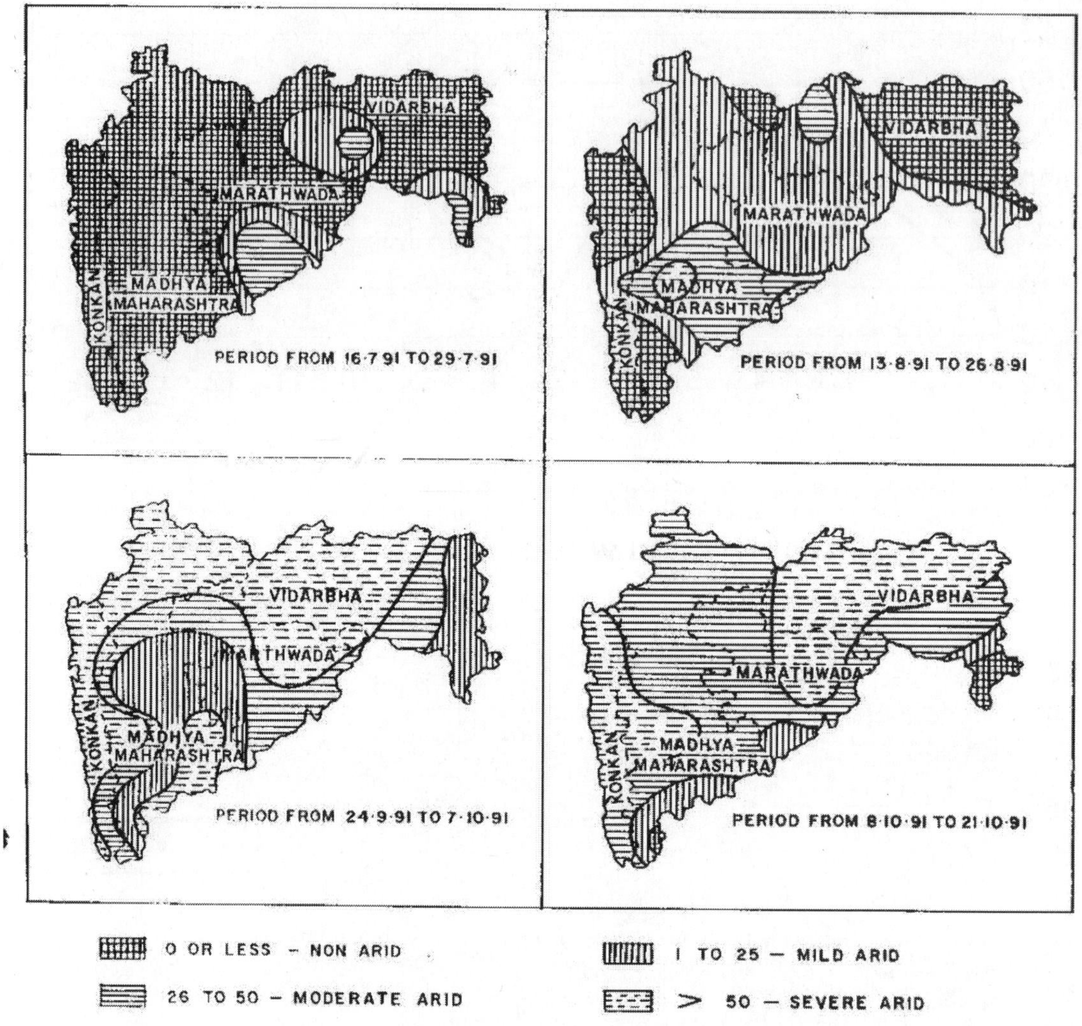

Fig. 2. Aridity anomaly map for Maharashtra

Deviation in rainfall from normalcy is recorded once in three years in India. Due to global warming, the trend may vary. (Table 3). Deep vertisols can sustain drought and sandy loam soils are more prone to drought. Drought tolerant crop species are found but, basically they are found poor yielders but it yields under abnormal year without heavy losses. Farmers can maintain his family for the season without starving for food. Vertisols are predominantly found in central plains but it has drawbacks. Frequent wetting and drying result in soil cracks and non-availability of nutrients due to root trimming. Cotton and pigeonpea are only the long duration crops that are suited to such unideal conditions of soil and therefore there is no other alternative for these crops in central plains. It can grow under moderate drought and yield is assured. Heavy rainfall for long duration is however detrimental and if it occurs at early seedling stage. the productivity is highly affected. The adopted cropping pattern includes, sorghum, Mung, soybean, cotton, pigeonpea, sunflower and pearl millet to satisfy the daily needs of the farmer. Technology adoption by a farmer is not satisfactory. The watersheds, contour bunding, BBF, organic recycling and live bunds are some of the solutions that came from exhaustive dryland research. Implementation has to be improved. Conventional cultivation is adopted by common farmer during last two decades. Though dramatic change is not seen, increased productivity indicates follow up of innovative ideas by few farmers.

Ranuzzi and Srivastava (2012), reported the impact of climate change on agriculture and food security. According to their report because of climate change, Indian agriculture is doubly vulnerable. India's backward districts as ranked by the Planning Commission are distinguished for the large-scale practice of rain-fed agriculture. In most agro-climatic regions, farmers have stopped cultivation of millets which are suitable to a particular agro-climatic region. Climate change is projected to have serious implications for these major crops especially wheat. The studies have already projected greater loss in Rabi season (e. g. in wheat yield) as compared to Kharif crops. Climate change will affect different parts of India in different ways. These differences are illustrated by the fact that, while large areas in Rajasthan, Andhra Pradesh, Gujarat, Orissa and Uttar Pradesh are frequented by drought, approximately 40 million hectares of land in the north and north eastern belt is flood-prone areas. India may also be exposed to a greater number of floods due to the intensification of the Indian monsoon.

Life exists where water prevails. *In situ* rain water conservation is only the solution over this acute problem to increase the productivity of dryland. Changing the minds of the people is highly difficult under democratic pattern and moreover to most illiterate farmer.

3. ANTITRANSPIRANT - CONCEPT AND TYPES

After having discussed the actual problems faced by the farmers, we shall now discuss what can best be the solution. Out of six concepts, we shall look into inventory planning of how can we save the standing crop under moderate drought situation. The concept of antitranspirant is developed in 1960. Any chemical that suppresses the transpiration without significantly affecting the growth and yield is called 'Antitranspirant'. Several chemicals are identified since 1960 and are classified as (1) Film forming types, (2) Radiation reflectant types, (3) Stomata closing types and (4) Wax coating types. These chemicals when sprayed or applied through soil can benefit the crop under adverse rainfall. Drought reduces the yield by 0-100% depending upon the severity. Normally it is reduced by 20-60%. Intermittant drought for 10-15 days at early or late stage is common under rainfed conditions. Crop productivity is dependant upon how fast a plant can recover after a stress of 6-10 days. Prolonged drought can drastically reduce the yield to zero level. Drought during critical phenological phase like flowering and grain development is highly detrimental. How do we prevent losses in yield?

AgriInfo.in. (2011), Some fungicides like phenyl mercuric acetate (PMA) and herbicides like Atrazine in low concentration serve as antitranspirants by inducing stomatal closing. These might reduce the photosynthesis. PMA was found to decrease transpiration than photosynthesis. Plastic and waxy material which form a thin film on the leaf surface and result into physical barrier. For example ethyl alcohol. It reduces photosynthesis eg. Tag 9; S - 789 foliate. They are white materials which form a coating on the leaves and increase the leaf reflectance (albedo). By reflecting the radiation, vapor pressure gradient and thus reduce transpiration. Application of 5 percent kaolin spray reduces transpiration losses. eg. Diatomaceous earth product (Celite), hydrated lime, calcium carbonate, magnesium carbonate, zincs sulphate etc. These chemicals reduce shoot growth and increase root growth and thus enable the plants to resist drought. They may also induce stomatal closure. Cycocel is useful for improving water status of the plant.

Transpiration is said to be unavoidable evil but it has several functions to attend in crop cycle. For one tonne of food, about 1000 tonnes of water are needed. Cereals and legumes need 400-500 litres of water/ Kg of grain while fruits and Vegetables require 1000 litres of water/Kg of food produced. The water use efficiency (WUE) of crops is different and ranges from 0.24 to 1.72 Kg/mm of water/ha. The WUE of sorghum is highest but that of cotton is lowest. Maturity period matters for such deviation. Cotton grows for six to seven months while sorghum grows for four months and it differs in nutritive value calorically.

Stomata occupy 1-2% of total leaf area and water losses are 80- 90%, 20% and 0.1% total through stomata, cuticle and lentical respectively. Various crops have different stomatal index and therefore their water requirement is varied depending upon the season (Table 4 & 5).

Table 4 : Water transpired by some crops/season/plant

Crop	Water transpired Litres/season/plant
Maize	200
Sunhemp	27
Trees (9-10 m height)	300-800
Forest of 400-600 trees	20,000 barrels/day (Barrel= 500 litres)
Cotton	8-10 litres/day
Citrus	100-120 Litres/day

Table 5 : Stomatal frequency and stomatal index in some crops

Crop	Stomatal frequency/cm2	Stomatal Index (%)
Sorghum/Maize	6800	28.6
Cotton	10300	15.8
Groundnut	21000	20.0
Tomato	13000	21.9
Pigeonpea	–	29.4
Mungbean	28000	25.0
Wheat	1400	32.0
Sunflower	28100	35.0
Pea	21600	30.0

4. WEATHER AND OPEN PAN EVAPORATION

Water losses/plant depends upon several factors but internal factors such as structure and its anatomy determine water requirement (Table 7). Drying capacity of air is dependant upon VPD and it primarily decides the magnitude of water loss. Stomatal frequency is genetic parameter but is found variable depending upon existing climate prevailing in the region. The open pan evaporation is an index of water losses and transpiration is almost 1.3 times higher than OPE. The graph on OPE and rainfall gives an idea about their relationship. The transpiration is controlled by light intensity and temperature during a day.

Vapour pressure of the air is dependant on relative humidity (RH) and drying capacity of air is mainly determined by VPD. The typical OPE of central plains are depicted in Fig. 3. The moisture use pattern of crop is dependent upon the leaf area index and phenological stages. The severity of intermittant drought of 6-10 days during critical stages of crop can reasonably be avoided by use of antitranspirants and crops can be saved.

Table 6 : Potential evaporation (mm/day) at Akola and Nagpur Districts in Maharashtra.

Month	Jan.	Feb.	March	April	May	June	July	Aug	Sept.	Oct.	Nov.	Dec.
Akola	6.1	7.9	10.0	12.3	15.3	12.3	8.4	6.3	6.6	7.2	6.2	5.3
Nagpur	5.1	7.9	11.1	16.0	11.6	6.7	5.8	6.3	6.7	5.9	4.5	

Source = Bathakal and Kaore (1980)

Table 7 : Water requirement of some crops and critical stages of growth

Crop	Water required (mm)	Critical period (Days after sowing)
Sorghum	400-500	28-30
Bajra	250-300	50-55
Maize	400-420	30-45
Wheat	400-450	60-65
Groundnut	400-450	40-45
Sunflower	300-350	30-35
Safflower	250-300	50-60
Gram	250-300	60-65
Cotton	550-600	80-90
Pigeonpea	500-550	90-100

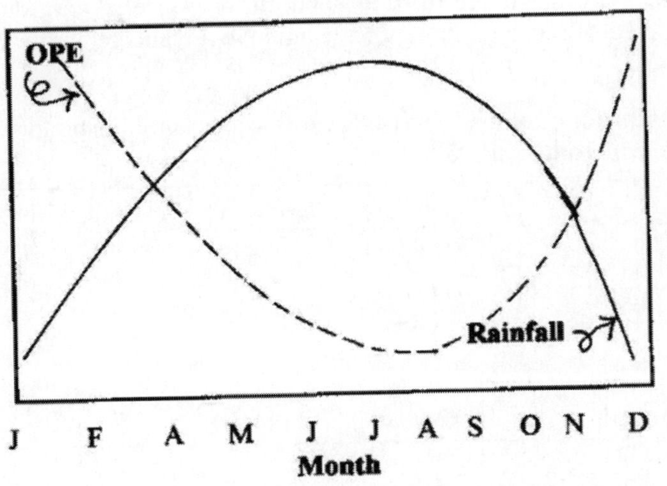

Fig. 3. Association of Open pan evaporation (OPE) and rainfall in central plains.

5. WHY ANTITRANSPIRANTS?

It is necessary to know why is there a need of antitranspirants. It can effectively be used to save the crop under water stress with proper planning. You may not wait drought to commence and then start emergency measures to save the crop. In long duration crops like, cotton, pigeonpea, antitranspirant can save the crop and maintain the productivity. In central plains (6-Deccan Plateau) where rainfed farming is predominant, there is no other alternative for cotton and pigeonpea. Soils are deep vertisol with high water holding capacity (>52%). Prolonged drought for 1015 days (RWC <70%) result in soil cracks at late flowering and lateral roots are trimmed of from main tap root. When functional roots are lost with heavy loss of root hairs, the crops experience water stress. Boll and pod development are drastically affected. Such situation can be avoided by sprays of antitranspirants, (AT) usually in Mid-Sept. This helps restoration of moisture in rooting zones by depletion of water loss through foliage (by about 20-30%) and make available the water at critical phenological stages of flowering boll and pod development. Fast decline in ASW can be prevented sprays of kaoline (1.25%) on cotton increased yield by 17.6% over control (Dhopte and Jamdar, 1995). The spray of Hico-emulsion-110 R increased yield by 15.3% in hirsutum cotton (Table 8). The ATs are required for stabilizing the yeild levels under erratic rainfall. This can be illustrated with average rainfall data given in Fig. 4 (Kayande *et al.,* 1991). The soil moisture decline is observed in Mid September in Maharashtra (central plains) when crop is at flowering or boll development. High leaf area index (LAI, >3.6) is recorded at this stage in cotton and pigeonpea. As per data of 1982-83, water requirement is increased but supply is limited due to receding soil moisture in Sept. onward. Suberization of root tissue due to secondary growth results impermeability to water and water absorption is arrested. The depletion in oxygen in soil is fast and root activity is highly affected. The water and mineral absorption are decreased and this is a stage when water is needed for higher productivity. Restoration of moisture in soil is therefore highly essential to maintain above process at optimum level and this can be achieved by AT spray like PMA or Kaoline. Transpiration is retarded by 20-30% with no significant reduction in photosynthesis. Growth can be retained stable for 20-30 days. Restored soil moisture maintains root activity and supply steady flow of water during critical reproductive stage. Persistance of AT is for 7-15 days and hence repeated sprays are inevitable to have optimum response in long duration crops. Rains may or may not be received. If received, it may have additive advantage and if not, the productivity can be optimised.

Table 8. Response of hirsutum cotton (AKH-081) to various antitranspirants. Seed cotton yield (kg/ha) (Dhopte and Jamdar, 1995)

Treatment	1986-87	Increase (%)	1987-88	Increase (%)
Control	773	–	638	–
Kaoline (1.25%)	914	18.3	118	12.5
Hico-110 R (1 ml/L)	889	15.0	735	15.3
Atonik + Kaoline	977	26.4	750	17.6
Ascorbic Acid + PMA (10 ppm)	906	17.1	710	11.4

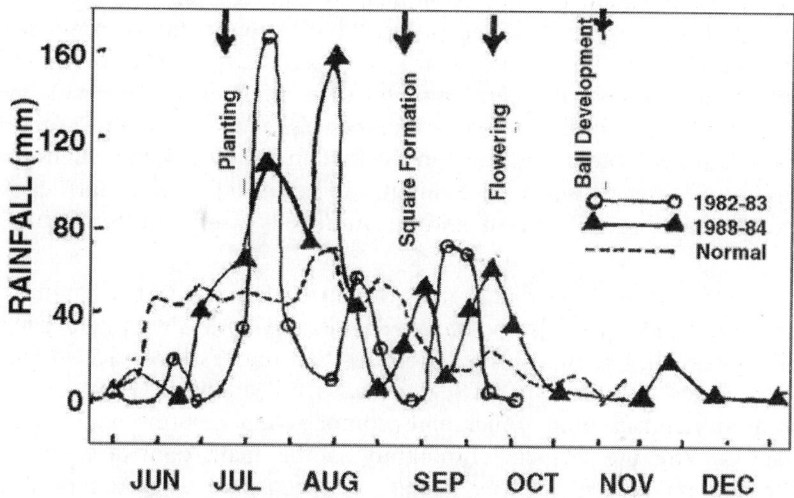

Fig. 4. Rainfall distribution pattern during 1982-83 and 1983-84 alongwith corresponding growth stages. Data Recorded at Dr. PDKV, Akola (MS)

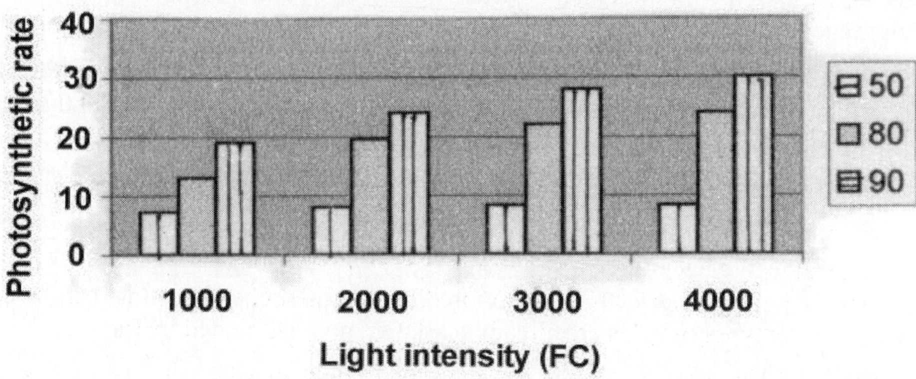

Fig. 5. Effect of RWC (%) on Photosynthetic rate (mg./dm^2/h)

Severe drought can lead to 70-80% losses in productivity but if plants are sprayed with AT, it can compensate the expected losses by about 20-30% with marginal expenses of Rs. 500/ ha. This is sensible reward to the farmer. In sorghum, drought during boot and grain filling stage results in no yield. If precautions are taken to spray the crop timely during expected drought, the yields are assured as against harvesting only fodder with no sprays. Award of grain yield becomes bonus in returns of the grain sown. Beneficial effects of AT are reported in various crops including fruits and vegetables and fodder grasses (Table 9). The irrigation frequency can be reduced in crops like banana, sugarcane and grapes with optimum yields.

Mohawesh *et al.* (2010), studied the effect of antitranspirant application on physiological and biochemical parameters of three orange cultivars grown under progressive water deficit, according to them the water is also lost to the atmosphere through transpiration and the potential for reducing transpiration water loss without significantly reducing photosynthetic rate is based on the premise that resistance to the movement of carbon dioxide in the mesophyll is greater than the stomatal resistance that limits water loss to the atmosphere. Application methods using antitranspirants have been proposed to reduce water loss and enhance the water status of plants. Agriculture in Jordan is restricted to areas that receive a respected amount of rainfall or above beside the irrigated areas. More than 90% of Jordan's 9 million hectares receives less than 200 mm of rainfall annually. Citrus species are grown under irrigated conditions in the Jordan valley within a subtropical climate. Even under irrigation, hot weather causes a huge amount of water loss that in turn causes an increase in irrigation intervals; therefore the use of antitranspirants is useful in agriculture to prevent such problems.

The yield optimization is highly important in rainfed farming rather having higher yields than the normal. The harvest index can also be improved with AT sprays with good seed quality. Drought during grain filling normally results shrivelled grains that fetches low price in the market. This can be confidently prevented with AT sprays if done with full attention. Alertness is therefore highly important in doing dryland farming. Quick and prompt action ensure rewards and benefit is always assured. Illiteracy among the farmers community is the main constraint in timely adoption of technology. The amount of Rs. 500 per hectare can be spent to insure productivity with cost-effectiveness. Solution is available but its adoption is very poor.

The ET is more during early stage of crop till 40 DAS but later it remains constant with slight variation. In general, the water requirement is increased with slight variation. In general, the water requirement is increased with increased growth (Fig. 1A & 6) and this is attributed to increased LAI. The plant spacing also does not significantly influence the moisture use pattern (Table 11).

There are various methods of decreasing the transpiration :

1. By increasing the leaf resistance to water vapour by applying material that close/ over the stomata.
2. By reducing the net energy uptake by leaves by increasing the leaf reflectance.
3. By reducing the top growth by growth retardants.
4. By increasing the air resistance to water vapour by using wind breaks

First three methods are effectively monitored by potential chemicals that can be categorised as antitranspirants. The types of AT's are already classified and mentioned in Table 12.

Table 9 : Percent stomatal closure in sunflower by CA and PMA (Dhopte and Aher, 1977)

Days after Treatment	Cetyl alcohol*			PMA (1 ppm)		
	C0	C1	C2	C0	C1	C2
3	25	31.6	47.6	22.7	52.5	51.2
5	23.5	47.4	55.0	21.1	47.6	55.0
6	25.0	47.6	57.1	20.0	50.0	52.0
7	20.8	47.6	46.7	23.8	50.0	52.6
9	18.7	32.3	27.3	14.0	54.8	63.7
11	10.0	40.0	30.0	12.5	52.4	52.6
14	17.6	54.5	56.3	16.7	69.2	56.3
19	29.4	42.9	46.7	13.3	57.1	47.1

* Cl = 30 Kg/ha soil applied, C2 = 60 Kg/ha

Plate 19.1. 1. Antitranspirant (Kaoline, 1.25%) sprayed cotton in intercrop after termination of monsoon (Multistoried cropping), **2.** Peanut sprayed (Left) with Paclobutrazol (50 ppm). Note more greenness of leaves and compact growth.

Plate 19.2.1. Narrow spaced cotton for higher yields (SRT-1), **2.** Soil cracking in vertisol trims lateral roots. Follow interculture and conserve water and avoid root trimming.

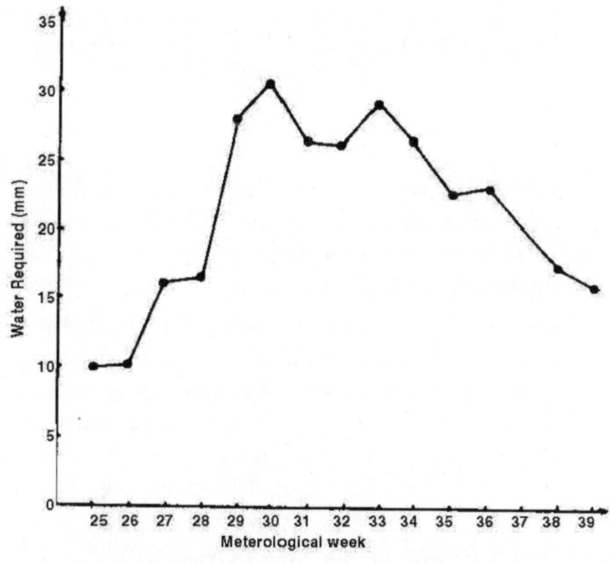

Fig. 6. Water Required for Sorghum (mm)

Antitranspirants	Concentration		
	C0	C1	C2
Cetyl alcohol	0.870	0.553	0.747
PMA	1.788	0.641	0.727

PMA. Cl = 1 ppm, C2 = 5 ppm, Sprays at vegetative stage.

Table 11 : Moisture use pattern under various spacing by hirsutum cotton under rainfed conditions (Dhopte *et al.*, 1994).

	Moisture use pattern (mm)			
	Days after sowing (DAS)			
Spacing (cm)	0-40	41-90	91-120	121-140
60 x 15	254	139	83	110
60 x 30	249	127	76	102
60 x 60	243	122	73	98

6. EFFECT OF ANTITRANSPIRANTS IN CROPS (AT) — A REVIEW

Adequate available water during the period from tillering through grain filling in cereals is necessary for the production of high yields in arid and semi-arid regions. The ATs may be used to retard the rate of soil water depletion and thus can extend the period over which precipitation may be received without sereous decline in yield (Gale and Hagan, 1966). Zhang et al. (2006), suggested that the ABA subsequently changes in gene expression that lead to rapid ABA biosynthesis and constitute the most important stress signal transduction pathway among all the plant responses to stresses. ABA-induced changes in gene expression and it plays an importance role in physiological regulation of plants. Considerable interest has been developed in recent years in the agronomic use of stomatal closing AT's.

Physiological Effects

The physiological effects are mostly centralised on photosynthesis and transpiration (TR) because of its direct effects on gas exchange. The ATs reduce the rate of TR by limiting the vapour exchange through the stomata. It is also found to have adverse effect on photosynthesis. Chemicals should be such that it should maintain the Ps normal while reducing TR losses.

The AT's that are now available are also reducing the Ps and some new AT's are being searched. Zelitch and waggoner (1962) found reduction in TR by application of hydroxysulfonates inhibitors of glycolic oxide applied to detached leaves in sunlight and are markedly reducing the TR through stomatal closure. They also found resistance in stomata with PMA. The TR was reduced more than assimilation. The PMA reduced TR in sunflower and tobacco, but reduction in Ps was less (Shimshi, 1963). A spray of cetyl alcohol (Adol-2), OED green, S-600 and PMA reduced TR. Low con. of PMA (10^{-4} and^{-5}M) was most effective in cotton. The PMA was found toxic to grasses (Davenport, 1963). It reduced TR by 7%. The 20% reduction in TR at $10^{-3.5}$ M con. of PMA was reported by him. The PMA reduced growth and water use in spring wheat (Brengle, 1968). The WUE was found increased with PMA treatment. Miller (1969) tried several AT's on water balance in corn. The PMA, N-decanyl succinic acid and hexadecanol resulted in reducing TR by 54.5, 51.5 and 34.9% respectively. The PMA also reduced TR by 30% in chrysanthemum when sprayed with wetting agent (Cocking and Tukey, 1970) without causing tissue damage. Higher concentrations were found to reduce the growth. In corn, it increased CO_2 compensation point and also resistance to diffusion of gas (Squire and Jones, 1971). In barley and tomato, the TR rate was lower due to PMA (Majernick, 1970; Cotter 1970). The experiments with Rb in TR studies indicated transport within the plant even though it was sprayed with AT' (Gale and Poljakoff, 1963). It can be concluded that the TR can be halved without harmful effects on leaf temperature and mineral nutrition. The undesirable effects can be relieved by AT's spray. In majority of crops it is useful in suppressing the TR and maintaining the vitality of crops under adverse conditions. The effect is however persisted for 8-10 days. The NAA and 2,4-D also cause stomatal closure (Zelitch and Waggoner, 1962).

In chrysanthemum, the reduction in TR was due to reduction in stomatal pore size (Miller, 1969; Cocking and Tukey, 1970). The ABA caused stomatal closure in tobacco (Jones and Mansfield, 1970) and *Xanthium*. The PMA had varied response in crops like, sorghum, maize, cotton and groundnut. Mishra and Pradhan (1972) found reduction in stomatal opening by 80% in tomato. Six days after spray, it reduced to 30-40%. Lower con. had better response in *Commelina communis* (Squire and Jones, 1971). The hexadecanol reduced plant height in corn (Peters and Roberts, 1963). The PMA spray showed some damage in sunflower (Shimshi, 1963). However, Dhopte and Aher (1977) reported reduction in transpiration by Cetyl alcohol and PMA application in sunflower in very low concentration. (Table 9 and 10). The CCC and PMA were found to cause early flowering in tomato (Mishra and Pradhan, 1972).

Some plant products are being explored by the researchers. The process is continued and no plant product has been reported so far as antitranspirants. Hopefully, we may have such products which are eco friendly and harmless in near future.

Table 12 : List of various types of antitranspirants (AT)

S.N.	Name of chemical	Type/Class	Mode of application
1	Adol-52 (Cetyl alcohol)*	Soil water binding	Soil
2	OED green (Octadecanol)*	–do–	–do–
3	S-800*	Plastic	Spray
4	Alkanyl succinic acid	Stomata	–do–
5	Sodium-L-Hydroxydecanol sulphonates	–do–	–do–
6	Monoethyl ether of decanyl succinic acid	–do–	–do–
7	8-Hydroxyquinoline sulphonate (8-HQ)	Stomata closing	–do–
8	Kaoline	Reflecting	–do–
9	Abscissic acid (ABA)	Stomata closing	–do–
10	Cycocel (CCC)	–do–	–do–
11	Ascorbic acid (AA)	–do–	–do–
12	Phenyl mercuric acetate (PMA)	–do–	–do–
13	Calcium bicarbonate	Reflecting	–do–
14	China clay	–do–	–do–
15	TIBA	Stomata closing	–do–
16	Waxol (Wax)*	Plastic	–do–
17	Mobileaf (Trade name)*	Plastic	–do–
18	Folicot (Trade name)*	Somata closing	–do–
19	Hico-110 R (Trade name)*	–do–	–do–
20	Polyetheline vinyl acrylate	Plastic	–do–
21	Phosphon	Stomata closing	–do–
22	Paclobutrazol	–do–	–do–
23	Methanol	–do–	–do–
24	24-D	–do–	Spray
25	Atrazine	–do–	–do–

* Some are trade names and active ingredients are not mentioned.

Kaya *et al.* (2006), pointed out that exogenous application of GA_3 improved the water stress tolerance in maize plants by maintaining membrane permeability, enhancing chlorophyll concentration, leaf relative water content (LRWC) and some macro-nutrient concentrations in leaves.

According to Sharp and LeNoble. (2002), the ethylene production is limited due to Abscisic Acid (ABA) and that interaction is involved in the effect of ABA status on shoot and root growth.

Gong *et al.* (2013), observed that with the application of Si, the leaves of stressed plants growing in pots were thicker compared to those without Si applied. Due to this it may have a beneficial effect by reducing the transpirational loss of water and maintain high RWC and water potential. According to them, the application of Si may be one of the available pathways to improve growth of the crop and increase its production in arid or semi-arid areas.

7. STOMATAL MECHANISM AND MODE OF ACTION OF ANTITRANSPIRANTS (AT)

In semi-arid tropics the AT's are found beneficial in WUE in various crops and restoration & conservation of soil moisture in rooting zone. The AT spray are always associated with increased leaf temperature due to increased stomatal resistances. In tomato and tobacco, the rise was observed to be 4 and 1.7°C with reduction in transpiration by 50 to 80% respectively. In other species, the rise in leaf temperature was insignificant with 30% reduced transpiration. The magnitude of TR reduction should not be 40% in order to have reasonable results without affecting growth considerably. The TR can be reduced to 50% without harmful effect. It should be remembered that the water stress occurs when TR exceeds the absorption. This undesirable effects can be relieved with AT sprays. (Kramer, 1983). The TR/Growth ratio must be maintained in plants and this can be achieved only when water losses are minimised without harmful effect on photosynthesis. Kastori *et al.* (1991) however reported reduction in CO_2 fixation from 25 to 66% with reduction in transpiration from 16 to 57% in various crops.

Table 13 : Response of various crops to antitranspirants (FOLICOT) on transpiration rate (Kastori *et al.*, 1991).

Crop	Reduction in Transpiration (%)
Sugarbeet	56.8
Phaseolus vulgaris	54.8
Chilli	48.4
Tomato	35.5
Maize	29.3
Grapevines	26.5
Sunflower	26.0
Tobacco	15.5

Stomatal opening is regulated by various sensors like, water, CO_2, light and hormones. The opening is strongly controlled by hydroactive mechanism while other sensors are hydropassive and is mediated through RWC of guard cells. In C_3 crops, the opening is slaved on guard cell chloroplast. Chemiosmotic mechanism predominantly regulates the stomatal opening. In C_4 crops, hormonal regulation is dominant and is highly sensitive to ethylene or ABA content. It is very efficient due to its special guard cells. In both types of crops, stomata closing types or reflectant types play significant role and TR is suppressed. The degree of stomatal closing is varied in different crops (Squire and Jones, 1971; Dhopte and Aher, 1977). The opening is partial and therefore the intake of CO_2 through stomata is not completely arrested. The photosynthesis (Ps) continues at slower space than the unsprayed plants (Gale, 1961). In various crops, the CO_2 enters through stoma during day through simultaneous loss of water (Except CAM plants).

Table 14 : Effect of methanol and urea on stomatal conductance and transpiration in groundnut in 1993 (data unpublished)

Treatment	Stomatal conductance (cm/s)	Transpiration (μg/ cm ^{2}/s)	Yield (kg/ha)
Control	0.23	6.48	1259
10%	0.13	4.93	1407
10%+Urea 2%	0.09	4.96	10.66
20%	0.19	6.08	1185
20% + Urea2%	0.05	2.07	1359
30%	0.20	6.17	1252
30% + Urea 2%	0.14	5.11	1344

Work at Dr. PDKV, Akola (MS) by Editor.

Table 15 : Effect of growth regulators two sprays in Rice Var. Co. 45 under light stress

Treatment (mg/l)	LAI	Chlorophyll (mg/g)	HI (%)	Yield (t/ha)
Control	3.40	1.82	38	3.56
CCC (500)	3.38	2.30	49	4.03
Salicilic acid (100)	4.42	2.37	45	4.62
Paclobutrazol (50)	3.55	2.26	48	4.06
Ascorbic acid (10)	3.55	1.97	42	3.90
Brassinosteroid (0.5)	4.43	2.4	46	4.76
CD 5%	0.48	0.45	7.0	0.38

Source : Maibangsa *et al.,* (1999), LAI = Leaf Area Index

Since stomatal opening is dependant on turgidity of GC, the RWC of GC must be higher. The Ps is higher at RWC 90%. If it remains between 80-90%, the process continues at slower rate and hence growth is affected. At severe water stress (RWC 70%), the growth is highly affected and productivity is reduced to half normal (Fig. 5). The water content monitors the Ps via hormonal regulations. The WUE is therefore the result of CO_2 fixed to the water lost through stomata. The efficient species in various crops need to be identified. A species or a variety that yields higher under low input of water are to be selected for dryland conditions. The ATs curtail the water requirement but it leads to proportionate reduction in Ps. Though the researches are being done since 1960, very few chemicals are found to have ideal properties of AT which are mentioned below (Gale and Hagon, 1966):

1. The chemical should be impermeabile to water and permeable to CO_2
2. It should be non-toxic to the plants and animals.
3. It should be have longer retention and greater specificity.

4. It should be degrade with solar UV radiation.
5. It should be relatively inexpensive and easily available.

Recently, methanol (10%) is shown to have AT property and reduces stomatal conductance and transpiration (Table 14). It is shown to be promising under dryland conditions and increases yield by 20- 30% in C_3plants (Nonomura and Benson, 1992). Very few chemicals have ideal properties of AT. The kaolin and lime water are found effective in some crops and are inexpensive. The only criticism the AT receives is that where to get huge quantity of water (500 L/ha) to spray the chemical under drought situation. This idea appears sloppy because to save or insure the crop from drought, some quantity of water is available. Two or three barrels of water can easily be managed through other resources if determinations are strong enough to safeguard the crop.

Wilson and Davies. (2006), reported in potted seedlings of four lines of maize and *Sorghum* differing in drought tolerance (subjected to a single soil drying cycle). They found that as plant water potential decreased in the unwatered plants of three of the lines investigated (Sorghum Piper and M35-1, V-4146 and maize Farz 27), endogenous levels of farnesol-like antitranspirants were increased. Closure of stomata correlated well with the increase in endogenous antitranspirant. In the fourth line (Sorghum M35-1, V-4184); stomata did not close as the level of plant water stress increased, although leaf diffusion resistance of even the well-watered plants of this line was quite high. In this line, there was no consistent relationship between plant water stress and antitranspirant level or between stomatal behavior and antitranspirant level. They also discussed the involvement of farnesol-like antitranspirants in the control of stomatal behavior in water-stressed plants.

Amor *et al.* (2010), found that the stomatal conductance and transpiration were reduced in sweet pepper, while the AT impaired photosynthesis at standard, but not at elevated CO_2 of fully irrigated plants. This effect was already apparent after 4 d of drought. Drought had a minor impact on chlorophyll fluorescence (Fv/Fm). Additionally, root respiration was increased at elevated [CO_2] but, after 8 d of drought, it was higher for plants treated with AT than for non-sprayed plants. Leaf water potential was affected more by drought at ambient compared to elevated [CO_2], and, especially after 8 d of drought. The AT minimized the reductions in leaf water potential. Leaf concentrations of proline and starch were affected by both [CO_2] and AT, especially after 8 d of drought. Moreover, increasing [CO_2] promoted the accumulation of starch, but led to decrease in the tissue concentrations of the soluble organic osmolytes, and hence diminished osmotic adjustment after 8 d of water withholding, relative to ambient [CO_2]. This study indicates that, in addition to the reported beneficial effect of elevated [CO_2] on drought stress, the AT could significantly improve drought tolerance in sweet pepper plants

According to Iriti *et al.* (2009), Chitosan (CHT) is a natural compound able to activate the plant own defence machinery against pathogen attacks and to reduce both transpiration and stomatal opening when applied as foliar spray. The data here reported show that CHT-induced antitranspirant activity in bean plants and is mediated by ABA, whose level raised over threefold in treated leaves, 24 h after foliar spraying.

8. CLASSIFICATION OF AT AND FIELD RESPONSES

The ATs are categorically classified on mode of action in following four classes.

I. Materials causing sotmatal closure

1. Herbicides like 2, 4-D, phosphon and Atrazine

2. Fungicides like PMA
3. Metabolic inhibitors like hydroxysulphonates, potassium metabisulphite.
4. Growth hormones like Cycocel, Ascorbic acid, Succinic acid, Ethrel, Abscissic acid.

II. Reflectant types

1. Kaoline, 2. China clay, 3. $CaHCO_3$, 4. Lime water

III. Thin film forming chemicals

1. Hexadecanol (Higher alcohols),
2. Cetyl alcohol
3. Triacontanol 4. Methanol 5. Paclobutrazol 6. Brassinollide 7. Resorcinol etc.

IV. Polyetheline material forming thick films

1. Mobileaf, 2. Folicot, 3. Waxol, (All trade names given by companies)

The trials are conducted on several crops and these ATs are shown to have beneficial effects in increasing biomass and economical yield (Table 19). The TR is basically a cooling process and if arrested it leads to increase in leaf temperature. It is therefore advised that measures to reduce the leaf area (Shoot growth) or increasing dessipation of solar energy by reflectants would be more appropriate in soil water conservation. Kaoline and lime water spray are advised after cesation of rains. In central plains, mid Sept or First week of Oct. can be suitable period to undertake this prophylactic measures. The beneficial response of hirsutum cotton to AT, Hicoemulsion - 110 R is reported by Dhopte and Jamdar, (1995). The data given in Table 16 A and 16B give the information. The purpose of AT is to maintain the growth and productivity under stress conditions and it is never recommended for higher productivity/unit area. It saves the crop and helps to get marginal yields when expectations are zero. Some products are released as AT in USA for better results. Some ATs can be tried through drip irrigation to save the frequency as well.

Table 16A : Response of dwarf hirsutum cotton to antitranspirants and growth regulators under rainfed conditions (Dhopte and Jamdar, 1995).

Treatment	Stom. Cond. (cm/s)	Dry matter prod. (g/plant)	Yield (Kg/ha)	
			1986	1987
Control	54.8	54.4	773	638
Kaoline (1.25%)	55.8	69.7	914	718
Hico-110R (1 ml/L)	48.5	46.7	889	735
Kaoline+Atonic+AA	43.8	56.2	977	750
PMA (10 ppm) + AA	50.8	51.9	906	710
CD = 5%	–	14.7	78.9	100

Table 16B: Response of cotton hybrid AHH-468 to antitranspirants, Hico-110 R (Dhopte and Jamdar, 1995).

Parameter	Control	Treated
Stomatal cond.(cm/s)	0.26	0.22
Transpiration rate (μg/cm^2/)s	8.32	7.34
Seed cotton yield (kg/ha)	950	1048 CD 5% = 73.8

Table 17 : Response of chickpea to various antitranspirants under limited irrigation (Anon, 1995-96). PCB = Paclobutrazol, Data of Author unpublished.

Treatment	Conc.	DMP (g/plant)	Seed yield (Kg/ha)	
			1995	1996
Control Water	14.8	1247	1315	
CCC	50 ppm	21.3	1203	1435
PCB	100 ppm	14.0	1250	1408
Hico-110R	100 ppm	12.4	1250	1306
8-HQ	100 ppm	24.4	1366	1435
Kaoline	2%	18.0	1181	1352
CD (5%)	–	4.36	NS	34

Table 18 : Response of rajma (*Phaseolus vulgaris*) to various antitranspirants (Rahangdale *et al.*, 1997) under limited irrigations

Treatment	Leaf temp (°C)	Transpiration (μg/cm^2/s)	Stom. Resi. (s/cm)	Yield (kg/ha)
Control	13.8	1.41	20.0	1000
PMA (10 ppm)	32.7	1.45	20.0	1270
8-HQ (10)	35.5	1.13	25.0	1290
Hico-110R (1 ml/L)	32.3	1.17	23.0	1400
CCC (10 ppm)	32.1	0.90	28.9	1290
Kaoline (1%)	31.4	0.57	66.6	1260
Succinic acid (0.1%)	30.7	0.85	27.8	1300
Paclobutrazol (20)	29.3	0.79	22.5	1460
CD (5%)	0.94	0.16	19.0	99

The productivity of rajma was significantly increased by 26.2, 23.6 and 15.4% over control by sprays with Hico -100 R, paclobutrazol and 8-HQ respectively under 6 limited irrigations as against 9 irrigations and saved water under rabi season. Thus, with limited irrigation potential, the crop of rajma can be taken with high remuneration. This is another advantage a ATs are having in dryland farming. The data in Table 17 indicate that the response is not significant during 1995 and this is due to more rainfall received in this season. Thus, positive responses during 1997 can be obtained only under adverse season.

9. CONSTRAINTS IN ADOPTION

The use of AT in agriculture is critcised for followed reasons :

1. Water is not available for spraying
2. Some ATs are expensive and are not cost effective.
3. Some chemicals are not available in shops.
4. Residual toxicity in soil and plant is not known.

The PMA can be used on crops like cotton, flowering plants to increase yield and shelf life of flowers. The turf grasses and lawns can be maintained green when water scarcity is found. Road side tree avenues can be maintained under severe drought. The non-availability of water can pose some problems but conservation of species under emergency is highly essential and water can be managed. Soil application may need little quantity of water. The chemical may bind water ten times of its weight. The atmospheric water can also be trapped with these chemicals and be made available to plants/tree. Some chemicals like triacontanol, paclobutrazol and resorcinol are imported and are definitely costly and it is said to be not easily available in market. The effective concentration of AT is very low and hence it can be cost effective. The positive response is dependant on climatic conditions. If rains are normal, the response may not be promotive but if drought occurs, it gives positive responses. This must be kept in mind and discouragement should be avoided. Timely spraying is, however essential after analysing the weather forecast. Constant watch is therefore necessary about rainfall and its probability for next 48 hours. Delayed action may not be fruitful. The advise can be obtained from experts. Discouragement is seen when positive responses are not experienced and this is due to failure of forecast. Prophylactic measures are always employed to avoid discouragement. The sprays are not costly, it should be done without waiting for adverse condition to occur. If it rains, you may loose Rs. 500 per hectare. In modern information technology, forecasting is highly accurate with internet satelite communication, website information. In such modern era, forecast prediction is 98% accurate. A farmer can plan such operations for better effects to accomplish targetted goals. He may become a member of such system through Agri. Universities.

The ATs are not some time available in market because of no demand. If demanded, supply can be assured. Chemical practiced in field is amply available in market. The effective use of chemical is not demonstrated to the farmer in past due to highly unpredictable weather. If sprayed, response may be positive or negative. This is for assurance of yield if adverse situation arises. Crop insurance is needed to avoid total failures. This is not properly understood by farmers.

Table 19 : Response of various antitranspirants under rainfed farming

Crop	Antitranspirants	Con.	Persistance (Days)	Response	Reference
I. Cereals:					
Wheat	PMA	$5x10^{-5}$	Wilting defered		Brengle (1968)
Maize	PMA	1 ppm	–	Reduced TR by 30%	Cocking and Tukey (1970)
	SA	0.1 ppm	–	Stomata closed	Roberts (1961)

Crop	Antitranspirants	Con.	Persistance (Days)	Response	Reference
	Hexadecanol	25-50 lbs/acre	–	Yield increased	Miller (1969)
	CCC	1000 ppm	–	TR reduced by 30%	Gale and Hagan (1966)
	Folicot	–	12	–do–	
II. Oil seed crops					
Cotton	Adol-52	–	–	TR reduced	Slatyer and Beirzuen (1964)
	AA	100 ppm	–	–do–	Verma (1976)
	PMA	10 ppm	–	–do–	Dhopte and Jamdar (1995)
	OED-green	–	–	–do–	Slatyer and Berzuen (1964)
	Kaoline	1.25%	–	Yield increased	Dhopte and Jamdar (1995)
Sunflower	PMA	1 ppm	10	TR reduced by 50%	Shimshi (1963)
	Cetyl alcohol	30 kg/ha	–	Yield increased	Dhopte and Aher (1976, 1977)
	Folicot	1:0	12	TR increased by 20%	Kastori *et al.*, (1991)
Peanut	Agent No. 1	75 g to seed	10	Yield increased by 16%	Liu et al. (1989)
	Kaoline	6%	15	Yield increased	Khan and Morey (198))
	Palcobutrazol	25 ppm	–	RWC and Chlorophyll increased	Anon (1996) and Anon (1995)
	Methanol	10%	–	–do–	Amaregouda *et al.*, (1994)
	8-HQ	10 ppm	10	–do–	De and Giri (1978)
Mustard	Kaoline	6%	–	Increased WUE by 10%	Patil and De (1978)
III. Legumes and Pulses					
Phaseolus vulgaris	Kaoline	6%	10	Improved water status	Kharnyan and Vikhera (1975)
	CCC	low	10	–do–	Giri and De (1976)
	Folicot	1:20		TR reduced by 55%	Kastori *et al.*, (1991)
Rajma	8-HQ	10 ppm	12	Yield increased	Rahangdale et al. (1991)
	PCB	20 ppm		TR reduced	Anon (1996)

Crop	Antitranspirants	Con.	Persistance (Days)	Response	Reference
	PCB	25 ppm	10	–do–	–do–
Chickpea	PCB	25 ppm	10	–do–	Bora and Mathur (1989) Dalvi et al. (1991) Chetti et al. (1994)
Cowpea	PCB	10 ppm		–do–	Bora and Mathur (1989) Shahine *et al.,* (1992)
IV. Flowering crops					
Chrysanthemum	PMA	2.6 ppm		TR reduced by 30%	Cocking and Tukey (1970)
Flowers	Waxon	–		Increased shelf life	Gale & Hagan (1966)
V. Fruits and Vegetables					
Tomato	PMA	100 ppm	6	Stom. closed by 80%	Cotter (1970)
Folicot	1:20	12	TR reduced by 36%		Mishra and Pradhan (1972)
	Hexadecanol	20-50 lbs/acre (Soil)		Yield increased by 80%	Gale & Hagen (1966)
Strawbery	8-HQS	–	–	Stom. closed 80%	Stodart and Miller (1962)
Citrus	Cetyl alcohol	0.1-10%		TR reduced 15-60%	Mishra (1962)
Grape	PEVA	Spray	–	Profitable	Gale & Hagen (1966)
	FOLICOT	1:20	12	TR reduced by 27%	Kastori *et al.* (1991)
Banana	PEVA	1:20	–	Good effects	Gale and Hagen (1966)
Sugarbeet	FOLICOT	1:20	12	TR reduced by 57%	Kastori *et al.,* (1991)
Chilli	Methanol	10%	15	Yield increased	Anon. (1997)
	FOLICOT	1:20	12	TR reduced by 47%	Kastori *et al.,* (1991)
Cherry	Waxol	–	–	Increased size	Davenport *et al.,* (1969)
Apple	–do–	–	–	–do–	-do-
Water melon	AT	–	–	WUE improved	Abdel and Naseer (1993)
VI. Cash crops	HS	0.1ppm	14	TR reduced	Zelitch and

Crop	Antitranspirants	Con.	Persistance (Days)	Response	Reference
					Waggoner
	PMA	10-4M		Yield increased	-do (1962)
	MOBILEAF	dip		–do–	Collins *et al.* (1970)
	ABA	0.1 ppm	12	Stom. closed	Jones and Mansfield (1970)
	Folicot	1:20	12	TR reduced by 16%	Kastori *et al.*, (1991)
VII. Forest sp.	PEVA	Spray	–	Increased survival	Berzilov & Khvalova (1970)

A literate farmer is reaping the advantage of ATs because he has paying capacity. A poor farmers is reluctant to spend even negligible amount. A Govt. motivation is therefore necessary to support his will power. There should be an urge to adopt full proof technology with proper monitoring to insure crop yields under aberrant weather. Farmer's reluctance is due to his self satisfaction about what he gets from his fields. He has never been greedy of money. All he wants is yearly ration for his family. This psychology needs to be changed in national interest and perspective logic needs to be inculcated in farmer's community because he produces not for himself alone but for the country as well.

10. COST EFFECTIVENESS AND PERSPECTIVES

Modern technology is developed in order to suit the present need for gallowping population. Every agrochemicals are deadly but it is being used to increase the productivity to feed a billion of population in India. We have no options. Had there not been a green revolution in past three decades, 1/4 of the population could have succomed to death. Thanks to breeders who have evolved hybrids in staple crops which made available the food for millions of people. Secondly, farmers are also equally credited for adopting of new seed material and improved technology. Today we are first in milk production, in the world with food security. Land remains same with simultaneous increase in population and this is the only major problem facing to policy makers. Deadly chemicals are being used to maintain the productivity. No chemical is 100% safe, including, drugs practised in allopathy. If not employed, the people may face starvation in days to come. Bio and eco friendly products are being developed to day with the reason that we have already enough spoiled in this sector. Useful flora, insects, soil degradation, pollution, endangering the environment are some of the results of indiscriminate use of poisonous chemicals in agriculture. Three decades from now on the global level, another 2.5 billion people need to be fed which amounts to 2/5 more than the existing today. India particularly may need to produce about 225 to 240 million tonnes of food grains in a period of next ten years (by 2010 AD). Todays challenge is made tougher by decline in the environmental resource base, water shortage, desertification, soil erosion, global warming and gallowping population. With heavy efforts through Agri. agencies, ICAR, state Agri. Universities and Departments, we could be able to achieve green revolution and increase our food basket from 50 million tonnes in 1950 to about 203 million tonnes in 1999. This has been radically achieved only through improved seeds and modern agrotechniques. There is no substitute for food, a dire need of man and animals. How do we cope up the growing demands of the day? If agro-chemicals are not employed, the production shall go down and it will lead to anarchy and lead to total deterioration of social health. If this be avoided, there is no way other than adopting precautionary preventive measures to increase food basket. Environmental damage is already done during past three decades. Another three decades may be needed to improve upon the present situation and who will feed growing population is sereous concern to the policy makers. Food security is needed by 2010 AD. Biodegradation, residual toxicity of ATs are considered only when stomach is full. All that man needs is food on first priority and therefore though it is essential, may not have first priority at national level.

The cost effectiveness of AT is found remunerative and most of the ATs being used in very low con. are profitable. The cost incurred normally is Rs. 400-500/ha and is easily in reach among average farmers. If rains continue after spray, responses may not be visible while it responds well under water stress conditions. Farmers are advised to adopt inexpensive chemicals, like CCC, Kaoline, Lime water, 8-HQ, china clay for higher productivity under dryland farming. The ATs are required to be timely sprayed at critical phenological stages for better results after keeping watch on weather and climate. The Extension agronomist at NARP/NATP must be consulted for proper (?) advise.

The rabi crops like wheat, mustard, chickpea, rajma, sunflower, safflower grown on residual moisture can equally be benefited and irrigation frequency can be reduced. Since persistance of AT is for 10-15 days, minimum two sprays are essential for demonstrable results. Most of above mentioned chemicals are biodegradable and will not have harmful effects on man and animals. The use of PMA should however be restricted to non-edible crops, ornamental crops, forest species and to maintain shelf life of cut flowers like roses, chrysanthemum, gladiolus, orchids, gerbera and merigold. This practice can be profitable to growers and exporters of such perishable articles. The use of waxol is already recommended in Mango, oranges, banana, apple, citrus, and papaya and the growers fetch remunerative prices for their produce in the export market. The following precautions are required to be necessarily followed:

1. Spray after rainfall stops.
2. Ensure two sprays of AT at an interval of one week.
3. Surfactants (Teepol 0.1%, Detergent powder, 1 g/l) should be mixed in spray for effective response.
4. Repeat spray of AT if rains follow after one hour of spray.

11. CONCLUSIONS

What is assured by AT application can be summerised below :

1. Yield levels can be optimised.
2. Better crop growth is assured.
3. Normal sized grain is obtained.
4. Seed quality is improved and produce can be used for seed purpose.
5. Crop life is saved with marginal crop productivity.
6. Irrigations can be reduced in post rainy season crops.
7. If applied through drip irrigation (Cetyl alcohol or Hexadecanol), the irrigation frequency can be minimised.

8. Some yields are assured when no yields are expected out of crop sown during severe drought.
9. Drought severity can be monitored.
10. Losses can be minimised with limited inputs.
11. Fast receding soil moisture can be arrested for better growth of rabi crops.
12. Very useful for farmers having minimum irrigation facilities.
13. Large nurseries can be saved when water is scarse in April & May.

The role of AT is important when crop life is in geopardy as in case of human being. A drug that can save the life is administered without considering its side effects. Many therapies and drugs against cancer are slightly harmful but life is saved without considering residual toxicity. Plant life is equally important for a farmer to sustain his family. Every drop of water received after waiting for eight months is precious and need to be effectively managed for human civilization to prevail. Life exists where water prevails. The avg. rainfall in India is 550 mm as against 500 mm in America, but water conserved on soil is 10% in India and 94% in America. It shows how poor we are in planning for future. Next generation will not excuse us for land degradation, nonmanagement of water, deforestation, pollution and dirty climate with high CO_2 for survival by 2025 AD. Rain water conservation should start at the point where it is received (Fields). No other strategy can be more profitable other than *in situ* conservation of water through watersheds and its effective utilization for crop growth with chemicals like antitranspirants. These chemicals can serve as boon to the frustrated farmers if rightly adopted with perspective vision to have food security. Miserable status of poor countries can be improved. Sustainable agriculture is the dream of many scientists to wipe out tears of poor farmers. Adoption of agrotechniques is the only solution for farmers of dryland to save millions of world population in millenium in developing countries.

REFERENCES

Abdel and Naseer, G (1993). Effect of antitranspirants on growth, yield , water content and water use of a squash plant. J. Agri. Sci. (Egypt) 24 (2) : 193-210.

Amaregouda, A., M.B. Chetti, K.C. Mandloi (1994). Effect of antitranspirants on stomatal resistance and frequency, RWC and pod yield in summer groundnut. Ann. Plant Physiol. 8 (1) : 18-23.

Anna Ranuzzi and Richa Srivastava. (2012). Impact of climate change on agriculture and food security. *ICRIER Policy Series, No. 16.*

Anonymous (1988, 1995, 1996, 1997); Ann. Rev. of Research Review Com. Report. Plant Physiology and Biochemistry. Dr. PDKV, Akola (MS), India.

Anonymous (2011). AgriInfo.in.Antitranspirants.http://www.agriinfo.in/?page=topic&superid=1&topicid=447

Bathakal, B.G. and S.V. Kaore (1980). Potential evaporation over the *Vidarbha* region. PKV. Res. J. 4 : 13-17.

Berzilov, V.F. and L.A., Khvalova (1970). Possibilities of reduction of plant transpiration during summer transplantation by means of antitranspirants. Inv. Acd. Nauk. Ser. Biol. 4 : 562-68 Quoted from Biolo. Abst. 1971 No. 118667

Bora, K.K. and S. Mathur (1989). Paclobutrazol as an antitranspirants in cowpea cultivar. Comp. Physiol. and Ecol. 14 (30) : 159-162.

Brengle, K.G. (1968). Effect of PMA on growth and water use of spring wheat. Agron. J. 60 (2) : 246-247.

Cengiz Kaya, A. Levent Tuna and Alfredo A. C. Alves (2006). Gibberellic acid improves water deficit tolerance in maize plants. *Acta Physiologiae Plantarum*, Vol. 28. No. **4. 331**-337.

Chema, S.S.; O. P. Malhotra, and J. Singh (1975). Effect of cycocel on barley under rainfed conditions. Indian. J. Plant Physiol. 8: 200-203.

Cocking, W.D. and H.B. Tukey (1970). Reduction in water loss from chrysanthemum sp. following foliar application of PMA. J. Amer. Soc. of Hort. Sci. 95 (3) : 382-384.

Collins, W.K.; J. Hawks; S.N. Kitreu, B.U. (1969). Yield of fluecured tobacco to an antitranspirant applied to transpirant. Tobacco Sci. 13 : 70.

Cotter, D.T. (1970). Phenyl mercuric acetate effects on water loss of tomato. J. Amer. Soc. Hort. Sci. 95 (6) : 774-776.

Dalvi, D.G., P.R. More and K.A. Nayeem (1991). Effect of antitranspirants on tranpiration rate, RWC, water saturation deficit and yield in chaffa chickpea. Indian J. Agri. Sci. 61 : 204-206.

Davenport, D.C. (1967). Effect of antitranspirants on transpiration and growth of grass. J. Expt. Bot. 18 : 332-347.

Davenport, D.C. and R.M. Hagen (1975). Role of antitranspirant in arid agriculture. In : Physiological aspects of dryland farming. By U.S. Gupta (Ed). pp. 315-323.

De, R. and G.Giri (1978). Effect of mulching and Kaoline foliar spray on mung (*Vignaradiata*), rapeseed, double cropping system under rainfed condition. J. Agri. Sci. (Camb.) 91 : 191- 197.

Dhopte, A.M. and S.B. Amarshettiwar and R.L. Kalane (1994). Linear models about soil moisture use for various leaf types in hirsutum cotton under competitive and water scarcity conditions. PKV. Res. J. 18 (1) : 68-72.

Dhopte, A.M. and S.B. Lall (1987). Relative efficacy of antitranspirant, growth regulators and mineral nutrients in control of leaf reddening in hirsutum cotton under dryland conditions. Ann. Plant Physiol. 1 (1) : 56 - 71.

Dhopte, A.M. and S.L. Jamdar (1995). Effect of antitranspirants and growth regulators on productivity of upland cotton. J. Cotton Res and Develop. 9 (2) : 136-141.

Dhopte, A.M. and S.S. Aher (1975). Effect of antitranspirants on water turn over, growth attributes and yield of sunflower. Oil Seed J. April/June :41-45.

Dhopte, A.M. and S.S. Aher (1977). Effect of antitranspirants on transpiration and stomatal dynamics of sunflower. Agri. Coll. Mag. Nagpur, India. 49 : 33-38.

Dhopte, A.M. and V.R. Zade (1980). Combined effects of ascorbic acid, CCC, PMA and nutrients on physiology of Desi and American cotton. PKV. Res. J. 4 : 24-27.

Dhruve, J. J., Jaydeep P Bhatt, Sarang Sapre, Avadh K Shah and K. B. Kathiria. (2011). Effect of Silixol as a silica source on quality and yield of chilli (*Capsicum Annum, L*). *International Journal of Pharma and Bio Sciences* **1**-39.

Francisco M. del Amor, Paula Cuadra-Crespo, David J. Walker, José M. Cámara and Ramón Madrid. (2010). Effect of foliar application of antitranspirant on photosynthesis and water relations of pepper plants under different levels of CO2 and water stress. *Journal of Plant Physiology* **167, 1232**–1238.

Gale, J. (1961). Studies on plant antitranspirants. Physiol. Plantarum. 14 : 777-786.

Gale, J. and M.A. Poljakoff (1963). Effect of antitranspirant sprays on uptake and transport of rubidium by plants. Plant Cell physiol. 4 (3) : 285-288 (Tokyo).

Gale, J. and R.M. Hagan (1966). Plant antitranspirants. Ann. Rev. Plant Physiol. 17 : 269.

Gale, J; B.R. Edwin and R.M. Hagen (1967). High alcohols as antitranspirants. Water Res. Rev. 3 (2) : 437-441.

Giri, G. and Rajat, De. (1976). Kaolinite-an antitranspirant for dryland crops. Science Reporter. 13 : 430-431.

Haijun Gong, Kunming Chen, Guocang Chen, Suomin Wang and Chenglie Zhang. (2013). Effects of silicon on growth of wheat under Drought. *Journal of Plant Nutrition* Vol. 26, No. **5, 1055**–1063.

Jianhua Zhang, Wensuo Jia, Jianchang Yang, Abdelbagi M. Ismail. (2006). Role of ABA in integrating plant responses to drought and salt stresses. *Field Crops Research* **97, 111**–119.

Jones, R.J. and T.A. Mansfield (1970). Suppression of stomatal opening in leaves treated with abscissic acid. J. Expt. Bot. 21 : 714-719.

Kastori, R.N.; Petrovic and Z. Stankovi (1991). Bioloski Vestnik. 39 (1-2) : 109-113 (Yugoslavia)

Kayande, K.S; R.M. Bhake; P.L.Narnaware and G.M. Bharad (1991). Dryland farming in Vidarbha (Marathi). Publ. of PKV, Akola (MS), India.

Khan, G.Z. and D.K. Morey (1980). Influence of kaoline spray on transpiration and water use efficiency in groundnut under different soil moisture regimes. J. Mah. Agri. Uni. 5 : 131-134.

Kharnyan, N.N. and V.N. Vikhera (1975). Fiziol. Raste. 22(4) : 806-809.

Kramer, P.J. (1983). Water relations of plants. Acad. Press, New York.

Liu, Q.S.; Y.S. Cao, X,Z, Wang; F.Z. Shi, W.D. Ding and Z.T. Zang (1989). Application of drought resistant agents to groundnut on different dates. Jianju Agri. Sci. 5 : 11-12.

Maibangsa, S; M. Thangraj and Stephen Roy (1999). Alleviation of low irradiance stress in rice (*Oryza sativa*) by growth regulators. Ann. Plant Physiol. 13(2): 133-142.

Majernick, O. (1970). Response of stomata of barley and maize to PMA. Biol. Plant. 12 (6) : 419-423.

Marcello Iriti, Valentina Picchi, Mara Rossonia, Stefano Gomarasca, Nicola Ludwig, Marco Gargano and Franco Faoro. (2009). Chitosan antitranspirant activity is due to abscisic acid-dependent stomatal closure. *Environmental and Experimental Botany* **66, 493**–500

Miller, N.A. (1969). The effect of several transpiration retardants on water balance, stomatal aperture and leaf temperature on field corn leaves. Diss. Abst. 29 (10) : No. 364013.

Mishra, D. and G.C. Pradhan (1972). Effect of transpiration reducing chemicals on growth, flowering and stomatal opening in tomato plants. Plant physiol. 50 : 271-274.

Nonomura, A.M. and A.A. Benson (1992). The path of carbon in photosynthesis. Improved crop yields with methanol. Proc. Nat. Acad. Sci. USA, Vol. 89. pp. 9794-9798. Agri. Sciences.

O.E. Mohawesh, K.M. Al-Absi and M.J. Tadros. (2010). Effect of antitranspirant application on physiological and biochemical parameters of three orange cultivars grown under progressive water deficit. *Adv. Hort. Sci.*, 24**(3): 183**-194.

Pasternak, D. and G.I. Wilson (1971). Some factors responsible for varying effectiveness of stomatal closing antitranspirants. Aust. J. Expt. Agri. 11 : 48-52.

Patil, B.P. and R.De (1978). Influence of antitranspirant on rapeseed plant under water stressed and non-stressed conditions. Plant Physiol. 57 : 941-943.

Peters, D.B. and W.J. Roberts (1963). Use of octadecanol as transpiration suppresants. Agro. J. 55 : 79.

Rahangdale, S.L.; A.M. Dhopte and K.S. Autkar (1997). Effect of antitranspirants on growth and yield of rajma. Ann. Plant Physiol. 11 (2) : 132-136.

Robert E. Shapr and Mary E. LeNoble. (2002). ABA, ethylene and the control of shoot and root growth under water stress. *Journal of Experimental Biology*, Vol.53, No.**366, 36-37**

Sadhu, M.K. (1975). Antitranspirants reduce water needs in agriculture. Sci. Reporter. pp. 980-990.

Shahine, S.H.; E.L. Desouky, S.A. Abdel Nayeem and A.L. Wani (1992). Response of pea (*Pisum sativum*) to foliar sprays of some growth regulators. Ann. Agri. Sci. 30 (2) : 755-775.

Shimshi, D. (1963). Effect of chemical closure of stomata on transpiration on varied soil and atmospheric environment. Plant physiol. 38 : 709-712.

Slatyer, R.O. and Bierhulzen (1964). The effect of foliar sprays on transpiration and water use efficiency of cotton plants Agri. Meteorol. 1 : 42-53.

Squire, G.R. and M.B. Jones (1971). Studies on the mechanism of antitranspirants, PMA and its penetration into mesophyll. J. Expt. Bot. 22 : 980-991.

Stodart, E.M. and P.M. Miller (1962). Chemical control of water loss in growing plants. Science. 137 : 224-225.

Verma, S.K. (1976). Control of abscission in cotton. (*G. hirsutum*). Proc. Indian Sci. Cong., 63 rd session. Agri. Sci. Section. Abst. 168 pp 100

Wilson, J.A. and W. J. Davies. (2006). Farnesol-like antitranspirant activity and stomatal behavior in maize and Sorghum lines of differing drought tolerance. *Plant, Cell & Environment* **2, 1: 49**–57.

Zelitch, I and P.E. Waggoner (1962). Effect of chemical control of stomata on transpiration of intact plants. Proc. Nat. Acad. Sci. US. 48 : 1297-1299.

Chapter - 20

PRE-SIGNS OF PHYSIOLOGICAL MATURITY FOR EARLY HARVEST

*N.G. Zode**

Seed Research Officer, Seed Technology Research Unit,
Dr. Panjabrao Deshmukh Krishi Vidhyapeeth, Akola - 444 104 (MS)
* Present address : Incharge & Senior Rice Breeder, Paddy Research Station,
Sakoli, Distt. Bhandara (M.S.)

1. INTRODUCTION

The deterioration of seed during the post maturation-Pre harvest period is a serious seed production problem. The frequent rains combined with high temperature result in rapid losses of viability and vigour of seed in standing crops.

The seed moisture content at physiological maturity ranges from 30-55% depending upon the species. In sorghum, it is reported around 30-40% while in cotton and groundnut it is noted around 50-55%. From physiological maturity there is a fast reduction in seed moisture content till it reaches the harvest maturity (below 20%). Climatic conditions during this post maturation pre harvest period have a great influence on quality of seed and grains harvested.

During Kharif 1987, the Maharashtra State Seed Corporation has record break seed production of hybrid sorghum CSH-5 (40,000 quintals) and CSH-9 (60,000 quintals). However, the seed production was caught in late September rains before harvest and resulted in failure of seed lots to the extent of 94-95% i.e. 5-6% seed lots only could achieve the germination up to minimum seed certification standard of 75%. Zode *et al.,* (1993) reported that sorghum seed caught in rain had poor germination and vigour. Now this phenomenon becomes a regular feature of the region, because of which whole sorghum seed production in Maharashtra State is shifted to Andhra Pradesh. Similarly the *Kharif* commercial sorghum is discoloured (blakenned) due to grain moulds. The black sorghum is not preferred for consumption & it has no market value. Government has to procure this produce, for protection of interest of farmers.

The cotton bolls picked from lower part of the plant, are poor in seed viability and vigour (Anonymous 1995). The produce from early picking of kharif cotton, is found inferior in quality as compared to later stages. The cotton bolls of early picking are mostly from the lower part of plant which were set early and are more in proximity of soil surface. These are exposed to high humidity and high temperature during developing and after boll burst causing severe damage to seed quality. Simpson and Stone (1935) reported losses of 20-30% in seed viability after only one weaks exposure to rainy conditions.

During Kharif 1997, due to continuous rains in months of Nov.-Dec. the cotton seed germinated in bolls on the plant. The high humidity or rains during cotton boll maturation and boll opening increase the free fatty acids content of seed which causes the seed viability and vigour loss, and affects the milling quality of seeds.

The soybean seed maturity during *Kharif* coincides with late rains during September -October months which are also characterised by high humidity and high temperature. These wheather conditions during post maturation preharvest period cause moderate to severe seed quality problems. Alternating periods of high humidity and high temperature accelerates seed deterioration. Yaklich (1987) studied the field weathering of pods and seed of soybean by determining moisture content. The results showed that pod and seed moisture changed daily during day and night. The largest single day increase in moisture content was 13.8% and 5.6% for pods and seed respectively. This diurnal change in moisture content leads to deterioration of seed quality.

The paddy seed/grain is also vulnerable to weather conditions. The harvested/ heaped paddy if caught in rains or high humidity and high temperature conditions lowers seed viability. During *Kharif* 1997, all the samples of paddy seed were caught in rains after harvest, lost germinability, however the samples from standing crop were least affected. The rice grain below the husk turns brown and reduces its market value.

A long term concern of farmers and seed companies is protection of seed or grain from the adverse environment. There is variation among crop species for resistance to weathering. Dormancy provides the natural protection against seed deterioration while seeds are still on the plant, however, it is not desirable in annual cropping system and man (including breeders) has selected for reduced dormancy in most major crop species.

Apart from the weathering of seeds, there are other factors also to be considered. During post-maturation-pre-harvest stage, the crop is always attacked by birds. The severity can be very high in crops like sunflower, which is severly damaged by parrots, if not cared. Another point is that there is always threat of theft. The produce which comes within reach shall be lost due to theft. If harvesting is delayed, there is possibility of grain shattering. It is true in case of crops like Soybean, Sesame, Pigeonpea etc.

2. HARVESTING AT PHYSIOLOGICAL MATURITY (PM)

Looking to the problems faced, harvesting the crop at physiological maturity or at an earliest after physiological maturing would be highly desirable.

Seed vigour and viability remains maximum at physiological maturity and it deteriorates progressively with delay in harvest because it is subjected to field weathering by changes in climatic conditions such as temperature and relative humidity (Delouche 1980). At physiological maturity, the seed attains maximum dry weight per seed, maximum viability and vigour, reduction in moisture content (Table 1) and initiation of seed shrinkage. The lower electrical conductivity of seed leachate at PM is also reported as indicator of superior function of cell membranes. The nutrient and water supply to the seed is "Cut off" by formation of abscission layer in between seed and mother plant; starting the rapid drying phase with initiation of seed shrinkage. In most of the agricultural crops this stage is attained at about 8-10 days earlier than harvest maturity.

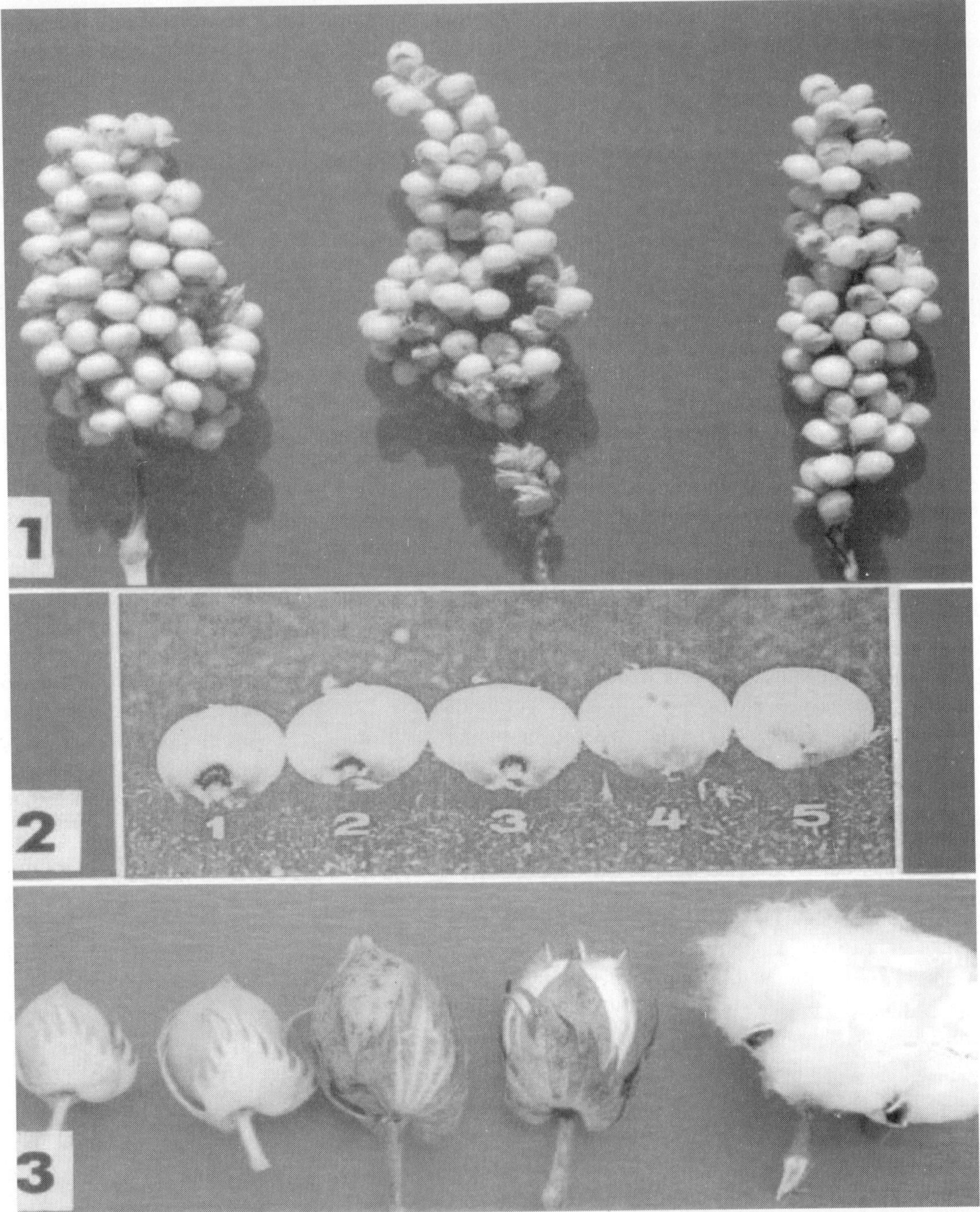

Plate 20.1. 1. Sorghum spike branch, left dough stage (immature), Physiological maturity (Middle) and stage of advanced maturity (Right). 2. Sorghum grains at various stages of (PM) maturity from most mature (1) to lest mature (5). Black layer is first visible in 3 and shows clearly in 2 and 1. 3. Maturity stages in cotton (PKV hybrid 4), 4th boll is mature (45 DAA) and drying onwards leading boll opening.

Plate 20.2. 1. Soybean cultivars with different maturity period. Note ununiform yellowness. 2. Sign of physiological maturity in soybean (JS-335), 4th pair with seed and pod wall (Dark yellow), Blackness shows advanced stage.

Table 1. Seed development and maturation characteristics in crops

S. No.	Soybean JS-80-21				Cowpea RC-19				Sorghum CSH-14				Cotton AKA-8410			
	01	02	03	04	01	02	03	04	01	02	03	04	01	02	03	04
1	28	74	21	00	08	83	05	15	10	62	07	07	20	80	26	00
2	31	73	23	00	09	81	06	14	13	53	14	32	27	69	34	02
3	34	68	38	44	10	71	11	26	16	42	17	44	34	61	48	16
4	37	64	49	60	11	66	25	52	19	40	19	46	41	48	61	25
5	40	59	58	73	12	57	35	64	22	39	26	48	48	26	62	62
6	43	47	74	80	13	51	53	60	25	33	27	56	55	09	63	66
7	46	31	82	82	14	46	65	81	28	29	27	74	58	07	63	66
8	49	17	88	84	15	45	71	84	31	28	28	84	–	–	–	–
9	52	11	89	83	16	38	73	87	34	20	28	84	–	–	–	–
10	–	–	–	–	17	30	75	89	37	15	28	86	–	–	–	–
11	–	–	–	–	18	26	76	89	–	–	–	–	–	–	–	–
12	–	–	–	–	19	12	76	94	–	–	–	–	–	–	—	–

01 = Days after anthesis 02 = Moisture content 03 = Dry seed weight mg/seed 04 = Germination percentage

Table 2. Vigour index for soybean varieties harvested at various dates.

S.N.	Treatments	Varieties			
		PKV 25	PKV 1	JS 35	Mean
1	5 days pre P.M.	1491.88	1941.67	1750.51	1728.02
2	P.M.	2002.02	2691.04	2347.09	2346.72
3	P.M. + 5 days	1883.64	2599.34	2259.38	2247.52
4	P.M. + 10 days	1856.37	2513.17	2190.83	2186.79
5	P.M. + 15 days	1823.30	2402.03	2058.22	2094.52
6	P.M. + 20 days	1735.22	2316.65	1988.79	2013.55
	Mean	1798.78	2410.65	2099.17	
		V	H	V x H	
	SE (M)	9.744	13.79	23.87	
	CD at 5%	26.738	37.81	65.49	

With a view to take advantage of early harvest, the crops can be harvested at physiological maturity with high vigour of seeds and following benefits (Table 2).

1. Seed weight attains maximum.
2. Seed development is completed.
3. Seed viability and vigour are maximum.

4. No reduction in yield.
5. No green seeds.
6. No weathering of seed.
7. No danger of rain damage.
8. Least damage from birds.
9. No seed shattering.
10. Nothreat of theft.
11. Field vacated 8-10 days early for next crop.
12. Increased time available for land preparation.
13. Can go for second crop.
14. Can take advantage of early marketing. for high prices.

Due to not knowing the advantages of harvesting at physiological maturing or for any other reason, if harvesting is delayed and seeds are left on the plant after physiological maturity, the producer can be put to following losses (Table 3)

1. Weathering of seed occurs
2. Danger of loss of seed quality if caught in rain.
3. Danger of seed sprouting while on the plant.
4. Danger of fungal invasion/insect infestation in field.
5. Increased yield losses due to bird damage.
6. Increased yield losses due to shattering.
7. Poor quality of seed with low germination and vigour.
8. Poor quality of grain - discoloration, grain becomes brittle, liable to breakage during threshing and processing.
9. Can not take advantage of early marketing.
10. Delayed land preparation.

The seeds are capable of germination much earlier than physiological maturity, in sorghum this capability is achieved within 10-15 days after anthesis (DAA), in groundnut at 28 DAA or just 6 DAA in cow pea. Considering this attribute, one may be tempted for early harvest even before physiological maturity. However, there is danger of yield and quality loss if harvesting is preponed to physiological maturity (Table 3) with following disadvantages.

1. Seed development remains incomplete.
2. Under sized shrivelled seeds are obtained
3. Seed moisture higher-difficulty in drying.
4. Seeds are troublesome to detach difficulty in threshing.
5. Seeds more soft-susceptibility to threshing damage.
6. Number of green seed very high-loss of market value.
7. Low vigoured seed-poor storability.
8. Reduced seed yield.

Table 3. Grain yield loss in % at various harvest dates in relation to physiological maturity of soybean

	Harvesting time	PKV 25	PKV 1	JS 335	Mean
1	5 days pre P.M.	21.49	13.73	15.15	16.32
2	P.M. + 5 days	3.50	1.27	2.42	2.26
3	P.M. + 10 days	3.97	1.49	3.03	2.67
4	P.M. + 15 days	29.00	1.01	3.63	8.76
5	P.M. + 20 days	33.77	16.41	9.09	14.38

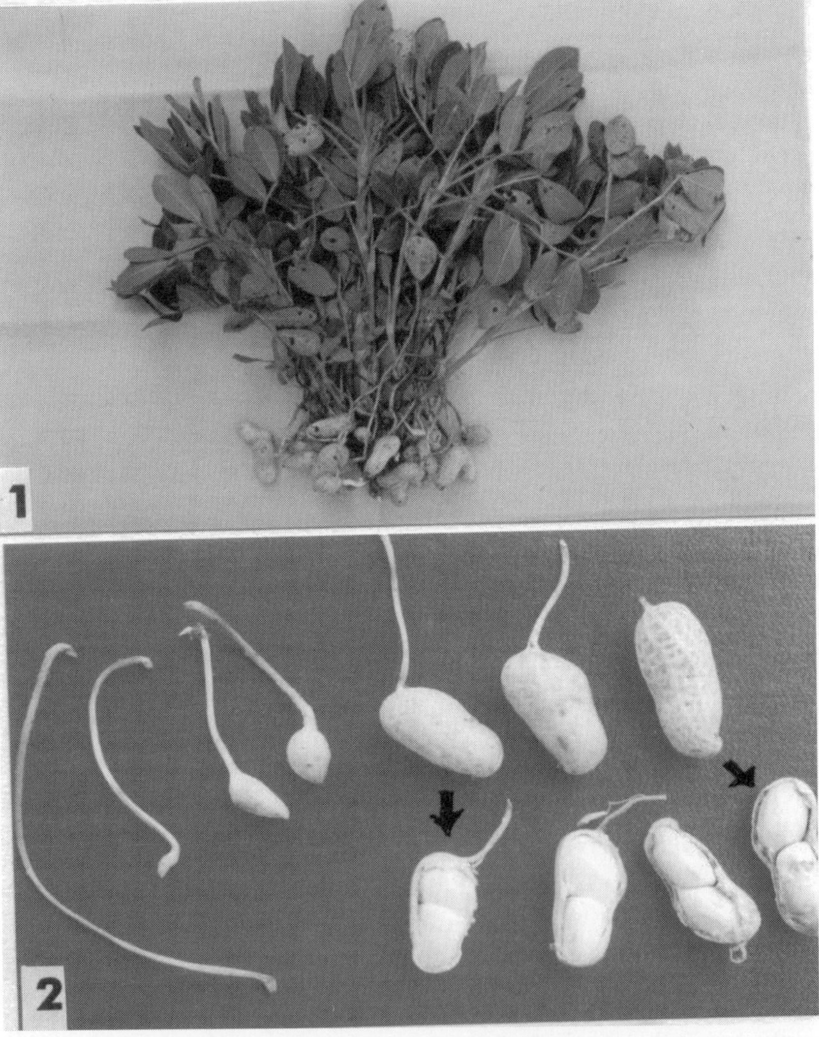

Plate 20.3. 1. Peanut plant with mature pods (TAG-24), Note synchronous maturity 2. Various stages of pod development (Peg to Pod-Left to right). 7th stage with wrinkled pod wall shows physiological maturity.

Plate 20.4. 1. Soybean crop (JS-335) with leaf yellowing due to waterlogging, 2. Pod shattering in soybean (JS-335) due to delayed harvest. Avoid this losses, 3. Over dried field of soybean.

3. MORPHOLOGICAL INDICES OF PHYSIOLOGICAL MATURITY

After physiological maturity is reached, the loss of moisture is accompanied by colour changes in the vegetative and reproductive plant parts including seed and fruit. The chlorophyll is lost and the leaf colour turns yellow, brown or black the range depending on the species. Therefore, the morphological characters like colour of pod, seed and seed parts and glumes etc. were studied to assess the physiological maturity in different crops. From farmers point of view, for harvesting crops at physiological maturity, it is essential to pin point specific morphological indicators coinciding with physiological maturity. Cropwise and if possible variety wise information on this aspect will be helpful to farmer in deciding the physiological maturity and plan the harvesting accordingly. The number of days after anthesis for physiological maturity and coinciding morphological markers are given in Table 4. The details of major crops like sorghum, maize, soybean, sunflower and groundnut are elaborated to understand the mechanism.

In sorghum and maize, Eastin *et al.,* 1973 reported that the basal region of placental zone of the kernels/grain can be used as a marker of physiological maturity. At this stage the vascular connection to the seed was broken by formation of a black coloured abscission layer which restricts the translocation of carbohydrates in maturing grain. This stage was also associated with disappearance of greenness and having firmness of the seed. The black layer formation in seed may be feasible approach for judging the end of grain filling period and stage of physiological maturity in sorghum and maize.

Attempts to describe physiological maturity in soybean have relied primarily on pod or leaf characters. Seed and pod colours are good indicators of physiological maturity. Yellow coloured seeds were obtained from yellow-brown or brown coloured pods to suggest that the appearance of pod can be used as an indicator of physiological maturity. Tekroney *et al.* (1979) reported that the physiological maturity of entire plant occurred when 26 per cent of the seeds were yellow and 35 per cent of the pods were yellow brown. However at this stage, the seed moisture content was above 54 per cent. They admitted in same article that the estimation was slight early. Dighe *et al.* (1995) reported 49 DAA in PKV-1 and JS-80-21. Shivankar (1999), reported physiological maturity at 34 days after anthesis in CV.PKV-25 whereas it was attained at 54-56 DAA, in JS-335 and JS-80-21. Physiological maturity was associated with 20 per cent green, 77 per cent yellow/pale-yellow and 2-3 per cent dry leaves coinciding with 23.66 per cent green and 76.66 percent yellow/pale-yellow pods. The seeds comprised 91.33 percent yellow/pale-yellow and only 8.66 per cent green seeds. The seed harvested at this stage were high in germination and strong in vigour.

Bhaskar Rao *et al.* (1993) reported physiological maturity at 30-40 DAA in morden, whereas in EC 68414, EC 68415, SPH-11 etc. it was 40-50 DAA. Whereas Kole and Gupta (1982) observed physiological maturity at 30 DAA in morden and 36 DAA in EC 68414. The physiological maturity was associated with head colour yellowish-green. Similar findings are reported by Mashakshetri *et al.,* (1992) and the yellow colour of head/capitulum shall be the better indicator of physiological maturity in sunflower.

In groundnut, Dighe *et al.* (1992) reported physiological maturity at 56 DAA in cv TAG 24. The PM in peanut was found at 63 DAA in JL 24 and SB XI and 70 DAA in ICGS XI (Anonymous, 1995). The stage was coincided with deep reticulations on pods, development of pink colour of kernel and turning placenta black. Saisawat (1980) also reported physiological maturity of peanut at 63-70 DAA with 35 percent kernel moisture content and pericarp colour as most satisfactory indicator of maturity. The appearance of crops is also found dull green/yellowish. If more than 80%

pods on one plant have attained physiological maturity, the crop is ready for harvest and black placenta colour is satisfactory indicator of maturity.

Table 4. Signs of physiological maturity (PM) in various crops.

	Crop	Days to PM DAA*	Physical indices of P.M.	Reference
1.	Sorghum *(Sorghum bicolor)*		Black layer	Eastin et al. (1973)
	Kharif	38-40	–	Dighe & Meshram (1982)
		25	–	Dighe et al. (1992)
			Dark closing layer in placental area near kernal attachment coincides with cut of radioactive assimilates.	
	Rabi	30	30% moisture content (MC)	Chaudhari & Bapat (1976)
		39-42	–	Sadashivamurthy (1982)
		35	–	Saibabu & Husaini (1984)
		35	–	Maiti *et al.* (1985)
2.	Fingermillet (*Eleusine coracana*)			
		30	Late cultivation 30% M.C.	Narayanaswamy & Jawaregouda, 1986
		27	Early cultivation 30% M.C.	
3.	Wheat (*Triticum aestivum*)			
	HD 2989	42	Withering of leaf blades and complete loss of green colour of flag leaf	Bharud & Patil, 1989
	NI 5439	46		
4.	Paddy (*Oryza sativa*)			
	BG 4001		25% M.C. at P.M. 17% M.C. at H.M. crop turns yellow in appearance	Shephard *et al.,* 1995
5.	Bajra *(Pennisetum typhoides*)	40	–	Vanagamudi, 1987
6.	Corn (*Zea mays*)	–	Black layer formation at P.M.	Daynard & Duncan, 1969
7.	Cotton (*G. hirsutum* and *G. arboreum*)			
	MCU4, MCU7, MCU8	55	Boll bursting	Thiagarajan & Ramaswamy, 1979
	Hy, CBS 156 and parents	50	–	Vanagamudi & Ramaswamy, 1979
	MCU 5 & Suvin	55	–	Thiagarajan & Ramaswamy, 1979
		55-60	–	Singh *et al.,* 1988
	AKH 081	48	Boll cracking	Changade *et al.,* 1990
	CAHH 468	52	Least electrical conductivity	Zade *et al.,* 1994
	AKA 8401 & LRA 5166	55	Boll cracking	
8.	Groundnut *(Arachis hypogaea)*			
		63-70	35% M.C. percarp colour most satisfactory indicator of maturity	Saisawat, 1980

	Crop	Days to PM DAA*	Physical indices of P.M.	Reference
		–	Mericarp turn black	Williams *et al.*, 1983
9.	Sunflower (*Helianthus annuus*)			
	Morden	30	Head colour yellowish green	Kole & Gupta, 1982
	EC 6814	36	Head colour yellowish green	
	KBSH 1	35	Head yellow, bracts deep brown	Mashakshetri *et al.*, 1992
	BSH 1	33	Head yellow, bracts deep brown	
	Morden EC 68414	30-40	–	Bhaskar Rao *et al.*, 1993
	EC 68415 SPH 11	40-50	–	
	–	35	Appearance of the first yellowish brown capitula in field	Adetunji, 1991
10	Soybean *(Glycine max)*			
	PKV 1 JS 8021	46	Greenish yellow Pod and yellow seed	Dighe *et al.*, 1995
11	Black gram (*Phaseolus mungo*)			
		13	–	Saha, 1987
12	Green gram (*Phaseolus aureus*)			
		25	–	Dharmalingan & Basu, 1988
13	Pigeon Pea (*Cajanus cajan*)			
		45-49 P.M.	Loss of green colour of pod	Kharb *et al.*, 1993
		51-55 H.M.		
14	Cowpea (*V. sinensis* and *V. unguiculata*)			
	V. sinensis	24	–	Palaniswamy *et al.*, 1986
	V. unguiculata			
	CV 152 & TUX 183609E	25	–	Ramaiah 1987
	5488	22	–	–
	RC 19	18	Pod colour yellow to brown	Dighe *et al.*, 1992
15.	Rice bean (*Vigna umbelata*)			
	17		Leaves, pods turn yellow, seeds whitish yellow	Zade *et al.*, 1993

* DAA = Days after anthesis. Readers are advised to refer plates.

4. PRECAUTIONS AND CONCLUSIONS

Though harvesting at physiological maturity is advantageous in terms of yield and quality, including other benefits, precaution is required at the time of harvest. The seed moisture content at physiological maturity ranges from 30-55 per cent, i.e., much higher than field maturity (below 20%), therefore the producer should have the artificial drying facility or the harvesting should be planned on clear sunny days so that natural drying shall be done without affecting seed quality. While harvesting at physiological maturity, the harvested produce should not be heaped, it should be spread in thin layer so that the drying shall be enhanced and seed quality be maintained.

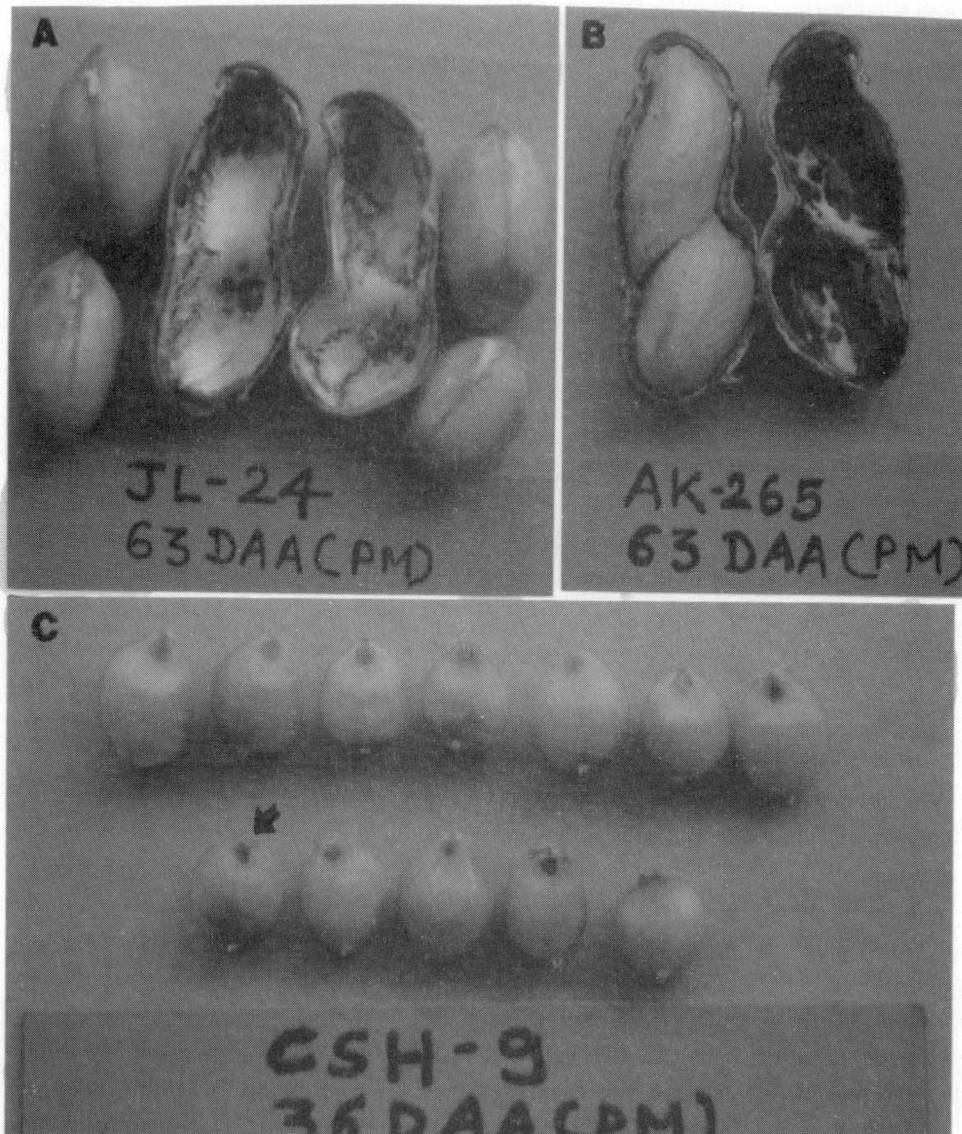

Plate 20.5. A&B. Note peanut pod maturity with shell and kernels in two varieties. C. Black layer formation in sorghum, a sign of PM. It means cut off stage of nourishing grains in CSH-9 hybrid 7th in first row and first in second row (36 DAA).

REFERENCES

Anonymous 1995. Bulletin on seed research highlights, Directorate of research, PKV, Akola.

Adetuniji, I.A. 1991. Effect of harvest date on seed quality and viability of sunflower in semi-arid tropics. *Seed Sci. and Technol.*, 19 : 571-580.

Bharud, R.W. and R.B. Patil 1989. Studies on physiological maturity in wheat (*Triticum aestivum* L.) *Ann. Pl. Physiol.* 3 (2) : 265 -268.

Bhaskar rao P, B.M. Reddy, K.G.R.S. Sai Babu and R. Ankaiah 1993. Effect of seed maturity on quality and dormancy in seven genotypes of sunflower. *Seed Res. Special vol.* 1 : 459 -466.

Changade S.P., V.N. Patil and N.G. Zode 1990. Studies on physiological maturity in cotton genotypes. *Ann. Pl. Physiol.* 4 (1) : 130-332.

Chaudhari S.D. and D.R. Bapat 1976. Studies on physiological maturity of grain sorghum, hybrids and varieties. *Sorghum News Letter* 19 : 56-57.

Daynard T.B. and W.G. Duncan 1969. The black layer and maturity in corn. *Crop. Sci.* 9 : 473-476.

Delouche, J.C. 1980. Environmental effects on seed development and seed quality. *Hort. Sci.* 16 (6) : 775-779.

Dighe, R.S and V.N. Patil 1979. Physiological maturity in sorghum. *Sorghum News Letter* 22 : 133-134.

Dighe, R.S., S.B. Lall and M.R. Patil 1988. Effect of stage of harvest on seed quality in grain sorghum. *Ann. Pl. Physiol.* 2 (1) : 82-86.

Dighe R.S., V.R. Zade, R.P. Chauke, N.G. Zode and S.P. Changade 1992. Studies on seed maturation and development in some crop species. *Ann. Pl. Physiol.* 6 (1) : 46-51.

Dighe, R.S., V.R. Zade, N.G. Zode and M.N. Asalmol 1995. Indices for proper stage of seed harvest in relation to seed quality of soybean. *Ann. Pl. Physiol.* 9 (2) : 142-145.

Dighe, R.S. and L.D. Meshram 1982. Optimum stage of harvest in grain sorghum, *PKV Res. J.* 6 (2) : 10-12.

Dharmalingam C. and R.N. Basu 1988. Seed quality in relation to protein of seed in the pod at different maturity periods in mungbean. CV. CO-3. *Seed Res.* 16 (2) : 168-172.

Eastin, J.D., J.H. Hulfquist and C.Y. Sullivan 19773. Physiological maturity in grain sorghum. *Crop Science.* 13 : 175-178.

Habib A.F., M.K. Sheikh, S. Sadaquat; G.K. Kulkarni and H.L. Nadaf 1985. Kernel filling as an index of maturity in bunch groundnut. *Seed Research.* 13 (2) : 142-144.

Kharb, R.P.S., B.P.S. Lather and D.P. Deshwal 1993. Studies on physiological maturity in early pigeonpea. *Seed Res.* 21 (1) : 45-48.

Khatra, S. and Gurmitsingh 1995. Indices of physiological maturity of seeds - a review. Seed Res. Vol. 23 (1) : 13-21.

Kole S. and K. Gupta 1982. The timing of physiological maturity of seeds of sunflower evaluation through multiple tests. *Seed Sci. & Technol* 10 : 457-467.

Maiti R.K., P.S. Raju and F.R. Bidinger 1985. Studies on germinability and some aspects of pre harvest physiology of grain sorghum. Seed Sci. and Technol 13 : 27-35.

Mashakshetri P.N., R.S. Dighe, V.R. Zade and R.T. Kausal 1992. Studies on determining harvest time for sunflower hybrids. *Ann. Pl. Physiol.* 6 (1) : 149-151.

Narayanswamy S. and S. Jawaregouda 1986. Studies on physiological maturity and germinability in finger millet (Regi). *Seed Res.* 14 (1) : 20 - 26.

Palaniswamy V., K. Vanagamudi and M. Jayabarathi 1986. Seed development and maturation in cowpea (*Vigna sinensis* (L.) Savi). Tropical Grain Legume Bulletin 33 : 24-26.

Patil V.N., and C.H. Andrews 1985. Development and release of hard seeded dormancy in cotton (*G. hirsutum (L.). Seed Sci and Technol* 13 (3) : 1-8.

Patil V.N., R.S., Dighe and J.D. Patel. 1987. Some aspects of seed maturation in cotton (*G. arboreum* (L.). Cotton Development 16 (3 and 4) : 45 - 46.

Ramaiah H. 1987. Studies on seed development, physiological maturity and their relationship on seed quality in cowpea (*Vigna unguiculata* L. Wals.) *Mysore J. Agri. Sci.* 20 (4) : 339-340.

Sadasivamurthy D.M. 1982. Studies on physiological maturity and factors affecting seed viability in some cultivars of sorghum. *(Sorghum bicolor* L. Moench) in Storage. *Thesis abstract.* 8 (3) : 271-272.

Sai Babu K.G.R.S. and H. Husaini 1984. Effect of maturity on seed quality in sorghum, *Seed Research*, 12 (2) : 51-55.

Saisawat P. 1980. Functional aspects of peanut (*Arachis hypogaea* L.) fruit and seed maturation, *Dissertation Abstract International.* 41 (3) : 768.

Savita Singh, G. Singh & H. Singh 1987. Assessing the physiological maturity of pigeon Pea (*Cajanus cajan* L. millsp) seed. *Indian J. of Pl. Physiol.* 30 : 304-307.

Saha A. 1987. Physiology of seed development in urdbean Indian *J. Pl. Physiol* 30 (2) : 199-201.

Shephard H.L., Naylor R.E.L. and T. Stuchbury 1995. The influence of seed maturity at harvest and drying methods on the embryo. α- amylase activity and seed vigour in rice (*Oryza sativa L.). Seed Sci. and Technol.* 23 : 487-499.

Shivankar, R.S. 1999. Effect of time of harvest on seed yield and seed quality of soybean (*Glyeine max* L. merril*). M.Sc. thesis* submitted to Post Graduate Institute Dr. P.D.K.V., Akola. (M.S.)

Simpson, D.M. and B.M. Stone 1935. Viability of cotton seed as affected by field conditions. *J. Agr. Res.* 50 : 435 - 447.

Singh A.R., S.B. Deshpande, and L.A. Deshpande 1988. Quality of seed as influenced by seed development and maturation in cotton. *Seed Research* 16 (2) : 131-135.

Thiagarajan C.P. and K.R. Ramaswamy 1979. Studies on seed maturation in two short duration cotton varieties, MCU -7 & PRS-72. Seed Research 7(2) : 67-91.

Thiagarajan C.P., and K.R. Ramaswamy 1979. Studies on seed maturation in MCU-4 and MCU-8 varieties of upland cotton. *Indian. J. of Agri. Sci.* 49 (11) : 844-850.

Thiagarajan C.P. and K.R. Ramaswamy 1982. Studies on seed maturation in MCU-5 and suvin cotton varieities. *Madras Agric. J.*. 69 (8): 529-534.

Thomson J.R. 1979. An introduction to seed Technology. *Leonard hill Pub.* 1-179.

Vanagamudi K. and K.R. Ramaswamy, 1979. Studies on boll and seed maturation in CBS - 156 Hybrid cotton and its parents. *Cotton Development* 9 (3) : 9-13.

Vanagamudi K. 1987. Studies on seed development and maturation in Bajra. *Madras Agril. J.*. 74 (3) : 132-134.

Williams E.J., J.S. Drexler and C.S. Kvien 1983. Effect of maturity and plant age on physical properties of pods and seeds. *Proceedings of American Peanut Research and Education Society.* 15 (1) : 25.

Yaklich, R.W. and P.B. Cregan 1987. A field study of moisture content after harvest maturity. *J. of Seed Tech.* 11 (1) : 62-68.

Zade, V.R., R.S. Dighe, N.G. Zode and S.P. Changade 1994. Seed development characteristics and stage of physiological maturity in hybrid cotton, CAHH 468 PKV Res. J. special issue on cotton PP- 122.

Zade, V.R., N.G. Zode, R.S. Dighe and S.P. Changade 1993. Seed development and maturation studies in rice rean (*Vigna umbelata Thumb). Ann. Pl. Physiol.* 7 (1) : 68 -72.

Zode, N.G., V.N. Patil and G.N. Kale 1993. Effect of different seed grades of sorghum (*Sorghum bicolor*) on vigour, growth and yield. Seed Res. Special Vol. 2 : 794-800.

Chapter - 21

SAFFLOWER — A SUCCESSFUL DRYLAND CROP

D.M. Hegde

Director, Project Coordinator (Safflower), Directorate of Oil Seed Research, Rajendranagar, Hyderabad -500 030 (AP).

1. INTRODUCTION

Safflower, *Carthamus tinctorius* L. isone of the important Rabi oilseed crops of India. The plant has been cultivated for centuries in the country either for its orange red dye (carthamin) extracted from its brilliantly coloured florets and/or for its much valued oil. With the advent of synthetic dyes and food colourants at the beginning of this century, the crop is grown mainly for edible oil obtained from the seeds. This has lead to disappearance of the crop from its traditional centres of cultivation in north and north-eastern parts of the country and is now mainly cultivated in the peninsular region. Currently, it is cultivated in an area of 707 thousand hectareswith aproduction of 432 thousand tonnes and average productivity of 611 Kg/ha. Maharashtra and Karnataka states account for about 95% of area and production of safflower in the country. It is also cultivated to a small extent in Andhra Pradesh, Madhya Pradesh, Orissa, Gujarat and Bihar.

Apart from its superior adaptability to scanty moisture conditions, safflower produces oil-rich in poly unsaturated fatty acids (linolenic acid 78%) which play an important role in reducing blood cholesterol level. The true potential of this crop as a dryland crop remained more or less obscure until about seventies. Extensive research carried out in different parts of the country over the past two decades under coordinated oilseeds and dryland projects clearly show that the crop which has been hitherto neglected, is more stable and remunerative than a number of traditional rabi crops.

2. ADOPTATION AND CLIMATIC NEEDS

Safflower shows a fairly wide adaptability within certain geographical limits, and may be found growing from the mountains of Afghanistan to the plains of the Ganges and Nile valleys. In the world, safflower cultivation is within latitudes 35 South to 45 North. In India, it is mostly cultivated between 14 and 22 North and 73 South and 79 East during rabi season (September/October to March/April). Saf-flower is normally grown below 900 to 1000 m and both seed yield and oil content fall with increasing height (Weiss, 1983). Altitude also affects the rate of growth and maximum height of the plant, and a given variety may be some 25% taller when grown below 300 m than when grown at 900 m.

Safflower is considered to be day-neutral but varieties may show adaptation to a specific photo-period when grown at the limits of their normal range. A short photo-period can prolong the rosette stage. Temperature is more important than day length. Good yields have been obtained in the plants' natural habitat with varying day-lengths.

Safflower is thermosensitive and does not favour extremes of either heat or cold. The main economic facets of temperature to be considered are those of tolerance to low temperatures as a seedling and susceptibility to high temperatures at flowering.

Tolerance to low temperatures depends on variety and stage of development. Seedling stage is most tolerant of low temperature while stem elongation, flowering and post-flowering stages are very sensitive. In seedling stage, it is reported to tolerate temperatures as low as -12 to -10ºC. When flower buds are being formed or flowering has commenced, temperatures below 0ºC may cause considerable damage in the form of sterile heads, even without any great visual effect on foliage. After flowering and when seeds have been formed, frost can reduce the quality and yield of seeds.

High temperature during early growth reduces the period spent in the rosette stage, and the transition can be rapid when there is no cool period after emergence. The effects of high temperature on safflower appear to be most severe at flowering, but such temperatures may also affect general plant growth.

There would appear to be a relationship between high temperature and available soil moisture for a shortage of the latter can exacerbate the deleterious effect of high temperatures. High temperatures also markedly reduce the rate of dry matter accumulation, and shorten the period between flowering on the main stem and subsequent branching. It also reduces the period to flowering, the size of flowers, seed set and 100 seed weight.

A temperature of 15ºC is highly favourable for rapid and even germination. A mean temperature of 20 - 21ºC favour optimum vegetative growth. At flowering, a day temperature ranging from 24 - 32ºC is congenial for obtaining higher yields.

Safflower has gained the reputation of being a drought resistant crop, but this is true only of its independence of rainfall. It is capable of obtaining moisture from levels not available to the majority of crops, and can thus tap sub-soil moisture or seepage. Climatic conditions during growth will affect the total moisture requirements and the rainfall necessary to mature a crop. In areas where there are no hot dry winds, and with adequate pre-planting soil moisture, 300 mm of rain will produce excellent yields if this falls mainly prior to flowering. In India, a rainfall of 650-1000 mm is considered optimum for obtaining maximum yields, depending on the soil type. Prolonged rainfall at flowering adversely affects pollination or after fertilization causes discoloured and small seeds. Persistant rain when heads are mature encourages seed to germinate in the head. The crop can not tolerate excess moisture or water logging at all stages of growth.

Excessive humidity during any stage of growth increases damage from fungal diseases. In areas where foggy or misty periods occur during flowering, head rots can be damaging. Persistent periods of high humidity when heads are mature encourages seed to germinate in the head, affecting yield and quality.

Young plants, particularly during the rosette stage, are little damaged by even quite severe hailstorms. Once stems have begun to elongate, it becomes more susceptible and hail in the period immediately prior to formation of flowers and secondary branches can do a great deal of damage. New buds and branches are often produced following such damage, but in regions with a short growing period, flowers from this new growth are often too late to mature seed. Mature plants are resistant to any but unusually severe storms.

The fully grown plant is extremely wind resistant, and even after seeds are mature, there are few losses from lodging or shattering. High speed winds cause damage to younger plants with more succulent stems. In areas where high winds are common, wider in the row spacing and narrower rows may be required.

3. SOILS

Safflower requires soils which are basically fertile, fairly deep and well drained. Highest yields are normally obtained on neutral soils but it is common to find crops in soils where pH value can range from 5 to 8. Irrespective of their fertility, shallow soils seldom produce high yields due to insufficient moisture. Dense subsoil will retard root growth, as will a layer of impermeable clay, and yields from rainfed or protective irrigated crop will be substantially reduced even though the sub-soil below the layer has adequate moisture. Low lying soils, acidic soils and shallow sandy soils are not suitable for safflower. Waterlogging due to poor drainage or prolonged rains even for relatively short periods predisposes the crop to damage from root rots and wilts and causes substantial fall in seed yield. Heavy soils with poor drainage must also be avoided for growing safflower, specially under irrigation.

In India, safflower is cultivated on a wide range of soils ranging from light to heavy soils. In traditional safflower areas of Karnataka and Maharashtra, safflower is grown on black cotton soils. In other states, it is cultivated on sandy loams, clay loams and alluvials both under irrigated and rainfed conditions. On heavy soils, it can follow an early *kharif* crop and may often be the single crop of the *rabi* season. On lighter soils, it is almost entirely irrigated and may be strip cropped with a cereal or legume (Chavan, 1960).

Safflower is considered to be salt tolerant, ranking just below cotton in its ability to produce profitable yields when grown in saline soils and is similar to barley in its tolerance to alkalinity when grown under rainfed conditions (Bernstein, 1964). It is especially tolerant of sodium salts, but less tolerant of calcium, magnesium or both. There is significant varietal differences in salinity tolerance (Gupta and Yadav, 1986; Janardhan *et al.* 1986). Salinity reduces seed size as well as seed oil content. Safflower was stated to be stimulated by an exchangeable sodium percentage level of 28.5 in the early stages of growth (Bains and Firman, 1964).

4. EFFICIENT CROPPING SYSTEMS

Traditionally, safflower is cultivated mainly in the states of Maharashtra and Karnataka and to a limited extent in Andhra Pradesh as an inter or mixed crop although the area under sole crop is on the increase in recent years. It is also cultivated in non-traditional areas like Gujarat, Orissa, Madhya Pradesh, Bihar etc.

5. POTENTIAL AREAS OF SAFFLOWER

Safflower is one of the efficient crops under rainfed conditions. Research carried out in All India Coordinated Safflower Improvement Project as well as in Dryland Agriculture have conclusively indicated that safflower, when grown as an entire crop, is more productive and remunerative than a number of conventional rainfed rabi crops not only in its principal belt but in a number of hitherto non-traditional areas representing diverse agro-ecological conditions in the country. A number of conventional dryland crops that can be profitably replaced by a sole crop of safflower, in different potential areas of its cultivation, are mentioned in Table 1. A shift or reorientation in the prevailing cropping patterns in the form of more efficient and productive crop like safflower would improve the aggregate returns from the drylands in these areas besides increasing the total oilseed production in the country.

Table 1. Low yielding dryland crops which could be replaced by safflower

S.N.	Region	Crops less remunerative than safflower
1	Parts of Maharashtra and Andhra Pradesh	Dryland wheat, linseed, and chickpea
2	Karnataka Bellary, Raichur and Chitradurga district Belgaum district	Wheat and coriander Wheat and linseed
3	Madhya Pradesh Malwa plateau, Chattisgarh region	Wheat, Chickpea and linseed, chickpea, linseed, urdbean (Black gram)
4	Coastal region (Bhal) of Gujarat	Dryland wheat
5	Uttar Pradesh Parts of eastern region Bundelkhand region	Wheat and linseed Wheat, niger, linssed and chickpea
6	South-eastern Rajasthan (Udaipur and adjoining areas)	Linseed, wheat, barley and mustard
7	Plateau region of Bihar	Linseed, lentil
8	Rice fallow medium lands of Orissa	Urdbean, horsegram

6. INTERCROPPING

In the traditional areas, safflower is generally raised as an intercrop with other rabi crops like sorghum, wheat, linseed, chickpea, coriander etc. Although sole crop of safflower is more profitable, it will be advantageous to intercrop safflower in a number of traditional crops otherwise it is low in their production potentials (Itnal *et al.,* 1980; Nikam *et al.,* 1987., Nikam *et al.,* 1988; Singh and Yadav, 1990; Autkar *et al.,* 1991; Rafey, 1993; Sawant *et al.,* 1993; Hegde and Koli, 1997). Some of the suggested intercrop combinations which are feasible, productive and profitable for different regions under rainfed conditions are listed in Table 2.

Table 2. Suggested intercropping systems in safflower in different regions

Region	Suggested intercropping system	Row proportion of main and intercrop
Maharashtra	Chickpea + Safflower	3 : 1 or 2 : 1
	Wheat + Safflower	3 : 1 or 2 : 1
	Linseed + Safflower	3 : 1 or 2 : 1
	Coriander + Safflower	3 : 1
Karnataka	Chickpea + Safflower	3 : 1
	Coriander + Safflower	3 : 1 or 2 : 1
	Wheat + Safflower	3 : 1 or 4 : 1

Andhra Pradesh	Chickpea + Safflower	3 : 1 or 2 : 1
	Wheat + Safflower	3 : 1 or 2 : 1
	Coriander + Safflower	3 : 1 or 2 : 1
	Linseed + Safflower	3 : 1 or 2 : 1
Madhya Pradesh	Chickpea + Safflower	6 : 2 or 4 : 2 or 2 : 1
	Linseed + Safflower	6 : 2 or 2 : 1
	Amaranthus + Safflower	6 : 2
	Toria + Safflower	6 : 2
	Mustard + Safflower	6 : 2
	Linseed + Safflower	3 : 1
Eastern Uttar Pradesh and Bundelkhad region	Chickpea + Safflower	3 : 1
	Barley + Safflower	6 : 2
	Toria + Safflower	1 : 2

To derive maximum benefits from the intercropping system, it is important to plant safflower at the optimum time recommended for each region. The seed rate of base crops must be same as per the recommendations for pure crop. For the intercrop of safflower, its seed rate must be adjusted based on its row proportion in the system. Avoid too wide a ratio between main crop and safflower. Seed safflower at closer interval of 2 : 1 or 3 : 1 or their multiples depending on seeding devices available in a region. Full dose of fertilizers recommended for main crop as well as intercrop of safflower, based on the area occupied by each crop in the system, must be applied to the intercropping system.

7. SEQUENCE CROPPING

Farmers rarely raise more than one crop in drylands. Safflower has deep and extensive root system with an ability to tap moisture from deeper layers, otherwise not available to a number of crops raised under similar situations. Hence, it offers an excellent opportunity for doubling the existing level of cropping intensity and thereby step-up returns from drylands in many conventional mono-cropped areas. A number of productive and remunerative cereal/legume-safflower sequences are identified for various agro-ecoregions of the country (Veeranna *et al.,* 1980; Joshi *et al.,* 1984; Nikam, *et al.,* 1985, Singh and Yadav, 1988; Raju and Reddy, 1990; Jha *et al.,* 1991; Tapadia *et al.,* 1991; Lomte *et al.,* 1993; Hegde, 1995) either as a contingent or regular practice which are given in Table 3. For successful double cropping in drylands involving safflower, the following points must be kept in mind:

1. Choose short duration high yielding hybrids/varieties of cereals and legumes to ensure timely planting of second crop of safflower when the soil moisture in the seed zone is adequate. Safflower performs better when it succeeds a short duration grain legume than a cereal.
2. Avoid deep or frequent tillage after harvest of Kharif crops as it promotes loss of moisture. Follow minimum tillage or if possible zero tillage to achieve satisfactory plant stand.

3. Take-up Kharif sowing immediately after the onset of monsoon so as to vacate the land well in time for safflower.
4. Avoid Kharif cropping in traditional rabi areas if the onset of south-west monsoon is delayed as it will hinder timely planting of safflower.
5. Adopt relay cropping wherever possible.
6. Adopt higher seed rates of 15 to 20 kg/ha to overcome the problems of inadequate moisture in the seed zone and the consequent problems of germination and emergence.
7. Place safflower seed in the moist zone by using appropriate seeding devices.

Table 3. Suggested crop sequences involving safflower in different regions

	Region	Crop sequence
1	Transitional tract of Karnataka (Dharwad, Belgaum and adjoining areas)	Mungbean - Safflower Groundnut - Safflower
2	Scanty rainfall areas of Karnataka (Medium deep black soils of Bijapur and Western parts of Bellary)	Mungbean - Safflower* Hybrid Sorghum - Safflower* Setaria -Safflower*
3	Northern Telangana of Andhra Pradesh (Parts of Ranga Reddy and Mahboobnagar districts, Adilabad, Medak and Nizamabad)	Mungbean - Safflower Maize - Safflower Hybrid Sorghum - Safflower* Sesame - Safflower
4	Assured moisture areas of Maharashtra (Khandesh tract, parts of Marathwada and Vidarbha)	Mungbean - Safflower Urdbean - Safflower Hybrid sorghum -Safflower* Groundnut - Safflower*
		Sesame - Safflower
		Sunflower -Safflower
		Soybean - Safflower
5	Drought prone areas of Maharashtra	Mungbean - Safflower Urdbean - Safflower
6	Malwa Plateau of Madhya Pradesh	Soybean - Safflower+ Maize - Safflower
		Groundnut - Safflower
7	Chattisgarh region	Upland rice -Safflower Soybean - Safflower
8	Bundelkhad region of Uttar Pradesh	Cowpea (Fodder) - Safflower Mungbean - Safflower Hybrid Sorghum - Safflower
9	Eastern Uttar Pradesh	Upland rice -Safflower+ Hybrid pearlmillet - Safflower Mungbean - Safflower Urdbean -Safflower
		Sesame - Safflower
10	Plateau region of Bihar	Upland rice -Safflower+ Maize - Safflower+
11	Medium and uplands of Orissa	Upland rice -Safflower

12	South-eastern Rajasthan (Udaipur and adjoining areas)	Mungbean - Safflower Urdbean - Safflower Maize (Fodder) - Safflower Sorghum (Fodder) - Safflower Cowpea (Fodder) - Safflower Cowpea (Vegetable) - Safflower

* As a contingent practice depending on the timely receipt and/or availability of favourable soil moisture conditions ; + As a contingent practice or alternatively with pre-plant irrigation if conditions warrant.

Table 3(A): Contribution of bracts in seed production in safflower (Mean of Nine varieties) Dhopte and Lall (1981)

Parameter	Stages		
	BS	PA	POA
Chlorophyll content of bracts (mg/g)	2.82	2.49	2.33
Range	1.72-3.86	1.83-3.22	1.74-2.88
Dry matter of bracts removed (mg)	115.8	221.9	258.9
Range	55-173	59-432	95-492
No. of bracts removed/capitulum	16.12	17.43	17.33
Range	12.7-22.0	12.7-22.0	13.0-23.3
No. of seeds obtained/capitulum	18.04	25.48	28.47
Range	11.7-31.3	19.7-37.7	15.3-44.7
Contribution	5.9	32.1	24.2
Seed yield/capitulum (g)	0.58	0.89	1.10
Range	0.31-1.13	0.47-1.20	0.75-2.0
Contribution (%)	61.72	40.06	32.2

BS=Bud stage, PA=Pre anthesis, POA=Post anthesis (Stages when bracts were removed)

Safflower is also emerging as an ideal sequence crop in irrigated deep soils such as in tail end areas of canals where water availability is limited and in other areas where irrigation sources are limited. With the availability of just 1 or 2 irrigations, safflower is generally preferred over other crops.

8. CROP ROTATIONS IN MONO-CROPPED AREAS

Continuous cropping of safflower encourages build-up of fungal diseases (wilts, root rots and foliar diseases) besides adversely affecting the long-term soil productivity and hence should be avoided. Farmers in the traditional rabi belt often fear low yields from Kharif/rabi crops grown immediately after safflower. In all the locations tested in the principal safflower belt, no such unfavourable influences on the succeeding crop could be detected (Anonymous, 1986). In fact, the yields of other crops were much better when grown after safflower. Therefore, safflower should be rotated with prevailing Kharif/rabi crops in the region without any fear or inhibition. However, cultivation of cotton after safflower or vice-versa should be avoided as it results in inefficient exploitation of nutrients because of their identical rooting habits. Further, this rotation increases build-up of Verticillium wilt (Weiss, 1983). Inclusion of a grain legume in the rotation should be preferred as it not only benefits safflower but also improves soil fertility.

9. CHOICE OF IMPROVED VARIETIES AND HYBRIDS

Adoption of appropriate variety/hybrid can have a large impact on productivity of safflower. Variety alone could contribute to the extent of 12 to 22% increase in productivity as observed in frontline demonstrations (Kiresur *et al.,* 1995). Breeding research since early 70's under the All India Coordinated Safflower Improvement Project has resulted in the identification and development of a number of high yielding varieties capable of yielding as much as 20 to 30 q/ha under favourable conditions of soil moisture and crop management in efficient areas of production. For the first time in India, two safflower hybrids viz., DSH-129 from Directorate of Oil seeds Research, Hyderabad and MKH-11 from Maharashtra Hybrid Seeds Company have been identified and released during 1997. These hybrids are capable of yielding about 20 to 22% higher seed yield and 29 to 32% higher oil yield than the best nationalcheck A-1.

Table 4. Recommended safflower varieties/hybrids for different regions

Region		Varieties/hybrids suggested
1	Andhra Pradesh	
	Normal sowings (October)	Manjira, DSH-129, MKH-11
	Late sowings (November)	Sagara Muthyulu (APRR-3) (Resistant to rust)
2	Maharashtra	
	Western Maharashtra	Bhima, DSH-129, MKH-11, Nira (NRS 209) under irrigation
	Khandesh tract	Bhima, Girna, DSH-129, MKH-11
	Marathwada	Bhima, Sharda, DSH-129, MKH-11
	Vidarbha	Bhima, DSH-129, MKH-11
3	Karnataka	
	Raichur, Bellary and Gulbarga	A-1, S-144, DSH-129, MKH-11
	All other areas	A-1, A-300, DSH-129, MKH-11
4	Madhya Pradesh	
	Malwa tract	JSF-1, JSI-7, DSH-129, MKH-11
	Eastern Madhya Pradesh	JSH-1, JSI-7, A-1, DSH-129, MKH-11
5	Gujarat	Bhima, A-1, DSH-129, MKH-11
6	Uttar Pradesh	HUS-305, T-65, DSH-129, MKH-11
7	Bihar	A-1, APRR-3, DSH-129, MKH-11
8	Orissa	A-300, N-62-8, DSH-129, MKH-11
9	Tamil Nadu	K-1, CO-1, DSH-129, MKH-11
10	West Bengal	HUS-305, DSH-129, MKH-11

The improved varieties and hybrids recommended for different regions are given in Table 4.

It is always important to use certified seeds for better impact on productivity instead of reusing the farmers own seed year after year.

Dhopte and Lall (1981) studied the contribution of bracts in nine varieties of safflower (NP-13, NP-23, NP-11, NP-19, UP-5, BC-32016, BC 32951, 892 and N-62-8) under rabi season at Parbhani in Marathwada region. They found wide variation in chlorophyll content of bracts, dry matter of bracts at various stages and number of seeds/capitulum obtained due to removal of bracts. There was reduction in yield by 61.7, 40.1 and 32.2% with reduction in seed number by 5.9., 32.1 and 24.2% when bracts were removed at bud stage, particularly at bud stage, heavy losses in yield can be observed (Table 3A). It is, therefore, necessary to keep capsules pest free in order to have maximum yield potential. Aphid infection is found common at reproductive stage. If plant protection is not followed and bracts are covered with black aphids that intercept the radiation and sucks the sap in large quantity. These losses can be avoided with proper prophylactic measures.

10. SOIL AND MOISTURE CONSER- VATION PRACTICES

Soil and moisture conservation practices form the basis of any sound crop production programme. This is particularly true in areas where safflower is raised as a monocrop on conserved soil moisture in large tracts of traditional safflower growing areas. In these areas, it is very important to check soil erosion and conserve maximum amount of rain water by following appropriate soil and moisture conservation practices for the region. Low infiltration rate of vertisols with their structural instability, often results in heavy runoff and soil losses besides creating waterlogged conditions which is further accentuated by the characteristic rainfall pattern in the safflower area. In all low to medium rainfall areas, graded bunds combined with interference in cultivation across slopes has been found to be quite effective in checking runoff and erosion and provide more opportunity time for rainwater to infiltrate into the soil. Formation of compartment bunds in the interbunded areas would be very useful for conserving maximum rainwater in low rainfall areas. Opening furrows every 3 m and tieing the furrow after every 10 m was found very effective in conserving moisture and increasing the yield of safflower in Telangana area (Anonymous, 1996). It is also necessary to make a provision for proper surface drainage to take out excess water.

The conventional harrowing for 3 to 4 times during the monsoon has been found to be as effective as deep ploughing in controlling weeds and conserving moisture in black soils (Anonymous, 1986).

10.1. Seed-bed Preparation

Safflower requires clod free seed-bed but not so fine that crusts after rain. The subsoil should be firm and there must be enough moisture for good germination and stand establishment. In monocropped black soils of *rabi* areas, harrowing 3 to 4 times during the monsoon is as effective as deep ploughing or subsoiling to keep fields weed free and to conserve moisture. To ensure greater weed control, the last harrowing should normally take place immediately before planting.

If the fields are infested with obnoxious weeds or if there is plough pans within the root zone, deep ploughing is recommended during summer as soon as the *rabi* crop is harvested. In areas where safflower is taken an intercrop the land prepared for main crop is good enough for safflower also.

In assured moisture areas, where safflower is to be planted after a short duration cereal or legume in *kharif*, restrict the tillage to the barest minimum. Too deep or frequent tillage after *Kharif* crop is not advisable as it leads to dry and unfavourable seed-bed. If conditions warrant, give just one or two harrowings after harvesting of grain legumes in black soil areas (Nikam *et al.,* 1989). In alluvial soils of Bundelkhand in Uttar Pradesh, one ploughing followed by cross harrowing was found adequate for safflower succeeding sorghum in *kharif* (Saxena, 1987). Similar tillage needs were found adequate for safflower after transplanted rice in West Bengal (Das and Ghosh, 1993). Proper control of weeds in the *kharif* crop helps greatly in minimising tillage needs of safflower during *rabi* season.

11. PRE-SOAKING & SEED TREATMENT

Pre-sowing seed hardening of safflower induces drought tolerance (Hegde and Kulkarni, 1974) although the practice has not found acceptance with the farmers. For eastern Madhya Pradesh, it has been recommended to pre-soak the seeds in pure water for 24 to 48 hours and shade dry for about 4 hours. Thereafter, seeds must be treated with appropriate fungicides before sowing.

Many of the safflower diseases are seed borne. Therefore, seeds should be treated with Thiram, Captan or Bavistin @ 3 g/kg seed before sowing to prevent losses from seed and soil borne diseases. It is always better to use fresh seeds every year. Certified seeds may be used which will enhance the yield levels by 12 to 22 % as compared to the use of own seeds by the farmers (Kiresur *et al.,* 1995).

12. SEEDING

12.1. Optimum seed rate and spacing

Safflower has considerable plasticity to seeding rates and row spacings because of its inherently superior branching potentials. Optimum plant population in different varieties seldom exceed 110000/ha even under ideal conditions (Rao, 1984). The seed rate to be used depends on various factors such as row spacing, depth of sowing, soil moisture status, previous crop etc. and may vary from 7 to 20 kg/ha in different situations (Nikam *et al.,* 1989). In both peninsular and northern regions, wider row spacings ranging from 30 to 45 cm to as much as 90 cm did not cause any striking yield differences (Rao, 1984). Row spacings can, therefore, be adopted depending on local convenience. Normally, row spacings of 45 to 60 cm were found to be quite ideal (Veeranna and Rudraradhya, 1980; Mane, 1983; Singh *et al.,* 1984). The optimum seed rates and spacings recommended for different regions to achieve desired plant stand is given in Table 5.

Table 5. Seed rates and spacings for safflower recommended in different regions

Region		Seed rate (kg/ha)	Spacing
1	Andhra Pradesh	7.5 to 10	45 x 20 cm
2	Maharashtra		
	Kharif fallows	10-12	45 x 20 cm
	Double cropped area	15	45 x 20 cm
3	Karnataka	7.5	60 x 30 cm
4	Gujarat	10-15	45 x 20 cm
5	Madhya Pradesh	20	45 x 20 - 25 cm

6	Orissa	20	30 x 15 cm
7	Uttar Pradesh and Bihar		
	Adequate moisture	12-15	45 x 20 cm
	Scanty moisture	15-20	45 x 20 cm
8	Tamil Nadu	10	45 x 20 cm

For higher seed yield, it is necessary to maintain the recommended plant stands. The seeding rate normally recommended for safflower often results in excess plant stand over and above those normally required whenever soil moisture conditions are favourable. Under such a situation, excess seedlings should be thinned within 10-15 days after emergence to maintain the desired plant to plant spacing.

12.2. Optimum time of seeding

Seeding time is a crucial non-monetary input for achieving adequate and uniform plant stand which can use the resources efficiently to give higher yields. The planting time will vary widely depending on a number of factors such as locations, time of withdrawal of monsoons, initial soil moisture at planting and the extent of pre-and post-planting rains. The crop planted at appropriate time suffers relatively less damage from aphids and diseases. Therefore, it is necessary to plant the crop at the recommended optimum time in different regions taking advantage of first rains received in September/October (Tiwari *et al.,* 1972; Saran *et al.,* 1980; Deokar *et al.,* 1984) which are indicated in Table 6.

Table 6. Recommended seeding time for safflower in different regions

Regions		Recommended seeding time
1	Andhra Pradesh	
	Coastal and Royalaseema areas	October
	Telangana	Late September to early October
2	Maharashtra	
	Drought prome areas of Western Maharashtra and adjoining areas of Marathwada	Second fortnight of September Last week of September to mid October
	Assured rainfall areas of Western Maharashtra, Marathwada and Vidarbha	Last week of September to first week of October
3	Karnataka	Second fortnight of September
4	Gujarat	Last week of October to first week of November
5	Madhya Pradesh	
	Malwa Plateau	Last week of September to first week of October
	Eastern Madhya Pradesh	Seed week of October to second week of November
6	Orissa	Second week of September to mid October
7	Uttar Pradesh	Second to third week of October
8	Rajasthan and Tamil Nadu	Early to mid October
9	Bihar	Second week of October

Wherever safflower is raised on conserved soil moisture, any delay in planting beyond the recommended time, would have adverse effect on plant growth and development and finally yields if the initial stored moisture is low and post-sowing rains are far and few. Under conditions of extreme moisture shortage, safflower performs best when seeded immediately after the receipt of first rains in September/ October. In case soil moisture conditions are extremely favourable as in seasons with extended monsoon or late rains or where facilities exist for pre-sowing irrigations, plantings of rainfed safflower can safely be extended by 1 to 2 weeks beyond the suggested optimum period for different regions.

12.3. Efficient seeding techniques

It is important to place safflower seeds in moist zone for good germination. Optimum depth of seeding has been found to be 5 cm (Daulay *et al.,* 1974; Veeranna *et al.,* 1976). The depth of planting must be even which ensures regular emergence and growth and the resulting uniform crop allows easier and more efficient field operations. Border method of planting in which one row is skipped after every two rows and opening a furrow in the skipped row was found to be better in scarcity zone of Maharashtra. It conserved the moisture better and gave higher yields of safflower. Anonymous, 1996).

Improper seeding device is often responsible for poor and uneven stands and low yields from drylands. For better stands, line sowing using improved seed drills or fertilizer-cum-seed drills available in different regions must be followed. Seeds can be sown behind plough also. Single row dryland deep-furrow seeder or corn planter with adjustable row opener may also be used specially in areas where the depth of sowing is to be increased to place the seed in the moist zone. Where soil is likely to dryout quickly following planting, leaving the surface fluffy, a light planting is recommended. In areas where 5 to 7.5 cm of soil is dry and loose, safflower must be planted into moist soil even if this means that seed is then 10-15 cm deep. If necessary, small furrows must be opened and seeds sown and the furrows should be kept half filled.

13. INTEGRATED NUTRIENT MANAGE- MENT

Safflower requires adequate supply of nutrients for good yields. The amounts of nutrients required to produce of a tonne of safflower seed is 38.8 kg N, 8.4 kg P_2O_5, 22.0 kg K_2O and 12.6 kg S besides significant amounts of other secondary and micronutrients (Hegde, 1998). The level of nutrient application depends on native soil fertility, soil moisture status, preceding crop and yield potential of cultivar. At a number of locations, economic response to N application was observed even beyond 60 kg/ha (Umrani *et al.,* 1981; Ahmed *et al.,* 1985; Zaman, 1988; Bohra *et al.,* 1989; Mane and Jadhav, 1993). In double cropped areas, the recommended dose of N to safflower can be reduced by 50 % amo-unting to 20 to 25 kg N/ha if the preceding crop is a grain legume receiving its full complement of fertilizers (Rao, 1984). In scanty rainfall areas, an additional basal/ top dressing of 20 to 25 kg N/ha is recommended if seasonal conditions turn out to be favourable such as high initial stored moisture at planting time and favourable post planting rains.

The phosphate requirement of safflower is moderate. Where the preceding crop was given large amount of this nutrient, it is seldom required for safflower (Weiss, 1983). However, safflower responds to adequate P in the seedling stage, and hence small-amounts band placed at planting time promote rapid growth in the seedling stage and improves general plant health. Safflower is deep rooted and can draw on a considerable volume of soil for nutrients. Its added P requirement in a particular area may thus be substantially smaller than more shallow rooted crops. Response to applied P was observed at a number of locations (Malewar, 1974; Veeranna and Channappa, 1981; Kummur *et al.,* 1984) and about 25 to 40 kg P_2O_5/ha was found to be optimum. However, it is always advisable to apply P based on soil test value.

Application of K fertilizer to safflower is usually unnecessary except in those areas where a serious deficiency of this nutrient exists, and it is also required for other crops. In general, the application of K to a preceding crop has been found sufficient for a following crop of safflower (Weiss, 1983).

Response to applied fertilizers depends on the method and time of fertilizer application. Under rainfed conditions, it is better to apply the entire quantity of fertilizers as basal by drilling. Fertilizers should be placed deep in the furrows about 10 cm below the soil and seed the crop adjacent to the fertilized rows. In mono-cropped areas of Maharashtra and Karnataka, application of fertilizers, 2 to 3 weeks prior to optimum planting time was found best for increasing the fertilizer use efficiency under receding soil moisture (Rao, 1984).

Recent studies in AICORPO (Safflower) have indicated good response of safflower to sulphur application in many areas. Application of 30 kg S/ha through single super phosphate or elemental sulphur before planting has been found to be quite ideal. If P is supplied through single super phosphate, sulphur requirement of the crop is taken care of adequately.

There has been no reports of trace elements deficiencies affecting safflower yield in the field.

There are some reports of beneficial effects of seed inoculation with Azotobacter (Rasal *et al.,* 1984), Azospirillum (Anonymous, 1998) and application of VAM (Janakirani and Manoharchary, 1993). However, field recommendations are yet to be made use of those biofertilizers. Integrated use of FYM (6 t/ha) and 20 kg N/ha gave 42% higher yield than use of 40 kg N/ha alone at Indore (Sharma, 1993). This also increased the water use efficiency and helped in alleviating the harmful effects of drought and frost. Recommended levels of manures and fertilizers for safflower in different regions are presented in Table 7.

Table 7. Manures and fertilizers recommended for safflower in different regions

Region	FYM (t/ha)	Fertilizer (Kg/ha)		
		N	P_2O_5	K_2O
1 Andhra Pradesh				
Rainfed	–	40	25	–
2 Maharashtra				
Western Maharashtra				
Rainfed				
Normal rainfall	–	50	25	–
Scarcity zone	–	25	12.5	–
Irrigated	–	60	30	–
Marathwada				
Rainfed	–	40	20	–
Irrigated	–	60	40	–
Vidarbha				

	Rainfed	5-10	25	25	–
	Irrigated	5-10	50	50	–
3	Karnataka				
	Rainfed	6	35	50	25
	Irrigated	6	75	75	35
4	Gujarat				
	Rainfed	–	25	10	–
5	Orissa				
	Rainfed	6	25	25	–
6	Uttar Pradesh				
	Rainfed	–	40	30	20
	Irrigated	–	60	30	20
7	Madhya Pradesh				
	Malwa tract				
	Rainfed	–	40	40	20
	Eastern Region				
	Rainfed	–	20-30	15-20	10-15
	Irrigated	–	50-60	20-30	20
8.	All other areas				
	Rainfed	–	30	20	–

14. WATER MANAGEMENT

Although safflower is raised mainly as a rainfed crop, the crop responds well to application of water through irrigation. The seasonal consumptive use of the crop seldom exceeds 250 to 300 mm. If the soil profile is fully saturated at planting which can accommodate about 250 mm of available soil moisture, the crop seldom responds to irrigation. However, under scanty moisture conditions in drylands, yield can be boosted by 40 to 60% by providing just one life saving irrigation (5 to 8cm) atcritical phases of crop growth (early stem elongation or flowering) or before soil moisture becomes limiting for crop growth. (Hegde, 1995). When raised as an irrigated crop, safflower can give almost double the yield obtained from rainfed crop. The crop is, however, very sensitive to excessive moisture. In poorly drained soils like heavy black soils, improper surface irrigation may lead to prolonged waterlogged conditions and the associated problems of wilts and root rots.

Give a light presowing irrigation if the soil moisture in the seed zone is not adequate for germination. Later, depending on the soil moisture status, give one irrigation at 35 days after planting in early elongation stage and another at 65-70 days during flowering. After this, no irrigations are generally required unless soils are too light with low water holding capacity. In light soils about 5-6 irrigations are needed including a preplanting irrigation amounting to 300 to 400 mm water.

Under irrigated conditions, it is advisable to plant the crop on broad beds with furrows laid at intervals of 1.35 to 1.8 m or preferably on flat-bed and then form irrigation furrows after every 2 or 3 rows at the time of first irrigation. Such a system of planting would greatly help to minimise contact of water with above ground parts. For getting maximum benefit from irrigation, the following precautions may be followed.

1. In soils that crack, apply irrigation well before cracks develop for better control of water. Do not delay irrigation till the crop suffers from moisture stress as it pre-disposes to wilts and root rots.
2. If there is provision for only one irrigation, apply it before soil moisture becomes very critical for crop growth.
3. During irrigation, avoid contact of above ground parts with irrigation water to minimise the chances of diseases.

There are no research reports on use of sprinkler and drip irrigation in safflower in India although sprinkler irrigation is used on large scale in California.

15. WEED MANAGEMENT AND INTERCULTURE

Normally weeds do not pose any serious problem in safflower except when frequent rains are received during initial phases of crop growth or when seed-bed preparation and tillage are improper. Safflower is very susceptible to weed competition during its rosette stage which lasts for about 25 to 30 days in peninsular India and 50 to 60 days or more in other parts with prolonged winter (Rao, 1984). Timely weeding and interculture during this period is a must for arresting weed growth and obtain full benefits from applied nutrients, rainfall and soil moisture. It is necessary to give one or two hand weedings and hoeings/ harrowings at 25 to 30 and 45 to 50 days after planting depending on the length of rosette period and the severity of weed infestation. In sorghum-safflower seque-nce, pre-emergence application of Atrazine 0.75 kg a.i./ha to sorghum followed by one weeding and hoeing at a later stage gave effective control of weeds in safflower also (Shelke and Bhosale, 1993). Pre-emergence application of Alachlor @ 1.5 (Kulaindavelu *et al.,* 1974) to 2 kg a.i./ha (Babu, 1979) or Trifluralin @ 1 kg a.i/ha (Panchal *et al.,* 1977) was very effective in controlling weeds in safflower. Pre-emergence application of Fluchloralin (1.5 kg a.i./ha) and Diuron (0.85 kg a.i./ha) were also effective in controlling weeds in safflower.

In black soil areas, cracks begin to appear from December onwards with the rapid loss of sub-soil moisture. To delay cracking, close them superficially with dust mulch as and when they appear and thereby minimise moisture loss. Give one additional interculture or hoeing using bullock drawn hoes/harrows/sweeps in December before the crop canopy is closed and the spines become problematic.

16. ROLE OF PLANT GROWTH REGULATORS

Plant growth regulators have been commercially used in many crops for different purposes such as amelioration of drought, control of vegetative growth or for induction of flowering etc. In safflower, there are some reports of use of gibberellins for induction of sterility for hybrid development. For improving the productivity of safflower under rainfed conditions, cycocel has been successfully used. Results of experiment conducted during 1995-96 and 1996-97 at Solapur clearly established the usefulness of cycocel spray @ 1000 ppm at flower initiation (Table 8).

Averaged over two years, there was more than 20% increase in the seed yield although the biological yield remained almost the same. The increased yield was brought about by marked increase in harvest index which increased from 27% in control treatment to 33% with 1000 ppm cycocel spray. Cycocel could translocate higher proportion of drymatter to the sink under rainfed conditions without affecting the total biological yield (Anonymous, 1996, 1998). However, this beneficial effect of cycocel was not observed under irrigated conditions (Anonymous, 1996).

Chopde and Dhopte (1970) studied the effect of two sprays of growth regulators (GA, IAA & MH) during reproductive stage at 100 and 200 ppm concentrations (Table 8A). Increase in seed yield by 20.9% is reported due to MH sprays and this increased yield was attributed to increase in number of capitula/plant and test weight. Thus, two sprays of MH (inhibitor) at flowering have beneficial role in raising yield by 21% in safflower, cv. N-62-8, a popular variety under dryland conditions which is recommended for Marathwada region. The inhibitor increased yield by monitoring the harvest index. Excessive growth is checked and dry matter partitioning to the sink is promoted in safflower. Growth promoters appear to have no positive role in safflower.

Table 8. Effect of cycocel on safflower under rainfed conditions (Mean data of 1995-96 and 1996-97)

S.N.	Treatment	Biological yield (kg/ha)	Seed yield (kg/ha)	Harvest index (%)
1	Cycocel @ 500 ppm at flower initiation	4971	1580	31.4
2	Cycocel @ 500 ppm at 50% flowering	4764	1449	30.4
3	Cycocel @ 1000 ppm at flower initiation	4836	1608	33.0
4	Cycocel @ 1000 ppm at 50% flowering	4830	1484	30.6
5	Water spray at flower initiation	4790	1338	27.9
6	Water spray at 50% flowering	4819	1305	27.1
7	Absolute control	4820	1308	27.0

17. INTEGRATED PEST AND DISEASE MANAGEMENT

Safflower is affected by a number of insect pests and diseases although only a few are of economic importance. Unless timely and appropriate measures are taken to manage them, there may be considerable yield losses.

Insect pests

Safflower aphid (*Uroleucon composilae* Theobald) is the most destructive pest in all safflower growing areas. Next important pest is the capsule borer (*Helicoverpa armigera* Hubner) which is a poly-phagous pest attacking many crops. In addition to these two pests, occasionally other insects such as cutworms, safflower fly, caterpillars, surface grasshoppers etc. also assume serious proportion and cause extensive damage.

Table 8A. Effect of growth regulators sprays (ppm) in safflower cv. N-62-8 (Two sprays during flowering) Chopde and Dhopte (1970).

Character		GA		IAA		MH	
	Control	100	200	100	200	100	200
Branches/plant	15.5	13.4	13.5	15.9	25.4	15.4	16.1
Capitula/plant	45.1	35.4	36.6	48.8	41.5	52.0	50.7
Seed number/ capit.	20.0	21.7	23.4	22.1	21.1	21.6	19.6
Test weight (g)	4.8	4.9	5.3	4.9	4.7	5.1	4.9
Seed yield/plant (g)	35.0	30.7	37.5	36.8	33.2	42.3	39.1
Pollen sterility (%)	4.8	1.5	4.4	9.2	4.0	5.3	5.1

Management

1. Improved varieties like Bhima, Sharda, Girna, Annigeri-1, HUS-305, JSF-1 are less susceptible to aphids and hence should be preferred. Hybrids are also less susceptible to aphids.
2. The crop should be planted as per the recommended planting time for the region as delayed plantings suffer more damage from aphids (Ghule *et al.,* 1987; Jagdale *et al.,* 1990).
3. In areas with high incidence of capsule borer, avoid intercropping of safflower in chickpea.
4. The caterpillars must be handpicked and destroyed wherever possible.
5. In case of high incidence of safflower fly, the infested flowers may be collected and destroyed and the light trapping may be followed to control it.
6. Spray Dimethoate (0.05%), or Methyl parathion (0.05%) or Monocrotophos (0.05%) or Chlorphyriphos (0.05%) or Endosulfan (0.05%) or alternately dust Quinolphos (1.5%) or Methyl parathion (2.5%) or Malathion (5%) or Endosulfan (4%) at 40 days and if necessary also at 60 days after sowing. Dimethoate (0.07%) and Endosulfan (0.07%) help to control capsule borer. Neem seed kernel extract (5%) spray at 40 days followed by the spray of any of the chemicals mentioned above also gives effective control of most of the pests. Use 500 litres of spray mixture and 20 kg of dust formulation/ha. In transitional tract of Karnataka, for control of safflower aphids, one spray of neem seed karnel extract (5%) a week after first incidence followed by the spray of recommended insecticide 15 days later was found as good as two sprays of insecticide (Anonymous, 1998).

Diseases

Alternaria leaf spot (*Alternaria Carthami* Chowdhary) is one of the most important diseases affecting the crop in all safflower growing regions in the country. Cercospora leaf spot (*Cercospora carthami* Sundaram and Ramakrishna) is of worldwide occurrence, but is important mainly, where the crop is grown in large pure stands. Ramularia leaf spot (*Ramularia carthami* Zaprometov) is mostly a problem in high soil moisture and irrigated areas. Rust (*Puccinia carthami* Corda) is a problem in some regions under extremely late sown conditions. Wilts and root rots are mainly problems in ill drained areas and under extensive wet weather conditions.

Management

1. Seeds should be collected from disease free plots for sowing.
2. Seeds should always be treated with fungicides like Thiram/Captan/Bavistin @ 3g/kg seeds before sowing as many diseases are seed borne.
3. The crop should be planted as per the recommendation for each region. Early planting encourages spread of Alternaria leaf spot and late planting rust.
4. Resistant high yielding varieties like APRR-3 must be used where there is rust incidence.
5. Safflower cultivationmustbe avoided in low-lying, waterlogged areas.
6. Continuous cropping of safflower in the same plot must be avoided and recommended crop sequences and rotations must be followed.
7. Wilt affected plants must be removed and destroyed as and when they are observed.
8. Spray Mancozeb (0.25%) immediately after disease symptoms of leaf spots are noticed. Repeat the spray 15 days later depending on the intensity of disease. Whenever rust is noticed, give one or two sprays of Calixin (0.05%) or Dithane M-45 (0.25%) at fortnightly interval.

The chemical control measures must be followed only after pest and disease symptoms are noticed. However, in endemic areas, where the problem is noticed every year, prophylactic measures must be followed.

18. HARVESTING AND THRESHING

All the currently available varieties come to harvest in about 115 to 140 days in the traditional safflower growing areas of Maharashtra, Karnataka and Andhra Pradesh. In Gujarat and Orissa, the crop takes about 145 to 150 days for maturity. In other non- traditional areas, 150 to 180 days are needed for maturity depending on weather conditions, time of seeding, soil moisture etc. The crop is ready for harvest when the leaves and most of the bracteoles except a few on last formed flower heads become dry and brown. The crop should be harvested preferably in the early hours when shattering would be minimum and spines relatively soft.

The plants must be cut with the help of sickles at the base or wherever possible uprooted (black soils) by pulling and stacked in the field in the form of small and well pressed heaps until they are fully dried. For easy handling and transport to the threshing yard, sets of plants are arranged in the heap in opposite directions with their cut ends overlapping and branched and spiny portions away. This arrangement provides fairly long and spine-free portion in the middle of stack/bundles. If necessary, improvised sacks and gloves made from used gunny bags may be used to protect legs and hands against spines. Threshing is done either by beating with sticks or with the help of bullock drawn stone-rollers or tractor and the resulting material is winnowed to clear seeds. The threshing and cleaning operation can also be made with power-operated threshers used for other crops such as wheat.

Combine harvesters used in rice and wheat could also be used for harvesting and threshing of safflower. In Vidarbha and Western Maharashtra where sole crop of safflower is grown, use of combine harvesters has become quite popular with the farmers for safflower.

18.1. Yield potential

If all the improved agro-production and protection technologies are followed, about 800 to 1200 kg/ha of safflower yield can be obtained under scanty moisture conditions and about 1500 to 2000 kg/ha under favourable moisture conditions. Under irrigated conditions, yield levels of 1800 to 2800 kg/ha could be achieved.

18.2. Seed production and Storage of Seed

The production of safflower seed for oil and production of seed for sowing require different techniques. The latter requires special attention, and costs of production are considered higher. More care is required to ensure above average weed and pest control since even slight damage from lygus and similar bugs will reduce the virility of the seed. Adequate fertilizers should be applied, and where nitrogen is required it is desirable to place this in the seed bed at planting, instead of split application as in normal practice under irrigated conditions. This will help reduce late branching and delayed maturity.

Seed production fields must not experience water stress. Fields should be carefully rogued for off types, sterile plants and those with deformed heads.

If combines are to be used for harvesting, it should be ensured that moisture-content in seed is not dropped below 8%. Before storing, seed should be carefully cleaned and dried to bring down the moisture content to 5%. Crops badly infested with a major disease such as rust, verticillum or Fusarium wilt should not be used for seed, but should be harvested and sold for oil.

Plate 21.1. Harvesting of safflower with combine. It facilitates cultivation on large scale laboures were reluctant to harvest due to leaf spinyness.

The safe moisture level for storage of safflower seed is 5%. Grain bins are quite suited for storage of safflower. The seed heats rapidly if not dried to at least 5% for bulk storage, and for safety should be scalped. Some dusting (Malathion 5%) may be made against storage pests before storing the seeds.

Storage for long periods of seeds intended for crushing is uneconomical and should be avoided whenever possible.

19. SAFFLOWER FOR ABERRANT WEATHER

Delayed rains

1. It is more profitable to go in for safflower whenever there are late rains in October/November.
2. Similarly, in several single cropped *rabi* areas, the occurrence of prolonged dry weather conditions immediately after planting coupled with insufficient moisture at seeding time result in poor and uneven plant stand and/or near failure of early sown *rabi* crops. In such a case, scrap the early sown *rabi* crop and resow with safflower taking advantage of favourable rains in October/early November.

19.1. Extended Monsoon

1. Large areas from otherwise traditionally *kharif* cropped soils are kept fallow in *rabi* for want of adequate moisture in seeding zone. Whenever favourable rains are received immediately after harvest of *kharif* crops sometimes in late September/ October and soil moisture adequate for second crop, take full advantage of the situation and raise two crops in place of a single crop by planting safflower immediately after harvest of *kharif* cereals (sorghum, pearlmillet, maize, Upland rice, setaria etc.), legumes (mungbean, urd bean, soybean etc.) and others (sesame).

2.Several light and medium textured soils in the potential safflower growing areas of Maharashtra, Karnataka, Andhra Pradesh, Madhya Pradesh, etc. which are normally cropped in *kharif* can be put to productive use by taking safflower either as a contingent crop after the harvest of *kharif* crops or as a compensatory crop in the event of total failure of *kharif* crops whenever near/above normal conditions prevail in *rabi*. In case plantings are delayed and/or soil depth is limited, use higher seed rate than normal to curtail excessive growth of safflower.

19.2. Insufficient rainfall and/or stored moisture in rabi

1. Whenever stored moisture is inadequate and post-sowing rains far and few, give either pre-planting irrigation from runoff water collected in dugout ponds or any other source, or alternatively one life saving irrigation at elongation/flowering or earlier if crop shows symptoms of moisture stress.
2. In situations where water levels in wells/tanks/reservoirs are limited and favourable and timely rains are received in *rabi*, plant safflower in place of other high water requiring crops like wheat to achieve efficient use of scarce water resources over a larger command area and obtain higher returns per unit area and unit amount of water.

20. MARKETING AND ECONOMICS OF CULTIVATION

Most of the marketing of safflower is through regulated markets. Oil millers and other wholesalers bid for purchase of produce fetching maximum price to the growers. Some times, NDDB and Marico Industries also enter the market for meeting their requirements.

The market prices will be lowest during March-April which coincides with harvesting season. If the farmers can store their produce at least for 3-4 months, they can get attractive prices.

Safflower is more remunerative than many of the traditional rabi crops under rainfed conditions. The cost of production of safflower varies with the region depending on the cost of inputs-mainly labour wages. However, under rainfed conditions, the cost of cultivation for one hectare comes between Rs. 4000 and 6000/ha and under irrigated conditions between Rs. 6000 to 8000/ha. The net profit works out between Rs. 4000 to 6000/ha under scanty moisture conditions, Rs. 10000 to 13000/ha under assured, favourable moisture conditions and Rs. 12000 to 18000/ha under irrigated conditions. Benefit-cost ratio ranging from 2 to 3 could be expected under different conditions of management and productivity levels.

REFERENCES

Ahmed, Z., Medekkar, S. and Mohammad, S. 1985. Response of safflower to nitrogen and phosphorus. Indian J. Agron., 30 (1) : 128-130.

Anonymous, 1986. Safflower Production Technology. Technologies for Better Crops, 29, ICAR, New Delhi.

Anonymous, 1994. Annual Progress Report. Safflower 1993- 94. Directorate of Oilseeds Research, Hyderabad.

Anonymous, 1996. Annual Progress Report, Safflower 1995-96. Directorate of Oilseeds Research, Hyderabad.

Anonymous, 1998. Annual Progress Report, Safflower 1997-98. Directorate of Oilseeds Research, Hyderabed.

Autkar, K.S., Vyas J.S. and Wanjari, K.B. 1991. Studies on chickpea based intercropping under irrigated conditions. *Ann. Plant Physiol.,* 5(1) : 103-109.

Babu, M.S.G. 1979. Studies on chemical weed control in linseed (*Linum usitatissimum* L.) and safflower (*Carthamus tinctorius* L.). M.Sc. Thesis, UAS, Bangalore.

Bains, S.S. and Firman, M. 1964. Effect of exchangeable sodium percentage on the growth and absorption of essential nutrients and sodium by five crop plants. *Agron. J.,* 56 : 432-435.

Bernstein, 1964. Salt tolerance of plants. *Agric. Inf. Bull. No. 283.* U.S. Dept. Agric., Washington D.C.

Bohra, J.S., Kumar, H. and Agrawal, R.K. 1989. Nitrogen management of safflower for different cropping sequences in drylands. Abstracts. Second International Safflower Conference, Hyderabad, India, January 9-13, 1989 : 57.

Chopde, P.R. and A.M. Dhopte, 1970. Effect of GA, IAA and MH on growth and yield of safflower variety, N62-8. Agri. Coll. Mag. Parbhani; Vol. 11 : 61-68.

Das, N.R. and Ghosh, N. 1993. Effect of number of tillage and N levels on yields of rainfed safflower (*Carthamus tinctorius* L.) after transplanted wet rice. In : Li, D and Han, Y. (Eds). *Proceedings of Third International Safflower Conference, Beijing,* China, June 9-13, 1993: 403-409.

Daulay, H.S., Ahuja, C.P. and Singh, R.P. 1974. Studies on the seeding depths and pre-sowing seed treatments on the seedling emergence of sunflowers and safflowers. *Ann. Arid Zone.* 13(3) : 232-236.

Deokar, A.B., Patil, N.D., Manke, B.S. and Munde, M.S. 1984. Response of safflower varieties to different dates of sowing. *J. Maharashtra Agric. Univ.,* 9(1) : 67-69.

Dhopte, A.M. and Lall, S.B. 1981. Effect of removal of bracts on seed yield of capitulum in relation to its chlorophyll content and dry matter in nine varieties of safflower. The PKV Res. J. 5(1): 6-10.

Ghule, B.D., Jagtap, A.B., Dhumal, V.S. and Deokar, A.B. 1987. Effects of sowing time on incidence of aphids on safflower. *J. Maharashtra Agric. Univ.* 12(2) : 259.

Gupta, I.C. and Yadav, J.S.P. 1986. Crop tolerance to saline irrigation water. *J. Indian Soc. Soil. Sci.* 34(2) : 379-386.

Hegde, D.M. 1995. Safflower : Improved Production Technology. Tech. Bull. Directorate of Oilseeds Research, Hyderabad, 26 pp.

Hegde, D.M. 1997. Safflower, In : Singh, R.P., Reddy, P.S. and Kiresur, V.(Eds) 1997. *Efficient Management of Dryland Crops in India* - Oilseeds. Indian Society of Oilseeds Research, Directorate of Oilseeds Research, Hyderabad-500 030, India. p. 273-298.

Hegde, D.M. 1998. Integrated nutrient management for production sustainability of oilseeds. *J. Oilseeds Res.,* 15(1) : 1-14.

Hegde, D.M. and Kulkarni, G.N. 1974. Presowing seed hardening to induce drought tolerance: Safflower (*Carthamus tinctorius* L). Farm J. 6: 8-12.

Itnal, C.J., Achar, M.S., Desai, G.S. and Sajjan, G.C. 1980. Inter-cropping of bengalgram and safflower pays rich dividends. *Curr. Res.,* 9(8) : 132-133.

Janakirani, S. and Manoharchary, C.1993. Field application of Vesicular arbuscular mycorrhizal fungi as benefactors in safflower. In : Li, D. and Han, Y. (Eds). *Proceedings of Third International Safflower Conference, Beijing. China.* June. 9-13. 1993 : 472-480.

Janardhan, K.V., Patil, B.N. and Raikar, D.S. 1986. Relative tolerance of safflower (*Carthamus tinctorius* L.) varieties to saline water irrigation. *Indian J. Plant Physiol.,* 29(2) : 118-124.

Jagdale, G.B., Dhoble, S.Y., Deokar, A.B. and Ajri, D.S. 1990. Effects of planting time on incidence of aphid on different safflower varieties. *J. Maharashtra Agric. Univ.,* 15(2) : 242-243.

Jha, K.P., Moorthy, B.T.S. and Mishra, A.P. 1991. Rice based cropping system for rainfed upland of eastern India. *Indian J. Agron,* 36(2) : 218-222.

Joshi, P.K., Raikhelkar, S.V. and Pawar, K.R. 1984. Crop sequence studies under dry farming conditions of Marathwada region. J. Maharashtra Agric. Univ., 9(1): 64-67.

Kiresur, V., Reddy, P.S., Prasad, M.V.R. and Reddy, B.N. 1995. Economic sustainability of improved oilseeds production technologies under real farm situations. Directorate of Oilseeds Research, Hyderabad, 32 pp.

Kumur, R.N., Sinha, M.N. and Rai, R.K. 1989. Response of safflower (*Carthamus tinctorius* L.) to levels and depth of phosphorus placement under two moisture regimes. *Haryana Agric. Univ. J. Res.,* 19(1) : 31-36.

Kulaindavelu, R., Thangavelu, S. and Sankaran, S. 1974. Effect of herbicides on nutrient removal by weeds in safflower (*Carthamus tinctorius* L.) and its impact on seed production. Pesticides, 8(9): 38-39.

Lomte, M.H. Kawarkhe, P.K. and Ateeque, M. 1993. Scope of safflower in sorghum based cropping system under dryland conditions. In : Li, D and Han. Y. (Eds). *Proceedings of Third International Safflower Conference,* Beijing, China. June 9-13, 1993 : 548-552.

Malewar, G.U. 1979. Effect of N and P fertilizers on the yield of safflower (*Carthamus tinctorius* L.). Res. Bull, MAU, Parbhani, 3(12) : 161-162.

Mane, V.S. 1983. Effect of spacing and fertilizer application on yield and quality of safflower (*Carthamus tinctorius* L.) variety N-62-8. *Madras Agric. J.,* 70(3) : 201-202.

Mane, V.S. and Jadhav, A.S. 1993. Effects of fertilizers and plant densities on growth and yield of irrigated safflower. In Li, D and Han, Y. (Eds). Proceedings of Third International Safflower Conference, Beijing, China, June 9-13, 1993 : 390-394.

Nikam, S.M., Mahajan, M.S., and Deokar, A.B. 1988. Improvement of monetary return of coriander (*Coriandrum sativum*) through safflower (*Carthamus tinctorius*) intercropping in dryland. Indian J. Agric. Sci., 58(3): 202-205.

Nikam, S.M., Mahajan, M.S. and Deokar, A.B. 1989. Effect of seed rates and seedbed preparation on the yield of safflower in the sequential cropping system in dryland. *J. Maharashtra Agrci. Univ.,* 14(1) : 108-109.

Nikam, S.M., Patil, N.Y. and Deokar, A.B. 1985. Performance of safflower based double cropping sequence under rainfed conditions. *Indian J. Agric. Sci.,* 55(3) : 160-166.

Nikam, S.M., Tendulkar, A.V. and Deokar, A.B. 1987. Production potential of safflower-chickpea intercropping under rainfed conditions. *Indian J. Agric. Sci.,* 57(3) : 151-156.

Panchal,V.C., Siddappa, L. and Rajanna, K. 1977. Chemical weed control in safflower. Abstracts, Weed science conference and workshop in India; 64.

Patil, B.C., Rao, M.R.G., Parama, V.R.R. and Viswanath, D.P. 1989. Relative salt tolerance of safflower genotypes based on yield, sodium, potassium and calcium contents. *Indian J. Plant Physiol.,* 32(1) : 90-94.

Rafey, A. 1993. Biological potential and economic feasibility of inter-cropping oilseeds and pulses with safflower (*Carthamus tinctorius* L.) in drylands. In : Li., D. and Han.,Y. (Eds) : Proceedings of Third International Safflower Conference, Beijing, China. June 9-13, 1993 : 77-82.

Raju, N.S. and Reddy, K.A. 1990. Studies on double cropping with Kharif pearlmillet based sequence under rainfed condition. *J. Research, APAU.,* 18(2) : 127-130.

Rao, M.R.G., Manjunath, M., Viswanath, D.P., Raikar, D.S. and Srinivasa, N. 1987. Screening of safflower genotypes for tolerance to saline water irrigation. *J. Oilseeds Res.,* 4(2) : 279-282.

Rao, V.R. 1984. Production practices for maximising productivity potential of safflower in scanty and assured rainfail areas. Lecture delivered on 10 October, 1984 to the participants of the training course "Maximizing the yields of safflower" organized by the National Communication and Training Centre on Oilseeds, Directorate of Oilseeds Research, Hyderabad.

Rasal, P.H., Patil, P.L. and Shingte, V.V. 1989. Effect of Azotobacter on safflower yield. Abstract. Second International Safflower Conference, Hyderabad, India, January 9-13, 1989 : 58.

Saran, G., Turkhede, B.B. and De, R. 1980. Influence of date of seeding and population density on the yield of safflower grown in drylands. *Indian J. Agric. Sci.,* 50(8) : 611-615.

Sawant, A.R., Deshpande, S.L. and Nadkarni, P.G. 1993. Safflower inter-cropping in gram and linseed. In : Li, D and Han. Y. (Eds). *Proceedings of Third International Safflower Conference, Beijing, China,* June 9-13, 1993: 646-651.

Saxena, J.P. 1987. Relative performance of tillage systems for production of sorghum and safflower crops under drylands of Bundelkhand region. In : The role of agril. engineering in dryland agriculture. Proc. 23 Annual Convention of the Indian Soc. of Agric. Engrs. Jabalpur 9-11 March, 1989 : 92 to 99.

Sharma, R.A. 1993. Response of safflower to integrated fertility management practices in black clay soils under dryland conditions. In: Li, D and Han, Y (Eds). *Proceedings of Third International Safflower Conference, Beijing, China, June 9-13, 1993* : 661-669.

Shelke, D.R. and Bhosle, R.H. 1993. Weed management in sorghum- safflower sequence cropping under shortage of labour on vertisols of Marathwada. In : Li, D and Han, Y. (Eds). *Proceedings of Third International Safflower Conference,* Beijing, China, June 9-13, 1993 : 670-673.

Singh, J. and Yadav, D.S. 1990. Studies on the wheat based intercropping systems under rainfed conditions. *Indian J. Agron.,* 35(3): 262-265.

Singh, S.B., Chauhan, Y.S. and Verma, G.S. 1989. Effect of row spacing and nitrogen level on safflower in partially reclaimed salt affected soils of eastern Uttar Pradesh. Abstracts. *Second International Safflower Conference,* Hyderabad, India, January 9-13, 1989 : 55.

Singh, S.B. and Yadav, R.D.S. 1988. Performance of rice based cropping systems in river flood plains. *Intern. Rice. Res. Newsletter,* 13(5) : 37-38.

Tapadia, B.B., Kale, V.C., Novhat, D.A. and Nagargoje, S.R. 1991. Sorghum based cropping system for irrigated areas of Marathwada region of Maharashtra. *J. Maharashtra Agric. Univ.,* 16(3) : 378-380.

Tiwari, K.P. and Mandloi, K.K. 1972. Optimum time of planting and plant population in safflower (*Carthamus tinctorius*). JNKVV Res. J., 6(2) : 172.

Umrani, N.K., Kale, S.P. and Bhoi, P.G. 1981. Response of safflower to application of nitrogen under rainfed conditions. *Indian J. Agron.,* 26(4) : 365-370.

Veeranna, V.S. and Channappa, K. 1981. Response of safflower (Annigeri-1) to nitrogen and phosphoric acid. *Curr. Res.,* 10(3) : 43-44.

Veeranna, V.S., Channappa, K. and Tippeswamy, 1976. The most advantageous depth of seeding for safflower is 5 cm. *Curr. Res.,* 5(12) : 201-202.

Veeranna, V.S., Jagannatha, B. and Gidnavar, V.S. 1980. Hybrid jowar in kharif followed by safflower in rabi is profitable under rainfed conditions. *Curr. Res.,* 9(1) : 5-6.

Veeranna, V.S. and Rudraradhya, M. 1980. 45 cm is the most advantageous row spacing for safflower. *Curr. Res.,* 9(7) : 111-113.

Weiss, E.A. 1983. Oilseed crops, Longman, London and New York, pp 216-281.

Zaman, A. 1988. Effect of nitrogen and phosphorus on yield attributes, Seed yield and oil content of irrigated safflower in laterite soil. *Ann. Arid Zone,* 27(1) : 37-40.

Chapter - 22

SOYBEAN — A REMUNERATIVE CROP FOR RAINFED FARMING

O.P. Joshi*, S.D. Billore, A. Ramesh and Ch. Bhardwaj

*Director, National Research Centre for Soybean, Indore - 452 001 (MP)

1. INTRODUCTION

Soybean has recently commanded an extensive area in India. As a commercial crop, it occupies premier position. It has brought socio-economic prosperity among farmers and earns our exchequer of over Rs. 24 billions annually. In addition to it, being a source of both high quality cheapest protein (40%) and oil (20%), it is likely to acquire a unique place in helping to alleviate wide spread pulse protein malnutrition of the Indian masses. The introduction of soybean as a major crop has long been of interest in tropical and subtropical countries, however, both agronomic and cultural difficulties have hindered serious development. Until recently, the varieties grown were primarily of temperate regions. Of late such obstacles have been largely overcome as a result of research break through in varietal development, production technology and pest control. Parallel growth of soya-based industries and great overseas demand for decorticated oil cake (DOC) have been instrumental in enhancement in area of soybean from 5 lakh hectaresto the present 60 lakh hectares within a short span of two decades (Table 1). Soybean has major contribution (8%) in achieving near self reliance in edible oil production of the country in addition to the socio-economic upliftment of Indian farmers.

This chapter covering most of the major aspects of soybean production would be of immense value to agricultural consultants and extension experts working on soybean in particular, and to the agricultural students and researchers in general.

2. ADAPTATION AND CLIMATIC NEEDS

The climatic requirements for cultivation of soybean are more or less similar to maize, sorghum and rainy season legumes. Being a remunerative cash crop and versatility in cultivation, it is replacing these traditional crops in Western Madhya Pradesh. Of late, it is also becoming popular in cotton dominating areas of Western Madhya Pradesh, Vidarbha region of Maharashtra and parts of Karnataka.

Table 1. Soybean Scenario in India

Year	Area (m ha)	Production (m.t)	Productivity (kg ha-1)
1979-80	0.50	0.28	570
1988-89	1.73	1.54	892
1989-90	2.25	1.80	801
1990-91	2.56	2.60	1015
1991-92	3.19	2.49	782
1992-93	3.79	3.39	894
1993-94	4.25	4.63	1089
1994-95	4.32	3.93	890
1995-96	5.03	5.09	910
1996-97	5.23	5.20	907
1997-98	5.60	5.30	946
1998-99*	6.20	5.70	919

Source : Directorate of Oilseeds Development, Govt. of India * Estimates

2.1 Temperature

Soybean is known to germinate at a soil temperature of above 10°C but the ideal temperature for germination and rapid seedling emergence is about 30°C (Delouche, 1963). The ideal temperature for further growth is also about 30°C (Brown, 1960). Germination and seedling emergence proceed slowly in cold soil and speed up as it warms. Vegetative growth, flowering and seed set are also affected by temperature. Warm temperature after seedling emergence results in vigorous plants and early flowering. Varietal differences in flowering time are mainly due to their response to day length. But the year to year variations in the time of flowering of one variety planted on same data can be attributed to prevailing temperature. Continuous temperatures of less than 24°C during crop growth will delay the onset of flowering. Temperature also affects the seed set and seed maturity. Floral induction and reproductive development in soybean are particularly sensitive to high temperature with the optimum being between 21°C and 27°C (Parker and Borthwick, 1943). Temperature below 22°C decrease pod initiation and no pods will be formed at temperature less than 14°C (Thomas and Raper, 1981).

2.2 Photoperiod

Most soybean çultivars respond to photoperiod as quantitative short day plants and are adapted for production in a narrow band of latitudes. In General, cultivars with the genetically controlled long juvenile trait are having wider adaptability. It has been reported by Kenworthy, Brown and Thibou (1989) that earlier maturing soybean strains are less sensitive to photoperiod than those maturing late. The varieties evolved for higher latitudes when taken towards equator experiencing photoperiod for short duration, they flower and mature early. There is also reduction in node number, height, leaf area and yield (Board and Settimi, 1988). The sub-optimal photo-period influences pod fill resulting there by in under developed seeds and ultimately into low yield. It has been reported that the critical photoperiod for flowering and pod fill is 12.5 hours.

2.3 Water

Rainfall is the principal source of water for soybean cultivation in the country. Only very few soybean farmers have irrigation facility to supplement the water during long dry spells often experienced during growing period of the crop. The critical periods for water management in soybean are planting to emergence, flowering and pod filling stage. A deficit or excess of moisture at these stages determines the yield.

The water content of the soil during this period should not exceed 85% or be less than 50% of potential available soil moisture (PASM) and the optimum being 75% PASM (Doorenbos and Kassam, 1986). A moisture stress at pod filling stage is more detrimental to yield than during flowering. Soybean can tolerate short periods of moisture stress at other stages as it has deep root system and a relatively long flowering period. Loss of early flower and pod may be compensated for by those produced later if moisture becomes available. To achieve optimum yield, an adequate supply of water must be made available during the critical growth period either through frequent rainfall, irrigation or stored soil moisture.

Based on soybean cultivation in India, five agro-ecological zones, namely central, southern, northern plain, northern hill and north eastern zone have been organised (Table 2).

Table 2. Soybean Agro-ecological Zones of India

S.N.	Zone	Area covered
I	Northern Hill Zone	H.P. and Northern Hills of U.P.
II	Northern Plain Zone	Punjab, Haryana, Delhi and North Eastern Plains of U.P. and Western Bihar
III	Central Zone	M.P., Bundelkhand Region of U.P., Rajasthan, Gujarat, North- Western Maharashtra and Orissa
IV	Southern Zone	Karnataka, Tamil nadu, A.P. Kerala and Southern parts of Maharashtra
V	North-Eastern Zone	Assam, West Bengal, Bihar, Meghalaya.

Source : Bhatnagar and Tiwari (1995)

3. PRODUCTION TECHNOLOGY

Studies conducted on optimization of input demand under the aegis of All India Coordinated Research Project on soybean, National Research Centre for Soybean and SAU's have led to the standardisation of production technology of soybean for different zones shown in Table -3.

3.1 Selection of field & land preparation

Soybean does well in a wide range of soils with textural range from sandy loam to clay. However, well drained and leveled field with near neutral pH i.e. around 7 (Table 4) and comparatively rich in organic carbon are ideal for soybean cultivation. Soybean poorly tolerates alkaline (8.5 pH), acidic (4 pH) and swampy soils.

One deep ploughing in summer followed by 2-3 cross harrowing or cultivation for breaking of soil clods will make ideal seed bed for a good crop of soybean. Gentle gradient, while levelling, should be provided to ensure good drainage. Unlevelled soil with clods with result in either too shallow or too deep placement of seed and will eventually lead to poor plant stand and ultimately lower yield. Since soybean is grown in deep vertisols which is difficult to work when wet, it is important to develop farming system where most soil tillage and cultivation operations can be executed in dry season. High intensity rains, a common phenomenon during *Kharif* in such soils adversely affect crop growth. A need thus arises to improve *in situ* soil and water conservation and at the same time provide proper drainage. Broadbed - and - furrow system has been found to satisfactorily attain these goals on deep Vertisols. The system consists of a relatively flat bed or ridge of 90 cm wide and shallow furrow of 60 cm wide. A gental grade should be laid out in order to give a slope of 0.4 to 0.8%. This system not only meets the requirements of conserving tillage, providing adequate surface drainage but also suits to fit cropping systems with widely different row spacing requirements.

Table 3. Appropriate technology of Soybean production for different Agro-climatic zones of India

S.N.	Input/ practices	North Hill Zone	Noth plain zone	Central Zone	Southern Zone	North Eastern Zone
1	Varieties	Bragg, PK 362, PK 308, PK 327, PK 416, VL Soya 1, VL Soya 2, Pusa 16, Pusa 24, Pusa 20, Shivalik	Bragg, PK 362, PK 308, PK 327, PK 564, SL4, SL-96, PK 416, Pusa 16, Pusa 24, Pusa 27, Pusa 22, PK 1024, PK 1042	Bragg, Durga, Gaurav, Gujarat, Soybean 1 & 2, MACS 13, MACS 58, Monetta, JS-2, JS 75-46, JS80-21, JS 335, Ahilya-1, Ahilya-2 Ahilya-3	Co-1, Hardee, KHSb 2, Monetta, PK 471, PK 1029, Pusa 40, MACS 57, MACS 124, ADT-1	JS 80-21, PK 472, Ankur Birsa Soybean - i, Bragg, PK 327
2	Planting time	Last week of May-June ending	Mid of June to 1st week of July	Middle of June to Middle of July (ii) 1st week October to Dec. (Rabi crp) (iii) 2nd fornight of Jan. (Summer crop)	(i) Middle of June to end of July (Kharif crop)	Middle of June to Middle of July
3	Planting	45 cm (row to row) x 5 cm (plant to plant)	45-60 cm x 5 x 8 cm	30-45 x 5 cm	30-45 x 5 cm	30-45 x 5 cm
4	Plant population	0.4 million plants/ha	0.4 million plants/ha	0.4 to 0.6 million plants/ha	0.4 to 0.6 million plants/ha	0.4 to 0.6 million plants/ha
5	Depth of seedling	3 to 5 cm	3 to 5 cm	3 to 5 cm	3 to 5 cm	3 to 5 cm
6	Fertilizers (kg/ha)	20 N : 80 P_2O_5 20, K_2O: 20 S	20 N: 60 P2O5, 20 K2O: 20 S	20 N: 60 P2O5, 20 K2O: 20 S	20 N: 80 P2O5, 40 K2O: 20 S	20 N: 80 P2O5, 40 K2O: 20 S
7	Seed treatment	Thiram + Bavistin (2:1) @ 3 g/kg seed				
8	Bacterial culture	About 500 g/ 75 kg of seed	About 500 g/ 75 kg of seed	About 500 g/ 75 kg of seed	About 500 g/ 75 kg of seed	About 500 g/ 75 kg of seed
9	Seed rate	75-100 kg/ha	75-100 kg/ha	75-100 kg/ha	75-100 kg/ha	75-100 kg/ha
10	Irrigation	Presowing and at flowering as well as pod filling stage, in case of drought				
11	Weed control	2 land weeding at 25 to 30 and 45 DAS* or Fluchloralin (Basalin) @ 10 kg a.i./ha in 750-800 of water as pre plant incorporation or Alachlor @ 2.0 a.i./ha in 750-800 l water as per emergence or Metalochlor (Dual) @ 1.0 kg a.i./ha in 750-800 l water as pre emergence.				
12	Plant protection	Furrow application of Thimet 10 g @ 10 kg/ha, followed by 1 or 2 sprays of either Quinalphos 25 EC (0.05%) or Endosulfan 35 EC (0.07%) or Monocrotophos 36 SC (0.04%) depending upon the insect pest and intensity of damage.				
13	Harvesting	When pods turn black, brown or golden, seed has about 17% moisture.				

14	Threashing	Operate thresher at low cylinder speed of 400 to 500 rpm at 14% seed moisture					
15	Seed drying and storage	At seed moisture of about 10%, store in moisture proof bags					
16	Inter cropping	Maize, Arhar, Sorghum	Maize, Arhar, Sorghum	Arhar, Sugarcane, Cotton, Bajra, Groundnut	Cotton Maize, Finger millet, Arhar, Groundnut, Wheat	Ragi, Paddy, Arhar, Temperate grass	

* DAS = Days after sowing

Source : Bhatnagar, P.S. & Tiwari, S.P. eds. (1995). Technology for increasing soybean production in India.Technological Bulletin No. 1, NRCS (ICAR),Indore PP. 49.

Table 4. Soil pH Dynamics suitable for soybean

pH Range	Type	Remarks
pH 9	Very strongly alkaline	Characteristic of arid and semi-arid regions, not suitable for soybean
pH 8	Strongly alkaline	Micronutrient deficiency are common in soybean particularly Mn and Fe
pH 6-7	Slightly acidic to neutral	Best range for soybean cultivation
pH 4-5	Very strong to strongly acidic	Mn and Al toxicity and Mo deficiency commonly observed in soybean - not recommended for soybean

Source : Walter and Samuel (1983)

3.2 Time of sowing

Soybean is predominantly grown in *Kharif i*n India. In suitable parts (Southern zone in particular) of the country, it can also be grown in *rabi,* spring/summer season. However, to avoid out break of rust caused by *Phakopsora pachyrhizi* Syd. It is advisable at present to cultivate soybean crop in one season during the year. For *Kharif* crop, pre-monsoon planting after irrigation in the second fortnight of June have been found to give good results in terms of desired plant population and good yields. This practice may not prove practical in swell-shrink soils which exhibits deep cracks during summer. When irrigation is not available, sowing should be effected with the commencement of monsoon rains. Sowing should be completed in last week of June or latest by first week of July because late planting leads to multiple problems like poor plant stand, pod bearing, seed filling and yield (Table 5). In case of delayed onset of monsoon, pre sowing irrigation can be practised to ensure optimum soil moisture for timely sowing.

Table 5. Influence of sowing dates on mean yield (t ha^{-1}) of soybean genotypes

Genotype	Sowing date					
	25th June	5th July	10th July	15th July	20th July	Mean
JS 335	2.30	1.84	1.63	1.61	1.37	1.75
JS 71-05	2.22	1.88	1.65	1.52	0.90	1.63
Ahilya 3	3.10	2.52	2.48	2.27	1.91	2.46
Ahilya 1	1.96	1.80	1.64	1.44	1.20	1.61
PK 472	2.12	1.77	2.04	1.68	1.41	1.80
Mean	2.34	1.94	1.89	1.70	1.36	

3.3 Plant population

One of the major reasons for poor soybean yields in India is sub-optimal plant population. A plant population of about 4.5 lakhs has been found to be optimum with a range of 4 to 6 lakhs plants per hectare. In general, comparatively higher plant population is desirable for determinate varieties. Use of quality seed and good seedbed are pre requisite to maintain the desired plant population. For ensuring minimum 70% seed germination, a 10 m x 10 m block should be test sown as per the recommended practices. Emergence of above 4500 plants shall meet the requirement. Results contrary to this will warrant change of seed or higher seed rate.

3.4 Depth of sowing

While sowing, the seed should be placed at the depth of 3-5 cm to ensure good germination and subsequent plant stand. Germination and emergence of seedling are severely affected if deep sowing is practiced. Formation of hard soil crust due to rains immediately after sowing followed by dry and sunny weather aggravates the problem of poor emergence. In such cases supplementing irrigation, breaking the crust without damaging the emerging seedling is advised. The shallower depths are recommended for soils that are wet and/or heavy and deeper depth are chosen for drier/or lighter soils. Since the bolder seed invariably shows rapid loss in seed viability and poor emergence, shallower placement is advocated. It is recommended not to plant deeper than 5 cm, because soybeans are epegial in emergence, by virtue of which elongating hypocotyl pulls the cotyledons to the soil surface. The required effort is greater with deeper planting.

3.5 Spacing, seed rate and seed treatment

Seed rate is dependent on seed index (SI) and germinability (Table 6). The required seed rate is generally 75 kg. ha^{-1}. In general, every gram increase in seed index increases seed rate approximately by 5 kg ha^{-1} viz.; suppose seed index is 13 g. ha^{-1}, it corresponds to 75 kg.ha^{-1}. If SI is 14 g the seed rate should be 80 kg.ha^{-1}. Sowing is generally done in rows of 30-45 cm apart. In Northern regions where the plants attain excessive vegetative growth, 60 cm row spacing is recommended. The ideal plant to plant distance is 4-5 cm. Soybean tolerates considerable variation in plant population and row spacing without significant changes in seed yield, but there is a tendency towards higher yields with narrower row spacing. In case of delayed sowing on account of late onset of monsoon, the yield reductions due to less vegetative growth coupled with early flowering can be compensated to some extent by narrowing the row to row spacing to 30 cm and increasing the seed rate by 25%.

Table 6. Plant population dynamics in soybean

Spacing inter row (cm)	Plant population (m ha-1)	Seed germinability (per cent)	Seed rate (kg ha^{-1}) based on seed index					
			10	11	12	13	14	15
30	0.66	100	66.0	72.6	79.2	85.8	92.4	99.0
		70	85.8	94.4	103.0	111.5	120.1	128.7
40	0.50	100	50.0	55.0	60.0	65.0	70.0	75.0
		70	65.0	71.5	78.0	84.5	91.0	97.5
45	0.44	100	44.0	48.4	52.8	57.2	61.6	66.0
		70	57.2	62.9	68.6	74.4	80.8	85.8

Nearly 20 or even more species of fungi are known to be seed borne affecting soybean plant stand, thus making it essential to resort to fungicidal treatment of seeds. Treating the seeds with Thiram + Carbendazim fungicide (2:1) at the rate of 3g. kg^{-1} seed has been found to prevent collar rot and ensure good plant stand. Seed treatment improves the germination of infected seeds, reduce the amount of seed borne inoculum and protects seed and seedling from soil borne pathogens.

To facilitate the effective nodulation and fixation of atmospheric nitrogen in root nodules of soybean plants, it is imperative that seed is inoculated with *Bradyrhizobium japonicum* culture @ 500 g per 75 kg seed. In recent years, application of phosphate solubilizing bacteria @ 500 g per 75 kg seed which is complementary to rhizobia is also recommended for improving P use efficiency for better yields. In the areas where soybean has been previously grown an enhanced level (2 to 5 times) of rhizobium is advisable during the initial years of soybean cultivation.

3.6 Germination and Dormancy

Checking for germination is an important aspect of soybean cultivation and management as the final yield depends on the total plant stand. A simple germination test involves placing of 100 seeds in between two newspaper covers and rolling them with a moist cloth. It has to be seen that the cloth is kept just moist through out the test. After fifth day the first count is taken and after the eighth day the second count is taken. The seed rates so prescribed are at 70% germination. Seed rate can be adjusted in seed lots showing germination upto 60%. Seed with germination lower than 60% should be replaced. Planting of soybean is not advisable with seed lots not subjected to germination/ emergence test prior to sowing.

4. VARIETAL SELECTION

A good soybean cultivar should be high yielding and should exhibit stable performance across a range of environments. Since the ideal cultivar has not been developed yet, farmers should choose the ones that most closely meet to their needs. In addition to it, aspects to consider include the size of the growing area to be planted, the length of the growing season, likely pest complex, cropping system and the desired probable conditions during season of growth. It is advisable to grow more than one variety with different maturity durations particularly when the planting area is quite large in order to increase production stability and to make an efficient use of harvesting equipment.

Importance of suitable varieties for long range of success in any crop has been well understood. The All India Coordinated Research Project on Soybean (AICRPS) by ICAR in 1967 and consequently establishment of National Research Center for Soybean (NRCS) at Indore, Madhya Pradesh in 1986 led to the release of 63 improved varieties which provide wide range for the farmer to choose from. The recommended varieties for different zones are listed in Table 7.

Plate 22.1. JS 335 : Since its release in 1994, this variety is dominating in major soybean growing areas in Madhya Pradesh, Eastern and Southern states. It is resistant to bacterial blight and stem fly. It matures in 95 to 100 days in Central Zone.

Plate 22.2. Ahilya 1 (NRC 2): Development at NRCS, Indore, through mutation and named after the famous ruler of Holkar regime - Ahilya Bai Holkar. It is suitable for cultivation in Central and North - Eastern Zones. It flowers profusely and bears two seed pods. It has good seed germiability.

Plate 22.3. Ahilya 2 (NRC 12): A mutant of exotic variety Bragg'' is resistant to major diseases and tolerant to defoliators, stem fly and girdle beetle. This is a case of land mark achievement of developing a variety through mutagenesis, with both insect tolerance and high yield.

Plate 22.4. Ahilya 3. (NRC 7): An early maturing (90-95 days), bold seeded with resistance to major diseases, insect pests and pod shattering. It has high oil content (21%). It is becoming highly popular in Madhya Pradesh, particularly among potato growers.

Table 7. Soybean varieties of India

Variety	Area of adaptability	Duration (days)	Yield potential (t ha-1)	Plant type
Alankar	Northern Plains	115-120	2.0-2.5	Determinate
Ankur	Northern Plains, Central zone	115-120	2.0-2.5	Determinate
ADT-1	Cauvery delta	85-90	2.0-2.5	Determinate
Birsa soy 1	Bihar plateau	106-110	2.0-2.5	Determinate
Bragg	Throughout India	112-115	1.5-2.0	Determinate
Clark - 63	Northern and Central zone	80-85	1.5-2.0	Determinate
Co-1	Tamil nadu	85-90	1.7-2.0	Determinate
Co-2	Tamil nadu	75-80		Determinate
Davis	Southern zone	110-115	1.5-2.0	Determinate
Durga	Central zone	102-105	2.0-2.2	Indeterminate
Gaurav	Central zone	104-106	2.0-2.5	Semi-determinate
Gujarat soy 1	Gujarat	90-95	1.5-2.0	Determinate
Gujarat soy 2	Gujarat	105-110	2.2-2.5	Determinate
Hardee	Karnataka	105-110	1.5-2.0	Determinate
Improved pelican	Southern zone	112-115	1.5-2.0	Determinate
Js 71-05	Central zone	90-95	2.0-2.4	Determinate
JS-2	Central zone	90-95	1.8-2.0	Determinate
JS 75-46	Central zone	100-106	2.5-3.0	Indeterminate
JS 76-205	Central zone	105-110	2.0-2.5	Determinate
JS 80-21	Central zone	90-109	2.5-3.0	Determinate
JS 335	Throughout India	98-101	2.5-3.5	Semi- determinate
JS79-81	Madhya Pradesh	104-107	2.5-3.0	Determinate
Kalitur	Madhya Pradesh	120-130	1.8-2.0	Indeterminate
KHSb2	Karnataka	115-120	2.0-2.5	Semi-determinate
Lee	Northern hill zone	105-115	2.1-2.5	Determinate
MACS-13	Central zone	90-100	2.5-3.0	Indeterminate
MACS-58	Central zone	90-100	2.5-3.5	Semi- determinate
MACS-57	Maharashtra	85-95	2.5-3.0	Semi- determinate
MACS-124	Southern zone	95-100	2.5-3.5	Semi- determinate
Monetta	Central and southern zone	98-102	2.5-3.0	Determinate
PK262	Northern hill and northern plain zones	120-125	2.5-3.0	Determinate
PK308	Northern hill and northern plain zones	110-115	2.0-2.5	Determinate
PK327	Northern hill and northern plain zones	100-105	2.0-2.5	Determinate

PK416	Northern hill and northern plain zones	115-120	3.0-3.5	Semi-determinate
PK472	Northern hill and northern plain zones	100-105	3.0-3.5	Determinate
PK471	Southern zone	90-95	2.5-3.0	Determinate
PK564	Northern plain zone	105-115	3.0-3.5	Determinate
PK1024	Northern plain zone	116-120	3.0-3.2	Determinate
PK1029	Southern zone	84-101	2.5-3.0	Determinate
PK1042	Northern plain zone	110-119	3.0-3.5	Determinate
Punjab 1	Central, northern plain and northern hill zone	95-100	2.0-2.2	Determinate
Pusa-16	Northern plain and northern hill zone	105-115	2.5-3.0	Determinate
Pusa-24	Northern plain and northern hill zone	110-115	2.5-3.0	Determinate
Pusa-20	Northern hill zone	105-120	2.5-3.0	Determinate
Pusa-22	Northern plain zone	105-110	2.5-3.0	Determinate
Pusa-37	Northern plain zone	105-115	2.5-3.0	Determinate
Pusa-40	Southern zone	110-115	2.5-3.0	Semi- determinate
Shilajeet	Northern hill and northern plain zone	100-105	2.0-2.5	Determinate
SL-4	Northern Plain zone	100-106	1.6-2.0	Determinate
SL-96	Punjab	110-112	1.8-2.0	Determinate
Type-49	Central zone	125-130	2.0-2.5	Indeterminate
VL Soya-1	U.P. plains and hills	110-113	2.0-2.5	Determinate
VL Soya-2	Northern hill zone	104-116	2.5-3.0	Determinate
Shivalik	Himachal Pradesh	120-125	2.5-3.0	Determinate
VLS 21	Northern U.P.	120-122	2.0-2.5	Determinate
Ahilya 1	Northern hill and central zone	90-95	2.5-3.0	Determinate
Ahilya 2	Central Zone	96-99	2.5-3.0	Determinate
Ahilya 3	Madhya Pradesh	90-99	2.5-3.5	Determinate
MAUS 1	Maharashtra	90-95	2.5-3.0	Determinate
MAUS 2	Southern Zone	105-110	2.5-3.5	Semi- determinate
KM 1	Tamil Nadu	95-100	1.2-1.5	Determinate
KB 79	Karnataka	85-93	1.7-1.8	Determinate

4.1 Soybean in cropping systems

Soybean is suitable for companion cropping/intercropping/mixed cropping due to its :

- Comparatively better tolerance to shade and drought (Lawn, 1982).
- Efficient light utilization due to lanceolate leaf type and compact canopy.
- Does not excessively compete for soil moisture (Nathanson *et al.,* 1984 and Wright *et al.,* 1988).

- It has better ability to fix atmospheric dinitrogen and tolerate high level of aluminium (Tanaka, 1983). Reports show that the succeeding crop benefits from the residual nitrogen fixed by rhizobium which is to the tune of 30 to 60 $kg.ha^{-1}$.
- Higher and remunerative land equivalent ratio (LER) makes soybean a better inter-crop/companion crop/mixed crop.

Plate 22.5. Soybean + pigeonpea intercropping: four rows of soybean with two rows of short/ medium pigeonpea has been found to be the most practical and remunerative intercropping for dry land agriculture.

Plate 22.6. Soybean + maize intercropping : A remunerative intercropping system for irrigated as well as dryland agriculture.

Soybean being a short duration and having deep rooted system, is best suited for intercropping with those crops having shallow and fibrous root system viz. maize, sorghum, finger millet, pearl millet etc. Studies at NRCS have shown the soybean + pigeonpea in 4:2 row ratio gave the highest returns.

For sustainability, it is generally advised to rotate legumes with non legumes. Soybean form excellent cropping sequence with cereals for effective utilization of residual nitrogen as well as to effectively control a larger spectrum of weeds and insects and pests. Further, on account of availability of 85 to 130 days maturity duration varieties, it fits well in traditional cropping sequences.

Remunerative cropping systems for different zones are shown in Table 8.

Table 8. Remunerative cropping systems for different zones

Zone	Inter/Mixed Companion-cropping	Sequence rotation	Suitable varieties
Central	Soybean+pigeonpea Soybean + sorghum Soybean + groundnut Soybean + pearlmiilet Soybean + cotton Soybean in mango/ guava orchards	Soybean-wheat Soybean-wheat-corn fodder Soybean-potato Early soybean-garlic Early soybean-rapseed mustard Soybean-pigeonpea Soybean-safflower Soybean-sorghum Soybean-chickpea	JS 335, JS 80-21, JS 71-05, PK 472, PK 416, PK 262, MACS 124, MACS 58, MACS 57, MACS 13, Durga, Punjab -1, Pusa 16, Pusa 20, MAUS 1, MAUS 2, Birsa Soya 1, Ahilya 1, Ahilya 2, and Ahilya 3.
Southern	Soybean+pigeonpea Soybean + Fingermillet Soybean + sugarcane Soybean + sorghum Soybean + groundnut Soybean + coconut/mango/ guava orchards Soybean in agro forestry Soybean + wheat (Rabi)	Wheat-soybean fingermillet-peas Oat-cowpea barley-soybean Soybean-fingermillet-beans Soybean-wheat groundnut	KHSb-2, MACS 124, MACS 58, Monetta, PK 472, PK 416, PK 262, PK 1029, JS 335, Pusa 16, Co-1, JS 80-21, MACS 13, Pusa 24
Northern Plain	Soybean + Pigeonpea Soybean + Corn Soybean + sorghum Soybean in mango guava orchards Soybean in agro forestry	Soybean - wheat Soybean - potato Soybean - chickpea	JS 335, PK 262, MACS 13, PK 327, Punjab 1, PK 472, PK 416, PK 1024, PK 1042, SL 4, SL 96
Northern Hill	Soybean + corn Soybean + pigeonpea	Soybean - wheat, Soybean - peas Soybean - lentil Soybean - toria	Shivali, VLS -2, Bragg, PK 262, PK 327, Punjab - 1, Ahilya -1
North east	Soybean + Fingermillet Soybean + rice Soybean + pigeonpea Soybean + roselle Soybean + temperate grass	Paddy - soybean Soybean - paddy	JS 80-21, PK 262, Bragg, PK 472, Pusa 16, JS 335, Ahilya 1, and Ahilya 3.

Source : Bhatnagar and Joshi (1998)

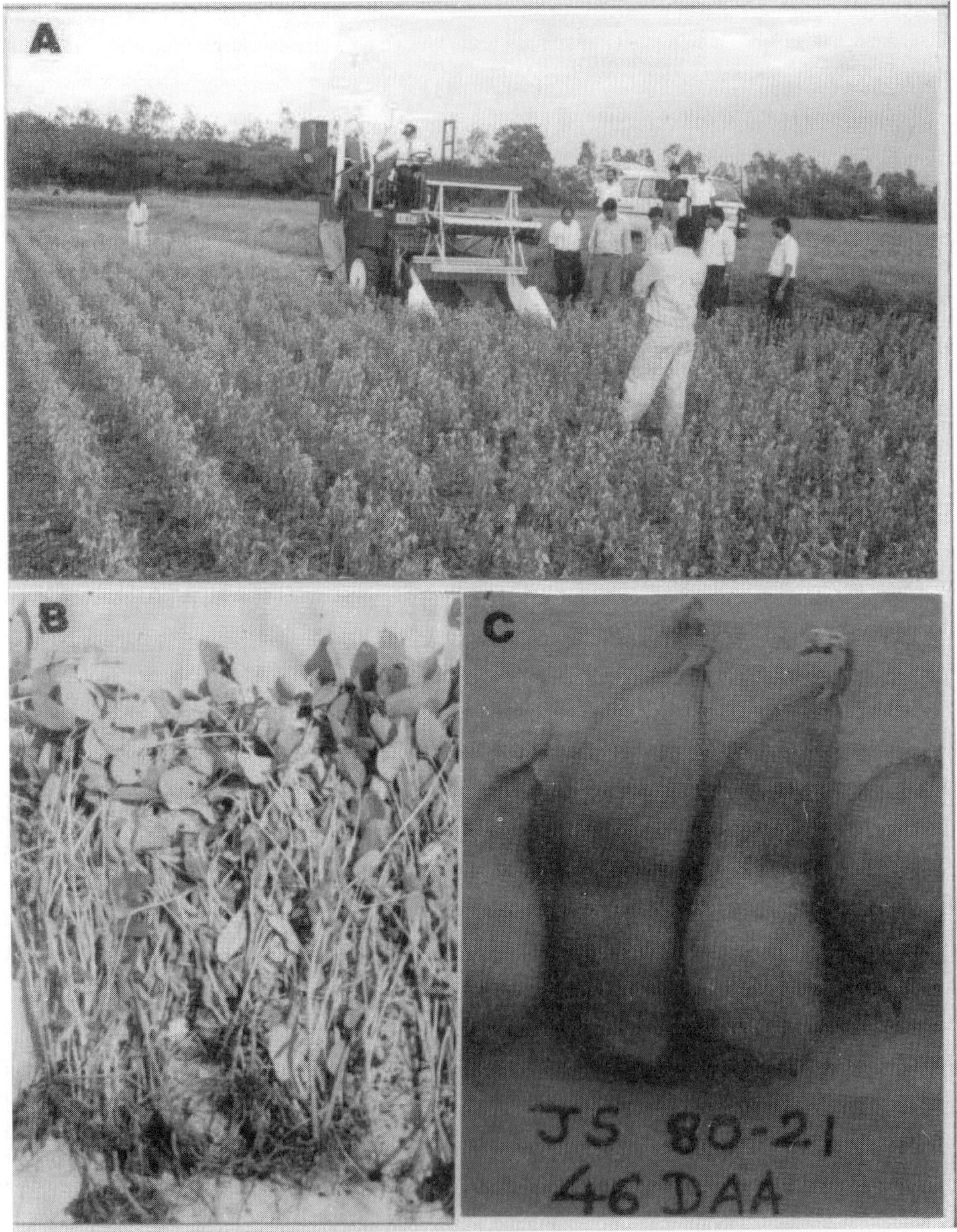

Plate 22.7. 1. **A**. Soybean harvest with combine, mechanical harvester. (Photo credit Dr. O.P. Joshi) **B.** Soybean cultivar, JS-80-21 and **C.** Pod at physiological maturity (46 DAA)

5. GROWTH KINETICS OF SOYBEAN

Soybean development is a continuous process that beings with seed germination and is completed with the maturity of seed. Variations in the relationship between vegetative and reproductive development is common in soybean due to the influence of planting date, variety, location and weather. There also exist a major difference in plant development between determinate, semi determinate and indeterminate varieties. In indeterminate varieties after the basic vegetative phase, the reproductive and vegetative development proceed simultaneously.

The description of the vegetative and reproductive phases in sequence (Fehr and Caviness, 1977) is given as below :

Notation	Stage	Description
Vegetative growth stages (V)		
VE	Emergence	Cotyledons above the soil surface
VC	Cotyledons	Unifoliate leaves unfold completely
V1	I node	Fully developed leaves at unifoliate node
V2	II node	Fully developed trifoliate leaf at node above the unifoliate node
Vn	n th node	n = number of nodes on the main stem to that before the one on which the flower has opened.
Reproductive growth stages (R)		
R1	Bloom initiation	One open flower at any node on main stem
R2	Full bloom	One flower at one of the two upper most nodes on the main stem with a fully developed leaf
R3	Pod initiation	5 mm long pod at one of the four upper most nodes on the main stem with fully developed leaf
R4	Full pod	2 cm long pod at as in R3
R5	Seed initiation	3 mm long seed in a pod at as in R3
R6	Full seed	Pod containing a green seed that fills the pod cavity at as in R3
R7	Maturity initiation	One mature pod on the main stem (Loss of green colour of the pod wall)
R8	Full maturity	95% of the pods have reached their matured pod colour

It has been generally noticed that after 5 to 10 days from R8 in dry weather brings pod moisture to 15%.

6. WEEDING AND INTERCULTURAL OPERATIONS

Weed density, type of weed and growing conditions are factors that affect the degree of competition between weeds and the crop. An yield loss of 35 to 70% has been reported depending on nature and intensity of weeds in soybean yield (Tiwari and Khurchania, 1990). It has been further observed that weed growing within a row are more harmful than those growing between rows on account of increased competitiveness.

The critical period for weed control (Table 9) in soybean is from sowing to 40 days after sowing (Prabhakar *et al.,* 1992). The most common weeds encountered in soybean ecosystem are *Cynodon dactylon* (Bermuda grass), *Cyperus rotundus* (Purple nut sedge), *Panicum* spp., *Dinebra arabica* (lona grass), *Commelina benghalensis* (day flower), *Elusine indica* (Goose grass), *Echinocloa crusgalli* (sanwa grass) and *Pasupalum* spp. (knot grass) in the monocots while *Boerhavia diffusa* (Khapra), *Celosia argentia* (white cockscomb), *Convolvulus arvensis* (field bind weed), *Chorcorus* spp. (wild jute), *Alysicarpus rugosus* (leaf flower) and *Phyllanthus niruri* (niruri) dominate in dicot weeds. Generally grasses are more competitive than broad leaved weeds.

Table 9. Yield erosion due to soybean - weed competition

Treatment	Seed yield (t ha-1)	Yield loss * (%)	Yield gain ** (%)
Weedy upto 20 DAS, then weed free	1.54	0.89	41.10
Weedy upto 40 DAS, then weed free	1.35	13.10	23.73
Weedy upto 60 DAS, then weed free	1.30	16.67	18.64
Weed free upto 20 DAS, then weedy	1.27	18.45	16.10
Weed free upto 40 DAS, then weedy	1.46	5.95	33.90
Weed free upto 60 DAS, then weedy	1.50	3.57	37.29
Pack sowing and weedy	1.26	19.35	14.83
Pack sowing + one HW at 30 DAS	1.50	3.94	36.78
Weedy check	1.09	29.70	
Weed free	1.56		42.37

* Yield loss as compared to weed free; ** Yield gain over weedy check. DAS - days after sowing; HW -hand weeding Source : Prabhakar et al., 1992.

Although two hand weedings, first at 25 to 30 days and second at 40 to 45 days after sowing are recommended, it shall be advisable to go for chemical weed control to ensure effective control even during incessant rains. For chemical weed control, Fluchloralin (45 EC) @ 1 kg ai.ha^{-1} or Trifluralin (48 EC) @ 1 kg ai.ha^{-1} as pre plant incorporation or Alachlor (50 EC or 10 G) @ 2 kg ai-ha^{-1} or Metolachlor (50 EC) @ 1 kg ai.ha^{-1} or Pendimethalin (30 EC) @ 1 kg ai.ha^{-1} as pre emergence application are recommended.

The water requirement for spraying the weedicides except granular formulations is 750 to 800 litres per ha. Herbicide should be applied on moist and pulverised soil only for better weed control. Use only flat jet nozzle and always follow weedicide rotations rather than using a same weedicide year after year.

6.1 Integrated weed control

Integrated weed control involves a combination of control methods in a view to minimise the use of pesticides and their adverse impact on the environment. It is always advantageous to combine whichever method used with cultural practices that favour soybean growth, making the crop more competitive against weeds.

In soybean, Ruedell (1990) observed that yields were higher in soybean when grown in rotation with wheat. Similarly mechanical control can be integrated with weedicide application. Pre-emergence application of herbicides, planting, and fertilization can be done simultaneously by indigenously mounting spray tank on the seed cum fertilizer drill and placing the spray nozzles at 15 cm above the soil. This system reduces the cost by 40 per cent. Deep ploughing in summer followed by recommended use of weedicide shall also substantially reduce the weed population. Manipulating the planting distance of soybean rows at 22.5 cm instead of 30 to 45 cm combined with spray of alachlor @ 1 kg ai.ha^{-1}. as pre emergence provides effective weed control and enhances yields.

The sequence of operations like deep summer ploughing, one inter-culture just before sowing, need based use of PPI or PE herbicides and one inter-culture at 25 to 30 days of sowing is recommended as integrated weed control measures in soybean.

7. IRRIGATION & WATER MANAGEMENT

As indicated earlier, the critical period for water requirement in soybean is planting to emergence, flowering and pod filling stage. A proper water management at these three stages is essential to optimise yield. Stress on account of excess or deficit of soil moisture would be detrimental to yield. A soil moisture of 75% PASM should be preferably maintained during the crop growth period. Seasonal water consumption is however dependent on radiation intensity, relative humidity of the air, wind velocity, water availability in soil and length of growing period. Only 25 to 30% of the total water consumed by soybean crop is used before flowering while reproductive stages account for 70 to 75% of water usage (Hiraoka *et al.,* 1986). Pod filling period consumed maximum water (Berlato and Bergamashi, 1978). Under conditions where water consumption exceeds the amount of natural water inputs, supplemental irrigation should be restorted to.

The soybean irrigation schedule of 0.6 IW/CPE ratio through double row furrow method increased water use efficiency (5.71 kg ha. mm^{-1}) and also showed high water saving (34.6%) under Coimbatore conditions (Elamathi and Singh 1998). Similarly, Patel *et al.,* (1998) and Kumawat and Dhakar (1998) also stipulated that soybean can profitably be grown by irrigating the crop at 0.8 IW/ CPE ratio under Gujarat and Rajasthan conditions, respectively.

8. MAJOR DISEASES AND INSECT PESTS AND THEIR CONTROL

Soybean is one of the most adoptable crop grown at almost all latitudes. Being a luxuriant crop with lush green, soft and succulent foliage is attacked by over 273 types of insects which include seed and seedling feeders, stem borers, foliage feeders, sap suckers, flower and pod feeders and storage insects. Similarly, more than 100 soybean diseases have been recorded and most of them are caused by fungi. Of these, about 35 are considered economically important. As soybean production expanded to new areas with different soil and climatic conditions, certain diseases of minor importance reached epiphytotic proportions like rust which caused epidemic in the plains of the India. Diseases and insects tend to be more severe in tropical and sub tropical environments where temperatures are high and rainfall is frequent and heavy during the growing season. The economic importance of these pests varies from year to year and region to region, depending on the climatic conditions, pest load and cultivar grown. For sustainable production of soybean, it is imperative to resort to the following control measures for insect pests and diseases :

Insect pest

A. Seed and Seedling feeders

Seed maggot (*Delia platura*), cut worms (*Agrotis* spp.), field cricket (*Gryllus* spp.) white grub (*Holotrichea consanguinea*).

Control measures

- Phorate 10 G @ 10 Kg. ha^{-1}
- Carbofuran 3 G @ 30 kg. ha^{-1}
- Light traps for white grub adults
- Poison bating for field crickets (1:1:10 BHC 0.5: Gur : Bran)

B. Stem borers

Girdle beetle (*Obereopsis brevis*, *Nupserha* spp.), Stem fly (*Melanagromyza sojae*, *Ophiomyia phaseoli*)

Control measures

- Mechanical removal of infested plant parts
- Soil application of phorate 10g @ 10 kg. ha^{-1}
- One/two sprays of 0.07% endosulfan 35 EC or 0.03% dimethoate 30 EC or 0.04% monocrotophos 36 EC.

C. Foliage feeders :

Green semiloopers (*Diachrysia orichalcea, Chrysodeixis ital*), Brown stripped semilooper (*Mocis undata*), leaf minor (*Bilobata subsecivella*), tobacco caterpillar (*Spodoptera litura*), bihar hairy caterpillar (*Spilosoma obliqua*), blue beetle (*Cneorane* spp.), grey weevil (*Myllocerus maculosus*).

Control measures

- light traps for adults
- Deep ploughing in summer
- Spraying 0.07% endosulfan 35 EC or 0.05% quinolphos 25 EC or 0.001% fenvelerate 20 EC.

D. Sap suckers

White fly (*Bemisia tabaci*), green jassids (*Empoasca terminalis*), thrips (*Caliothrips indicus, Thrips tabaci, Ayyaria chaetophora*), green stink bug (*Nezara viridula*).

Control measures

- Apply phorate 10g @ 10 kg ha^{-1}/ Carbofuran 3g @ 30 kg ha^{-1} at the time of sowing.
- Spray of monocrotophos @ 0.04% / endosulfan 0.07% / methyl parathion/dimethoate @ 0.04%

E. Flower and pod feeders

Blister beetle (*Mylabris pustulata, Epicauta hirtipes, E. mennerheim*), pod borer (*Cydia ptychora*).

Control measures :

- Spray of endosulfan/malathion @ 0.1% / quinalphos @ 0.05%.

Major diseases of soybean and their control

A. Seedling/stem/root diseases :

Charcoal rot (*Macrophomina phaseolina*) and collar rot (*Sclerotium rolfsii*).

Control measures :

- Grow resistant varieties.
- Follow crop rotation.
- Seed treatment with thiram + carbendazim (2:1) @ 3g kg^{-1} seed.

B. Foliage/pod diseases :

Bacterial pustule (*Xanthomonas compestris* pv. *glycines*)

Control measures :

- Grow resistant/tolerant varieties (Bragg, PK 416, PK 472, PK 564).
- Seed treatment with carboxin @ 0.2%.
- Spray of mancozeb @ 0.2%.

Frog eye leaf spot (*Cercospora sojina* Hara), pod blight/anthracnose (*Colletortrichum dematium*), Rhizoctonia aerial blight (*Rhizoctonia solani*) and Myrothecium leaf spot (*Myrothecium roridum*).

Control measures :

- Use healthy seeds.
- Use of tolerant/resistant varieties.
- Seed treatment with thiram + carbendazim (2:1) @ 3 g kg^{-1} seed.
- Spray of Zineb @ 0.2%.

Rust (*Phakopsora pachyrhizi*)

Control measures :

- Grow resistant/tolerant varieties (Ankur, PK 1024, PK 1029, and JS 80-21).
- Grow early maturing varieties (JS 71-05, Ahilya 3, Monetta).
- Spray of hexaconazole / propiconazole/ triadimefon @ 0.1%.

C. Viral diseases :

Soybean mosaic (SMV) and soybean yellow mosaic (YMV).

Control measures :

- Grow resistant/tolerant cultivars (PK 416, PK 472 and PK 564).
- Keep down weed and white fly population.

9. NUTRIENT MANAGEMENT

Nutrient management is integral part of yield stabilization. Soybean being a leguminious crop does not require high doses of nitrogen. Nevertheless, a starter dose of 20 kg N ha^{-1} is adequate. Application of 60-80 kg P_2O_5 ha^{-1} and 20 kg K_2O ha^{-1} is recommended. These levels can be adjusted to the need based on soil test values to provide balanced nutrition. Being an oil yielding legume, sulphur forms the crux of integrated nutrient management programme. Twenty kilograms sulphur per hectare is advocated. Joshi and Billore (1998) reported 10 to 22 kg S ha^{-1} to be the economical optimum dose for soybean depending on the variety sown. For better soybean production, application of 10 t ha^{-1} farm yard manure is recommended in the basal ploughing. Soybean is highly sensitive to fertilizer injury. Seedling emergence in the field is greatly reduced when fertilizer is placed in contact with seed. It should always be placed just one and half inches besides and below the seed, while using a seed cum ferti-drill. Sowing of seed mixed with fertilizer should be avoided.

Plate 22.8. Lowland patches of Soybean crop (JS 335) affected with collar rot (*Sclerotium rolfsii*) due to water-logged condition of grain development stage. (Sindhi Area of Wardha dist in Maharashtra). Grow tolerant genotypes

Application of FYM along with balanced nutrition to soybean, wheat, maize fodder was documented to provide sustainability. Joshi and Billore (1998) reported that the recommended levels of fertilizer along with 10 t FYM ha^{-1} to each crop in soybean -chickpea/wheat cropping sequence application gives stabilised and sustainable production.

To resort integrated nutrient management for soybean, the components are application of organic manures, use of bio fertilizers for N-fixation and phosphorus solubilization, efficient control of weeds to check nutrient drain and balanced application of fertilizer on soil test basis. This schedule not only supplement micronutrients but also provides sustainability to soybean cultivation.

10. HARVESTING, THRESHING AND SEED STORAGE

Timely harvesting and proper handling are important for enhancing quality and quantity of soybean production. Delayed harvesting leading to pod-shattering is one of the major causes of reduced yields in soybean. Besides cultivation of shattering resistant varieties, timely harvesting also minimizes the yield losses. It also preserves the seed quality. Moisture content of the seed is the criterion for seed harvest. Generally the seed moisture at harvest should be 14 to 16%. In most of the varieties change of pod colour to golden yellow indicates the stage of harvest. Harvesting can be done by cutting the stalks to the ground level by sickles or by tractor driven reapers or combines. The harvested plants should be left on the threshing floor for 2-3 days for drying. The dried produce can be threshed by operating mechanical threshers at a low cylinder speed of 400-500 rpm at seed moisture levels of 14% and at a speed of 300-400 rpm at a seed moisture of about 13%. Seed cracking and seed splitting are seen if the seed moisture is below 13% while seed bruising is seen if the seed moisture is above 15%. Threshing can also be done by beating the harvest with wooden sticks.

10.1 Seed treatment

Fungicide treatment of soybean seed is beneficial particularly if the seed is infested with diseases like pod and stem blight. However, once the seed is treated with fungicide, it can not be put to any other use than planting. Thus, the treated seed which is not sold for any reason, i.e., drop in germination percentage, lack of demand etc. will have to be destroyed. Therefore, to avoid financial risk, seed treatment with fungicide is recommended prior to sowing.

10.2 Seed packing and Storage

If the seed moisture has been brought down to 9% or less, water proof bags should be used otherwise jute bags are recommended. Soybean seed being hygroscopic in nature, absorbs moisture from atmosphere or loses moisture till the equilibrium is reached. Hence precaution should be taken to see that relative humidity is kept as low as possible and any chance of absorbing moisture is avoided.

Aeration during storage is important, particularly when the moisture content is not low to the desired extent. Proper aeration helps in conditioning the seed, equalising the temperature within the desired mass, cools the seed to ambient temperature. Generally a relative humidity of 65% or less is preferable. For longer storability of the seed, the relative humidity should be less than 50%. Temperature in the store room has profound effect on seed viability, seed germination and vigour of seedlings. Ideal temperature for maintaining the quality of the seeds for 8-9 months is recommended to be 20ºC at a relative humidity of 50%. For long term storage needed to maintain valuable seed material of genetic lines, cultivar collections or limited resources of breeders seed the storage temperature should be 10ºC and relative temperature should be 50%.

11. SOYBEAN SEED PRODUCTION

Availability of high quality seed is a pre-requisite for higher seed yields and exploitation of crop potential. Unlike other self-pollinated crops, soybean seed production is a bit difficult as special care is essential at all stages of crop growth. Both genetic and physical purity of seed can be maintained through systematic seed production and rigorous quality control measures. Seed certification standards for soybean seed production are given in Table 10.

Table 10. Seed certification standards

Item	Type	Foundation	Certified
Pure Seed	Minimum	97%	97.0%
Inert matter	Maximum	3.0%	3.0%
Other crops	Maximum	Nil	0.05%
Weed seed	Maximum	5 seeds/kg	0.1%
Variety of same crop	Maximum	0.1%	0.1%
Germination	Minimum	70%	70%
Moisture	Maximum	12%	12%

In seed production programme harvesting and threshing are very crucial because :

- Harvesting at most appropriate time will minimise the effect of field weathering, seed deterioration and seed yield erosions on account of pod shattering.
- Proper cleaning and method of threshing will ensure high physical purity of seed.
- Proper handling during harvesting and threshing will decrease seed injury and the loss of viability.
- Proper handling during harvesting and threshing will decrease seed injury and the loss of viability.

12. COST BENEFIT RATIO FOR SOYBEAN PRODUCTION WITH IMPROVED PRODUCTION TECHNOLOGY

Soybean being a high returns low input annual cash crop, quickly got popularised with the farmers. An estimation of cost of production and C:B ratio affirm the statement mentioned in Table 11.

Table 11. Cost effectiveness of soybean production

Sl. No.	Item	Cost (Rs ha^{-1})
	Expenditure	
1.	Land preparation (Two harrowing)	900.00
2.	Phorate application (10 kg ha^{-1})	500.00
3.	Seed (75 kg ha^{-1})	1275.00

4.	Seed treatment and culture (Thirum + carbendazim) +*Bradyrhizobium japonicum* culture	130.00
5.	Farm yard manure (10 t ha^{-1})	1000.00
6.	Chemical fertilizer (kg ha^{-1})	
	20 kg N (Urea)	150.00
	80 kg P_2O_5 (SSP)	1300.00
	20 kg K_2O (MOP)	140.00
7.	Sowing charge (Fertilizer + seed)	400.00
8.	Chemical weed control	900.00
9.	Interculture (Dora or Kolpa)	150.00
10.	Plant protection (Two sprays)	500.00
11.	Harvesting and threshing	800.00
12.	Land value @ Rs 5000 per year	2500.00
	Total	**10645.00**
	Returns	
	Seed 2000 kg ha^{-1} @ Rs 10 kg^{-1}	20000.00
	Straw	2000.00
	Gross returns	22000.00
	Net returns	11355.00
	C:B ratio	**1:2.07**

13. A CASE OF VALUE ADDITION

Besides using soybean in vegetable oil industries, its constitution of 40% protein and 20% oil makes, it a potential crop for value addition as well as to meet the protein calorie requirements of the masses. Full potential of soybean in combating pulse-protein deficiency and provision of nutritious food have not been exploited in India and concerted efforts are required. This shall minimise the external dependency of soya farming and industries. Soy-based food products are also suitable to diabetics as they contain less carbohydrates (Table 12). Soy-protein is also good for those allergic to animal protein or milk. Soybean gives 2 to 3 times more protein yield than pulses or animal products, hence is the most economical protein source in the world. Soymilk is palatable to lactogen intolerant people.

Table 12. Soyba-based food products

Product	Processed ingredients	Uses
Deflated flake	Protein concentrate	Meat additive, bakery
	Flour and grits	Edible : Bakery odles, meal maker, sausages etc.
		Industrial : Adhesive for plywood, insecticidal spray; nutrient for beer, yeast

	Soy milk	Feeds, protein concentrates
	Extracted soy protein	Fudarative, stimulated meat, soy milk, and various industrial uses.
Mill feed		Vitamins, drugs, carrier
Natural full fat	Baked soybeans	Soy sprouts, feed
products	Soyflour	Bread, gruel, candy
	Roasted beans	Soybutter, Soycoffee,
Oil	Refined oil	Margarain, cooking oil, mayonnaise, salad oil
	Soylecithin	Emulsifier, stabilizer, antifoamer, disperser & wetting agent.

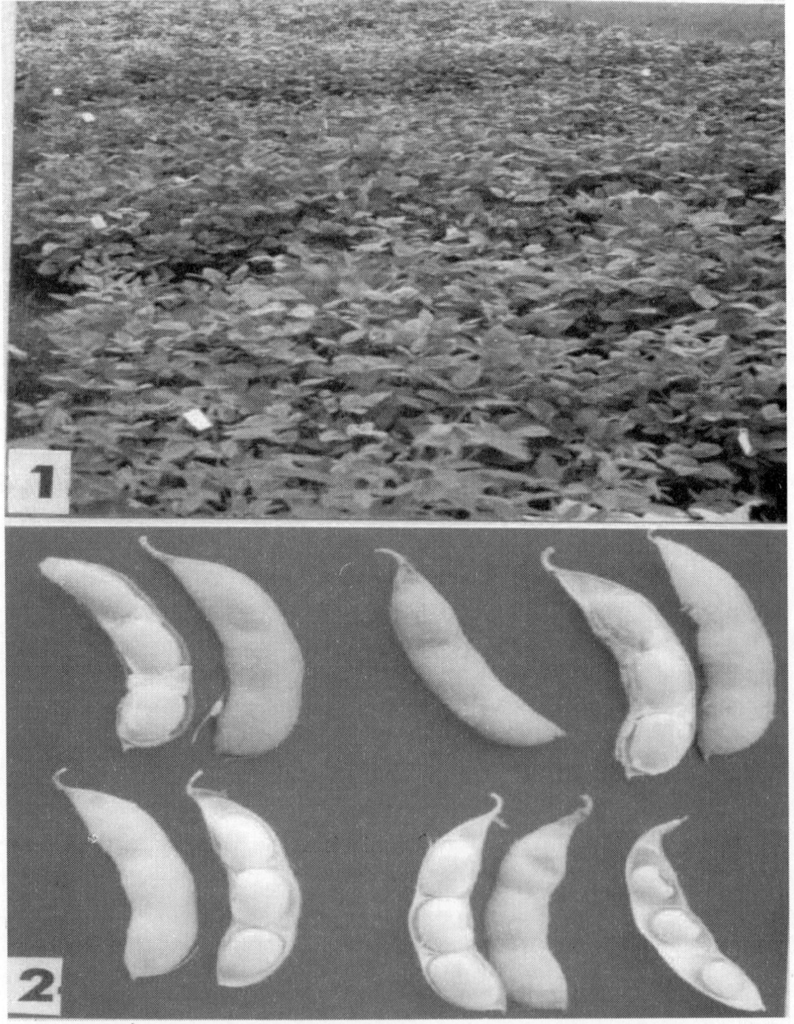

Plate 22.9. 1. Leaf yellowness is the criteria of maturity in soybean (JS-335). It sometime indicates forced maturity due to nutrient deficiency or leaf virus 2. Pods with seed development and 4th pair fully mature with yellow pod wall.

Broadly soybean products can be classified as whole soybean, defatted soyflour, soy protein isolates, and soy oil.

Raw soybean contains some anti nutritional factors which if not eliminated or destroyed reduces the nutritional value. Most commonly found tripsin inhibitor is heat labile and easily destroyed by moist heat of about 100°C.

14. TROUBLES ENCOUNTERED DURING SOYBEAN PLANT GROWTH

Problem	Probable causes
Irregular plants stand	Clogged planter, dry soil
Rotted seed/seedlings	Disease. (Pythium/Phytophthora/Fusarium), excessive moisture.
Abnormal seedling	Herbicide injury, soil compaction.
Broken seedlings	Bird, insect damage.
Wilted plants	Rot causing pathogenic fungi.
Stunted plants and disclosured leaves	Nematodes, grubs, herbicide injury, nutrient deficiency.
Spotted leaves	Diseases
Girdle on stem/petiole	Girdle beetle
Damaged leaved	Insect attack
Deformed / abnormal cause	Stink bug, purple seed stain

15. FUTURE PRODUCTION STRATEGIES - QUESTIONS TO BE ANSWERED

Some important problems which research workers have to answer to augment soybean yields are :

- How to increase photosynthetic efficiency of soybean ?
- How to alter source-sink relationship to decrease the blossom drop?
- How to eliminate genetically the trypsin inhibitor?
- How to decrease linolenic acid thereby decreasing in rancidity on storage?
- How to stabilize yields?
- How to improve the ecological balance of soil while maintaining the population of useful soil microorganisms.
- How to improve efficiency of nitrogen fixation ?
- How to improve the soybean plant through biochemical and cytogenetic studies?
- How to increase fertilizer and water use efficiency ?
- How to remove photosensitivity ?

Of late, bio-technology has come in a big way to circumvent some of the above said problems. Breeding for drought resistance, rust resistant type selection, gene pyramiding and for efficient weed control was achieved by producing transgenics. There is seldom an immediate dividend to research but has tremendous impact on the future of maintaining the sustainability of the crop and meeting the ever increasing demands of food production.

REFERENCES

Bhatnagar, P.S. and Joshi, O.P. 1998. Soybean in cropping systems in India. *FAO Series on Integrated Crop Management, Crops and Grassland Series, Plant Protection Division*, Food and Agriculture Organisation of United Nations.

Bhatnagar, P.S. and Tiwari, S.P. 1995. Technology for increasing soybean production in India. *NRCS. Tech. Bull.* 1. (Revised). pp. 10.

Berlato, J.M. and Albrecht, S.B. 1984. Drought and flooding effects on N2 - fixation, water relations and diffusive resistance of soybean. *Agron. J.,* 735-740.

Board, J.E. and Settimi, J.R. 1988. Photoperiod requirement for flowering and flower production in soybean. Agron. J., 80 : 518- 525.

Brown, D.M. 1960. Soybean ecology. I. Development - temperature relationships from controlled environment studies. Agron. J., 52 : 493-6.

Delouche, J.C. 1953. Influence of moisture and temperature levels on germination of corn, soybean and watermelons. Proc. Ass. Offic. Seed anal., 43 : 117-126.

Doorenbos, Y. and Kassam, A.H. 1986. Yield response to water. FAO Irrigation and Drainage, Paper No. 33. Rome, FAO.

Elamathy, S. and Singh, S.D.S. 1998. Effect of irrigation regimes, methods of management practice for water use efficiency in soybean Paper presented : In : Intl. Agronomy Cong.: Agronomy, Environment and Food Security for 21st century - 23 to 27th November, 1998, organised by ISA, IARI, New Delhi. pp. 111- 12.

Fehr, W.R. and Caviness, C.E. 1977. Stages of soybean development. Special report 80. Ames, IA, USA, Iowa State University.

Hiraoka, H., Sasiprapa, V. and Piyawongsomboon, W. 1986. Water consumption of maize and soybean. In : Tech. Bull. Trop. Res. Center, 20 : 9-16.

Joshi, O.P. and Billore, S.D. 1998. Economic optima of sulphur fertilizer for soybean. Indian. J. agric. Sci. 68 (5) : 244-46.

Joshi, O.P., Billore, S.D. and Bundela, V.P.S. 1998. Fertilizer management in soybean -chickpea crop sequence. J. Oilseeds. Res. 15 (1) : 118-122.

Kanwar, J.S., Kampen, J. and Virmani, S.M. 1982. Management of Vertisols for maximum crop production - ICRISAT experience. In : Vertisols and rice soils of the tropics -symposia Paper II. 12th Intl. Cong. of Soil Sci. New Delhi, 8 - 16 February. pp. 94-120.

Kenworthy, W.J., Brown, A.H.D. and Thibou, G.A. 1989. Variation in flowering response to photoperiod in perennial Glycine spp. Crop Sci. 29 : 678-682.

Kumawat, S.M. and Dhakar, L.L. 1998. Effect of different nitrogen levels and irrigation regimes on productivity and economics of soybean. *Paper presented : In : Intl. Agronomy Cong. : Agronomy, Environment and Food Security for 21st Century-* 23 to 27th November, 1998, organised by ISA, IARI, New Delhi. pp, 268-69.

Lawn, R.J. 1982. Response of four grain legumes to water stress in Southern Queensland. I. Physiological response mechanisms. Aust. J. Agril. Res. 33 : 487-496.

Nathanson, K., Lawn, R.J.; De Jabsun, P.L.N. and Byth, D.E. 1984. Growth, nodulation and nitrogen accumulation by soybean in saturated soil culture. *Field crops Res.* 8 : 73-92.

Parker, M.W. and Borthwick, H.A. 1943. Influence of temperature on photoperiod reactions in leaf blades of Biloxi soybean. Bot. Gaz., 104 : 612-629.

Prabhakar, Tiwari; S.P. and Bhatnagar, P.S. 1992. Production technology for augmenting soybean productivity in India. J. Oilseeds. Res. 9 (1) : 1-13.

Patel, Z.G., Chavan, S.A. and Arvadia, M.K. 1998. Influence of irrigation, fertilizer and mulching on yield performance of hot weather soybean. Paper presented: In : Intl. Agronomy Cong. : Agronomy, Environment and Food security for 21st Century - 23 to 27th November, 1998, organised by ISA, IARI, New Delhi. pp. 268- 69.

Ruedell, J. 1990. Efeito do manejo do solo e da rotacao ded culturas sobre a populaco de plantas daninhas e a productividade das culturas. In : *C. Crovetto, ed. Primeras Jornadas Nacionales de cero Labranza chequen,* 29-31 Oct., p. 169-182. Concepcion, Chile, Instituto de Investigationes Agropecvarias (INDIA).

Tanaka, A. 1983. The physiology of soybean yield improvement. In : Soybean in Tropical and Sub-tropical systems. *Proc. symposium TSUKUVA,* Japan, pp. 323-331.

Thomas, J.F. and Raper, C.D., Jr. 1981. Day and night temperature influence on carpel initiation and growth in soybean. *Bot. Gaz.,* 142 : 183-187.

Tiwari, J.P. and Khurchania, S.P. 1990. Survey and management of weeds in soybean ecosystem in Madhya Pradesh. *Indian J. agric. Sci.* 60 (10) : 672-676.

Walter, O.S. and Samuel, R.A. 1983. Fertility management. In : Modern soybean production. Pub. S. and A publications, Inc. pp. 229.

Wright, G.C., Smith, C.J., and Nilson, J.B. 1988. Growth and yield of soybean under wet soil culture and conventional furrow irrigation in South Eastern Australia. *Irr. Sci.* 9 : 127- 142.

Chapter - 23

SESAME — AN IDEAL CROP FOR DIVERSE AGRO-ECOLOGICAL SITUATIONS

B.N. Reddy and A.R. G. Ranganatha

Directorate of Oilseed Research, Rajendranagar, Hyderabad-500 030 (AP)

1. INTRODUCTION

Sesame (*Sesamum indicum* L.) is an important and ancient oilseed crop of India grown mainly under a system of small scale peasant agriculture mostly by small and marginal farmers. The seeds are rich source of oil, protein, calcium and phosphorus. The oil is extensively used for culinary purpose while cake serves as manure or cattle feed, the protein in cake contains relatively high quantity of sulphur containing amino acid methionine and serves as useful protein for human consumption (Rajan, 1967). The oil is of high quality, resistant to oxidation and rancidity, even when stored under ambient temperature.

The Ethiopian area is generally accepted as the origin of cultivated sesame. India and Japan are considered as the secondary centres. Sesame (*Sesamum indicum* L. Syn. *Sesamum orientale* L.) is a member of the order Tubiflorae, family Pedaliaceae.

Sesame is currently grown (1995-96) in an area of 1.92 million hectares with a production of 0.55 million tonnes having national average yield as low as 288 kg/ha. The crop contributes nine per cent of area and four per cent of production to the total oilseeds in the country. The crop has witnessed substantial increase in production (6.53 lakh/tonns) during 1985-94 from 4.69 lakh tonnes (1965-84) which was possible only due to 43% gain in productivity. During 1985-94, the area expansion was marginal but there was increased growth rate in terms of production (5.2%) and productivity (4.06%).

There is an export demand for white seeded sesame. During 1996-97, 85,000 tonnes of seed valued at Rs.238 crores has been exported. The limiting factor for export of sesame is reported to be the pesticide residues observed in the seed. The major sesame producing countries in the world are India, China, Burma, Sudan, Kenya, Nigeria, Mexico, Ethiopia besides Venezuela. India is the world's major producer with a third of the world's acreage and approximately quarter of the total production. (Ashri, 1985).

Sesame is grown in almost all parts of the country viz., Northern India (Punjab, Haryana and Himachal Pradesh states), North Western India (Rajasthan and Gujarat), Himalayan foot hills and Indogangetic plains (Uttar Pradesh, Bihar and West Bengal), Central India (Maharashtra, Madhya Pradesh and Orissa) and Southern India (Andhra Pradesh, Karnataka and Kerala states). In a limited area, it is also cultivated in north eastern states.

Sesame seeds are used for oil extraction (77.6%), for edible purpose (20.1%) and for seed purpose (2.3%). Sesame oil is used for edible purpose (70%) as salad and cooking oil for non-edible purpose like domestic and toilet (13.0%) and for manufacture of margarine (17%). Sesame oil forms the basis of most of the fragrant and scented oils and in perfumaries it is used as fixtures.

2. ADAPTATION AND CLIMATIC NEEDS

Sesamum indicum is basically considered a crop of the tropics and subtropics but the cultivation can be extended to temperate zones through the development of suitable varieties. The diversity of local ecotypes well adopted to the particular locality is an indication of the plant's potential. Sesame's main distribution is between 25 S and 25 N but it is found growing upto 40 N in China, Russia and U.S.A., 30 S in Australia and 35S in South America. It is normally found below 1250 m. altitude, however, some varieties may be adapted upto 1500 m. It is grown in the Himalayas upto 1250 m and in Nepal upto 2000 m. Experimental crop has been grown in Kenya and South America upto 1800 m. The high altitude types are usually small, quick growing, relatively unbranched and frequently have only two flowers per leaf axil and low seed yields. The oil content is also high at low altitudes in the same varieties (Weiss, 1983).

2.1. Temperature

Sesame normally requires fairly high temperature during its life cycle. Temperature for optimum growth from seedling emergence to flowering and fruiting has been found to be in the range of 27-33°C. Considerable genotypic variation in germination response to temperature has been reported. A temperature of 25-27°C encourages rapid germination, initial growth and flower formation. If temperature falls below 20°C for a length of time, germination and seedling growth are delayed and these processes are inhibited at temperatures below 10°C. High temperatures, particularly high night temperatures, promote stem growth and leaf production. Temperature above optimum i.e. 40°C or above at flowering can seriously affect fertilization and the number of capsule set.

2.2. Radiation

The rate of net total dry matter production per unit of ground area is related to the daily amount of photosynthetically active radiation intercepted by the crop. Low yields in kharif season (summer rainfall areas) could result from low radiation levels caused by heavy cloud cover resulting in reduced radiation input or from low plant density rendering suboptimal interception by crop canopy. Reduced solar radiation decreases dry matter production and seed yield.

2.3. Photoperiod

Sesame is basically a quantitative short day plant, and with a 10 hour day will normally flower in 40-45 days, but, many varieties have become locally adapted to various light periods. When sesame varieties are introduced to other areas which have a similar day length but with different rainfall or temperature patterns, there is considerable variation in growth and yield from that in their original place. This is because of interaction of photoperiod with factors such as light intensity, rainfall and temperature. Light intensity has a significant morphogenic effect influencing yield and oil content. Considering yield obtained at the optimum planting period as maximum, then as the difference from the optimum sowing time increases then yield from sowings after this period decrease. Rainfall has a major modifying influence on optimm time of planting relative to photoperiod.

The local adaptability of sesame varieties have been utilised in countries like India with distinct growing seasons. Varieties adapted to one season give uneconomic yield if grown in other season because of photoperiod and light intensity adaptability. The relationship of time of planting to maximum yield is generally well appreciated, although less understood.

2.4. Rainfall

Sesame has great adaptability with regard to rainfall. It is possible to produce an excellent crop with a rainfall of 500-650 mm but down to 300 mm and upto 1000 mm will also produce a crop under certain conditions, particularly under irrigation from newer varieties. In India and Pakistan, sesame is grown as a rainfed crop in kharif season when 500 mm rainfall is normal, and in rabi or spring/summer seasons as an irrigated crop. In all seasons and areas, the locally adapted variety gives the maximum yield with 600-1000 mm rainfall.

For maximum yield, precipitation should be distributed over the period of plant growth as follows : germination to first bud formation 35%, bud formation to main flowering 45%, flowering to maturity not more than 20%. Heavy rain at flowering will drastically reduce yield. Rainfall when plants are ready for harvest also reduces yield by increasing susceptibility to diseases and prolonging the period required for drying of capsules. Sesame is extremely susceptible to waterlogging, heavy continuous rains at any time during growth will greatly increase the incidence of diseases.

2.5 Humidity

Sesame will have lower net photosynthesis and possibly lower yield potential when grown in an arid environment than when grown in areas which have a higher humidity. This is because at high humidity gradient between leaf and air there is reduction in stomatal aperture. Large humidity gradient may cause mid-day closure of stomata and depression of photosynthesis. There is also increase in leaf resistance at high temperature. However, the highest yields are obtained under irrigation in arid regions.

2.6. Soil moisture

Sesame is considered to be a drought resistant crop. It is capable of withstanding a high degree of water stress than many other cultivated plants. However, during the plant establishment phase, it is extremely susceptible to moisture shortage. Once established, the crop will grow almost entirely on stored soil moisture and with only occasional shower of rain in the early stages good yields are obtained. This ability to produce a crop under adverse conditions makes sesame an important crop under semi-arid conditions. Waterlogging is highly detrimental to the crop.

Soil

Sesame grows well on a wide range of soils, from highly sandy soils to the black cotton and clay soils, but it thrives best on well drained, moderately fertile soils of medium texture. Shallow soils with impervious sub-soil are not suitable. The soils on which sesame is grown range from sandy soils to clay soils. Soils with a neutral reaction are preferred, although good results have been obtained on slightly acid and slightly alkali soils. Sesame does not thrive on acid soils. It will grow in soil of pH 5.5 to 8.0 but at higher pH soil structure becomes increasingly important.

3. EFFICIENT CROP ZONES

The sesame crop is grown in different regions of the country in different seasons and also in more than one season in the same region. Thus, the crop has different varieties and crop production practices for different regions and growing conditions (seasons). In northern India the crop is grown as a rainfed kharif crop. The crop is also grown as a semi-rabi crop (September-January) in parts of Madhya Pradesh, Maharashtra, Gujarat and Andhra Pradesh. Sesame crop has also gained popularity, as a summer crop (January- May) after potato or late paddy in Bengal, Bihar and Orissa. However, in Southern India, the crop is taken in three seasons (i) Kharif/ Semi-rabi : July/August-November/ December under rainfed conditions, (ii) Rabi : November-January/February and (iii) summer irrigated crop : February/March -May/June. Thus, the sesame growing periods and conditions in India can be summarised as follows (Table 1).

Table 1. Sesame growing periods and conditions in India

Season time	Planting time	Harvesting time	Stand	Water source	% of area
1. Rainy (Kharif)	June-July	Oct.- Nov.	Mixed	Rainfed	70
2. Mid-rainy (Semi-rabi)	Aug-Sept.	Dec.- Jan.	Sole	Rainfed	20
3. Summer	Jan-March	April- June	Sole	Irrigated	10
4. Winter	Oct-Nov.	Feb - March	Sole	Irrigated	Rare

4. EFFICIENT CROPPING SYSTEMS

Sesame by virtue of its short duration and wide adaptability fits well into a number of intercropping and crop sequences in different agro-ecological situations. The availability of high yielding varieties coupled with remunerative market prices has become attractive for the farmers to include sesame in different crop sequences under irrigated as well as rainfed conditions.

4.1. Crop sequences

The promising crop sequences involving sesame in different sesame growing regions are given in Table 2.

Table 2. Crop sequences involving sesame

State	Crop sequence
Rajasthan	Sesame - Wheat-mung Sesame - wheat/Barley
Gujarat	Sesame - Wheat/mustard
Maharashtra	Sesame - safflower Sesame - rabi sorghum Sesame (early) - Ragi Cotton + sesame - wheat
Madhya Pradesh	Sesame - wheat Sesame - chickpea Cotton + sesame - wheat

Uttar Pradesh & Punjab	Sesame - chickpea/mustard/lentil/pea/Barley
Haryana	Sesame - wheat/mustard
Andhra Pradesh	Paddy - sesame
Tamil Nadu	Paddy - sesame Groundnut - sesame Sesame - Ragi/sorghum
Karnataka	Sesame - Horsegram Sesame - chickpea
West Bengal	Potato - sesame Paddy - sesame
Orissa	Paddy - sesame Potato - sesame
Bihar	Early paddy - potato - sesame Maize - gram/wheat-sesame Paddy - wheat - sesame

Inter/mixed cropping

Inter/mixed cropping helps in achieving greater stability in production under dryland conditions besides increasing the total returns. Sesame is usually raised as mixed or intercrop with different cereals, pulses and oilseeds traditionally. However, by following proper planting geometry of compatible crops, the efficiency of intercropping systems can be raised substantially. Yields and monetary returns from sesame are more when it is raised as sole crop rather than as a mixed/inter crop. However, in large parts of the country sesame is usually raised as a mixed crop with one or the other crops. Some of the remunerative intercropping systems which are popular in different states are presented in Table 3.

5. SOIL AND MOISTURE CONSERVATION PRACTICES

Soil and water are basic resources to get higher and stable yields of sesame in rainfed environment. Sesame grown on rainfed soil suffer from various degradation process like limited soil depth, stoneness, low water holding capacity, poor infiltration etc. Hence, adoption of appropriate soil and water conservation measures are essential for obtaining stable yields besides other improved crop management practices. The soil and water conservation practices for sesame can be grouped into two categories : (A). In-situ moisture conservation and (B). mechanical conservation techniques.

(A) In-situ moisture conservation techniques

I. Alfisols

(a). Contour farming : In situations where the soil slope is less than 1%, contour farming alone can serve the purpose of soil and water conservation measure. Contour ploughing acts as barrier to water flow in the bund which helps in recharging the soil profile. contour seeding further increases the moisture recharge.

Table 3. Recommended mixed/intercropping systems involving sesame in different states

State	Mixed/Intercropping
Rajasthan	Mixed cropping with guar, greengram, mothbean, pearlmillet, Sesame+ mothbean (3 : 1), sesame + millet (3 : 1)
Gujarat	Mixed cropping with pigeonpea, cotton and castor
Madhya Pradesh	Mixed cropping with maize, pigeonpea and greengram, sesame + greengram (1 : 1), sesame + soybean (2 : 1 or 3 : 1).
Uttar Pradesh	Mixed cropping with maize, pearlmillet, sorghum and pigeonpea, sesame + greemgram (1 : 1), sesame + pigeonpea (2 : 1)
Tamil Nadu	Mixed cropping with cotton, pearlmillet and pigeonpea
Karnataka	Mixed cropping with pigeonpea, pearlmillet and castor
West Bengal	Intercropping with spring/summer groundnut.
Orissa	Intercropping with spring/summer groundnut.

(b) Off-season tillage : The practice of off-season tillage coinciding, with pre-monsoon showers helps in increasing moisture conservation and efficient weed control.

(c) Dead furrow : opening of dead furrow at an interval of 3.6 m across the slope serves as mini water storage channel in the land when intense rain occurs. Slow infiltration of water helps in moisture recharge in the soil profile.

II. Vertisols

In case of sesame where it is grown on black soils, provision for removal of excess water during monsoon season is an integral part of in situ moisture conservation. The excess water could be collected in small farm ponds and could be used for life saving irrigation during the drought period. Broad-bed and furrows are effective in safe disposal of run off where water logging is a problem.

(B) Mechanical measures

In well drained alfisols where rainfall is less than 650 mm, contour bunds with 0.5 - 2 m section are recommended. Diversion drains wherever required are formed prior to bunding works. The alignment of bunds should be in the permissible deviation limit of 15 cm above the contour and 30 cm below the contour. Graded bunds of 0.5 m^2. section in alsifols and 0.65 m^2 in vertisols are recommended with 0.2% grade. Graded border strips are also useful where water disposal is required in areas receiving rainfall of 650-900 mm per annum. Bench terracing is recommended for very sloppy lands (5%) for effective soil and water conservation.

6. SEED BED PREPARATION

The land selected for growing sesame should be well drained or provision of drainage should be made. The soil should be well pulverised and brought to fine tilth to ensure good germination and plant growth. The field should be kept free from weeds and perfectly levelled (Joshi, 1961). Generally, for kharif sesame, soil is ploughed deep in summer and with the onset of monsoon 2 to 3 harrows in crosswise direction are done with blade harrow. For good germination and growth of the crop moderate compaction of soil by planking after second harrowing is desirable. For semi-rabi sesame under fallow land situations, ploughing in summer and fortnightly harrowing in rainy season is required to keep the soil loose and weed free, Semi-rabi sesame under double cropping following kharif legumes or maize requires ploughing to a depth of 15-20 cm followed by 3 to 4 harrowings so as to attain smooth, fine and friable seed bed.

7. VARIETAL RECOMMENDATION FOR DIFFERENT REGIONS

In the light of the advent of superior plant types in major crop species, the situation warranted the sesame breeders to come in aid to the farming community. The efforts in reconstruction of sesame plant types led to the isolation of desirable types for stress intensive agriculture. The features of the plant types recommended for stress intensive agriculture are as follows : Medium/long duration (90-110 days), however, very late is undesirable; branching types (4-6 branches); high degree of tolerance to leaf spots; powdery mildew and moderate tolerance to phyllody; moderate tolerance to leaf roller/shoot webber and gall fly; high regenerative capacity after moisture stress; tolerance to heavy rains/ water logging; medium/tall types; less number of immature/abortive capsules; fair synchrony in maturity and better adaptability.

The presently grown local types are misfit for input intensive agriculture due to low response to higher levels of management like fertilizers, irrigation and plant protection (Ranganatha, 1989). The agronomic features of the plant types for input intensive agriculture are as follows : Short duration (75-85 days); dwarf plant type; photo and thermo insensitivity; higher response to added inputs and irrigation; short internodel length; higher number of effective capsules; synchrony in maturity and high harvest index.

The sesame improvement and varietal recommendation details for different regions/states are as follows :

Maharashtra

In the old Bombay state, breeding work on sesame started at the Agricultural Research Station, Niphad and at Nagpur from 1936-37. The research work at Jalgaon was intensified with the commencement of ICAR project in 1967. The early work resulted in the release of varieties D-7-11-1 (from Dhule in 1962), N-58-2 (from Niphad in 1960), No. 41 and No. 128 for kharif and No. 8 for semi-rabi (from Nagpur). The new varieties recommended are as follows for Maharashtra.

Phule Til-1 : Released in 1978; Selection from local material, duration 90-95 days, seed yield 650 kg/ha, oil content 51%. Bold and light brown seed.

No. 8 : Recommended for semi-rabi areas. Selection from local material, duration 120 days, seed yield 500 kg/ha, oil content 50%, seed brown.

JLT-7 (Tapi) : Evolved from N-58-2 x C-50. Yields higher than Phule Til-1; duration 80-85 days, seeds white, released for Maharashtra, evolved from Jalgaon, average yield 700 kg/ha, oil 51%, seed white, released in 1986 for kharif and semi-rabi.

JLT-26: Released in 1990. Recommended for Khandesh and adjoining areas of Vidarbha. Average yield 700 kg/ha, duration 85-90 days.

TC-25 variety is also recommended for Vidarbha areas of Maharashtra.

Orissa

Vinayak : Released in 1972. Selection from local material. Duration 80 days, seed yield 500 kg/ha. Recommended for all seasons.

Kanak (BS-6-1) : Evolved from Vinayak x T-4. Released in 1979, for kharif and summer, duration 78-80 days, seed yield 600 kg/ha, oil content 47% , highest yield of 1350 kg/ha was realised.

Kalika (BM-3-7): Mutant of Vinayak. Suitable for kharif and summer. Released in 1979. Duration 82 days, seed yield 600 kg/ha, oil content 48% , highest yield of 1350 kg/ha was realised.

Uma (OMT-11-6-3): Released in 1990, average yield 600-800 kg/ha, duration 80-85 days.

Usha (OMT-11-6-5): Released in 1992. Average yield 700 kg/ha, duration 85 days in kharif and in summer 75 days, seed brown.

TKG-21 and TKG-22 : These varieties are also recommended.

Andhra Pradesh

Gouri : Suitable for coastal areas of Andhra Pradesh. Recommended for kharif -rainfed and rabi-summer irrigated areas. Evolved from Yellamanchilli in 1974. seed brown and medium in size, medium duration (85-95 days). Average yield 800 kglha, oil content 44 per cent.

Madhavi : Suitable for coastal areas of Andhra Pradesh. Recommended for kharif, rabi and summer seasons. Developed from Yellamanchilli in 1978, seed light brown and medium in size, short duration (75-80 days). Average yield 800 to 850 kg/ha, oil content 46 per cent.

Rajeshwari : Released in 1989, recommended for Telangana areas for late kharif and rabi seasons. Seed yield 800-900 kg/ha. White and bold seeded, duration 90- 95 days, oil content 50%.

YLM-1(Varaha) : Recommended for kharif, rabi and summer seasons. Released in 1993 from Yellamanchilli. Seed yield 950 kg/ha, duration 80-85 days, brown seed, oil content 53%.

YLM-2 (Gautam) : Recommended for kharif, rabi and summer seasons. Released during 1993 from Yellamanchilli. Seed yield 850 kg/ha, duration 80-85 days. Light brown seeded, oil content 52%.

Swetha Til (JCS-96) : Released in 1997 for kharif from Jagtial, duration 85-95 days. Recommended for Telangana, white seeded, oil 52%, seed yield 600 kg/ha.

RT-54 : Released in1992, seed yield 950 kg/ha, duration 80 days, oil 48%, seed brown.

RT-103 : Variety is also recommended.

Karnataka

E-8 : Selection from Dharwad local. Recorded 39% higher yield than local material, duration 100-110 days, oil content 53%, seed yield 500 kg/ha. Recommended for Northern Karnataka.

DS-1 : Released in 1997, Average yield 500 kg/ha, medium duration, Non branching, seed white.

TMV-3 and Co-1 varieties are also recommended.

Kerala

Kayamkulam-1 : Duration 90-100 days, black seed, released in 1980. Recommended for rice fallows, seed yield 500 kg/ha, oil content 46%.

Kayamkulam-2 (Thilothama) : Evolved from PT-58-35 x kayamkulam-1, duration 80 days and multi capsuled, seed brown, released in 1982. Seed yield 600 kg/ha.

Soma : Released in 1986 for summer rice fallows, Selection from culture 38, multi loculed (8 locules), seed white, duration 83-85 days, seed yield 700 kg/ha, oil content 51%.

Surya : Released in 1986 for upland semi-rabi situations. Selection from culture 42-1, multi loculed (8 locules), seeds grey, duration 87-90 days, seed yield 400 kg/ha, oil content 50%.

Plate 23.1. Field view of sesame crop of variety Madhvi (Top) and YLM 17 (Bottom)

Tamil Nadu

TMV-3 : Evolved from south Arcot local x Malabar wild type , Duration 80-85 days, yield 400 kg/ha rainfed and 800 kg/ha irrigated, oil content 52% .

TMV-4 : Pureline selection from sattur local, Duration 85 days, seed brown, suitable for summer irrigated areas. Released in 1977, seed yield 650 kg/ha, oil content 52% .

TMV-5 : Short duration (75 days). Suitable for kharif, seeds brown. Released in 1978 for Trichi areas. Seed yield 400 kg/ha, oil content 54% . Drought tolerant.

TMV-6 : Released in 1979. Suitable for Salem areas, seeds brown. Duration 85-90 days, Seed yield 700 kg/ha, oil content 54%, drought tolerant.

CO-1 : Cosmopolitan variety. Released in 1983, duration 90 days, Rainfed yield 450-650 kg/ha, Irrigated yield 750-800 kg/ha, oil content 50%, Black seed.

TSS-6 (SVPR-1) : Released in 1994 for Tamil Nadu, seed yield 800 kg/ha, duration 80 days, oil 54%, seed white.

Paiyur-1 : Released in 1997, seed yield 650 kg/ha, duration 90 Days, oil 50%, seed black.

Uttar Pradesh

T-4 : Evolved from T-10 x 3 A, late in maturity (100 days), seeds white, recommended for alluvial tract, oil content 52% , seed yield 600 Kg/ha.

T-12 : Evolved from Kaushani local, duration 85 days, seeds white, seed yield 500 kg/ha, recommended for central and western areas, oil content 53% .

T-13 : Selection from Hamirpur local, medium duration (90 days), seeds white in colour, suitable for Bundelkhand tract and for poor and marginal soils, seed yield 450 kg/ha (upto 700 kg/ha), oil content 50% .

T.78 : Released in 1994, seed yield 700 kg/ha, duration 90 days, oil 50% , seed white.

RT-46 variety is also recommended.

Rajasthan

Pratap (C-50) : Evolved from local material of Kota, suitable for Kota, Jhalawar, Bundi, Sawai Madhopur and other areas of heavy soils, plants unbranched (1-2 rarely), multiloculed (4,6,8, and 12), seeds dull white, seeds are long, oil content 48%-50%, released in 1968, seed yield 500 kg/ha, duration 100-105 days.

TC-25 : Evolved in Punjab from EC 7 x TS 24-1. Released in 1978, recommended for light soil areas, duration 90 days, seeds white, oil content 50%, seed yield 500 kg/ha.

RT-46 : Released in 1990, derived from T-12 x Pb Til-1, duration 80 days, seed yield 600-800 kg/ha, oil 49%, seed white.

RT-125 : Released in 1994 seed yield 900 kg/ha, duration 80 days, oil 49% , seed white.

T-13 variety is also recommended.

Madhya Pradesh

N-32 : Released in 1970. Pure line selection from Chhatarpur local, multi capsular, seeds white, seeds bold, duration 95 days, seed yield 750 kg/ha, oil content 53%, suitable for kharif and semi-rabi sesame areas, single stemmed.

JT-7 : Duration 85 days, seeds white, seeds bold, recommended for all parts including Bundelkhand and Baghelkhand areas, seed yield 800 kg/ha and upto 1200 kg/ha, oil content 54%, released in 1980-81 for all kharif areas.

TKG-21 : Released in 1993, seed yield 950 kg/ha, duration 75-80 days, oil 55% , seed white.

TKG-22 : Released in 1994, seed yield 950 kg/ha, duration 85 days, oil 53% , seed white.

RT-46, OMT-11-6-3 and Imp. Sel. -5 (Rama) varieties are also recommended.

Gujarat

Purva-1 : Suitable for late kharif/ semi-rabi, multi-capsular, duration 90 days, seed yield 500 kg/ha, oil content 50%, released in 1968 for whole of Gujarat.

Mrug-1 : Released in 1968 for Saurashtra region, duration 90 days, seed yield 500 kg/ha, seeds white and oil content 53% .

Gujarat Til-1 : Selection from local material of Amreli, early in maturity (85 days), seeds light brown and bold, seed yield 550 kg/ha, oil content 51%, recommended for whole of Gujarat, released in 1979-80.

Gujarat Til-2 : Released in 1994, seed yield 700-800 kg/ha, duration 85 days, oil 51% , creamy white seed.

RT-103 : Released in 1994, yield 800-900 kg/ha, duration 80 days, oil 48%, seed white.

Punjab and Himachal Pradesh

Pb. Til. 1 : Selection from Kangra material, released in the year 1966, early in maturity (80 days), seeds are white and bold, suited to sandy loam and loamy soils, seed yield 500 kg/ha, oil content 50% .

TC-25: Released in 1986, seed yield 600 kg/ha, it is recommended for whole of Punjab and Himachal Pradesh, evolved from EC 7 x TS 24-1, duration 85-90 days, oil content 50%.

TC-289 : It is released in 1986, seed yield 700 kg/ha, suitable for whole of Punjab and Himachal Pradesh, Evolved from Pb. Til. 1 x EC 4619, duration 95 days, bold and oil content 52%.

Haryana

Haryana Til No. 1 : Selection from local material, suitable for whole of state, seed bold and white, duration 85 days, oil content 50%, released in 1979.

RT-46 variety is also recommended.

Jammu and Kashmir

Sesame is grown in Kathua, Jammu and Udhampur districts. The varieties Pb. Til. 1, TC-289 and TC-25 may be cultivated in this state, because these are doing better in adjoining Punjab and Himachal Pradesh states.

Bihar

B-67: Selection from local material, duration 80 days, seeds black, seed yield 650-750 kg/ha, oil content 48%, released in 1974 for whole of Bihar.

Kanke white : Selection from local material. Recommended for eastern parts (plateau region), duration 85-90 days, seed dull white and medium in size, seed yield 500 kg/ha, oil content 50% .

Krishna : Released in 1986, seed yield 750 kg/ha, duration 85-90 days, oil 46%, seed black, suitable for kharif, semi-rabi and summer seasons.

TKG-21, TKG-22 and Uma (OMT-11-6-3) varieties are also recommended.

West Bengal

B-67 : Suitable to grow after potato crop, duration 80 days, seeds black and oil content 50%, released in 1974 for whole of West Bengal, seed yield 750 kg/ha, suitable for summer also.

Imp. Sel. 5 (Rama) : Released in 1989, seed yield 700 kg/ha (kharif) and upto 1100 kg/ha, duration 85 days, oil 50%, seed brown.

TKG-21 and TKG-22 varieties are also recommended.

Assam

ST-1683 : Released in 1981 for whole of Assam, branching type, duration 85- 90 days, seeds black and oil content 50%, seed yield 650 kg/ha.

RS-1 : Recommended for kharif for whole of Assam, white seeded, oil content 50%, seed yield 800 kg/ha.

RT-1 : Recommended 'for kharif for whole of Assam, duration 90 days, seeds black, oil content 50%, seed yield 950 kg/ha.

SP-1181 : This is another variety recommended for Assam. It is performing well. The above variety gives yield upto 1300 kg/ha.

Gauri, Madhavi and OMT-11-6-3 varieties are also recommended.

NEH and other States

For other states like Meghalaya, Manipur, Arunachal Pradesh, Mizoram, Tripura, Nagaland, Andaman-Nicobar islands, the varieties B-67 and TKG-21 are recommended. Also varieties recommended for Assam are suitable for the adjacent areas. Similarly for Pondichery CO-1 variety is recommended.

8. TIME OF SOWING

The sesame is grown in different regions of the country in different seasons and also in more than one season. The crop is raised as rainfed kharif, semi rabi and as irrigated crop during winter and summer.

In northern India, the crop is grown as a rainfed kharif crop. It is also grown as a semi-rabi crop (September-January) in parts of Madhya Pradesh, Maharashtra, and Gujarat.

The crop has gained popularity as a summer crop after potato or late paddy in West Bengal, Bihar and Orissa. However, in Southern India, crop is grown in three seasons. Time of sowing has greater influence on growth and yield of crop. Sowing at the onset of monsoon has invariabily found advantageous during kharif season. However, there are reports to show that early sowing resulted in increased vegetative growth and incidence of phyllody. Late sowing resulted in poor growth and seed set. The recommended dates of sowing of sesame in different states as reported by Sharma and Reddy (1983) are furnished in Table-4

9. SEED RATE, SPACING AND SOWING METHODS

Sesame is usually grown as broadcast. However, for easy interculture and higher yields, line sowing is recommended. A seed rate of 4-5 kg/ha is adequate to get the required plant stand.

Wherever seed drills are used as in the case of Maharashtra, the seed rate may be reduced to 2.5 to 3 kg/ha. In mixed stand, seed rate would seldom exceed 0.5 kg/ha. Spacing between and within the rows depends on the specific variety chosen, its plant type (branching pattern), specific season etc. The spacings recommended for different states are reported by (Sharma, 1998). In the line sowing spacing should be maintained at 30 to 45 cm between row and 10-15 cm between plant to plant. Seed should be treated with thiram at the rate of 3 g/kg in order to prevent seed borne diseases. As the seed of sesame is small it should be mixed with dry earth or sand or well sieved farm yard manure to ensure even distribution of seed in soil and to minimise the crowding of seed. Seed should be placed at an appropriate depth (3-5 cm) under assured soil moisture condition. Deep placement of seed will not germinate and seed placed at shallow will be eaten away by ants.

Table 4. Optimum time of sowing sesame in different regions

State	Season	Sowing time
Rajasthan	Kharif	Late June-Early July
	Kharif	Last week of June to second fortnight of July
	Semi rabi	Mid September
	Summer	January - February
Maharashtra	Kharif	Second fortnight of June-first week of July
Madhya Pradesh	Kharif	First week of July
	Semi-rabi	Late August-early September
Uttar Pradesh	Kharif	Second fortnight of July
Andhra Pradesh		
– Coastal	Kharif	First fortnight of May
	Summer	Second fortnight of January
– Telangana	Kharif	Second fortnight of July
Tamil Nadu	Kharif	Second fortnight of May to second fortnight of July
	Rabi	November -December
	Summer	Second fortnight of January
Karnataka		
– Northern region	Kharif	June-July
– Southern region	Kharif	April - May
West Bengal	Kharif	End of May to first week of June
	Rabi	September - October
	Summer	February -March
Orissa	Kharif	June -July
	Rabi	September - October
	Summer	February

10. NUTRIENT MANAGEMENT

One of the constraints for low productivity of sesame is owing to its cultivation in marginal lands without any nutrient or input management under rainfed conditions. Under the existing cropping situations, just by adding manure/fertilizers in a limited extent, the yield of sesame can be improved significantly. Sesame crop yielding 1200 kg/ha has been found to remove 62 kg N, 24 kg P, 64 kg K, 14 kg S, 45 kg Ca, 19 kg Mg, 202g Zn, 952 g Fe, 138 g Mn and 140 g Cu. Depending on the level of production, the values of nutrient uptake vary. The response to application of fertilizer varied depending on the variety and availability of soil moisture. The studies conducted by Agricultural Finance Corporation, Hyderabad revealed that nearly 61.8% of farmers in Maharashtra, 76.3% in Gujarat, 64.5% in Orissa and 62.5% in Rajasthan used below optimum dose of fertilisers. Thus, there is a great scope to increase the productivity through balanced fertilization.

Table 5. Fertilizer recommendations for sesame in different states

State	Recommended N : P : K (kg/ha)	Remarks
Rajasthan		
Light soils	20 - 20 - 0	With less than 350 mm rains
Light soils	40 - 25 - 0	With more than 350 mm rains
Gujarat		
Kharif	30 - 25 - 0	
Semi-rabi	25 - 25 - 0	
Maharashtra	50 -0 -0	Half at 3 weeks after sowing and half at 6 weeks of sowing
Madhya Pradesh	40 -30 -0	Apply 25 kg zinc sulphate once in 3 years in zinc deficient soils
Uttar Pradesh	20 - 10 - 0	
Andhra Pradesh		
Coastal region	40 - 40 - 20	
Telangana region	30 - 30 - 20	
Tamil Nadu Irrigated	35 - 23- 23	Apply full dose of NPK as basal
	25 - 15 - 15	Seed treatment with 600 g Azospyrillum
Karnataka	37.5 -25 -25	
Kerala	30 - 15 - 30	
West Bengal		
Irrigated	50 - 25 - 25	No fertilizer if sown after potato
Rainfed	25 - 13 - 13	
Orissa	30 - 20 - 20	

Although, it is recommended to apply 10 t/ha of farm yard manure a month before sowing, farmers are not able to apply adequate quantities of either F.Y.M. or compost due to the non-availability. The varieties to produce high yields require 40 to 60 kg N/ha at different locations. It is desirable to apply 50% N at sowing and 50% N at 30 days after sowing. Full dose of P_2O_5 and K_2O may be applied at the time of sowing. The top dressed nitrogen should be incorporated into the soil by hoeing. Adequate moisture should be present at the time of incorporation. Statewise fertiliser recommendations are given in Table 5.

Integrated nutrient management has been found to be beneficial for balanced nutrition, high yield and sustaining soil fertility. Application of nitrogen and FYM in conjunction with Azatobactor or Azosporillum significantly enhanced the seed yield over nitrogen or FYM alone. Under prolonged dry conditions, foliar spray of urea 2 per cent is economic and safe at all the stages of the crop growth (Gaur, 1969). There is positive interaction between application of nutrients and irrigation. Studies conducted at Berhampore during pre-kharif 1995 revealed that providing three irrigations to sesame and application of 75 kg N/ha gave the highest yield in sesame. The seed yield and oil content of sesame has been found to increase with addition of sulphur and magnesium to NPK dose. At Vriddhachalam, application of gypsum @ 200 kg/ha recorded the highest seed yield which was comparable with that obtained under $ZnSO_4$ application (Bala Subramanian *et al.* 1994).

11. WEED MANAGEMENT

Depending on the time of sowing, composition and intensity of weed flora the yield losses due to weeds in sesame range from 30-70 per cent. Sesame has shy growth in early stage and the critical period of crop weed competition was up to 30 days after sowing (Kondap and Rao, 1978). Usually, two hand weedings first at 2-4 week stage and another 5-6 week stage are necessary for good yield of sesame. Two to three hoeings by hand or bullock drawn hoe may be carried out for satisfactory weed control. Effective control of weeds with herbicides is also reported by (Gaur and Tomer, 1978, Vishnu Shukla 1984). Application of herbicides like Alachlor @1.5 l/ha pre-emergence or Fluchloralin @ 1 kg/ha pre-plant incorporation is equally effective in checking weed growth. Integrated weed management is most efficient and acceptable approach to combat weed problems (Padmaja *et al.,* 1994). Studies conducted at Mandore revealed the effective control of weeds with alachlor granules + one hand weeding at 30 days after sowing.

12. WATER MANAGEMENT

The sesame crop is generally not irrigated during kharif. However, when the crop is grown in semi-rabi/rabi or summer season, irrigation is required. Protective irrigation may also sometimes needed in rainfed kharif crop during prolonged drought. The work of Rao and Raju (1991) clearly demonstrated that moisture stress (severe) during flowering, pod initiation and development period spanning over 25 days (36-60 days from sowing) caused significant reduction in leaf area, capsule number, test weight and seed yield. Studies conducted at Hyderabad on sandy loamsoil revealed that 40% Available soil moisture depletion (ASMD) gave higher yield and increasing the depletion from 40 to 80% ASMD caused significant reduction in sesame yield. According to Arunachalam and Venkatesan (1984) under Tamil Nadu conditions on sandy loam soils, irrigating sesame at 0.30 or 0.45 IW/CPE ratio during vegetative and reproductive phases (3 to 4 irrigations) recorded consistently higher mean seed yields (714 kg/ha) in all the three years. Ghosh and Biswas (1984) also observed that under West Bengal conditions, sesame yields increased when the number of irrigations were three. Irrigation at 0.6 IW/CPE with 50 mm depth gave highest seed yield and water use efficiency as compared to 0.4 or 0.8 IW/CPE in summer sesame at Kharagpur. In vertisols of

Parbhani, Rao *et al.* (1994) observed increased yield of sesame when the irrigation is scheduled at 0.9 IW/CPE during summer. At Raipur, irrigation at 0.7 IW/CPE was found to be the best for summer sesame (Chandrakar *et al.,* 1994). The water requirement depending on the variety and climate varied between 25.6 to 51.4 cm at different locations in India. There is positive interaction between irrigation and nitrogen.

13. INTEGRATED PEST AND DISEASE MANAGEMENT

Sesame is attacked by several insect pests at various stages of growth. Of all the insect pests, sesame shoot webber and pod borer (*Antigastra catalaunalis*), is the serious pest and attacks the crop all round the crop stage i.e. vegetative, reproductive and seed filling stage. This is the most destructive pest which naturally causes maximum damage. The gall fly (*Aspondylia sesami*) is another serious pest which affects flowers and results in no seed formation. In addition to these two major pests, there are number of other pests like sphingid moth, hairy caterpillars, jassids, thrips, budfly and beetles cause damage sometimes to a considerable extent. Whenever, there is economic damage by the insect pests at any of the crop growth stages, integrated pest management approach may be effectively followed.

1. Resistant variety : In the integrated management of sesame pests, varieties which have been found resistant, after through screening against major pests of sesame should be recommended to the farmers. For gallfly: OMT-26, RT-108, RT-125, RT-57 and RAUSS-17-4. For leaf roller and capsule borer : Padma.

2. Cultural Methods : (i) After harvesting of kharif crops the field should be ploughed to expose pupae for predation. (ii) Soon after harvesting, straw and stubbles should be removed and utilised as animal feed or composted so that the carryover of the pests is avoided. (iii) Early planting (in the lst week of July) and use of early varieties like OMT-11-6-3 and JT-7 for 80-85 days and the proper use of balanced doses of fertilisers to give a philip to the crop growth make the crop to pass the vulnerable stage before gallfly build up the population above economic threshold level.

3. Mechanical methods : (i) Light traps are helpful to monitor local populations of some pests for forecasting/ warning systems and also as tools for their control. They are useful against leaf roller and capsule borer, hairy caterpillar and effective. (ii) collection and destruction of sphynx caterpillar. (iii) Collection and destruction of egg masses of *Spillosoma obliqua* and early instar of caterpillars which they are in congregation. (iv) Collection and destruction of leaf rolls, shoots, infested pods and gall fly infested buds. (v) Pearching: Birds rapidly eat the caterpillars when they find them and help to check them when they are numerous.

4. Chemical control : (i) Three rounds of spraying of endosulfan 0.07% will effectively control *Antigastra* and gall fly. It gave the cost benefit ratio of 1 : 3.86. (ii) Two rounds of dusting with endosulfan 4% or phosalone 4% or carbaryl 5% dusts @ 25 kg/ha for effective control or *Acherontia styx* and other leaf feeding caterpillars. (iii) Spraying with Dimethoate 0.03%, or monocrotophos 0.04% or endosulfan 0.07% at bud initiation stage has been found effective against *A. sesami.*

Integrated Disease Management

Sesame is affected by a number of fungal, bacterial and viral diseases. Among all, phyllody is most destructive disease capable of causing yield losses over 50%. The disease if occurs in the early stage of the crop results in complete sterility. In addition, leaf spots by *Cercospora* and *alternaria*, bacterial leaf spots, stem and root rot, seedling blight and powdery mildew may also damage the crop to a considerable extent. The fungal leaf spots can be managed by spraying Dithane M-45 @

0.2% and powdery mildew by spraying 0.2% wettable sulphur. The bacterial leaf spots may be controlled by spraying Agrimycin @ 250 ppm. The seed treatment with Thiram or Captan @ 0.25% minimises seed borne and seedling diseases to a considerable extent and it is cost effective also. The plant protection for the disease management may be taken up whenever there is economic damage and considering the cost benefit ratio in the cropping system. Further, proper adoption of recommended agro production practices like optimum date of sowing, cultural practices like weed management, deep ploughing, proper inter and sequence cropping systems are required. This approach will help in maintaining optimum ecological balance for natural bio control agents for sustainable production in different agro-ecological situations of sesame.

14. HARVESTING AND THRESHING

The sesame crop should be harvested when the leaves turn yellow and start dropping while the capsules are still greenish. Kaushal *et al.* (1969) reported that harvesting of sesame should be done when the capsules have turned yellow and leaves have dropped as it gives highest seed oil content. Further, delay in harvesting causes shattering of the seed without increase in oil content. According to Lee (1980) the optimum time of harvesting will be started when the moisture content in the capsules drops below 70 per cent, leaf senescence reaches upper-nodes and 50 per cent of capsules lose green colour. At maturity, leaves and stems tend to change from green to yellowish and then reddish tint. If the harvesting is postponed and the crop is allowed to completely dry there is loss in yield due to bursting and shattering of capsules. Harvesting is done by cutting plants with sickle and bundles stacked around bamboo stake. A week drying is enough and then by letting the bundles upside down, the seeds are separated and cleared. The threshing is also done by gentle beating of well dried plants with sticks. The seed is dried to bring the moisture level to 8-9 per cent.

Marketing

Since the cultivation of sesame is scattered in India, the produce is available in small lots. The cultivator brings the produce to the local market or *Mandis* where from it is marketed after cleaning and packing the seed. The oil from village *Ghani* is sold by the local traders at the place of production itself or at the nearest weekly market.

15. SEED PRODUCTION AND SEED STORAGE

The studies indicated that use of quality seed of improved variety increases the yield levels by 20-30% in sesame. Though sesame is a self pollinated crop, there are around 5% cross-pollination report by insects. Further, though there is large quantity of breeder seed produced, there is no continuity in the foundation and certified seed production chain, resulting in less availability of improved seed to the farmers. Hence, state seed corporations need to maintain the seed chain. An isolation distance of 100 m. for foundation and 50 m. for certified seed production is required. To produce good quality seed, fertile soil, better weather (Rabi-summer) and better management are required, in addition to proper rogueing and isolation requirements. In general, seed production problems are less in sesame, seed rate is low and seed multiplication rate is very high. hence, considering these desirable features, state seed corporations should produce good quality seed to increase seed yield levels of sesame in farmers' fields.

Normally sesame has better seed viability, however appropriate post harvest processes like threshing, drying, cleaning and proper storage are essential. If the moisture levels are high and storage is not proper, the seed loses the viability. At 7% moisture, viability was good. Hence, precautions are needed for better storage for higher longevity of seed and optimum germination, plant stand to realise the superior genetic potentials of the improved variety. Seed should be completely dried and made air tight for effective storage.

16. IMPACT OF IMPROVED TECHNOLOGIES AND ITS ECONOMICS

The development of improved varieties coupled with agro-production and protection technologies in different agro-ecological and crop growing situations have shown that a well managed sesame crop can yield 900 to 1000 kg/ha under irrigation and 500 to 600 kg/ha under rainfed conditions. Under the All India Coordinated Research Project on Oilseeds (AICORPO) frontline demonstrations were conducted on farmers' field in an area of 0.4 ha under improved practice and adjoining plot of the same farmer with prevailing practices as local check/control plot. The results of whole package demonstrations in sesame under kharif, rabi/summer season of 1996-97 were reported by Virupakshappa *et al.* (1998). The results of rainfed demonstrations conclusively proved the superiority of improved technology (IT) over prevailing farmer's practices (FP). The improved technology registered an yield advantage ranging from 34 to 48 per cent and the B : C ratios varying from 1.63 to 2.92 for IT and 1.04 to 4.55 for F.P. (Table 6). The highest productivity recorded in IT and FP plots ranged from 373 to 1383 kg and 250 to 1013 kg respectively, while the corresponding mean yield levels varied from 285 to 1200 kg and 200 to 894 kg/ha. Under irrigated conditions at Mandore, the yield increase was 40 per cent with net returns to the tune of Rs. 4325/ha. Similarly, during rabi/summer irrigated conditions at Bhubaneswar, on an average, the IT yielded 737 kg/ha as against 605 kg/ha in F.P. thus providing an yield advantage of 22 per cent, with Benefit cost Ratios of 4.35 and 4.55. If improved practices are followed by the farmers, it is possible to attain additional 7.23 lakh tonnes at national level.

Table 6. Impact of improved technologies on the productivity potentials and economics of sesame under real farm situations

Centre	No. of demo.	Highest yield (kg/ha)		Mean yield (kg/ha)		% increase in yield	Cost of cultivation (Rs/ha)		Gross returns (Rs/ha)		B : C ratio	
		IT	FP	IT	FP		IT	FP	IT	FP	IT	FP
Kharif												
Amreli	5(R)	1383	1013	1200	894	34	8275	7727	24154	17994	2.92	2.33
Mandore	2(I)	688	450	594	425	40	5954	4593	10279	7513	1.73	1.64
	4(R)	375	250	285	200	43	3085	2013	5019	3318	1.63	1.65
Mauranipur	11(R)	415	295	322	217	48	2778	1849	5484	3360	1.97	1.82
Tikamgarh	3(R)	640	310	607	287	111	5616	3920	10920	5160	1.94	1.04
Rabi												
Bhubaneswar	8(1)	950	720	737	605	22	3394	2654	14769	12100	4.35	4.55

IT = Improved Technology; FP= Farmer's practice; I = Irrigated, R = Rainfed

REFERENCES

Arunachalam, L. and Venkateshan 1984. Water management studies in sesamum, Madras Agricultural Journal 71 : 339-340.

Ashri, A. 1985. Sesame and safflower; Status and potentials, Proc. Expert Consultation, FAO Paper 66, FAO, Rome.

Balasubramanian, T.N., Sivanatham, M. Alagesan, V.A and Palaniappan, S.P. 1994. Gypsum increases irrigated sesame yield. In Prasad, M.V.R. et al (Ed). Sustainability in oilseeds, Indian Society of Oilseeds Research, Hyderabad, pp. 300-303.

Chandrakar, B.L., Sekhar, N., Tuteja, S.S. and Tripathi, R.S. 1994. Effect of irrigation and nitrogen on growth and yield of summer sesame (*Sesamum indicum* L.), Indian Journal of Agronomy 39 : 701-702.

Gaur, B.L., 1969. A note on optimum concentration of urea for foliar spray of sesamum, Oilseeds Journal 1(4) : 22.

Gaur, B.L. and Tomer, D.S. 1978. Chemical weed control in sesame, Indian Journal of Agronomy 23(1) : 71.

Ghosh, D.C. and Biswas, S.K. 1984. Influence of irrigation and straw mulch on the growth and yield of sesamum grown in summer season. Indian Agriculture 28: 275-279.

Joshi, A.B. 1961. 'Sesamum' Monograph, Indian Central Oilseeds Committee, Hyderabad.

Kaushal. P.K., Sawani, A.R. amd Dhagat, N.K. 1969. Oil content in sesame at various stages of maturity, JNKVV Research Journal 3(1) : 53-54.

kondap, S.M., Chandrasekhar Rao, J. 1978. Weed control in sesamum (*Sesamum indicum* L.) In abstract of papers, ISWS/TNAU, Weed Science, Conference, Coimbatore.

Lee, H.J. 1980. Sesame seed maturity, determination of harvesting time and effect of chemical desiccation, Agron. Abst. 101.

Padmaja, D., Satyanarayana, V. and Chandrasekhara Rao, P. 1994. Increased productivity of sesame through Integrated Weed Management Systems. In : Prasad, M.V.R. et al (Ed) Sustainability in oilseeds, Indian Society of Oilseeds Research, Hyderabad, pp.294-296.

Rajan, S.S. 1967. Sesamum high productivity potential, Indian Farming 17(9) : 14-18.

Ranganatha, A.R.G. 1989. Sesame. In : Agriculture in Andhra Pradesh, Vol.2, Crops, S.S.A.A., Hyderabad pp. 151-158.

Rao, V.P. and Raju, C.H.S. 1991. Moisture sensitive growth stages of sesame (*Sesamum indicum* L.) and optional sequencing of evapotranspiration deficits. Journal of Research, APAU, 19 : 203-209.

Rao, V.P., Raikhelkar, S.V. and Sondge, V.D. 1994. Effect of irrigation and fertilization on yield and its components in sesame (*Sesamum indicum* L.) Indian Journal of Agricultural Science, 164 : 93-100.

Reddy, B.N. and Sharma, S.M. 1983. Agronomic Research on Sesame in India, Oilseeds Journal 13 (1-4) : 47-52.

Sharma, S.M. and Reddy, B.N. 1983. Research on sesame makes head way. Indian Farming : 32 (12) : 3-10.

Sharma, S.M. 1998. Sesame. In : Singh R.P. et al (Ed). Efficient management of Dryland crops in India - Oilseeds, Indian Society of Oilseeds Research, Hyderabad, pp. 193-227.

Virupakshappa, K., Ramanjaneyulu, G.V. and Padmaiah, M. 1998. Technology transfer in oilseeds (Annual Report on frontline demonstration in Oilseeds, 1996-97), Directorate of Oilseeds Research, Hyderabad, India.

Vishnu Shukla 1984. Chemical weed control in sesame, Pesticides 18 (3) : 13- 14.

Weiss, E.A. 1983. Oilseed Crops : Longmen Group Limited, London.

Chapter - 24

SUITABLE MEDICINAL AND AROMATIC CROP PLANTS FOR DRYLAND FARMING

B.V. Hudge* , V.S. Hudge, M.R. Shobhane, M.V. Dhoble and A.M. Dhopte

*College of Agriculture, Dongarshelki Tanda, Udgir-413 517 (M.S.) India

1. INTRODUCTION

The Indian system of medicine (ISM), *Ayurveda* and Homoeopathy (H) uses plant based raw materials in most of preparations. In all, about 1100 medicinal plants are estimated to find use in ISM and H and 500 out of these are commonly used in ISM. For untold centuries, the area under forest has been the source for crude drugs for medicinal plants (MP). In last few decades, the area under forest has been decreasing and there is increase in demand for medicinal plants. Due to heavy exploitation of MP, number of species have become vulnerable to extinction due to lack of cultivation. It is therefore essential to promote the cultivation of MP and aromatic plants (AP) to create awareness amongst farmers about its economic value. With this strategy, the precious plants can be conserved for centuries to come.

In general, following advantages may be exhibited by cultivation of MP and AP:

1. The MP and APs are valuable as it is demanded all over the world.
2. Plants can be grown successfully in available agroclimatic conditions.
3. These plants have wider adaptability to adverse climatic conditions.
4. Some crops are best accomodated into mix cropping, relay cropping etc.
5. Such plants need least care due to rare incidence of pests and diseases.
6. Plants can be planted on field borders and stray animals may not be attracted.
7. The agrotechnology for cultivation of these plants is now available.

2. IMPROVED DRYLAND AGROTECHNOLOGY

Dryland agriculture is characterised by low crop yield due to uncertain monsoon. Inspite of irrigation facilities likely developed in future, 75% land shall remain as Dryland and it needs to be exploited. In India, 42% of countries food grains are produced in drylands and over 108 m ha of cropped area is under rainfed farms. These land produces 1/5 of what it is capable of. Cultivation practices are developed for MP and APs. Improved varieties of MP and APs are now available in centres of AICRP, NBPGR, New Delhi and CIMAP, Lucknow and other related regional centres in India (Table 1(a) and Table 1(b)).

3. SCOPE OF MEDICINAL AND AROMATIC PLANTS

In India, the plant based system of health care is practiced since centuries. Synthetic drugs although used extensively, could not completely replace the medicines of traditional system. The plants have wonderful curative properties without any side effects. Some diseases are cured with plant based drugs. The research is on way to find drugs for Cancer, AIDS, some chronic diseases and viral infections. These plants appear to have excellent source of bioactive compounds. A worldwide interest has developed rapidly for phytomedicinal sector.

At present, .the population of India is over 1.2 billion and 75% lives in villages. There are about six lakh villages in India. Two third of rural population is illiterate and proportion of women folks is quite high. The 1/3 of population is living under poverty line. Infant mortality rate is also high. Among cities 21% population lives in unhygienic shelters. Most of the people are poor and are below social economic conditions. The country offers great potential of cultivation of MP and APs. In recent past, there has been renewed interest for plants as a source of medicines, perfumes and cosmetics because it is harmless. This may be due to:

- Little or no side effects
- Relatively cheaper
- Readily available in rural area
- Environmental friendliness
- Some diseases are cured only with herbal medicines
- Creates job opportunities in rural area

In India with a 8000 medicinal plant species and 1200-2500 species of aromatic group has become global destination for pyto-molecules and products. Several technologies are now available for commercial utilization by small and medium enterprises and international market is ready to accept quality herbal products from India (Rajeswara Rao *et al*. 2012)

There are about 700 naturally occurring MP used in Ayurveda system of medicine. Numerous formulations have derived from these plants. At present, 1000 single drugs and 8000 compound formulations are in use. In India, 2300 plants have been recorded with medicinal properties and it is documented. The herbal medicines are now popular in western countries. The developed countries like USA, Germany, UK, Europe and Russia have realised the importance of herbal medicines and many pharmaceutical companies are using plants as source of medicines. Business of millions of dollars is being performed annually on plants after hectic research. China, Japan, France, Italy, Germany and many European countries are fast capturing the drug market based on plants.

According to world Health Organization (WHO) the international market of herbal products in around \$6.2 billion, which is poised to grow to \$5 trillion by the year 2050 (Ramesh Kumar and Janagam 2011).

4. LIMITATIONS OF PHARMACEUTICAL INDUSTRIES

India could not utilize the full potential of plant drugs as a trade. *Ayurvedic* and *Unani* medicines are prepared from about 3000 plant species. There are organised whole sale and retail traders. Crisis of raw material exists today because of increasing demands. This increased the prices of many plant medicines because of non availability of raw material. *Asparagus, Gulvel, Hirada, Behada, Aloe, Kuda, Ringni Sarpagnadha* have become rare due to heavy deforestation. Cheaper and inferior substitutes are being used as raw material or in some cases falsified adulterated version of genuine raw material is used in the brand name of herbal medicines. This results in ineffectiveness of drugs in curing the diseases.

Table 1(a). Improved varieties of Medicinal plants and Aromatic plants and their source of availability in India

Crop	Variety	Institute where developed / available
1. Isabgol	Gujarat Isabgol - 1	M and AP centre, Agri. Inst. Anand
	Gujarat Isabgol - 2	(Gujarat)
2. Senna	Tinneveley senna, ALFT -2	TNAU, Coimbatore and -do-
3. Opium popy	Jawahar Apheem - 16	JNKV, Bahaduri Res Farm, Mandasaur (MP)
	Trishna (IC-2)	M and AP Centre, New Delhi
	40-285 (Chetak)	Raj . College of Agriculture, Udaipur
	NOP-4 (Kirtiman)	M and AP Centre, Deptt. of Horticulture and Agri. University, Faizabad (UP)
	NBRI-3	Plant Breeding Lab. Nat. Bot Res. inst. Lucknow (UP)
	Sweta (GS-2)	CIMAP, Lucknow
4. Khasi kateri (*Solamun viarum*)		
	Glaxo Variety, Divn. of M and AP IIHR, Bangalore	
	Arka-sanjeevani	
5. Steroidal yam	FB(C)-1, Ark'a upukar	-do-
6. Pyrethrum	Hansa (SPS-387)	CIMAP Reg. Res. Centre, Kodaicanal (TN)
7. Rauvolfia	RI-1	M and AP Unit, JNKVV, Agri. College, Indore (MP)
8. Japanese	MAS-1	CIMAP, Lucknow
Mint	MAS-2	
	MAHYB-77	
	Supriya and MCAS-2	
9. Lemongrass	OD-19	Aromatic and Medicinal Plant Res. Sta. PO. Oddakalli, Dist. Ernakulam (Kerala)
10. Java-Citronella (*C. winterianus*)		
	Manjusha, Mandakini, Bio-13	CIMAP, Lucknow at SZC, Bangalore
11. Vetiver	Hyb-8	NBPGR, New Delhi /Mandasaur
	Sugandha	CIMAP, Lucknow (UP)
	ICS-1 & KS-2	
12. French Jasmine (*J. grandiflorum*)		Deptt. of Horticulture, TNAU, Coimbatore
	Pitchi (Co-1)	

13. Rose Geranium (*Pelargorium graveolens*)		
	PG-7	Hort. Res. Station, TNAU, Kodaikanal (TN)
14. French Anise	(*Dimpinella anisum*)	
	EC-22091	M and AP centre, JNKVV, Agri. College, Indore (MP)
15. Davana	Common culture	M and AP unit, Deptt. of Hort. UAS, Bangalore
16. Celery	(*Apium graveolens*)	New Crops Unit, Dept, of Agronomy
	Ann. Common culture	Punjab Agri. University, Ludhiana, (Punjab)
17. Lavender	(*Lavendula officinalis*)	
	Sher -e-Kashmir	CIMAP, Lucknow (UP)

M = Medicinal, AP = Aromatic plants
CIMAP = Central Institute for Medicinal and Aromatic Plants

Table 1(b). Improved varieties of Medicinal plants and Aromatic plants developed at various research centres

Sr.No.	Crop	Variety	Developed at	Year of Release
Medicinal Plant				
1	*Chlorophytum borivilianum (Safed musli)*	JS405	Mandsaur	2004
2	*Cassia angustifolia (Senna)*	Anand Late Selection	Anand	1989
3	*Diascoria floribunda*	FB (C)-1	Bangalore	1974
4	*Diascoria floribunda*	Arka Upakar	Bangalore	1980
5	*Digitalis lanata (Foxglove)*	D-76	Solan	1991
6	*Glaucium flavum (Yellow Horned poppy)*	H-47-3	Solan	1991
7	*Glycyrrhiza glabra (Liquorice)*	Haryana Mulhatti-1	Hisar	1989
8	*Hyoscyamus muticus (Egyptian Henbane)*	HMI-80-1	Indore	-
9	*Lepidium sativum (Cress)*	GA-1	Anand	1998
10	*Rauvolfia serpentina (Sarpagandha)*	RI-1	Indore	-
11	*Papaver somniferum (Opium poppy)*	Jawahar Aphim 16	Mandsaur	1984
12	*Papaver somniferum (Opium poppy)*	Kirtiman	Faizabad	1990
13	*Papaver somniferum (Opium poppy)*	Jawahar Aphim 16	Mandsaur	1997
14	*Papaver somniferum (Opium poppy)*	Jawahar Aphim 16	Mandasur	1998
15	*Papaver somniferum (Opium poppy)*	Chetak Aphim	Udaipur	1994
16	*Papaver somniferum (Opium poppy)*	Trisna	Delhi	-
17	*Piper longum (Long pepper)*	Viswam	Trichur	1996
18	*Plantago ovata (Isabgol)*	Gujarat Isabgol-1	Anand	1976
19	*Plantago ovata (Isabgol)*	Gujarat Isabgol-2	Anand	1983
20	*Plantago ovata (Isabgol)*	Haryana Isabgol-5	Hisar	1989

21	*Plantago ovata (Isabgol)*	Jawahar Isagol-4	Mandsur	1996
22	*Solanum laciniatum*	NH-88-12	Solan	1991
23	*Solanum Viarum (Khasi Kateri)*	Arka Sanjeevani	Bangalore	1989
24	*Solanum Viarum (Khasi Kateri)*	Arka Mahima	Bangalore	1992
25	*Withania somnifera (Aswagandha)*	Jawahar Asgand-20	Mandsur	1989
26	*Withania somnifera (Aswagandha)*	Jawahar Asgand-134	Mandsur	1998
Aromatic Plants				
1	*Cymbopogon flexuosus (Lemon Grass)*	NLG-84	Faizabad	1984
2	*C. martinii Var. Motia (Palmarosa)*	Rosha Grass-49	Hisar	1989
3	*C. martinii Var. Motia (Palmarosa)*	CI-80-68	Indore	-
4	*Jasminum grandiflorum (Jasmine)*	Arka Surabhi	Bangalore	1993
5	*Mentha spicata (Spearmint)*	Punjab Spearmint-1	Solan	1991
6	*Valeriana jatamansi (Mushakbala)*	Dalhousi Clone	Solan	1994
7	*Vetiveria zizanioides (Vetiver)*	Hyb-8	Mandsaur	-

It is, therefore, suggested to adopt cultivation of genuine species so that superior quality indigenous medicines could be manufactured and exported. The following points have to be considered if this trade has to be promoted in future :

1. How to ensure sustained supply of quality raw herbal material for phytomedicines ?
2. How to accord respectful position to phytomedicines.?
3. How to involve rural and tribal folks in collection and cultivation of plant material on sustainable basis?
4. How to exploit the tremendous export possibilities of phytomedicines?

5. CULTIVATION AND FUTURE PROSPECTS OF MEDICINAL PLANTS

There are several medicinal and aromatic plants which can be grown successfully at various places other than traditional farm area. Plants like *Adulsa, Aloe, Neem, Tulsi, Sadafuli* etc. are normally seen in and around village area. In forest area also several plants naturally grow well. *Chitrak, Bell, Ashok, Hirda, Behda, Gulvel, Sarpgandha* may be grown in the gardens. Some plants like *Shatavari, Satap, Citronella grass, Tulsi* etc. can be grown in earthen pots. Other places like school, college, office area should also be used for growing medicinal and aromatic plants (Hudge *et al*. 2007).

Cultivation of medicinal plants is scanty though agrotechnology is developed. Some plants are being cultivated in India. Some examples are of Isabgol, (Psylium), Ahiphen (Popy) and *Swarnpatri* (Senna). Some plants are cultivated on small scale such as *Ashwagandha. Adulsa, Pangara, Khair, Arrarot, Ambehalad, Shatawari, Gulwel, Kal-lawi, Gokarni* are mostly found in Konkan regions of seashore. Some plants like *Haladi (Curcuma longa), Adraka (Zingiber offininale), Coriandrum, Ajawan, Kalimirch* etc are grown as spices which are used indirectly in medicines. A market survey revealed that the drugs such as *Bhringraj, Ashok, Withania, Anantmool* etc. occur partially or wholy in adulterated or in substandard quality. It is therefore necessary to think about agrotechnology of plants and its synthetic components for further cultivation.

Plate 24.1. 1. *Vinca rosea* (Sadaphuli), 2. *Asparagus resemosus* (Shatawari)

Merits of drugs obtained through cultivation:

1. Supply of authentic and uniform quality drug by companies etc.
2. Good price for raw material collected through cultivation rather than wild collection.
3. Close link between farmers, traders, industries etc for mutual benefit.
4. Assured supply of drugs and raw materials.
5. Protection and preservation of rare plants.
6. Maintenance of standard quality drug materials and promote export potentials.

Research and Development in Medicinal and aromatic plants

Contributions have been made by Central Institute of Medicinal and Aromatic Plants (CIMAP), Lucknow (UP). India, on research and promotion of medicinal plants. Cultivation and promotion are gaining momentum day by day as they offer great potential for creation of job opportunities especially in rural sector. This institute provides every thing that is needed for promotion like :

1. Current awareness service like promotion etc.
2. Commercial statistics like marketing export import
3. Technical enquiry service about all plants
4. Consultancy service like Agrotechniques
5. Bibliography service like plant information
6. Extension service like training to farmers, industries, doctors etc.
7. Crude drugs analytical service like identification of crude drugs.
8. Publication of farmers bulletins in local languages
9. Publication of books, booklets, monographs and news
10. Extension of Agrotechnology and supply of seeds, slips, suckers, cuttings and seedlings.

6. MEDICINAL AND AROMATIC CROP PLANTS

A. Medicinal Plants

1. *Sapogenin-bearing yams :* Various species of Dioscorea are today one of the important medicinal plants of economic value throughout the world. It is basic raw material of steroidal drugs. Most of plant material is collected from Jammu and Kashmir, Himachal Pradesh and Uttar Pradesh. Mexican species are cultivated. The steroid drugs are imported from abroad. Plant is diocious.
2. *Tropane-alkaloid bearing plants : Atropa belladona*, various species of Datura and Duboisia are source of these alkaloids – atropine which finds use in medicines due to antispasmodic actions. Technology for a belladona is developed by CIMPA, Lucknow. Duboisia contains 3% alkaloides.
3. *Chrysanthemum cinerariaefolium (Pyrethrum) :* Flowers are source of pyrethrines, the safest insecticides for the control of household insects, and in protection of food grains. Improved technology for cultivation has been developed in Kashmir valley. It is also being grown in Tamil Nadu. It is produced on limited scale in India. It flowers in 10-15 days and does not suffer from frost, disease and pests.
4. *Plantago ovata (Psyllium/ Isabgol)* : India is major supplier of seeds and husk that is used for bulk laxative. It relieves intestinal irritation. It is cultivated in Gujarat and adjoining area of Rajasthan. 80 to 85% of produce is exported abroad earning substantial amount of foreign exchange annually.
5. *Cassia aungustifolia (Senna) :* Dried leaves and pods of senna are frequently used as stimulant laxative. It is grown extensively in south India and some districts of Andhra Pradesh and Maharashtra. The area under senna is >2500 ha, yielding 3000 to 5000 tonnes of leaves and pods annually. It is exported to Europe.
6. *Commiphora wightii (Guggul) :* It is an important MP grown on rocky hills of Rajasthan and Kathiawar. The gum is extremely used in arthritis, diabets and obesity. It is not regenerated, and sources are depleting. At present, the requirement is met from wild plants.
7. *Withania somnifera (Ashwagandha) :* It is long time used as folk medicine. The roots are valued today as tonic in various preparations of *Ayurvedic* and *Unani* medicines. It has

antiarthritic and antinflammatory activities. It is cultivated on 2000 ha in Neemuch and Mandsaur area of MP where from it is supplied to all over the country. It is important source of medicine and needs to be cultivated on large scale.

8. *Rauvolfia serpentina (Serpangandha):* Dried root of this plant yields several alkaloids which are used as tranquillizer and as antihypertensive agent. Mostly it is obtained from wild collection. The agrotechnology is developed by CIMAP and they distribute quality seeds for cultivation. It has good scope in future.

B. Aromatic Crops

1. *Mentha arvensis (Japanese mint):* Japanese mint oil, obtained from distillation of herb is a rich source of natural menthol used in pharmaceuticals, flavors and cosmetical industries. Before 50 years, it was imported totally but now 5000 tonnes of mint oil and 2500 tonnes of menthol valued at Rs. 75 crores are being produced in the country and some of it is exported abroad. It is being grown in menthol producing belt. Somaclones are also developed by CIMAP and released for cultivation.
2. *Cymbopogon winterianus (Citronella):* Citronella java oil is obtained by steam distillation of leaves. It is important source of perfumery chemical. Till 1973-74, the oil was imported to meet the requirement of country. After release of a variety by CIMAP in 1962-63, (Jorthat), 600 tonnes of citronella java oil worth about 8 crores of rupees is produced annually. The plant is grown in North eastern states and in South India. It has great potential in UP, MP and Maharashtra where good irrigation facilities are available. Manjusha, Mandakini, CIMAP Jeeva and Bio-13 are newly released varieties by CIMAP.
3. *Cymbopogon flexuosus (Lemongrass):* The oil is obtained from leaves and is used in soap perfumery. It is source of citral used in flavor industry. It is raw material of Vitamin A. About 600 tonnes of oil is being produced annually for internal consumption as well as for export. The oil has very good demand in world market.
4. *Vetiveria zizanoides (Vetiver) :* The oil is obtained from roots and it is used as high grade perfumes. Wild species are used for oil extraction. KS-1, CIMAP Hy-1, Khas-15 and Khas-22 are the varieties developed by CIMAP that contains 0.7 to 0.8% oil. A tetraploid variety that yields 1.4% oil is also developed for general release.
5. *Artemisia pallens (Davana) :* Davana is a traditional herb in south India prized for its fruity fragrance. The oil distilled from flowering plants of Davana is used in food and flavour industries. Technology is developed at CIMAP, Bangalore. Two tonnes of oil is being produced and exported, mainly to USA and France.
6. *Jasminum Species :* There are four varieties of scented jasmine which are cultivated in India. It is used for perfumery blends. Presently jasmine concrete and absolute are being exported from the country. The main buyers of these items are USA, France, UK, Japan and Holland. The CIMAP has developed technology for jasmine concrete and absolute from jasmine flowers.

Farmers are showing increasing interest in cultivation of medicinal and aromatic crops due to increased demand for Ayurvedic drugs at national and international levels. Crops like *Kasturbhendi* (*Abelmoschus moschatus*) has immense demand for use in preparation of high grade perfumes, scents and cosmetics. It is also used in tobacco flavouring, in liquors and some food flavours. In medicine, seeds are used as tonic, aphrodisiac, antispasmodic and carminative. Growth regulators like NAA at 25 and 50 ppm concentration levels when applied to *A. moschatus*, increased number of fruits per plant, total biological yield and seed yield per plant (Hudge *et al*. 2000).

Plate 24.2.1. Aloe vera (Korphad), 2. Jatropha sp. 3. Trachispermum sp. (Ajwan). 4. *Vitex negundo* (Nirgudi)

7. IMPROVED VARIETIES OF MEDICINAL AND AROMATIC PLANTS AND MAIN SOURCES

Great deal of work has been done on MP and AP at various institutes in India and better varieties are now available for cultivation. Main contribution is from CIMAP, Lucknow which pioneered the work on medicinal plants. Processing technology has been developed for many MP and APs and many consultancies are now available for all those who are interested in cultivation, conservation, promotion and pharmaceuticals industries. The seeds, seedlings, suckers, rhizomes and cuttings are made available to the needy persons. Other institutes have also contributed for promotion of MP and AP in India including agricultural universities. Some MPs are very slow in growth and therefore, tissue culture techniques are employed for propagation of saplings. The protocol has been developed for plants like Aloe, *Aswagandha, Serpgandha, Shatawari, Mentha and Safed musali* which face problems in general propagation under field conditions. It is necessary to propagate the saplings for large cultivation because conventional sources are tremendously declined. Forest reserves are heavily exploited by the rural folks and some miscreants and some species of MPs are extinct now or unless proper methods are not employed, we shall loose these biodiversity. Reserve forests were the sources for MPs in past but now there is rare chance of getting important species. Fortunately, some forests are still unattended by human folks and probability of having these species can not be ruled out. Thanks to wild animals who helped preserve the precious plant species. Who knows, one day we will find the plant that shall have property to cure Cancer or AIDS in future. All that is needed is to search. Cultivation practices are mentioned in Table 2(a), Table 2(b) and Table 3 for use of readers.

8. COST-EFFECTIVENESS OF CULTIVATION

(A) Medicinal crop plants

1. ***Sarpgandha*** : The plants raised from seeds give the maximum yield of roots varying widely from 0.1 to 4.0 Kg/plant. If planted on 60 x 30 cm with 80% survival, yield of 1,175 Kg/ha is obtained as compared to 175 Kg in plants raised from stem cuttings and 345 Kg in plants raised through root cuttings. An irrigated two year old plantation on sandy clay loam yields 2200 Kg and three year old plantation 3300 Kg air dried roots/ha. The average yield is 1000-1200 Kg/ha. With Rs. 45-50 per kg, net profit of Rs. 25,000-30,000 in two years can be obtained.
2. ***Ashwagandha*** : Average fresh root yield is 500 Kg/ha and of dried root is 150 Kg/ha. Dried root record is 300 to 400 Kg/ha. Commercially, a root of 6-15 mm diameter and of 7-10 cm length are better. Alkaloid in root is 0.13 to 0.31%. Cost is Rs. 50 per kg.
3. ***Guggul*** : Considering the rate of gum @ Rs. 200/Kg, the gross profit comes out to be Rs. 6,40,000 over 10 years. Therefore a gross profit of Rs. 64,000 can be estimated annually. Deleting the cost of cultivation as Rs. 10,000 to 12,000 per year/ha, a profitability of Rs. 50,000/year/ ha can be obtained.
4. ***Senna*** : The average crop yield is about 10 q/ha of dried leaves and 5 q/ha of pods under good management practices. The yield under rainfed condition is about 7 q leaves and 4 q of pods. Cultivation expenses is about Rs. 10,000/ ha under satisfactory management. Prices fluctuate widely depending upon demand of buyers abroad. Current price is around Rs. 3000 to 3500 for leaves and Rs. 3500 for pods per quintal of average fair quality.
5. ***Pyrethrum*** : Pyrethrin is obtained from dried flower heads, The average annual yield is about 400-500 kg of dried flowers/ ha. Pyrethrum content of clones is about 1%. More profit can be obtained by getting high pyrethrin content in flowers.

(B) Aromatic crop plants

1. ***Mint*** : The plant is harvested at flowering stage for maximum yield of spearmint. Under local conditions, Japanese mint and bergamot mint do not bloom and therefore it is difficult to determine the proper stage of harvest. Crop should be harvested in bright sunny weather. Oil yield depends upon crop growth. A good crop can give about 40 to 45 tonnes of herb yield and 175 to 200 Kg oil yield/ha from two harvests irrespective of species.

2. ***Citronella*** : The oil yield of java citronella is 100 to 130 Kg in first year and 250 to 300 Kg in second and third year. Oil yield declines to about 200 to 225 Kg in fourth year. Plantation becomes uneconomical after fourth year. A net profit of Rs. 30,000 to 40,000/ha/year can be obtained from commercial cultivation of java citronella.

3. ***Lemon grass*** : An yield of 10-12 tonnes/harvest and recovery of 0.3% oil from fresh grass is considered normal. The oil yield from second year onward should be about 100-120 Kg/ha. A profit of Rs.10,000 to 12000 can be earned with this crop.

Plate 24.3.1. *Ocirnurn basilicum* (Tulsi), 2. *Ocimum* sp (Sabja), 3. *Datura alba* (Datura)

4. ***Palmarosa*** : It can be grown in various agroclimatic conditions as mentioned below :

(a) *Perrenial crop under rainfall conditions:* It grows well under well distribution of rainfall. Closer spacing is recommended. Fertilizer doses should be reduced by 1 / 3 and dust or organic mulch should be applied to take maximum benefit from natural precipitation by conservation of moisture. The balance sheet is given below under very good condition of growth and yield :

Year	Cost of Cultivation (Rs)	Oil yield (Kg)	Selling price (Rs)	Profit (Rs)
I	12000	80	40,000	28,000
II	8000	100	50,000	42.000
III	8000	100	50,000	42,000
IV	8000	100	50.000	42,000
Total	36.000	380	1,90,000	1.54,000

In general, there is average profit of Rs. 38.000 per year. Under second grade management of rainfed farming, it may be Rs. 30,000 per year/hectare.

(b) *Annual crop under rainfed conditions :* With scanty irrigation facility, it can be taken in *rabi* season with spacing of 30 x 30 cm. Cost of cultivation is observed as Rs. 7000 with returns of Rs. 20,000 and net profit of Rs. 13000 in a period of six months/ha.

(c) *Cultivation on saline/sodic soils or saline water :* The palmarosa can be grown in saline/sodic soils having salinity up to 12 ds m^{1} and alkali soils having pH up to 9.5. It can also be grown under saline irrigation water having conductivity up to 16 d sm^{1} without affecting herbal oil yield. This is additional advantage of this crop. Neglected land and an area of sodic soils can be brought under cultivation of this medicinal plant. Purna valley tract of Akola district (Maharashtra) and adjoining area of about 50 km can be effectively used for growing this crop with better remuneration. The farmers of this region may think of utilizing their such land for cultivation.

5. ***Vetiver*** **:** Depending upon climate, soil and management practices, root yield vary from 10 to 50 q/ha. In light textured soil, yield is less as compared to heavy and waterlogged soils. The oil recovery ranges from 0.5 to 0.8%. Thus, a minimum of 5 Kg and maximum of 25-30 Kg oil/ha can be produced. Considering average yield of 10-15 Kg of oil/ha, a grower may hope for Rs. 40,000 to 50,000 return/ha for 18 to 24 month crop period.

6. ***Davana*** **:** The yield of fresh herbage from the main crop and the ratoon crop is about 12 t/ha which on shade drying and distillation yields about 8 to 10 Kg of oil. Appropriate maturity of crop and proper shade drying of herbage are important factors affecting the quality and yield of oil. In large scale distillation, an average yield of 0.2% from a material dried for about 2 days may be considered satisfactory. The oil is used in costly perfumes.

7. ***Patchouli*** **:** (*Pogestemon patchouli*) : The yield of oil is varied from 2.5 to 3.5% on shade dry basis of the leaves and an average yield of 2.5% may be considered satisfactory in commercial distillation. The yield of fresh leaves/ha/year from the harvests is about 8000 Kg which on shade drying reduces to 1600 Kg and on distillation yields about 40 Kg of oil.

8. ***Lavender*** **:** The flower yield of 60 to 80 q/ha can be obtained from third year. The flower contains 1.2% essential oil which yields 72 to 96 Kg oil/ha. Considering the oil rate of Rs. 800/Kg, a gross profit of Rs. 57,600 to 70,800/ha can be estimated. Deleting the cost of cultivation and distillation for Rs. 25,000 ha, a net profit of Rs. 32,600 to 51,800/ha can be obtained.

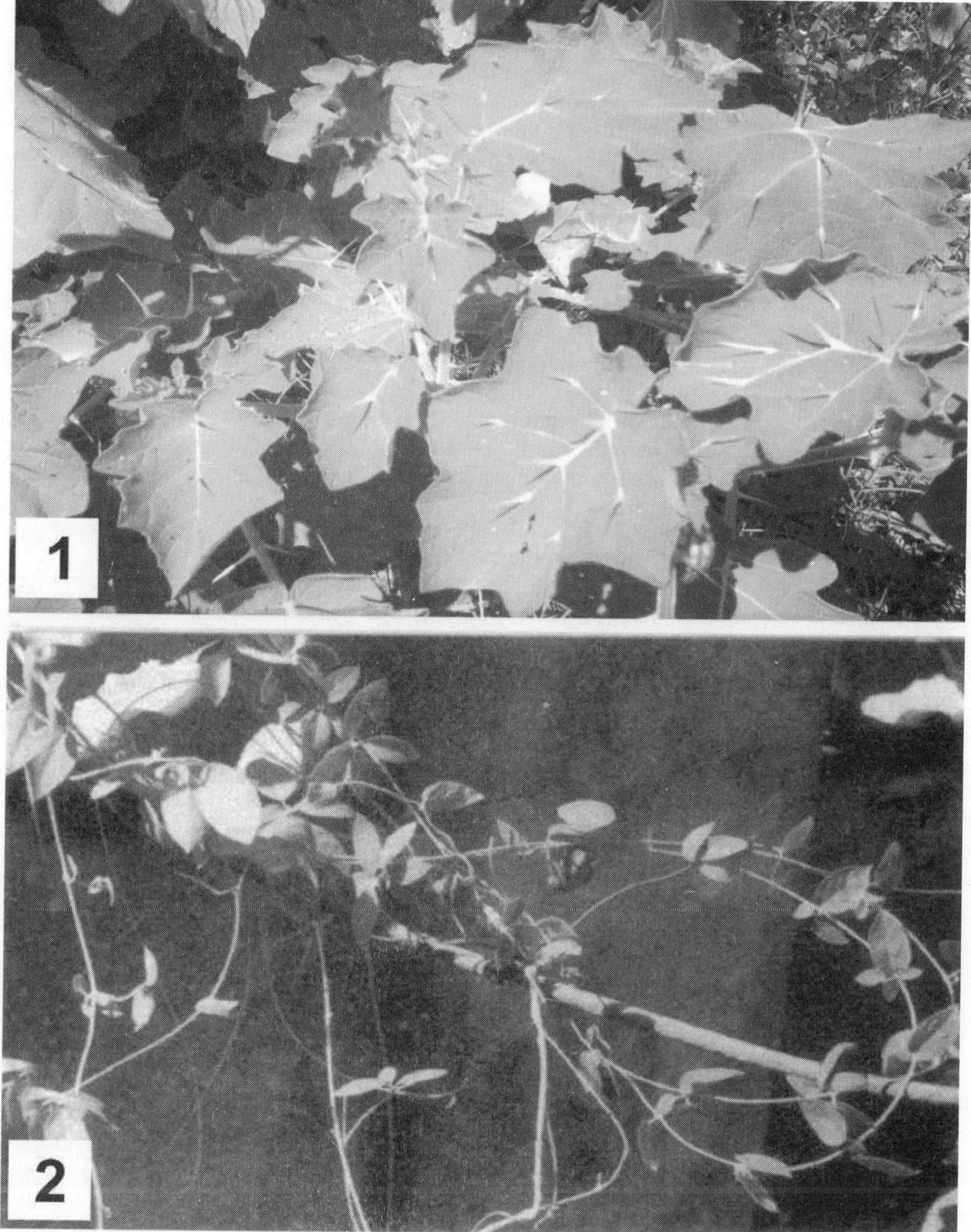

Plate 24.4. 1. *Solanum xanthocarpum* (Ranwangi). 2. *Gymnema sylvestris* (Gudmar)

9. **Geranium :** Normally, with good soil and management, an yield of 25 to 30 t of fresh herb is obtained. The herb is distilled fresh and the oil content on commercial scale is

0.1%. Thus, a total of 45 to 50 Kg oil/ha can be produced. The oil rate is Rs. 2500 per Kg. A gross profit of Rs. 1.12 to 1.25 lakh is expected. Deducting the cost of cultivation and distillation (Rs. 40,000) a net profit of Rs. 70,000 to 80,000 can be fairly expected from the crop.

10. ***Korphad (Aloe vera)*** It is a xerophytic succulent crop and its growth is very slow under natural conditions. It can be profitably cultivated on light soils under rainfed farming. Thick leaves contain glycosides, aromatic oils, fructose and useful enzymes. All these ingredients are useful in medicines. Its cultivation is done on large scale in Gujrat, Madhya Pradesh and Rajasthan. The waste lands can be used for cultivation and good remuneration can be assured. It is neglected species and is almost at extinction. Its propagation has been started. Its growth is found best under polynet house conditions (Dhopte *et al.*, 1999). Five to eight leaves of good size is obtained per plant under good management and 15 to 20 tonnes of fresh leaves/ha can be obtained. The cost of fresh leaves is Rs. 1500 to 2000/tonne. After deducting the expenses on cultivation, a net profit of Rs. 20,000 to 25,000 per hectare can be obtained. The gel is used in cosmetics and it costs Rs 200 litre and remnant can be sold at the rate of Rs. 50-100/kg. It is a crop of one year duration. The oil is used in cosmetics. For multiplication of the species, polynet house condition is found most suitable.

Table 2(a). Cultivation of medicinal crops (Hudge *et al.*, 2003)

Particulars	Crops				
	Ashwagandha-*Withania Somhifera*	Shatawari- *Asparagus racemosus*	Safed musli- *Chlorophytum borivilianum*	Jestmadh- *Glycyrrhiza glabra*	Guggul- *Commiphora wightii*
Season	*Kharif*	*Kharif*	*Kharif*	*Kharif*	*Kharif*
Soil	Medium, well drained	Light, medium well drained	Black & well drained	Medium, well drained	Light to medium
Seed/ seedlings/ ha	1kg seed /ha seedling, 60x30 cm	1kg seed/ ha seedling, 1.5x1 m	1kg seed/ ha seedling, 30x15 cm	1kg seed/ ha cutting, (3-4 buds,) 60x45 cm	1kg seed /ha planting 2x2m
Fertilizers kg ha	N.P.K. 15:15:15 & N:40 after 40 days	N.P.K. 15:15:15 every after 3 months	N.P.K. 15:15:15	N.P.K. 40:40:0 N:30 every year	N.P.K. 15:15:15
Harvesting & yield/ha	After 6 months 12-15 q dry roots/ ha or 100kg seed	After 1.5 to 2 yrs. 12-15 q dry roots/ ha	After 8-10 months 15-20 q/ ha wet OR 3-4 q ha dry roots	After 1.5 up to 3 years 25-30 q /ha	After 7 years 500-800 g. Gum /plant
Use	Tonic, Cough, Asthma, Arthritis, Inflammatory	Tonic, More milk, Narayan tel	Diabetic, Tonic, Ayurvedic preparation	Ulcers, Cough, Asthma (Laxative, expectorant)	Arthritis, Cough, Blood pressure, Obesity, Heart problems, Diabetes
Price	Rs. 50 /kg dry roots	Rs. 50/ kg dry roots	Rs. 400-400 /kg dry roots	Rs. 60/ kg dry roots	Rs. 300 /kg gum

Table 2 (b). Cultivation of aromatic crops (Hudge *et al.*, 2003)

Particulars	Crops				
	Java citronella-*Cymbopogon citrates*	Roshagawat-*Cymbopogon martinii*	Vala-*Vetiveria zizanioides*	Lemongrass-*Cymbopogon flexuosus*	Kustur bhendi-*Abelmoshcus moschatus*
Season	*Kharif*	*Kharif*	*Kharif*	*Kharif*	*Kharif*
Soil	Medium, well drained	Light, medium, well drained	Light to medium or medium drained	Light well drained	Medium well drained
Seed/ slips/ ha spacing	25000 slips, 75x 45 cm	2.5 kg seed 75x45 cm or 60 cm	8-10 kg seed 60 or 75x 30 cm	2kg seed/ ha 75 x 45 cm	2-3 kg seed
Fertilizers kg/ha	N:P:K 20:60:50 N:30 after each cutting	N:P:K 20:40:40 N:20 after every cutting	N:P:K 60:30:30 N:20 after 45 days	N:P:K 30:40:40 N:40 after 45 days	N:P:K 40:40:40 N:40 after 40 days
Harvesting & yield/ha	4-5 months after 3 months 15-20 t leaves 150-200 kg. oil/ha	4 months (when flowering) 20-25 t leaves/ha 60-80 kg. oil/ha	15-18 months 30-40 q root 15-20 kg. oil	4-5 months after 3 months 30-40 t leaves /years 100-125 kg. oil/year	3 months onwards (2 months periods) 8-10 q seed (1-2% oil)
Use	Aromatic oil, perfumery chemicals, cosmetics, food preservations. Odomoss	Soap, Skin diseases, perfumery, Cosmetic and flavour industries	Cool, cough, Attar, High grade cosmetics and perfumes	Cosmetic, Perfumery, Vit. A, flavour, insecticidal	Aromatic, Pan masala, Tobacco products, cough
Price	Rs. 250-350 kg oil	Rs. 400-600 kg oil	Rs. 4000-5000 kg oil	Rs. 300 kg oil	Rs. 30-40 kg seed

Table 3. Cultivation details of tree medicinal plants (Paturde *et al.*, 1998)

Name	Spacing (m)	Flowering	Productivity/tree (Kg)	Life of tree
Amla	7x7	June/July	50-150	–
Bibba	-do-	July/Aug.	5-10 Kg	25-30 years
Hirda	-do-	Dec.	40-50 Kg	-do-
Behada	-do-	Dec./March	-do-	30-40 years
Bel	6x6, 8x8	June-Dec.	20-40	25-30 years
Shikakai	2x1, 2x2	-	2-4	-do-
Ritha	7x7, 10x10	Oct.	30-40	-do-
Shevga	3x3, 4x4	-	1000-3000 Fruits	-

Plate 24.5. 1. Medicinal plant *Adulsa*, (Photo credit, Dr. V.S. Hudge), 2. A dwarf variety of castor (*Ricinus communis*), 3. A species of wild Sabja (*Ocimum sp.*)

9. AVAILABILITY OF SPECIES AND PRECAUTIONS

The most useful and suitable medicinal and aromatic plants for rainfed farming are given in Table 4. The rates of improved varieties of planting materials are mentioned in Table 5 and are available at CIMAP, Lucknow, (UP) India. The medicinal plants are cultivated in Central India at

Table 4. List of medicinal and aromatic plants suitable for rainfed farming system

Local name	Scientific name	Classification	Plant type
Juhi	*Jasminum auriculatum*	AP	Vines
Datura	*Datura stramonium* & *D. alba*	MP	Shrub
Bhringraj	*Eclipta prostrata*	MP	Plant
Musali (White)	*Lawsonia intermis* and *L. alba*	MP	Root crop
Mogra	*Arabium jasmine*	AP	Plant
Shikakai	*Acacia rugata*	MP	Shrub

Sadaphuli	*Vinca rosea* and *V. alba*	MP	Shrub
Jatropha	*Jatropa* sp.	MP	Herb
Bel	*Aegle marmelos*	MP	Tree
Tulasi	*Ocimum basilicum* (Shyam)	MP	Plant
Jakhamjudi	*Tridax procumbense*	MP	Plant, creeper
Korphad	*Aloe vira & A. barhadensis*	MP	Succulent
Dhotra	*Argemone mexicana* (Yellow)	MP	Plant
Gulvel	*Tinospora cordifolia*	MP	Vines
Ashwagandh	*Withania somnifera*	MP	Plant
Shatawarl	*Asparagus resemosus* (Wild)	MP	Shrub
Jamun	*Syzygium cumini*	.MP	Tree
Kali musali	*Curculigo orchiodes*	MP	Root plant
Amala	*Phyllanthus emblica*	MP	Tree
Camphor	*Camphora officinarum*	MP	Plant
Kaiphal	*Myrica nagi*	MP	Tree
Khair	*Acacia catechu*	MP	Tree
Guggul	*Balsamodendron mukul*	MP	Tree
Ranwangi	*Solanum viarum*	MP	Plant
Jayphal	*Myristica fragrans*	MP	Tree
Dikamali	*Gardenia gumnifera*	MP	Tree
Nirgudi	*Vitex negundo*	MP	Herb
Behada	*Terminalia belerica*	MP	Tree
Bibba	*Semicarpus anacardium*	MP	Shrub
Hirda	*Terminalia chebula*	MP	Tree
Tylopha	*Tylopha aesthematica*	MP	Shrub
Kalimire	*Piper nigrum*	MP	Vines
Sabja	*Ocimum* sp.	AP	Plant
Gudmar	*Gymnema sylvestris*	MP	Vines
Tambul	*Piper betel*	MP	Vines
Madhumalti	*Hiptago madablota*	MP	Vines
Sagargoti	*Caesalpinia bonducella*	MP	Shrub
Pudina	*Mentha spicata*	MP	Plant
Sarpgandha	*Rauvolfla serpentina*	MP	Plant
Isabgol	*Plantago ovata*	MP	Plant
Mehandi	*Randia dumetorum*	MP	Herb
Bawchi	*Psorolea corylifolia*	MP	Plant

Nagarmotha	*Cyperus rotundus*	AP	Sucker
Vekhand	*Acorus calamus*	MP	Tree
Ritha	*Sapindus trifoliatus*	MP	Tree

Table 5. Rate list of improved varieties of planting materials

Sr. No.	Materials		Rates (Rs.)
A. Slips/Suckers/Cuttings/seedlings/shoots			
1	Bergamot mint (*Mentha citrate*) cv: kiran:	all material	50/kg
2	Brahmi cv: CIMAP Jagriti:	Fresh planting material	200/kg
3	Bursera:	Fresh cutting	3/cutting
4	Citronella (*cymbopogon winterianus*) cv: Majusha, Mandakini, Bio-13, Manjari, CIMAP Jeeva: (Minimum 40 slip will be sold)		75/ for 100 slips
5	Eucalyptus citroda:	Rooted sprouts	10/plant
6	Garden mint (*menthe virdis*) cv: Supriya, Ganga:	Runners	50/kg
7	Geranium cv: CIMAP Pawan, Borbon, CIM BIO-171:	Fresh cutting/	1/cutting
8	Ghrit kumara (*Aloe vera*) cv: CIMAP Sheetal:	Rooted sprouts	3/sprouts
9	Guggul (*Commiphora mukul*) cv: Marusudha:	Fresh cutting	3/cutting
10	Herbal plants in polybags for kitchen garden/small beds: 1. Aloe vera 13. Meethi neem 2. Ashwagandha 14. Patchouli 3. Brahmi 15. Patharchoor 4. Citronella 16. Pipli 5. Coleu 17. Rosemary 6. Geranium 18. Sadabahar 7. Giloy 19. Sarpgandha 8. Guggul 20. Satavar 9. Kalmegh 21. Stevia 10. Lavender 22. Tulsi 11. Lemon grass 23. Vach 12. Mandukparni 24. Vetiver		10/plant
11	Jasmine (*Jasminum grandiflorum*):	Fresh cutting	3/cutting
12	Lavender: Fresh cutting		3/cutting
13	Lemongrass (*Cymbopogon flexuosus*) cv: Pragati, Praman, Krishna, Nima , Chirharit, CIMAP Swarna, Kaveri (Minimum 40 slip will be sold)		75/ for 100 slips
14	Liquorice/Mulethi (*Glycyrhiza glabra*) cv: CIMAP Mishri:	Fresh cutting	3/cutting
15	Mandukparni (*Centella*)	Fresh planting material	400/kg
16	Mentha species cv: CIMAP Patra:	Sukers/runners	100/kg
17	Menthol mint (*Mentha arvensis*) cv: Kosi, Saksham, Kushal, Saryu, etc.	Sukers	50/kg

18	Menthol Mint:	Shoots	
19	Nursery raised seedling/plantlets of important MAPs (Tulsi, Kalmegh , African Marigold, Chamomile, Mint, Satavari etc.)		25/100 seedling
20	Patchouli cv: CIMAP Shreshtha, CIMAP Samarth:	Fresh cutting	1 cutting
21	Peppermint (*Mentha piperita*) cv: Kukrail, Madhuras, Tushar, Pranjal	Runners/Suckers	50/kg
22	Peppermint cv: CIMAP Indus:	Runners/Suckers	100/kg
23	Pipli	fresh planting material	500/kg
24	Rosemary cv: CIMAP Hariyali:	Fresh cutting	1/cutting
25	Safed musli (*Chlorophytum borivillianum*) cv: CIMAP-Oj	Fresh planting material	500/kg
26	Scented Rose (Rosa damascene) cv: Noorjahan, Ranishahiba:	Fresh cutting Rooted plant	2/ cutting 25/plant
27	Scotch spearmint (*Mentha cardiac*) cv: MCAS-2:	Suckers	50/kg
28	Serpent wood (*Rauvolfia serpentine*) cv: CIMAP Sheel	Root cutting	2/cutting
29	Spearmint (*Mentha spicata*) cv: Arka, Neera, MSS-5, Neer Kalka	runners/Suckers	50/kg
30	Stevia cv: CIMAP Mithi & CIMAP Madhu:	Fresh rhizomes	200/kg
31	Vach (*Acorus calamus*) cv: CIMAP Balya:		200/kg
32	Vetiver (*Vetiveria zizanoides*) cv: KS-1, Dharini, CIMAP Hy 1, Kesari, Gulabi, CIMAP Vriddhi, CIMAP khus-15, CIMAP khus-22 Minimum 10 slips will be sold (Minimum 40 slip will be sold)		75/- for 100 slips

B. Seeds

1	African Marigold (*Tagets minuta*) cv: Vanphool	1000/kg
2	Ammi majus and Ammi visnaga	500/kg
3	Ashwagandha (*Withania somnifera*) cv: Poshita, NMITLI-118, Chetak, Pratap	500/kg
4	Black henbane (*Hyoscyamus niger*) cv: Aela	500/kg
5	Chamomile (*Chamomilla recutita*) cv: Vallary, Prashant, CIMAP Sammohak	1500/kg
6	Clarysage cv: CIMAP Chandani	1000/kg
7	Coriander (*Coriandrum sativam*) S-33	500/kg
8	Eucalyptus citriodora	500/kg
9	Hazardana (*Phylanthus*) cv: CIMAP-Jeevan	6000/kg
10	Isabgol/Psyllium (*Plantago ovata*) cv: Mayuri, Niharika	500/kg
11	Kalmegh /(*Andrographis paniculata*) cv: CIMAP Megha	2000/kg
12	Kewanch (*M.puriances*) cv: CIMAP Ajar	500/kg
13	Milk Thistle (*Silybum marianum*)	1000/kg

14	Opium Poppy (*Papaver somniferum*) cv: Shyama, Sampada, Rakshit (To be grown with the license only)	500/kg
15	Palmarosa (*Cymbopogon martini*) cv: PRC-1, Trishna, Tripta, Vaishnavi, Harsh	1500/kg
16	Pyrethrum cv: Avadh	5000/kg
17	Sadabahar/Periwinkle (*Catharanthus*) cv: Dahwal, Nirmal, Prabal	10000/kg
18	Satavar (*Asparagus recemosus*) cv: CIMAP Shakti	1500/kg
19	Senna (*Cassia angustifolia*) cv: Sona	350/kg
20	Sweet Fennal (*F. vulgare*) cv: CIMAP Sujal	1300/kg
21	Tulsa (*Ocimum basilicum*) cv: CIMAP Saumya	1500/kg
22	Tulsi (*Ocimum sanctum*) cv: CIMAP Ayu & CIMAP Angna	3000/kg

* Prices are subject to change as per notification

Source : Director, Central Institute of Medicinal and Aromatic Plants. PO, CIMAP, Lucknow-226 015 (UP). India

Kuhi, Dist. Nagpur (MS) on 25 Acres farm and one processing unit as *Harihar Herbal* Industries, has been launched in Sept. 2000. The interested farmers can contact the concerned person. A herbal garden of medicinal plants has been developed at Ramtek, near Nagpur.

The cultivation of these medicinal trees can be conveniently undertaken on waste lands. Some patches of a field are not feasible for normal cultivation due to several reasons. It can be either used for agroforestry or for growing MP conveniently with little care. The species are hardy and can grow well on rain water. The care should, however, be taken during seedling stage. The saplings should be protected from scorching heat during summer by various kinds of mulches. Once the trees are established in one season, there is need to fertilize properly so that it can yield profitably. The produce can fetch good price in the market. The species like Jamun can be grown for good remuneration. If the land is too sloppy, it can bunded with contour trenches so that the rain water can best be conserved and soil erosion can be stopped. Thus, it has dual purpose. It checks the soil erosion and degradation is avoided.

The medicinal plants are also available at Nagarjun Medicinal Garden, Dr. PDKV, Akola 444 104 (MS) and some one interested, correspondence should be made before actual steps are taken to ensure the availability of saplings or propagation materials (seeds, seedlings, rhizomes, cuttings, suckers, saplings) etc. To begin with, few plants can be procured and grown and then its multiplication can be done for large scale cultivation. The tissue culture saplings of very rare species are also available in some Agriculture Universities or Central Institutes of India on reasonable price. Very rare species are costly. Few plantlets can be purchased and its multiplication can be done on own with expert help. Consultancies are now available for those who are interested in cultivation of medicinal plants.

The tissue culture laboratories at national level are functioning for multiplication of saplings and these are genuine sources of plant material. Various species of a genera are available and only those which are in demand should be brought under cultivation to have better price by pharmaceutical companies. In central India, an International cargo facility shall be available at Nagpur in near future and it can be exported to the needy country. Reliability and genuineness are important in this sector and reputation needs to be developed for genuine raw material of medicinal plants. Corporate sectors can be launched in this regard and farmers should form their own organization involving cultivation, expert guidance, transport, erecting plant for crude extracts and national and international contacts with companies interested in purchase. A society can be registered by a group of farmers interested in cultivation and a new endeavour can be launched for the benefit of farmers. "Who to bell the cat" is the riddle of today's problems. Some one has to come forward in larger interest of the farmers in this alarming sector of medicinal plants. An honest approach only shall bring new hopes and better future for the farmers.

Several medicinal plants are adapted for dryland conditions. The MPs like, wild brinjal, datura, dhotra, nirgudi, jatropa, sabja, korphad, tulsi, shatawari, gudmar, gokhru, pipri, bawchi, nagarmotha, bhendi, and bhringraj can be well cultivated with little attention. It is always convenient to approach pharmaceutical company within nearby area for easy sale out of the produce. Out of above listed MPs, few can be selected for cultivation on large scale and the produce can be sold as such or after processing in crude form. The requirement of the company and sustained supply are highly important. Renowned pharmaceutical companies like, Baidyanath, Unza, Dabar, etc. can be contacted and contract must be assured and signed so that produce can be easily collected for processing. Transport facilities are likely provided by the company. The herbal medicines are prepared either from roots, leaves, flowers or branches. The economical parts can be made available in saparate cartons and care should be taken in transport of the produce. Some produce like korphad leaves (Aloe) is in much demand by beauty parlours and it is used as raw material without processing.

The purchaser can be contacted by approaching the concerned patronage. There can be little more efforts in the beginning but once the contacts are established, further it becomes the routine for the buyer and the producer. The rates can be negotiable on the basis of demand and supply. In near future, the raw material requirement is going to be more. Presently, most of the produce is exported abroad after processing. The western countries have more requirement since they are now aware of side effects of allopathy drugs. The indigenous people are also now interested in herbal medicines and it has now become tedious to meet the requirement of indigenous as well as foreign countries. This situation created the need for systematic cultivation of herbal medicines. Primarily, forests were the major sources of raw material but with increased population and demand all over the world, it can not satisfy the raw material needs.

Small processing units are being erected by private agencies and the crude form of product is being sold to the pharmaceutical companies. Consultancy can be had from various central institutes (Addresses are given for the use of readers in Table 1). It can be a good enterprise if it is done systematically by the growers of medicinal plants. The planning is however needed. Normally it is observed that the people start growing the MPs without surveying the requirement and then it ends up in loss. Such situation can be avoided after studying the status and present requirement at indigenous and global level. Frustrations in this kind of business can be prevented with systematic approach.

Several MPs are listed in this book but few of them which have demand can be highly remunerative to the farmer. Several books on medicinal plants are now available and some doctors and pharmaceutical companies are in need of growers for furtherance of industry. Benefit can be had of such situation with proper planning. The economics of cultivation should not be ignored. It is the cost effectiveness that can make the entrepreneurship profitable to the grower.

In future, the herbal medicines shall have remarkable scope and people will prefer herbal drugs for routine ailments instead of planning for allopathic drugs having side effects. Every pathy has its own advantage. Promotion of herbal medicines may in no case be harmful to allopathic drugs. It has its own merit and it is certainly life saving. The merit of herbal medicine can also not be ignored. The farmer should be more tactful in cultivation of MPs and APs after watching the demand. The price list is provided to have comparative idea about the importance of each MP. The readers are suggested to seek guidance in planning the cultivation.

10. CONCLUSIONS

Ayurveda incorporates ancient plant remedy for different kind of ailments and it is now steered up at global level because of its unique approach without side effects. Considerable countries are interested in herbal medicines and it is supposed to have bright future. Few medicinal and aromatic plants can be cultivated under rainfed conditions on neglected shallow soils. Every field has small piece of land which remains uncultivated due to poor soil properties' and water holding capacity. Presently, it is either occupied by agroforestry or is kept fallow for grasses to the cattle. *Small herb like Citronella, Aloe, Mentha. Senna, Sagargoti, Reetha, Serpagandha, Isabgol. Guggul, Dikamali, Shatawari, Behada. Hirda. Gokarni, Bhringraj, Belladona and Lemongrass* ; can be grown under natural conditions with minimum attention and dry farming can be made remunerative. Two objectives can be achieved. One of utilization of land for additional gain and another for conserving the important plant species which are now neglected and under extinction due to deforestation. Future generation can be benefited by this kind of action. Some species are at the verge of extinction due to heavy exploitation by pharmaceutical industries. Deforestation resulted in loosing important treasure of unique medicinal plants. It needs now to be cultivated for harmless medicines (Plate 24.1 to 24.5).

Biodiversity of regional flora needs to be conserved for the reason that one day it may come to our rescue to find suitable drugs for non curable diseases like cancer and AIDS. All that is needed is to search from thousands of plant species. Several thousand of plant species are growing wild in reserved forests and remained unattended by human beings. Several useful herbal drugs are today being used effectively to cure the diseases. Several foreign tourists have taken away the plant species from India for further exploration. Plants are at the service of human beings and cattle life as well. We depend heavily upon plants and never the plant on us. Plants are autotrophs. Nature takes care of them but we owe much for them because we get every livelihood from plants and trees. This concept should not be ignored. Let us hand over this environment and plant treasure to our successors in better form that may last longer for the benefit of mankind in days to come. Let us help preserve and maintain the precious wealth of nature together hand in hand.

The information on medicinal plants and aromatic plants can be had from Director, Medicinal and Aromatic Plants, CIMAP, Lucknow (UP), and other Central Institutes of India or from Officer In charge, Nagarjun Medicinal Garden, Dr. PDKV, Akola – 444 104 (MS) India. Other institutions in India as listed in Table 1(a) and Table 1 (b) can also be consulted for related information and problems. Since this approach has geared up recently, many farmers are interested in cultivation of medicinal plants for good remuneration. It is honestly suggested that they should contact reliable institutes for genuine information, get the saplings / seed from reliable source, find the required demand by the pharmaceutical companies, get involved in contract and then should plan for cultivation to avoid frustration. After assuring regular market for the produce, the cultivation can be thought of.

Cultivation should not necessarily be tried on waste land only if sizeable demands are obtained for produce. It can be promoted on solo cultivation of one or two species depending on the climate and soil type. The selection of species and plant variety should certainly be based upon the available resources, irrigation facilities and known limitations. This shall be a wise step to start with cultivation of medicinal and aromatic plants. An extraction plant for essential oils can be designed and instead of selling the raw produce (Leaves, root or flowers), crude form of extract can be brought into the market to have better profit to make the dryland more sustainable.

REFERENCES

Anonymous, (1994). Processing technologies available at CIMAP, Lucknow. News Letter 21 (2) : 24-25 Pub. CIMAP, Lucknow - 226 015, India.

Anonymous (1994). Price list of seeds and planting materials of medicinal and aromatic plants. News Letter 21(2) : 26-28 Pub. CIMAP, Lucknow – 226 015, India.

Anonymous (1998). Pub. Director, CIMAP, Lucknow - 226 015, India.

Anonymous (1995). Source of Breeder and nucleus seed of new improved varieties of medicinal and aromatic plants in India. Commu. Deputy Director of Cocoa, Arecanut and spices development; Ministry of Agri. Govt, of India. Calicut – 673 005, Kerala, India.

Anonymous (1995). Medicinal plants and economic value having regular trade in raw materials Comm. Depty. Director,. Directorate of Cocoa, Arecanut and spices development. Min. Agri. Govt. of India, Calicut – 673 005, Kerala, India.

Anonymous (1998). Technology available for medicinal and aromatic plants development in India. Comm, by Project Coordinator, Medicinal and Aromatic plants, NBPGR, New Delhi- 110 12.

Anonymous (1992). Cultivation of medicinal and aromatic plants (Special issue). *Baliraja* Mag. (Marathi), Sadashiv Peth, Pune 411 001 (MS), India.

Anonymous (1994). Current research on medicinal and aromatic plants (CROMAP), CIMAP, Lucknow – 226 015, India Vol. 16(4).

Anonymous (1999). Medicinal Trees (Ed) (Marathi). Pub. Director, Institute of Medicinal and Aromatic Plants. 120/9, Pashan flats, Sus Road, Pashan, Pune 411 021 (MS). India

Akhtar Hussain (1994). Essential oil plants and their cultivation (Ed) Pub. CIMAP, Lucknow – 226 015, India.

Atal, B.K. and B.M. Kapu (1989). Cultivation and utilization of medicinal and aromatic plants Vol I and II. (Ed), CSIR, Pub. by RRL, Jammu and Kashmir, India.

Chadha. M.L. and Rajendra Gupta (1995). Advances in Horticulture. Vol II. Medicinal and aromatic plants.

Dhopte, A.M., R.D. Deotale and C.N. Chore (1999). Accelerated growth and sucker formation in *Aloe barbedensis* under polynet house conditions. J. Soils and Crops 9 (2) : 275-277.

Hudge V. S., Godawari Pawar., R.H. Bhosle., M. R. Shobhane., V. D. Patil and D. G. Kulkarni (2000). Effect of phytoharmones on yield of *Kasturbhendi* (*Abelmoschus moschatus*) J. of medicinal and aromatic plant sciences vol. 22/4A and vol. 23/1A Oct 2000 to March 2001 pp. 331-332.

Hudge V. S., M. R. Shobhane., B. M. Koreppa and D. V. Kalburge (2003). Cultivation of medicinal and aromatic crops proceedings of first national interactive meet on medicinal & aromatic plants. Editors : A.K. mathur, S. Dwivedi, D.D. Patra, D Bagchi, N. S. Sangwan, A Sharma and S P S Khanuja, CIMAP, Lucknow (India) © 2003: pp 174-175.

Hudge V.S., M.R. Shobhane and I.A. Madrap (2007). *Aushadhi va sugandhi vanaspatichi lagwad* (Marathi). Booklet No. 24/2007. Published by Marathwada Agricultural University, Parbhani-401 402 (MS), India.

Jain, S.K. (1994). Medicinal plants. Pub. National Book Trust A-5 Green Park: India New Delhi- 110 016.

Joshi, B.P. (1994). Cultivation of medicinal and aromatic plants (Ed). Continental Pub. Vijay nagar. Pune – 411 030 (MS), India.

Kirtikar, K.R. (1975). Indian Medicinal plants (Ed). Vol I to IV.

Paturde, J.T., S.G. Wankhade and Pradeep Khode (1998). Cultivation of medicinal plants. *Shetkari* Mag. (Marathi). Nagarjun Medicinal Plant Deptt. Dr. PDKV, Akola – 444 104. (MS). India pp 29-31.

Rajeswara Rao., B. R. Rajput., D. K. Nagaraju and G. Adinarayana (2012). Scope and potential of medicinal and aromatic plants products for small and medium enterprises. Journal of pharmacognosy vol. 3, issue 2, 2012 pp-112-114

Ramesh Kumar and D. Janagam (2011) Export and Import pattern of medicinal plants in India. Indian J. Sci. Technol. Vol. 4 issue 3 March 2011 pp 45-248

Sharma. Devendra (1994). Compendium on phytomedicines. (Ed.). CSIR. Anusandhan Bhavan Rafi Marg. New Delhi- 110 001.

Sawant. Y. (1969). Unique Medicinal plants of Maharashtra. (Ed.). (Marathi). Maharashtra Uni. Book Prod. Board, Law College Square, Nagpur – 440 001 (MS), India.

Thakre, R.D. and P.P. Khode (1992). Useful Medicinal and Aromatic plants (Marathi). (Nagarjun Medicinal Garden), Pub. Director of Extension, Dr. PDKV, Akola- 444 104 (MS). India.

Chapter - 25

DROUGHT TOLERANT CROP GENOTYPES ANDSTRATEGIC APPROACHES FOR RAINFED FARMING

V. Rajagopal[1] and A.M. Dhopte[2]

Principal Scientist and Head, Plant Physiology, Biochem. and PHT,
Central Plantation CropResearch Inst. (ICAR), Kassaragod -671124 (Kerala)
2. Retd. Prof. of Botany, 49, Vivekanand Nagar, Khamla Road,Nagpur -440015 (MS)

1. INTRODUCTION

Population and agricultural production paralleled since independence until 1977 when food basket was short to feed gallowping population in India. Later with introduction of improved varieties and hybrids and irrigated area resulted self sufficiency and moreover matched with increased population. Rainfed areas that support 40% human population and 65% cattle heads showed higher productivity due to addition of new improved seeds. Credit goes mainly to the breeders who brought revolution in agricultural productivity. The hybrids in cereals, namely maize, bajra & Sorghum, satisfied the growing food demands and saved country from begging bowl. Major crops like cereals (90%), oilseeds (80%) and cotton (65%) are produced in rainfed conditions. With recent advancement in rain water management, the productivity is almost stabilised. Today we are one billion in population with limited land and resources and if possible steps are not initiated for effective water management, we may not be self sufficient in food and this is a point of sereous concern. Yield levels are platea used since a decade inspite of better seeds. New genotypes may be introduced with advanced gene transfer technology in various crops in near future that may probably show hopes of self sufficiency .

Today we have hybrids in sorghum, bajra, maize, rice in cereals, cotton, sunflower, safflower in oilseeds; pigeonpea in legumes and vegetable crops as well. The area under hybrid seed cultivation has been tremendously increased from 1970 to 1999 and positive results are in focus. The hybrids are highly responsive to fertilizers and irrigation and its potential is only realised under ideal field conditions (Not seen every year). Aberrant weather and erratic rainfall bring down the productivity. Some varieties do not withstand abiotic stresses. Some hybrids are susceptible to various stresses.

2. CAUSES OF LOW PRODUCTIVITY

Today we find heavy degradation of land, paucity of required soil fauna and polluted environment, salinity, alkalinity and poor water holding capacity of soil. These are some of causes of low productivity which constitutes 90% of total cultivable area. What therefore we need today is the crop genotype that can withstand above limitations and sustain productivity. Dramatic results can not be achieved in degraded soils. It will take its own time for improvement. Soils and environment are corrected only with long term planning of 10-15 years. For this, we have to have tolerant species / genotypes which may sustain productivity of drylands under aberrant weather and soil conditions. Productivity of such genotypes is certainly less than improved varieties but it can positively yield marginal under unpresidented weather conditions. These genotypes have potential to yield under all odds and yield is definitely assured. This is what we need under present crisis. Primarily, attempts are needed first for conservation of water and soil that will insure crop productivity and secondly, evolution of tolerant genotypes shall add to maintain the productivity for sustainable agriculture. Sustainability is important rather having random bumper yields once in a while.

Seed, soil and water are basic inputs in agriculture and provision of best seed material to the farmer shall be highly satisfactory if not much remunerative. The performance of hybrid is best under assured rainfall zone & heavy soil. The arid regions with low avg. rainfall of 200-300 mm and minimum rainy days (50 days) are generally occupied with lesser and pearl millets. This crop is tolerant to drought. Sorghum is a crop of tropics but is highly sensitive to water and temperature stress, imposed during panicle initiation and floret development and grain fill period. Cereals are staple food of most developing countries. Evolution of hybrids made radical revolution in agri. productivity and helped maintain the civilization. The oilseed crops like peanut, sesamum, sunflower, safflower and soybean are highly susceptible to water stress and yield poor when drought occurs at critical period of growth. Drought tolerant crop genotypes in all crops are now known after many years of research by the scientists (Rajagopal, 1997) and assures marginal productivity under harsh environment. These genotypes grow well where hybrids fail under semiarid environment.

3. BIOTECHNOLOGY AND CROP IMPROVEMENT

Tissue culture (Plate 25.1) and gene transfer technology is capable of evolving genotypes to suit our needs of aberrant weather but it is under development. Many genes have been identified for nitrogen fixation, drought tolerance etc. but it may take another 10 years to evolve suitable varieties. At present, known sources of tolerance are being used for human welfare. Following paragraphs deal with drought tolerant genotypes for benefit of farmers and consultants. The Agri. Universities and Research Institutions have played significant role in identifying these genotypes with suitable techniques and it is extensively reviewed by Rajagopal (1997). Some tolerant genotypes are still under cultivation because of its inherant potential proved since many years. Farmers shall have access to these varieties if they are interested. It must be remembered that these varieties are not high yielders but certainly help assuring yield under incidental low rainfall, short term water stress, waterlogging, salinization or high temperature to sustain his family in epidemics. Poverty is elevated in some farmers in hopes of growing high yielding improved genotypes without necessary inputs. Farmers of dryland areas must understand their limitations to avoid gross failure in production. Marginal failure can provide food rather hoping for bumper yields. This risky preposition can certainly be avoided. Farmer's choice for variety/ hybrid or crop necessarily depends upon what kind of land and resources he has to make the farm more economical and profitable enterprise. Crops are to be cultivated on degraded land with low nutrient availability under conditions of poor resources. Limitations in availability of water at desired time and quantity in irrigated areas also face problems of water deficit. The rabi crops also often experience water stress coupled with high temperature during reproductive phase. Thus, interaction of various stresses is an emerging problem to which we have to pay attention. Though tolerant genotypes are poor yielders, it is an important asset of gene pool for further improvement in crop productivity under harsh environment.

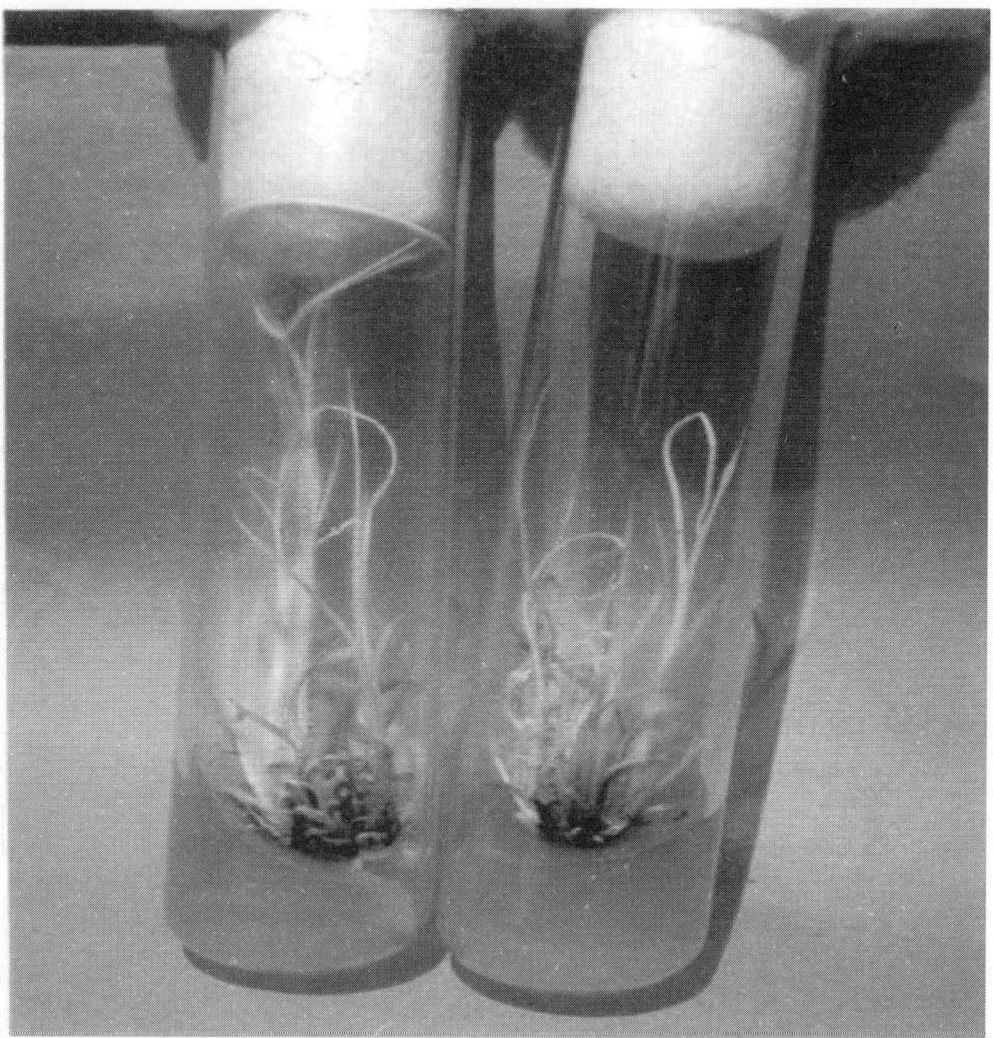

Plate 25.1. 1. Microropagation of plantlets of sugarcane with tissue culture

An attempt has been made by Sehgal *et al.* (1990) to identify the agro-ecological regions of India and the map of India depicting the various regions is given in Map A. Following are the agro ecological regions which are included in arid, semiarid, hot climate and their boundaries are depicted in map of India. Only those which are referred as dryland farming in this book are mentioned below for the use of readers.

Among the various abiotic stresses, the most common is drought. Water is pivotal factor in growth of crop. If it is deficit, other stresses follow in sequence. Metabolic imbalance occurs and leads to further anomalies in growth. The term tolerant means withstanding drought with certain amount of loss. The following paragraphs Agro Ecological Regions of India (Map A) provide the information on suitability of drought tolerant varieties/genotypes for specific areas. From this list, crop genotypes may be selected for the specified area or region. If we have to sustain in crop productivity, we have to select tolerant genotype not with expectations of high yields but to find the solution to survive in all odds.

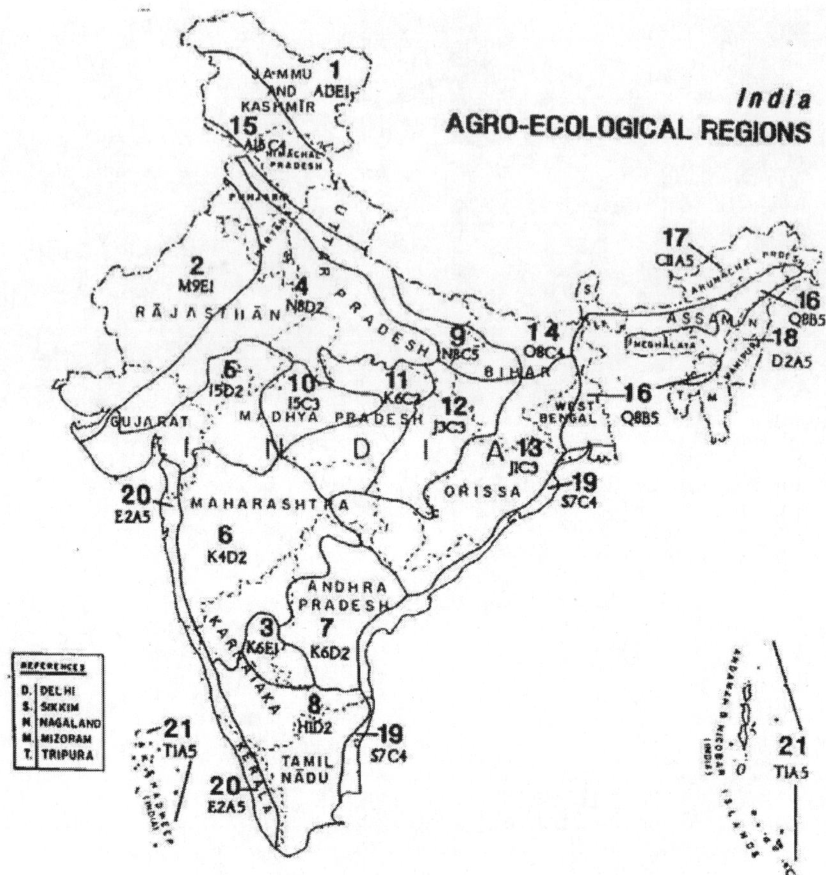

Map A: Agro Ecological Regions of India (Sehgal et al., 1990)

LEGEND FOR AGRO-ECOLOGICAL REGIONS OF INDIA (MAP - A)

1. Western Himalayas, cold arid ecoregion with shallow skeletal soils & length of Growing Period (GP) 90 days
2. Western Plain and Kutch Peninsula, hot arid and ecoregion with desert & saline soils & GP 90 days.
3. Deccan Plateau, hot arid ecoregion with mixed red and black soils & GP 90 days.
4. Northern Plain and Central Highlands, hot semi arid ecoregion with alluvium derived soils & GP 90-150 days
5. Central (Malwa) Highlands and Kathiawar Peninsula, hot semi- arid ecoregion with medium & deep black soils & GP 90-150 days.
6. Deccan Plateau, hot semi-arid ecoregion with shallow & medium (inclusion of deep) black soils & GP 90-150 days
7. Deccan Plateau and Eastern Ghats, hot semi-arid ecoregion with red & Black soils & GP 90-150 days.

8. Eastern Ghats (TN uplands) and Deccan Plateau, hot semi-arid ecoregion with red loamy soils & GP 90-150 days.
9. Northern Plain, hot subhumid ecoregion with alluvium derived soils & GP 150-180 days.
10. Central Highlands (Malwa & Bundelkhand), hot subhumid ecoregion with medium & deep black soils & GP 90-150 days
11. Deccan Plateau and Central Highlands (Bundelkhand), hot subhumid ecoregion with mixed red and black soils & GP 150-180 days.
12. Eastern Plateau (Chhatisgarh), hot subhumid ecoregion with red and yellow soils & GP 150-180 days
13. Eastern (Chhota Nagpur) Plateau and Eastern Ghats, hot subhumid ecoregion with red loamy soils & GP 150-180 days.
14. Eastern Plain, hot subhumid ecoregion with alluvium derived soils & GP 180-210 days
15. Western Himalayas, warm subhumid (inclusion humid) ecoregion with brown forest and podzolic soils & GP 180-210 (+) days
16. Assam and Bengal Plains, hot humid ecoregion with alluvium derived soils & GP 210 days
17. Eastern Himalayas, warm perhumid ecoregion with red and lateritic soild & GP 210 days
18. North Eastern Hills (Purvachall), warm perhumid ecoregion with red and lateritic soils & GP 210 days
19. Eastern Coastal Plains, hot subhumid ecoregion with alluvium derived soils & GP 150-210 days.
20. Western ghats & Coastal Plains, hot humid perhumid ecoregion with red, lateritic and alluvium-derived soils & GP 210 days
21. Islands of Andaman - Nicobar and Lakshadweep, hot perhumid ecoregion with red loamy and sandy soils & GP 210 days

4. CROP VARIETIES TOLERANT TO STRESSES IN CEREALS

4.1. Sorghum

The 94% of sorghum is grown in rainfed conditions in India. Maharashtra and MP occupy 2.7 and 1.71 m. ha area under cultivation respectively (Fig. 1). Approximately, 1/4 of total area of sorghum is under shallow soils in Maharashtra state and average productivity is 541 Kg/ha. Normal yield potential under deep soil is 3 to 4 t/ha. Tolerant genotypes can maintain the yield levels under rainfed farming. Gujar et al., (1995) reported Selection -3 to be better than M35-1 for cultivation in shallow soil under dryland farming. This is selected from local bedar. They reported 44% more yield than M35-1 under multilocation trials. Studies on yield stability of five hybrids (CSH1, CSH5, CSH9, CSH14 and SPH388) under rainfed conditions (Table 1) following the method of Eastin et al., (1988), revealed R-263, R-12 & SPH388 with high stability (Dhopte et al., 1994 & 1995). Sorghum is often exposed to intermittent drought during either PI or grain fill stage due to erratic monsoon and productivity is adversely affected. Drought probability is more in Vidarbha region. No genotype has been found with high tolerance for both these stages and research is still needed (Jordan and Sullivan, 1982).

Category No.	Region
2	Western plain and kutch peninsula, hot arid and ecoregion with desert and saline soils and GP <90 days
3	Deccan plateau, hot arid ecoregion with mixed red and black soils and GP<90 days
4	Northern plain and central high lands, hot semiarid ecoregion with alluvium-derived soils and GP 90-150 days
5	Central (Malwa) Highlands and Kathiawar peninsula, hot semiarid ecoregion with medium and deep black soils and GP 90-150 days
6	Deccan plateau and hot semiarid ecoregion with shallow and (inclusion of deep) medium black soils and GP 90-150 days
7	Deccan plateau and eastern ghats hot semiarid ecoregion with red and black soils and G90-150 days.
8	Eastern ghats (TN uplands) and deccan plateau, hot semi arid ecoregion with red loamy soils and GP 150-180 days
10	Central highlands (Malwa and Bundelkhand) hot subhumid ecoregion with medium and deep black soils and GP 90-150 days
11	Deccan plateau and central (Bundelkhand) high lands, hot sub humid ecoregion with mixed red and black soils and GP 150-180 days
12	Eastern plateau (Chatisgarh), hot subhumid ecoregion with red and yellow soils and GP 150-180 days
13	Eastern (Chota Nagpur) plateau and eastern ghats, hot subhumid ecoregion with red loamy soils and GP 150-180 days
14	Eastern plain, hot sub humid ecoregion with alluvium derived soils and GP 180-210 days
16	Assam Bangal plains, hot humid ecoregion with alluvium derived soils and GP>210 days
19	Eastern coastal plains, hot subhumid ecoregion with alluvium derived soils and GP 150-250 days.

Source : Sehgal *et al.,* (1990). Agro-ecological regions of India. GP = Growing period.

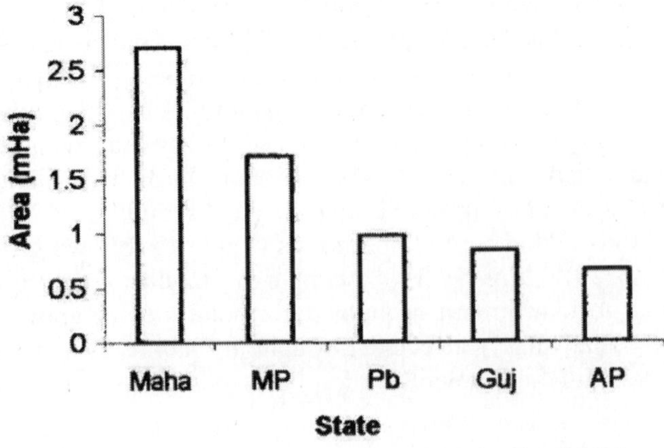

Fig. 1. Area under sorghum, Total 15.9 mha (1997)

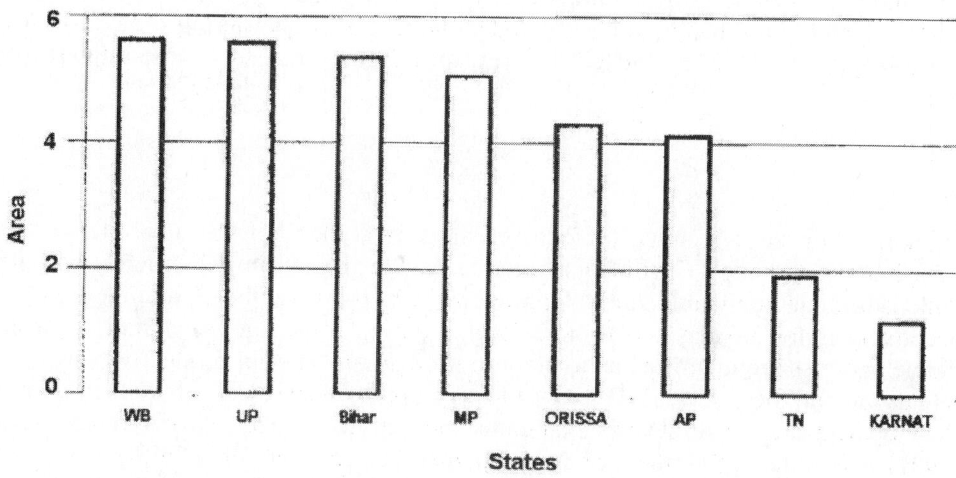

Fig. 2. Area under Rice cultivation in India. Total 41.9 m ha (1997)

Table 1. Evaluation of sorghum genotypes for yield stability under rainfed conditions (Grain yield g/plant). Dhopte et al. (1994)

Genotype	1991			1992		
	IR	RF	YSI	IR	RF	YSI
CSH1	50.0	27.0	0.53	19.4	10.3	0.53
CSH14	89.0	30.5	0.34	27.5	21.0	0.76*
CSH9	77.0	29.0	0.38	17.8	9.3	0.52
CSH5	64.0	21.0	0.33	19.8	5.2	0.26
SPH388*	76.5	50.5	0.66	25.1	16.5	0.66
R-150	79.5	36.5	0.46	9.3	7.9	0.85
R-263*	29.5	30.5	1.03*	12.6	12.8	1.62*
R-12*	28.0	22.0	0.78*	4.73	6.9	1.45*
R-208	69.0	19.5	0.28	21.1	9.2	0.43

* Genotypes were found tolerant to water stress at panicle initiation Data were collected in field. IR=Irrigated, RF = Rainfed, YSI = Yield stability index It was calculated dividing yield under stress with yield under normal condition. The PI is a stage when drought normally is experienced in central plains.

Sorghum is protoandrous where pollens mature first. Receptivity of stigma remains for 2-3 days. It is oftenly self pollinated crop. The ms lines are fertilized with cross pollination of R line. Sowing before July, 15th protects the crop from shoot borer. Most sensitive stages for drought are panicle initiation (PI = 40 DAS) and floret differentiation (FD = 80 DAS) (Dhopte *et al.*, 1995, and Dhopte and Eastin, 1999). Dry spells during this period do not affect yield much in tolerant varieties. Dhopte *et al.* (1987) studied the drought tolerance in relation to stomatal factors of 40 exotic lines and wild species of sorghum. Drought tolerant varieties yield 3-4 t/ha under rainfed conditions. Temperature tolerance of some hybrids is studied by Dhopte and Eastin (1988) and they found yield reduction by 30-60% when stress was imposed at PI and FD stages. The recovery from drought of short duration is better and it has high harvest index. Greeness of leaf during drought is the special character of tolerant species. A hybrid CSH-14 escapes drought due to earliness by 10 days than other hybrids (Shekar *et al.,* 1993). It yields 4 t/ha under rainfed farming. Under stress conditions, the yields of 3-4 t/ha are assured (Plate 25.3).

4.2 Maize

Morphology of maize is peculiar. Male tassels are seen at the top while a cob (female florets) is almost in the centre (7th node). Fertilization leads to seed setting when pollen falls upon the silk of the cob. Pollens become non viable due to high temperature (40°C) and dehydration during drought. Stigma receptively is for 3-5 days. If drought occurs during this stage, seed setting/cob is highly reduced. Some varieties are identified as tolerant to drought and sustain productivity avoiding these anomalies (Tunwar and Singh, 1985). Out of 14 leaves, upper 5 leaves contribute their assimilates to the cob development and lower leaves maintains the activity of root during reproductive stage (Dhopte, 1981). The water requirement of corn is higher than sorghum though it is a crop of tropical region. It is grown world wide in subtropical zone. It can tolerate day temperature to the extent of 35-40°C. Better cob development is only observed when adequate moistures available at boot stage and milk stage. It helps better seed setting and grain filling (Table 3).

Table 2. Drought tolerant varieties/hybrids of sorghum for various regions in India

Variety/Hybrid	Type	Duration	Adaptability	Reference
E-36-1	Tall	Medium	7-Deccan plateau	Rajagopal (1997)
DJ1195	Tall	–do–	–do–	–do–
IS 2037	–do–	–do–	–do–	
IS2280	–do–	–do–	–do–	–do–
R-263	Dwaf	115-120	Maharashtra	Dhopte, et al., (1994)
R-12	–do–	–do–	–do–	–do–
Pariyur-I	Tall	145-150	8-Deccan plateau	Rajagopal, (1997)
NTJ-1	–do–	100-105	Tamil nadu	–do–
NTJ-2	–do–	100	7-Deccan plateau	–do–
N-14	–do–	110-115	Eastern ghat	–do–
M35-1	Tall	125-135	Rabi sorghum tract	–do–
Selection -3	–do–	–do–	–do–	Gujar et al. (1995)
SPH-388	–do–	110-115	Hot semiarid Kharif growing area in the country	Dhope et al., (1995)

Table 3. Drought tolerant varieties of maize in India and adaptability

Variety	Release	Duration	Adaptability	Reference
Aswani	1988	105-110	7-Deccan plateau, Hot semiarid	Rajagopal, 1997
Varuna	1990	110-120	Andhra Pradesh	–do–

Harsha	1988	95-100	–do–	–do–
Ganga-5	1968	90-95	Peninsular India in most states	–do–
Agetic	1982	85-90	All maize growing areas, TN, Karnataka & Rajasthan	–do–
Vikram	1967	96-100	Northern plains	–do–
Navin	1979	86-90	Uttar Pradesh	–do–

Table 4. Drought tolerant varieties in Rice with course grain quality in India.

Variety	Type	Duration (Days)	Adaptability	Reference
Chanda-2	Tall	Medium	Maharashtra, AP, UP & MP	Rajagopal, 1997
Sakoli-6	–do–	–do–	–do–	–do–
Ambika	–do–	110-115	–do–	–do–
N-22	–do–	90-100	–do–	IRRI, Phillipines
N-27	–do–	–do–	–do–	–do–
Tuljapur-1	–do–	105-115	Maharashtra	Rajagopal, 1997
MTU-17	–do–	100-110	Orissa & AP	–do–
Jaldi-7	Dwarf	90-100	Orissa & MP	Anon, 1997
Jaldi-3	–do–	–do–	–do–	–do–
Kalinga-3	–do–	90-110	–do–	Rajagopal, 1997
IR-60080-42	Erect	80-90	UP	–do–
PMK-1	Tall & Erect	110-115	TN	–do–
NLR-145	–	130-140	AP	–do–
Pushkala	–	100-105	AP	Rajagopal (1997)
Erramallelu	–	120-130	Warangal (AP)	–do–
Birsadhan-101	–	80-85	Bihar	Tunwar and Singh,1985
Suhasini	–	100-125	Maharashtra	–do–
Surya	–	120-122	–do–	–do–

4.3 Rice

Rice is a staple food of many countries and 58% of total area (42 m ha) of upland paddy is under rainfed farming in India. Major growing states are West Bengal, UP, Bihar, MP, Orissa and AP. The states like Maharashtra, TN and Karnataka have less than 2 m ha area each under rice (Fig. 2). Drilled paddy is taken in high rainfall (1000 mm) zones of India. Drought tolerant genotypes are identified for different regions and is documented by Rajagopal (1997) from sources of various national research centres. Some varieties are given in Table 4. and it yields 1 t/ha under rainfed conditions. Most drought sensitive stages are anthesis and post anthesis. High temperature during anthesis is detrimental for seed setting. Pollen viability is lost and fertilization is affected. Tip sterility is of common occurrence in rice which leads to empty florets. It is mostly self pollinated crop. Humid climate during reproductive stage is ideal for better seed setting. (Plate 25.2).

Plate 25.2. 1. Upland paddy (Jaldi 3), 2. Spikes of rice, Jaldi-8 (Left) and Jaldi-3 (Right)

4.4 Pearl Millet

The bajra is protogynous, stigma matures first and then the anthers. It is highly hardy crop and is adapted to arid and semi-arid regions of subtropics. Drought tolerance is attributed to maintenance of better harvest index and ability to grain fill under conditions of terminal drought in Hyderabad region. The variety, PCB yields 1.08 t/ha and Mukta. 1.7 t/ha under rainfed conditions. The ICMS 7703 has potential of 1.75 t/ha under drought conditions. The ICMV 221 can tolerate 38°C maximum temperature under Rajasthan conditions. It has suitability in pearl millet tract under varied soil types. Most sensitive stage is vegetative and preanthesis (Rajagopal, 1997).

4.5 Wheat

Wheat is a *rabi* crop grown on residual soil moisture. It is a major food of developing countries. Maximum area is observed in UP (8.70 m ha) followed by MP (3.35 m ha) and Punjab (3.30 m ha Fig. 3). It yields 4 to 4.5 t/ha under no stress condition but yields 1.5 to 2.5 t/ha under stress conditions. This crop has better recovery after stress. Late sown crop suffers due to high temperature at anthesis in central plains. Thermoinsensitive varieties are developed to have satisfactory yields. Leaves show inward rolling during water stress. Leaf tip burn is also observed when night temperatures are lowered down to 10°C at boot stage. Early maturity is seen in late sown crop (Dec. 15th) and normally results in shrivalled grains. The area under purely rainfed condition is very meagre.

Land is kept fallow if wheat is required to be sown and it is sown in late Sept. or first week of October. Most sensitive stage for drought is anthesis and post-anthesis. Ear initiation commences at 50 DAS and it is emerged out at 80-85 DAS. The spiklet initiation is most critical stage when floret number is set. Critical phenological stages are crown root initiation, tillering, boot, flowering, milk stage and dough stage of grain in the spikes. With limited three irrigations (21, 42 and 65 DAS) the crop can be successfully cultivated. Normally it needs six irrigations during above mentioned phenological stages for satisfactory yields. None of the varieties are specifically released as drought tolerant to any area for 4-Northern plains for Hot semi-arid climate (Jobner, Rajasthan). The drought tolerant varieties are presented in Table 5B.

5. CROP GENOTYPES WITH TOLERANCE TO STRESSERS IN LEGUMES AND PULSES

5.1. Genotypes

Pigeonpea, being long duration crop is best suited to rainfed farming with deep vertisol that is generally observed in penninsular India. It blooms 2-3 times in a season and serves as assurance for yield even if short duration dry spell prevails. Soybean area is reached to 12 lakh hectares in last 5 years in Maharashtra (1995-2000) and it has dual purpose property of oil and pulses. It is highly remunerative to dryland farmers who have realised the importance of the crop. Conventionally, greengram and blackgram are grown as intercrops and these crops being of short duration escapes drought but are normally affected by waterlogging during flowering stage. Very rarely it suffers from drought when dry spells continue for 6-10 days. Warm and humid climate is ideal for these crops. Chickpea is cultivated in rabi on residual moisture and 85% of total area is mostly under rainfed farming. It has capability of using atmospheric moisture in cold season due to salt depositis (Rahangdale *et al.,* 1995). Some of the drought tolerant genotypes are listed in Tables 6 & 7 for the use of readers. It can be cultivated on alfisol and vertisol soils. These crops are very much remunerative and fetch superior attractive prices in the market. The productivity of these crops is found slightly low but can respond to better management and offer crop insurance.

These genotypes yield 1.0 to 1.5 t/ha under rainfed conditions with average rainfall of 752-800 mm. The probability of crop insurance is high because of its perennial habit. Out of total area under cultivation in India (3.2 m ha), 96% is grown under dryland conditions.

5.2 Soybean

Out of 26.42 mha area under cultivation in India, 87.7% (22.11 m ha) is cultivated in Madhya Pradesh and 8% in Maharashtra (Fig. 4A). Because of double advantage for oil and pulse, it has become most popular among farmers in short period due to high remuneration with very less attention. The soybean area is increased from 1989 to 1998 (Fig 4B). It yields 0.6 to 1.0 t/ha under drought conditions. The occurrence of yellow mosaic virus (Plate 25.4) on JS-335 (Most popular in Maharashtra and MP) in recent years (1996-2000), the area under soybean may be reduced. Drought results in poor seed setting and small seed size. Sensitive stage for drought is shown to be post anthesis. Another stress experienced by the crop is heavy rains during flowering and fruiting and this affects harvesting. Pod shattering is serious problem if harvesting is delayed by about 6-10 days (Plate 20.5). If harvesting is not timely attended, heavy losses can be faced.

Table 5A. Drought tolerant varieties in Pearl millet and Adaptability (Rajagopal, 1997)

Variety	Type	Duration	Release	Adaptability
ICMS-7703	Tall	85-90	1985	7-Deccan plateau, Hot semiarid tropics
ICTP-8203	–do–	–do–		–do–
MP-124	–do–	– 1988	Maharashtra and AP	
PCB-138	–do–	75-80	1989	Punjab
Mukta	–do–	–do–	1985	Mehboobnagar Dist. of AP
HS-1	–do–	–do–	–	Haryana
X-5	–	90-95	1983	Tamil Nadu
RHR-1	–	85-90	1985	Rainfed area of Haryana
BK-560	–	–do–	1979	Kharif sowing for all India

Table 5B. Drought tolerant varieties of wheat and its adaptability in India (Thermotolerant)

Variety	Type	Duration	Adaptability	Reference
C-306	Tall, erect	120-140 4-	Northern Plains	Rajagopal, 1997
WT-245	–do–	–do–	–do–	–do–
HW-517	–do–	116	Palani hills of Southern zone	Tunwar & Singh, 1985
Mukta (HI-385)	Tall	120-125	Dryland areas of MP, Gujarat, UP and Rajasthan	–do–
Sujata (HI-617)	–do–	135-140	Central zone	–do–
MACS-9	–do–	112-120	Penninsular zone	–do–
AKW-381*	Dwarf	90-95	–do–	Dr. PDKV Diary, 1997
AKW-1071*	Tall	100-105	–do–	–do–
N-59 (Purna)	–do–	115-120	Central zone, and MP	Tunwar & Singh, 1985
NI-5439	–do–	105-110	Maharashtra and MP	–do–

Table 6. Drought tolerant genotypes in pigeonpea (Rajagopal, 1997)

Variety	Release	Type	Duration (days)	Adaptability
ICPL-87	1982	Determinate	120-125	7-Deccan plateau, Hot semiarid
ICPL-87119	1992	Spreading	150-160	Penninsular India

Madhira-66	1993	Semispreading	160-180	Andhra Pradesh
CAUT-87-17	–	–	–	Rajkot (Gujarat)
GAU-82-99	–	–	–	–do–
C-11	–	Spreading	160-180	Maharashtra (Personal Observation by Dhopte)

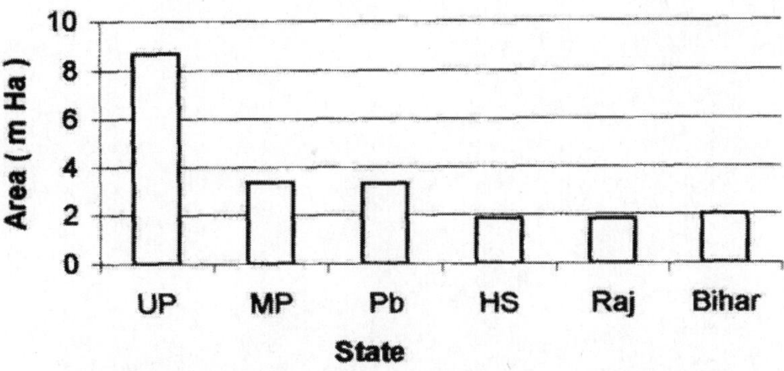

Fig. 3. Area under wheat cultivation in India (1997)

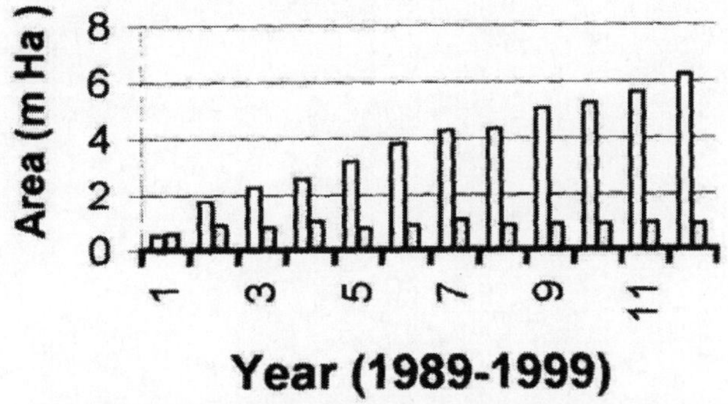

Fig. 4A. Area and Avg. Yield of Soybean in India

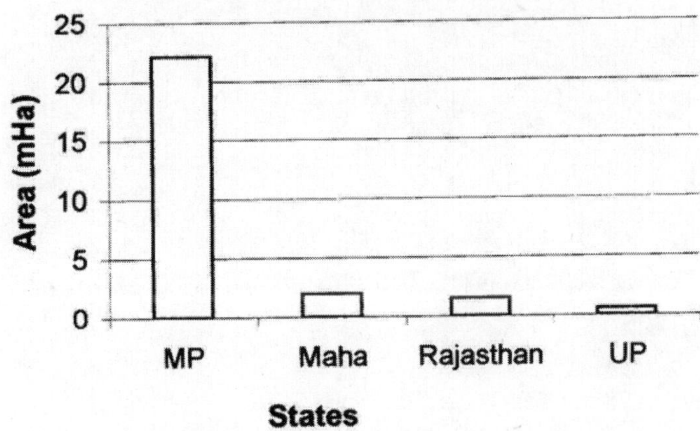

Fig. 4B. Area under cultivation of Soybean in India (1997)

Plate 25.3. 1. Sorghum hybrid CSH-9 (Left) and early (CSH-14) (Right) 2. Pigeonpea (ICPL-87119), 3 Chickpea (Chaffa), 4. Soybean (MACS-13)

Plate 25.4. 1. Soybean virus affected plant (JS-335), 2. Greengram waterlogged tolerant (TAP-7)

Non-shattering types with determinate habit are developed by the breeders. Harvesting, threshing, stacking the produce, non viability of seed are major problems regarding this crop. Timely harvesting is therefore essential to avoid the losses. The germinability is lost if it is cought in rains. A variety with black seed coat is preferred in pennicillin production in pharmaceutical industries. (Plate 25.3).

Flood Tolerance in Soybean

The flooding at flowering did not have much influence on WLTI of cultivars but at vegetative stage, the variety, MACS-58 was better while water logging for 8 days at pod development stage, the cultivar PKV-1 was found tolerant to flooding (Table 7B). Thus, cultivars differed in tolerance to waterlogging. The root growth was not much affected in PKV-1 when flooding occurred at vegetative stage (Table 7C). The flooding at pod development stage, however, had least effect on root growth of cultivar PKV-1. Normally, flooding is experienced during pod development stage in central plains and it is suggested that the cultivar PKV-1 can be a good preposition for such flooding conditions because it showed least damage either at vegetative stage or at pod development stage regarding pod yield and root growth. This is evident on the basis of water logged tolerance index calculated for different stages. Thus, this response supports the concept that selection for one variety is always risky and multiline approach shall be more profitable under rainfed farming (Plate 20.4).

Table 7A. Drought tolerant genotypes in soybean (Ann. Rep. of CRIDA, Hyderabad, 1990-91)

Variety	Type	Duration (Days)	Adaptability	Reference
VMC-3068	Bushy	110-120	9-Northern Hilly, UP	Rajagopal, 1997
Himso-352	–do–	–do–	–do–	–do–
Gujarat-Soya-I (J-2,3,1)	Tall	–do–	Gujarat	–do–
LBS-2	–do–	110-115	Karnataka	Tunwar & Singh, 1985
MACS-58	–do–	–do–	MP	Rajagopal, 1997
MACS-13	–do–	–do–	–do–	–do–
JS-81-355	–do–	–do–	Maharashtra	–do–
Gaurao	–do–	–do–	MP	–do–

5.3 Greengram and Blackgram (Mung and Urd)

This crop is of very short duration and very rarely it is subjected to drought. Dry spells of 6-10 days during early stage may affect the crop. Under drought situation, the tolerant varieties (Table 8) yield 0.7 to 0.8 t/ha as against 1.1 t/ha under normal rainfall. Main problem is waterlogging and it is mostly harvested when rainfall is highest in the season (Aug.). Waterlogging reduces yield by 18-40% depending upon duration of flooding (4-8 days) at 40 DAS. The varieties tolerant to waterlogging are also identified by Dr. PDKV, Akola (Table 9).

5.4. Chickpea

All India hectarage is 6.9 m ha (1997) and 85% is under rainfed conditions. Central plains have 2.3 m ha followed by UP (1.27 m ha) and Rajasthan (1.2 m ha Fig. 5A). This is a crop of rabi season and duration ranges from 75-170 days. Crop is mainly grown on residual moisture with protective irrigation on vertisol and alfisol soils. The most sensitive stage to drought is pre anthesis and pod development. Few chickpea varieties were screened for drought tolerance by Rahangdale *et al.* (1994). using zero irrigation and irrigation (3) at Dr. PDKV, Akola and they found AKG-46 to be tolerant. Data are given in table 10. along with yield stability.

Plate 25.5. 1. Peanut drought tolerant cultivar JL-24, 2. Peanut field cv. Girnar.

6. DROUGHT TOLERANT CROP GENOTYPES IN OILSEEDS

6.1 : Groundnut

Asia is a larger producer of peanut. It is major oilseed crop occupying over 8 m ha of which 85% is rained. India ranks first in area in world but avg. yields are low (1 t/ha in rainy season and 2 t/ha in post rainy season). Major growing states are Gujarat, AP, TN, Karnataka and Maharashtra among 80% of total area. Agrotechnology has to be developed for higher yields (Kak, 1995). The most sensitive stages for drought are flowering, pegging and pod development. Tolerant varieties yield 1.0 to 1.5 t/ha under drought situation. Six genotypes were evaluated for drought tolerance (Dhopte *et al.,* 1991) following the method of sowing crop on two sites (Deep vertisol and shallow vertisol) during kharif season. Drought tolerant index was calculated dividing yield under shallow soil with yield under deep soil. Data are presented in Table 12.

Field methods are employed for knowing the tolerance (Plate 25.5).

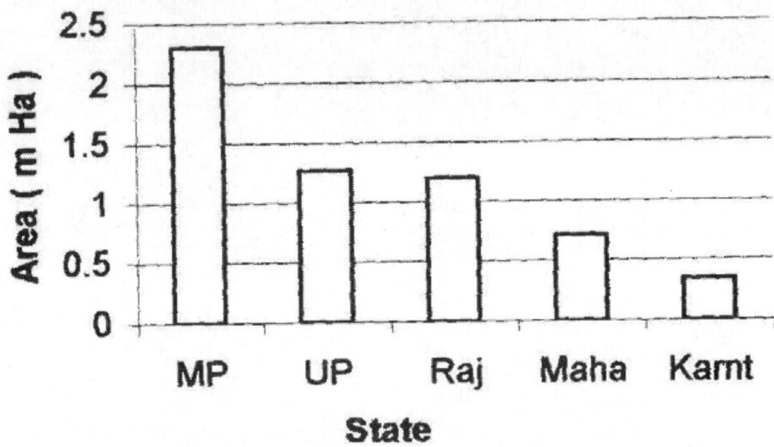

Fig. 5A. Area under chickpea in India (1997)

Table 7B. Evaluation of waterlogged tolerant varieties in soybean (Sorte et al., 1996) Seed yield/plant (g)

Cultivars Treat.	Vegetative stage		Flowering stage		Pod development Stage	
	Control	WL-8	Control	WL-8	Control	WL-8
JS80-21	24.3	17.4	25.1	17.5	26.0	19.9
WLTI	–	0.72	–	0.70*	–	0.77
MACS-58	22.6	18.1	21.3	14.3	22.0	21.0
WLTI	–	0.80*	–	0.67*	–	0.81
PKV-1	22.2	15.2	18.3	11.3	23.1	21.8
WLTI	–	0.68	–	0.62	–	0.94*

WL-8 = Waterlogging for 8 days, WLTI=Waterlogged Tolerant Index *Waterlogged tolerant cultivar

Table 7C. Effect of waterlogging on Root volume (cm^3) in soybean varieties (Sorte *et al.,* 1996)

Cultivars Treat.	Vegetative stage		Flowering stage		Pod development Stage	
	Control	WL-8	Control	WL-8	Control	WL-8
JS80-21	16.9	8.8	28.3	21.8	28.4	19.5
WLTI	–	0.52	–	0.77*	–	0.69

MACS-58	15.1	9.5	25.6	17.5	25.9	13.9
WLTI	–	0.63	–	0.68	–	0.54
PKV-1	11.1	7.7	25.3	17.8	23.0	17.5
WLTI	–	0.68*	–	0.70	–	0.76*

* Tolerant to flooding for 8 days. WLTI=Waterlogged Tolerant Index

Table 8. Drought tolerant genotypes in greengram (Mung) Ann. Report of CRIDA, Hyderabad

Variety	Duration	Release	Adaptability	Reference
Greengram (Mung)				
Madhira -295 (MGG-295)	65-75	1993	7-Deccan plateau Andhra Pradesh	Rajagopal, 1997
KGG-373	–do–	–do–	Kovilpatti	–do–
PDM-54	–do–	–do–	–do–	–do–
G-860612	–do–	–do–	–do–	–do–
MCH-26	70-75	–do–	Hissar	–do–
Blackgram (Urd)				
APK-1	65-70	–do–	Tamilnadu	–do–
KBG-512	–do–	–do–	–do–	–do–
KB-51	–do–	–do–	–do–	–do–

Table 9. Waterlogged tolerant varieties in greengram (Mung) Ann. Report of Dr. PDKV, Akola, 1998-99.

Variety	Type	Duration	Adaptability	Reference
AKM-8803	40-45 cm	65-70	Maharashtra	Anon. 1998-99
TAP-7	–do–	–do–	–do–	-do-

Table 10. Screening of chickpea varieties for drought tolerance (Rahangade *et al.,* 1994)

	Seed Yield (Kg/ha)		
Genotype	Control (IR)	Stressed	Yield stability (%)
Chaffa*	36.9	32.0	86.7
AKG-1	30.0	26.2	87.3
AKG-46*	35.7	31.4	87.9
AKG-50	36.3	31.4	86.5
AKG-52	36.4	27.2	74.7
AKG-58	37.2	29.6	79.6

*Tolerant to drought

Table 11. Drought tolerant varieties in chickpea

Variety	Type	Duration (days)	Adaptability	Reference
ICC-4958	Semispreading	–	AP	Rajagopal, 1997
AKG-1	Dwarf	103	Maharashtra	Rahangdale et al., 1994 and 1995
AKG-46	–do–	130-150	–do–	–do–
ICCV-2	Spreading	75	–do–	Rajagopal, 1997
Vijay (Phule G-81-1)	–do–	105-110	–do–	–do–
Chaffa	–do–	Medium	–do–	–do–
H-208	–do–	–do–	Hissar	–do–
ICCC-37	–do–	–do–	–do–	–do–
GNG-16	–do–	150-170	Rajasthan	Tunwar & Singh, 1985
Pant-G-114	–do–	140-155	All tracts	–do–
Pusa-209	–do–	140-160	Rajasthan, UP & Bihar	–do–
Pusa-212	–do–	130-150	Gujarat, MP	–do–
S-26	–do–	135-140	Punjab	–do–

Table 12. Evaluation of peanut genotypes for drought tolerance (Dhopte *et al.,* 1991)

Genotype	Yield (Kg/ha)		Drought tolerance
	Deep soil	Shallow soil	Efficiency (%)
Girnar-1*	806.7	473.3	59
TAG-24	800.0	206.7	26
JL-24*	842.2	566.7	67
SB-XI	824.4	35.6	43
ICGS-11	704.4	166.7	24
AK-9003	764.4	364.4	48

* Drought tolerant

6.2 Mustard (*Brasica juncea*)

It is a crop of rabi season and adapted to cold climate but it is shown that it can grow equally well under high temperature zone of central plains with one or two irrigations. The range of ideal temperature is from 35.2ºC (Max.) & 5.2ºC (Min.). It yields 0.5 to 1.5 t/ha during drought conditions. Most sensitive stage is post anthesis. In central plains of high night temperature, Chitanwis *et al.* (1989) reported that the variety *Kranti* has shown high drought tolerance (Table 14) with marginal yields of 70-90 Kg/ha under no irrigation. Even with limited irrigation it may not yield more than 500 Kg/ha.

The mustard is mostly grown with protective irrigation in central plains and it is not a crop of dryland but following varieties may optimise the yield when irrigation facility is limited. The spray of gibberllic acid at 50 ppm con. can provide compensation for cold environment and yield levels can be maximised under high night temperature. This crop is widely grown in UP, Hissar, Himachal Pradesh, Punjab where night temperature during rabi season ranges from 3-8°C. Drought tolerant varieties are given in Table 15 along with its suitability.

Some varieties like Bio-109, are evolved through Biotechnology (Tissue culture) and they have found better under central plains of high temperature. Cold temperature requirement is highly necessary for this crop and yield potential is as high as 1.5 to 2 t/ha. Due to global warming, the night temperatures are 12°C in central plains. In some areas, it goes down to 3°C in Nasik district.

6.3 Sunflower

This crop is grown round the year because of its day neutral behaviour. Area under cultivation is 2.37 m ha in India (1997). Karnataka shares 1.29 m ha and 0.52 and 0.35 m ha are found in Maharashtra and Karnataka (Fig. 5). Negligible area is found in other states. It is very much remunerative crop and has become most popular in penninsular India due to its multiple cropping advantage. It is grown on alfisol and vertisol mainly in kharif, rabi and summer where limited irrigation facilities are available. Lift and canal irrigation are main sources of irrigation. Most sensitive stage for drought is flowering and grain filling. It can tolerate the day temperature up to 40°C in south India. The fatty acid synthesis is highly variable due to season. Drought tolerance of 14 varieties were evaluated by Jaipurkar *et al.* (1994) and some of them are mentioned in Table 16. for the use of readers (Plate 25.6).

The farmers are advised to grow short duration variety for better crop assurance. Hybrids are available but it is grown well only under high inputs and water. If rains are scanty, the seed filling is incomplete in the capitulum and yield is drastically affected. Total leaf area/plant is highest at flowering and it is a crop of high water requirement. The LAI ranges from 4-6 and hence ET is higher than other crops. Some farmers should organise the planting of sunflower in nearby locality to avoid bird damage. Isolated patches are very much affected and yields are reduced. The above mentioned varieties can tolerate terminal drought or high temperature at post anthesis.

Table 13. Drought tolerant genotypes in groundnut and suitability for the region in India

Genotype	Type	Duration	Adaptability	Reference
ICGV-87885	Erect	90-100	7-Deccan plateau	Rajagopal, 1997
ICGV-86707	Spanish Bunch	110-120	3-Deccan plateau AP	-do-
JL-24	Spreading	90-109	6-Deccan plateau Maharashtra	Dhopte et al. 1992
Girnar	–do–	90-100	-do-	-do-
TVG-4	Erect bunch	100-105	-do-	Rajagopal, 1997
TVG-5	–do–	-do-	-do-	-do-
Tirupati-1	–do–	100	8-Deccan plateau (AP)	-do-
BG-2	Virgina	115-120	Bihar	Tunwar & Singh, 1985
DH-3-30	Spanish	100-120	Karnataka	-do-
JSG-21	–	-	Gujarat	Rajagopal, 1997

Field methods are employed for knowning the tolerance (Plate 25.5).

Table 14. Screening in mustard varieties for drought tolerance (Chitanwis *et al.*, 1989)

Cultivar	Seed Yield (Kg/ha)		Drought tolerance Efficiency (%)
	Irrigated	Rainfed	
Waruna	201	74	36.8
Pusa bold	303	74	24.4
Pusa Barani	239	83	34.7
Kranti*	196	81	41.3*
Krishna	190	19	10.0
Seeta	211	74	35.1
Local	197	13	6.6

Table 15. Drought tolerant varieties in Musratd *(Brasica juncea)*

Variety	Type	Duration	Adaptability	Reference
RH-819	Tall spreading	145	2-Western plains Hot semiarid Hissar	Rajagopal, 1997
RH-30	-do-	140-142	–do–	–do–
Vaibhav (RK-1418)	-do-	120-125	UP and MP	Tunwar & Singh, 1985
Kranti	-do-	120-125	Maharashtra	Chitanwis *et al.*, 1989

Table 16. Evaluation of sunflower genotypes for drought tolerance using biochemical parameters. (Jaipurkar *et al.*, 1994)

Variety/Hybrids	Yield (t/ha)	Leaf Proline content (mg/g)
Surya	1.50	13.22
Morden	1.26	14.66
MSFH-8	1.52	14.95
PKVSH-28**	2.27	14.95
PKVSH-32*	2.07	16.68

* Hybrids evolved at Dr. PDKV, Akola (MS) were drought tolerant on basis of proline accumulation and yield under limited irrigations

Table 17. Drought tolerant varieties of sunflower in India

Variety	Release	Type	Duration	Adaptability	Reference
KBSH-1	–	Hybrid	Medium	8-Deccan plateau Karnataka	Rajagopal, 1997
62-B	1992	Semidwarf	115-120	Andhra Pradesh	–do–
APSH II	1957	Hybrid	95-100	Karnataka	–do–
Morden	1979	Semidwarf	80-90	–do–	Tunwar & Singh, 1985
ESH-2	–	–do–	–	–do–	CRIDA, Hyderabad
PKVSH-28	–	Hybrid	115-120	Maharashtra	Jaipurkar et al., 1994

Table 18. Evaluation of Drought tolerant genotypes in safflower (Bharad *et al.,* 1995)

Variety	Yield (kg/ha)	Chlorophyll stability index	Capsule/plant	Osmotic potential (Bars)
JLSF-88	1981	0.06	44.9	-22.8
JLSF_228	1910	0.08	50.0	-21.9
AKS-96	1710	0.08	46.5	-17.1
AKS-146*	2290*	0.16	46.5	-21.2
CTV-212	1280	0.18	54.3	-22.7
A-1*	1520	0.09	55.5	-23.6
Bhima*	1970	0.09	41.9	-26.5

* Drought tolerant genotypes 14 Genotypes were evaluated. Avg. Rainfall = 800 mm, Tolerant genotype had high Osmotic potential

Table 19. Drought tolerant genotypes in safflower and adaptability in India

Variety	Release	Type	Duration	Adaptability	Reference
A-300	1957	Semi spreading	125-135	8-Deccan plateau Hot semiarid tropics	Rajagopal, 1997
A-1	1969	–do–	–do–	Karnataka	–do–
S-144	1976	–do–	–do–	Raichur,	Bihar, Tunwar & Singh, 1985
Bhima	1982	–do–	–do–	Maharashtra	–do–
AKS-146	–	–do–	–do–	-do-	Bharad, et al., 1995

Table 20. Screening of cotton genotypes for drought tolerance under filed conditions (Yield. g/plant). After Dhopte and Livera (1989)

Variety	1986 (Rigorous stress)			1987 (Moderate stress)		
	Normal	Stressed	DTE (%)	Normal	Stressed	DTE (%)
AKH-081	1013	44.8	4.4	335	237	85**(2)
LRA-5166	122	67.8	5.6**(3)	440	265	60
DHY-286	633	32.7	5.2	470	360	77
L-147	647	56.8	8.8**(1)	221	290	131**(1)
AKH-4*	797	30.6	3.8	640	490	77
AKA-5*	847	27.3	3.2	693	355	51
AKA-8401	1450	29.5	2.0	820	620	76
SRT-1	773	39.3	5.1	328	250	77
Red Pig.	377	29.0	7.7**(2)	200	160	80

DTE = Drought Tolerant Efficiency; * Tolerant; ** Highly tolerant

Plate 25.6. 1. Sunflower drought tolerant hybrid, PKVSH-28, 2. Safflower cv. A1, 2. Cotton cv. L-147 (Left), Red pigmented (Right)

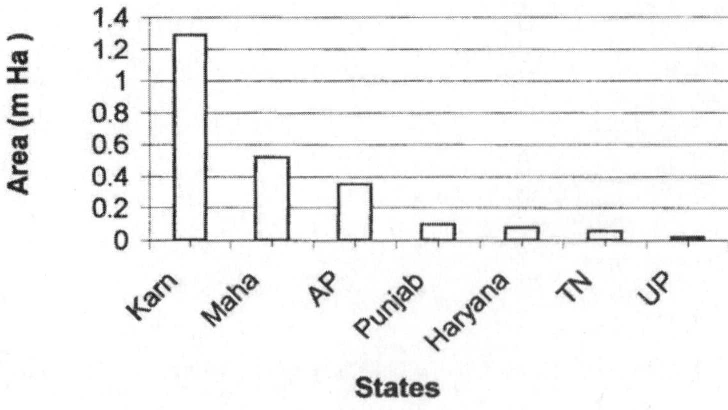

Fig. 5. Area under sunflower, Total 2.37 m Ha (1997)

6.4 Safflower

It is highly drought tolerant crop grown in rainfed conditions in rabi season. Maximum area under cultivation is in Maharashtra (0.58 m ha. followed by Karnataka 0.17 m ha) and Andhra Pradesh (0.02 m ha) out of 0.80 m ha in India. It grown well on residual moisture with average rainfall of 750 mm. Tolerance to day temperature is 41°C. High rainfall during flowering is detrimental for crop growth and is affected by wilting. The incidence of aphids at vegetative and preflowering stage lowers the productivity. The drought and wilt tolerant genotypes are evolved (Table 19). It yields 0.3 to 0.4 t/ha under drought condition. Bharad *et al.* (1995) screened 14 genotypes for drought tolerance under high temperature zone of *Vidarbha* on the basis of biochemcial characters (Table 18). Crop is harvested with harvesters in parts of Marathwada and Vidarbha in Maharashtra. Sensitive stages for drought are branching and pre-flowering. Winter rains some time occur during peak flowering and crop is severely affected. Thus, there is a need to evolve the variety for waterlogging (Plate 25.6).

6.5 Cotton

Cotton area is largest in central plains under rainfed farming. The area under various states is mentioned in Fig. 6. It is a duration crop of 5-6 months and it can tolerate few dry spells during entire growth period. Long dry spells of 8-15 days at flowering affect crop growth and yield is reduced. Extensive work on drought tolerance is done at Dr. PDKV, Akola (MS) to identify drought tolerant varieties (Dhopte and Livera, 1989; Dhopte *et al.,* 1988; Dhopte *et al.*, 1995, and Dhopte *et al.,* 1999) under AICCIP project during 1984-1989 Drought tolerance indices have been identified for arboreum and hirsutum cottons (Dhopte *et al.,* 1989).

Cotton is grown on 91 m ha land in India (1999) and 70% is under rainfed farming. Maharashtra has maximum area (3.2 m ha) followed by Gujarat (1.69) and Andhra Pradesh (1.0 m ha) (Fig. 6). Intermittent dry spells usually occur and growth is affected. Root growth is not affected due to water stress at flowering (Dhopte *et al.,* 1988). Deep vertisols in Maharashtra have maximum water holding capacity and receives average rainfall around 800-900 mm. Soil is subjected to frequent wetting and drying and therefore results in cracks which leads to trimming of lateral roots. This creates stress situation unless proper intercultural operations are not adopted.

Various genotypes were tested for drought tolerance under field conditions during 1985-1988 and variety L-147 and AKH-081 were found to be drought tolerant. These varieties are characterised by deep root system with more laterals, leaf hairyness, low stomatal index and conductance, good fruit setting and high harvest index. It yields 0.8 to 1.0 t/ha under drought conditions. Most sensitive stages for drought are seedling, flowering and boll development. Dhopte *et al.* (1995) studied ideal plant spacing and plant type for rainfed farming and they reported 60 x 15 cm (2,24000 plant/ha) with 2.6 LAI to be more suitable for dryland conditions. The broad lobed hairy leaf character can assure yield of 1.05 t/ha with narrow spacing of 60x15 cm. High LAI harvests more solar energy and accounts for higher yield as compared to Okra, red pigmented and pilose leaf types. The Okra, red pigmented, and pilose leaf types yield 31,30 and 13% less than normal broad lobed hairy type plant. The red pigmented plant types however, have higher water use efficiency.

Drought predisposes the photo-synthetic system to photoinhibition. Photosystem II is agravated by the presence of Mg. The water stress markedly increases the extent of photoinhibition of photo-synthesis in mesophyll chloroplast in spinatch and pea (Raghvendra, 1991). The selection for drought tolerance was based on rigorous testing under totally dryland with normal operations and fertilizer doses. It was not irrigated even acute symptoms of wilting was observed. The plants were raised on deep vertisol and very shallow vertisol. The yield levels are very poor because of thie situation. The trial was designed only for the purpose of screening and the data are reliable for DTE. The drought index was calculated dividing yield under stress with yield under normal conditions. The CV ranged from 15-20%. Only those genotypes having inherent potential of productivity under extreme drought condition could be identified and these are presented in Table 20 for the use of readers (Plate 25.6).

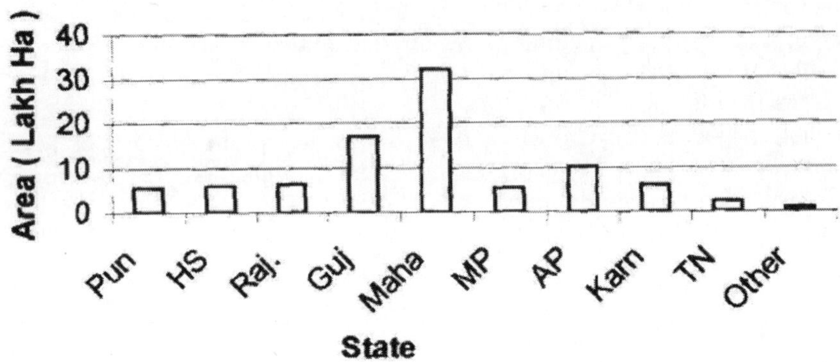

Fig. 6. Area under cotton in India Total 91.12 Lakh Ha. (1998-99).

Dryland farming is not only subjected to water stress but it is also oftenly subjected to low light intensity during vegetative stage. This phenological stage parallels with August when sky is covered with clouds and the light intensity is reduced to half than normal. (Bright sunshine intensity is 1000 W/m^2, cloudy sky had 500 W/m^2 and this is below light compensation point). Growth in cotton is very slow because of deficit solar radiation. The soils are flooded and soil aeration is highly reduced. Some genotypes are also screened for photo efficiency and it is reported in Table 21.

The effect of waterlogging during flowering for 6 days was studied by Dhopte *et al.* (1990). Three varieties and two hybrids were tested under pot culture experiment during kharif, 1985. Flooded plants increased stomatal conductance and TR without affecting leaf temperature. Flooding exhibited chlorosis in terminal leaves. The DMP was increased due to flooding. The plants were recovered in 11 days. The H-4 plants however recovered in 7 days and it was found tolerant for waterlogging during flowering indicating its responsiveness to more water. The AKH-081 was found tolerant to drought but it was susceptible to flooding at flowering. Thus, it can respond to low soil moisture but is affected by saturated soil conditions. The responses of genotypes are varied for different environments. The application of CCC was shown to relieve the effect of waterlogging in gram (Krishnamurthy *et al.,* 1983). The application of nitrogen could ameliorate the effect of waterlogging in cotton (Hodgeson and Leod, 1987). The crop pigeonpea is also highly sensitive to waterlogging (Singh 1986). Bennet and Albrecht (1984) reported that flooding of soybean for 14 days had little effects and plants recovered quickly after termination of flooding. It is reviewed by Yoshida and Tadano (1988).

7. STRATEGIC APPROACHES FOR DRYLAND FARMING

7.1 Know your soil and conserve rain water

Soil variation regarding texture, depth, organic matter and physical properties is evident in every field. Some piece may be eroded and may contain less quantity of clay and silt. Slopey land may be subjected to contour bunding and vitiver, subabhul, or grass (Citronella) should be grown for better rain water conservation. According to slope and topography, watershed should be planned to conserve runoff losses in a corner of field. If catchment area is more, the reservoir may be bigger. Expertise may be obtained for this planning. This will help increase of soil water table and slow release of soil moisture for crop growth. Evaporation losses from pond may be checked with appropriate treatment. Consultation of an expert is needed. Hiwase *et al.* (1994) suggested use of polymulches in cash crops for water efficiency.

7.2. Prepare the compost pit

A pit may be dug at suitable place to dump all field organic matters (Weeds, leucina loppings, subabhul, post harvest plant residue, glyrisydia leaves, Dhaincha leaves, cow dung, urine and animal waste etc.). Trichograma culture can be used for early decomposition. A pit of 2 x 5 x 1 m size can be comfortable for operation. The layers of SSP, DAP, soil every after 60 cm layer may be added in a pit to have better farm yard manure. This, though old technique, is not adopted by the farmer. It can minimise the cost of manures and fertilisers and maintain the physical property of soil high for better absorption of nutrients. There is no substitute for organic matter compost for better soil health. Very low organic matter in our soil is major constraint in productivity. If this is achieved, 20% yield can be assured. Our soils contain <1% organic matter while in developed countries, it is more than 2%. The practice of burning farm waste should be immediately stopped. It can be easily converted in precious manures and a dramatic change in productivity can be experienced. Broadcasting the FYM should be avoided. It should be applied at the hill of plant. The doses of chemical fertilizers can be lowered down without affecting the yield. This is throughly discussed in chapter 6.

Table 21. Screening of cotton genotypes for photoefficiency in India (Dhopte and Livera, 1989)

Genotype	Light Intensity (%)			Photoefficiency Index	
	100	80	400	Mod. stress	Sev. Stress
L-147	78.9	46.9	13.8	0.51	0.17
AKH-081	80.6	59.0	21.5	0.73*(3)	0.26
DHY-286	42.9	23.7	10.3	0.55	0.24
LRA-5166	171.1	120.2	11.5	0.70	0.07
MCU-5	50.3	39.8	18.3	0.79*(2)	0.36*(2)
AKA-8401**	93.7	25.9	10.8	0.28	0.12
AKH-4**	92.1	48.8	6.1	0.53	0.66*(1)
AKA-5**	106.3	39.9	8.9	0.38	0.28*(3)
SRT-1	123.2	37.5	26.0	0.30	0.21
Red Pig.	51.5	47.4	29.6	0.92*(1)	0.57

** Arboreum types adopted to dryland conditions. Tents were specially designed for the study.

1. Grown without cover = 100% Light intensity, Roughly 600 w/m^2 in cloudy sky.
2. Covered with thick cloth tent (1 x 2 x 1 m tent) = 80%
3. Covered with muslin cloth tent (1 x 2 x 1 m) = 40% MS = Moderate stress
4. Light compensation point is 400 w/m^2 SS = Severe stress In all 14 genotypes were evaluated during investigations.

Note : Some genotypes are only available in germplasm and it may not be available for commercial cultivation eg. L-147 as drought tolerant, AKH-081 can be available at Dr. PDKV, Akola (MS) 444 104

Table 22. Screening of cotton varieties for waterlogged tolerance (Dhopte *et al.*, 1990)

Variety		Leaf area (dm^2)	DMP (g/plant)	Stom. Cond (cm/s)	Seed cotton yield (g/plant)	FTI
AKH-081	C	29.7	37.9	1.11	36.6	1.06
	WL	33.6	44.8	1.21	38.8	–
SRT 1	C	26.6	34.1	0.56	19.3	1.40
	WL	30.8	41.4	0.99	27.0	–
AHH-468	C	28.4	46.0	0.61	18.7	1.47*
	WL	49.5	66.6	0.99	27.4	–
H-4	C	21.2	28.3	0.72	21.2	1.82*
	WL	43.3	49.9	0.82	38.7	–

* Waterlogged tolerant during flowering (70-76 days) C= Control; WL = Waterlogged,
FTI = Flood Tolerance Index Plants were waterlogged continuously for 6 days and then relieved from stress

Table 23. Drought tolerant varieties in cotton and its adaptability

Variety	Release	Type	Duration	Adaptability	Reference
AKA-5	1981	Erect	180-190	Maharashtra	Dhopte and Livera (1989)
L-147	1970	Erect	190-200	–do–	–do–
AKH-081	1985	Dwarf	150-160	–do–	–do–
Paiyur-1	–	Erect	145-150	Tamil nadu	Rajagoptal, 1997
LK-861 (Amravati)	1993	Erect	160-170	AP	–do–
NHH-390	1991	Hybrid	150-160	–do–	–do–
Eknath	1956	Erect	170-180	Gujarat	Tunwar and Singh 1985
Bharati (MCU-6)	1970	Hybrid	150-160	TN	–do–

7.3. Know rainfall pattern

Try to get the data of rainfall from nearest meteorology lab or Panchayat office for last ten years. General information can be obtained from elders but it is explained in *Nakshtras.* The horoscope also predicts the likely pattern of rainfall. The modern information technology that is based on satellite communication is highly reliable. Further planning of crops can be decided on the basis of predictions. Conventional cropping pattern is adopted by the farmers to avoid the risk of failure because this information is passed from father to son. Any new concept is hard to digest. Its adoption on experimental basis can be suggested. Agrotechniques are demonstrated by the village level workers and green revolution that occurred in past is the result of this devoted efforts. Farmers can adopt variety, hybrids and technology if they are monitarily benefited. Wider adaptability of hybrids in most of the crops by farmers is certainly the result of untiring efforts made at government level during last two decades.

7.4. Selection of variety/Hybrid of crops

Planning and selection of a variety should be based on rainfall pattern. Agroclimatic zones are identified and crops are normally fixed. What remains to be selected is the variety. Generally, farmers sow the variety which is adopted to the region. Selection for one variety in a crop is not strategic approach under rainfed farming because if climate is not favourable, the loss is total. A model approach should be to select 2-3 varieties with different duration (Early, medium or late). If rains are erratic, the selection should be accomodative of climate. If one fails, two are assured. If two fail, atleast one is assured. The chances of failure is 1/3 of total. Gross failure can be certainly avoided with this strategy. Rainfall may some time be favourable for all the varieties or hybrids and higher productivity can be assured. For erratic rains, random selection can be profitable rather sowing one variety/hybrid and experiencing gross failure. This tendency should be totally stopped to have assurance in productivity. The illeterate and poor farmer adopts the strategy of a progressive farmer without considering his own limitations. Better inputs and favourable climate result higher productivity. First input in agriculture is best seed for higher productivity. Drought tolerant varieties can definitely offer assurance to a farmer and gross failure can be avoided.

Agri. price policy also decides the cropping pattern. A crop with high price is preferred first for higher remuneration without consideration of rainfall and climate limitations. This approach is suicidal. Better crop management is not generally attempted by common farmer and therefore leads to frustration. Selection of a variety should necessarily be dependant on soil, climate, rainfall and nutrient management affordable by the farmer. It should not be based upon other's selection (Neighbour). Sequence and relay cropping are profitable for better remuneration. Multistoried approach of intercropping can have best utilization of land resources for higher rewards with minimum inputs.

7.5. Seedling raising, Seed Pilleting and Nursery management

The onset of monsoon is late once in three years and such situation results in late sowing of specified crop. Late sowing reduces yield by 30-50% because appropriate climate (Light, RH and Temperature) is not obtained by the crop for full growth and expression of potential. Nursery seedling raising in polybags or seedling trays can avoid this precarious situation (Plate 25.7). A container of 28.5 x 52 cm can provide 288 seedlings. Different sizes of containers are available in the market. Thousands of seedlings can be made available in a small area with limited water under polynet house conditions. Any bomboo shade can be used to protect the seedlings from high heat of May. (42- 45ºC). The company "COMPACKT" (Address : COMPACKT, 16A, South end Road, Bangalore - 566 004 (Karnataka), India) has been meeting the needs of professional growers since 1984 providing a full time top quality growing containers at affordable prices. The design of seedling trays not only meets biophysical specifications but also technical ease of handling and automation. These trays have following advantages:

- Easy to fill manually and mechanically
- Easy to wash and disinfect after use.
- Easy to stack and transport and for storage
- Robust root system is developed
- Trays are easy to handle
- Plants are easy to extract from trays
- Uniformity in crop stand can be maintained
- Plants can be hardened to our need

Plate 25.7. 1. Seedling trays for seedlings, 2. Seedling production on large scale in polynet house shelter, 3. Tray with seedlings raised in net house, (Cheap technology) 4. Transport system of seedlings. 5. Glass house shades for raising seedlings under ideal conditions (Costly technology)

Due to high cost of seeds, every seed must emerge into a plant. Hybrid seeds of vegetables and flowers are very costly and using seedling trays can be a proper solution to have plants from all seeds under intensive care. Tray system provides ideal conditions of media and water for better growth and it has thereforem become profitable in crop production particularly, tomato, chilli, brinjal, cabbage and cauliflower, It can suitably be used for Kharif crops like cotton. pigeonpea, maize, sorghum, soybean, sesame, castor, sunflower, greengram and blackgram This system is being used for tissue culture plants and flowering plants. Costly seeds of watermelon, squash, cucumber, gourd and papaya can be grown for healthy saplings. Some hybrid seeds are very costly and if sown in field may not germinate under abnormal conditions. Desiccation of seedling by low soil moisture is common. Germinated seedling may die if rains are not received 20 days after germination. To avoid this mortality, seedling trays are highly beneficial. Some initial cost is incurred but can last longer for 2-3 seasons. This strategy should not be adopted for enitre field but needs to be reserved for abnormal conditions to insure losses in productivity. Fine tilth is never prepared for sowing and large seed quantity goes waste in cracks and crevices and results in poor crop stand (See Chapter 7).

For growth and vigour and timely sowing, this practice can be followed. It is feasible on large scale. Paddy, chilli and brinjal transplantings are done routinely with nursery saplings. A. "CROW BAR" techniques is followed in cotton and *in situ* seedlings are raised in last week of May with very limited water. The productivity is 2.5 times higher than the normal. In paddy, the seedlings are transplanted after stagnating water in field. The germinated seedlings are broadcasted in field under *Dapok* system. Seed pilleting is also a phenomena recently followed for better vigour and growth. Ideal crop stand is necessary for required productivity. This is explained in Chapter 7. This can be accomplished with seedling raising in trays. If agriculture has to be made sustainable, these key factors should not be ignored.

In China, maximum productivity of rice is 6 t/ha as against 1-2 t/ha in rest of Asia. This is achieved with use of tissue culture saplings of hybrids obtained through systematic approach by Govt. agencies. Saplings are made available to the farmer. Large area (55%) is brought under cultivation of hybrids following these packages. The *"Dapok"* system of seedling raising has been attempted in 1970's in rice. These examples are adequate to understand the impact of agrotechnology in agriculture. Nothing is impossible. All that is needed is determination, devotion, commitments and nationalism to accomplish the target. Unfortunately, very few people have these qualities in developing countries. The president of United States, John F, Kennedy said :

"Don't think what country can do for you but think what you can do for the country"
– John F. Kennedy

Plate 25.8. A. Black polynet cover to protect seedlings from high solar intensity in off season and B. storage of Poly pipes and other irrigation pipes during off season after through cleaning. Maintenance is essential for longevity.

Plate 25.9. A. Drought tolerant cultivar of pigeonpea, ICPL-87119 at flowering and B. Mustard cultivar adopted to high temperature zone in *Vidarbha* region

This essence of living is only observed in countries like Japan, UK, Vietnam, Korea, Germany, USA, Russia, China and Australia. A farmer is a food producer not only for himself but for the nation as well. Right policies, ideal infrastructure, determination and sacrifice are lostessence of life in people of third world and therefore are found behind the time schedule. Several factors are responsible for such ill-mentality and this is not the forum to discuss all those here. A Govt. agencies can think of subsidies on this system for farmer for yield assurance. Farmers are given grant-in-aids when crop is failure. These funds can be diverted for such kind of innovative fruitful ideas which may result in dramatic change.

7.6. Water management

The old practice of furrow irrigation needs to be stopped and drip or sprinkler irrigation should be adopted. Though it is promoted by Govt. during last 10 years offering subsidies to the grower, the results are not in focus. Huge funds are allotted but are gone in vain without success. We face lower aqua layer in soil due to heavy exploitation under furrow irrigation. Sprinklers are best solution for farmers having meagre sources of irrigation to save 40-60% of water with ideal productivity. Ground water is now understood as Govt. property and it has to be lawfully handled. Maharashtra state has made legislation and notification is issued. If this is not executed properly, one person out of three person shall face drinking water problem by 2025 AD in Asia as per World Bank Report.

Indiscriminate use of ground water need to be stopped on top priority. Water under ground is fluid and flows as per geological conditions. Crops are saved and productivity is assured with minimum expenses. The conserved water in watersheds can be used by lift irrigation under critical crop stage of drought. The *in situ* rain water conservation is the only solution to solve this acute problem of drought and it helps to raise the water table in nearby area. The crop life is saved under adverse conditions. Probability of crop failure can be minimised. This is well illustrated in Chapter 2 by Dr. Bhaskar. The readers are suggested to make use of this valuable information. It can be well followed with minimum efforts. A farmer should follow these practices to reap the advantages of assured yields in dryland farming.

Various strategic approaches are already dealt by different scientists of eminence in various chapters. The phase wise follow up of corrective measures for improvement in dryland productivity should enable average farmer to achieve assured yields in rainfed farming. Biginers may have some problems but constant follow up shall bring new hopes of better living and no more time shall be needed to achieve desired goal of improved productivity. If China can do, we are no way inferior to them. Success may be achieved trying over and over until we accomplish the targeted goal of new millennium. Let us wake up and join hands together for building nation and maintaining nation self sufficient in food always.

REFERENCES

Anonymous (1998-99) Ann. Report of Plant Physiology and Biochemisty. Res. Rev. Comm. Dr. PDKV, Akola (MS), India.

Bennet, J.M. and S.L. Albrecht (1984). Drought and flooding effect on N_2 fixation, water relations and diffusive resistance of soybeans. Agro. J. 76 (5) : 730-740.

Bharad, G.M; M.A. Japurkar; A.M. Dhopte and G.N. Satpute (1995). Evaluation of sunflower genotypes for physiological variability and stability under rainfed conditions. Ann. Plant. Physiol. 9 (1) : 34-37.

Chitanvis, A.G.; G.R. Fulzele; A.M. Dhopte and P.B. Ghavghave (1989). Performance of mustard cultivars and their degree of drought tolerance efficiency under high temperature zone of Vidarbha. Ann. Plant Physiol. 3 (1) 84-87.

Eastin, J.D; P.K. Verma; A.M. Dhopte; M.D. Witt; D.R. Kreig and J.L. Hatfield (1988). Stress screening and sorghum improvement. Proc. Int. Conf. Dryland Agri.; Bushland, Texas, USA.

Dhopte, A.M.; S.L. Rahangdale and S.L. Jamdar. (1987). Physiological evaluation of forty exotic and wild sorghum in relation to stomatal factors involved in drought resistance. Ann. Plant. Physiol. 1 (2) : 143-150.

Dhopte, A.M; N.N. Potkile and N.R. Shastri (1988). Pattern of root growth and growth analysis in *hirsutum* and *arboreum* cotton in response to short term water stress imposed during reproduction. Ann. Plant. Physiol. 2 (1) : 166- 175.

Dhopte, A.M; M.A. Tayyab; S.L. Jamdar and M.B. Wadodkar (1989). Value of leaf characteristics in selection for drought tolerance in upland cotton. J. Cotton Res. and Develop. 3 : 52-63.

Dhopte, A.M. and S.L. Jamdar, (1991). Association of physiological parameters with cotton productivity under dryland conditions. J. Cotton Res. and Develop. 5 : 135-146.

Dhopte, A.M; S.D. Ramteke and S.G. Thote (1991). Water use and drought tolerance efficiency of peanut genotypes in relation to canopy growth under field conditions. Ann. Plant. Physiol. 5 : 202-212.

Dhopte, A.M; V.B. Shekar; R.S, Wadhokar and N.R. Potdukhe (1994). Field evaluation of sorghum genotypes for yield stability under soil moisture deficit during panicle development. PKV Res. J. 18 : 181-185.

Dhopte, A.M.; S.L. Jamdar and S.L. Rahangdale (1990). Physiological responses of hirsutum cottons. to six days root submersion during flowering in vertisol. Ann. Plant Physiol. 4 (2) : 159-160.

Dhopte, A.M. and M. Livera - M (1989). Few tests for various stresses in crop plants. In. Useful Techniques for Plant Scientists. Pub. Forum. for plant physiologist, Akola (MS). India.

Dhopte, A.M.; S.B. Amarshettiwar and R.L. Kalane (1995). Interaction of leaf types with leaf area index on physiological responses in hirsutum cotton isolines under dryland conditions. J. Cotton Res. and Develop. 9(2) : 209-215.

Dhopte, A.M., V.B. Shekar; R.Pandrangi; S.K. Wankhade and S.L. Rahangdale (1995). Variation in leaf amino acid contents in drought tolerant and susceptible genotypes and relationship of plant traits with yield stability in grain sorghum. Ann. Plant Physiol. 9(1) : 28-33.

Dhopte, A.M. and J.D. Eastin (1998-99). Influence of temperature stress on reproductive physiology of field crops. In. Agro's Ann. Rev. Plant Physiol. by S.S. Purohit. Vol. 4 pp. 148-190.

Dhopte A.M. and S.B. Lall (1981). Contribution of upper and lower leaves in cob development of maize. P.K.V. Res. J. 5 : 113-116.

Gujar, A.R., M.R. Shinde and S.D. Ugale (1995). Selection 3- A rainfed rabi sorghum for shallow soils. Ann. Plant Physiol. 9(1) : 124-129.

Hodgeson, A.S. and Mac Leod (1997). Effect of foliar applied N fertilizer on cotton waterlogged in eraking grey clay. Aust. J. Agri. Sci. 38 (4) : 681-688.

J.N.Uttarwar and C.K. Tijare (1994). Effect of drip irrigation and polymulches on leaf physiology and yield of *Solanum melongena*. Ann. Plant Physiol. 8 (2) : 140-145.

Jaipurkar, M.A.; A.M. Dhopte and G.N. Satpute (1994). Evaluation of sunflower genotypes for physiological variability and stability under rainfed conditions. Ann. Plant Physiol. 8(2) : 129-134.

Jordan. W.R. and C.Y. Sullivan (1982). Reaction and resistance of grain sorghum to heat and drought. Sorghum in Eighties ICRISAT (Ed), Patancheru, (AP) India.

Kak, R.D. (1995). Groundnut in India. Directorate of Publ. and Information, New Delhi. India

Krishnamoorthy, H.N; G.L. Goswami and Jay Dayal. (1983). Effect of CCC and B-Nine on water logged gram. Indian J. Plant Physiol. 26 (3) : 258-263.

Rahangdale, S.L., A.M. Dhopte and K.B. Wanjari (1994). Evaluation of chickpea genotypes for yield stability under soil moisture deficit. Ann. Plant. Physiol. 8 (2) : 179-184.

Rahangdale, S.L., A.M. Dhopte and K.B. Wanjari (1995). Alteration in osmoregulation, dew formation and salt deposits in leaves of chickpea genotypes under soil moisture stress. Ann. Plant Physiol. 9(1) : 17-23.

Raghvendra, A.S. (1991). Photosynthesis under water stress. Nat. Symp. Recent Adv. Drought Res. Kottayam (India). Dec. 10-13 pp. 6 (Lead Paper).

Rajgopal, V. (1997). A handbook of drought tolerant crop species of India (Ed). Publ. Soc. Plant Physiol. and Biochem. New Delhi - 11012, India.

Shekar, V.B.; R.S. Wadhokar, SK, Jillani and N.R. Potdukhe (1993). A new grain sorghum hybrid SPH-468 (CSH-14) with better performance and high adaptability at international level. Ann. Plant Physiol. 7 (1) : 136-139.

Singh, D.N. (1986). Influence of waterlogging on growth and yielding ability of early maturing pigeonpea genotypes. Ann. Report. Divn of Pulse Res. (ICAR) Kanpur. pp. 87, India.

Sorte, N.V., R.D. Deotale, J.H. Meshram and M.A. Chanekar (1996). Tolerance of soybean cultivars to waterlogging at various growth stages. J. Soils and Crops 6(1) : 68-72.

Tunwar, N.S. and S.V. Singh (1985). Handbook on cultivars. Publ. Central Seed. Comm. Dept. of Agri. and Coop. Ministry of Agri. and Rural Develop. New Delhi. India.

Yoshida, S. and T. Tadano. (1988). Adaptation of plants to submerged soils. In. Crop tolerance to suboptimal land conditions. By G.A. Fung (Ed). Amer. Soc. Agro. Madison, WI, USA pp. 233-254.

Chapter - 26

APPLICATIONS OF REMOTE SENSING FOR DRYLAND AGRICULTURE

Shyam. L. Deotale* and Sapna. S. Deotale**

* Assistant Professor, Extension Education Section, College of Agriculture, Nagpur - 440 001 (Maharashtra) and Ex-Scientific Associate, Maharashtra Remote Sensing Applications Centre, VNIT Campus South Ambazari Road, Nagpur (Maharashtra)

** Associate Scientist, Maharashtra Remote Sensing Applications Centre, VNIT Campus, South Ambazari Road, Nagpur (Maharashtra)

1.0. INTRODUCTION

Agriculture plays a crucial role in ensuring food security and accounts for a significant share of, India's Gross Domestic Product (GDP). It engages almost two-thirds of the workforce in gainful employment. Several industries such as sugar, textiles, jute, food and milk processing etc. depend on agricultural production for their requirement of raw materials. On account of its close linkages with other economic sectors, agricultural growth has a multiplier effect on the entire economy. In India out of the total cultivated area of 143 m. ha the area under drylands is about 85 m. ha, which comes to 60% and dryland agriculture accounts for nearly two-thirds of total cropped area and generates nearly half of the total value of agricultural output. The major factors those determine the dryland farming land use are landforms, hydrology, climate and soils etc.

Dry lands, besides being water deficient, are characterized by high evaporation rates, exceptionally high day temperature during summer, low humidity and high run off and soil erosion. The soils of such areas are often found to be saline and low in fertility. As water is the most important factor of crop production, inadequacy and uncertainty of rainfall often cause partial or complete failure of the crops which leads to periods of scarcities and famines consequently the life of both, human beings and cattle in such areas become difficult sustain and insecure too.

Growing of crops entirely under rainfed conditions is known as dryland agriculture. Depending on the amount of rainfall received, dryland agriculture can be grouped into three categories:

a. Dry farming: It is cultivation of crops in regions with annual rainfall less than 750 mm. Crop failures is most common due to prolonged dry spells during the crop period. These are arid regions with a growing season (period of adequate soil moisture) less than 75 days. Moisture conservation practices are necessary for crop production.

b. Dryland farming: It is cultivation of crops in regions with annual rainfall more than 750 mm. In spite of prolonged dry spells crop failure is relatively less frequent. These are semi arid tracts with a growing period between 75 and 120 days. Moisture conservation practices are necessary for crop production. However, adequate drainage is required especially for vertisols or black soils.

c. Rainfed farming: It is crop production in regions with annual rainfall more than 1150 mm. Crops are not subjected to soil moisture stress during the crop period. Emphasis is often on disposal of excess water. These are humid regions with growing period more than 120 days.

United Nations Economic and Social Commission for Asia and the Pacific distinguished dryland agriculture mainly into two categories: dryland and rainfed farming. The distinguishing features of these two types of farming are given below.

Table 1. Dryland vs. rainfed farming

Constituent	Dryland farming	Rainfed farming
Rainfall (mm) >800	<800	>800
Moisture availability to the crop	Shortage	Enough
Growing season (days)	<200	>200
Growing regions	Arid and semiarid as well as uplands of sub-humid and humid regions	Humid and sub-humid regions
Cropping system	Single crop or intercropping	Intercropping or double cropping
Constraints	Wind and water erosion	Water erosion

1.1. Importance of Dry farming in Indian Agriculture

About 70% of rural population lives in dry farming areas and their livelihood depend on success or failure of the crops. Dryland agriculture plays a distinct role in Indian agriculture occupying 60% of cultivated area and supports 40% of human population and 60 % livestock population. The contribution (production) of rainfed agriculture in India is about 42 percent of the total food grain, 75 per cent of oilseeds, 90 per cent of pulses and about 70 per cent of cotton. The contribution of drylands will have to be more than 60 percent to provide adequate food to 1210 million people. (as per the 2011 census). Hence tremendous efforts both in the development and research fronts are essential to achieve this target. More than 90 per cent of the area under sorghum, groundnut, and pulses is rainfed. In case of maize and chickpea, 82 to 85 per cent area is rainfed. Even 78 percent of cotton area is rainfed. In case of rapeseed/mustard, about 65.8 per cent of the area is rainfed. Interestingly, but not surprisingly, 61.7, 44.0, and 35.0 percent area under rice, barley and wheat, respectively, is rainfed. At present, 3 ha of dryland crop produce cereal grain equivalent to that produced in one ha irrigated crop. With limited scope for increasing the area under plough, only option left is to increase the productivity with the modern technology and inputs, since the per capita land availability which was 0.28 ha in 1990 had to decline 0.19 ha in 2010.

Thus, while the strategic importance of dry-land agriculture cannot be ignored, its state in India is very precarious, to say the least. The multifaceted problems associated with dryland and rain-fed agriculture severely impoverish the rural communities and populations dependent on them. Specifically, several factors exacerbate the problems of vulnerability and rural poverty among such dry-land farmers: a) poor/low soil fertility; b) inappropriate soil/water management practices; c) lack of improved/suitable varieties to counter uncertainty and diversity in climatic conditions; d) devastating pest/disease attacks that ruins the crops and/or drastically reduce the yields; e) declining land-man ratio; and f) resource-poor rural communities (that are unable to meet even minimum standards of health, nutrition and hygiene).

The vulnerability and poverty of these farmers dependent on rain-fed agriculture is further accentuated by low cropping intensity, high cost of cultivation, poor adoption of modern technology, low productivity and uncertainty in output, inadequate public investment in water/ market/ related infrastructure, lack of appropriate and quality institutional credit (post-harvest, especially), decreasing returns, increasing indebtedness and related factors. The net result of all of this is that these resource-poor farmers become poorer and vulnerable, with each passing day.

As it is observed that these dryland/rainfed areas are often affected by periodic droughts, floods, soil erosion, instability of crop production, drinking water scarcity etc. In order to improve food production in such areas development of resources area judicious management of available natural resources are imperative. There exists potential to enhance the crop production to meet the growing population. There are several ways of augmenting the food production such as: increasing physical area of cultivation, expansion of irrigation facilities, stepping up production of unit area etc.

Development of irrigation potential is though indicative of the progress in agriculture, the experiences from several irrigation projects in the country with respect to crop production is not very encouraging. Faulty irrigation practices and improper drainage in several command areas have lead to wastage of precious water, water logging, salinity and alkalinity etc. The productivity of grains already showed a plateau in irrigated agriculture due to problems related to nutrient exhaustion, salinity build up and raising water table. Therefore, the challenges of the present millennium would be to produce more from drylands while ensuring conservation of existing resources. Hence, new strategies would have to be evolved which would make the fragile dryland ecosystems more productive as well as sustainable. In order to achieve evergreen revolution, we have to make grey areas (drylands) as green through latest technological innovations. Drylands offer good scope for development of agro-forestry, social forestry, horti-sylvi-pasture and such other similar systems which will not only supply food, fuel to the village people and fodder to the cattle but forms a suitable vegetative cover for ecological maintenance.

There exists potential to enhance the crop production to meet the growing population with judicious conservation and management of land and water resources coupled with adoption of improved crop production technology. It may not be difficult to achieve the set target, provided, the information related to available resources, the nature, extent, spatial distribution etc. are available for planning and management of the natural resources.

Scientific study of dryland farming was initiated by the Government of India in 1923. Early research focused more on improving crop yields. The important practices included: (i) bunding, to conserve soil and water; (ii) deep ploughing, once in three years for better intake and storage of water; (iii) use of farmyard manure, to supply plant nutrients; (iv) use of a low seed rate; and (v) inter-cultivation for weed and evaporation control etc. These practices resulted in, approximately about 15 to 20 percent increase over the base yields.

2.0. DRY LAND/ RAINFED AGRICULTURAL SCENARIO

Dry land agriculture is largely practiced in arid, semi-arid and sub- humid regions in the country. With about 68 per cent of rural population (Kumar et al., 2009), these regions are also home to 81 per cent of rural poor (Rao *et al.*, 2005). In rainfed regions, the annual precipitation is lower than the evapo-transpiration demand particularly in arid and dry semi-arid zones. Coarse cereals (85%), pulses (83%), oilseeds (70%), and cotton (65%) are the predominant rainfed crops grown in India (CRIDA, 2007). Dry Land agriculture is considered as a gamble with monsoon while soils in these regions are not only thirsty but also hungry. Rainfed agricultural scenario is influenced by both bio-physical and socio-economic factors and their interaction.

2.1 Bio-physical Parameters

1. **Rainfall:** India is blessed with a monsoon regime that is more or less regular in its cycle of onset, spread and withdrawal over the country. The southwest monsoon experienced by India is a part of the larger Asian monsoon circulation and provides a major portion of the annual rainfall of the country. India on an annual basis receives about 4 x 10^3 km^3 (400 M ha m) of precipitation (FAI, 1994) out of the 5 x 10^5 km^3 precipitation received globally (Lal, 1994). India's share, thus, is about one percent of the global precipitation. The rainwater availability in different monsoon periods indicates major contribution from the southwest monsoon (74%) as compared to northeast monsoon (10%). Of late, there is an evidence of erratic behaviour of rainfall like increased frequency of high intensity rains, reduced number of rainy days and shift in rainfall. Most future scenarios indicate that there may not be a significant change in the total rainfall in the country in the years to come but the distribution could become more erratic with high intensity storms followed by long dry spells. The challenge is to harvest the monsoon rains during excess rainy events and re-use efficiently during dry spells for improving the yield and income per drop of rainwater.
2. **Soils:** Alfisols and Vertisols of peninsular India and Aridisols of extremely dry climates are the principal soil orders in dry areas, although Entisols and Inceptisols also occur, especially in topo-sequences. Vertisols, Alfisols, Entisols and associated soils are the major soil orders extensively are under rainfed crops. About 30% of dryland area is covered by Alfisols and associated soils while 35% by Vertisols and associated soils having vertic properties and 10% by Entisols of the alluvial soil regions (Virmani et al., 1991). These soils are extremely prone to rainwater induced erosion. As per the harmonized dataset, 73.3 M ha of arable land is affected by water erosion resulting in soil loss of more than 10t/ ha/ annum (Anonymous 2008). Apart from soil and nutrient losses, poor nutrient management is leading to multi-nutrient deficiency of essential nutrients, which is posing a threat to rainfed agriculture. Due to increased rainfall intensities, land degradation is likely to increase in future. The gap between nutrient supply and demand is likely to widen further by 2030 to 50 M tons of NPK. With phosphate and potash raw materials reserves depleting world wide, the availability of chemical fertilizers may become increasingly difficult. Biomass recycling and legumes in the crop rotation are to be encouraged in order to sustain soil fertility in rainfed areas.
3. **Livestock:** India has world's largest livestock population accounting for about 55 and 16 percent of the world's buffalo and cattle populations, respectively. Rainfed areas account for two-thirds of total livestock population. India ranks second in goats (124.5m out of world's 764.5 m) while third in sheep (59m out of world's 1028.6m) but of late there is a significant change in the livestock composition. There is a steep fall in bullock's population. India is ranked number one in milk production too. Most of the milk production comes from the rainfed regions where livestock population is large. Currently there is deficit in demand and supply of dry and green fodder which stands at 12% and 63% respectively. This gap of demand and supply of fodder is likely to be further widened with increase in demand in milk and meat products and change in cropping pattern, like cotton at the cost of coarse cereal. Hence there is a need to devote at least 10% of the net sown area for fodder production.

4. **Production systems:** Historically rainfed farmers practice high diversity in cropping systems with livestock integration which is an inbuilt risk management strategy. The cropping patterns have evolved based on the rainfall, length of the growing season and soil types. However, due to changed consumer preferences and market demand, farmers are now rapidly shifting to crops and cropping patterns which are more remunerative.
5. The change in cropping pattern will have implications on the resource use. Continuous mono-cropping increases vulnerability of farmers to weather risks, depletes soil fertility, ground water and leads to build up of pests and diseases. This issue has to be dealt both through technology and policy. For example, we have to evolve management practices for farmers' choice of remunerative crops without degradation of the natural resource base and also define agro-ecological zones where such cropping systems can be adopted sustainably. Simultaneously, need based policy incentives are required to encourage farmers adopt agro-ecology- compatible cropping systems so that the farmers' income is maintained and the natural resource base of the country is not degraded.
6. **Biodiversi**ty: India's natural ecosystems have immense richness of agricultural biodiversity including diversity in crops, plants, wild plants, livestock, aquatic species, below ground biota, microbes, etc. India has ranked seventh in the world in number of species contributing to agriculture and animal husbandry. Rainfed areas have a distinct superiority over irrigated in terms of diversity in flora and fauna. The entire forest area (6,92,027 Km^2 i.e. 21.05% of geographical area as per Forest Report 2011 Govt. of India) is rich in biodiversity and it is mostly rainfed. Apart from livestock and fishes, the contribution of forest areas to livelihoods of poor is substantial in rainfed regions. The forest area is major source of water and fodder. Climate change is likely to impact biodiversity in rainfed regions significantly. This impact will be complex with differential responses between interacting species, such as, crop, pest and pathogen. Certain genotypes may be preferred over others. This may lead to loss of useful gene pool and hence *ex-situ conservation* in identifying hot spots becomes important.

2.2 Socio-economic Parameters

Rainfed areas show poor socio-economic status such as limited irrigation facility, population pressure, fragmentation of land holdings, tenancy farming, low investment capacity, low productivity of crops and income, credit, pricing policy, marketing and wage rates are the major socio-economic issues that impact rainfed agriculture. A major concern is the continuous decline in land holding size which is a major hindrance in mechanization. Though significant proportion of the rural population is expected to move to cities by 2030, still substantial numbers will remain in rural areas.

Uncertain rainfall, growing water scarcity, rising input costs and stagnant output prices have rendered many of the current crop and livestock enterprises in rainfed areas non-remunerative. Further, as the labour force is not finding enough work in the villages, they are migrating to long distances in search of work; as rainfed areas are mostly mono-cropped and provide work for four to five months only. The interventions thus need to be designed in such a way that farmers and rural youth are retained in farming round the year, may be through alternate livelihood options. One of the ways is agro-forestry coupled with livestock on watershed mode besides other enterprises which could even be outside agriculture, meeting the local needs.

Under these circumstances, it becomes not only necessary but is essential to adopt proper land and water management practices, to conserve basic resources and make best use of the conserved resources not only to increase in agriculture production but also to develop, exploit and utilize the resources already developed in a most scientific manner. Although, agriculture research has achieved many milestones towards sustainable development, the expected results are still not achieved so far, it seems that the research/technology has not reached the farming community despite several attempts at various levels. Unless and until the rural masses are not convinced properly then one cannot expect the desired development of the rural India. The extension machinery still has to work hard to achieve the desired rural development.

Optimal utilization of available natural resources is of utmost importance. Now a day the world is experiencing erratic climate behavior and rainfed agriculture in such circumstances is a big challenge in front of the farming community.

3.0. GLOBAL CLIMATE CHANGE

The climate of earth is dynamic which is also promoting the evolution of various living forms and also changing the structure and chemical composition of the atmosphere. Over the past few decades, acceleration in the human-induced changes in the climate of the earth has become the focus of scientific and social scrutiny. The gaseous composition of the atmosphere has undergone a significant change mainly through increased industrial emissions, fossil fuel combustion, widespread deforestation and burning of biomass as well as changes in land use and land management practices. These anthropogenic activities have resulted in an increased emission of harmful gases, e.g., carbon dioxide, carbon monoxide (CO), methane etc. popularly known as the 'greenhouse gases'.

Global warming, in turn, leads to regional changes in climate related parameters such as rainfall, soil moisture, and sea level. The extensive and frequent occurrence of climatic extremes such as droughts, heat and floods in the last decade in many parts of the world may be the fallout of this process. The sea level has risen by 10-20 cm with regional variations. Similarly, snow cover is also believed to be gradually decreasing.

3.1 Impact of climate change

More floods, frequent droughts and forest fires, decrease in agricultural and aquacultural productivity, displacement of coastal dwellers by sea level rise and intense tropical cyclones, and the degradation of mangroves may be some of the likely consequences of climate change in Asia. Such consequences could considerably affect the food supply and access through their direct and indirect effects on crops, soils, livestock, fisheries and pests. On the other hand increase in temperature, can reduce crop duration, increase crop respiration rates, affect the equilibrium between crops and pests, and hasten nutrient mineralization in soils, decrease fertilizer use efficiencies, and increase evapo-transpiration among others.

Uncertainty in precipitation is causing droughts and floods have been responsible for many famines, rural poverty and migration despite development of impressive irrigation potentials. These environmental changes, particularly temperature increase and sea level rise, could also affect fisheries directly and indirectly through changes in the availability of feed. Similarly, by increased temperatures the changes in fodder and water availability may affect production of meat and milk. Indirectly, there may be considerable impact on agricultural land use due to snow melt, availability of irrigation, frequency and intensity of inter- and intra-seasonal droughts and floods, soil organic matter transformations, soil erosion, decline in arable areas (due to submergence of coastal lands), and availability of energy.

All these changes would have tremendous adverse impact on agricultural production and, hence, on the food security of any region. Several important socio-economic determinants of food supply such as government policies, capital availability, prices and returns, infrastructure, land reforms, and inter- and intra-national trades are also expected to be influenced and altered by environmental changes.

3.2 Need for indigenous efforts in quantifying greenhouse gas (GHG) emission

Alarmed by the possible impact of global climatic changes on the quality of life of human beings, as well as animals; there has been a serious concern all over the world in understanding the processes and developing strategies to mitigate the negative impact. It has been realized by all that a change in environmental quality affects all aspects of life. Efforts are, therefore, needed to reduce the emission of greenhouse gases, which are mainly responsible for atmospheric warming. The solution to such environmental issues is closely linked with the issues of socio-economic problems. Each country or region addresses the problem with their own individualistic perception on what is best suited for their economic development and not necessarily what is best for the health of the world as a whole. Remote Sensing plays a significant role in monitoring the climate changes by temporal analysis of various earth and environmental resources.

Another very important factor which influences dryland agriculture is weather. The timely and accurate information regarding weather is very crucial for proper crop production planning process.

4.0. WEATHER FORECASTING AND AGRICULTURE PLANNING

Weather plays an important role in agricultural production. It has a profound influence on crop growth and yields; on the incidence of pests and diseases; on water needs; and on fertilizer requirements. This is due to differences in nutrient mobilization as a result of water stresses, as well as the timeliness and effectiveness of preventive measures and cultural operations with crops. Weather aberrations may cause physical damage to crops and soil erosion. The quality of crop produce during movement from field to storage and transport to market depends on weather. Bad weather may affect the quality of produce during transport, viability and vigour of seeds and planting material during storage. Thus, there is no aspect of crop culture that is immune to the impact of weather. Weather factors contribute to optimal crop growth, development and yield. Weather also influences on the incidence and spread of pests and diseases.

Weather factors show significant spatial and temporal variations at a given place and time. In strategic planning of crops and cropping practices, current and historical climatic data plays a crucial role.

4.1 Usefulness of weather forecasts

Occurrences of erratic weather are beyond human control. It is possible, however, to adapt to or mitigate the effects of adverse weather if a forecast of the expected weather can be obtained in time. The farmers have been keen to know in advance the likely weather situations for crop operations from time immemorial.

The first step in familiarizing weather forecasters with the weather warning requirements of farmers is the preparation of crop guides for forecasters, which should give the times of occurrence and duration of developmental phases from sowing to harvest of major crops in the regions of their forecast interest, and specify the types of weather phenomena for which weather warnings and forecasts are to be issued in the different crop phases. Such guides can be used by forecasters to prepare calendars of agricultural weather warnings with a breakdown by periods and regions.

In the crop guide for forecasters, normal values of important weather elements in the crop season should also be given for the short period adopted at the national level for agro-meteorological work; this guide should also be made available to the farming community so that any farmer will know immediately what the normal features of weather will be for a given crop and season at his location.

4.2 Need of Weather Forecast

Following points should be kept in mind while thinking of importance of weather forecast in agriculture crop production-

- Basically weather has many social and economic impacts in a place.
- Among different factors that influence crop production, weather plays a decisive role.
- Rainfall is the most important factor among the required forecast, which decides the crop production in a region and ultimately the country's economy.
- The planning for moisture conservation under weak monsoon conditions and for flood relief under strong monsoon condition is important in a region.
- A reliable weather forecasting when disseminated appropriately will pave way for the effective sustainability.
- One can minimize the crop damage, which may be caused directly or indirectly by unfavourable weather.
- The recurring crop losses can be minimized if reliable forecast on incidence of pests and diseases is given timely, based on weather variables.
- It helps in holding the food grain prices in check through buffer stock operations.
- This means that in good monsoon years when prices fall, the government may step in and buy while in bad years when price tend to rise, it may unload a part of what it had purchased.
- Judicious use of water can be planned in a region depending up on the weather forecast.

4.3 Weather calendar

In order to provide the farmers with an efficient weather service, it is essential that the weather forecaster should be familiar with the crops that are grown in a particular agro-climatic zone. The type of forewarnings to be given depends on the stages of the crop. In case of farmers, they should become familiar with weather bulletins and learn how to interpret. To meet the above requirement, the detailed information collected from the agricultural departments has been condensed by the IMD and presented in a pictorial form known as crop weather calendar.

The weather calendar has three parts as follows -

a. Bottom part

b. Middle part

c. Top part

Bottom part provides the activities related to crop or information related to phenological stages of the crop and the months.

Middle part gives information regarding normal weather condition required for active crop growth. It is divided into different sections according to rainfall, rainy days, minimum temperature, maximum temperature, pan evaporation and sunshine hours.

Top part gives information related to the weather abnormalities or to take precautionary measures. Top part is divided into different sections according to dry spell length, high wind, heavy rainfall and cloudy weather etc.

In short weather forecasting plays a very significant role in farm production. Farmers can be immensely benefited by the weather forecasting for timely decision making regarding crop production. It is very important to provide timely weather information to the farmers so as to minimise the inherent risks in dryland agriculture.

The concept of sustainable development is very essential to increase the agricultural production in general and dryland farming in particular; better results can be achieved by understanding resources limitations as well as their potential. For this purpose, information about the existing resources are necessary for planning and making certain decision making policy. Currently, these are being achieved through conventional techniques which are inadequate, time consuming and many a times inaccurate too. In order to speed up the process, latest survey and monitoring technique such as 'Remote Sensing technique plays a crucial role in acquiring the desired information quickly, economically and accurately.

Remote sensing is largely concerned with the measurement of electromagnetic energy which is reflected, scattered and emitted by the object receiving and returning the energy from the sun. Different objects of earth surfaces such as soil, vegetation, water, buildings etc., return different amounts of energy in different spectral bands of Electro Magnetic Spectrum (EMS). Detecting and identifying the differences enables identification of objects of interest under study. Besides considering the spectral variations, spatial & temporal information is also utilized in the remote sensing data interpretation and analysis for agriculture planning purposes.

5.0. REMOTE SENSING - OPTICAL AND INFRARED

By definition "Remote Sensing" is the science and art of acquiring information about an object, area or phenomena from a distance by a device or instrument. The device or instrument can be sensor like camera or scanner which are placed on the aircraft or space platform for surveying the object on the surface of earth. The data obtained can be in the form of a photograph or as a multispectral data both in image and digital format.

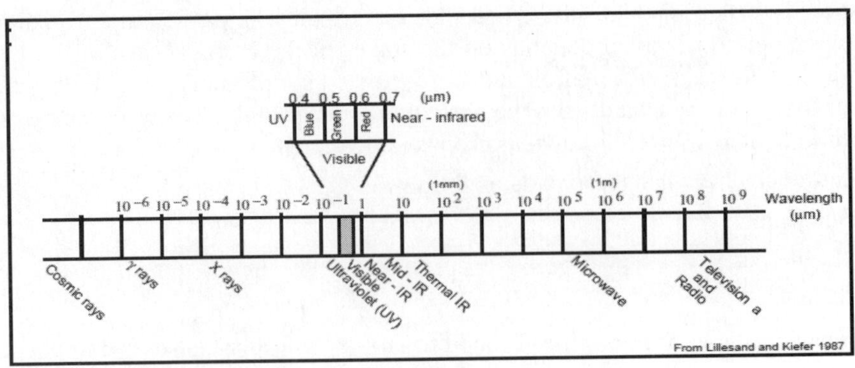

Figure 1. Electromagnetic Spectrum

Remote sensing is done in several bands of electromagnetic spectrum. This technique ranges through the spectrum from the very short wavelengths at which gamma rays are emitted to the comparatively long wavelengths at which radar operates. In this way, much more information about an area can be secured and compared to that obtained with conventional photography which is limited to the visible band of the electromagnetic spectrum.

5.1 Signatures in Remote Sensing

The underlying principle, on which remote sensing technique is developed, is that all objects on the surface of earth have characteristic spectral signatures. A spectral signature of an object or a ground surface feature is a set of values for the reflectance or radiance of the feature, each value corresponding to the reflectance or radiance arranged over a different, well defined wavelength interval. In other words, it is distinctive set of distinguishable characteristics.

The knowledge of spectral signatures is essential for exploiting the potential of remote sensing techniques. This knowledge enables one to identify and classify the objects. Interpretation of satellite data/imagery is carried out visually or using digital techniques.

5.2 Spectral Signatures

The spectral signatures produced by wavelength-dependent absorption provide the key in discriminating different objects in images of reflected solar energy. The property used to quantify these spectral signatures is called ***spectral reflectance***; the ratio of reflected energy to incident energy as a function of wavelength. The spectral reflectance of different objects can be measured in the laboratory or in the field, providing reference data that can be used to interpret images.

5.3 Passive Sensing and Active Sensing

The sun provides a very convenient source of energy for remote sensing. The sun's energy is either reflected, as it is for visible wavelengths, or absorbed and then re-emitted, as it is for thermal infrared wavelengths. Remote sensing systems which measure energy that is naturally available are called passive sensors.

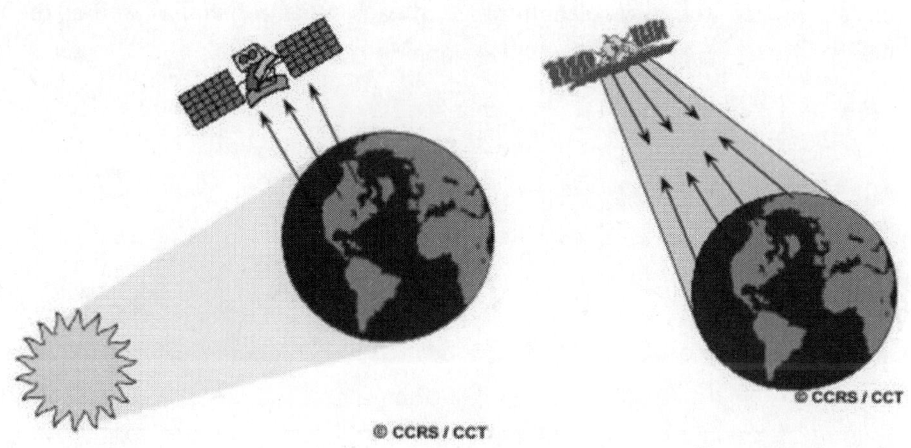

a. ***Passive sensors*** can only be used to detect energy when the naturally occurring energy is available. For all reflected energy, this can only take place during the time when the sun is illuminating the Earth. There is no reflected energy available from the sun at night. Energy that is naturally emitted (such as thermal infrared) can be detected in a day or at night, as long as the amount of energy is large enough to be recorded.

b. ***Active sensors,*** on the other hand, provide their own energy source for illumination. The sensor emits radiation which is directed toward the target to be investigated. The radiation reflected from that target is detected and measured by the sensor. Following are the advantages of active Remote Sensing; The ability to obtain measurements anytime, regardless of the time of day or season and they can be used for examining wavelengths that are not sufficiently provided by the sun, such as microwaves, or to better control the way a target is illuminated. However, active systems require the generation of a fairly large amount of energy to adequately illuminate targets.

5.4 Properties of Remotely Sensed Data

The property of remotely sensed data most critical to their utility is their resolution. It refers to an imaging system's capability of resolving two adjacent features or phenomena. There are four types of resolution for remote sensing imagery: spatial, spectral, radiometric and temporal.

1. Spatial Resolution

Also called ground sampling distance, spatial resolution of imagery refers to its ability to distinguish two spatially adjacent objects on the ground. Spatial resolution is the equivalent of the spatial dimension of scanning on the ground during image acquisition.

2. Spectral Resolution

Spectral Resolution refers to the ability of a remote sensing system to differentiate the subtle difference in reflectance of the same ground object at different wavelengths. Spectral resolution is determined by the number of spectral bands used to record spectrally split radiative energy received from a target. It is related to the wavelength range of each spectral band, as well as the wavelength range of all bands.

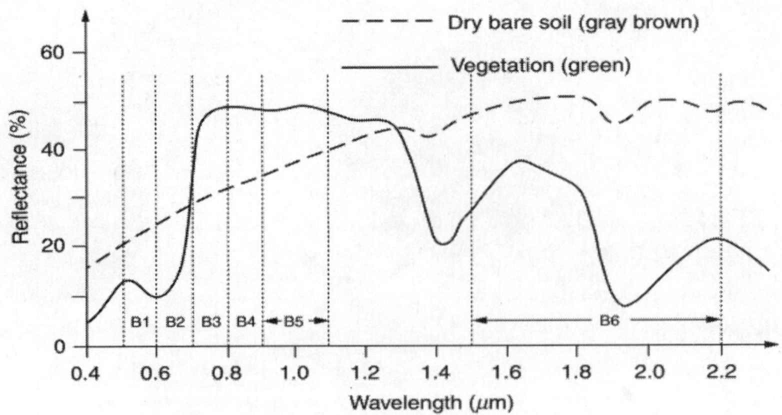

Figure 2. Spectral Resolution of imagery.

It is defined as the width of a spectral band. As illustrated in this figure, band 6 has the coarsest spectral resolution against band 1 and 2. Since the reflectance curves of most ground objects vary with wavelength (Fig. 2), in general, the finer the spectral resolution, the more information about the target is captured.

3. Radiometric Resolution

Radiometric Resolution refers to the ability of a remote sensing system to distinguish the subtle disparity in the intensity of the radiant energy from a target at the sensor. It is determined by the level of quantizing the electrical signal converted from the radiant energy.

4. Temporal Resolution

Also known as revisit period, temporal resolution refers to the temporal frequency at which the same ground area is sensed consecutively by the same sensing system. Since remote sensing satellite are revolving around the earth 24 hours a day and 365 days a year, the temporal resolution is directly related to the satellite orbital period. Among the four image resolutions, temporal resolution bears no direct relationship to the other three resolutions. Although spectral and spatial resolution are independent of each other, both are tied closely to radiometric resolution

5.5 Remote Sensing Platforms

The base, stationary or moving, on which remote sensors are mounted, is called a remote sensing platform. The choice of a particular platform is mainly guided by the choice of sensors which in turn is guided by the objectives of the programme. However, other key factors in the choice of the platform are the payload, operating height, operating range, time, and cost.

5.6. Analysis and Interpretation Techniques

Visual interpretation of remote sensing data with appropriate field checking is perhaps the best way to begin understanding what the imagery reveals for agricultural applications. This process utilizes the human computer (your brain) to derive information through interpretation of the remote sensing data. The other interpretation option is to use computer classification

i) Visual interpretation of pictorial image data has long been the work horse of remote sensing. Visual techniques make use of the excellent ability of human mind to quantitatively evaluate spatial pattern. Recognizing targets is the key to interpretation and information extraction. Observing the differences between targets and their backgrounds involves comparing different targets based on any, or all, of the visual elements of tone, shape, size, pattern, texture, shadow, and association.

Tone refers to the relative brightness or colour of objects in an image.

Shape refers to the general form, structure, or outline of individual objects. Shape can be a very distinctive clue for interpretation.

Size of objects in an image is a function of scale. It is important to assess the size of a target relative to other objects in a scene, as well as the absolute size, to aid in the interpretation of that target.

Pattern refers to the spatial arrangement of visibly discernible objects.

Shadow is also helpful in interpretation as it may provide an idea of the profile and relative height of a target or targets which may make identification easier.

Association takes into account the relationship between other recognizable objects or features in proximity to the target of interest.

ii) Computer aided digital classification/interpretation is complex and often involves expensive software. Supervised and unsupervised classifications are the two main types of computer classification for data interpretation. In supervised classification the computer is used to classify the image according to what the user has identified as training data. In unsupervised classification, the computer will aggregate the digital image according to the number of categories that the user has specified. Two other computer applications are also important: **Geometric Rectification and Ratioing.**

The **Rectification** process of the imagery allows relatively accurate positions to be determined so that area measurements can be made. Latitude and longitude points are located on the computer - displayed imagery, and a software algorithm then transforms and redisplays the image to remove much of inherent distortion. Rectification also allows the digital imagery to be incorporated into a GIS.

Ratioing involves a mathematical manipulation of the digital numbers (spectral reflectance). A well-known ratio is the Normalized Difference Vegetative Index (NDVI), which is computed as follows: NIR band – red band divided by NIR band + red band. The resulting image emphasizes differences in vegetative "greenness" and is often used with digital imagery to determine plant condition or status.

Recent years have witnessed phenomenal development in sensor technology and emergence of a wide variety of remote sensing satellites. Now it is possible to acquire satellite images with a spatial resolution as fine as sub meters or comparable to that of airborne photographs. The wide and easy availability of remote sensing data from various sensors in the digital format creates an ideal opportunity to process and analyze them automatically. Satellite data are routinely analyzed using various image processing systems to fulfill different application needs. The functionality and sophistication of image analysis have evolved considerably over the last two decades, thanks to the incessant advances in computing technology

Remote sensing of the earth resources, using the airborne and space borne sensors, is being now considered as a major component of the natural resources monitoring and management system. Though most of the data used today is from the optical sensors, in the recent past sensors operating in the millimeter wave and centimeter wave regions of the electromagnetic spectrum have also been extensively used for a variety of applications (***hyperspectral sensor***). With the advancement of the technology of signal processing, the resolution achieved at these frequencies is comparable to that of the optical sensors, even from the spaceborne platforms. Further they have the advantage of propagating through the clouds; vegetation and soil cover and are highly sensitive to moisture in the soil and vegetation. They are now being experimented extensively in the fields of agriculture, geology, hydrology, oceanography, meteorology and environmental studies. The ability to collect data at any time of the year makes the **microwave sensor** highly suitable for agricultural resources survey.

6.0. MICROWAVE REMOTE SENSING

Microwave sensors have certain unique capabilities due to the wavelengths used. The unique capabilities are : propagation through cloud cover, independent of sun's illumination, sensitive to moisture and penetration of soil and vegetation. Table 2 gives the unique capabilities of microwave sensors and compares them with that of the optical sensors for different applications.

Microwave sensors can be classified into two broad categories viz., passive and active sensors. The passive sensors measure the microwave radiation emitted from the target while the active systems provide their own illumination and measure the scattered signal.

Several studies have been conducted using both passive and active sensors for a variety of applications in the field of agricultural sciences. It has been fairly well established by these studies that measurement of soil moisture could be effectively carried out by microwave sensors and optimum radar parameters for this application.

7.0. HYPERSPECTRAL REMOTE SENSING

The third trend in data acquisition that emerged about a decade ago is termed hyperspectral remote sensing because it involves sensing the target in hundreds of spectral bands simultaneously in sharp contrast to tens of spectral bands that have been the norm of multispectral remote sensing. These hyperspectral bands cover roughly the same wavelength range as that of multispectral bands. Hence, each hyperspectral band is much narrower in its bandwidth. The hyperspectral sensor is either spaceborne or airborne.

Hyperspectral data are ideally applied in fields in which the subtle spectral variation among ground targets at some wavelengths needs to be identified and differentiated. Usually, such identification and differentiation are almost impossible with standard multispectral data, such as mapping of soil salinity, accurate mineral exploration, better predictions and assessments of crop yield, and better containment mapping.

8.0. LIDAR REMOTE SENSING

LIDAR (Light Detection and Ranging) remote sensing has the unique advantage of providing three-dimensional data through direct and indirect retrievals with unprecedented accuracy. It has enormous potential because it directly measures the physical attributes of vegetation canopy structure that are highly correlated with the basic plant community measurement. This technology is new and is restricted to only few countries of the world and has not yet properly reached to many countries including India.

Parallel to the advances in sensing technology is the process in pertinent geo-computational fields such as positioning system and geographic information systems. The geospatial data from these sources not only enrich the sources of data in digital image analysis, but also broaden the avenues to which digitally processed results are exported.

9.0. GLOBAL POSITIONING SYSTEM

GPS stands for **G**lobal **P**ositioning **S**ystem. GPS is a satellite-based system that uses a constellation of 24 satellites to give a user an accurate position. It is important at this point to define accurate. To a hiker or soldier in the desert, accurate means about 15m. To a ship in coastal waters, accurate means 5m. To a land surveyor, accurate means 1cm or less. GPS can be used to achieve all of these accuracies in all of these applications, the difference being the type of GPS receiver used and the technique employed. GPS was originally designed for military use at any time anywhere on the surface of the earth. Soon after the original proposals were made, it became clear that civilians could also use GPS, and not only for personal positioning (as was intended for the military). The first two major civilian applications to emerge were marine navigation and surveying. Nowadays applications range from in car navigation through truck fleet management to automation of construction machinery.

10.0. GEOGRAPHIC INFORMATION SYSTEM (GIS)

Geographic Information System (GIS) can be defined as - A computer based information system which involves collecting/capturing, storing, processing. manipulating, analyzing, managing, retrieving and displaying data (information) which is, essentially, referenced to the real-world or the earth (i.e. geographically referenced).

10.1 Components of GIS:

There are five essential components which make a complete Geographic Information System. These are: **Hardware, Software, Data, Method, and People**

10.2 Data in GIS

Broadly, the basic data for any GIS applications can be categorized as

- **Spatial Data** consists of maps which have been prepared either with the help of field surveys or with the help of interpreted remotely sensed data (RS). Remote sensing data is a classic source of data on natural resources for a region and provides a record of the continuum of resource status because of its repetitive coverage. Remotely sensed data in the form of satellite imageries can be used to study and monitor land features, natural resources and dynamic aspects of human activities for preparation of thematic maps.
- **Non Spatial Data** is the attribute data as a complimentary to the spatial data and describes what is a point, along a line or in a polygon and as a socio-economic characteristics from census or other sources. The attributes of a soil category could be the depth of soil, texture, erosion, drainage etc. and for geological category could be rock type, its age, major composition etc. The socio-economic characteristics could be the demographic data, occupation data for a village etc.

The GIS package offers efficient utilities to handle both these data sets and also allows for the spatial database organization along with non-spatial database organization. It is also capable to transform as well as integrate these two different kinds of information.

10.3 Advantages of GIS:

- Though collection of data is difficult once it is collected, it can be stored safely in digital form (in computer) forever without damage.
- Once a map is prepared accurately it can be printed in any scale without any redrawing.
- Large scale maps at smaller units (Beats) level can be joined for creating maps at divisions or district level, very accurately within few minutes.
- Changes can be incorporated very easily in computer and it doesn't require any redrawing.
- Same map can be generated for many times.
- New things can be known after overlay of different themes and new types of maps can be generated.

- Work can be done faster than manual practice.
- The work of decision makers and end users will become easier.

As explained, the capabilities and applications of remote sensing and GIS techniques play a vital role in resources development viz. agriculture (dry land and irrigated), forestry, geology, geomorphology, soils, urban and rural sectors etc. The mingling of spatial and non-spatial data can achieved with these technologies.

11.0 APPLICATIONS OF REMOTE SENSING IN AGRICULTURE

Remote sensing in association with GIS and GPS has found a number of myriad of applications in agriculture field also thereby benefiting the common man and also the scientific community. These tools are already widely used in rainfed agriculture, particularly in planning and monitoring of integrated watershed development projects and land use planning. With expansion of the technology and availability of high spatial and temporal resolution satellite data, many novel applications are emerging. Monitoring the status of natural resources over time, seasonal drought monitoring through Normalized Difference Vegetation Index (NDVI), in-season crop health monitoring over large areas for timely interventions on irrigation and pest management and locating potential water harvesting structures are some such applications where remote sensing is being used effectively. Under Mahatma Gandhi National Rural Employment Guarantee Scheme (MGNREGS), substantial investments are being made on land and water development works. Use of RS and GIS will facilitate better location of such ponds and contribute to drought proofing. Water harvesting structures like check dams and percolation tanks can be more efficiently planned at mini basin level if RS and GIS techniques effectively. Increasing use of these tools are found in land degradation, soil crop planning in rainfed agriculture.

Applications of Remote Sensing in various areas -

- Crop acreage estimation.
- Crop production estimation.
- Identification of crop pests and diseases.
- Soil mapping.
- Tree census.
- Crop damage estimation.
- Agricultural zonation
- Groundwater prospecting.
- Monitoring of dynamic characteristics like crop growth.
- Studying climate change.
- Precision farming.
- Building up database on soil, climate, land characteristics.
- Monitoring deforestation.
- Rainfall prediction
- Weed identification.
- Creation of global soil map

- LiDAR for farmers.
- Vegetative analysis.
- Bio-prospecting.
- Crop information system
- Watershed development and management
- Land use Land cover Mapping
- Wasteland Mapping
- Change detection studies

Some of the applications of remote sensing are explained in brief as under-

11.1 Crop Acreage Estimation

State and Central Government (Department of Agriculture), usually undertake initiatives to estimate the area under different crops on district, state and national level each year, which is very important prerequisite for crop production estimation. Govt. of India have been undertaking various projects like Cotton Acreage and Condition Assessment (CACA), Crop Acreage and Production Estimation (CAPE), Forecasting of Agriculture outputs through Satellite, Agro meteorology and Land based observations (FASAL), wherein dominant crops of the area are identified with multi-temporal data using visual or digital techniques of Remote Sensing and further the data is used for production estimation using various modules prepared for production estimation of different crops.

Crop Acreage and Production Estimation (CAPE) project is most successful Remote Sensing Applications Mission (RSAM) initiated in 1986 as a joint venture between Department of Space (DOS) and Ministry of Agriculture & Textile. The mission has reached its maturity and is operational in all the states of India covering dominant crops of the region. Remote Sensing Applications Centre has been participating in the CAPE mission since 1993-94 to forecast pre-harvest acreage and production estimation. Studies carried out in the mission have graduated from IRS 1A/1B LISS II of 36 m resolution through IRS 1C/1D LISS III of 23.5 m resolution to IRS P6 LISS III and AWiFS sensors, thus providing reliable and timely results.

Since cotton is a well-established commodity for domestic and international trade and also being a dominant cash crop, due attention is paid to its pre-harvest acreage and production estimates by a number of trading bodies, private as well as government corporations, which is an important input in decision making on procurement, pricing policy planning and export programme of the country.

The major objectives under the gamut of CAPE / FASAL -

- Tahsil/District wise acreage estimation using sample segment / complete area enumeration approach
- Condition assessment using Vegetation Index (VI)
- Production forecasting using spectral and meteorological models
- Linked with resources management activities by monitoring crop type of the region.
- Preparation of stratification plan

The digital analysis using customized software CAPEMAN developed by Space Application Centre (SAC), Ahmedabad was used. Selected sample segments are extracted from full scene of LISS II / III / AWiFS and joined together to form a synthetic image in order to reduce the volume of data for further processing. This mosaic is subjected to maximum likelihood classification based on the real time ground truth data. The segment-wise area statistics is generated for district-wise acreage aggregation and complete area enumeration approach is being adopted.

11.2. Crop Production Estimation:

This is extremely important for national level policy planning. For this, the crop production forecast of the state / nation should be made available to the policy planners. All over the country different crops yield estimation are predicted under projects like CAPE (Crop Acreage and Production Estimation). The crops taken up were Wheat, Paddy, and Rabi Sorghum etc.

Rabi sorghum is a semi-arid tropical crop cultivated extensively in Central India. Rabi sorghum grown in Western Maharashtra and Marathwada region of Maharashtra State alone contributes to about 67 % of all India acreage and 55 % of all India rabi sorghum production. During rabi season, sorghum is the major crop and is cultivated in large contiguous areas. Monitoring of crop condition, development of crop growth profile and prediction of yield based on remotely sensed multi-spectral data holds great promise.

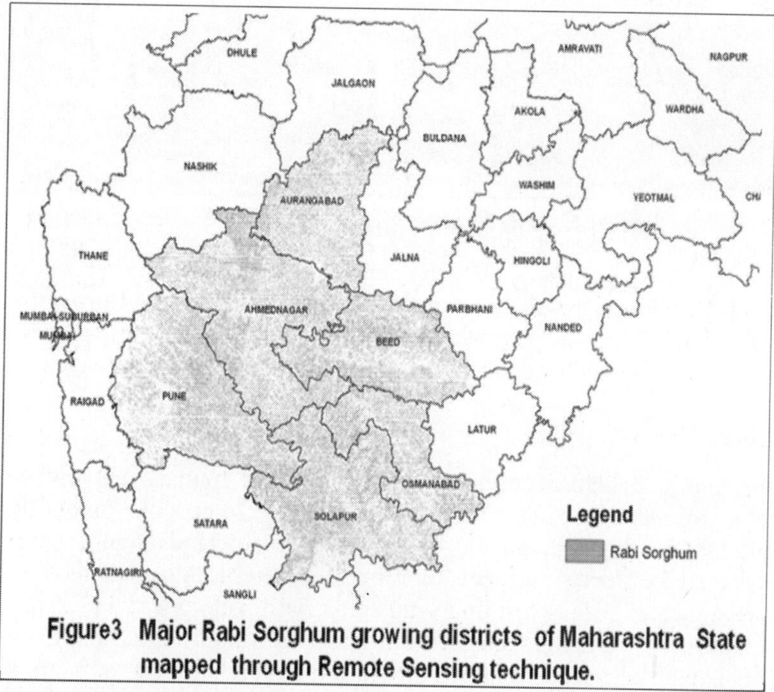

Figure3 Major Rabi Sorghum growing districts of Maharashtra State mapped through Remote Sensing technique.

Rabi Sorghum, is a staple food of the people which forms a vital part of the socio-economic setup in Marathwada region of Maharashtra State. Fig. 3. shows the major rabi sorghum growing districts of Maharashtra State mapped under CAPE project. The yearly acreage and yield forecasting of the rabi sorghum is of immense use to the State Department of Agriculture and Dry land crop Research Institute in perspective planning of the region.

11.3 Crop damage estimation

Crop damage studies are carried out for agricultural crops by comparing pre and post events satellite images to arrive at qualitative assessment. Homogeneous crop areas can be easily picked up by visible and infrared bands of electromagnetic spectrum used in the sensors attached to these satellite are recorded as a smooth pattern on satellite data, which facilitate the rapid and accurate enumeration of crop acreage and crop damage. Remote Sensing Applications Centre can analyse the crop damage by using pre-post satellite images of the events and generated the village-wise crop damage extent by overlaying the geo-referenced cadastral maps on the satellite image. The inputs thus, generated on extent of crop damage in the village using multi-temporal satellite images, against the quantum of damage reported by the concerned authority, facilitate to assess and ascertain the extent and severity of crop damage. Geo-referenced cadastral map overlay helps in extraction of parcel level information for devising a compensation package and / or to provide a financial assistance.

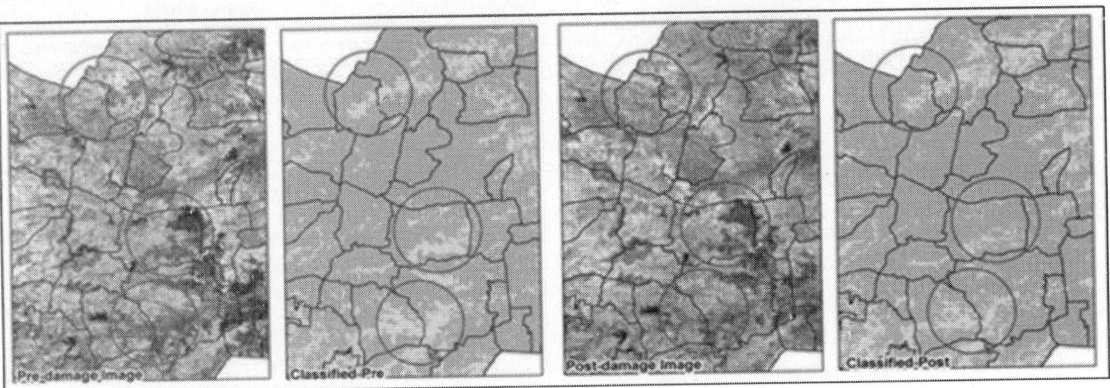

Figure 4. Crop damage Assessment due to hailstorm

The first two pictures in Fig.4 shows the remote sensing and interpreted data of the area in which the circled area are showing crops in good condition whereas the third and fourth pictures are the pictures of hailstorm affected area in which the circled area of the same places as that of first two pictures are showing crop damages occurred due to hailstorm at the same places.

11.4 Land use Land cover Mapping

The growing biotic pressure coupled with increasing human demands exerts tremendous pressure on the available land resources all over the country. In order to optimally use the land, it is necessary to have the information on the existing land use/ land cover, which permits a better understanding of the land utilization aspects on cropping pattern, fallow lands, forests, grazing lands, wasteland and surface water bodies forming vital areas for developmental planning. One of the basic requirements of planning is to have accurate and timely information on land use within the shortest possible time. This can be achieved from satellite based high resolution remotely sensed data on cost effective basis. This data, therefore, can be integrated with satellite data to evaluate the agriculture and urban land use, forest and plantation, extent of wastelands and pasture lands, land suitability for optimum utilization besides season wise availability of surface water for irrigation prospects etc. Digital image analysis techniques have been found more useful in land use / land cover inventories since segregates of different land use / land cover classes is feasible due to its synoptic coverage, near real time base line information, multi-spectral mode of data acquisition and its relative economy over other methods of survey.

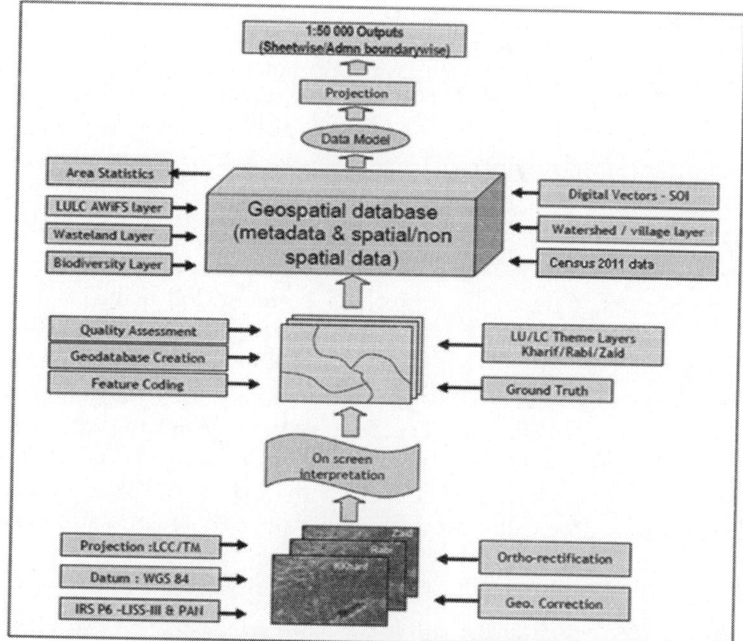

Figure 5. Broad Project Approach

Fig. 5. gives the broad methodology adopted to execute the Nationwide land use/ land cover mapping project on 1:50,000 scale for the year 2011-12. In the project three season data i.e. Kharif, Rabi and Zaid (summer) are mapped to prepare the maps. Fig. 6. shows land use/ land cover map prepared with the same methodology for Aurangabad District of Maharashtra State.

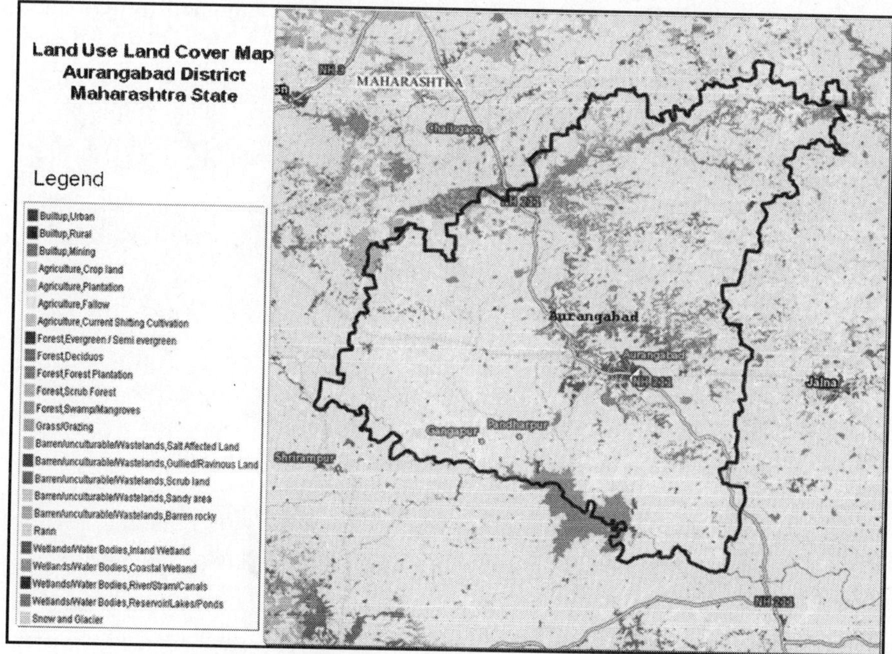

Figure 6 Detailed Land Use / Land Cover Map

11.5 Identification of crop pests and diseases:

Pests and diseases continue to affect the overall production, with losses ranging from 18-30% of production. Every year about 8.5 billion US $ are lost in India due to pests and diseases. Under the climate change scenario, this is expected to increase at least four fold (Pandey, 2012). Pests and diseases adversely limit the crop productivity, increase the crop management costs, restrict the free trade and threaten the biodiversity in the natural ecosystems. This is a major challenge to be addressed to contain the losses by the adoption of the rapid advancements occurring in the applications of Science and Technology.

Pests and diseases affecting various crops can be identified using remote sensing and GIS technology. Assessment of the health of a crop, as well as early detection of crop infestations, is critical in ensuring good agricultural productivity. Stress associated with, insects and fungal infestations must be detected early enough to provide an opportunity for the farmer to mitigate. This process requires that remote sensing imagery be provided on a frequent basis (at a minimum, weekly) and be delivered to the farmer quickly, usually within two days. The potential risk of pest and disease damage to rice crops was studied at the province level using a geographic information system (GIS) to monitor the relationship between environmental and climatic variables.

11.6 Soil mapping

One of the most important applications of RS and GIS in the field of agriculture is preparing soil maps in various themes like fertility, land use / land cover, extent of erosion, slope map, vegetation cover map, physiography map which is an important application of GIS, Groundwater availability map etc for formulating effective land use plans for the area under study. The Soil Resource Management and Conservation Division under the Dept. of Agriculture is instrumental in preparing soil maps, soil erosion maps, land use maps, agro-eco maps. The division is utilizing geospatial technologies including land surveying, GPS and satellite imagery to produce the imagery. The Soil Moisture and ocean Salinity (SMOS) Satellite has been launched by Russia which will be mapping soil moisture in large scale.

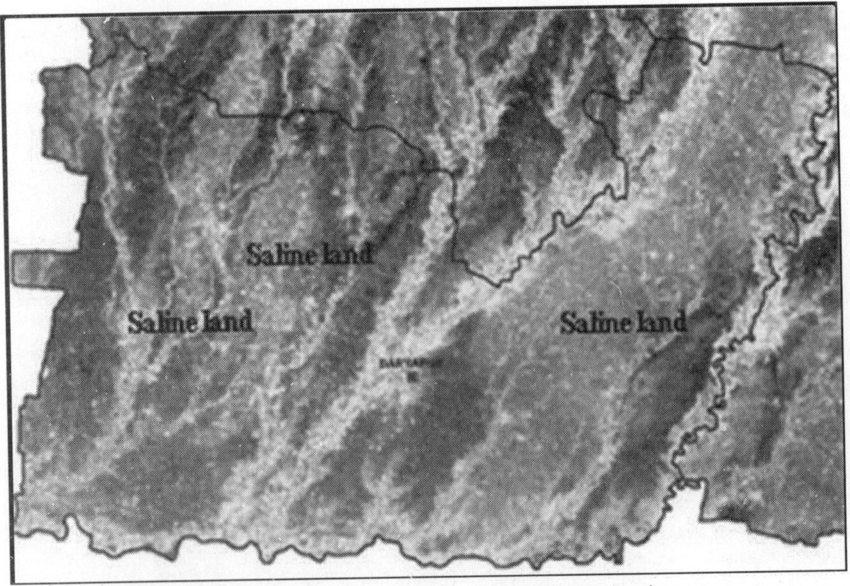

Figure 7 Saline Soils Mapping and monitoring

11.7 Tree Census

Various types of trees growing in a forest can be identified with the help of remote sensing technology by studying the NDVI. We have now incorporated tree census / inventory with latest technologies like GIS, GPS which makes the results more accurate. Each project involves physical survey and recording of individual trees in the defined area with relevant details like latitude, longitude, botanical name, common name, age, height, girth, and canopy. Terracon has recently launched proprietary, user friendly software, 'Vruksh Sharad" for managing tree inventory and related analytics. The software can provide quick search results, analytical tools and act as knowledge bank of trees and other forms of vegetation. The customized GIS based software has the added feature of plotting trees on maps with location accuracy. The software is also beneficial to clients and users in obtaining information on species composition in an area and facilitates species diversity management. Also helps in taking decisions on tree planting and removal planning and prioritization. The software will help the researchers, students and planners (landscape, ecological and restoration) to study and develop plans.

11.8 Watershed development and management

Sustainable development of natural resources is primarily based on maintaining fragile balance between productivity functions and conservation practices through identification and systematic monitoring of problem area in various resources and development sectors. It calls for application of alternate agricultural practices or land use, crop rotations, energy-efficient farming methods, reclamation of under-utilized and wastelands, efficient water use practices and conservation of water resources. Optimal exploitation of the resources (both renewable and non-renewable) with proper mechanism calls for a hydrological approach for a region as whole. The synoptic view provided by satellite remote sensing, offers technologically suitable method for studying the land and water resources with reference to agro-climatic zones and identifying constraints / ecological problems at micro-level. Effective use of remotely sensed data base merged with collateral socio-economic and meteorological data helps in arriving at locale specific prescriptions to achieve sustainable development of natural resources of a given area.

Department of Space, Govt. of India along with Remote Sensing Application Centers of the concerned states in the year 1988 had initiated pilot studies in 21 perennially drought affected districts of the country to generate locale-specific action plans by integrating thematic information on land and water resources derived using data from Indian Remote Sensing satellites along with socio-economic information and meteorological data.

In Maharashtra, Ahmadnagar district was identified as a study area, since large amount of data were available in various forms and many on-going projects related to drought management were being undertaken in Ahmadnagar district. The integrated project was primarily aimed to generate thematic data base and evolving a tractable package for the management of agriculture / land resources, water resources and fodder resources. In the process various thematic maps showing present land use / land cover, types of wastelands, forest cover types, surface water resources, drainage pattern, hydro-geomorphological situation, soils were generated using data from Indian Remote Sensing satellite adopting tahsil as a unit on 1:50,000 scale. These maps (spatial data) and other collateral information (attribute data) were integrated to identify a set of coherent micro-level land units which are unique in their resource potential and problems.

Taking into account the available resources integrated land use plans were prepared for three tahsils (Pathardi, Karjat and Rahuri) of Ahmednagar district. Before arriving at a concrete development plan, the ground realities were also assessed by conducting limited ground truth in the area under study. Interactions with the local farmers were also held so as to get the correct feedback regarding various problems in the area. While preparing these specific developmental plans the views of the experts from various state / central developmental departments, agriculture universities research institutions were also considered so as to ensure the technical feasibility and cultural acceptability of such action plans. A small watershed covering an area of 12000 ha forming part of the watershed GV-130 had been selected for actual implementation of action plans for optimum management of land and water resources.

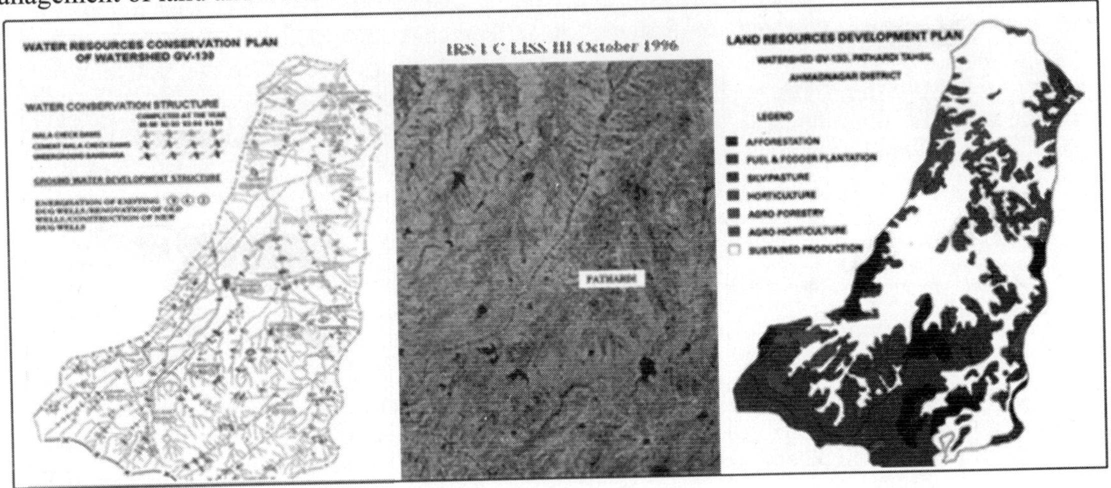

Figure 8 Watershed development and management

The suggestions implemented as per the Land and Water resources maps prepared for watershed GV-130 were again studied after the gap of five years and the result for development of Land and Water Resources were found to be very encouraging even though the rainfall was deficit in those years. The suggestion had surely increased the surface and ground water level and the effect was clearly seen on the cropping pattern of the watershed GV-130. Fig. 8. shows the Land and Water resources development plan prepared with the remote sensing and GIS technology for Watershed GV-130, which was implemented with the help of local authorities and yielded very good results helping the farmers of the dryland areas.

11.9 Change detection studies

As multi-date data is easily available, change detection studies for any area, be it a district, tahsil, village or any particular crop within the specified region , can be monitored effectively e.g. monitoring natural resources, their degradation, change of agricultural land into wasteland and other categories etc. and the results can be declared.

Remote Sensing provides authentic, accurate and timely information regarding available natural resources for various aspects of agricultural development planning process and these becomes the input data for the various Web-Portals. Apart from Remote Sensing information, easy availability of other relevant database is also needed, this database/information can be acquired through various Web-Portals.

Fig. 9. shows change detection in land use/ land cover in the area in a decade. The changes can be seen very clearly in the western part.

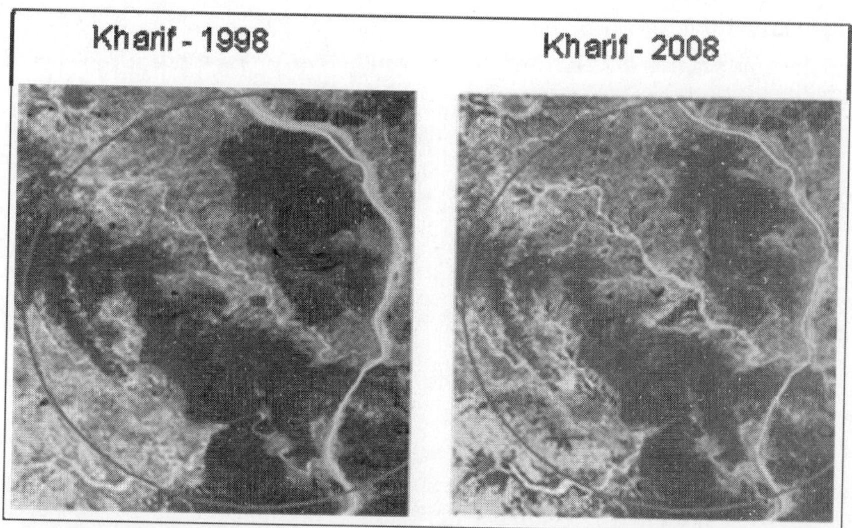

Figure 9 Detection of changes in land use and land cover over space and time using remote sensing technique

12.0. WEB PORTAL FOR AGRICULTURE

The database generated with the help of Remote Sensing and GIS techniques is very authentic, unbiased and real time data of natural resources and can be effectively used for creating the relevant Web-Portals. The database thus generated for agricultural sectors is used to create Web Portal for Agriculture by Govt. of India, State Govt., Agricultural Universities, Agriculture Department etc.

The Web Portal caters to the information needs of farmers, scientific community, administrators, planners, private sectors, Non Governmental Organizations (NGO) etc. The Web Portal facilitates the farmers to access the requisite information in different Indian languages. There are various web portals which are involved in agriculture development. One of the examples of such web portals developed by Ministry of Agriculture, Govt. of India is, the Central Agricultural Portal (CAP) which is involved in various functions such as, Content Management, Pesticides Registration, e-Certification etc.. All the services requests to these applications initiated using different delivery channels (Govt. Departments, KCC, Private Kiosks, Mobile, IVRS, Agri-Clinics and agri-business centres etc.) It will also be integrated with existing applications such as AGMARKNET, SEEDNET, etc.

12.1 Major objectives of Web Portal for Agriculture –

- To provide **one stop, single window solution** to all stakeholders, especially farmers for accessing information and services in agricultural sector eliminating the need to remember multiple URL and to navigate multiple web sites and applications.
- To provide **localized** and **personalized** services for agriculture related information.
- To act as a **gateway for delivery of information and services** in Agricultural sector this will be accessed using various delivery channels.

- To deliver agriculture related government services in an integrated manner.
- To **complement and enhance service delivery channels** which may already exist at the central government level.
- To build a **comprehensive Knowledge Management system** for Indian Agriculture by providing platform to collaborate and share information.
- To act as a **bridge between farmers and Government bodies** at the highest decision making level by streamlining and simplifying their interactions with Government entities, reducing service cycle time and providing enriching experience.
- To provide mechanism for **grievance redressal and management** in agriculture sector.
- To provide **Expert Advisory services** to the farmers during each stage of the entire crop cycle, while on the other hand, to provide the platform for the experts to receive and address specific queries of the farmers.
- To build **National Farmers' database** to understand the information requirements of the farmers and to provide relevant services through various delivery channels.
- To provide **anytime, anywhere**, single window delivery of government services to **agri-business industry** comprising of dealers, wholesalers, retailers of the agricultural inputs and exporters, importers, traders of the agricultural products.
- To streamline and simplify their interactions with Government entities reducing service cycle time and providing enriching experience.
- To **minimize gaps** between Service providers(Government entities) and service consumers (farmers, Agri-businesses entities etc.)
- To build **platform for Government departments to present information, applications and services** in a single consolidated browser view.
- To provide a secure and individualized view of multiple online resources and interactive services.

12.2 List of Web Portals for agriculture development:

1. www.indiaagristat.com- It is a comprehensive source for Indian agriculture statistics which is regularly updated. It provides authentic statistical information on sectors like agricultural education, agricultural export, agriculture census, agriculture prices, agricultural insurance, animal husbandry, agricultural marketing, horticulture production, agricultural wages and all other relevant agricultural statistics of India. It also provides the agriculture related news.

The statistics available on this web site is widely used by Indian agriculture research institutes, Indian agriculture sector, agricultural economists and Indian agriculture universities for Indian agricultural information, agriculture research and agricultural production data.

2. **www.isapindia.org** - The site is all about the Indian society of agribusiness professional (ISAP) which works for helping the farmers' community and address the rural- urban income divide. It is a network of agriculture and allied sector professionals in India and developing countries.

3. www.carrittmoran.com - This site provides, information regarding tea and coffee- statistics (on production and sale), catalogues (of area- wise sale of different varieties of tea and coffee in India), market reports and TASI.

4. www.fciweb.nic.in - This is the site by Food Corporation of India. It works for effective price support, food security, price stabilization and distribute food grains through public distribution system for the benefit of both farmers and consumers.

5. www.fredisurti.com - Fredisurti is a flower company specializing in flower seeds. This site offers garden consultancy and distributor for all kinds of seeds.

6. www.indiancommodities.com - Information on cotton, rice, wheat, oilseeds, pulses, spices, coffee, and tea. This site requires registration, login and payment.

7. http://www.isapindia.org - Indian society of agribusiness professionals (ISAP) is a non-government, non-profit organisation. It is a network of agriculture and allied sector professionals in India and developing countries. Its vision is to instill economic security and stability among farming community particularly small and marginal farmers through holistic development of agriculture and rural sector.

8. http://agricoop.nic.in/- This website on agriculture will lay stress on the agricultural sector, employment, opportunities, industrial sector and infrastructure. The site also displays the plant protection information network, Rashtriya Krishi Vikas Yojana, and National Food Security Mission.

9. http://www.apeda.com- agricultural and processed food products export development authority (APEDA) is mandated with the responsibility of export promotion and development of the scheduled products like: floriculture, fruits and vegetables, processed foods, organic foods, animal products and cereals. In addition to this, APEDA is also responsible to monitor export of some non - scheduled items such as basmati rice, wheat, and coarse grains and also import of sugar.

10. http://fert.nic.in- The website Department of fertilizers, Govt. of India, gives information on different aspects of fertilizer.

11. http://mofpi.nic.in/ - Ministry of food processing industries, is the main central agency of the Government responsible for developing a strong and vibrant food processing sector; with a view to create increased job opportunities in rural areas, enable the farmers to reap benefit from modern technology, create surplus for exports and stimulating demand for processed food.

12. http://www.nationalfertilizers.com/- National fertilizer ltd. is the second largest producer of nitrogenous fertilizers in the country. They produce urea and fertilizers which are beneficial for all types of crops.

13. http://www.fertindia.com/ - This website gives in detail about fertilizer statistics in India, fertilizer prices and fertilizer marketing in India.

14. http://www.upagriculture.org/- This agricultural site shows new schemes related to farmers, facilities given to them, latest techniques, new agri policies, quality control, useful machines, insects/diseases of crop and many useful information's related to agriculture.

15. http://www.krishi.net/GovtLinks.asp#IO- This websites has a collection of various sites related to agriculture and have links to Govt. institution, agricultural universities, state links, and international organizations.

16. http://www.indg.in/agriculture/- This site aims to disseminate useful information about improved technology to the farming community and service providers in the rural areas. It aims to create a platform for different levels in the rural agricultural landscape - farmers, cooperatives and professional bodies, farm machinery vendors, fertilizer and chemical companies, insurance regulators and agronomists, consultants, and farm advisors.

17. www.kissankerala.net/ - This website is a complete agri - info system for Kerala. Information on farming practices, fertilizers & pesticides, planting material are available.

18.http://www.sciencedaily.com/releases/2012/01/120119143328.htm - The site consists of a regular trickle of information and Expert comments from the International group of Scientists writing in the Science Magazine, regarding a road map for ensuring food security in the face of Climatic Change.

13.0. CONCLUSION

The world is facing many challenges in agriculture development. The major challenge in front of the farming community is to feed the ever increasing world population and improving their quality of life. This challenge is more complex in the developing countries like India. For overall development of any country there has to be an integrated approach by the constituents who are involved in this development process. For initiating this process, the first and foremost prerequisite is to have an accurate knowledge about various resources with the planners, the resources may be natural resources, socio-economic resources, infrastructural potential etc. For initiating this developmental process one must know the strengths as well as weaknesses of the given area and accordingly the development planning process can be initiated. Remote sensing and GIS technology can help immensely in this regard.

The remote sensing technology plays a very pivotal role in the field of agriculture for the benefit of the scientific as well as farming community in knowing the present and past situations and ultimately planning for the future strategies for sustainable agriculture development. The technology can be effectively used in monitoring various natural resources, infrastructural potential studies, weather studies, fisheries, oceanographic studies, crop acreage and condition assessment, crop production estimation, forestry, mineral exploration, geological studies, soil mapping, environmental impact assessment studies, wasteland assessment studies etc .

In dryland agriculture, Remote Sensing and Geographical Information System techniques have immense potential in the areas like assessing surface and ground water potential, estimation of land degradation, identifying suitable lands for its alternate use, weather forecasting etc. Now a days it is playing a significant role in dryland agriculture, watershed development, land use planning forest mapping, mineral exploration, urban development and many more areas.

In earlier days, acquiring information related to various natural resources like soil mapping involved huge manpower, time and money. With the advent of remote sensing technology such huge tasks have become more accurate, less exhaustive involving comparatively very less time and cheap too. This technology has many advantages over the age old conventional practices, e.g. it gives the information in spatial form, information can be sought with desired repeativity (temporal data) as and when required, which helps the planners, administrators tremendously in executing various agricultural development plans precisely.

The remote sensing technology has progressed tremendously; it is up to the various constituents of the society, like administrators, planners, scientific community, social groups, social scientists etc. as to how well they can use this technology for agricultural development in general and dry land agriculture in particular, ultimately helping the farming community for their overall wellbeing to lead a good quality of life.

REFERENCES

An Asian Journal of Soil Sciences Vol. 5 No.2 (December,2010): Pages, 419-425

AGRO 102 (New) Dryland Farming and Watershed Management, Lecture Notes prepared by Dr. T Yellamanda Reddy and Dr. G. Prabhakara Reddy, Department of Agronomy, Acharya N.G. Ranga Agriculture University.

Compendium of Remote Sensing & GIS Application Projects., (MRSAC, Nagpur) January 2011.

CRIDA 2011. VISION 2030, Central Research Institute for Dryland Agriculture, Santoshnagar, Hyderabad, India, pp: 31, July, 2011 edition.

CRIDA. 2007. CRIDA - Perspective Plan, Vision 2025. Central Research Institute for Dryland Agriculture, Hyderabad, Andhra Pradesh.

Current Science, Vol. 83, No.11, 10 December 2002, Pages: 1320-1328.

Digital Analysis of Remotely Sensed Imagery by Jay Gao. McGraw-Hill Companies Inc.(2009)

GPS Satellite Surveying by Alfred Leick (John Wiley and Sons.) (2004)

Introduction to Hyperspectral Imaging by Randall B. Smith, September 2011(TNTmips).

Introduction to GPS (Global Positioning System), Version 1.0 by Leica

International Journal of scientific research and management (IJSRM),Volume -1, Issue -8, Pages: 421-425 (2013) Website: www.ijsrm.in ISSN (e): 2321-3418.

Journal of Arid Environments 70 (2007) : Pages, 329-343.

"Microwave Remote Sensing for Agricultural Applications" by O P N Calla, Space Applications Centre (ISRO), Ahmedabad.(1985)

National Mission for Sustainable Agriculture, Strategies for Meeting the Challenges of Climate Change , Department of Agriculture and Cooperation, Ministry of Agriculture, New Delhi (August 2010).

National Remote Sensing Agency (NRSA) (2006), "Manual of National Land use Land Cover", Dept. of Space, Hyderabad, Government of India.

National Water Policy (2012) , Minister of Water Resources, Government of India.

"Remote Sensing and Image Interpretation", Lillesand, T. and Kiefer, R., Wiley, New York.(2004)

Satellite Remote Sensing and GIS Applications in Agriculture Meteorology, Proceedings of a Training Workshop held 7-11 July 2003 in Dehra Dun, India, Edited by M.V.K. Sivakumar, P.S. Roy, K. Harmsen, and S.K. Saha. pp. 263-289.

Site http://en.wikipedia.org/wiki/Dryland_farming for dryland farming

Site http://dacnet.nic.in INTRADAC-G2E e-Governance model for Department of Agriculture & Cooperation

Site http://dacnet.nic.in/AMMP/CAPSRSVer1.0.pdf

Site http:// www. microimages.com for Geospatial Analysis software

Site bhuvan.nrsc.gov.in for land use/land cover and crops.

Site http://agropedia.iitk.ac.in/ for list of Indian Agricultural portal.

Site http://www.iari.res.in for Global Climate Changes.

Pib.nic.in/newsite/erelease.aspx, India State of Forest Report, 2011, Press Information Bureau, Govt. of India, August 2014, for forest cover in India.

Chapter - 27

AMELIORATION OF ADVERSE EFFECT OF DROUGHT WITH PLANT GROWTH REGULATORS AND MINERAL NUTRIENTS

R.D. Deotale

Dr. P.D.K.V. Akola, Botany section, College of agriculture, Nagpur

Stress is an altered physiological condition caused by factors that tend to disrupt the equilibrium. Strain is any physical and chemical change produced by a stress (Gaspar *et al.,* 2002). The term stress is used with various meanings, like the physiological definition and appropriate term as responses in different situations. The flexibility of normal metabolism allows the response initiation to the environmental changes, which fluctuate regularly and are predictable over daily and seasonal cycles. Plants are frequently exposed to many stresses such as drought, low temperature, salt, flooding, heat, oxidative stress and heavy metal toxicity, while growing in nature.

Drought is a meteorological term and is commonly defined as a period without significant rainfall. Generally, drought stress occurs when the available water in the soil is reduced and atmospheric conditions cause continuous loss of water by transpiration and evaporation. Water deficit and salt stresses are global issues to ensure survival of agricultural crops and sustainable food production (Jaleel *et al.,* 2007b-e; Nakayama *et al.,* 2007). Tolerance to abiotic stresses is very complex, due to the intricate of interactions between stress factors and various molecular, biochemical and physiological phenomena affecting plant growth and development (Razmjoo *et al.,* 2008). High yield potential under drought stress is the target of crop breeding. In many cases, high yield potential can contribute to yield in moderate stress environment (Blum, 1996).

Drought stress is considered to be a moderate loss of water, which leads to stomatal closure and limitation of gas exchange. Desiccation is much more extensive loss of water, which can potentially lead to gross disruption of metabolism and cell structure and eventually to the cessation of enzyme catalyzed reactions (Smirnoff, 1993; Jaleel *et al.,* 2007d). Drought stress is characterized by reduction of water content, diminished leaf water potential and turgor loss, closure of stomata and decrease in cell enlargement and growth. Severe water stress may result in the arrest of photosynthesis, disturbance of metabolism and finally the death of plant (Jaleel *et al.,* 2008c).

Water stress inhibits cell enlargement more than cell division. It reduces plant growth by affecting various physiological and biochemical processes, such as photosynthesis, respiration, translocation, ion uptake, carbohydrates, nutrient metabolism and growth promoters (Jaleel *et al.,* 2008; Farooq *et al.,* 2008). In plants, a better understanding of the morpho-anatomical and physiological basis of changes in water stress resistance could be used to select or create new varieties of crops to obtain a better productivity under water stress conditions (Nam *et al.,* 2001; Martinez *et al.,* 2007). The reactions of plants to water stress differ significantly at various organizational levels depending upon intensity and duration of stress as well as plant species and its stage of growth (Chaves *et al.,* 2002; Jaleel *et al.,* 2008b). Understanding plant responses to drought is of great importance and also a fundamental part for making the crops stress tolerant (Reddy *et al.,* 2004; Zhao *et al.,* 2008).

Drought is one of the most important abiotic stress which affects crop growth and yield (Lutts *et al.,* 2004, Wahid, 2004; Luo *et al.,* 2005; Parida and Das, 2005; Mahajan & Tuteja, 2005; Castillo *et al.,* 2007; Khan *et al.,* 2010; Bakht *et al.,* 2010). The stage at which the drought stress occurs plays a major role in the final yield of the crop. Numerous physiological and biochemical changes occur in response to drought stress in various plant species. Many defense mechanisms i.e., osmoregulation, ion homeostasis, antioxidant and hormonal systems are invoked in tolerant plants which allow the plants to stay alive and properly develop prior to reproductive stages (Reddy *et al.,* 2004; Sairam and Tyagi, 2004; Mahajan and Tuteja, 2005; Ashraf, 2010). The alteration of protein synthesis or degradation is one of the fundamental metabolic processes that may influence drought tolerance. Both quantitative and qualitative changes of proteins were detected during water stress (Riccardi *et al.,* 1998). Evidence is increasing in favour of a relationship between the accumulation of drought-induced proteins and physiological adaptations to water limitation (Riccardi *et al.,* 1998; Han and Kermode, 1999). Research into the plant response to water stress is becoming increasingly important as most climatic change scenario suggest an increase in aridity in many areas of the globe. Agriculture is the major user of water resources in many regions of the world. With increasing aridity and growing population, water will become an increasingly scarce commodity in the near future. Even though in viable agriculture, severe water deficits should be a rare (but catastrophic) event, a better understanding of the effects of drought on plants is vital for improved management practices in breeding efforts in agriculture and for predicting the fate of natural vegetation under climatic change (Passioura, 2002).

Different Ways of Viewing and Defining Drought

Which type of drought people mean when they talk about "drought" often depends on who they are, the kind of work they do, and the perspective that gives them.

Farmers and ranchers are most often concerned with agricultural drought, for example, agricultural drought is also the type of drought that worries people in the grocery and meat business or people in farm communities that depend indirectly on agricultural income for their livelihoods. Urban planners usually mean hydrological drought when they talk about drought, because water supplies and reserves are key components in managing urban growth. The most common use of the term "drought" refers to meteorological drought, because that is the drought condition most familiar to the general public and the one most easily identified.

What are Causes of Drought?

Drought is one of the main environmental factors limiting plant growth and crop production. Water stress causes visible injuries to leaves, induces stomatal closure, leaf rolling and osmotic adjustment. Water stress causes a faster decline in chlorophyll and protein content, altering both the structure and function of membranes (Wang, 1999).

Drought can be caused by too little precipitation (rain and snow) over an extended period, as most people assume, but drought can also be caused by increased demand for the available supply of usable water even during periods of average or above average precipitation.

TYPES OF DROUGHT

There are three conditions that are generally referred to as drought:

- **Meteorological drought** : This type of drought is all about the weather and occurs when there is a prolonged period of below average precipitation, which creates a natural shortage of available water.
- **Agricultural drought** : This type of drought occurs when there isn't enough moisture to support average crop production on farms or average grass production on range land. Although agricultural drought often occurs during dry, hot periods of low precipitation, it can also occur during periods of average precipitation when soil conditions or agricultural techniques require extra water.
- **Hydrological drought** : This type of drought occurs when water reserves in aquifers, lakes and reservoirs fall below an established statistical average. Again, hydrological drought can happen even during times of average or above average precipitation, if human demand for water is high and increased usage has lowered the water reserves.

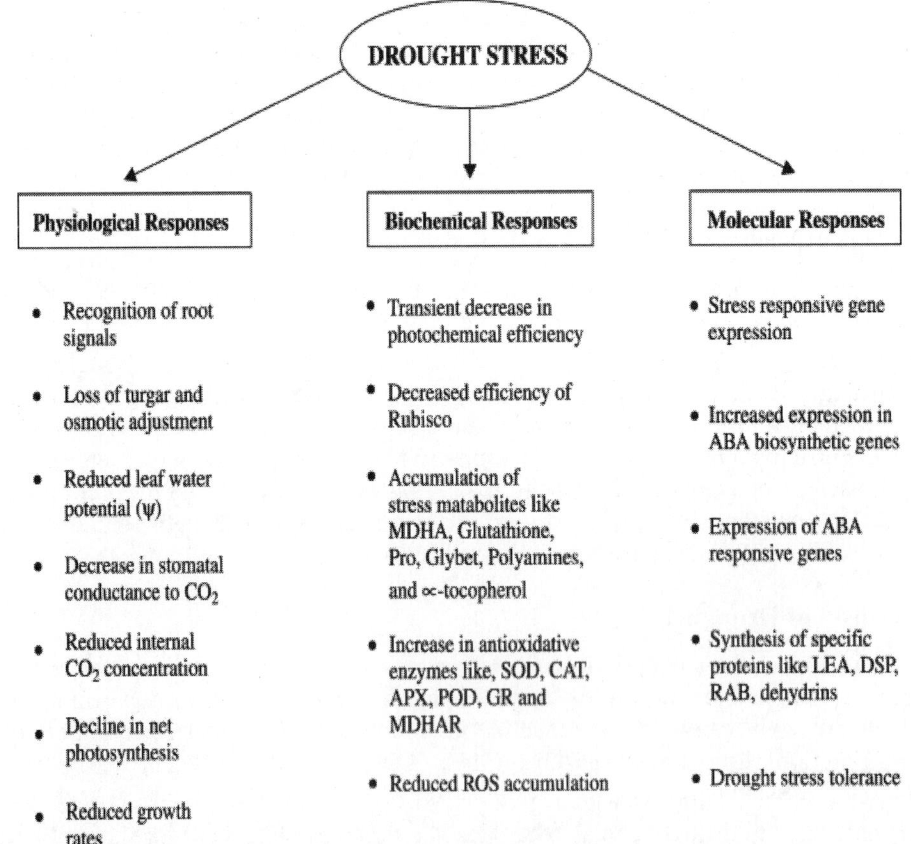

Fig. 1. Physiological, biochemical and molecular basis of drought stress tolerance in plants [Reprinted with permission from 'Shao *et al.* (2008). *Comptes Rendus Biologies*, 331:215-225.' ©2008, Elsevier Limited, UK].

PLANT RESPONSES TO DROUGHT STRESS

Plants growing in deserts or high salinity habitats are all exposed to more or less constant water stress. To survive such conditions, plants have developed growth strategies such as increased water use efficiency with C_4 or CAM metabolism (Keeley and Rundel, 2003), succulent growth and extensive root systems (Henry *et al.*, 2011). These strategies are good in a dry environment, but in more "favourable" conditions at least some of these plants may, due to lower growth rates, more easily be out competed by other less drought tolerant plants. Other adaptations to plant life in dry environments are thick cuticula and wax layers, depressed stomata and high density of trichomes. Thick cuticula and wax layers reduce extra stomatal transpiration, and depressed stomata and trichomes create a thicker boundary layer outside the stomata, where the humidity gradient is more gradual, thereby reducing the stomatal transpiration. Plants living in saline environments (e.g. beaches) commonly keep a low osmotic potential in their cells, which facilitates water uptake. They usually also have the ability to exclude or excrete salt from their cells to avoid too high salt concentrations. A variety of perennials commonly avoid water stress during the winter by entering dormancy and often shedding leaves (deciduous woody species) before the onset of the harsh conditions when water is unavailable due to frost. However, plants keeping the leaves on through the winter commonly face water stress in the spring when air temperatures are high while the soil is still frozen. Even if they do not live in particularly dry places, most plants will occasionally encounter water stress for shorter or longer periods of time. Most of these plants do not have many of the adaptations of desert plants and must respond to the water stress in other ways. When these plants are exposed to water stress, such as drought or saline conditions, to survive, they must be able to retain as much water as possible. If the plants are not able to cope with the water stress, they will not be able to survive. The sensitivity and response time to drought differs between different species and slow growing species have been found to be more sensitive (Aasamaa and Sober, 2011). Repeated drought encounters increases the sensitivity to environmental changes that induce stomatal closure, while the sensitivity to changes that induce stomatal opening is decreased (Aasamaa and Sober, 2011). In response to water stress, plants have developed several different mechanisms that increase the desiccation tolerance and water retention. These responses can be divided into short term and long term responses.

2.1 Long term responses

During prolonged water stress, plants must be able to survive with low water content and maintain a minimum amount of water, through water uptake and retention. To cope with prolonged drought stress, plants respond with energy demanding processes that alter the growth pattern, chemical content of the plants and the up or down regulation of genes.

Short term response

When plants suddenly encounter drought it, is important to respond as quickly as possible. A faster drought response means that less water is lost and the survival rate of the plants is increased. The most important quick response is stomatal closure. Stomata consist of two guard cells surrounding the stomatal pore. When the stomata are open water is transpired and CO_2 enters the leaf through the stomatal pore. During water stress, the stomatal pore can be closed to reduce water loss. By closing the stomatal pore the water use efficiency is increased (Farooq *et al.*, 2009), reducing the amount of water lost per CO_2 molecule assimilated. Several mechanisms work together to close the stomata, such as hydro passive closure and chemical signals from the plant stress hormone, ABA. Increased levels of ABA also cause increased hydraulic conductivity in the roots and xylem, enabling the plants to transport more water and thereby recover more rapidly after water stress (Kudoyarova *et al.*, 2011)

Biochemical changes

When the water availability is reduced, plants change the biochemistry to be able to retain as much water as possible and take up whatever water they can. During water stress, plants produce and accumulate compatible solutes such as sugars, polyols and amino acid to lower the osmotic potential in the cells to facilitate water absorption and retention (Xiong and Zhu, 2002). Some of the compatible solutes also contribute to maintaining the conformation of macromolecules by preventing misfolding or denaturation (Xiong and Zhu, 2002). Plants also produce higher levels of the plant stress hormone ABA during water stress and this affects their growth pattern and stress tolerance.

Effects of water stress on plants

Drought, as an abiotic stress, is multidimensional in nature, and it affects plants at various levels of their organization. In fact, under prolonged drought, many plants will dehydrate and die. Water stress in plant reduces the plant-cell's water potential and turgor, which elevate the solutes' concentrations in the cytosol and extracellular matrices. As a result, cell enlargement decreases leading to growth inhibition and reproductive failure. This is followed by accumulation of abscisic acid (ABA) and compatible osmolytes like proline, which cause wilting. At this stage, overproduction of reactive oxygen species (ROS) and formation of radical scavenging compounds such as ascorbate and glutathione which further aggravate the adverse influence. Drought not only affects plant water relations through the reduction of water content, turgor and total water, but it also affects stomatal closure, limits gaseous exchange, reduces transpiration and arrests carbon assimilation (photosynthesis) rates. Negative effects on mineral nutrition (uptake and transport of nutrients) and metabolism leads to a decrease in the leaf area and alteration in assimilate partitioning among the organs. Alteration in plant cell wall elasticity and disruption of homeostasis and ion distribution in the cell has also been reported. Synthesis of new protein and mRNAs associated with the drought response are another outcome of water stress on plants. Under the water stress, cell expansion slows down or ceases, and plant growth is retarded. However, water stress influences cell enlargement more than cell division. Plant growth under drought is influenced by altered photosynthesis, respiration, translocation, ion uptake, carbohydrates, nutrient metabolism, and hormones.

PLANT GROWTH REGULATORS IN RELATION TO DROUGHT

Phytohormones such as IAA, gibberellic acid (GA_3), and kinetin are known to be involved in the regulation of plant response to the adverse effect of stress conditions (Walker and Dumbroff, 1981; Hamdia , 1998; Kaya *et al.* 2009). The plant hormones can be defined as organic substances that are produced in one part of plant and translocated to another parts, where at very low concentration, they stimulate physiological response. Plants are usually subjected to environmental factors such as drought or water salinity.

Abscisic acid (ABA) plays an important role in abiotic stress (Bakht *et al.,* 2006; Bakht *et al.,* 2011). Root-originated xylem sap, the ABA can move to crop reproductive structures and accumulate there to a high level under drought conditions in wheat crop (Lie *et al.,* 2003). This elevated ABA content in the crop reproductive structures had been thought to be involved in controlling kernel pod^{-1} abortion, presumably *via* inhibition of cell division in the young ovaries (Setter and Ammgam, 2001; Lie *et al.,* 2003). In addition, exogenous application of ABA to developed maize ovaries inhibits cell division in the embryo and endosperm, and this effect is probably due to a depression of cell cycle gene expression by high levels of ABA (Setter and Ammgam, 2001). These studies suggested that drought induced increase in xylem sap (ABA) might affect expansion growth of crop reproductive structure resulting in a weak sink intensity, which fails to attract assimilate from source organs and eventually leads to abortion.

Foliar application of ABA had a significant ($p<0.05$) effect on pod yield of groundnut. Drought stress reduced 46% pod yield when compared with control. There was an increase of 11% in pod yield under drought stress when ABA was applied at 41-60 DAS compared with drought stressed plants alone. Wheat and other grain crops under water deficit during grain filling substantially affects grain weight (Rahman and Yoshida, 1985) due to early plant senescence, cessation of grain filling (Hossain *et al.,* 1990) and shortening of the grain filling period (Royo *et al.,* 2000).

Abscisic acid (ABA), an apocarotenoid synthesized from cleavage of carotenoids, is involved in the signal transduction pathway regulating several genes that are expressed at specific developmental stages or as a result of an environmental stresses. The ABA accumulates in vegetative cells in response to water deficit, salinity, cold temperature, and light variations was thought to act as a signal for the initiation of acclimation to these stresses (Xiong *et al.,* 2002 and Swiatek *et al.,* 2003).

Jasmonic acid (JA) is a naturally occurring growth regulator found in higher plants. Several physiological roles have been described for this compound (or related compound, methyl jasmonate) during plant development and in response to biotic and abiotic stresses by inducing the biosynthesis of defense proteins and protective secondary metabolites (Creelman and Mullet, 1995).

Jasmonates (i.e., JA or its methyl ester) can modulate many physiological events, such as resistance responses to pathogens and insects, pollen development, root growth and senescence (Sasaki *et al.,* 2000). These substances can also activate the expression of several genes, leading to the accumulation of their products, which are referred to as jasmonate-induced proteins (Benedetti *et al.,* 1998). The best studied jasmonate induced proteins include proteinase inhibitors (Lawrence and Koundal, 2002), thionins (Vignutelli *et al.,* 1998), vegetative storage proteins (Benedetti *et al.,* 1995), lipoxygenases, ribosome inactivating proteins, enzymes of phenylpropanoid metabolism, and others (Devoto and Turner, 2003).

The combination of JA and ABA contents were increased by 56.55% and 33.83%, respectively in response to the same treatment (40% WHC) as compared with control values. These results are in a good agreement with those of Lehmann *et al.* (1995) who showed that treatment of *Hordeum vulgare* with sorbitol, mannitol, sucrose, glucose and the neutral desiccation agent polyethylene glycol (PEG) caused an increase in endogenous JA and ABA. Similar results were obtained by Xin *et al.* (1997) working on maize plants. Also, Finch-Savage *et al.* (1996) proved that the concentration of JA, MeJA and ABA progressively increased during drying in both cotyledons and axes of whole seeds of *Quercus robur.*

The plants grown under drought stress conditions exhibited pronounced increases in the detectable amounts of zeatin, zeatin riboside and total cytokinins over the control value (80% WHC) by 23.35%, 76.23% and 61.58%, respectively. Similar results were obtained by El-Meleigy *et al.* (1999) who found that kinetin can partially counteract the adverse effects of drought stress in *Arachis hypogaea* plant. With regard to ethylene content, it has been found in the present work that this content was increased by 11.59 fold in drought stressed plants (40% WHC) compared with the control value (80% WHC). These results are in a good agreement with those of Mcjeon *et al.* (1982) and Chrominski *et al.* (1988). Therefore, the increase in jasmonic acid, abscisic acid, ethylene and cytokinin contents are adaptive responses and safeguard mechanisms of soybean plants to cope with drought stress.

Cycocel is used as plant growth regulator because it has a low toxicity and affects many crops. Both roots and shoots can absorb it (Banerjee *et al.*, 2012). It is an anti-gibberellin agent that prevent Trans - Geranyl pyro phosphate in Mevalonic acid routes and leads to lower levels of corn (gibberellic acid precursor synthesis) and thus reduces gibberellin levels in plant and plant height is reduced (De *et al.*, 1982).

It has been reported that plants treated with cycocel in Hana (Khandewel *et al.*, 2002), spring and winter barley (Ma & Smith, 1992) and wheat (De *et al.*, 1982), with reduced apical dominance, resulted in the growth of vegetative buds and formation of side branches and ultimately increased the number of branches and the productivity. Research shows that cycocel spraying on plants enhanced performance of barely (Humphries *et al.*, 1965) and rained wheat (De *et al.*, 1982) .Function in plants treated with cycocel due to root growth, water potential and stomatal resistance in leaves ultimately increased the efficiency of water use through increasing root activity and reduction of plant transpiration (Starman & Williams, 2000).

Phytohormones or poly-amines have multiple roles in improving drought tolerance of maize. These functions are improved with cell water status and alleviation of oxidative damage on the biological membranes. This suggests that maintenance of water economy through stabilized cellular structure is an important mechanism of drought tolerance in maize. Establishment of similar roles of polyamines are likely to be a great step in improving drought tolerance in high water requiring plant species.

Polyamines are now being regarded as plant growth regulators and secondary messenger in signaling path-ways (Davis, 2004; Liu *et al.* 2007; Kusano *et al.* 2008), and play an array of physiological roles in plant growth and development (Ali 2000). Although, they induce tolerance against several abiotic stresses in plants (Nayar *et al.* 2005; Yang *et al.* 2007; Farooq *et al.* 2007), mechanisms of their action during exogenous application in modulating physiological phenomena and improving drought stress tolerance are not fully under-stood. Thus, the effect of exogenous treatments with some phytohormones and polyamines have counteracting the adverse effects of drought on growth, photosynthetic pigments, mineral contents and yield, production of the maize (Shaddad *et al.* 2009).

The application of putrescine, spermine or mixture of putrescine and spermine polyamines or any one of 10 ml (200 μ. Mol) the polyamines (20 μ. Mol) or one of the phytohormones, IAA, GA3, Kinetin and mixture of these phytohormones resulted. Polyamines alone or in combination increased the dry matter yield at 50% field capacity over those of the relative control values, the level which reduces dry matter yield to about 50% in only drought stressed plants. The IAA promoted the dry matter production in plants irrigated with 50% field capacity by about 56% over those of related control value. Kinetine and mixture phytohormones induced a slight increase in these values at 50% field capacity (about 27%). GA3 alleviated the inhibitory effect of drought at 50% field capacity. In general Polyamines seemed to be the superior up to 50% field capacity, while phytohormones were superior at 30% field capacity. Most of them nearly alleviated the inhibitory effect of drought stress.

Plant growth regulators like salicylic acid, cytokinin, gibberrellins and abscisic acid amend the plant responses towards drought stress (Farooq *et al.,* 2009). Foliar application of salicylic acid ameliorates the adverse effects of water stress and enhances the restoration of the growth process in wheat (Sakhabutdinova *et al.,* 2003). Putrescine, a new plant growth regulator, belonging to polyamines, is reported for its beneficial effects on plant growth under environmental stresses (Perez-Adamor *et al.,* 2002). Salicylic acid and Putrescine are economical, non-hazardous and bring stress resistance through metabolic defense mechanisms leading to better plant growth and yields.

The growth regulators like salicylic acid and Putrescine were applied @ 10^{-5}mol/L as foliar spray 3 days after drought induction. Drought stress significantly reduced leaf relative water content (LRWC), chl *a,* chl *b,* carotenoids and soluble proteins but augmented the leaf proline, seed glucosinolates and oil erucic acid (C22: 1) contents. Growth regulators were highly effective in ameliorating the adverse effects of drought stress on both the canola cultivars. The applied growth regulators maintained the water budget of canola plants, augmented the accumulation of osmolyte proline and protected photosynthetic pigments from adverse effects of drought stress. The SA was effective to reduce the drought induced accumulation of glucosinolates and erucic acid in canola oil and both the growth regulators overcame the drought induced decrease in oleic acid (C18:1). It is inferred that SA is economical and environment friendly alternative and can be implicated to improve the plant growth and oil quality of canola in current scenario of drought and climate change.

Polyamines are small ubiquitous nitrogenous compounds. Increased polyamine levels in stressed plants are of adaptive significance because of their involvement in the regulation of cellular ionic environment, maintenance of membrane integrity, prevention of chlorophyll loss and stimulation of protein synthesis, nucleic acids and protective alkaloids (Bouchereau et al. 1999; Kusano et al. 2008). Stabilization of membranes and minimization of water stress of various kinds of cells are of the several known physiological effects of polyamines (eg. putrescine) in the plant system (Liu et al. 2007; Goyal and Asthir 2010).

Cytokinins are important group of plant bioregualators which play important role in greater partitioning of photosynthates towards reproductive sink thereby improving the harvest index (Sharma et al. 2008). They protect cell membranes against degradation by preventing oxidation of unsaturated fatty acids and hence increase drought tolerance in plants. Further, cytokinins inhibit formation and speed up break down of free radicals such as superoxide (O_2^-) and hydroxy radical (OH^-) that otherwise oxidizes membrane lipids (Werner and Schmulling, 2009).

The role of polyamines and cytokinins in imparting drought tolerance has been reported in wheat (Sharma 1999; Gupta et al. 2000), but their comparative tolerance to drought has not been explored. .They have studied the effect of foliar application of putrecine and benzyladenine on various physio-biochemical and yield parameters of wheat under water stress conditions. Efforts have also been made to know that which one acts more efficiently to improve drought tolerance in wheat.

Plant growth regulators act as modulators of plant responses to water stress. In recent years, polyamines are also being considered as growth regulators and secondary messengers in signaling pathways (Kusano et al. 2008). Maintenance of favourable water status is of paramount importance in plants. Water stress conditions reduced the relative water content significantly but putrescine (put) and benzyladenine (BA) played pivotal role in maintaining higher water status. Further, the percent increase was higher with foliar spray of putrescine (24.39 %) over benzyladenine (18.90 %) under water stress conditions. It is suggested that this increase in relative water content as a result of putrescine and benzyladenine treatment might be due to enhanced water absorption. Gupta et al. (2000) reported that exogenous application of cytokinin increased water absorption in wheat and thus yield was enhanced. Sharma (1999) reported that polyamines prevented the water stress induced decline in leaf relative water content in pea.

MINERAL NUTRITION IN RELATION TO DROUGHT

Water stress also affects plant mineral nutrition and disrupts ion homeostasis. Calcium plays an essential role in structural and functional integrity of plant membrane and other structures. Decrease in plant Ca_2+ content was reported in many plants. For example, approximately 50% decrease in Ca_2+ in drought stressed was found in maize leaves, while in roots Ca_2^+ concentration was higher compared to control. Potassium is an important nutrient and plays an essential role in water relation, osmotic adjustment, stomatal movement and finally plant resistance to drought. Decrease in K+ concentration was reported in many plant species under water deficient plants which is conditioned mainly due to membrane damage and disruption in ion homeostasis. The K+ deficient plant has lower resistance to water stress. Nitrogen metabolism is the most important factor that influences plant growth and performance. Disruption in N metabolism is crucial in X-plant injury under the water deficit conditions. Some studies showed the reduction of nitrate uptake and decrease in nitrate reductase activity under water stress.

1. Nitrogen

Nitrogen is the mineral element that plants require in the largest amounts and is a constituent of many plant cell components, including amino and nucleic acids. Therefore, nitrogen deficiency rapidly inhibits plant growth. An early study of the relationships between water availability and the N fertilizer responses from Smika *et al*. (1965) demonstrated that fertilizer N will not increase yield without sufficient water being available to the plant, and increasing soil-water availability will not increase production without adequate N supply. A recent study on wheat grown in a sandy soil under different moisture-stress levels (Geesing and Schmidhalter, unpublished) showed similar interactive effects of N supply and drought stress. Application of additional N enhanced wheat yield only when the drought was not severe. While drought conditions may reduce soil-N mineralization, thus lowering the N availability (Bloem *et al*., 1992), a reduced crop N uptake may also be attributed to a decreased transpiration rate to transport N from roots to shoots (Tanguilig *et al*., 1987).

2. Phosphorus

Phosphorus is a constituent of nucleic acids, phospholipids, phosphoproteins, dinucleotides, and adenosine triphosphate. Hence, P is required for processes including the storage and transfer of energy, photosynthesis, the regulation of some enzymes, and the transport of carbohydrates. The data of the interactive effect between P nutrition on plant growth in the literature are frequently contradictory. However, the addition of P or the strategies employed to increase P uptake in most cases seem to be more important in drought conditions.

An important approach for increasing P uptake involves taking advantage of the symbiosis between the roots and mycorrhiza, the latter of which enhances both the growth or resistance of plants subjected to drought, and also the uptake of P, Zn, Cu, Mn, and Fe (e.g., Bagayoko *et al*., 2000). Ajouri *et al*. (2004) reported that priming seeds with solutions containing the limiting nutrients under drought conditions (such as P and Zn) can improve barley establishment. Smith (2002) suggested that strategies for increasing nutrient uptake by over-expressing genes encoding for high-affinity P transporters are likely to be an important strategy in the future, especially in light of the increasing problems caused by P-deficient soils of the semi-arid tropics.

3. Potassium

Potassium is an essential factor in protein synthesis, glycolytic enzymes, and photosynthesis; an osmoticum mediating cell expansion and turgor-driven movements; and a competitor of Na+ under saline conditions (Marschner, 1995). Because both drought and salinity affect plant growth similarly through water deficit, the K+ is equally important for maintaining the turgor pressure in plants under either stress. Additionally, higher K+, Na+ ratios will also improve the resistance of the plant to salinity (Asch *et al*., 2000).

The availability of K+ to the plant decreases with decreasing soil water content, due to the decreasing mobility of K+ under these conditions. Kuchenbuch *et al*. (1986) reported that low levels of soil moisture reduced root growth and the rate of potassium inflow in onion plants in terms of both per unit of root growth and per unit of root length. Under drought conditions, wilting in plants suggest possible K+ deficiency (Beringer and Troldenier, 1978). This differs from salt-stressed plants, which are characterized by the absence of wilting at the same water potentials as that for drought stress (Shalhevet and Hsiao, 1986). Numerous studies have shown that the application of K fertilizer mitigates the adverse effects of drought on plant growth (e.g., Andersen *et al.,* 1992; Studer, 1993; Sangakkara *et al*., 2001). Potassium increases the plant's drought resistance through its functions in stomatal regulation, osmoregrulation, energy status, charge balance, protein synthesis, and homeostasis (Beringer and Troldenier, 1978; Marschner, 1995). It also maintains turgor pressure (Mengel and Arneke, 1982) and reduces transpiration under drought conditions (Andersen *et al*., 1992). In plants coping with drought stress, the accumulation of K^+ may be more important than the production of organic solutes during the initial adjustment phase, because osmotic adjustment through ion uptake like K^+ is more energy efficient (Hsiao, 1973). Working with wheat, Morgan (1992) showed that the lines displaying high osmotic adjustments had a high accumulation of K^+ in their tissues. In these lines, K^+ accounted for about 78% of all osmotica. By contrast, amino acids, which were the only other important contributor, constituted only about 22%. A study by *Ma et al*. (2004) showed that K^+ accumulation in the expanding leaves in three genotypes of *Brassica napus* oilseeds accounted for about 25% of drought-induced changes in osmotic adjustment.

High Na^+ concentrations in the external solution cause a decrease in both K^+ and Ca^{2+} concentrations in the tissues of many plant species (Hu and Schmidhalter, 1997). These decreases could be due to the antagonism of Na^+ and K^+ at uptake sites in the roots, the effect of Na^+ on K^+ transport into the xylem (Lynch and Läuchli, 1984), or the inhibition of uptake processes (Suhayda et al., 1990). Data from a study by Hu and Schmidhalter (1997) illustrate that, although salinity decreases K^+ accumulation in wheat leaves, the increase in macronutrients that occurs at an optimal K^- fertilization level does not alter the K^+ concentration for a given salinity level. At a deficient level of K^+, however, K^+ accumulation under salinity was significantly increased with an increase in the macronutrient content. However, Bernstein *et al*. (1974) reported that, despite the beneficial effects of K^+ fertilization on increasing the ratio of K^+/ Na^+ within maize plants, it did not reduce the adverse effects of salinity on plant growth.

Reasonable amounts of both K^+ and Ca^{2+} are required to maintain cell-membrane integrity and its function (Wei *et al*., 2003). In glycophytes, which comprise most crop species, exclusion is the predominant strategy for plant resistance to salt stress. Glycophytes restrict both the uptake of toxic ions like Na+ by the roots from the soil and also the movement of the toxic ions to the shoot by controlling their influx into the root xylem from the root cells. A high K^+/ Na^+ ratio in the plants is also maintained by the roots, showing a selectivity of K^+ over Na^+ and by a preferential loading of K^+ rather than of Na^+ into the xylem. Toxic ions like Na^+ can also be transported away from the cytoplasm into the vacuole of cells (intracellular compartmentalization).

4. Calcium

Calcium plays a vital role in regulating many physiological processes that influence both growth and responses to environmental stresses. Included among these are water and solute movement as influenced through the effects of Ca^{2+} on membrane structure and stomatal function, cell division and cell-wall synthesis, direct or signaling roles in systems involved in plant defense and repair of damage from biotic and abiotic stress, and rates of respiratory metabolism and translocation (McLaughlin and Wimmer, 1999). Despite its obvious importance, the low mobility of Ca^{2+} makes the rates of its uptake and distribution limiting processes for many key plant functions. Furthermore, the general lack of recognition of the limiting role of Ca^{2+} is due, in part, to the fact that some important plant functions are controlled by changes in very small physiologically active pools of Ca^{2+} within the cytoplasm. As such, whole-leaf Ca^{2+} levels might not reflect any potential limitations (McLaughlin and Wimmer, 1999).

Although Ca uptake, as for other elements, is decreased under drought conditions, overall Ca accumulation is depressed only slightly compared with P and K. In mature maize plants grown in dry conditions, for example, component accumulations of P, K, and Ca were 40%, 71%, and 91%, respectively, of those found in mature plants grown in well-watered conditions (Jenne *et al.*, 1958). Oertli (1991) also reported that sufficient Ca^{2+} should be present in arid soils with high P^{H}. However, the Ca^{2+} status of more acidic tropical soils in semi-arid regions can be more problematic.

5. Other elements

Little information is available on the effect of drought on Mg and S nutrition of plants. However, drought reduces both Mg and S uptake because Mg deficiency is occurring more frequently as the cropping intensity increases and because the decrease in atmospheric S during the past decade has been associated with the appearance of S deficiency symptoms (Scherer, 2001) and has had severe consequences for S nutrition and crop production. Although many studies have analyzed plant tissues for Mg^{2+}, few salinity nutrient studies have directed any attention to Mg^{2+} nutrition as affected by salinity (Grattan and Grieve, 1999). A study by Hu and Schmidhalter (1997) revealed that the Mg^{2+} concentration in wheat leaves was reduced, but could be increased for a given salinity level by increasing the Mg^{2+} concentration in the nutrient solution. A correlation between wheat yield and Mg^{2+} accumulation in the plant under saline conditions was also noted, suggesting that Mg^{2+} supplementation could play a role in increasing plant salt resistance when this nutrient is available at low levels.

Because the transport of micronutrients to the plant roots occurs *via* diffusion, low soil moisture content, as with P uptake, will reduce micronutrient uptake. However, because plants require much smaller quantities of micronutrients, the effects of drought stress are not as great as for P. Temporary B deficiency is common under drought and is related both to low mineralization, because much of the B is in the organic matter, and also to the lower subsoil B in some areas (relative to the surface soil), because water uptake occurs predominantly from the subsoil under dry conditions. By contrast, excessive rainfall can leach some of the available soil B in sandy soils. Low soil moisture can also induce deficiencies in Mn, Mo, Fe, and Zn. On the other hand, Mn and Fe become increasingly available under moist conditions because of its conversion to reduced and more soluble forms (Havlin *et al.*, 1999). Several recent studies have shown that an increased Si accumulation in plants can improve growth under drought conditions, due to a reduction in transpiration (Epstein, 1999; Ma, 2004).

Naik (2004) conducted field experiment to study the physiological responses of groundnut to hormone and nutrients. He observed that the soil application of ½ RDF with foliar spray of NAA (50 ppm) and KNO_3 (0.5%) at 25 and 45 DAS gave highest yield of 25.11 q ha^{-1}which was significantly higher by 5.61% respectively over RDF and their B:C ratio was 2.65. and therefore there is a economy in ½ RDF.

Table: Effect of hormone and nutrients on yield and yield contributing parameters in groundnut

Treatments	No. of pods $plant^{-1}$ (g)	100 kernel weight (g)	Shelling (%)	Pod yield ha^{-1} (q)
T_1 RDF	18.43	29.33	56.33	23.77
T_2 ½ RDF	17.20	28.33	44.66	21.55
T_3 ½ RDF + NAA (50 ppm)	17.23	28.33	48.00	22.44
T_4 ½ RDF + $FeSO_4$ (0.5%)	17.80	29.00	53.66	23.66
T_5 ½ RDF + KNO_3 (0.5%)	18.76	30.00	57.00	24.00
T_6 ½ RDF + $ZnSO_4$ (0.5%)	17.23	28.33	50.00	22.88
T_7 ½ RDF + $MgSO_4$ (0.5%)	17.63	28.66	53.33	23.33
T_8 ½ RDF + NAA (50 ppm) + $FeSO_4$ (0.5%)	18.96	31.33	59.66	24.88
T_9 ½ RDF + NAA (50 ppm) + KNO_3 (0.5%)	19.73	32.66	61.33	25.11
T_{10} ½ RDF + NAA (50 ppm) + $ZnSO_4$ (0.5%)	18.83	30.33	58.00	24.55
T_{11} ½ RDF + NAA (50 ppm) + $MgSO_4$ (0.5%)	17.76	28.33	53.33	23.44
SE (m) ±	0.193	0.364	0.271	0.097
CD at 5%	0.568	1.073	0.800	0.288

RDF=Recommended dose of fertilizer 25:50:0 NPK kg ha^{-1} (Source: Naik R.W., 2004)

Basole (2002) carried out an field experiment to study the physiological responses of soybean to hormone and nutrients. He observed that soil application of ½ RDF and foliar spray of NAA (50 ppm) and $ZnSO_4$ (0.5%) at 50 and 90 DAS gave highest yield of 11.66 q ha^{-1} which was significantly higher by 4.76% over a RDF and their B : C ratio was 2.0 and therefore there is a economy in ½ RDF.

Table : Effect of hormone and nutrients on yield and yield contributing parameters in soybean

Treatments	No. of pods $plant^{-1}$ (g)	100 seed weight (g)	seed yield ha^{-1} (q)
T_0 RDF	28.0	10.06	11.13
T_1 ½ RDF	17.6	8.30	5.74
T_2 ½ RDF + NAA (50 ppm)	24.6	9.53	8.52
T_3 ½ RDF + $FeSO_4$ (0.5%)	21.0	8.70	6.73
T_4 ½ RDF + KNO_3 (0.5%)	22.6	8.90	7.00
T_5 ½ RDF + $ZnSO_4$ (0.5%)	24.3	9.03	7.19
T_6 ½ RDF + $MgSO_4$ (0.5%)	23.0	9.23	7.07
T_7 ½ RDF + NAA (50 ppm) + $FeSO_4$ (0.5%)	28.0	9.60	8.91

T_8 ½ RDF + NAA (50 ppm)+ KNO_3 (0.5%)	32.3	10.46	11.37
T_9 ½ RDF + NAA (50 ppm) + $ZnSO_4$ (0.5%)	36.0	10.90	11.66
T_{10} ½ RDF + NAA (50 ppm) + $MgSO_4$ (0.5%)	32.0	10.06	11.22
SE (m) ±	0.5504	0.0535	0.0507
CD at 5%	1.6191	0.1575	0.1491

RDF=Recommended dose of fertilizer 30:75:30 NPK kg ha^{-1} (Source- Basole V.B., 2002)

Suke (2009) conducted field experiment to study the physiological responses of nutrients and biofertilizers in maize. He observed that soil application of RDF with *Azotobactor* and PSB gave highest yield of 66.55 q ha^{-1} which was significantly higher by 30.18 % and their B : C ratio was 1.94.

Table : Effect of RDF, *Azotobactor* and PSB on yield and yield contributing parameters in maize

Treatments	Grain yield $plant^{-1}$ (g)	Grain yield $plot^{-1}$ (kg)	Grain yield ha^{-1}
T_1 RDF(120 kg N + 60 kg P_2O_5 + 40 kg k_2O ha^{-1})	33.53	4.46	46.45
T_2 RDF + *Azotobactor*	37.40	4.81	50.17
T_3 RDF + *Azotobactor* + PSB	38.88	5.28	54.99
T_4 RDF + *Azotobactor* + PSB	46.46	6.38	66.55
T_5 90 % RDF + *Azotobactor* + PSB	43.20	5.82	60.65
T_6 80 % RDF + *Azotobactor* + PSB	42.77	5.72	59.58
T_7 70 % RDF + *Azotobactor* + PSB	40.25	5.44	56.73
T_8 60 % RDF + *Azotobactor* + PSB	33.33	4.55	47.72
T_9 50 % RDF + *Azotobactor* + PSB	27.91	3.48	36.28
SE (m) ±	3.7670	0.4831	5.0316
CD at 5%	11.192	1.4355	14.9503

RDF=Recommended dose of fertilizer 30:75:30 NPK kg ha^{-1} (source- Suke S.N., 2009)

Table : Effect of growth regulators on different crops.

S.N.	Name of growth regulator	Concentrations used	Results	Name of scientist
1.	Ethrel	20 ppm	Increased the straw and grain yields of wheat.	Alam *et al.* (2002)
2.	Ethrel	250 ppm	Increased plant height, protein and oil content, 100 grain weight and grain yield ha^{-1} in maize.	Hamed (2003)
3.	Ethrel	250 ppm	Increased leaf area and dry weight, number of pods	Saxsena *et al.* (2007)

			plant $^{-1}$, 100 grain weight, seed yield and harvest index in chickpea.	
4.	Ethrel	50 ppm	Increased the leaf area, dry weight of plant, AGR, RGR and NAR, grain yield in urdbean.	Aucharmal *et al.* (2008)
5.	Ethrel	200 ppm	Increased number of primary branches plant^{-1}, leaf area number of pods^{-1}, number of seeds pod^{-1} and seed yield ha^{-1} in cowpea.	Chovatia *et al.* (2010)
6.	Ethrel	200 micro l^{-1}	Enhanced leaf area index, net photosynthetic rate, nitrate reductase activity and plant dry mass, number of pods plant $^{-1}$, 1000 seed weight, seed yield ha^{-1} in mustard.	Lone *et al.*(2010)
7.	Ethrel	200 ppm	Highest plant height, number of branches, dry weight plant^{-1}, leaf area plant^{-1}, protein content, oil content, number of pods plant^{-1}, 100 seed weight, yield ha $^{-1}$ and harvest index over control in soybean.	Devi *et al.* (2011)
8.	kinetin homobrassinolide	kinetin @ 5 ppm and homobrassinolide @ 1 ppm	Increased seed yield by 20.9% in chickpea.	Vijayakumar *et al.*(2013)
9.	Humic acid with brassinosteroid	0.1 % humic acid and 1 ppm brassinosteroid	Increased Chlorophyll, total chlorophyll and yield in green gram.	K. Ananthi and Mallika Vanangamudi (2013)
10.	Brassinosteriods	1 ppm	Increased plant height, number of branches, leaf area, chlorophyll, oil content, number of capsuleplant^{-1}, 1000 seed weight (g), yield plant^{-1} and harvest index in sesame.	Prakash *et al.* (2008)
11.	Putrescine	20 ppm	Increased plant height, leaf area, Number of pods plant^{-1}, number of seeds plant^{-1}, 100 seed weight, and seed yield in green gram.	Thavaprakash *et al.*(2006)

12.	Putrescine	100 mg L^{-1}	Increased seed yield by 21.3% in chickpea.	Aman *et al.* (2013)
13.	Salicyclic acid	100 ppm	Increased plant height, number of branches, leaf area, Number of pods plant^{-1}, 100seed weight and grain yield in soybean.	Vaiyapuri *et al.* (2012)
14.	TIBA, GA , AA (Ascorbic acid), IAA, NAA, Cytokinin, Triacontanol and CCC (cycocel)	500 ppm, 50 ppm, 50 ppm, 50 ppm, 50 ppm, 5 ppm, 10 ppm and 4000 ppm	All growth regulators significantly increased plant height, number of branches plant^{-1}, number of leaves plant^{-1}, leaf area, total dry matter production, relative growth rate, net assimilation rate), total chlorophyll content, protein and yield, whereas TIBA and CCC reduced plant height in pea.	Tripathi and Kumar (2006)
15.	Mepiquat chloride	2000 ppm	Increased stem diameter, leaf area, dry matter, number of seeds plant^{-1}, head diameter and seed yield, whereas plant height was decreased in sunflower.	Doddamani *et al.* (2010)
16.	TIBA, Mepiquat chloride and Cycocel	50 ppm, 100 ppm and 250 ppm	Increased number of seeds plant^{-1} and seed yield in soybean.	Pankaj Kumar *et al.* (2001)
17.	Atonik andTIBA	0.25 % and 100 ppm	Increased the growth parameters viz., plant height, leaf area and yield plant^{-1} whereas TIBA reduced plant height in cotton.	Djanaguiraman *et al.* (2005)
18.	TIBA, NAA, GA3 and CCC	100 ppm, 50 ppm, 50 ppm, 500 and 1000 ppm	Increased number of branches, leaf area, grain yield (q ha^{-1}) in soybean.	Salunkhe *et al.* (2008)
19.	TIBA 75 ppm, Cycocel and Mepiquat chloride	75 ppm, 500, 1000, 1500 ppm and 1000, 1500, 2000 ppm	Increased seed yields in sunflower.	Kashid *et al.* (2010)
20.	TIBA	50 ppm	Increased total dry matter weight, number of branches plant^{-1}, leaf area, 100 seed weight and seed yield in cowpea.	Ganiger *et al.* (2002)

Table : Effect of plant nutrients on different crops.

Sr. No.	Name of Nutrients	Concentrations used	Result	Name of scientist
1.	NPK and chelated micronutrients	20 ppm	Increased yields of wheat.	Manivannan *et al.* (2002)
2.	NAA + KNO_3	NAA 20 ppm + KNO_3 0.5%	Increased the dry matter production, seed yield of red gram.	Reddy *et al.* (2004)
3.	Potassium nitrate, urea and muriate of potash	2% potassium nitrate, 2% urea and 2% muriate of potash	Increased in the number of flowers and bolls plant^{-1} and potassium nitrate increased the N and K contents by 4.2 and 31.6%, while urea enhanced the N and K content by 11.3 and 21.1% in cotton leaves, respectively.	Brar and Brar (2004)
4.	pre-sowing seed treatments (water, KCl, KH_2PO_4 and $(NH_4)_6MO_7O_{24}$), rhizobium inoculation (*Bradyrhizobium japonicum*) and foliar spraying of nitrate salts KNO_3 and $Ca(NO_3)_2$)	0.50% KNO_3 and 0.203 and 0.406% $Ca(NO_3)_2$	increase in grain yield in green gram	Sarkar and Pal (2006)
5.	Nitrogen fertigation level and foliar application of chelated calcium	160 kg N fed^{-1} and chelated calcium at 1000 ppm	increased the leaf area, plant height, fresh and dry weight, total chlorophyll, nitrogen, phosphorus, potassium and calcium in leaves and yield characters ear length, ear diameter, weight of 1000 seed, unhusked ear, husked ear and total yield in sweet gram.	El-Yazied *et al.* (2007)
6.	Potassium nitrate	0.4%	pod growth rate (PGR), crop growth rate (CGR) and significantly increased grain yield by 18.4 per cent in green gram.	Vekaria *et al.* (2012)

REFERENCES

Ajouri, A., H. Asgedom, and M. Becker, 2004. Seed priming enhances germination and seedling growth of barley under conditions of P and Zn deficiency. J. Plant Nutr. Soil Sci. **167**: 630–636.

Alam, S.M, A. Shereen and M.A. Khan, 2002. Growth response of wheat cultivars to NAA and Ethrel. Pak. J. Bot. **34**(2): 135-137.

Ali R.M., 2000. "Role of Putrescine in Salt Tolerance of Atropa Belladonna Plant," Plant Sci., **52** (2): 173-179.

Amin, A.A., F.A. Gharib, H. F. Abouziena and M. G. Daeood, 2013. Role of Indole-3-butyric Acid and Putrescine in Improving Productivity of Chickpea (*Cicer arientinum* L.) Pak. J. Bio. Sci., **16**:1894-1903.

Ananthi, K. and M. Vanangamudi, 2013. Foliar application of humic acid with brassinosteroid on Chlorophyll content and yield of greengram [*vigna radiata* (L.) Wilczek]. Legume Res., **36** (3): 241 – 244.

Andersen, M.N., C.R. Jensen, R. Losch, 1992. The interaction effects of potassium and drought in field grown barley. 1. Yield, water-use efficiency and growth. Acta Agric. Scand. Sect. B Soil Plant Sci., **42**: 34–44.

Asch, F., Dingkuhn, M., Miezan, K., Dorffling, K. 2000: Leaf K/Na ratio predicts salinity induced yield loss in irrigated rice. Euphytica **113**: 109–118.

Ashraf, M., 2010. Inducing drought tolerance in plants some recent advances. Biotech. Adv., **28**: 169-183.

Aucharmal, S.M., S.G., Bhosle, H. Pawar and K. Jadhav, 2008. Growth analysis of urdbean under different growth regulators. J. Food Leg., **21**(3): 204-205.

Bagayoko, M., E. George, V. Römheld and A.B. Buerkert, 2000. Effects of mycorrhizae and phosphorus on growth and nutrient uptake of millet, cowpea and sorghum on a West African soil. J. Agric. Sci. **135**: 399–407.

Bakht, J., A. Bano and P. Dominy. 2006. The role of abscisic acid and low temperature in chickpea *(Cicer arietinum)* cold tolerance: II. Effects on plasma membrane structure and function. J. Exp. Bot., **57**: 3697-3706.

Bakht, J., M. M. Shafi, Yousaf and M.A. Khan. 2010. Effect of Irrigation on Physiology and Yield of Sunflower Hybrids. Pak J. Bot. **42**: 1317-1326.

Bakht, J., Y. Jamal, M. Shafi and H. Sher. 2011. Response of maize *(Zea mays* L.) to seed priming with NaCl and salinity stress. Spanish J. Agric., **9**: 252-261.

Banerjee A, J. Datta, and N. Kumar Mondal, 2012. Changes in morpho-physiological traits of mustard under the influence of different fertilizers and plant growth regulator cycocel. J. Saudi Soci. Agric. Sci., **11**: 89–97.

Bar, M.S.and A.S. Brar, 2004. Foliar nutrition as a supplement to soil fertilizer application to increase yield of upland cotton. Ind. J. Agric. Sci., **74**(9): 472-475.

Basole, V,B. 2002. Physiological responses of soybean to hormone and nutrients unpublished thesis submitted to Dr. P.D.K.V., Akola (Maharashtra).

Benedetti, C.E., C.L. Costa, S.R. Turcinelli and P. Arruda, 1998. Differential expression of a novel gene in response to coronatine, methyl jasmonate and wounding in the coi1 mutant of *Arabidopsis.* Plant Physiol., **116**: 1037-1042.

Benedetti, C.E., D. Xie and J.G. Turner, 1995. Coi1-dependent expression of an *Arabidopsis* vegetative storage protein in flowers and siliques and in response to coronatine or methyl jasmonate. Plant Physiol., **109**: 567-572.

Beringer, H. and G. Troldenier, 1978. Influence of K nutrition on the response to environmental stress, in: Potassium Research Review and Trends. Proceedings of the 11th Congress of the International Potash Institute. Internat. Potash Inst., Bern, pp. 189–222.

Bernstein, L., Francois, L. E., Clark, R. A. 1974. Interactive effects of salinity and fertility on yields of grains and vegetables. Agron. J. **66**: 412–421.

Bloem, J., P.C. Deruiter, G.J. Koopman, G. Lebbink and L. Brussaard, 1992. Microbial numbers and activity in dried and rewetted arable soil under integrated and conventional management. Soil Biol. Biochem. **24**: 655–665.

Blum, A., 1996. Constitutive traitsaffecting plant performance under stress. In G.O. Edmeades, M. Banziger, H.R. Mickelson and C.B. Pena-Valdivia (eds.), Developing Drought and Low N Tolerant Maize, pp. 131–135. Proceedings of the Symposium, Mexico D.F. Cimmyt.

Bouchereau A, A. Aziz, F. Larher and J. Martin Tanguy, 1999. Polyamines and environmental challenges: recent development. Plant Sci., **140**: 103–125.

Castillo, E.G., T.P. Tuong, A.M. Ismail and K. Inumbushi. 2007. Response to salinity in rice: Comparative effects of osmotic and ionic stresses. Plant Prod. Sci., **10**: 159-170.

Chaves M.M., J.S. Pereira, J. Maroco, M.L. Rodriques, C.P.P. Ricardo, M.L. Osorio, I. Carvatho, T. Faria and C. Pinheiro, 2002. How plants cope with water stress in the field photosynthesis and growth Ann. Bot., **89**: 907–916.

Chovatia R.S., T.R., Ahlawat, S.V., Mepa and G. Jat, 2010. Response of cowpwa (*Vigna unguiculata L.*) cv. GUJ-4 to the foliar application of plant growth regulating chemicals. Veg. Sci., **37**(2): 196-197.

Chrominski, A., R.B. Bhat, D.J. Weber and B.N. Smith, 1988. Osmotic stress-dependent conversion of 1-aminocyclopropane-1-carboxylic acid (ACC) to ethylene in the halophyte, *Allenrolfea occidentalis.* Environ. Exp. Bot., **28**: 171-174.

Creelman, R.A. and J.E. Mullet, 1995. Jasmonic acid distribution and action in plants: regulation during development and response to biotic and abiotic stresses. Proc. Natl. Acad. Sci., **92**: 4114-4119.

Davies P. J. 2004. "The Plant Hormones: Their Nature, Occurrence and Function," In: P. J. Davies, Eds., Plant Hormones, Biosynthesis, Signal Transduction, Action, Kluwer, Dordrecht.

Dea R, G. Giria, G. Sarana, R.K. Singha and G.S. Chaturvedia, 1982. Modification of water balance of dryland wheat through the use of chlormequat chloride. J. agric. Sci., **98**(03): 593-597.

Devi, K.N., N.G. Vyas, and M.S. Singh, 2011. Effect of bioregulators on growth, yield and chemical constituents of soybean *(Glycine max)*. J. Agric.Sci. **3**(4): 151-159.

Devoto, A. and J. Turner, 2003. Regulation of jasmonate-mediated plant responses in *Arabidopsis.* Annu. Bot., **92**: 329-337.

Djanaguiraman, M., J.A. Sheeba, D.D. Devi and U.Bangarusamy, 2005. Response of cotton to Atonik and IBA for growth, enzymes and yield. J. Biol. Sci., **5** (2): 158-162.

El-Yazied, A.A. Ragab, M.E. Rawia, E.I. El-Wafa, 2007. Effect of nitrogen fertigation levels and chelated calcium foliar application on the productivity of sweet corn. Arab Uni. J. Agric. Sci., **15**(1): 131-139.

El Meleigy, E.A., R.A. Hassanein and D.Z. Abdel Kader, 1999. Improvement of drought tolerance in *Araches hypogaea* L. plant by some growth substances 1. Growth and Productivity. Bull. Fac. Sci., Assiut Univ., **28**: 159-185.

Epstein, E. 1999. Silicon. Annu. Rev. Plant Physiol. **50**: 641–664.

Farooq M., S. M. A. Basra, M. H. Rehman and B. A. Saleem, 2007. "Incorporation of Polyamines in the Priming Media Enhances the Germination and Early Seedling Growth in Hybrid Sunflower (*Helianthus annuus* L.)," Int. J. agric. Bio., **9**(6): 868-872.

Farooq, M., A. Wahid and D.J. Lee. 2009. Exogenously applied polyamines increase drought tolerance of rice by improving leaf water status, photosynthesis and membrane properties. Acta Physiol. Planta., **31**: 937-945.

Farooq, M., S.M.A. Basra, A. Wahid, Z.A. Cheema, M.A. Cheema and A. Khaliq, 2008.Physiological role of exogenously applied glycinebetaine in improving drought tolerance of fine grain aromatic rice *(Oryza sativa* L.*)*. J. Agron. Crop Sci., **194**: 325–333.

Finch Savage, W.E., P.S. Blake and H.A. Clay, 1996. Desiccation stress in recalcitrant *Quercus robur* L. seeds results in lipid peroxidation and increased synthesis of jasmonates and abscisic acid. J. Exp. Bot., **47**: 66 1-667.

Ganiger, T.S., S.R. Kareekatti and B.C. Patil, 2002. Effect of plant growth regulators on yield and yield components in cowpea Karnataka J. Agric. Sci., **15** (4): 693-695.

Gaspar, T., T. Franck, B. Bisbis, C. Kevers, L. Jouve, J.F. Hausman and J. Dommes, 2002. Concepts in plant stress physiology. Application to plant tissue cultures. Plant Growth Regul., **37**: 263–285.

Govindan, K and V. thirumurugan, 2000. Response of green gram to foliar nutrition of potassium., J. Mah. Agric. Uni., **25**: 302-303.

Grattan, S.R., Grieve, C.M. 1999. Mineral nutrient acquisition and response by plants grown in saline environments, in Pessarakli, M.: Handbook of Plant and Crop Stress. Marcel Dekker, New York, pp. 203–229.

Gupta N.K, S. Gupta, A. Kumar, 2000. Exogenous cytokinin application increases chlorophyll and cell membrane stability index in wheat (*Triticum aestivum* L.). Cereal. Res. Comm., **28**: 287–291.

Hamdia, M.A 1998 " The counteraction effect of GA_3 or .IAA with Endogenous Ethylene of Wheat Plants under Salt Stress Conditions," Sixth Egyptian Botanical Conference Cairo University, **1**: 97-110.

Hamed, M. F., 2003. Performance of two maize hybrids under irrigation intervals and ethrel treatments. Ann. Agric. Sci., **41**(2): 625-634.

Han, B. and A. R. Kermode. 1999. Dehydrin like proteins in caster bean seeds and seedlings are differentially produced in responses to ABA and water stress related stresses. J. Exp. Bot., **47**: 933-939.

Havlin, J.L, J.D. Beaton, S.L. Tisdale and W.L. Nelson, 1999. Soil Fertility and Fertilizers: An Introduction to Nutrient Management. Prentice-Hall, Inc., London, pp. 406–425.

Hayat, S. and A. Ahmed. 2007. Salicylic acid: A plant hormone. Springer, Dordrecht, The Netherland, pp. 1-14.

Hossain, A.B.S, R.G. Sears, T.X. Cox and G.M. Paulsen. 1990. Dessication tolerance and its relationship to assimilate partitioning in winter wheat. Crop Sci., **30**: 622-627.

Hsiao, T.C. 1973. Plant responses to water stress. Annu. Rev. Plant Physiol. **24**: 519–570.

Hu, Y., Schmidhalter, 1997. Interactive effects of salinity and macronutrient level on wheat. 2. Composition., J. Plant Nutr., **20**: 1169–1182.

Humphries E.C, P.J, Welbank and K.J. Witts 1965. Effect of CCC (*Chlorocholine Chloride*) on growth and yield of spring wheat in the field. Ann. Applied Biol., **56** (3): 351-361.

Jaleel, C.A., B. Sankar, P.V. Murali, M. Gomathinayagam, G.M.A. Lakshmanan and R. Panneerselvam, 2008e. Water deficit stress effects on reactive oxygen metabolism in *Catharanthus roseus;* impacts on ajmalicine accumulation. Colloids Surf. B: Biointerfaces, **62**: 105–111.

Jaleel, C.A., P. Manivannan, B. Sankar, A. Kishorekumar, R. Gopi, R. Somasundaram and R. Panneerselvam, 2007b. Water deficit stress mitigation by calcium chloride in *Catharanthus roseus;* effects on oxidative stress, proline metabolism and indole alkaloid accumulation. Colloids Surf. B: Biointerfaces, **60**: 110–116.

Jaleel, C.A., P. Manivannan, B. Sankar, A. Kishorekumar, R. Gopi, R. Somasundaram and R. Panneerselvam, 2007d. Induction of drought stress tolerance by ketoconazole in *Catharanthus roseus* is mediated by enhanced antioxidant potentials and secondary metabolite accumulation. Colloids Surf. B: Biointerfaces, **60**: 201–206.

Jaleel, C.A., P. Manivannan, B. Sankar, A. Kishorekumar, S. Sankari and R. Panneerselvam, 2007e. Paclobutrazol enhances photosynthesis and ajmalicine production in *Catharanthus roseus.* Process Biochem., **42**: 1566–1570.

Jaleel, C.A., P. Manivannan, G.M.A. Lakshmanan, M. Gomathinayagam and R. Panneerselvam, 2008c. Alterations in morphological parameters and photosynthetic pigment responses of *Catharanthus roseus* under soil water deficits. Colloids Surf. B: Biointerfaces, **61**: 298–303.

Jaleel, C.A., P. Manivannan, P.V. Murali, M. Gomathinayagam and R. Panneerselvam, 2008d. Antioxidant potential and indole alkaloid profile variations with water deficits along different parts of two varieties of *Catharanthus roseus.* Colloids Surf. B: Biointerfaces, **62**: 312–318.

Jaleel, C.A., R. Gopi and R. Panneerselvam, 2008b. Growth and photosynthetic pigments responses of two varieties of *Catharanthus roseus* to triadimefon treatment. Comp. Rend. Biol., **331**: 272–277.

Jaleel, C.A., R. Gopi, B. Sankar, M. Gomathinayagam and R. Panneerselvam, 2008a. Differential responses in water use efficiency in two varieties of *Catharanthus roseus* under drought stress. Comp. Rend. Biol., **331**: 42–47.

Jenne, E., H. Rhoades, C.I. Yien and O. Howe, 1958, Change in nutrient element accumulation by corn with depletion of soil moisture. Agron. J. **50**: 71–80.

Kashid, D. A., M. B. Doddamani, M. B. Chetti, S. M. Hiremath and B. N. Arvindkumar, 2010. Effect of growth retardants on morpho-physiological traits and yield in Sunflower. Karnataka J. agric. Sci., **23** (2): 347-349.

Kaya C., A. L. Tuna and I. Yokas, 2009, "The Role of Plant Hormones in Plants under Salinity Stress," Book Salinity and Water Stress, **44** (1): 45-50.

Khan, A.J., F. Azam and A. Ali. 2010. Relationship of morphological traits and grain yield in recombinant inbreed wheat lines grown under drought conditions. Pak. J. Bot., **42**: 259-267.

Khandewal, K.S., N.K. Gupta and M.P. Sahu, 2002. Effect of plant growth regulators on growth, yield and essential oil production of henna (*Lowsonid inermis* L.). J. Horticultural Sci. Biotech., **77** (1): 67-72.

Kuchenbuch, R., N. Claassen and A. Jungk, 1986. Potassium availability in relation to soil moisture. 1. Effect of soil moisture on potassium diffusion, root growth and potassium uptake of onion plants. Plant Soil **95**: 221–231.

Kusano T., T. Berberich, C. Tateda and Y. Takahashi, 2008. "Polyamines: Essential Factors for Growth and Survival," Planta, **228** (3): 367-381.

Lawrence, R.K. and K.R. Koundal, 2002. Plant protease inhibitors in control of phytophagous insects. Plant Biotech., **5**: 10-29.

Lehmann, J., R. Atzorn, C. Bruckner, S. Reinbothe, J. Leopold, C. Wasternack and B. Parthier, 1995. Accumulation of jasmonate, abscisic acid, specific transcripts and proteins in osmotically stressed barley leaf segments. Planta, **197**: 156-162.

Lie, F., M.N. Anderson and C.P. Jensen. 2003. Loss of pod set caused by drought stress is associated with water status and ABA content of reproductive structure in soybean. Funct. Plant Biol., **30**: 271-280.

Liu J. H, H. Kitashiba, J. Wang, Y. Ban and T. Moriguchi, 2007. "Polyamines and Their Ability to Provide Environmental Stress Tolerance to Plants," Plant Biotech., **24** (12): 117-126.

Lone, N.A., M.R. Mir, M.A. Bhat, Haleema Ashraf, K.A. Bhat, R.Rashid, Asma Hassan, Nasier Ahmed, S.Akhtar, J.A Bhat, M.Habib, 2010. Effect of ethrel and nitrogen on nitrate reductase activity, photosynthesis, biomass, and yield of mustard (*Brassica junicea* L.) Rec. Res. Sci.Tech., **2**(2): 25-26.

Luo, Q., B. Yu and Y. Liu. 2005. Differential sensitivity to chloride and sodium ions in seedlings of *Glycine max* and *G. soja* under NaCl stress. J. Plant Physiol., **162**: 1003-1012.

Lutts, S., M. Almansouri and J.M. Kinet. 2004. Salinity and water stress have contrasting effects on the relationship between growth and cell viability during and after stress exposure in durum wheat callus. Plant Sci., **167**: 9-18.

Lynch, J., Lauchli, A. 1984: Potassium transport in salt stressed barley roots. Planta **161**: 295–301.

Ma B.L. and D.L. Smith 1992. Chlormequat and ethephon timing and grain production of spring barley. Agron. J., **84**(6): 934-939.

Ma, Q.F., D.W. Turner, D. Levy and W.A. Cowling, 2004. Solute accumulation and osmotic adjustment in leaves of *Brassica* oilseeds in response to soil water deficit. Aust. J. Agric. Res. **55**: 939–945.

Mahajan, S. and N. Tuteja. 2005. Cold, salinity and drought stresses: An overview. Arch. Biochem. Biophys., **444**: 139-158.

Marschner, H. 1995. Mineral Nutrition of Higher Plants. Academic Press, London.

Martinez, J.P., H. Silva, J.F. Ledent and M. Pinto, 2007. Effect of drought stress on the osmotic adjustment, cell wall elasticity and cell volume of six cultivars of common beans *(Phaseolus vulgaris* L.*)*. European J.Agron., **26**: 30–38.

Mcjeon, T.A., N.E. Hoffman and S.F. Yang, 1982. The effect of plant hormone pre-treatments on ethylene production and synthesis of 1-aminocyclopropane-1-carboxylic acid in water stressed wheat leaves. Planta, **155**: 437-443.

McLaughlin, S.B. and R. Wimmer, 1999. Physiology terrestrial ecosystem processes. New Phytol. **142**: 373–417.

Men gel, K. and W.W. Arneke, 1982. Effect of potassium on the water potential, the pressure potential, the osmotic potential and cell elongation in leaves of *Phaseolus vulgaris.* Physiol. Plant. **54**: 402–408.

Morgan, J.M. 1992. Osmotic components and properties associated with genotypic differences in osmoregulation in wheat. Aust. J. Plant Physiol. **19**: 67–76.

Naik, R.W. 2004. Physiological responses of groundnut to hormone and nutrients unpublishar thesis submitted to Dr. P.D.K.V., Akola (Maharashtra).

Nam, N.H., Y.S. Chauhan and C. Johansen, 2001. Effect of timing of drought stress on growth and grain yield of extra-short-duration pigeonpea lines. J, Agric. Sci., **136**: 179–189.

Nayyar H., S. Kaur, S.S. Kumar, K.J. Singh and K.K. Dhir, 2005. "Involvement of Polyamines in the Contrasting Sensitivity of Chickpea (*Cicer arietinum* L.) and Soybean (*Glycine max* L.)," Botanical Bulletin Academia Sinica, **46** (4): 333-338.

Oertli, J.J. 1991. Nutrient management under water and salinity stress, in: Proceeding of the symposium on nutrient management for sustained productivity. Depart. Soils Punjab Agric. Univ.. Ludhiana, India, pp. 138–165.

Pankaj Kumar, S.M. Hiremath and P.S. Deshmukh. 2001, Studies on foliar application of growth regulators on biomass production, harvest index and yield of soybean (*Glycine max* (L.) Merrill). Annals agric. Res., **22** (2): 221-224.

Parida, A.K. and A.B. Das. 2005. Salt tolerance and salinity effects on plants a review. Ecotoxcol. Environ. Safe., **60**: 324-349.

Passioura, J.B. 2002. Environmental plant biology and crop improvement. Funct. Plant Biol., **29**: 537-546.

Perez Adamor, M.A., J. Leon, P.J. Green and J. Carbonell. 2002. Induction of the arginine decarboxilase *ADC2* gene provides evidence for the involvement of polyamines in the wound response in Arabidopsis. Plant Physiol., **130**: 1454-1463.

Prakash, M., S. Suganthi, J. Gokulakrishnan, T. Sabesan, 2007. Effect of Homobrassinolide on growth, physiology and biochemical aspects of sesame. Karnataka J. Agric. Sci., **20** (1): 110-112.

Rahman, M.S. and S. Yoshida. 1985. Effect of water stress on grain filling in rice. Soil SM. Plant Nair., **31**: 497-5I1.

Razmjoo, K., P. Heydarizadeh and M.R. Sabzalian, 2008. Effect of salinity and drought stresses on growth parameters and essential oil content of Matricaria chamomile. Int. J. Agric. Biol., **10**: 451–454.

Reddy, A.R., K.V. Chaitanya and M. Vivekanandan, 2004. Drought induced responses of photosynthesis and antioxidant metabolism in higher plants. J. Plant Physiol., **161**: 1189–1202.

Reddy, J, N. Rao and B.K. Mahalakshmi, 2004. Effect of different chemicals on growth, yield and yield attributes of pigeonpea in vertisol. Ann. Plant Physiol., **17**(2): 120-124.

Riccardi, F., P. Gazeau, D.V. Ienne and M. Zivy. 1998. Protein changes in responses to progressive water deficit in maize. Plant Physiol., **117**: 1253-1263.

Royo, C., M, Abaza, R., Blanco and D.G.F. delMorat. 2000. Triticate grain growth and morphometry as affected by drought stress late sowing and stimulated drought. Aust. J. Plant Physiol., **27**: 1051-1059.

Sairam, R.K. and A. Tyagi. 2004. Physiology and molecular biology of salinity stress tolerance in plants. Curr. Sci., **86**: 407-421.

Sakhabutdinova, A.R., D.R. Fatkhutdinova, M.V. Bezrukova and F.M. Shakirova. 2003. Salicylic acid prevents the damaging action of stress factors on wheat plants. Bulgar. J. Plant Physiol., **21**: 314-319.

Salunke, T. E., A. V. Khairnar, H. M. Patil and S. D. Shinde, 2008. Influence of plant growth regulators on yield attributes and quality of soybean. International J. agric. Sci., **4** (2): 621-623.

Sangakkara, U.R., M. Frehner and J. Nösberger, 2001. Influence of soil moisture and fertilizer potassium on the vegetative growth of mungbean (*Vigna radiata* L. Wilczek) and cowpea *(Vigna unguiculata* L. Walp). J. Agron. Crop Sci., **186**: 73–81.

Sarkar, R.K. and P.K. Pal, 2006. Effect of pre-sowing seed treatment and foliar spray of nitrate salt on growth and yield of greengram. Ind. J. Agric. Sci., **76**(1): 62-65.

Sasaki, Y., E. Asamizu, D. Shibata, Y. Nakamura, T. Kaneko, K. Awai, T. Masuda, H. Shimada, K. Takamiya, S. Tabata and H. Ohta, 2000. Genome wide expression monitoring of jasmonate responsive genes of Arabidopsis using cDNA arrays. Biochem. Soci. Transactions, **28**: 863-864.

Saxsena D.C., Abbas and R.K., Sairam, 2007. Effect of ethrel on reproductive efficiency in chickpea. Ind. J. Pl. Physiol., **12**(2): 162-167.

Scherer, H.W. 2001. Sulphur in crop production Invited paper. Eur. J. Agron. **14**: 81–111.

Schmidhalter, U., Z. Burucs, S. Hu. von Tucher and R. Gutser, 1999. Foliar fertilization applied to droughted and salinized wheat and maize seedlings, in: Proceedings of the 2nd International Workshop on Foliar Fertilization. Bangkok, Thailand, pp. 345–358.

Setter, T.F. and B.A. Ammgam. 2001. Water deficit inhibits cell division and expression of transcripts involved in cell proliferation and end or duplication in maize endosperm. J. Exp. Bot, **52**: 1401-1408.

Shaddad M. A. K., M. A. Hamdia and H. T. Mohammed, 2009, "Drought Tolerance of Some *Zea mays* L. Genotypes at the Early Growth Stage," Assiut University J. Bot., **38** (2): 135-147.

Shalhevet, J. and T.C. Hsiao, 1986. Salinity and drought – a comparison of their effects on osmotic adjustment, assimilation, transpiration and growth. Irri. Sci. **7**: 249–264.

Sharma K.M., D.D. Sharma, K.B. Shukla and B. Upashyay, 2008. Growth, partitioning and productivity of wheat as influenced by fertilization and foliar application of bioregulators. Indian J. Plant Physiol **13**: 387–393.

Sharma M.L. 1999. Polyamine metabolism under abiotic stress in higher plants: salinity, drought and high temperature. Physiol. Mol. Biol. Plants, **5**: 103–113.

Smika, D., H. Haas and W. Power, 1965. Effects of moisture and nitrogen fertilizer on growth and water use by native grass. Agron. J. **57**: 483–486.

Smirnoff, N. 1993. The role of active oxygen in the response of plants to water deficit and desiccation. New Phytol., **125**: 27–58.

Smith, F. W. 2002. The phosphate uptake mechanism. Plant Soil **245**: 105–114.

Starman T.W and M.S. Williams, 2000. Growth retardants affect growth and flowering of Scaevola. Horticultural Sci., **35** (1): 36-38.

Studer, C. 1993. Interactive effects of N, P, K nutrition and water stress on the development of young maize plants. Ph.D. Thesis, ETHZ, Zurich, Switzerland.

Subramani M., A. Solaimalai and A. Velayutham, 2002. Effect of plant population and methods of fertilizer application on yield attributes and yield of irrigated blackgram. Madras Agric. J., **89**: 305-306.

Suhayda, C.G., J.L. Giannini, D.P. Briskin and M.C. Shannon, 1990. Electrostatic changes in *Lycopersicon esculentum* root plasma membrane resulting from salt stress. Plant Physiol. **93**: 471–478.

Suke, S.N. 2009. Physiological responses of nutrients and biofertilizes in maize unpublishar thesis submitted to Dr. P.D.K.V., Akola, (Maharashtra).

Swiatek, A., A. Azmi, E. Witters and H. Van Onckelen, 2003. Stress messengers' jasmonic acid and abscisic acid negatively regulate plant cell cycle. Bulg. J. Plant Physiol., special issue, 172-178.

Tanguilig, V.C., E.B. Yambao, J.C. 'Toole and S.K. Dedatta, 1987. Water-stress effects on leaf elongation, leaf water potential, transpiration, and nutrient uptake of rice, maize, and soybean. Plant Soil **103**: 155–168.

Thavaprakash, N., K. Velayudham, M. Djanaguiraman, P. Subramanian, S. Panneerselvam and C. Prabakaran, 2006. Influence of plant growth promoters on assimilate partitioning and seed yield Of green gram (*Vigna radiata* L.). Leg. Res., **29**(1):18-24.

Vaiyapuri, K., M. Amanullah and K. Rajendran, 2012. Growth and yield of soybean as influenced by nitrogen and growth regulators. Madras Agric. J., **99**: 79-80.

Vekaria, G.B., M.M. Talpada, G.S. Sutaria, and K.N. Akbari, 2013. Effect of foliar nutrition of potassium nitrate on growth and yield of greengram. Leg. Res., **36**(2): 162-164.

Vignutelli, A., C. Wasternack, K. Apel and H. Bohlmann, 1998. Systemic and local induction of an Arabidopsis thionin gene by wounding and pathogen. Plant J., **14**: 285-295.

Vijayakumar, G., P.V. Ramesh and K. Kumar, 2013. Effect of foliar sprays of brassinosteroids and kinetin on biochemical parameters and yield of chickpea under drought stress. Indian J. Dryland Agric. Res. & Dev. 2013 28(2): 52-56.

Wahid, A. 2004. Analysis of toxic and osmotic effects of sodium chloride on leaf growth and economic yield of sugar cane. Bot. Bull. Acad. Sin., **45**: 133-141.

Walker M. A. and B. Dumbroff, 1981. "Effect of Salt Stress on Abscisic and Cytokinin Levels," Zeitschrift für Pflanzen Physiologie, **101**: 661.

Wei, W.X., P.E. Bilsborrow, P. Hooley, D.A. Fincham, Lombi, E., B.P. Forster, 2003. Salinity induced differences in growth, ion distribution and partitioning in barley between the cultivar Maythorpe and its derived mutant golden promise. Plant Soil **250**: 183–191.

Werner T, T. Schmulling, 2009. Cytokinin action in plant development. Curr Opin Plant Biol. **12**: 527–538.

Xin, Z.Y., X. Zhou and P.E. Pilet, 1997. Level changes of jasmonic, abscisic and indole acetic acids in maize under desiccation stress, J. Plant Physiol., **151**: 120-124.

Xiong, L., K.S. Schumaker and J. Zhu, 2002. Cell signalling during cold, drought and salt stresses. The Plant Cell, **14**: 163-183.

Yang J., J. Zhang, K. Liu, Z. Wang and L. Liu, 2007. "Involvement of Polyamines in the Drought Resistance of Rice," J. exp. Bot., **58** (6): 1545-1555.

Zhao, C.X, L.Y. Guo, C.A. Jaleel, H.B. Shao and H.B. Yang, 2008. Prospects for dissecting plant-adaptive molecular mechanisms to improve wheat cultivars in drought environments. Comp. Rend. Biol., **331**: 579–586.

Zhu, J.K., 2002. Salt and drought stress signal transduction in plants. Annu. Rev. Plant Biol., **53**: 247–273.

Other Related Books

Agricultural Drainage Engineering

Hiteshkumar V. Parmar

The book on "Agricultural Drainage Engineering" deals with the problem of drainage, drainage investigation, salt balance, surface drainage systems, subsurface drainage systems, reclamation of degraded land, designs and installation of drainage system, etc.

This book will assist in imparting better skills and understanding to Under Graduate (UG) and Post Graduate (PG) students, research scholars and professionals in the field of Agricultural Engineering.

Contents

1 Introduction; **2** Drainage Investigation; **3** Hydraulic Conductivity; **4** Surface Drainage System and Design; **5** Subsurface Drainage System and Design; **6** Drainage Materials; **7** Installation and Maintenance of Drainage Systems; **8** Special Drainage Systems; **9** Supplementary Protection Structures; **10** Reclamation Of Salt Affected Soils; **11** Soil and Water Quality; **12** Water and Salt Balance of Agricultural Land; **13** Glossary; References

ISBN: 978-81-7233-859-6 | 2014 | 238 pp

Agricultural Drainage Engineering : Field & Laboratory Manual

Hiteshkumar V. Parmar

Agricultural drainage system is planned to remove excess water from the crop land, whether coming through irrigation system, rainfall or runoff, before it creates waterlogging or high salinity condition. The Field and Laboratory Manual will assist in imparting better practical skills and understanding to Under Graduate (UG) and Post Graduate (PG) Students of Agricultural Engineering and Technology faculty. The manual will be helpful to the students as a reference for preparation of competitive examinations like GATE/ JRF/ SRF/ NET/ ARS, etc.

Contents

1 In-situ Measurement of Hydraulic Conductivity; **2** Determination of Drainage Coefficient; **3** Installation of Piezometer and Observation Well; **4** Preparation of ISO-bath and Isobar Maps; **5** Measurement of Hydraulic Conductivity and Drainable Porosity; **6** Design of Surface Drainage System; **7** Design of Subsurface Drainage System; **8** Determination of Chemical Properties of Soil And Water; **9** Fabrication of Drainage Tiles; **10** Testing of Drainage Tiles; **11** Determination of Gypsum Requirement for Land Reclamation; **12** Installation of Subsurface Drainage System; **13** Operation and Maintenance of Surface and Sub-surface Drainage System; **14** Design of Drainage Envelope or Drainage Filter; **15** Leaching Requirement for Reclamation of Soil; **16** Cost Analysis of Surface and Sub-surface Drainage System; References

ISBN: 978-81-7233-863-3 | 2014 | 107 pp

For more details visit : www.scientificpub.com